I would like to dedicate this educational book to my late father, Tang, Xian-Zhou (1935–1979), a very well-accomplished high school Chinese language teacher, who dedicated his life to educate growing-up youths in rural China. He was very passionate about telling illustrative stories from classic Chinese novels, the Water Margin, the Dream of the Red Chamber, Romance of the Three Kingdoms, and the Journey to the West, not only to his pupils but also to thousands of people beyond school boundaries. He was truly a Chinese classic literature expert. Meanwhile, his enthusiastic interest in Western classical music and art, his skills in mastering musical instruments, such as organ, accordion, and violin; and to sports, such as basketball and soccer, as well as serving as a swimming couch and lifeguard in his school; all of this inspired me to journey this far to accomplish this book.

**Tao Tang, Ph.D., GISP at Clarence Center, New York, USA**

# Contents

# Preface

In my 26 years of teaching and research career path in Geographic Information Systems (GIS), I always felt that there was a lack of connections of theories and concepts that were gradually growing into the current stage and the Earth's surface data processing and analysis routings that we practice daily using software or in the software environment. Such misconnections are very obvious when professors teach their GIS classes. After many years of struggling to balance teaching theories and concepts and presenting myself as a "software super-genius" to demonstrate my students "How to do?" in ArcInfo, ArcView, ArcGIS, and then ArcGIS Pro environments, I realized that a textbook is needed to cover both sides and make clear connections.

Therefore, this textbook is my attempt to cover both theories and approaches in the GIS field and demonstrate them in the ArcGIS Pro environment. The same strategy can be applied to other GIS software platforms as well. The objective of this textbook is to avoid being either software-centric or theory-centric. This attempt is the first of its kind and definitely a preliminary one. Whether the tutorials of software and data analysis match with the illustrations of the theories and concepts in this book, the readers will judge, and they can make suggestions to improve. Why is the balance of learning the theories or concepts and GIS software or analytical skills important? I used to tell students in that theories and concepts are the forest, and GIS software and analytical skills are the trees. If a person continuously flies over a forest and looks through its beauty without working on trees, the entire adventure could be like talking the talk. However, if you dive into an individual tree to work on it, either trying to grow the tree or to harvest its lumber to be used for other purposes, you will either confuse yourself sooner or later or lose interest altogether in doing it. I hope that this textbook can help the readers and learners to find both the forest and trees.

Dr. Tao Tang, Ph.D., GISP, Professor of GIS and Remote Sensing,
State University of New York (SUNY) – Buffalo State University

## Acknowledgment

First, I would like to thank two emeritus professors at the University of Wisconsin-Milwaukee for inspiring me. My Ph.D. dissertation advisor, Dr. Michael J. Day at Geography Department, for his unwavering, longtime support on my academic endeavors. Professor Williams Huxhold at the Urban Planning Department for his GIS class teaching to me during my Ph.D. graduate program. His textbook *GIS County – User Guide* inspired me to apply all my knowledge of physical geography and human geography into real-world practices.

Second, I would like to thank my two sons, younger son Rockwell and elder son Derrick, for their help revising the early draft for the manuscript and for being inspiration to me. Third, I would like to thank Mr. Jack Dangermond, CEO and founder of ESRI Company, and his wife Laura for their support and inspiration during my early years of studying GIS. Jack and Laura met me in person at the ESRI campus in Redland, California, to briefly discuss the future development of ArcInfo and ArcView, then the ArcGIS software platform when I went for a job interview.

Fourth, I would like to thank Dr. John R. Jensen, Emeritus Professor at the Department of Geography, University of South Carolina, and Dr. David Mark, late Emeritus Professor at the Department of Geography, University at Buffalo, for their support in my academic endeavors. I also would like to thank Dr. Ling Bian, Professor at the Department of Geography, University at Buffalo, and Dr. A-Xing Zhu, Professor at the Department of Geography, University of Wisconsin, Madison, for their wonderful exchange of academic ideas and interactions with me. Last, but not least, I would like to thank my former colleague, late Emeritus Professor and former chair at the Department of Geography and Planning, Buffalo State University, Dr. Raymond Waxmonsky, for the inspiration through his strategic vision on GIS and disciplinary development of geography and urban planning. Many thanks also go to the SUNY-Buffalo State University for granting me a sabbatical leave to write this textbook, to Mrs. Lisa Matthies-Wiza, the Director of Geographic Information Services, Erie County, New York, for covering my classes during my sabbatical leave and sharing data with me. I am also very grateful to the people establishing and managing the New York State spatial data sharing program, https://data.gis.ny.gov/ to provide the public spatial data available to explore in research and teaching. Many friends at the New York State GIS Association supported me all along the way.

I would also like to express my sincere thanks to the editors of this textbook at the John Wiley and Sons Company for their help in publishing this book.

Tao Tang, Ph.D.,
GISP at Clarence Center, New York, USA

## About the Author

Dr. Tao Tang is a professor at the State University of New York (SUNY) – Buffalo State University. He earned his bachelor of science degree from Northwest University in China, his first master's degree from the Institute of Geographic Sciences and Natural Resources Research, Chinese Academy of Sciences (CAS), second master's degree from the University of Guelph, Ontario, Canada, and Ph.D. degree from the University of Wisconsin – Milwaukee. All his academic degrees are in Physical and Environmental Geography. Dr. Tang started his career as a professor at SUNY – Buffalo State University in 1998. Dr. Tang's major research and teaching interests include Geographic Information Systems (GIS), Remote Sensing, and Physical and Environmental Geography. In his 40-plus years of development in the field of Earth's surface sciences, physical geography, and then GIS, in particularly during his 26 years of teaching and research at SUNY – Buffalo State University, Dr. Tang conducted fieldwork and traveled extensively on four of the five continents worldwide. Dr. Tang conducted field research in the countries Cambodia, China, Mexico, and the United States, to name a few.

Dr. Tang views that the comprehensive approach of GIS is the new era of digital geography and Earth's surface sciences. This approach is beyond mapping although mapping and cartographic processes are fully included. Growing up academically with a strong theoretical background of knowledge, Dr. Tang has always been searching for methods to apply the theories and concepts he learned to serve the society where he lives and to humankind worldwide. In his 26 years of teaching and research practices, Dr. Tang believes that academic research and professional services to government agencies and private industries go hand in hand. Inbreathing the digital Earth's surface sciences and Earth's surface large data computational processing, Dr. Tang is interested in sharing his continued learning experiences both within the discipline and across different disciplines through writing more text or learning books.

## About the Companion Website

*Fundamentals of Geographic Information Systems (GIS) Analysis* is accompanied by a companion website:

**https://www.wiley.com/go/tanggis**

The website includes:

- Tutorial Datasets—Ch01 to Ch12

# Part I

# Foundations of Geographic Information Systems (GIS)

# 1

# Introduction to Analyses in Geographic Information Systems (GIS)

## Concepts and Theories

Why do geographic information systems (GIS) matter? As its name suggests, GIS contains the concepts and theories of Geography and Locational Science. The historical development of geography and geographical ideas can be traced back to the early stages of human civilizations. Locational or geospatial science is and has been essential for almost all human activities because it is the foundation of both science and social science studies that involve locations on the Earth's surface. In other words, GIS is the science of studying where objects or phenomena on the Earth's surface are and attempts to answer the essential question "Why are these objects there?" The issue of geographic location is so important, it is embedded in many problems that our society needs to solve, and in many objectives or goals that our society needs to achieve. Here we can present some examples of GIS analyses:

1. Local governments need to manage geographic locations and the ownership of parcels of the land by its residents for various services and property taxation (Figure 1.1). GIS road and street spatial databases help the local or state government maintain and develop infrastructure in the region the government manages. National and state-level governments always need to match people with their locations for the issues of information management and services from time to time.
2. Private companies involving deliveries of goods and services, such as Federal Express (FedEx) or United Parcel Services (UPS), need to use GIS to analyze the locations of their customers in order to plan their transportation routes to reach their clients effectively and efficiently (Figure 1.2).
3. Civil engineering and construction companies need to utilize GIS to study the natural and human factors at site locations that impact their project. GIS are applied by these companies to study the impacts of construction on the surrounding environment as well.
4. Utility supply and transportation companies, such as the National Grid electricity supplier or the Chicago Transportation Authority (CTA), utilize GIS to manage their utility network or transportation network daily in order to provide geographically distributed services to their customers (Figure 1.3).
5. Retail stores and banks, such as grocery stores or fast-food chains, can utilize GIS to locate their customers and access their store needs relating to customer locations (Figure 1.4).

*Fundamentals of Geographic Information Systems (GIS) Analysis: Applying ArcGIS Pro*, First Edition. Tao Tang.

**Figure 1.1** Digital map of parcel polygons for land ownerships and streets that are connected to attribute database in Buffalo, NY.

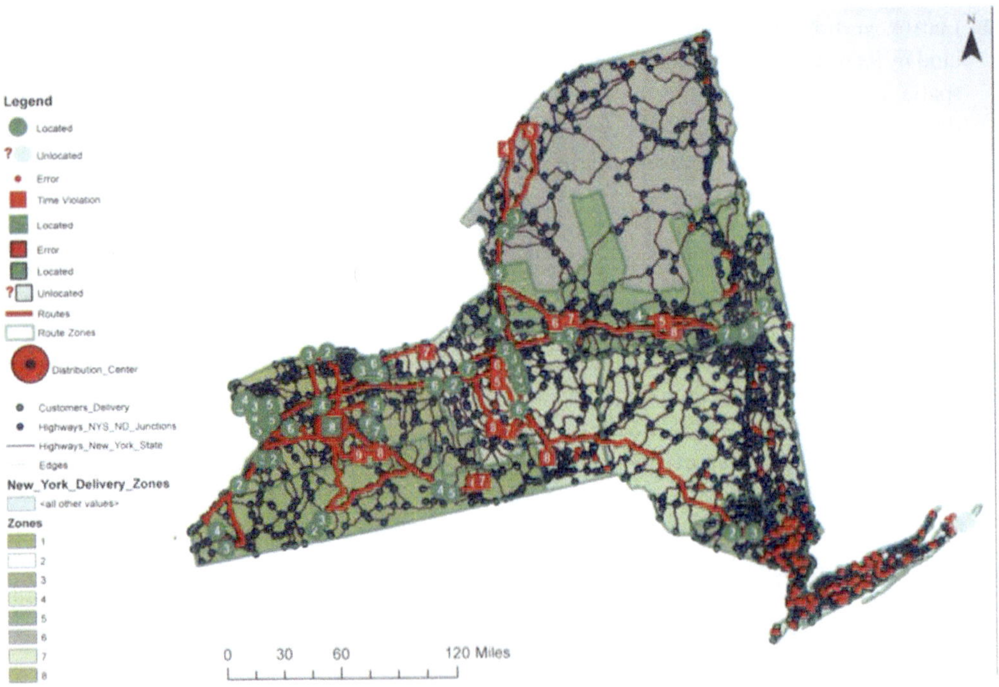


**Figure 1.2** Example of delivery route map and route planning in GIS, New York State.

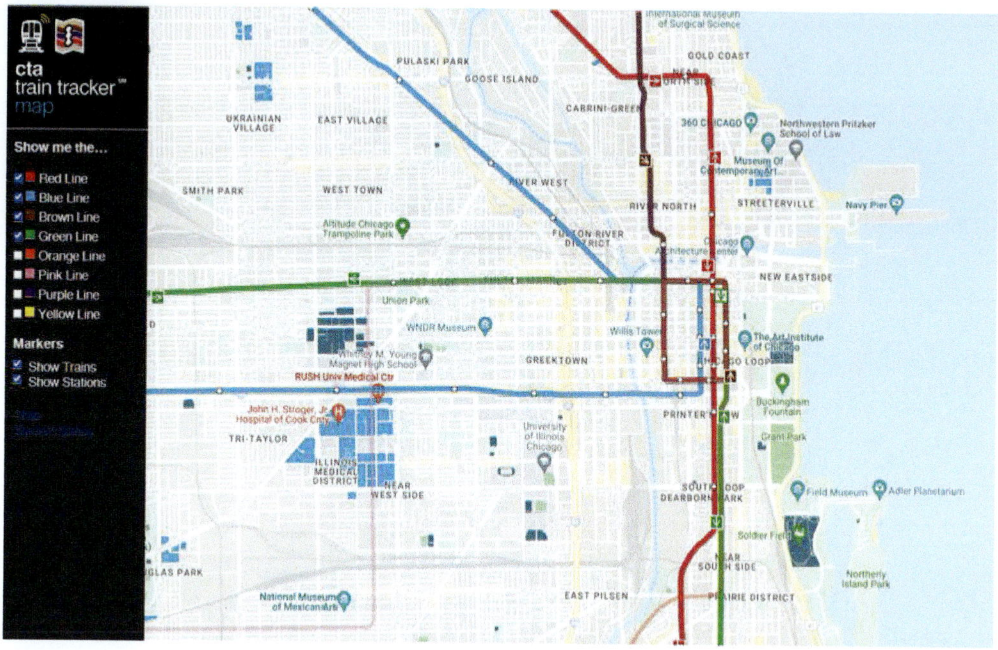


**Figure 1.3** Example of online subway train tracking digital map at CTA website powered by GIS. *Source:* Chicago Transit Authority/https://www.transitchicago.com/maps//last accessed on November 20, 2024.

6. Emergency response organizations either public or private in nature, such as firefighters, police, and the US Federal Emergency Management Agency (FEMA), can utilize GIS to access, manage, and mitigate the effects of natural disasters and societal emergencies by their locations and areas.
7. Health care systems and health management organizations, such as the Center for Disease Controls (CDC) can use GIS to visualize and analyze the distribution of infectious diseases, such as COVID-19. GIS can also help health providers study the regional distributions of diseases and answer questions about environmental causes of diseases.
8. Human living environments are by nature locationally registered or geographically distributed. Therefore, GIS provides an effective utility platform that can be used in a wide range of environmental science and environmental studies, from air pollution to water pollution to other types of environmental management issues.

As Longley et al. (2015) indicated, almost everything that happens, happens somewhere. Knowing where something happens can be critically important. In essence, all the activities or events that occur in either the natural environment or in human-made society can be located in a three-dimensional (3D) geographic space without considering the time issue. If the time scale and time range are considered, we can identify a four-dimensional geographic cube to locate all the various events and activities. The significance of GIS is that we can utilize modern computing power and information system technology to explore the answers to questions such as: where are events happening or where will they happen? How big is the geographic area? Why does it happen there? In addition, GIS can facilitate the simulation of events in a four-dimensional geographic or spatial-temporal cube to answer the question of what will happen under a particular set of conditions.

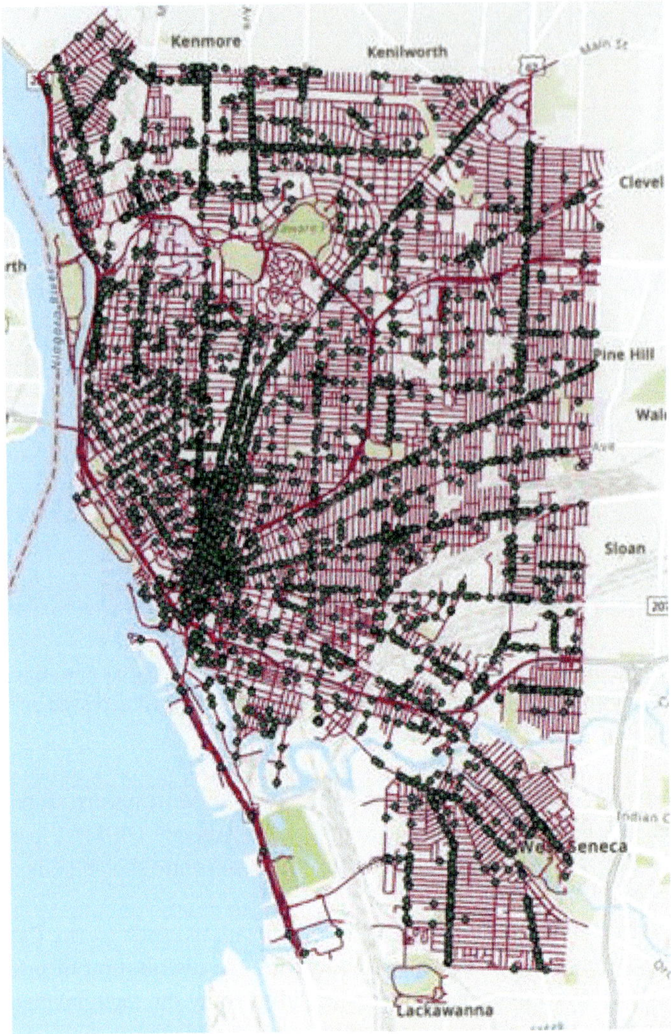


**Figure 1.4** Geocoded business locations with varieties of services in the City of Buffalo, NY. *Source:* Dr. Tao Tang.

## What Is an Information System?

An information system is an integrated set of components for collecting, storing, and processing data to provide information, knowledge, and digital products. The purpose of processing data and building an information system is to facilitate decision-making for various objectives. Information systems in the modern era are rooted in computer science or the study of computation, automation, and information (The University of York 2020). The study of computer science includes the theoretical and algorithmic foundations of hardware and software, and computer hardware or software used for processing information.

## What Does Geography Study?

According to a philosopher in the 1800s, Kant (1755) (Elden 2009), there are three different categories of understanding: (i) the understanding of objects or phenomena, such as in the studies of physics; (ii) the understanding of characteristics of objects through time, such as in the studies of archeology; and (iii) the understanding of the distributions of objects across the space, such as in the studies of geography. Therefore, geography is the study of spatial distributions and patterns of objects that are related to the human living environment on the Earth's surface (Figure 1.5).

## What Is a GIS?

GIS are special categories or special cases of information systems that combine digital map layers with database to capture (Figure 1.6a), analyze (Figure 1.6b), and model (Figure 1.6c) spatial or geographic features, objects, or Earth's surface processes (events).

In practice, geographic information analysts need to capture spatial objects and their characteristics first. Hundreds of unique methods have been developed in various disciplines of science or social science to obtain location-stamped data or spatial data on the Earth's surface. Meanwhile, geographic information analysts cannot practice all these different scientific or social science data collection methods simply because of too many of them. Therefore, utilizing the datasets collected by other scientists or social scientists for geographic information analysis is a common practice in the discipline especially with the exploration of data captures and dataset publications on the internet. Several traditional surveying and remote sensing technologies have been widely used by geographic information analysts to capture GIS data. Some of these are satellite remote

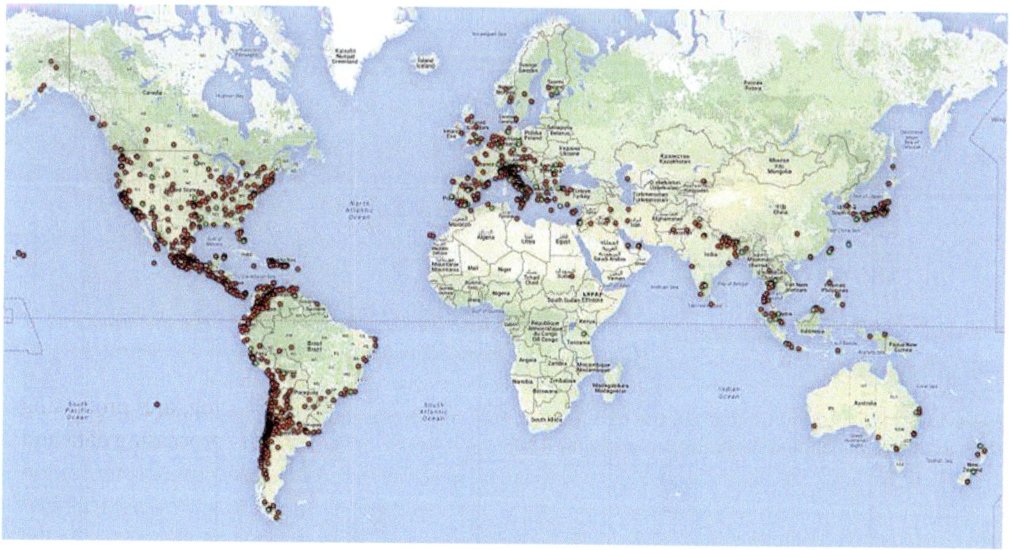

Figure 1.5 Spatial distribution of earthquakes worldwide. *Source:* Blue8111/Wikimedia Commons/ CC BY SA 4.0.

(a)

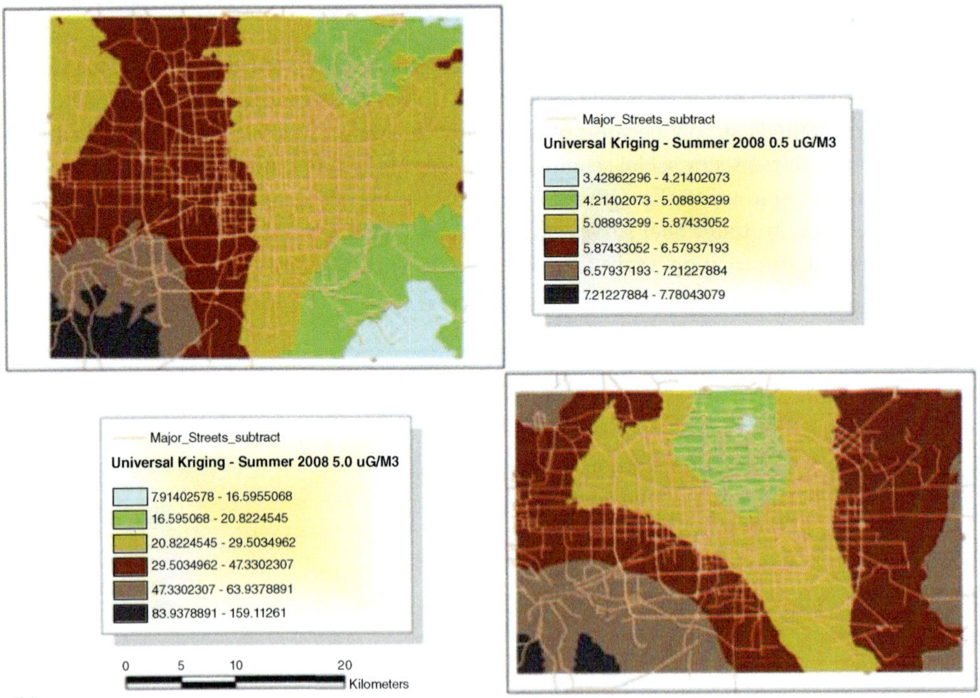


(b)

**Figure 1.6** (a) Capturing GIS data in the field using LiDAR and drone equipment. *Source:* Dr. Tao Tang. (b) Analyzing and mapping air pollution applying GIS.

*(Continued)*

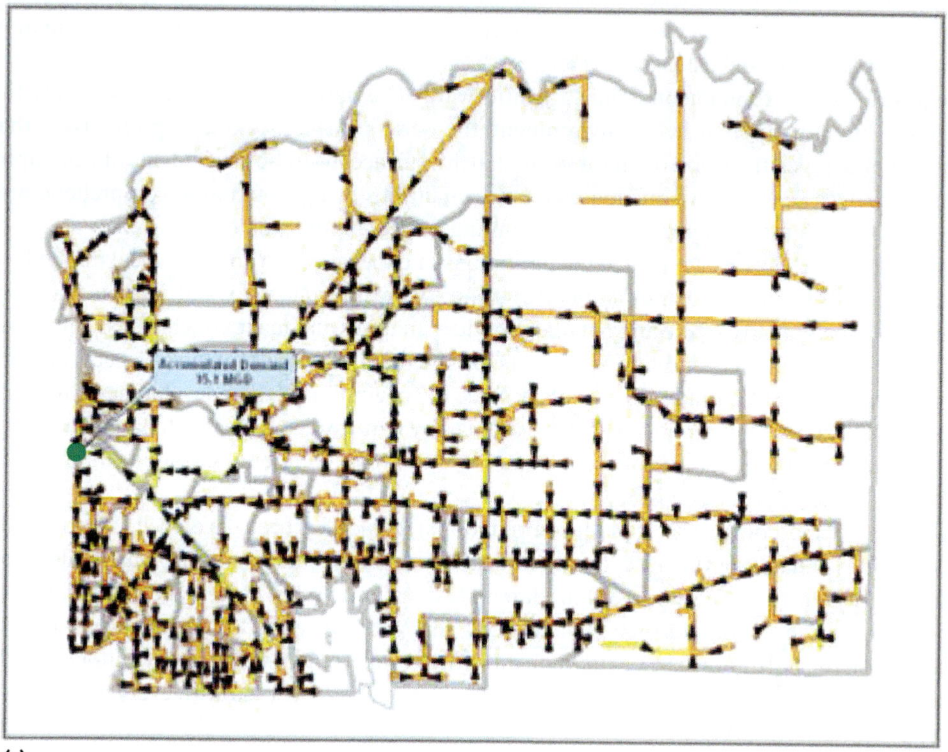

(c)

**Figure 1.6 (Cont'd)** (c) Modeling water main pipelines for household water supplies applying GIS in the Towns of Amherst and Clarence, NY.

sensing, aerial or drone remote sensing, photogrammetrically enabled survey, and Global Positioning System (GPS) survey.

Analysis of the spatial data involves many different levels of data processing, from overlaying map layers to geocoding tabular data, to mathematical or statistical interpolations of grid cell surfaces, to conducting geographical regressions to test the relations of two or more spatial variables in a region. Modeling geographic objects, features, or Earth's surface events is considered more advanced data processing than that of analysis. The goal of modeling is mainly to answer the question "What if?." For instance, if we can simulate a certain amount of rainfall in a short period of time, what kind of flooding would result? How many human-made buildings, houses, and infrastructures would be destroyed? The significance of modeling and simulating in GIS is to provide possible measures to avoid such kinds of events happening in the real world in the first place. If such disastrous events do happen, GIS-based modeling and simulation will help to mitigate the losses.

What does computational science or information technology have to do with geography? Equally speaking, why is geography a discipline that needs to use computational science or information technology intensively and extensively?

The answer is very fundamental and straightforward. The nature of geography is to study the relations of human and their living environment, the Earth's surface. Taking a specific issue that either occupies a specific region or on the entire Earth's surface to study, the students of geography or Earth's surface science have encountered large quantities of information to comprehend and complex information to analyze since the beginning of geographic studies in the early stages of human civilization. However, owing to the limitations of data gathering and analytical methods in the past, students of both geography and Earth sciences were forced to use small datasets that were within their capability to capture and obtain to conduct analyses and make conclusions. Therefore, it is natural for geographers and Earth scientists to seek the utilization of various aspects of computational science and information technology since the beginning of their development. In other words, the disciplines of geography and earth sciences by nature need to deal with large quantities of data. Therefore, it is necessary for these studies to utilize computational science or information technology intensively and extensively.

On the other hand, computer science or computational science has been developed for processing large quantities of data, converting data into valuable information, and analyzing data for decision making. In this regard, we need to ask a fundamental question: "What percentage of the data that computational science and information technology deal with every day are geographic or related to locations on the Earth's surface?" Although an accurate answer to this question cannot be easily found, it is evident that the majority of data we deal with or are concerned with every day are geographic or related to locations on the Earth's surface. Therefore, GIS is not only essential but also significant among the science and social science studies.

Since the 1960s, the development of GIS as a scientific discipline and its practice as a set of professional skills in the workplace has been fraught with difficulties, uncertainties, and a lot of debates. It is clear that many of the significant contributions that paved the way for the success of GIS today in industrial and professional settings were not made in academic areas of geography or by academic geographers. Popular uses and analyses of geographic information for various practical purposes by non-geographers contributed a great deal of cases, concepts, and theories in GIS. Meanwhile, GIS approaches encountered several misconceptions or misinterpretations. First, since GIS utilizes digital maps and digital graphics, many people consider it to be a specific type of computer graphic design. While GIS does make a large quantity of maps and utilize graphic design tools in the computer environment, its goal is to analyze the spatial data and present the analytical results and data on digital maps. The purpose of GIS is not just for designing better graphs or maps although all these tasks are included. Secondly, since cartography has been a subdiscipline in geography long before the computer was invented, many practitioners attribute GIS as a branch of cartography or call it "computer mapping" or "computer cartography." In the late 1990s, the American Association of Geographers hosted a widely involved professional debate on the theme "Is cartography swallowed by GIS?" in one of its annual national academic conferences. After that debate, the general consensus in the association is that GIS has "swallowed" the cartography as a subdiscipline. Meanwhile, many new subdisciplines, such as 3D visualizations and network analysis have been developed in GIS as well since then.

Since GIS have heavily utilized computer hardware, software, and new technologies for its own development, many people might be most familiar with GIS as a computer software package or one of the applications in computer science. People with this kind of understanding have overlooked the significance of the data side of GIS approaches. In real-world practice, data and data captures are equally important as software implementations and utilizations. Toward the theoretical and conceptual end, a newly established science discipline called Geographic Information Science has been developed since the late 1980s. As one of the leading researchers in the field, Dr. Michael Goodchild illustrated that Geographic Information Science is a discipline to study "the generic issues that surround the use of GIS technology, impede its successful implementation, or emerge from an understanding of its potential capabilities" (Goodchild 1992).

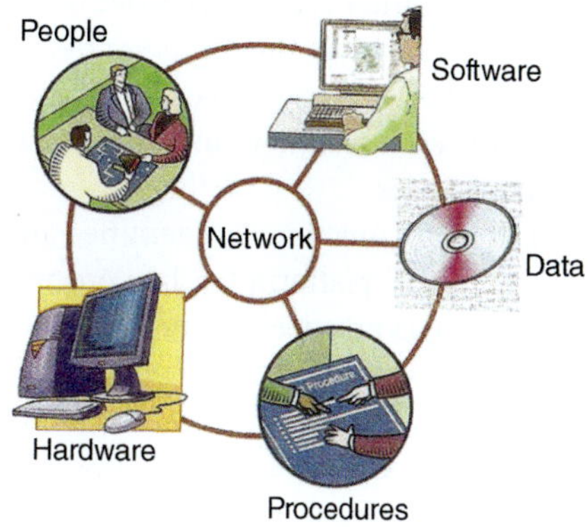


**Figure 1.7** ESRI illustration of GIS approach. *Source:* Paul A. Longley et al., 2005/John Wiley & Sons.

It is evident that GIS is a unique and special approach that is different from many other computer software applications. Summarizing the current practices of GIS in various sectors in our society, scientists and geographic information analysts at Environmental Systems Research Institute (ESRI) company illustrated (ESRI 1995) that there are six major components in GIS: data, software, hardware, applications (procedures), network, and people (Figure 1.7).

The goals of GIS in practice are mainly to analyze the geographic or spatial features or events on the Earth's surface. Therefore, conceptualizing geographic features in the computational environment or in computer systems is essential to our approaches. To this date, there are two major methods to conceptualize geographic features or objects in the computational environment or computer storage spaces. One is the object-oriented interpretation method that treats each geographic feature on the Earth's surface as one object in a computing environment. This method of geographic feature illustration is called vector data structure, vector data format, or vector data illustrations in GIS. The objects or features are then classified into three geometric categories: points, lines, and polygons or areas. The fundamental characteristics of these vector objects or features are: (i) Each of these features or objects occupies a location (or a range of or area of locations) on the Earth's surface. (ii) Each of the features or objects carries its attributes. In other words, the recorded attributes are characteristics of the geographic features. (iii) Geographic features can be classified into feature classes by their commonality of major characteristics or attributes. From a computational perspective, geographic features of points, lines, and polygons can be stored in the computer database with their own locational information and other attributes. One feature is one record in the computational database. Meanwhile, features or objects can be drawn on the computer screen using computer graphics tools, which one feature class can be easily interpreted as one digital map layer.

The second method to illustrate the geographic objects on the Earth's surface is the field-oriented or grid cell-based interpretation. This method is called raster data type of illustration. Raster data structure or raster data format decomposes the study area by grid cells. In general, the size or spatial resolution of the grid cells of a specific dataset is predetermined according to the objectives of the studies and data capture processes. The geographic locations of the grid cells are assigned to either each or a cluster of these cells. From a computational perspective, each of the grid cells is stored in the computer database originally and separately. The shape and characteristics of real-world geographic objects can still be recognized by their common values. The boundaries of feature locations and their occupation on the Earth's surface can be represented by grid cell boundaries.

GIS can be functionally identified into two major technical components: data and software. Software is the platform for data processing to solve practical problems, while data is always the "blood" of GIS practice and analysis. As time goes on and human social and economic development pushes forward, we always generate and gather new data for analysis in general and GIS analysis in particular. Therefore, we can explore the chronological development of GIS along these two lines. GIS as a discipline of studies in science and social science was not established until the 1960s. Previously, several subdisciplines or new data-gathering technologies and applications were either embedded or adopted in the studies of geography for large volumes of data acquisitions and presentations. Cartography has always been a major component in geographic studies and served as a method of presenting results in geographic studies. During the World War II, aerial survey and aerial reconnaissance were rapidly developed. Since then, both aerial photogrammetric and ground geodetic surveys have been developed as basic methods for large quantities of geographic data gathering and collection. According to scientific studies (Tomlinson 1999; National Academies of Sciences, Engineering, and Medicine, 2007; Tomlinson 2012; ESRI 2017), the first functioning GIS system was created in Canada in 1962. Dr. Roger Tomlinson, an educated physical geographer who specialized in geomorphology for his PhD degree program led a group of scientists and technical staff to develop the Canadian Land Inventory System for the government of Canada. Dr. Tomlinson later named this system the "Canadian Geographic Information System." Thus, the term "geographic information system" was first introduced, marking the beginning of the development of GIS. As Dr. Tomlinson illustrated in creating the Canadian Geographic Information System, the original and fundamental idea is to use computers to ask questions about maps and to render useful information from them. Therefore, maps had to be digitized into computer formats. This led to the idea that many digital maps could be stitched together to represent the entirety of Canada and that the maps could be linked intelligently to digital databases for statistical analysis, such as with census data of the Federal Government of Canada. Cultivatable land and potential farming populations in the proximity of the areas could then be analyzed.

In 1964, the Urban and Regional Information Systems Association (URISA) was established by a group of architects, urban planners, and urban landscape designers. The purpose of this association is to utilize computers to gather information and conduct graphic design to solve problems related to urban planning and the human living environment. In 1965, an established architect and urban designer, Howard T. Fisher founded the Harvard Laboratory for Computer Graphics and Spatial Analysis at the Graduate School of Design, Harvard University. This laboratory pioneered early integrations of cartography, computer science, and architectural graphic design. Such efforts lead to the initiation and development of software for GIS. Among the graduate students who were trained in this laboratory, Mr. Jack Dangermond, a California native, established the company called Environmental Systems Research Institute (ESRI) Inc. in 1969 after graduation from the Harvard University. Since then, ESRI Inc. developed a series of GIS software, from ArcInfo, to ArcView, to ArcGIS, and ArcGIS Pro (Figure 1.8). Currently, ArcGIS software occupies approximately 43% of the global GIS software market. It is the most popular GIS software worldwide. The estimated annual revenues are approximately US$1.1 billion.

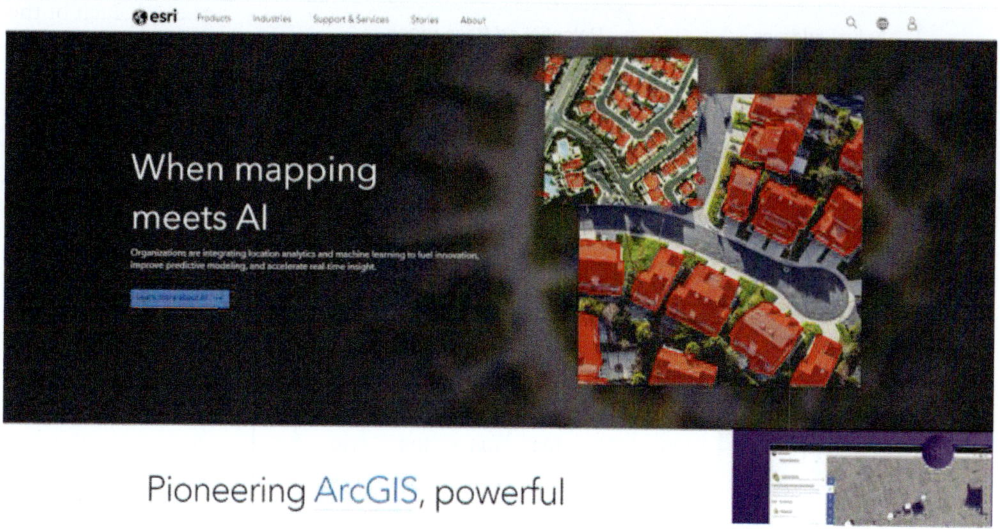


**Figure 1.8** The website of ArcGIS software company. *Source:* ESRI, Inc.

In addition to this type of global flagship GIS software, we can also find and access many other GIS software, for instance, Hexagon Geomedia, QGIS, MapInfo Professional, Maptitude, Google Maps, and SuperMap GIS. Some GIS software is free for users, such as QGIS. Others may just be online for everyone to use, such as Google Maps. Some others are more specialized for either a particular task or a particular country.

One of the conventional software operations in GIS is called "map overlay," which is a process to overlay digital map layers in a computer environment to pinpoint the overlapping areas or locations of two or more geographic objects on the Earth's surface. Map overlay spatial data processing technique was first implemented in a computerized environment by the Federation of Rocky Mountain States, a not-for-profit organization in 1979 (Federation of Rocky Mountain States 1978). In 1977, the organization obtained a grant from the US Fish and Wildlife Service (FWS) to develop a spatially enabled information system for environmental impact and habitat mitigation studies for an after-strip-mining wildlife recovery project. As part of the results of this study, the Federation of Rocky Mountain States published the digital map overlay method.

Meanwhile, the structural database or smart spatial database was also developed since the 1960s. Based on their practical needs, the US Bureau of the Census at the Department of the Interior developed the Dual Independent Map Encoding (DIME) system in 1967. Attempting to convert analog maps into numerically encoded renderings using computer data storage, the bureau's mathematician James Corbett then created the DIME representation system. The concepts of map topology were first introduced in this approach. The bureau later improved the DIME file and renamed it as Geographic Base File (GBF), then the Topologically Integrated Geographic Encoding and Referencing (TIGER) database system.

The TIGER spatial data system is the first vector database structure that interprets the elements on a digital map into the computer database by applying topology. In this illustration system of relational features or geographic objects, point features were called 0-cells, serving as fundamental building blocks. Line features were called 1-cells, serving as walls of buildings. Finally, lines were

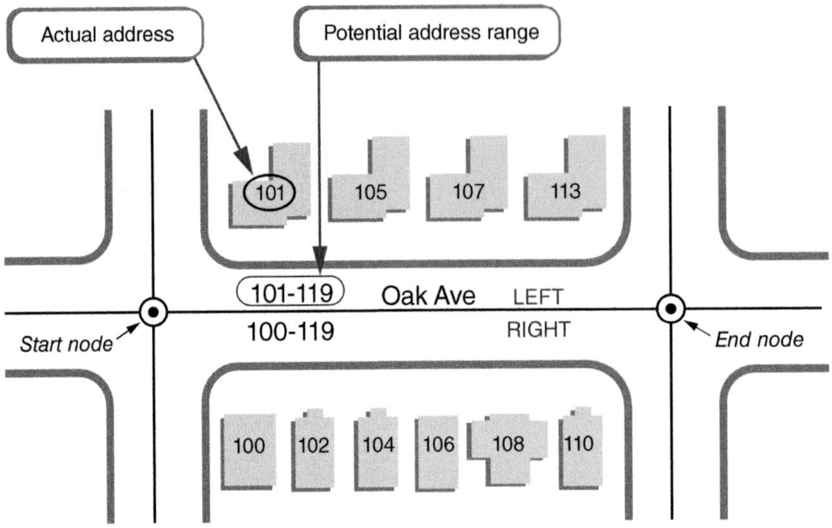


Record Type 1 contains separate data fields for both the start and end of each address range.

| Record type 1 | | | | Address range | | | |
|---|---|---|---|---|---|---|---|
| | | | | Left side | | Right side | |
| | | | | Start | End | Start | End |
| RT | TLTD | FENAME | FETYPE | FRADDL | TOADDL | FRADDR | TOADDR |
| 1 | 0007654320 | Oak | Ave | 101 | 119 | 100 | 118 |

**Figure 1.9** US Census TIGER file referencing database system. *Source:* 2000 TIGER/Line Files /U.S. Census Bureau/Public Domain.

put together to form polygons that were called 2-cells, similar to building houses or structures in the digital or computational environment. The principal rules of this approach are: (i) the building blocks on a digital map are hierarchical, and the points are fundamental elements to build lines and polygons in the way that bricks are used to build walls and houses. (ii) Lines are identified by starting point (node) and ending point (node) to ensure connectivity to other lines. (iii) Left address numbers and right address numbers of starting and ending nodes of a line or street segment are recognized and stored in the computational database (Figure 1.9). (iv) Left or right regions or polygons are identified. This approach set up the basis of relational vector feature geodatabases that we have used up to today for both digital databases and mapping. It is the foundation of the modern vector GIS database.

In order to facilitate decades of popular movements in developing GIS and its applications, the National Science Foundation of the United States founded research, training, and innovation centers for GIS at three universities in 1988. These centers are part of the National Center for Geographic Information and Analysis (NCGIA), and they are hosted at the University of California, Santa Barbara; the State University of New York at Buffalo; and the University of Maine. Both the centers at the University of California, Santa Barbara and the State University of New York at

Buffalo are housed in the Department of Geography, while the center at the University of Maine is hosted at the Civil Engineering Department. Currently, educational programs up to PhD degrees are offered by these centers. As GIS course offerings are increasingly popular at the university campuses across the United States, the University Consortium for Geographic Information Science (UCGIS) was established in 1994. The primary missions of the UCGIS were to advance research in the field of Geographic Information Science; expand and strengthen Geographic Information Science education, and build scholarly communities and networks to foster multidisciplinary GIS research and education. As GIS gradually became a digital infrastructure of the nation for both government functions and private industrial operations in the United States, Executive Order 12906 – the National Spatial Data Infrastructure was signed by President Bill Clinton on 11 April 1994. The objective of this initiative is to coordinate geographic data acquisition and access, and to share, distribute, and improve the utilization of geographic datasets.

Currently, the global market size of the GIS industry, including software, data, hardware, and services, was projected to reach roughly US$10.1 billion in 2021 and US$21.1 billion by 2027 (IMARC 2021). From a daily application perspective, the GIS industry can be further identified into the following: (i) the GIS data or spatial data industry, (ii) the GIS software industry, (iii) the GIS service industry, and (iv) the online and web-based geographic information service industry. GIS is data intensive and extensive. Therefore, it is significant for conducting spatial data acquisitions, converting data from invisible to visible formats, and visualizing the spatial data for many other operations and for decision-making in both private sector and government agencies. For instance, the US-based Maxar, Inc. is one of industrial leaders to utilize satellite or spacecraft remote sensing technology to collect geographic information for anywhere or everywhere on the Earth's surface. Software serves a platform to process data. Integration of software to solve problems and the development of new software to fulfill a specific task or use of geospatial data are always key activities in the practice of GIS.

Naturally, many of the private industries and three levels of government in the United States need to process and analyze spatial data in order to make decisions and provide services and products. For instance, FedEx, UPS, and Pizza Hut, Inc. need GIS analysis to ensure effective deliveries. Walmart, Inc. and MacDonald, Inc. need to utilize GIS to effectively select the new locations of their retail stores. Local city or county governments need to utilize GIS to manage land parcel ownership databases for the purpose of property taxation. Since not every organization has the resources to readily establish one professional office or department to conduct these kinds of GIS tasks, GIS service companies have been developed following a business-to-business model. Online and web-based GIS information are vital to every person. Web-based geographic information retrievals are important for the decision-making of individual people and businesses. One example of a successful web-based geographic information service is Google Maps. The Google Maps GIS system is maintained and powered by Google, Inc. In summary, it is apparent that geographic information gathering, analysis, and presentation are important to ensure daily operations of our human society by its own purposes and practices.

## References

Elden, S. (2009). Reassessing Kant's geography (PDF). *Journal of Historical Geography* 35 (1): 3–25.

Federation of Rocky Mountain States – Information Systems Technical Laboratory (1978). WELUT – 02 Project: report 2200: Logical capabilities of the GIS: management summary version. Bureau of Land Management Library Publication, Fort Collins, Colo.

Imarc. (2021). https://www.imarcgroup.com/geographic-information-system-market#:~:text=Market%20Overview%3A,13.1%25%20during%202022%2D2027.

Kant, I. (1755). *Universal Natural History and Theory of the Heavens.* Berlin Academy Publication, Berlin, Germany.

Longley, P.A., Goodchild, M.F., Maguire, D.J., and Rhind, D.W. (2005). *Geographic Information Systems and Science.* 2e. John Wiley & Sons.

National Academies of Sciences, Engineering, and Medicine. (2007). *National Land Parcel Data: A Vision for the Future.* Washington, DC: The National Academies Press. https://doi.org/10.17226/11978.

Tomlinson, R. (2012). *Origins of the Canada Geographic Information System.* ESRI Press, Redlands, CA. https://www.esri.com/news/arcnews/fall12articles/origins-of-the-canada-geographic-information-system.html. ESRI Press.

Tomlinson, R. and Toomey, M.A.G. (1999). "GIS and LIS in Canada," chapter 15 in Mapping a Northern Land: The Survey of Canada 1947–1994, Gerald McGrath and Louis Sebert, eds. (McGill-Queen's University Press).

The University of York, What is Computer Science? Computer Science. www.cs.york.ac.uk. Retrieved June 11, 2020.

## Tutorial in ArcGIS Pro Environment

ArcGIS Pro is the most advanced desktop-based GIS software among the various software platforms that the ArcGIS software family provides up to this date. ArcGIS Pro also has some internet mapping capabilities. ArcGIS software family is produced and supported by the Environmental Systems Research Institute (ESRI), Inc. ESRI is a private company that was established in 1969. In the early stage, the company mainly conducted consulting projects for geographic analysis, urban planning, and urban infrastructure construction applying computer databases. Meanwhile, the company developed its own software which was called ArcInfo. Strategically, the software company kept a clear vision and objectives since its beginning. "ArcInfo" by name reflects what the company has been doing up to this date. Arc means the curves of lines on a digital map. Info means computer databases. People can comprehend what the software does by the illustrations of its name. For years, this company has endeavored to combine digital maps and computer databases to make sense of information systems (IT) in various practices in our society. Over time, the name of the software was changed from ArcInfo to ArcGIS, then ArcGIS Pro, while the main contents have been always kept the same.

The structure of the ArcGIS Pro software, inherited from the ArcGIS software package, has three major functional components. These are: (i) mapping and visualizing data, (ii) geoprocessing and analyzing data, and (iii) managing and editing data. In the background of the software design, it has been strongly believed that digital or computerized datasets are mostly geographic or spatial. Geography not only matters but is also significant. This is the reason that the company declared recently on its website that ESRI is a company for exploring the "science of where." In practice in the software environment, people can strongly feel the reflections of these three components. In ArcGIS, three separated modules or windows were originally developed to match with the three conceptual components. These are ArcMap, Arc-Toolbox, and ArcCatalog. ArcMap is mainly for data visualization and presentation. Arc-Toolbox is mainly for geoprocessing of data and data analysis. ArcCatalog is mainly for managing spatial databases. The major structural improvement of ArcGIS Pro from the ArcGIS package is that the new software combined all three windows into one platform in the Microsoft environment. Now, all the databases, the digital map layers, and the software tools for geoprocessing and data analysis are organized in one window with one management file or one big organizer called a "project." The purpose of doing this is to put everything under one roof in this comprehensive software environment. Connecting to real-world practices, the project file or project window reflects various projects that geographic information systems (GIS) professionals carry out every day. Project-oriented software design has a lot of advantages in practice. However, the disadvantage is that even just to read or explore the databases that a GIS analyst obtained without intending to do a specific GIS project, one must create a new project file with a specific name in order to do so. The alternative way is to use a previously created project file that may have nothing to do with this newly obtained database that the GIS analyst is looking into to perform data exploration and inspections.

GIS is both software-intensive, and database intensive and extensive. In our tutorials in this textbook, we will use geographic database of Erie County, New York State which is called Erie-County-Spatial as one of the major tutorial databases. Other databases collected are also to be utilized for GIS tutorials and student exercises (assignments). From Chapter 3 onward, students studying GIS will be given assignments. These assignments will apply various datasets of the United States. However, digital geographic datasets or spatial data can be generated or collected anywhere worldwide on the Earth's surface. Users are encouraged to utilize the skills and methods learned in this textbook to explore your local spatial data, conduct local GIS projects, and solve local problems as well.

Before we start exploring the ArcGIS Pro software platform, we need to copy our datasets to the hard drive of our local computer. First, let's open the Microsoft Windows - File Explorer . Then, copy the "Chapter01-Tutorial" folder to the computer hard drive that you are going to use.

Now, let's open the ArcGIS Pro by clicking on its icon on the desktop, or from ArcGIS software folder at the Windows start menu . Then, the following opening dialog window opens.

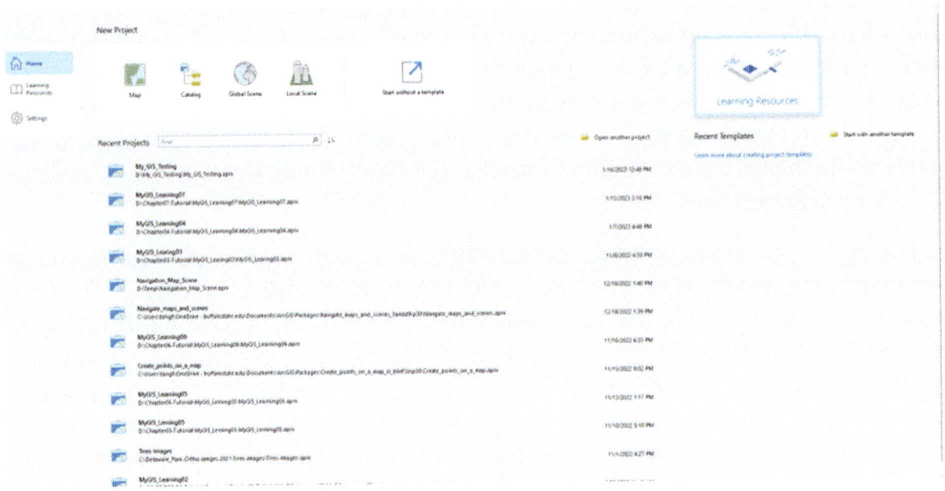


ArcGIS Pro 3.0 or later version

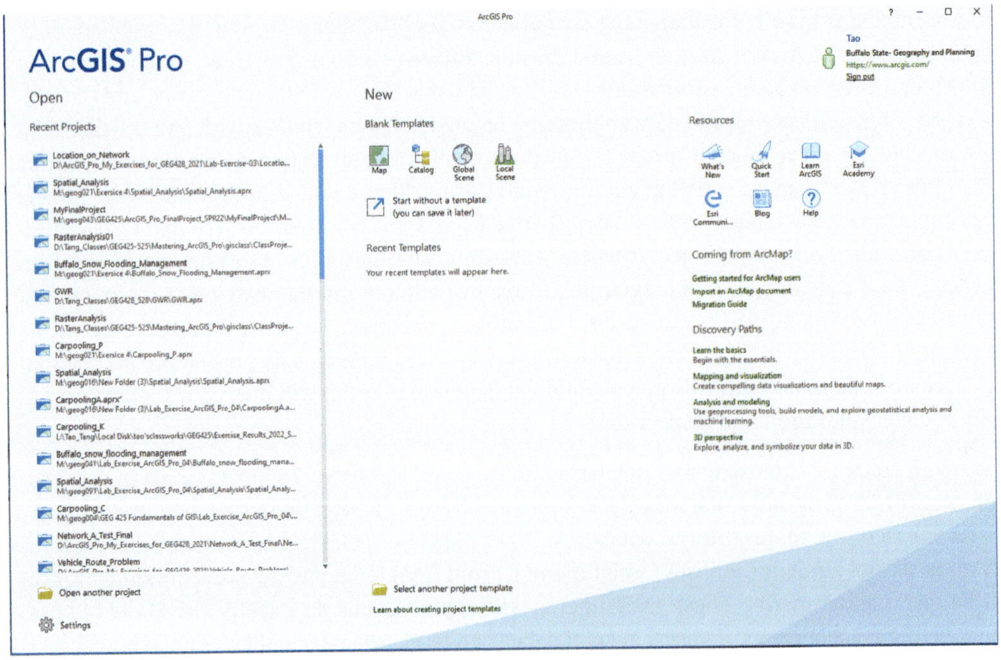


ArcGIS Pro 2.0 or later version

Each time we open the ArcGIS Pro, we have two main choices. One is to "Open another project." In this case, you can click either on one of the links on the left side under "Open" category to open, or to click on the icon "Open another project" and navigate to a folder to open a project file (.aprx). This is the standard method to open an existing GIS project file (.aprx). The other way is to create a new project file ("New Project") under the "New" menu on the top or in the middle of this dialog (depending on the ArcGIS Pro version that you use). Under the "New" menu, we have four choices of selection of the template icons that we can use to create (open) a project window.

1. Create a project file with a "Map" window in the center of the APP 

.
2. Create a project file with a "Catalog" window in the center of the APP 

.
3. Create a project file with a "Global Scene" window in the center of the APP 

.
4. Create a project file with a "Local Scene" window in the center of the APP 

.

Both the global scene and local scene are digital three-dimensional (3D) geographic visualization windows. The major difference is that the global scene can handle the curvature of the Earth's surface in a large geographic region while the local scene focuses on a small region, such as a city simplifying the Earth's surface as flat. Of course, one can select another template than these four or no template at all to create a new project file. Normally, we initiate an ArcGIS Pro project with a "Map" template. All other templates or windows can be easily added later in addition to the "Map" template.

Name the project file as "MyGIS_Learning01," make sure the "Create a new folder for this project" is checked, and save it inside your Chapter01-Tutorial folder as shown in the following.

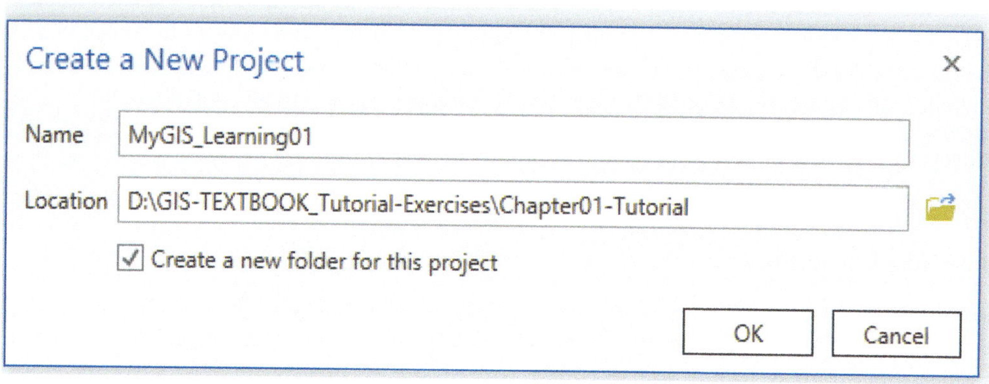


Click OK to start the ArcGIS Pro window. The new ArcGIS Pro project window opens.

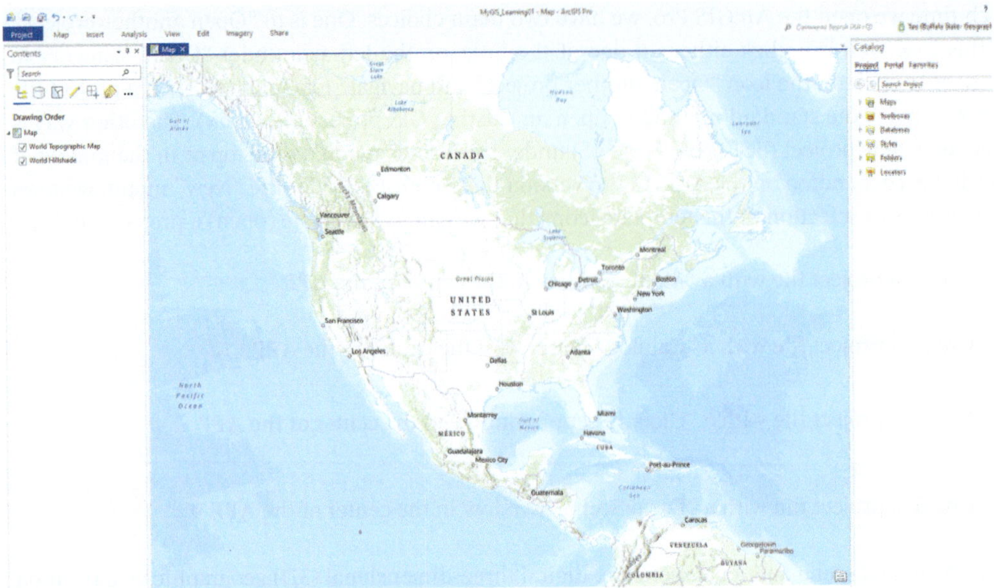


In the middle of the ArcGIS Pro software platform is the mapping template or mapping frame. The background worldwide map is provided and embedded in this software for your reference. The background map is not the local data. The Table of Contents (Content Pane) of the digital maps and data shown in the "Map" window is on the left-hand side. The Catalog Pane (equivalent to ArcCatalog in ArcGIS) for the database viewing and management of visualization windows is on the right-hand side. The title of the project is presented in the middle of the top margin of the software APP window.

Now, let us add the geographic (spatial) databases that are called "Geodatabase" in ArcGIS Pro. The spatial database management module here is the Catalog Pane on the right-hand side. Normally, we use the "Databases" category on the list of Catalog Pane to add and connect to geodatabases only, and the "Folders" category is to add and connect to data folders. After the geodatabases and data folders are added on the list of Catalog Pane, we can use them in the ArcGIS Pro.

Now, let's add the "Chapter01-Tutorial" folder on the list under "Folders" category.

To accomplish this task, we can right-click on the "Folders" icon to activate the contact menu, "Add Folder Connection" 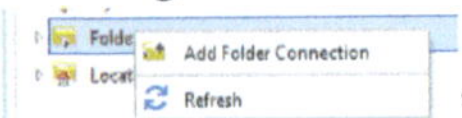

, then navigate in the opening window to the folder location on computer network. Alternatively, we can activate the "Insert" ribbon on the top of the software window and click on the "Add folder" icon to navigate to the folder location.

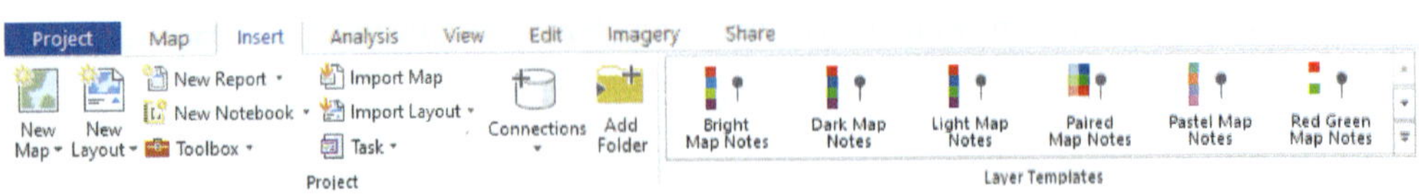


Click OK while the folder is highlighted to add it. You may notice that the folder is added on the Catalog Pane.

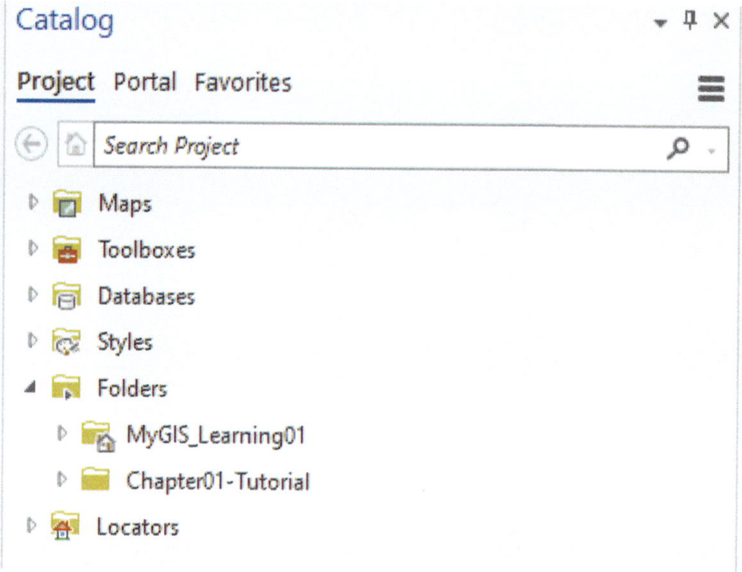


Now, let us add a few map layers or feature classes on the "Map" window. Expand the "Chapter01-Tutorial" folder, and you will see three geodatabases in this folder. These are namely, Erie_County_Spatial_Data.gdb, USA.gdb, and World_Countries.gdb. A suffix of "gdb" stands for geodatabase and the icon of a grey can in ArcGIS Pro environment is the geodatabase icon.

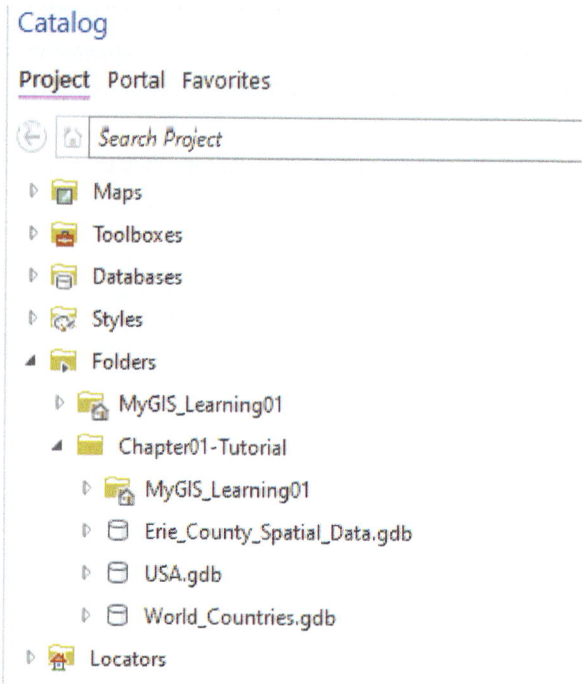


Expand the World_Countries geodatabase, and you can see there are a total of four map layers (feature classes) in this geodatabase.

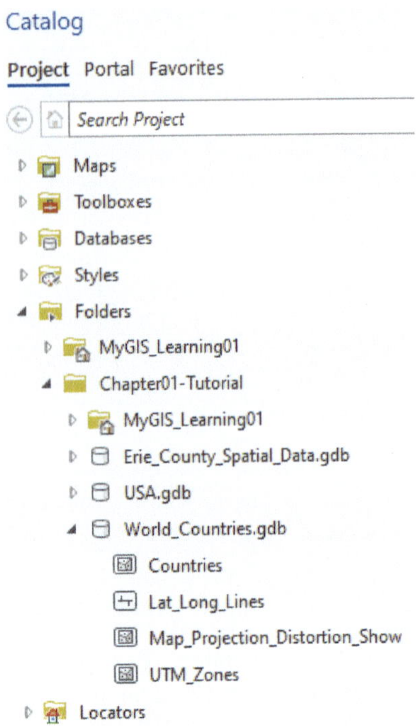


- Generally speaking, one map layer in a mapping frame window represents one feature class in a geodatabase. GIS Analysts use the terms "map layer" or "feature class" interchangeably.

The icon of a map layer also represents its contents. A polygon icon Countries indicates the dataset contains polygon features. A line icon Lat_Long_Lines indicates it contains the lines or arcs. A point icon Cities shows that it contains point features.

We can either right-click on the icon of a map layer to activate the contact menu then click on "Add to Current Map" like below, or just click and drag the map layer or feature class to the "Map" window.

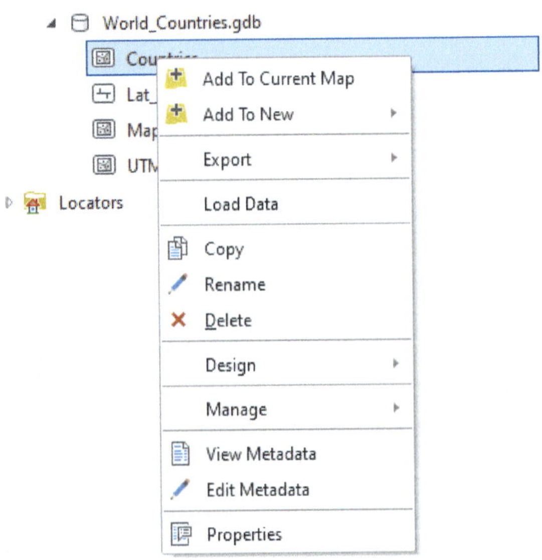

You may notice the map layer is added. In addition, add the "Lat_Long_Lines" in the map composition ("Map" window) as well. Each of the map layers in the Table of Contents on the left-hand side can be clicked to check-on or check-off. The GIS Analyst has the control to show any of map layers in the "Map" window from the table of contents.

The general rule of the arrangement of map layers in the Table of Contents is that the map layer (feature class) whose title is listed on the top, is presented in the front in the "Map" window in the center of the software platform. Whichever map layer on the bottom of the Table of Contents is, shown in the back of the window. The ranking of the title of the map layers in the Table of the Contents reflects the map layer overlay from the front to the back. For each of the map layers or feature classes, we can use the "Symbology" menu to open the Symbology Pane for coloring and symbolizing. Now, let's right-click on the title of "Countries" in the Table of Content to obtain the contact menu. Then, click on Symbology on the contact menu. The Symbology Pane opens on the right-hand side on top of the Catalog Pane.

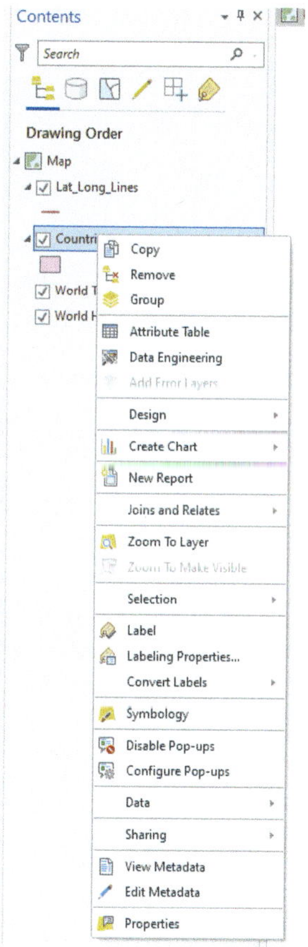


Right now, we are at the single symbol mode. Change the drop-down name list to "Unique Values." For "Field 1" select the "Name" (Name of countries). Answer the pop-up question "Yes." Then, we get a colorful map of world countries by their names. Since we have more than 100 countries worldwide, the color of countries could repeat. In this case, it is fine. The digital map should look like the following. You can click to practice checking on or off the "Countries" map layer as well.

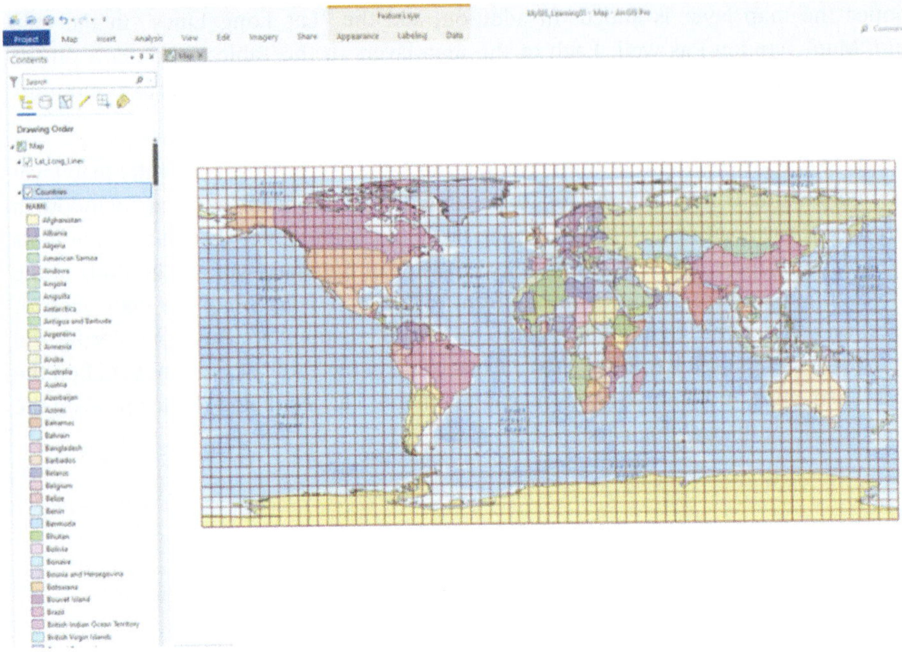

Now, we can practice clicking and dragging the "Countries" map layer on to the top of the list in the Table of Contents. This can only be accomplished when the "Drawing Order" is activated on the Table of Contents ···. The result should look like the following. This time, the countries map is on the front.

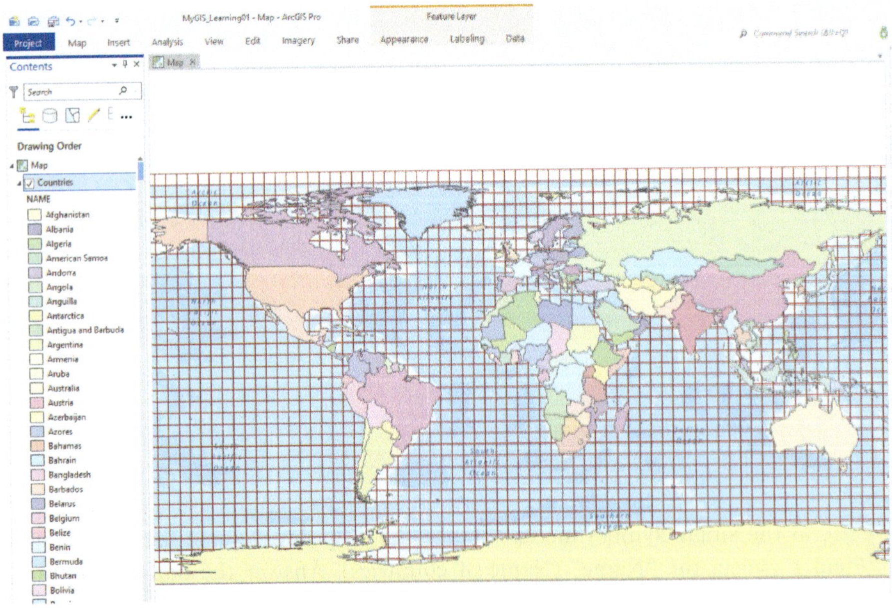


We can conduct a very simple approach for labeling the countries by their names on this digital map. First, highlight and right-click on the "Countries" map layer title in the Table of Contents to obtain the contact menu.

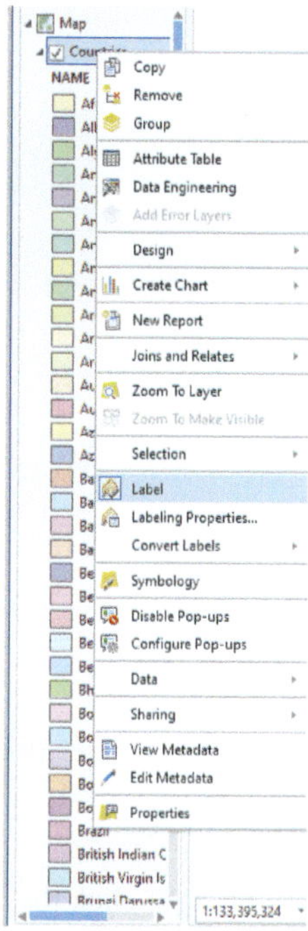


Click on "Labeling Properties" first to check the properties. Make sure the dialog has "$feature.NAME" in the expression window. If the expression is not this, double-click the "NAME" on the list of "Fields" to get it. "Fields" is a list of applicable attributes (columns) in the attribute table of the "Countries" map layer or feature class.

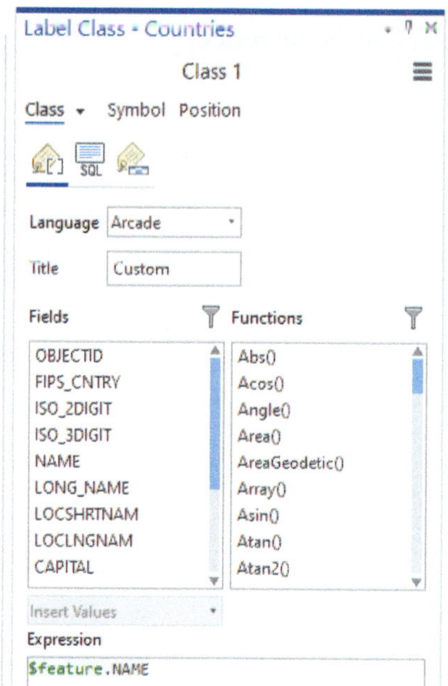


Now, activate the "Symbol" category, select the size of symbols as 8 points (pt) on this Label Class dialog and apply. Close this dialog.

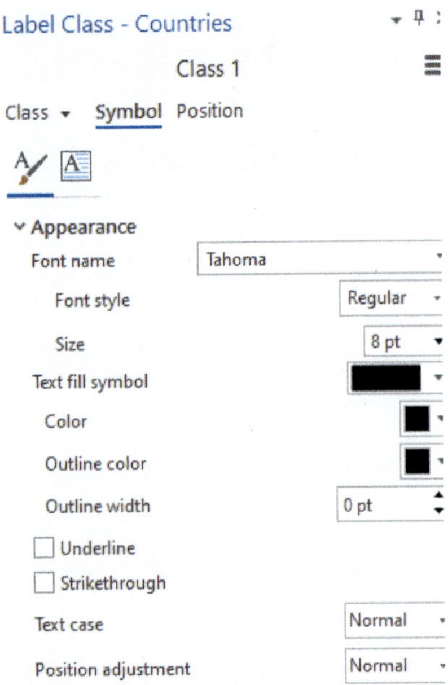


Now, we can activate the "Labeling" menu on the top to check the "Label" icon or right-click the map layer to check it.

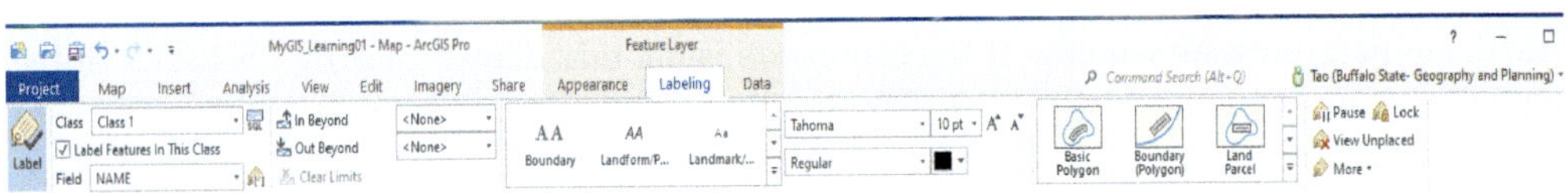

The digital map window should look like this now with labels of all the countries.

Moving to the right-hand side of the ArcGIS Pro window, we can explore the database management tool, Catalog Pane now. We normally have a total of six categories as showing in follows when the "Project" theme is activated on the top.

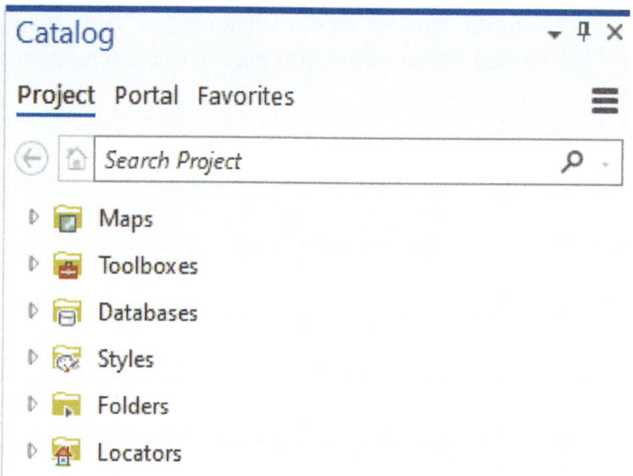


"Maps" is a category to manage windows of "Map," "Local Scene," "Global Scene" for spatial data visualizations. We can create new mapping frames and new ArcGIS scene frames using this icon (right-click). All the mapping and ArcGIS Scene frames under this project or saved in this project file are listed here, no matter whether or not it is active and shown in the big window in the center. "Toolboxes" is a category that handles the local toolboxes. You can create a new toolbox, and then create new tools or new models in the newly created toolboxes. You can also add the toolboxes on your list to be available to use. Always, when you create a project file, you automatically create one toolbox that is named the same as your project title. "Databases" is a category that you can use to create a new geodatabase, such as file geodatabase, mobile geodatabase, or add all these types of geodatabases as well. You can also connect to many other types of computational databases to explore and analyze as a geodatabase. Again, when you created a new project file in ArcGIS Pro, you always create a new geodatabase with the same name and automatically put it as the default geodatabase. "Styles" category allows you to create new style of map symbols or colors. You can add these items from your local collections or import some styles from other software environments.

We can utilize the "Folders" category to connect ("Add folder connections") various data folders on the computational network that we use. This will make the data in that folder available for ArcGIS Pro to use. Meanwhile, when we create a new project file, we also automatically create a new folder of the same name with the newly created geodatabase in it. The folder is recognized as the default folder automatically in ArcGIS Pro. We can also create a new folder under the folders that are connected using this icon. Finally, the "Locators" category is the category to provide Geographic locations of your geographic features stored in the databases. We can either use the ESRI company supplied locator file worldwide, or XY coordinate system worldwide to convert the tabular database to digital maps in ArcGIS Pro. Alternatively, we can create a locator file ourselves using a local base-map for converting the tabular database (such as Excel) to a digital map layer or feature class.

Now, let's expand the "Maps" category. The mapping frame "Map" that we are using now is on the list in this category. We can change the title of this mapping frame from default to a new name called "World Map." Double-click the name of it in the Catalog Pane, and we will be able to change the text. You may notice the title name on the mapping window is changed as well to follow this change. Now, right-click on the "Maps" to activate the contact menu. Click on the "New Map" to create a new mapping frame. Rename it as "United States Map." You may notice the mapping window has automatically switched to the new mapping frame.

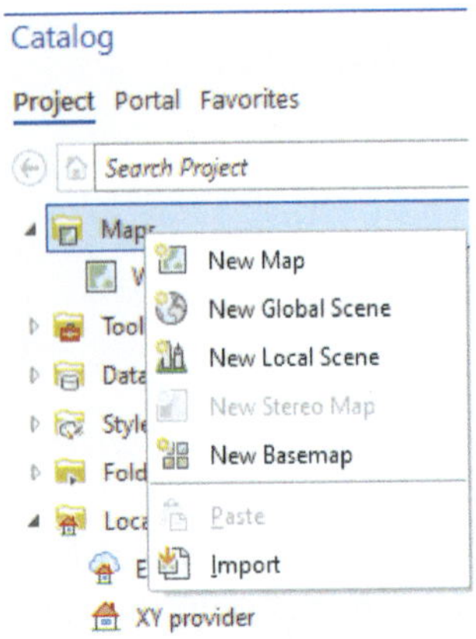


Now, we can add the spatial data from geodatabases in order to visualize it. Collapse the "Map" category. Expand the "Folders" category, then "Chapter01-Tutorial." Let's explore the USA.gdb geodatabase in this folder now. Expand the USA.gdb, and there are three geodatabase feature datasets in this geodatabase. In this case, the feature classes or map layers are stored in each of these feature datasets. Feature datasets are very good organizers that can store the map layers or feature classes with the same general theme, such as "transportation" or "natural features" together.

In the "Transportation" feature-dataset, drag and drop the "Interstate_Highway" to the mapping frame. In the "Political_Subdivisions" feature-dataset, drag and drop "Counties" to the mapping frame. In the "Natural_Features" feature-dataset, drag and drop both "Lakes" and "Rivers" to the mapping window. Since each of the feature classes is shown by randomly assigned colors, we need to adjust or change the colors and symbols using Symbology Pane as the following:

1. Lakes: Ultra blue (dark blue).
2. River: Cretan blue and 2-points line width.
3. Interstate highway: Mars red and 2-points line width.
4. Counties: Tzavorite green (in filling of polygons).

The result should look like the following.

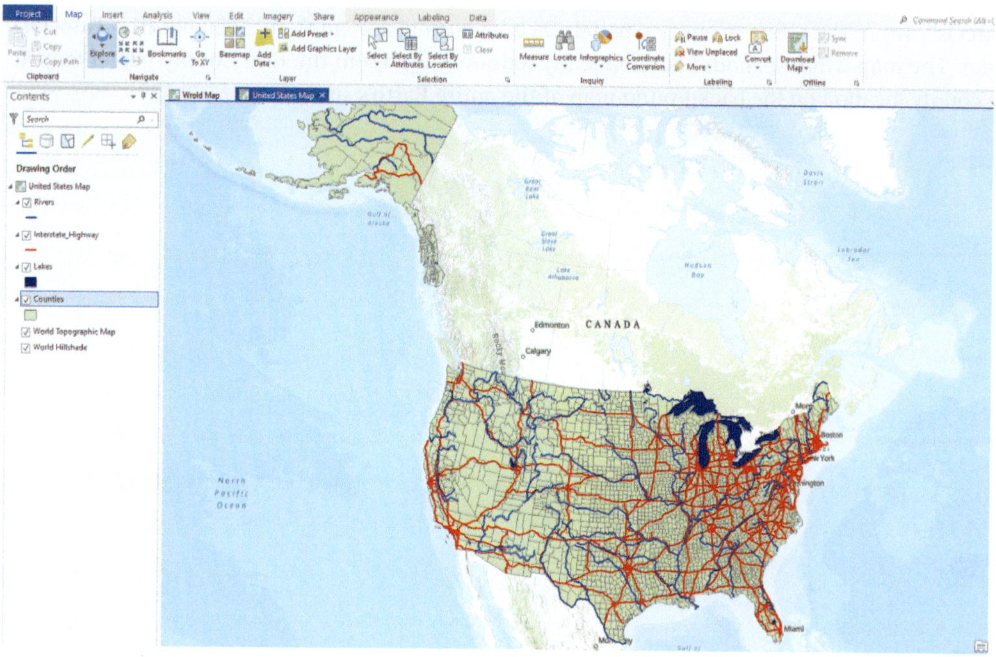


Up to this stage, we have created two mapping frames (Maps) in this project. Let us practice saving the project file. In essence, ArcGIS Pro project file is an organizer type of file. When you save it, you will save all the settings, symbols, colors, and layouts of maps. However, this process does not save or change any original datasets in the geodatabase that we use. When you open a project after it is saved, you will restore all the settings and layouts that you updated most recently. We can go to the "Project" menu on the top of the software window to use either "Save" or "Save As" to open a dialog window to navigate to a location we desire to on the computational network to save our project. There are differences in detail between these two routes in this APP compared to that in the MS-Word environment. If you choose "Save As," the file name will be empty, and it will expect you to type in a new name. Otherwise, if you would like to just save the project and update it to the current status, just use "Save." Click on Project -> Save to save this project. If you would like to take a break, you can always save and close the project, and exit the software platform. When you are ready to resume, you can open the project again with ArcGIS Pro.

If the project "MyGIS_Learning01" is closed, open ArcGIS Pro software and the project. Otherwise, continue. Our third task is to create another mapping window and present some local geographic or spatial data for Erie County, New York. Erie County is one of the counties in the New York State. We would like to use Erie County spatial data to show the application needs of GIS by local government agencies. First, we create another (new) mapping frame at the "Maps" category on the Catalog Pane and call it "Erie County Map." Then, we add the map layer or feature class of "Parcels" from the "Erie_County_Spatial_Data" geodatabase(.gdb) in "Chapter01- Tutorial" folder. The map window should automatically redraw and present the Erie County parcel map at the county location on the background map of the entire Earth's surface.

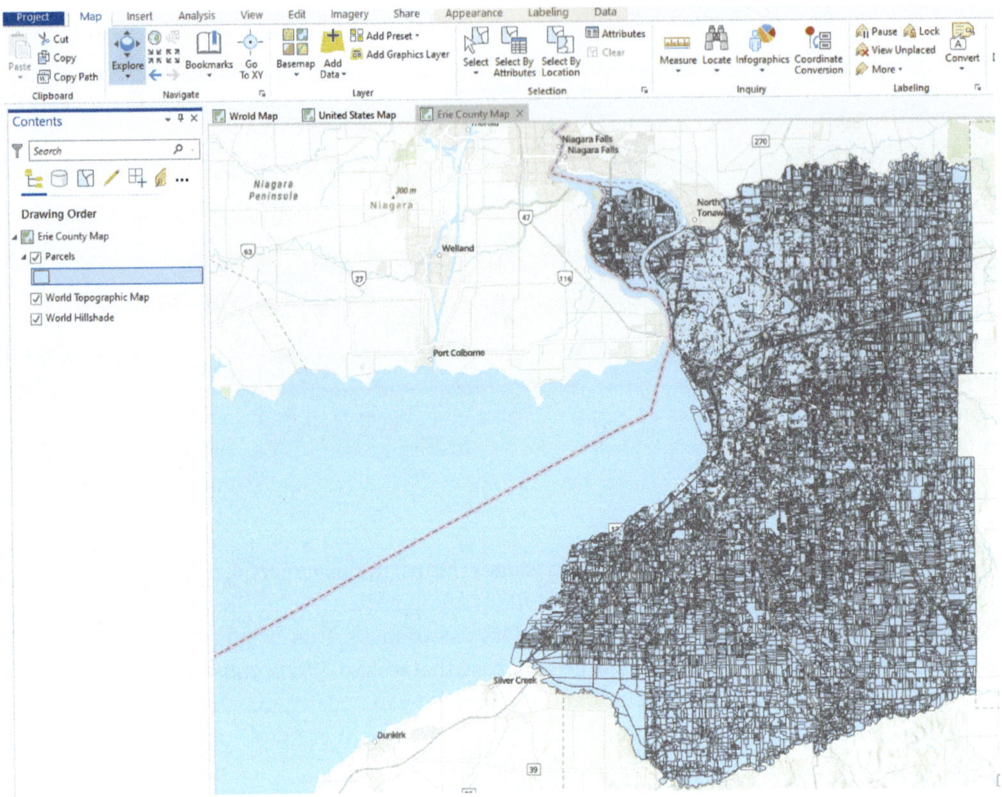


A parcel of land is an unambiguously defined unit of land within which a bundle of rights and interests are legally recognized in a community. A parcel encloses a contiguous area of land for which location and boundaries are known, described, and maintained, and for which there is a history of defined, legally recognized interests (National Academies of Sciences, Engineering, and Medicine 2007). Geographically, a land parcel is the smallest subdivision of land or regions in the terms of society. In legal terms, land parcels indicate parcels of land located in the United States wholly owned by a Borrower or any Borrowing Base Subsidiary that are unencumbered by any Lien or Liens.

Land parcels have the following characteristics:

1. The value: It is the worth of the land area and the properties or buildings on the land. The value can be the selling price on the market or the assessed value by local government. In practice, the local governments of counties, cities, or towns collect property tax from the owners of land parcel in the United States as their major revenue.
2. The ownership: It is the information that describes who owns or manages the land. The ownership records and detailed map or drawing of the land parcels are stored registers of deeds offices in county or local government.

3. Land parcels are subject to land use and zoning regulations established by a local government agency.
4. A land parcel has an address in the country. The addresses can be further identified as site address or mailing address. Although these two addresses are the same in most cases, they can be different ones. The address is unique that represents the relative location of the land parcel unit on the Earth's surface. In essence, land parcel is a geographic term. Therefore, managing the land parcels is one of the major GIS applications of local government in the United States. Land parcels and their property conditions are the base of property taxation that is maintained by county or local governments.

Now, we would like to convert one old spatial data file that is called "shapefile" to a feature class in the geodatabase. Then, we will load the feature class or map layer into the "Erie County Map" window. Expand the "Chapter01- Tutorial" folder as needed. We notice that one individual map layer, "Erie_Streets.shp" is listed outside of geodatabases. This is a shapefile, a file type that many professional GIS organizations still use. Historically speaking, the default data file development on the ArcGIS software platform went through three stages since its beginning. The first-generation data file format is called ArcInfo coverages. One dataset of ArcInfo coverage can host several related map layers. ArcInfo software was originally designed to run on the mainframe computers in the 1970s and 1980s since the mainframe computers were the most popular computational tools then. In the late 1980s to 1990s, personal computers were getting more and more popular for data processing. ESRI company developed PC-ArcInfo first to cope with changes in the computational environment. Then, the company developed the ArcView GIS software running in the personal computing (PC) environment to view and present the spatial data as well as conducting data analysis. The shapefile file format was created as the second-generation data file in the ArcView GIS environment. ArcGIS was then developed in the early 2000s building upon these previous software and database environment. To this date, professionals practicing GIS still need to convert or use these old GIS data file formats, such as coverages or shapefiles. We strongly encourage GIS practitioners to convert these old data file format to geodatabase format in order to explore the full functionalities of data analysis and geoprocessing in ArcGIS Pro.

To practice the steps to access ArcGIS Pro Toolboxes and to review various toolboxes and tools in processing and analyzing data, we click on the "Analysis" menu on the top of software APP. In the "Analysis" ribbon, we click on the "Tools" icon Tools to open the Geoprocessing Pane.

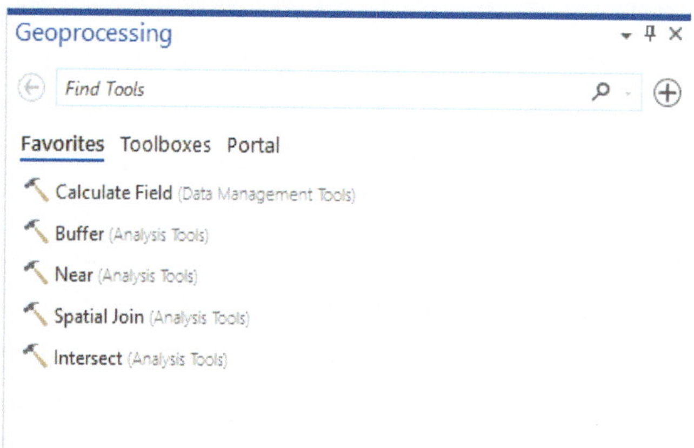

You may notice the Geoprocessing Pane is overlaid on top of the Catalog Pane. We can switch the rank at any time. The Geoprocessing Pane is a module in ArcGIS Pro to support geographic data analyses and data processing, while the Catalog Pane is a module to manage and maintain the geographic or spatial databases in the computational systems and computer storage. Now, let's activate the "Toolboxes" category on the top. You will see the entire list of toolboxes.

On the list of toolboxes, expand the "Conversion Tools" box, then "To Geodatabase" toolset.

Click and select the "Feature Class to Feature Class" tool like the following.

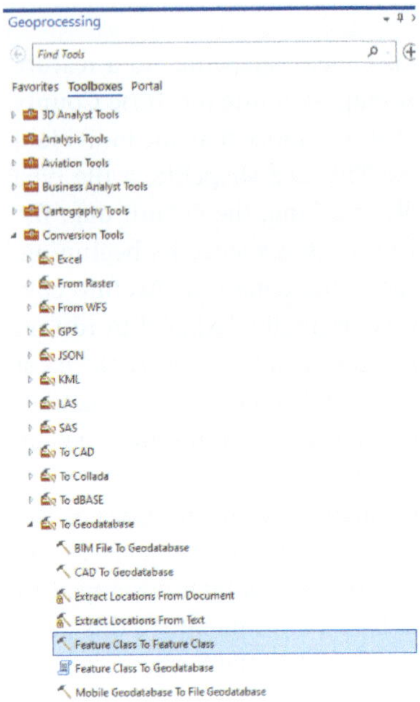


You may notice that the Geoprocessing Pane is changed to the dialog of this tool. If you accidentally clicked a different tool, you could always click the back arrow to return to the previous dialog.

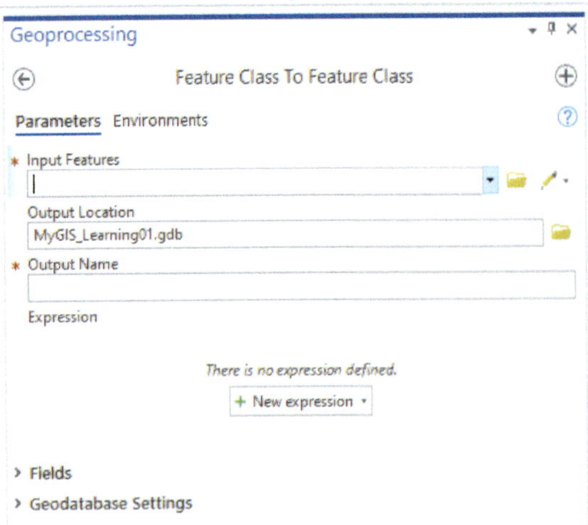

We need to fill in all the criteria of input and output files in order to click "Run" to run the tool and convert the file. For "Input Features," navigate to your "Chapter01- Tutorial" folder and select the "Erie_Streets.shp" file. For "Output Location," select Erie_County_Spatial_Data.gdb in the same folder. For "Output Name," type in "Streets." Then, run this tool to convert the shapefile to a geodatabase feature class.

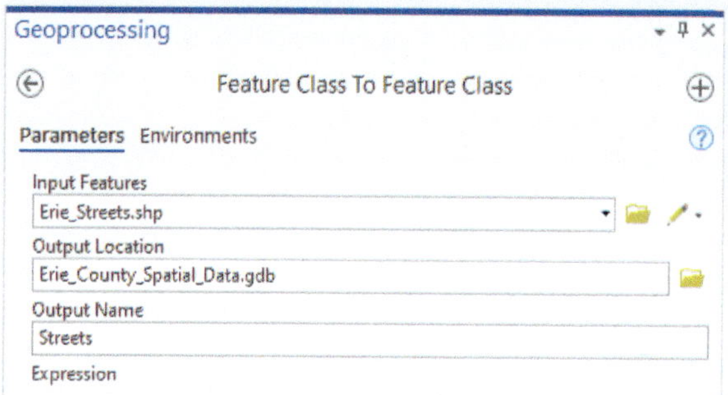


When the process is finished, you can close the Geoprocessing Pane.

Now, we do have a "Streets" map layer or feature class in Erie_County_Spatial_Data geodatabase. Add the data into the mapping window.

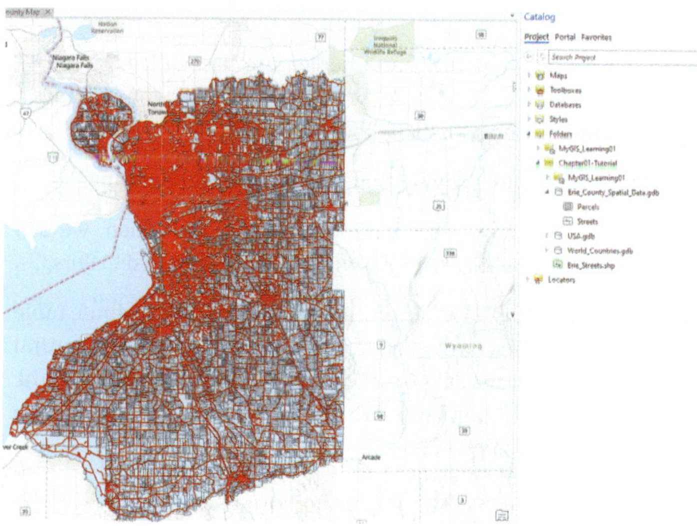


Alternatively, we can also right-click on the Erie_County_Spatial_Data.gdb to get the contact menu. Then, select "Import" and "Feature Class" to open the same tool.

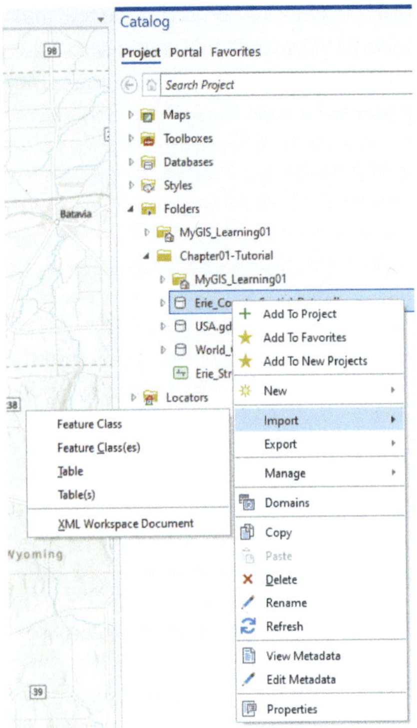


You can practice this process again to convert the shapefile to another feature class called "Street01" in the same geodatabase. In ArcGIS Pro software environment, we can always try right-clicking to obtain a contact menu. We can also accomplish one task using different routes.

This is really a learning environment like the popular saying that "All Roads Lead to Rome."

We can use the Symbology Pane to change the color of "Streets" to "Mars Red." Then, we can change the parcel polygon symbology method from the "Single Symbol" to "Graduated Colors."

In the "Field" drop-down list, select "COTAX." "COTAX" is a column or field in the attribute table that comes with the digital map layer. It means the county tax. Keep the "Method" as the "Natural Breaks (Jenks)" and change the "Classes" to 6. For the "Color scheme," select cyan to purple on the drop-down list. Change the symbols of the Parcels map layer. We can zoom into a specific area on the digital map. It looks like this. Close the Symbology Pane.

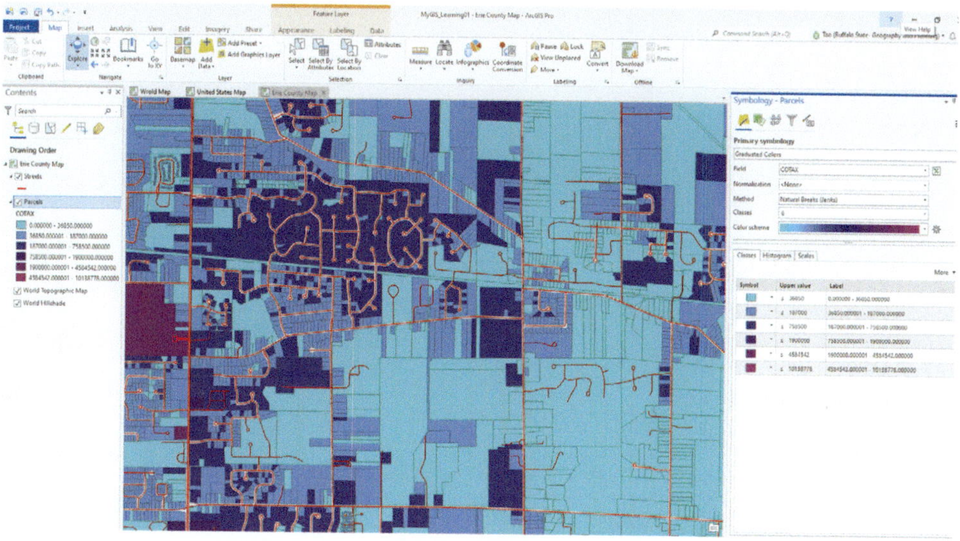

Now, let us open the attribute table of the "Parcels" feature class on this map.

Right-click on the "Parcels" in the Table of Contents to obtain the contact menu. Then, click on "Attribute Table" icon to open the attribute table. Now, we have both features of parcels are presented on the digital map and the attributes of each feature in the "Parcels" feature class are presented in the table. It should look like the following.

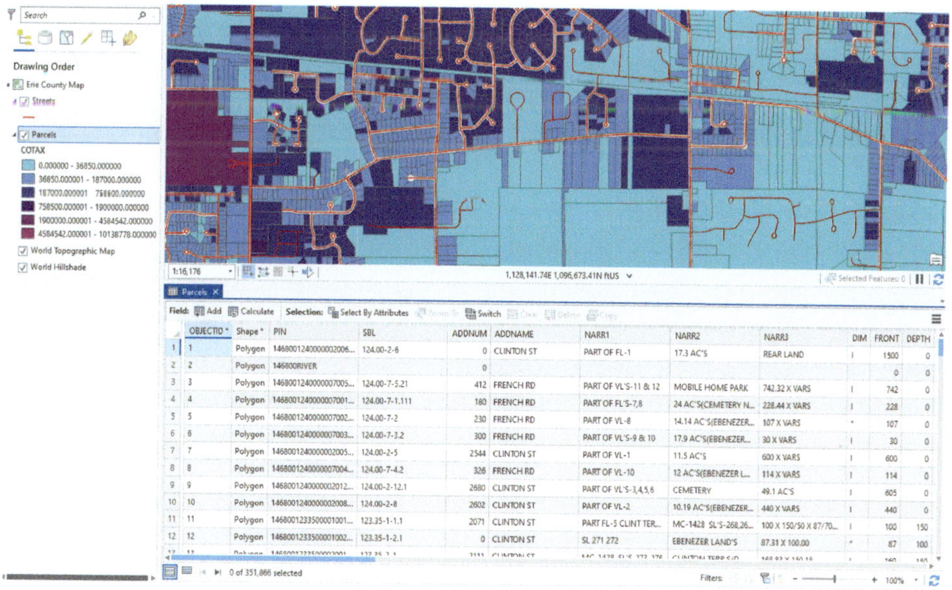


Save your project file as "MyGIS_Learning01" using the "Project" tab (ribbon) then "Save" on the main menu.

In summary, we created a project file (.aprx) first. Then, we explored three different geodatabases in three different mapping frames. We also explored the "Analysis" ribbon, "Geoprocessing Pane," and the Toolboxes on the geoprocessing pane.

2

# Digital Data Models Representing Geographic Information

## Concepts and Theories

Digital representation of geographic features or events in the computational environment, without considering the time issue, currently can be classified into three different formats or models. These are vector models, raster models, and three-dimensional (3D) models. The first two types are originated from two-dimensional (2D) geographic space and mainly represent geographic features in a 2D digital environment on a map. The third type of representation inherits characteristics of both vector and raster data illustrations while the main objective is to illustrate geographic features and events on the Earth's surface with their original 3D properties in the digital environment. In essence, the digital data formats or models are the methods of representing real-world geographic (spatial) features in GIS.

Geographic phenomena by nature can be studied or analyzed either as a set of features, objects, or events occupying the locations on the Earth's surface. This approach is called the object-based or vector type of data illustrations. Alternatively, geographic phenomena can be studied or analyzed as a group of fields, or small (micro) regions called grid cells. This approach is called a field-based or raster type of data interpretation. The first data illustration method is to utilize conceptualized features to populate the geographic space on the Earth's surface. The latter illustration is to decompose geographic space into small grid cells or fields. Maps either analog or digital are in essence the abstract representations of a geographic area and the geographic features in that area.

A core part of any GIS platform is data management and compilation. ArcGIS Pro uses the geodatabase to manage geographic data. The geodatabase stores spatial and nonspatial data and organizes them into datasets. Examples of the types of datasets that might be maintained in a geodatabase include tables, feature classes, raster grids, relationship classes, topologies, and utility networks. An enterprise geodatabase stores data in a relational database and leverages its capabilities to support storing large datasets as well as provide multiuser access to data efficiently. It also allows users to share data as services at the designated portal or online.

## Vector Data Format

The concept of vector data is to illustrate geographic features on the Earth's surface as objects in the digital environment. Then, these geographic objects are further classified theoretically into three geometries or shapes. These are points, lines, and areas or polygons. Each of the objects or geographic features is individually stored in the computational system with its attribute data or characteristics. In this method, the computational system does not deal with geographic areas

*Fundamentals of Geographic Information Systems (GIS) Analysis: Applying ArcGIS Pro*, First Edition. Tao Tang.

or regions. Rather it deals with these objects or geographic features that are connected to their attribute dataset. Each feature is separately geo-referenced with its location on the Earth's surface in the computational database. Therefore, these types of geographic or spatial feature illustrations are also called object-based data structures or object-oriented data models. The so-called object-based data structure means the computational database is centered with geographic objects or features and their attributes, one object is one database item.

Vector databases in modern computational systems are also designed in the topological structured format. It utilizes the starting and ending points of a line with registered locations both on the digital map and in the real world in building lines in spatial database. Then, the lines are used to build polygons or enclosed areas in spatial database. This approach of structure building increases the connectivity of different features and enhances the effectiveness of tracing and querying the features on a complex digital map or in a spatial database.

## Points

A point is a spatial object with no dimension. Points are located with x, y coordinates on a digital map or x, y, z coordinates in 3D geographic space. In other words, a point in GIS must have its location property registered on the Earth's surface. This location property is represented by one of the absolute coordinate systems that we will discuss later. Examples of absolute coordinate systems are the Geographic Coordinate System (latitude and longitude), the Universal Transverse Mercator Coordinate System, the State Plane Coordinate System, etc. In ArcGIS Pro, the points are also called nodes or vertices. Since ArcGIS Pro vector feature class of geodatabase utilizes points to build a line feature, the end points of one line are called nodes, and the points in the middle of a line are called vertices. Many of the real-world features can be illustrated in points in GIS, such as traffic lights on a street, water wells, fire hydrants, or individual trees (Figure 2.1).

**Figure 2.1** Digital map of point features of fire hydrants in Buffalo, NY with streets in the background (in black).

## Lines

A line is a spatial object with one dimension. A line carries the property of length and location and is elongated from one point to another point. The length can be measured in real-world values in meters, kilometers, feet, or miles. Lines are also called polylines or arcs in ArcGIS or ArcGIS Pro software environment. From a database storage and management point of view, a real-world line, such as a street, is stored as connected line segments. Each of the line segments is one record in the database. Examples of real-world features that can be illustrated into lines, polylines or arcs are rivers, streets, highways, water pipelines, oil pipelines, or high-wattage electricity transportation lines (Figure 2.2).

## Polygons or Areas

A polygon or area is a spatial object with two dimensions. A polygon is bounded by the connected lines to form an enclosed region on a digital map. Polygon carries the property of area, perimeter, and size. The area of a polygon can be measured in square meters, square kilometers, square feet,

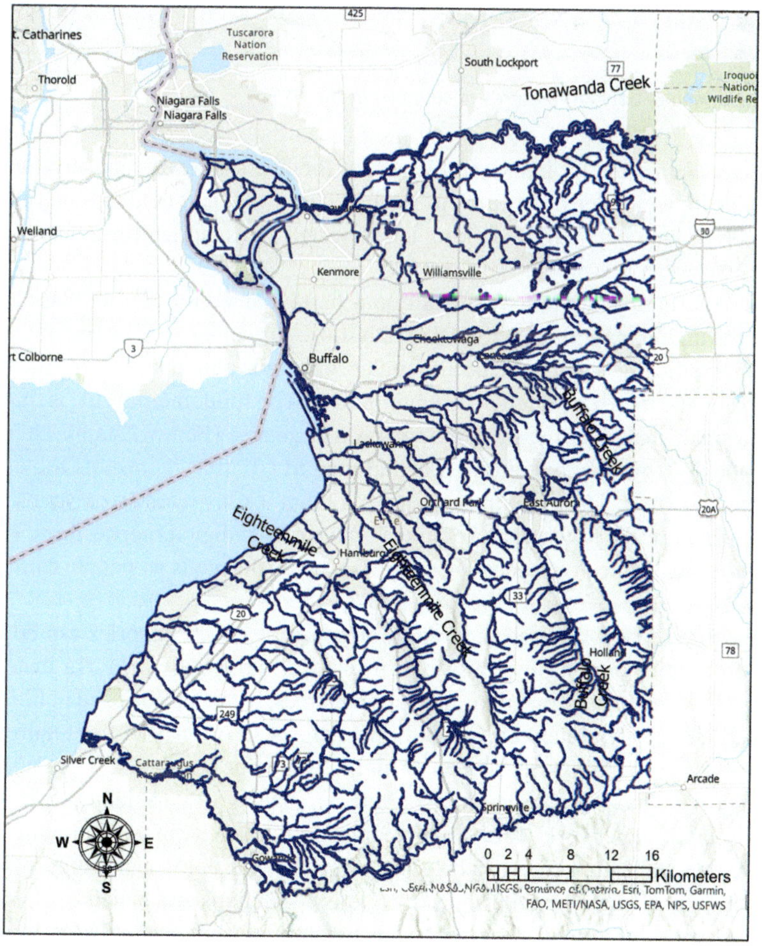


**Figure 2.2** Digital map of line (polyline) features of river line segments in Erie County, NY.

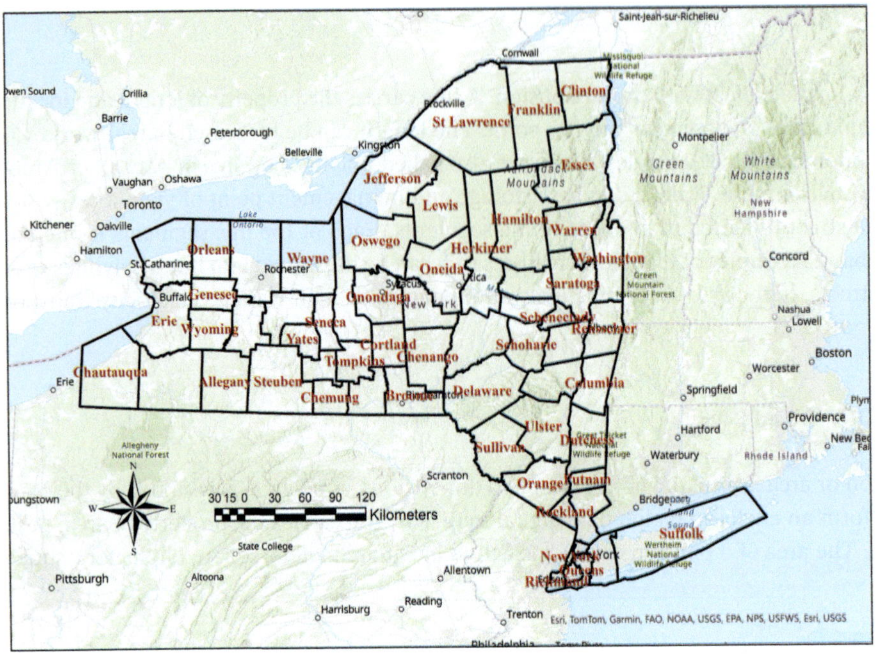


**Figure 2.3** Digital map of polygon features of counties in the NYS.

square miles, etc. The perimeter is the total length of the line segments that serve as the boundaries of the polygon. Examples of features in the real world that can be illustrated into polygons are lakes, parcels of land ownership, parking lots, forested areas, agricultural farms, counties, or states (Figure 2.3).

Illustrations of certain geographic features into a particular vector feature class type are also influenced by the digital map scale that a GIS analyst uses or constructs. In general, digital maps follow the cartographic rules on map scale although people can zoom in to enlarge the map scale or zoom out to shrink the map scale in the GIS environment. The fundamental rule is that large-scale maps portray a small area. Small-scale maps portray a large area (Figure 2.4a,b). Therefore, a city or a town can be illustrated as a point on a digital map in GIS that represents the entire United States with a small map scale. By contrast, a city or a town is illustrated as polygons on a county map with a large map scale. Rivers may be shown on a national digital map as lines. However, the same rivers can be illustrated on a local or regional map as polygons so people can measure the drainage areas.

Each of these three conceptualized geographic objects or feature categories can be used to represent physically existing or non-physically existing objects in the real world. For instance, a point feature class can be used to illustrate naturally growing trees. It can also be used to illustrate occurrences of infectious diseases, such as points of infections of COVID-19 virus (Figure 2.5). A line feature class can be used to illustrate streets or highways by line segments. It can also be used to illustrate a network of goods and signal transportations, such as a TV cable network, high voltage electricity transportation network, water pipeline transportation networks, or aircraft flying routes of an airline. A polygon feature class can be used to illustrate the political or administrative subdivisions. It can also be used to illustrate areas or regions of religions and beliefs.

In vector GIS data models, we recognize the topological characteristics of the geographic objects or features. Topology is one of the subdisciplines of mathematics. Topology studies the

(a)

(b)

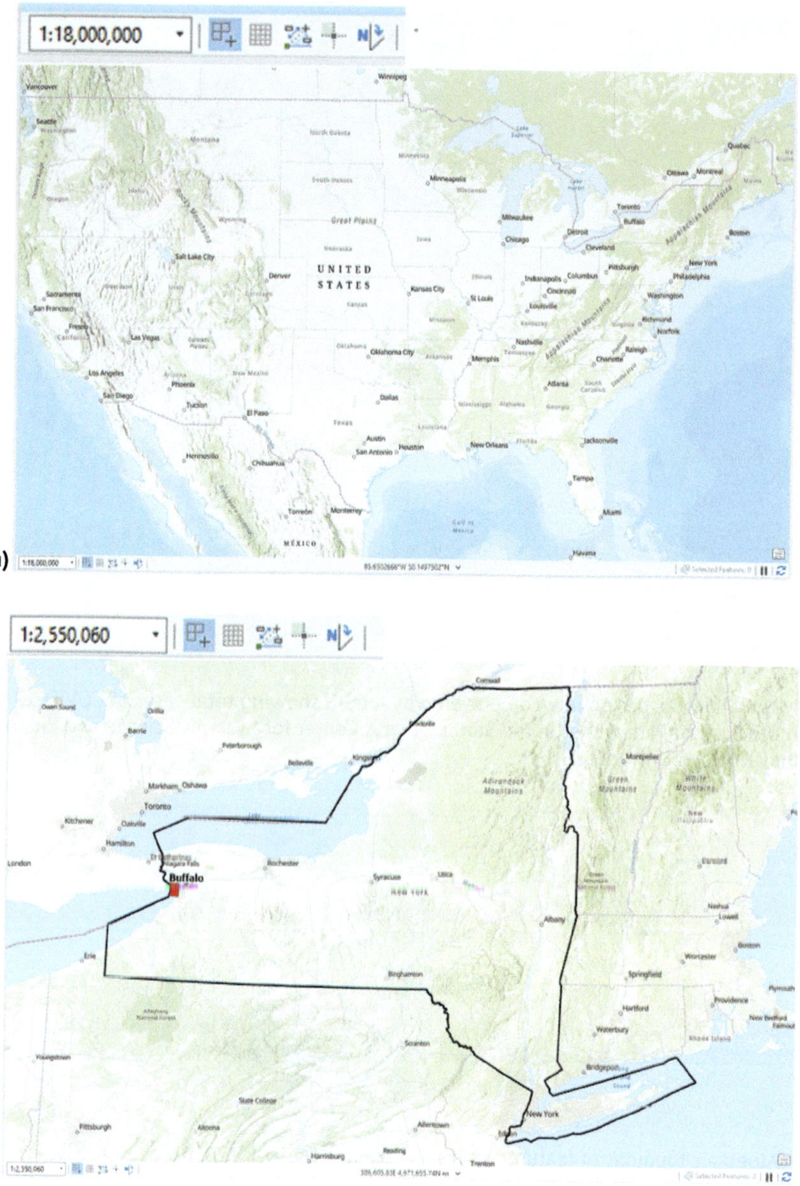


**Figure 2.4** (a) Small scale map portrays large area – entire United States. (b) Large-scale map portrays small area, NYS (the City of Buffalo can be illustrated in a polygon at this map scale).

properties of a geometric object that are preserved under continuous deformations. It is sometimes called "rubber-sheet geometry" because the objects can be stretched and contracted like rubber (Figure 2.6). One of the major principles in topology is that the geometric relationships of the objects being studied never change if the study space is changed from four-dimensional (4D) to 3D or 2D. In topology, the objects cannot be broken in the study space (University of Waterloo 2020). In GIS, we mainly focus on the rules of topological relationships of geographic objects or features in geographic space.

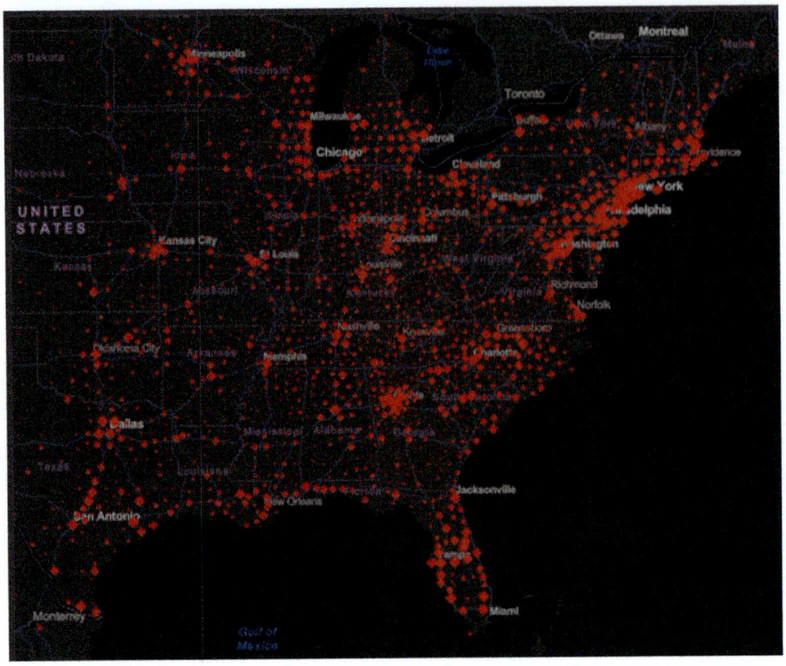


**Figure 2.5** Digital dashboard online map powered by ArcGIS showing total cases of COVID-19 virus infections in the Eastern Part of the United States. *Source:* Center for Systems Science and Engineering (CSSE) at Johns Hopkins University.

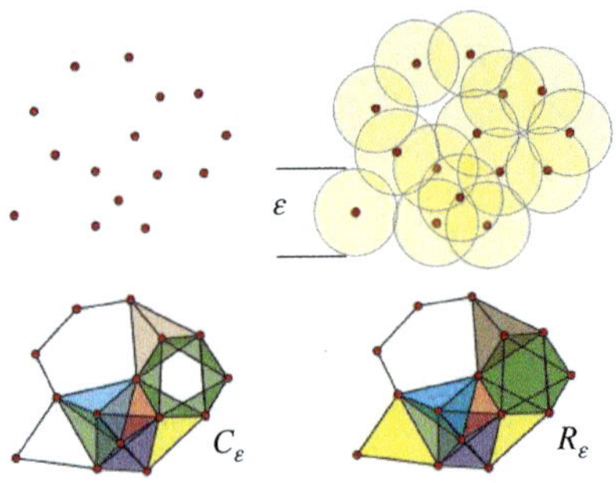


**Figure 2.6** Algebraic topology of features on graphs. *Source:* Ghrist R., 2016/Robert Ghrist.

In vector GIS systems, the topology is mainly identified into three applicable categories. These are connectivity (Figure 2.7), adjacency (Figure 2.8), and containment (Figure 2.9). Connectivity is the way in which geographic features in GIS data are attached or linked to one another functionally or spatially or both. Features in the same feature class are normally interpreted with connectivity both functionally and spatially, such as a water or oil pipeline network. Features in different feature classes are mainly ensured by the spatial connectivity of geometric locations on a digital map. Adjacency is the topological relationship in which two or more features, in particular the polygons, share a side or boundary. Closely related to adjacency, proximity is also a concept that identifies the neighborhood of spatial features without attachment to each other. Containment is a topological relationship in which a point, line, or polygon feature or set of features is enclosed completely within a polygon or a set of polygons.

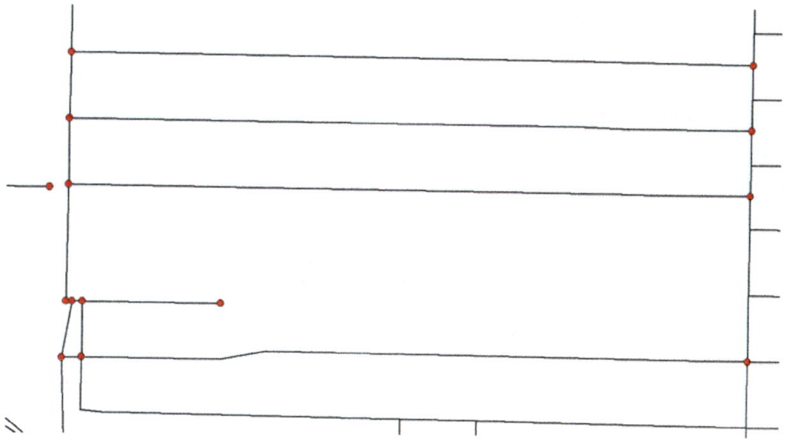

**Figure 2.7.** Digitized digital street segment map showing the connectivity of streets in Buffalo, NY.

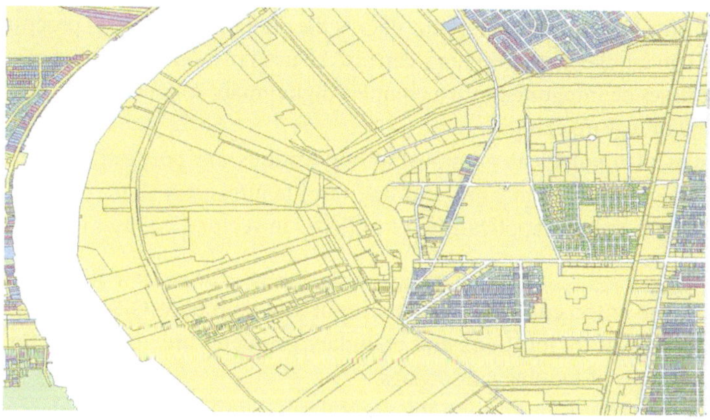

**Figure 2.8** Parcel polygons in the Town of Tonawanda, NY showing the land ownership adjacency.

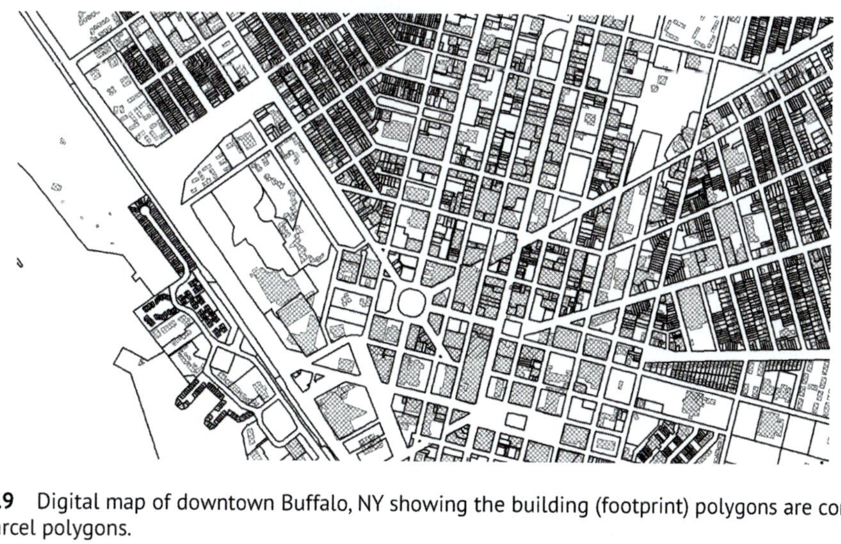

**Figure 2.9** Digital map of downtown Buffalo, NY showing the building (footprint) polygons are contained in the parcel polygons.

Fundamentally, topological relationships and rules in vector GIS datasets enable us to more accurately model the geometries of the geographic features and to precisely pinpoint their locations on a digital map. Topology is implemented through a set of rules in GIS that define how features may share a geographic space. ArcGIS Pro created a set of editing tools that work with features that share geometry in an integrated fashion. Topological rules are stored in an ArcGIS Pro geodatabase as one or more relationships that define how the features in one or more feature classes share geometry. The features participating in a set of topological rules are still stored in simple feature classes, rather than modifying the definition of the feature class, topological rules in ArcGIS Pro serve as a description and regulation of how the features can be spatially related.

Topology has long been a key requirement for vector GIS data management and integrity. In general, a topological data model manages spatial relationships by representing spatial objects of point, line, and area features as an underlying graph of topological primitives—nodes, faces, and edges. In essence, topology is used to ensure the data quality of the spatial relationships and to aid in data compilation. Topology is also used for analyzing spatial relationships in many situations, for instance, dissolving the boundaries between adjacent polygons with the same attribute values or traversing a network of the elements in a topological graph. In particular, connectivity tracing of street line segments on a base map or reference data is the core of a geocoding process to create a new map layer based on tabular data. Topology can also be used to model how the geometry from a number of feature classes be integrated into a complex digital map. Some GIS analysts refer to this as vertical integration of feature classes.

Important examples of topological relationship applications are polygon (area) features can share boundaries (adjacency relationship), and line (arc) features can share endpoints (connectivity relationship). In general, the applications of topology in vector databases can be identified in the following three areas:

1. Topological rules are enforced for vector data input and editing in GIS. Errors are automatically corrected, and feature locations and boundary geometry are automatically changed to fulfill the requirements of topological settings.
2. Artificial intelligence (AI) is built upon topological relationships in large datasets from local regions to a state or a nation for tracing, matching, or detecting a specific or a group of specific geographic features. These tasks are accomplished by the establishment of topological rules and relations of connectivity, adjacency, and containment in the "smart" database otherwise would not be able to do.
3. Topology can help GIS practitioners to construct complex objects or features in vector data structure. It also enables GIS analysts to quantify precisely the shared geographic space by multiple features on the Earth's surface. Many practical operations in a GIS software environment, such as geocoding addresses, map overlay, and point or line in polygon analysis, are logically based on topological reasoning.

Topological principles of vector objects were first defined by the scientists and GIS professionals working at the US Census Bureau, and the Department of Commerce in the late 1970s (Corbett 1979). Based on the topological relations of the digital objects and features, the bureau developed the "Topologically Integrated Geographic Encoding and Referencing (TIGER)" vector data structure in the computational database. This is considered as a benchmark on contributions of vector GIS data development. This approach also sets the standards for modern vector GIS data structure design and implementation.

In the TIGER data structure, points are called 0-cells and are considered the lowest level of building blocks in the hierarchy in constructing of a complex digital map representation of

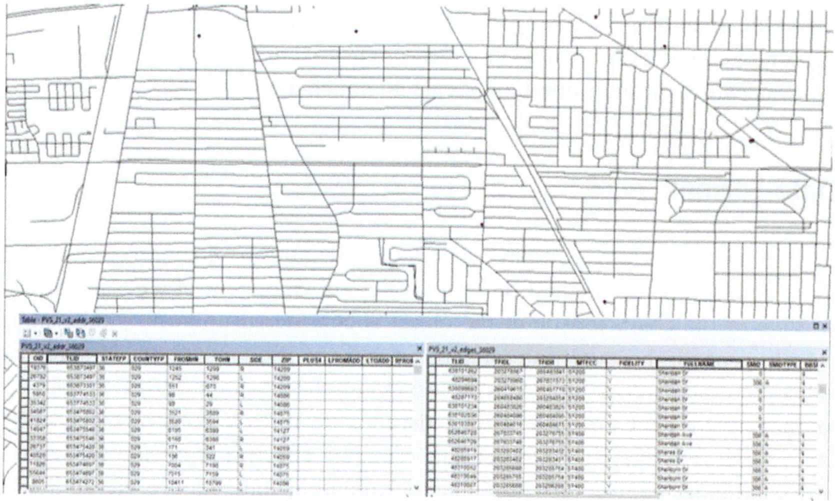

**Figure 2.10** Currently published TIGER line graph spatial database showing streets in Buffalo, NY by US Census. TLID stands for Tiger Line Segment ID; FROMHN stands for the house number from (starting node); and TOHN stands for house number to (ending node).

geographic features. Points are referenced and registered by x, y coordinates in a specific geo-reference system. Lines are called 1-cell and constructed with starting and ending nodes (points). Relative geo-reference or locational system, street address, is matched at the starting and ending points of a line parallel to the absolute coordinate system registered. Since lines are illustrated in this case with directions, the left-side or right-side polygons, such as counties or postal code regions, are also recognized and recorded in the attribute table of line feature datasets. Polygons in this illustration are called 2-cells, which are constructed by digitally connected lines. An example of TIGER line graph illustration is shown in Figure 2.10.

Since vector data structure is to utilize the conceptualized geographic objects and features to construct digital map layers in a GIS environment, the scale of a map layer and the objects that comprise the map layer is very important in this case. Map scale is a fundamental concept in cartography or map-making process. Map scale is the ratio of the distance measured on a map to that measured in the real world on the Earth's surface. The original map scale format is called a representative fraction, which is a fraction number without any unit. For instance, 1/1,000,000 map scale demonstrates that one unit measured on this map equals to one million of the same units on the ground at the Earth's surface. The significant rule is that we need to use the same distance unit to measure on the map and in the real world, no matter what type of unit we use—centimeters, inches, kilometers, or miles. The large number of real-world values in a small distance unit can then be converted to a suitable real-world distance unit later. In practice, two other map scale formats are also widely used on hard copy and digital maps. These are: (i) graphic scale or bar scale, and (ii) verbal statement scale. The most popularly used map scale on digital maps is the graphic scale.

In the GIS software environment in a mapping frame, we can zoom in and zoom out in the window. The map scale is constantly changing as you change the extent of the mapping area and presentation. The map scale always matches the current mapping extent and presentation details. Again, the general rule governing the map scale of geographic feature presentations is that the large-scale map portrays a small geographic area and the small-scale map portrays a large geographic area. Although geographic features and feature classes on a digital map in a mapping frame in ArcGIS Pro can be enlarged or shrunken, the original characteristics and properties of the features at their input map scale do not change.

## Raster Data Format

Raster data structure illustrates geography or geographic features based on the fundamental units of grid cells. A grid cell is the smallest unit for a value measurement of a geographic feature or event. The value of the characteristic of the object is considered uniform within one grid cell (Figure 2.11). In general, the raster interpretation treats the geographic area of study as a field. Then, it utilizes the standardized grid cells to decompose the geographic area on the Earth's surface. By contrast, vector data structure utilizes identified features to construct a geographic area on the Earth's surface. In this regard, raster data are more effective to illustrate geographic phenomena with continuous nature, such as rainfall precipitation, temperature, elevation or altitude, and soil texture.

Raster grid cells in GIS are arranged in rows and columns on a digital map. They are also called tessellations (Figure 2.11). Ideally, the grid cells can be in any geometric shape as long as they decompose the study area without gaps. In practice, there are mainly two types of grid cells either squares or rectangles. The square shape of grid cells is the most popular format in GIS and related scientific disciplines of studies. Each of the raster grid cells carries a unique value of the objects on a map layer. Grid cells are referenced to geographic locations on the Earth's surface using the same coordinate systems or geo-referencing systems that vector data use (Figure 2.11). Since one of the popular data sources of raster format is remotely sensed satellite or aerial images, raster grid cells are also often called pixels. Pixel is a term for one individual dot on an image or picture that carries a color value.

Figure 2.12 is a real-world example of a raster digital map and spatial database of agricultural cropland and developed land cover in the NYS. Each of the agricultural crop species and its distribution across the state is stored and presented on this digital map. The legend of the digital map is simplified from the original database varieties. The enlarged view that focuses on the Finger

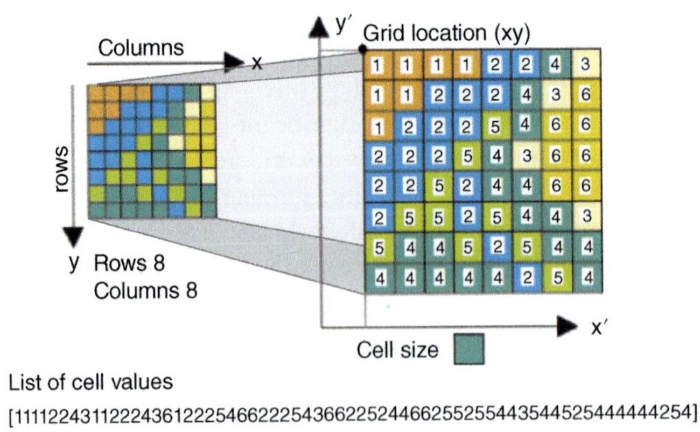


**Figure 2.11** Sample of details on a raster digital map showing the grid cells. *Source:* Adapted from ESRI, Inc.

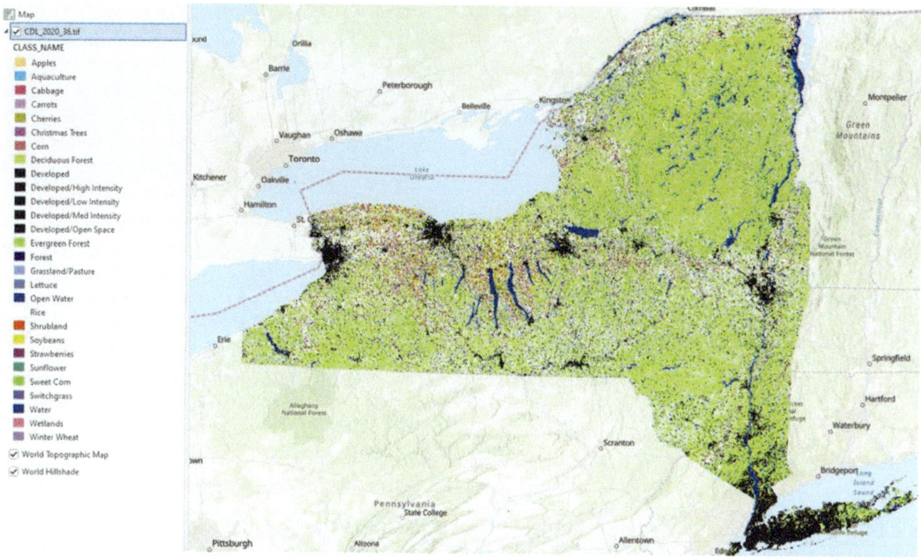


**Figure 2.12** Agricultural production land by crop species and the developed land across the NYS.

Lake region is shown in Figure 2.13. This raster database can help farming companies and US Department of Agriculture agencies to determine where a particular crop is produced currently. This raster spatial database can also be used to help locate new production fields if the production increase is required.

In mathematics, the levels of measurement for the attribute data can be identified into four categories. These are: (i) Nominal: the data can only be categorized; (ii) Ordinal: the data

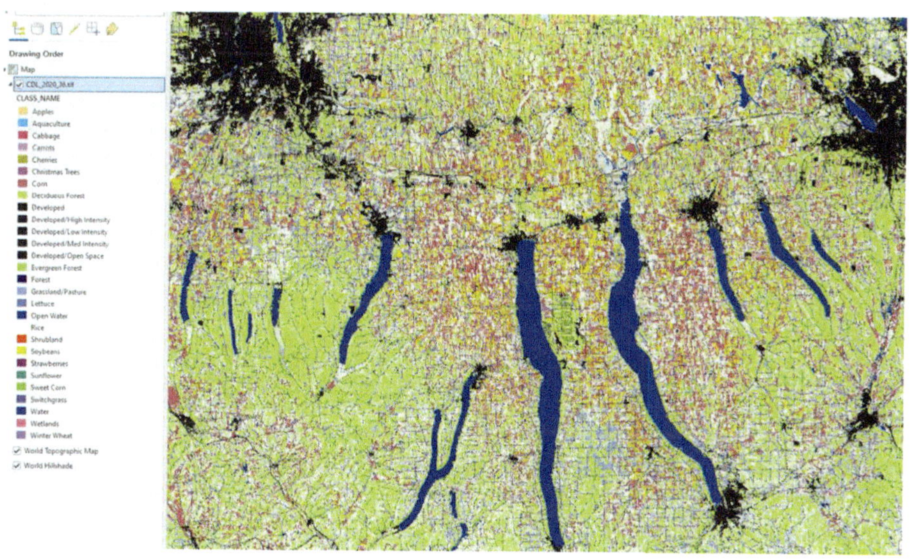


**Figure 2.13** Agricultural production land by crop species and the developed land in the Figure Lake Region in NYS.

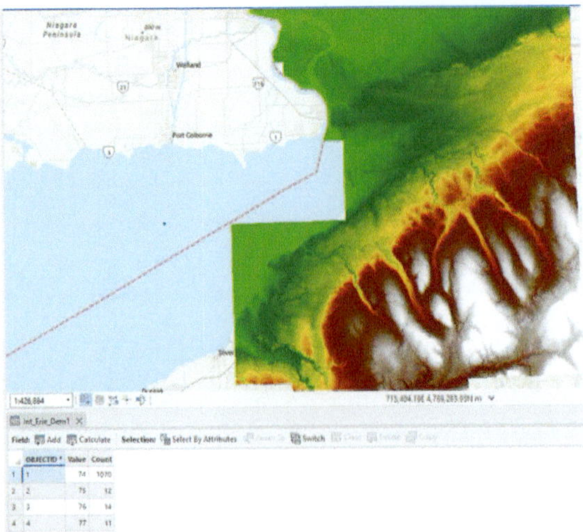

**Figure 2.14** Simplified digital elevation models (DEM) in integer values in Buffalo, New York are presented with their value attribute table (Count stands for the number of grid cells that have the value on the digital map).

can be categorized and ranked; (iii) Interval: the data can be categorized, ranked, and evenly spaced; and (iv) Ratio: the data can be categorized, ranked, evenly spaced, and has a natural zero. Values or characteristics of grid cells can be in any data format, such as nominal or ratio type.

However, each of the raster digital map layers can only present one value of the geographic phenomena owing to that a grid cell can only take one attribute value. Therefore, the attribute table of a raster digital map layer is called its value attribute table. Raster data values can be integer or floating numbers. Currently, only raster digital map layers with integer values are associated with a value attribute table in ArcGIS Pro (Figure 2.14). Grid cells present other types of values with decimal numbers that do not have a viewable attribute table that is associated with the map layer.

In raster GIS data format, we also recognize the spatial resolution of a dataset or a digital map layer. Spatial resolution in essence is the size of a grid cell in the real world or on the Earth's surface. For example, 5m spatial resolution means one grid cell on the map covers an area of 5m by 5m on the Earth's surface. The smaller the dimensions of a grid cell on the ground, the more details it represents. Dataset of smaller grid cell dimensions is called higher spatial resolution data (Figure 2.15). Relating to the digital map scale, the higher the spatial resolution, the larger the map scale that the dataset can present.

High spatial resolution raster data carries more details of features and objects in the geographic space. However, such kind of a dataset may occupy large computational storage space since each of the grid cells is separately stored in the computational systems. Therefore, raster dataset compressions have long been the focus of GIS and computational science to handle this type of data. Many methods of raster data compression have been developed over the years. We introduce two most popular methods that we use in GIS. These are run-length encoding (RLE) and quadtree method.

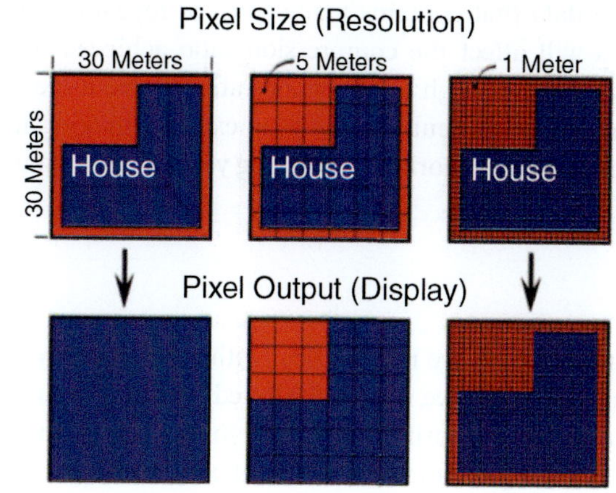


**Figure 2.15** Spatial resolution of grid cells or pixels. *Source:* Adapted from Jensen, 2007.

## Run-length Encoding

RLE is an algorithm of lossless data compression that is supported by most bitmap file formats in raster data handling, such as TIFF, BMP, and PCX. RLE works by reducing the physical size of a repeating string of grid cells or pixels. This repeating string is called a run. One run is typically encoded into two bytes. The first byte represents the number of grid cells in the run and is called the run count. The second byte is the value of the grid cell in the run, which is in the range of 0–255 in colors and hues in the raster imagery case and is called the run value. An encoded run of remotely sensed raster data may contain 1–128 or 256 colors or hues as its value. In practice, the run value usually contains the number of colors or hues of the imagery raster data minus one, since 0 is also considered as a value of colors and hues (Figure 2.16).

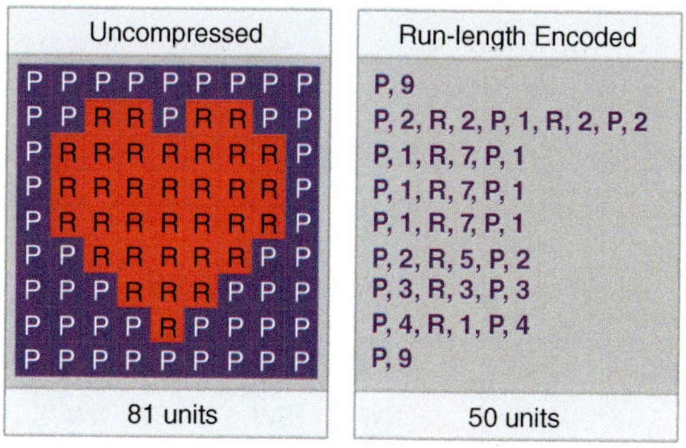


**Figure 2.16** RLE for compression of data on a raster digital map. *Source:* Adapted from GITTA- Geographic Information Technology Training Alliance.

RLE is most efficient on data that contains many runs, or repeating values of grid cells. The content of the raster data will affect the compression ratio achieved by RLE. Although most RLE algorithms cannot achieve the high compression ratios of the more advanced compression methods, RLE is both easy to implement and quick to execute, making it a good alternative to either using a complex compression algorithm or leaving your image data uncompressed.

## Quadtree Method

Since in raster spatial data structure, we need to deal with large quantity of grid cells that might have the same value across the space, we would like to reduce the computational storage requirements and improve access speeds by storing values only for homogeneous regions rather than for every grid cell. Quadtree encoding is a method of building a structure of raster data based on the successive reduction of homogeneous cells. It recursively subdivides a raster dataset into quarters. It produces a multilayer index system. If one-quarter of four grid cells have homogenous value, the upper-level indexing layer will only record one value in the database. The subdivision process continues until each of the original cells is classed. Quadtree reduces raster storage and retrieval requirements in computational systems (Figure 2.17).

Quadtrees are only used on a 2D space now in raster GIS data. In particular, quadtrees are used in image data compression, where each node contains the average color of each of its four children. The deeper you traverse in the tree, the more the detail of the image. Quadtrees are also used in searching for nodes or grid cells in a 2D geographic area. For instance, if we would like to find the closest grid cell to a given pair of coordinates, we can search using quadtrees. In ArcGIS Pro, a quadtree is called a pyramid. One only needs to build the pyramid for a raster dataset the first time they use it. This will improve the efficiency of computational storage and retrieval of the data.

The major difference between raster and vector data is that the raster data represents the geographic space as cells or grid matrix while vector data represents the geographic objects using sequential referenced points or points, lines, and polygons. Raster data is continuous data whereas vector data is discrete data. In the digital data processing and GIS analyses, different types of data structures always present some pros and cons. In practice, it is up to the GIS analyst to select a

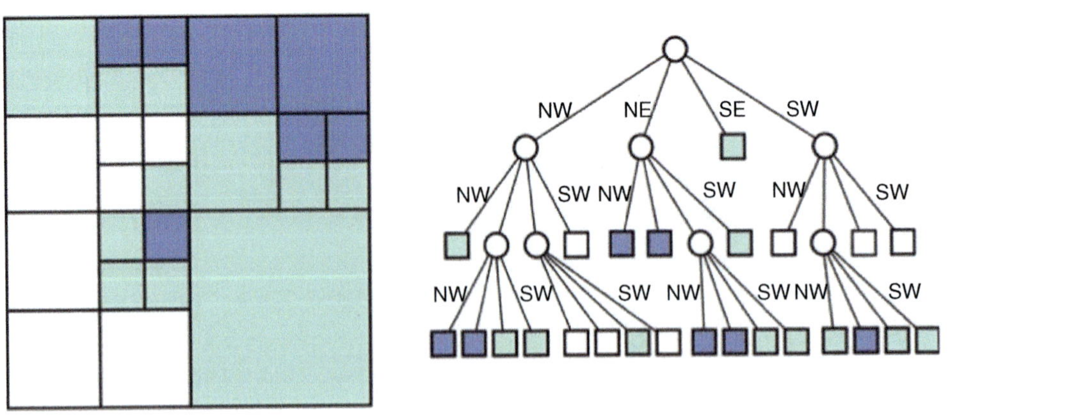


**Figure 2.17** Quadtree method of raster data compression. *Source:* Adapted from GITTA- Geographic Information Technology Training Alliance.

**Table 2.1** Characteristics of vector and raster data models.

| Vector data model | Raster data model |
|---|---|
| Represent geographic features on the Earth's surface as objects of points, lines, and polygons | Represent geographic environment on the Earth's surface by grid cells |
| Discrete data | Continuous data |
| Simple spatial data structure | Complex spatial data structure |
| Effective to represent geographic features with clear boundaries | Effective to represent geographic features without clear boundaries |

more suitable data model or potentially use both type of models to accomplish one GIS project. Table 2.1 is the summary of characteristics of vector and raster data models.

Table 2.2 is the comparison of advantages and disadvantages of vector and raster data models.

Inputting spatial data into the vector or raster format has been always the frontier of both practices and research in the GIS field since the beginning of its development. Naturally, raster data can be readily obtained in digital or computational format through the conventional aerial survey or remote sensing technologies. In the modern era, analog aerial survey and photogrammetry have been transformed to digital remote sensing and image processing. However, two outstanding data collection and processing issues still exist. First, the majority of remotely sensed data collection methods in the field yield raster data than vector data; while raster data by its nature is more difficult to be integrated with attribute datasets of the geographic objects. Second, inputting vector spatial data into the computational environment is both time-consuming and labor-intensive

**Table 2.2** Comparison of advantages and disadvantages of vector and raster data models.

| Data structure | Advantages | Disadvantages |
|---|---|---|
| Vector data | a. Simple discrete geometry and logical data structure<br>b. Easy to edit<br>c. One geographic feature or object can carry many attributes or characteristics<br>d. Can be scaled to any size in GIS environment without losing data quality<br>e. File size in computational system is relatively small even portraying a large geographic area on the Earth's surface<br>f. Analytical techniques are well developed in GIS | a. Difficult to create, lack of data collection method in the field on the Earth's surface<br>b. Difficult to represent continuous geographic features and objects<br>c. Inaccuracy of locations of geographic features may create big errors and uncertainty for the GIS analytical results |
| Raster data | a. Easy to create, can be produced directly by data collection technology, such as remote sensing<br>b. Easy GIS workflow and analysis<br>c. Easy to represent continuous geographic objects or features<br>d. Easy to apply mathematical modeling | a. Relatively large storage file size in computational systems in particular for high spatial resolution and detail-oriented datasets<br>b. Difficult to represent clear boundaries of geographic features<br>c. Grid cell scale up and summary will lose data quality and original values |

although the vector data structure can be easily integrated with object-centered attribute data sources that are gathered from other means for various analyses.

Traditionally, vector objects are digitized or traced either from hard copy maps or from remote sensing satellite or aerial images. Hard copy maps and satellite or aerial images can be viewed as the raster data in analog format. While the hard copy maps can be scanned into a computerized environment, and satellite or aerial images can be readily provided in the computer file format, they are both in raster data types that are made by pixels or grid cells. Therefore, the real daily practice and challenge in vector GIS data input is to extract features from the raster data. Extracting vector features from a raster grid cell data or a remotely sensed raster image can be classified into roughly three approaches. These are: (i) tracing or digitizing, (ii) extracting points and lines, and (iii) extracting polygons (Warner 2017).

## Tracing or Digitizing

Digitizing in GIS is the process by which coordinates from either a hard copy or digital map, image, or other sources of data are converted into a digital format. This is the most popular method for converting the raster data or hard copy analog map to the vector feature classes. Digitizing can be further identified into manual digitizing and heads-up digitizing.

In manual digitizing, the digitizer uses a digitizing tablet (also known as a digitizing table) to trace the points, lines, and polygons of a hard-copy map (Figure 2.18). This is done using a special magnetic pen, or a digitizing puck (mouse-like tool) that feeds information into a computer to create an identical, digital map. Manual digitizing is still a useful technique because of its ability to accurately copy maps in poor condition.

**Figure 2.18** Manual digitizing of a hard copy map using a digitizing table. *Source*: Dr. Tao Tang.

Heads-up digitizing involves scanning a map or inputting an image into a computer. The GIS analyst who are conducting the digitizing then trace the points, lines, and polygons on the computer screen using digitizing software in GIS (Figure 2.19). This method of digitizing has been named "heads-up" digitizing because the focus of the user is up on the screen, rather than down on a digitizing tablet. This method has largely replaced manual digitizing because of its speed and accuracy. Since the tracing is done on a computer, lines can be set to snap together, and polygons can be programmed to share an edge, thus removing accidental sliver polygons. Heads-up digitizing also reduces or removes the need for digitizing tables.

**Figure 2.19** Heads-up digitizing. *Source*: Dr. Tao Tang.

## Extracting Vector Objects in Points, Lines, or Polygons from Raster Data

Human tracing or digitizing geographic objects on the Earth's surface from satellite or aerial images, or scanned maps to vector GIS database is very time consuming and labor intensive. Automatic extraction of points, lines, or polygons that represent the Earth's surface objects are always the main challenge for raster to vector data conversion in particular, and GIS data input in general. Recently, some specific feature extraction software has been developed for converting imagery objects to readable vector GIS format for GIS applications, such as ENVI image processing software. GIS analysts can take advantage of this type of software to convert raster data to vector data automatically. Researchers (Tang and Wang 2011) have tried to develop a specific computational model to extract line features from remote sensing images to the vector format database for GIS analyses. Some remote sensing or imagery processing software provide tools that GIS analysts can utilize to automatically extract polygon map layers with attribute values. Meanwhile, programming codes written in free GIS software environments, such as GRASS can be used to convert raster datasets to polygons as well.

Up to this date, no fully developed feature extraction module from raster data format have been implemented in either ArcGIS or other major GIS software environment. Recent breakthrough of imagery pattern recognition applying AI provides GIS field the new direction in developing the automatic vector data extraction or creation modules. After the automatic extraction method is succeeded, the human digitizing method can be put to the rest.

## 3D GIS Data Models or Data Format

Conventionally, GIS mainly deals with 2D geographic environment on digital maps. With the improvement of computational systems and capabilities, GIS is more capable to handle and analyze 3D features and objects in the digital world. Meanwhile, the real world is actually in the 3D in nature. Therefore, the 3D GIS data models are not only necessary but also important. 3D versions of representations of both vector and raster data models were developed in GIS. Currently, the popularly used vector 3D data model is called the triangulated irregular network (TIN) and that of raster data is called the digital elevation model (DEM). Meanwhile, several new 3D vector models were developed, such as voxel layers. One newly developed 3D vector GIS data in the last decade is called light detection and ranging (LiDAR) data or laser scan point cloud. This type of dataset consists of 3D point clouds and each of the points in the cloud carries its own real-world 3D or x, y, z location. Point cloud datasets are directly collected by laser scan equipment, a new remote sensing technology. LiDAR or laser scan is the first remote sensing technology that can directly collect vector GIS data from the real world and in 3D. It revolutionized the GIS data collection methods.

## Triangulated Irregular Network

A TIN, which originated in computer graphics, is a representation of a continuous surface consisting entirely of triangular facets or an irregular triangle mesh. The vertices of these triangles are created from field-recorded spot elevations through a variety of means, including surveying through conventional techniques, Global Positioning System (GPS), photogrammetry, or some

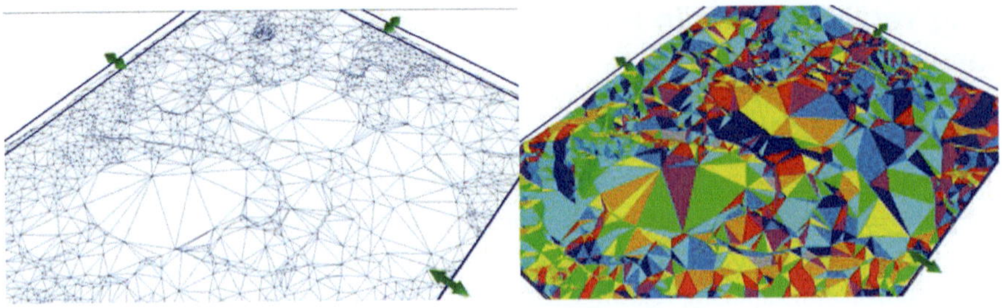

**Figure 2.20** TIN 3D models.

other means. Associated 3D (x, y, z) data and topography, TIN models are useful for the description and analysis of surfaces of terrain in a 3D geographic space (Figure 2.20).

Construction of TIN is based on a Delaunay triangulation. Delaunay triangulations maximize the minimum of all the angles of the triangles in the triangulation; they tend to avoid sliver triangles. The triangulation is named after Boris Delaunay for his work on this topic published in 1934 (Delaunay 1934). The first TIN model in GIS was published by Randolph Franklin in 1973 (Franklin 1973).

## Digital Elevation Models

A DEM is a representation of the bare topographic ground (bare earth) of the Earth's surface excluding all the natural or human-built objects, such as trees or buildings (USGS 2018). Previously, DEMs were mainly created from hard copy or digital topographic maps. Recently, the new productions of DEMs are being systematically replaced by using high-resolution LiDAR and IfSAR satellite (Alaska only) data (Figure 2.17). In essence, DEMs are raster grid cells that carry the value of elevation or altitude in the attribute tables. Therefore, all the characteristics of 2D grids, such as spatial resolution also apply to DEMs.

When discussing DEMs, we also need to illustrate two related concepts: the digital surface model (DSM) and digital terrain model (DTM). In contrast to DEM, DSM captures and includes both natural and human-built features in the environment. DSM, by its meaning, represents the current surface on the Earth, no matter how it is covered by trees or human-made buildings. In practice, the interpretations of DTM are more complicated. Some GIS analysts consider DTM as also including natural features of rivers and mountain ridges but similar to DEM excluding trees, vegetation, and human-made buildings. Others think DTM is similar to DEM, and it is just a different name (Figure 2.21). A third group considers DTM to be similar to DSM, particularly in the natural environment.

## LiDAR Point Cloud

LiDAR stands for light detection and ranging. It is a newly developed remote sensing and digital data collection technology. In practice, the technology is called 3D laser scan. 3D laser scanning is the controlled deflection of invisible laser beams generated by a laser source device. Before the laser scanning technology was used in GIS and remote sensing fields, it had been widely used for designing and rapid prototyping for 3D printing in various industries. While many other areas of application still call the technology as 3D laser scan, the application of

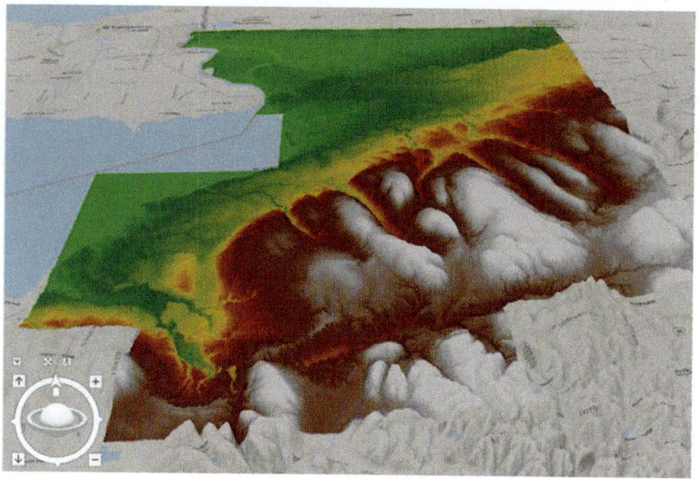

**Figure 2.21** DEM in Buffalo, NY are presented in ArcGIS Pro 3D local scene.

laser scanning in GIS and remote sensing has been named as LiDAR. LiDAR detection of surrounding environment in 3D point clouds is not only very useful and important to GIS, remote sensing, and digital mapping, but also very significant in supporting the automatic-driving technology of cars and trucks.

The data produced by LiDAR or laser scanning are 3D point clouds. Each of the points in digital database carries a 3D coordinate location (x, y, z) in standard real-world coordinate system (Figure 2.18). In this regard, the point is a true digital representation of a particular location of the object in the geographic space on the Earth's surface. Viewing from the real-world perspective, an object's surface can be represented by a list of 3D coordinates (x, y, z) points, referred to as a "Point Cloud." Therefore, a point cloud is naturally in 3D from its collection time (Figure 2.22).

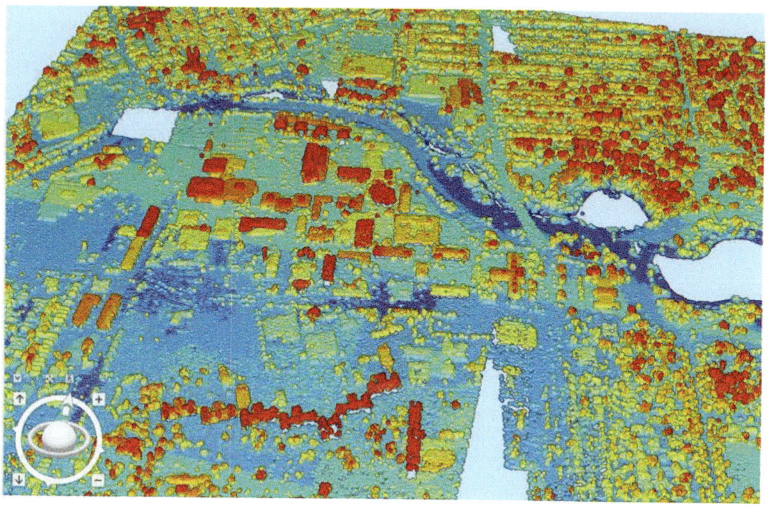

**Figure 2.22** Point cloud of LiDAR data showing the SUNY– Buffalo State University campus area in 3D in a local scene.

Since LiDAR surveying technology was invented after the development of GIS and remote sensing, we are still in the process of determining the most effective and accurate ways to incorporate this type of data into GIS analyses.

## References

Cook, J.D. (2016). Next areas of math to be applied. https://www.johndcook.com/blog/2016/01/27/next-areas-of-math-to-be-applied/.

Corbett, J.P. (1979). Topological principles in cartography. *United States. Bureau of the Census*, 48: p. 50 Technical paper. Washington, D.C.: U.S. Department of Commerce, Bureau of the Census.

Delaunay, B. (1934). Sur la sphère vide. bulletin de l'académie des sciences de l'URSS. *Classe des Sciences Mathématiques et Naturelles* 6: 793–800.

Franklin, R. (1973). *Triangulated Irregular Network Program*. Accessed 30 Oct 2008. http://wrf.ecse.rpi.edu/wiki/Research/tin73.tgz.

Jensen, J.R. (2007). *Remote Sensing of the Environment – An Earth Resource Perspective*, 2e. Upper Saddle River: Prentice Hall, Inc. NJ 07458, p. 592.

University of Waterloo. (2020). "What is Topology?." https://uwaterloo.ca/pure-mathematics/about-pure-math/.

USGS (United States Geological Survey). (2018). https://www.usgs.gov/.

Warner, T. (2017). *Three Ways to Convert Raster Images to Vector for CAD/GIS*. https://www.safe.com/blog/2017/08/3-ways-to-convert-a-raster-image-to-vector/.

## Tutorial in ArcGIS Pro Environment

Copy or download the "Chapter02-Tutorial" folder from the textbook data-site to your working location (either the C: drive of your computer, your portable drive, or a share of a network drive).

Open ArcGIS Pro and create a new project called "MyGIS_Learning02" with Map template. Save the project folder in the Chpter02-Tutorial folder.

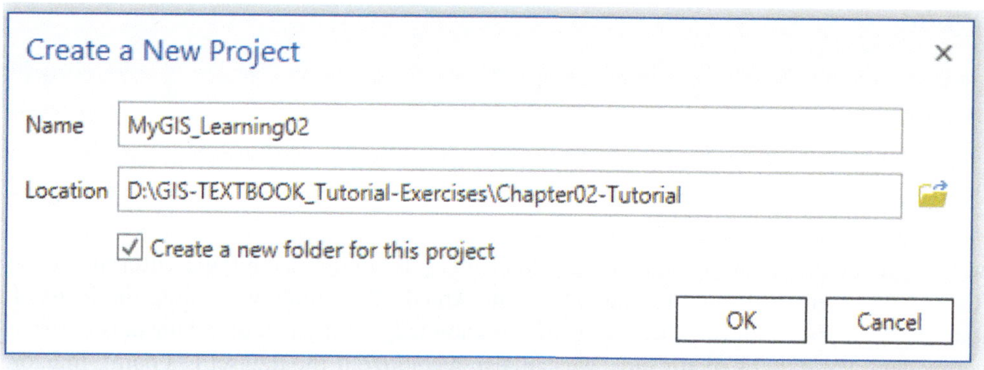


Under the category of "Folders" on the Catalog Pane, add the entire folder of "Chapter02-Tutorial." Now we can see all the spatial databases and files.

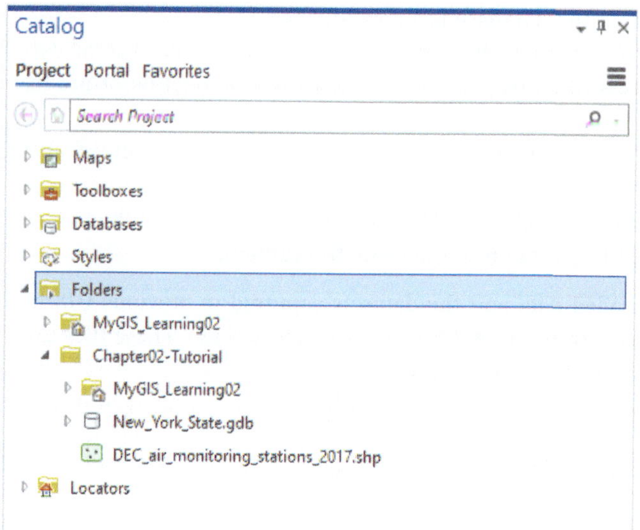


Different users of ArcGIS Pro may have different ways of connecting spatial or geographic databases. We can also practice adding the "New_York_State" geodatabase (.gdb) under the "Databases" category by right-clicking on its title on the Catalog Pane, obtaining the contact menu , and navigating to the folder. However, any spatial datasets outside of a geodatabase will be invisible under this category. This is the reason that it is more popular to add data folders than to add geodatabases.

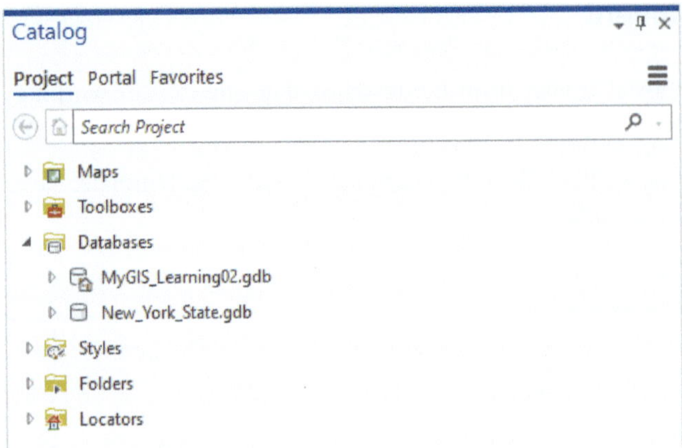


Now, let's expand the "New_York_State" geodatabase (.gdb) in the "Chapter02-Tutorial" folder on the Catalog Pane at the right-hand side of the ArcGIS Pro window. Reading the icons of spatial or geographic data files, we can gain the knowledge if a particular dataset is a raster or vector data. For data of vector type, we can also recognize if the dataset is a point, line, or polygon type, which is called a feature class. First, a geodatabase can store both vector and raster data. For vector datasets in a geodatabase, the feature classes can be organized under different feature datasets or can be directly stored underneath a geodatabase. By contrast, raster datasets and map layers can only be stored individually directly within a geodatabase. Secondly, a can-like icon indicates that the database is a geodatabase. New_York_State.gdb : A grid cell-like icon indicates that it is a raster dataset t_6875068000_2017 . For vector datasets, an icon of a point feature class is like this Cities ; the icon of a polyline or arc feature class is like this Intersate_Highway ; and the icon of a polygon feature class is like this Counties . In order to organize the vector data, one can also store several feature classes into one geodatabase feature dataset NYS_Social_Economic_Data . This structural organization can help GIS analysts to put the geographic feature classes with similar characteristics into a same feature dataset, like filing them into the same file cabinet.

Meanwhile, many externally downloaded data and the datasets captured in previous GIS formats, such as shapefiles or ArcInfo coverages are not hosted by a geodatabase. These files exist in the data folders. The example of a shapefile is like this DEC_air_monitoring_stations_2017.shp . The old ArcGIS data format files and the non-native ArcGIS Pro files, either downloaded or from field surveys and either in raster or vector data models can be easily identified by its individual file locations and the color of icons. The graphic patterns of these icons are normally the same.

In the "New_York_State" geodatabase (.gdb), all the vector data feature classes cover the extent of entire New York State, while raster datasets in the geodatabase only cover a small geographic area in the New York State. It is called the Holliday Valley in the Town of Ellicottville, NY. The simple reason for this is that raster datasets occupy or consume large computational storage space. Normally, our computational storage space cannot afford to store the detail raster datasets for a large area like New York State. Let's explore the vector data first.

Add the following feature classes from the New_York_State.gdb. We can either drag the feature class from the Catalog Pane to current mapping frame, or right-click on the targeted feature class to get the contact menu. Then, click on "Add To Current Map" icon.

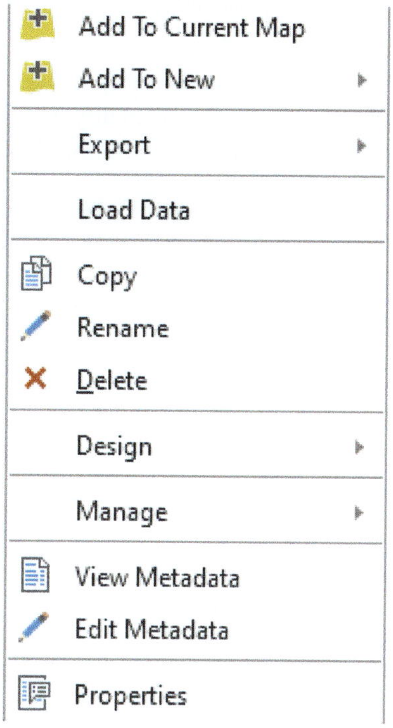


1. Counties – symbolize it in “No color” for filling and “Black” and 2 points of width for the boundary.
2. Cities – symbolize it using a “House” symbol with a pink color and font size of 14 points.
3. Interstate_Highway – symbolize it in red color with 2 points of width.

Also expand the NYS_Water_Bodies feature dataset to load the NYS_Rivers feature class and symbolize the rivers in blue color. In ArcGIS Pro, a feature class is a vector spatial or geographic database name, such as NYS_Rivers feature class. Meanwhile, a feature class in a mapping frame (Map) is shown as one map layer. Therefore, it can also be called a map layer, such as NYS_Rivers map layer. The mapping product should look like this.

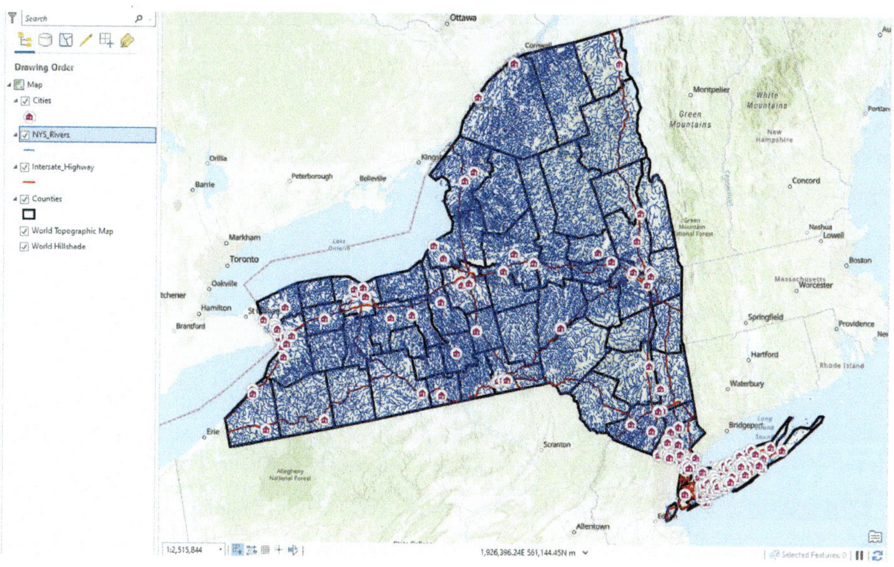

The background map in ArcGIS Pro "Map" window is called the base map. The base map comes with the ArcGIS Pro software. A base map is a layer that helps users of ArcGIS Pro to orient and locate their own data layers worldwide on the Earth's surface. The base map is different than your local map layers. It is provided by the software company, ESRI, Inc. Users cannot modify base maps but can change from one type to another. We can practice changing the base map now. On the "Map" tab of the main menu, click on the "Basemap" icon 

to open the selection menu as follows.

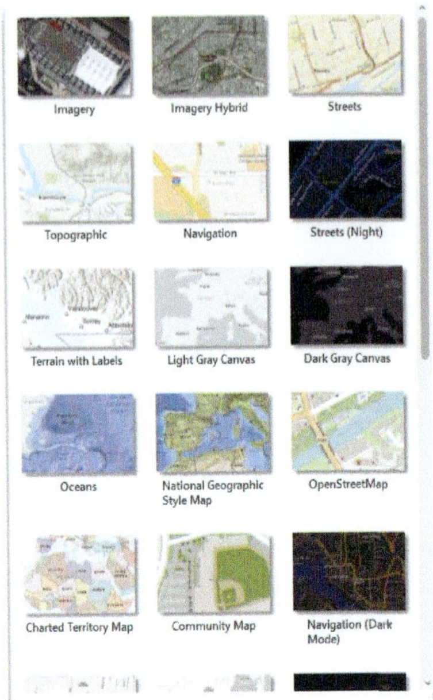


Change the default "Topographic" base map to "Imagery Hybrid." It would look like this.

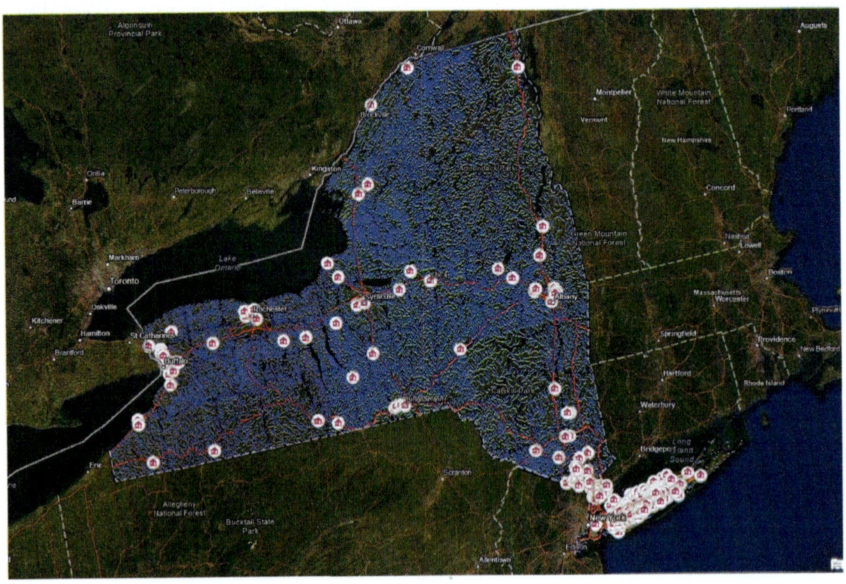

Try different base map styles. Then, change back to "Topographic."

We can show and query the spatial or geographic feature classes both on the map and in the attribute table. Both vector feature classes and raster grid cells can be added from the Catalog Pane on the right-hand side to the main window of "Map," "Local Scene," or "Global Scene" whichever is appropriate. As soon as the spatial data is inputted or added in the mapping window as a map layer, it is listed on the "Table of Contents" on the left-hand side. Next to the title of each map layer in the Table of Contents, there is a check icon. We can turn the map layers on or off by checking on or off each map layer in the Table of Contents. Practice turning on and off map layers by checking them on or off.

The group of icons on the top of the Table of Contents show the method that the data layers are handled with . The first is called "List by Drawing Order." That means whichever map layer is listed on top of the Table of Contents, that map layer would be shown in the front in the mapping window. Whichever map layer is listed on the bottom of the Table of Contents would be presented in the back of the mapping window just in front of the base-map. We can change the order of drawing for a map layer by dragging it up or down. The second icon is called "List by Data Source," which shows the computational storage location that a particular map layer is at. The third is called "List by Selection," which is related to the select-ability of map layer set-ups. The fourth one from the left is "List by Editing." This will put the map layers in editing mode on the top. While "List by Drawing Order" is activated, we can drag a particular map layer up or down on the list of the Table of Contents. This means that we can rearrange the map layers in the mapping window.

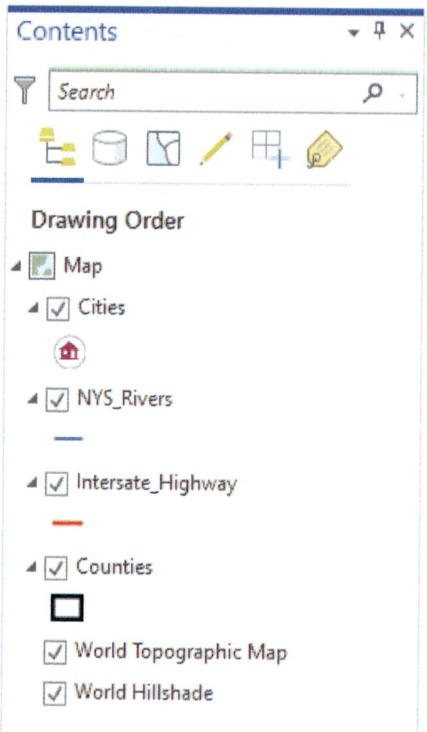

We can zoom in on the mapping window to view the details in any locations on your digital map. For the zoom in or zoom out process, we can use the middle roller on the mouse of the computer. Pointing your computer cursor on a particular location that you would like to zoom in, push the middle roller on the mouse forward to zoom in, and push backward toward you to zoom out. Another way to zoom in and out on a digital map in your mapping window is to click on the zoom-in or zoom-out icon in the "Navigate" group under the "Map" ribbon on the main menu. You can drag the digital map around by depressing the left button while drag the mouse cursor. In practice, it is not suggested for you to use the zoom to full extent icon in obtaining the full view of your mapping area. If you do click on this icon, the map extent will be zoomed to the entire world according to extent of the base map. We can use any of the map layers as a reference to return to the entire mapping area of your local study region. Open the contact menu of a map layer that you use and click on "Zoom to the Layer" to return our full extent of the mapping area.

The "Bookmarks" icon Bookmarks in the "Navigate" group on the "Map" tab is the icon to save and manage the bookmarks of the digital map that we use. We can zoom into the Manhattan region in the New York City area on the digital map.

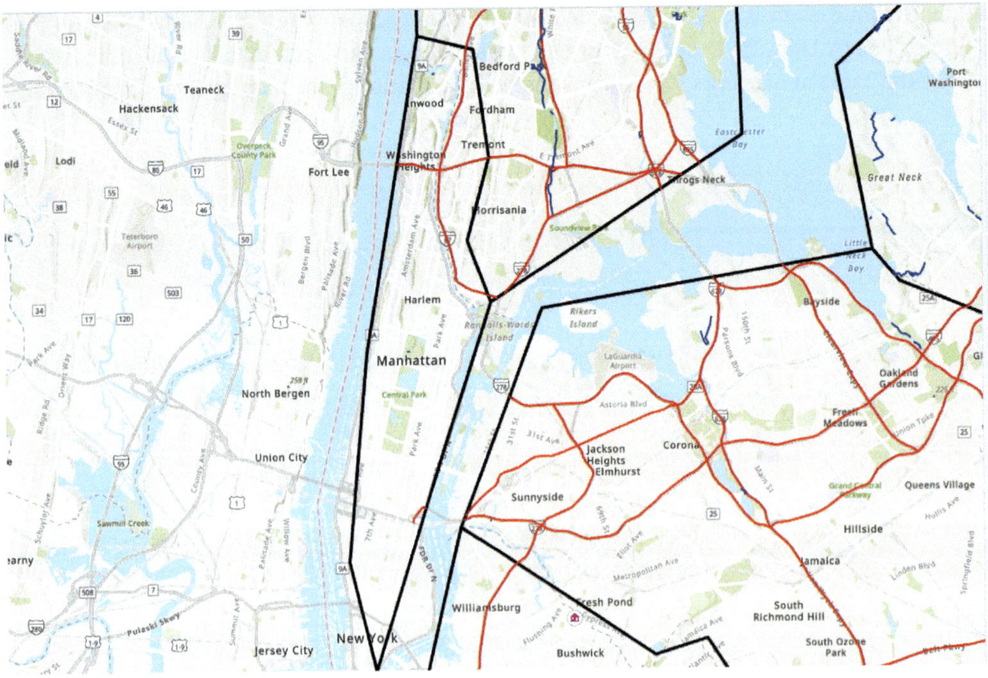


Open the bookmark function list to create a new bookmark called "Manhattan."  You notice that a bookmark icon is created on the list. Now, use the "Counties" map layer to return to the full extent of local mapping area across the New York State. We can practice making the direct jump by clicking on the "Manhattan" bookmark.

Each of the map layers in the digital map frame is also a spatial database. Therefore, we can query each of these map layers. The graphic icons to query the geographic or spatial datasets in the mapping window are collected in the "Selection" group on the "Map" ribbon (tab) of the main menu 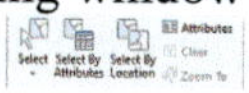. The "Select" icon is to use a drawing graph on the digital map to select the features. The drop-down list is as follows. You can select to draw a polygon; or rectangle to select the geographic features that you would like to get.

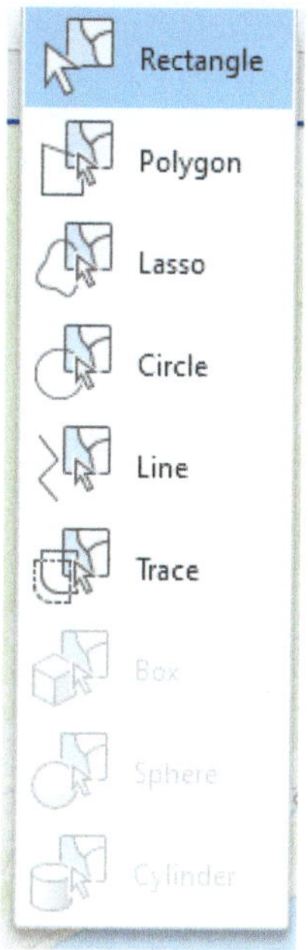


Clicking on the "Select by Attributes" icon will open a structured query language dialog for you to build a query to select features. Query is the standard way to extract spatial features and database according to a specific set of characteristics of the items.

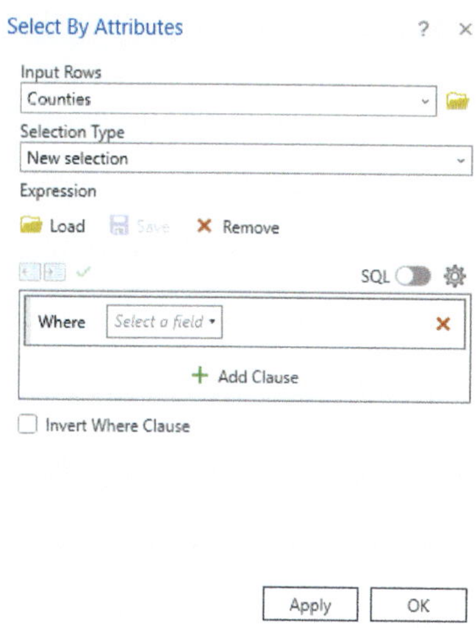

The "Select by Location" icon is the geographic or spatial query that is unique in the GIS software environment only. This query will select the features according to their geographic relationships and relative locations to each other. There could be many different geographic relationships between features on the different feature classes. The following is a list of some of the major geographic relationships.

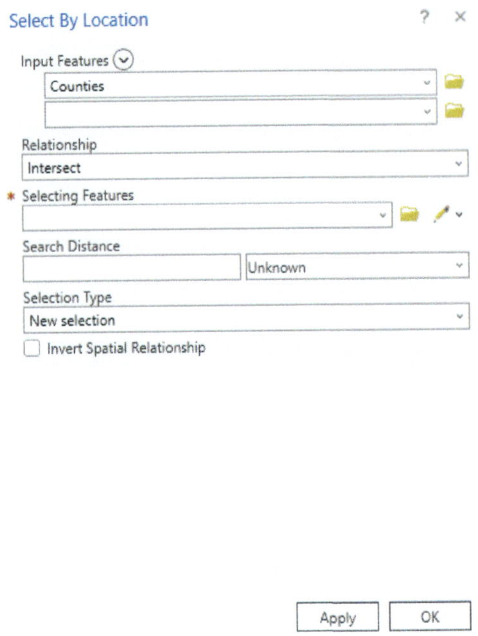


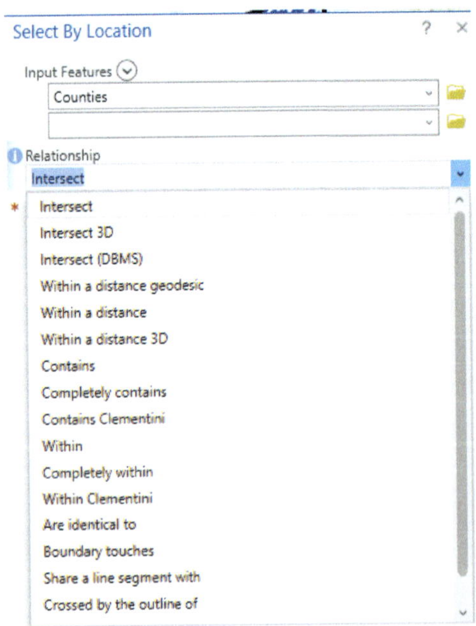

Now, let us use the "Select by Attributes" tool to build a query and select on the "Counties" feature class. Click the icon to open the dialog. For "Input Rows," select "Counties." For "Selection Type," choose "New Selection." Build the query as "Where" "NAME" "is equal to" "Cayuga" (NAME = "Cayuga"). The dialog should look like the following. Click on "Apply" and "OK" to process the query. You notice that the Cayuga County that is on "Counties" map layer in the New York State is selected. The default selection color of selected feature is light blue.

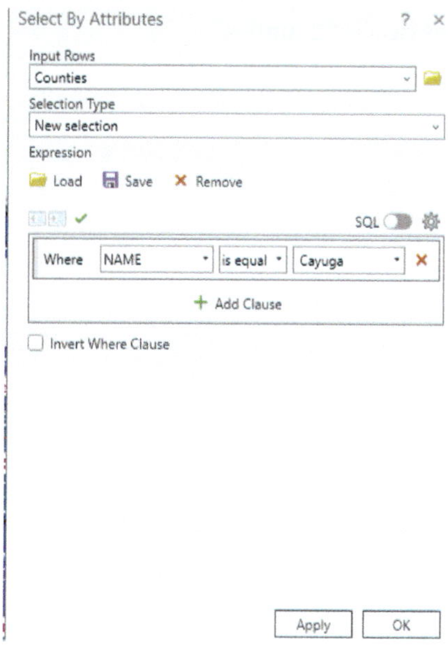


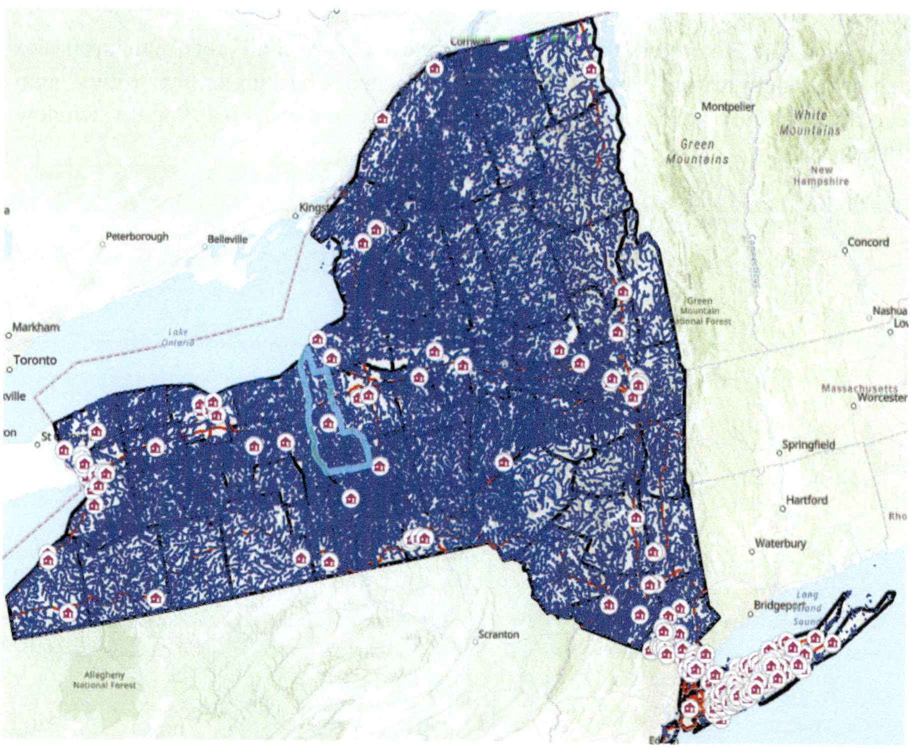

We can use the "Select by Location" (Geographic Query) icon to select other features, such as rivers, interstate highways, and cities in Cayuga County now. Click the icon to open the dialog of "Select by Location" (Geographic Query). For "Input Features," we can fill in "Interstate_Highway," "NYS_Rivers," and "Cities." For "Relationship" (Geographic relationship), we keep the default as "Intersect." That means we will select the features that intersect with the county feature on the digital map. For "Selecting Features," we choose "Counties." The fundamental rule in ArcGIS Pro is that the software APP only works for selected features if we selected previously. If there is no selection before, it will use the entire map layer, which in this case is "Counties."

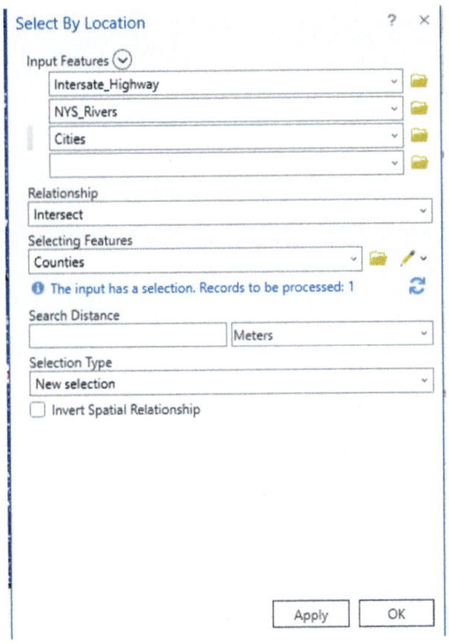


Click on "Apply" and "OK" to process the query. Now, we can check off all other map layers and check on each of the input feature classes to see what happened. Checking on the "Cities" map layer, only one city is selected. We can point to this city and click on it to open a "Pop-up" window to read its attributes.

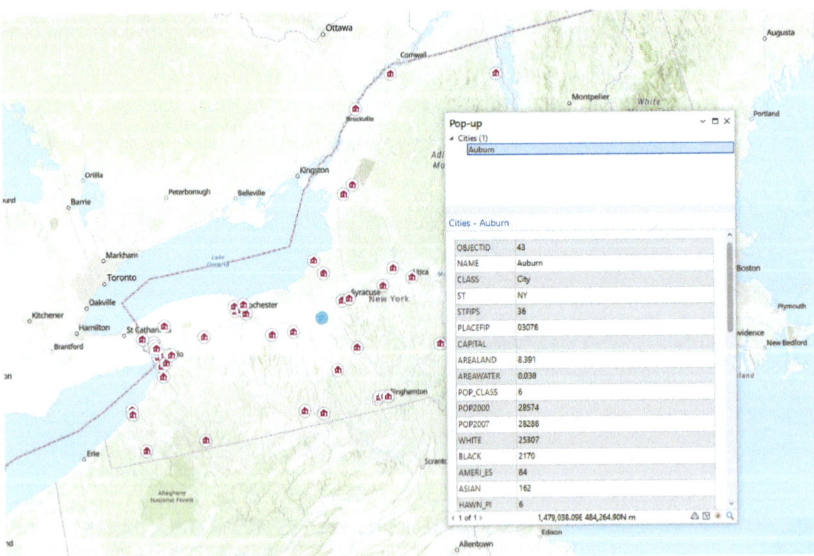

What is the name of this city or town?

Now, we can close the "Pop-up" window and check off the "Cities" map layer.

Check on the "NY_Rivers" map layer and zoom into the area of selected features. We see many rivers within the county boundaries are selected.

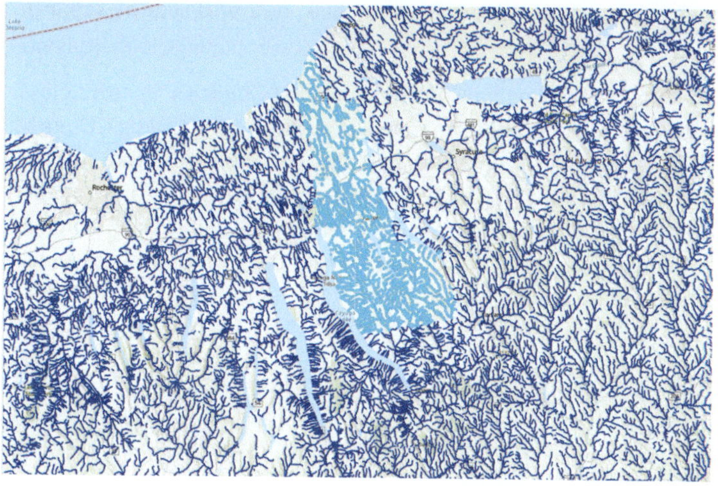

Now return to the local full extent using the "Previous extent" icon.

Check off the "NY_Rivers" map layer and check on the "Interstate_Highway" map layer. You notice roughly the entire I-90 interstate highway is selected. This is because the entire I-90 is one feature or line segment that intersected with the feature of Cayuga County.

Now, we can check on all the four feature classes or map layers in the "Map" window. Another important process is that the selected features can be exported (saved) to a separate feature class as well. In practice, this is one of the methods that GIS analysts use often to create new geographic or spatial dataset from the existing datasets. Highlight and right-click on the title of the "NY_Rivers" map layer to get the contact menu. Click on "Data" to get the second contact menu like the following.

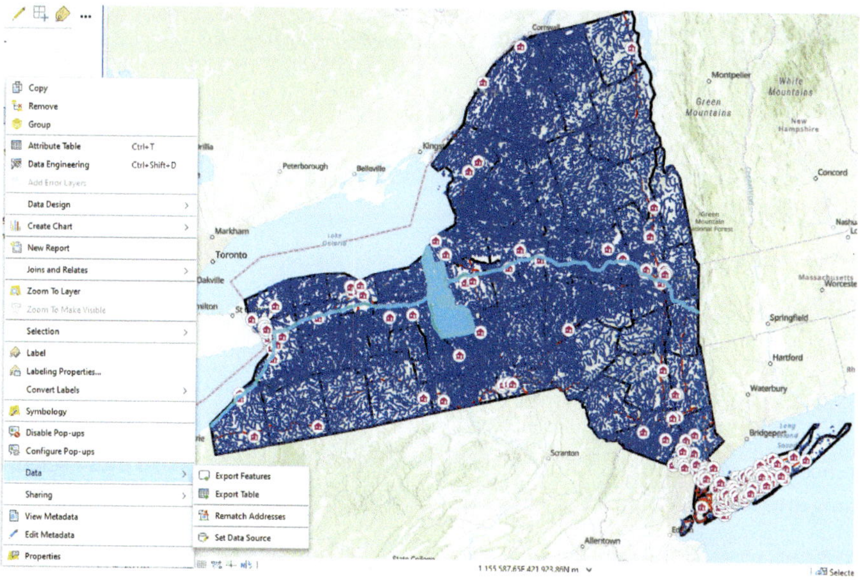

Choose "Export Features" on this contact menu, and the dialog opens. Rename the "Output feature Class" as "Cayuga_County_Rivers," which reflects what it is. The export in ArcGIS Pro only works for selected features unless you deselect all the features. The default storage location of the new feature class should be in the MyGIS_Learning02.gdb geodatabase that we created for this project. You notice that the new feature class or map layer is added on the Table of Contents. If you check off the NYS_Rivers feature class, you will see the separate map layer.

Now use the same steps as above to export Cayuga County from the "Counties" feature class. Call it "Cayuga_County." Now, we have two new separated map layers for Cayuga County, New York that are saved in the MyGIS_Learning02.gdb geodatabase. We would like to clear the selections now for all the feature classes. Normally, there are two different ways to clear the selected features. These are: (i) click on the clear icon Clear in the "Selection" group at the "Map" ribbon of the main menu on the top of APP, and (ii) right-click on the title of a map layer in the Table of Contents to get contact menu, then "Selection" menu and click on "Clear Selection."

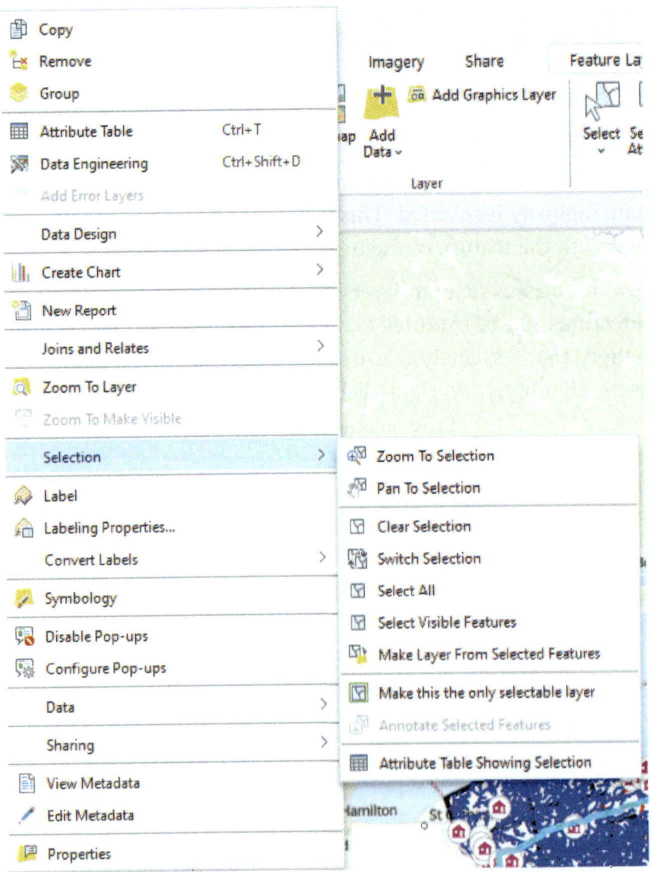


The major difference of these two routes is the level of control. The first one will clear every selection in the mapping area for every map layer. The latter one will only deselect the selected features on that particular map layer. Selections on other map layers will not be touched. Let's right-click on the "Counties" feature class title in the Table of Contents, then click on the clear selection command. You notice that Cayuga County is deselected. In order to clear all the selections on various map layers in this mapping window, click on the "Clear" icon on the "Map" ribbon. This action clears all the selections.

We can work with both digital map layers and the attribute tables that come with these spatial datasets. Let's check off the river classes (such as NYS_Rivers) on the Content Pane (Table of Contents). This will make it easier to visualize and focus on other feature classes. Now, we can right-click on the "Counties" title in the Content Pane to activate the contact menu and open the attribute table by clicking on the menu. The attribute table of "Counties" opens on the bottom half of the main window.

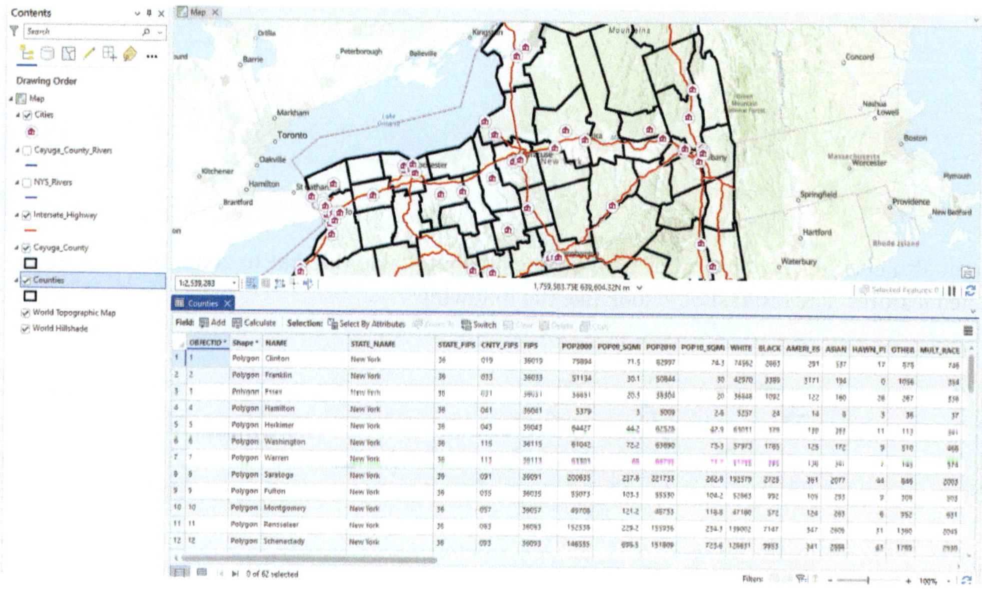


This table view is called the "Content View" of the attribute table. We can pull the lower gray bar to the right to view all the different columns or fields of features. We can pull the right-hand side gray bar to view the attributes of each of the features. For vector feature class, each row in this table is the representation of characteristics of one feature on the digital map. Each column or field in this table is the collection of a particular characteristic or called a particular attribute for all the features in this feature class. We can query the database to select the items that we would like to get. Meanwhile, the related features will be automatically selected on the digital map as well. This is simply because the features are connected to their characteristics in the attribute table. Click on the "Select by Attributes" on the title bar of the attribute table. The query dialog opens. Build the new selection as "POP2000 (Population in 2000) is greater or equal to 1,000,000." This will select those counties where the population is higher than 1,000,000. Click on "Apply" and "Ok" to process. You notice that 6 out of a total of 62 counties are selected. At the lower task bar of the attribute table, click on and highlight the "Show Selected Records" icon 

.
The table now shows only selected records in blue.

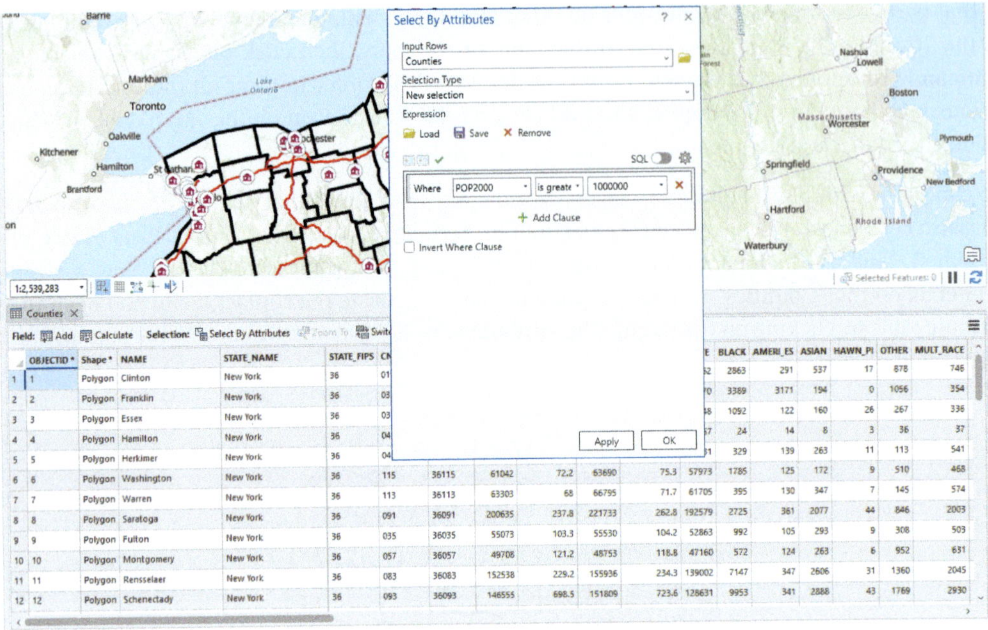


We can also click on the "Zoom to" icon on the attribute table title bar to zoom the map to the selected features. The result should look like the following.

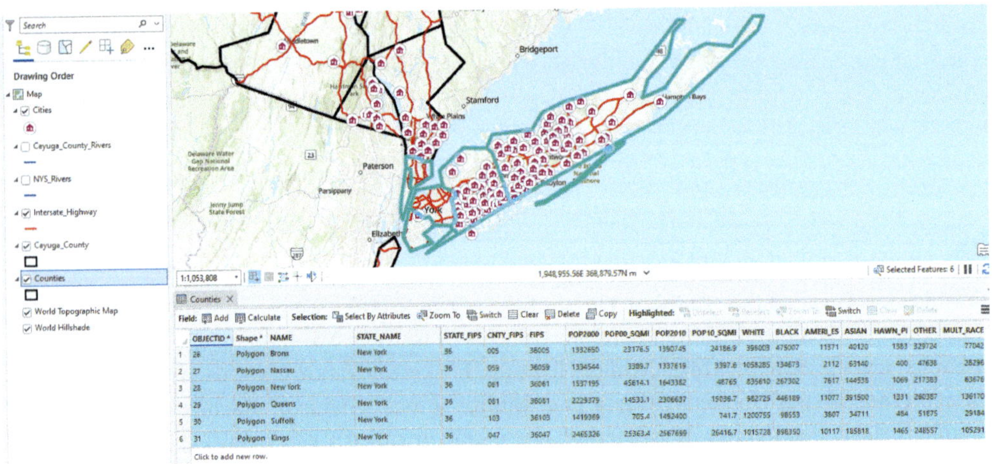


The six counties in New York State where the population are either higher or equal to one million are: Bronx, Nassau, New York, Queens, Suffolk, and Kings. This process demonstrates that geographic or spatial datasets that we use and analyze in GIS software environment have both database aspects and locational characteristics. Now, we can click on the clear icon under "Map" ribbon on the main menu to clear all the selections. Click on the "Show All Records" icon at the lower task bar of the table to see all the records in the attribute table again.

In ArcGIS Pro, GIS database can be managed by both the "data content view" and the "fields view." Click on the "Options" icon on the task bar at upper right-hand side of the attribute table to open the contact menu. Then, select "Fields View" Fields View. "Fields: Counties" tab is added on the upper task bar and the field view window is added on top of the data content view window. In the "fields view" window, the columns (fields) in data content view are changed to rows. We can change some settings to make our working environment more convenient here. The rank of fields from top to bottom here is the reflection of columns from left to right in the data view. Let's drag the "CNTY_FIPS" (means Federal Information Processing Standards [FIPS] code of counties) field just below the "Name" field. Since we may not use many of the attribute table columns or fields in our GIS project, we can always check (uncheck the "Visible") those off. Check the following fields off from the bottom up. "HISPENIC_PER," "AVG_SALE07," "CROP_ACR07," "AVG_SIZE07," "NO_FARMS07," "FHH_CHILD," "MHH_CHILD," "MARHH_NO_C," and "MARHH_CHD." In the "Alias" column, change the "NAME" (county name) to "COUNTY NAME." Now, we close the "Fields View" tab and answer "Yes" to the message of "Save all the changes?" You notice that the fields named above disappeared in the table in data view. The original field "NAME" is now "COUNTY NAME." County FIPS code follows the county name field now. We successfully modified the attribute table for our use. We can always come back to modify either to the previous status or a new status. FIPS codes are unique identification code (unique ID) in the computational environment to identify the geographic regions of political subdivision or administrative entities. In this case, The FIPS code of New York State is 36. The FIPS code for Erie County is 029. Therefore, the combination of FIPS code for Erie County, New York is 36029.

We can conduct a simple statistical and geometric analysis on each of the attributes (fields) in the table. Highlight the "POP10_SQMI" by clicking its column (field) title, then right-click on it to get the contact menu. "POP10_SQMI" means population in 2010 per square mile. Select "Statistics."

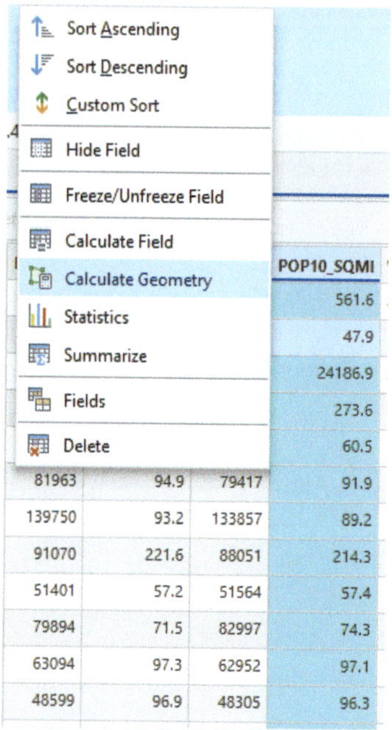

You will get a simple statistic chart and summary report of the values of this column (field).

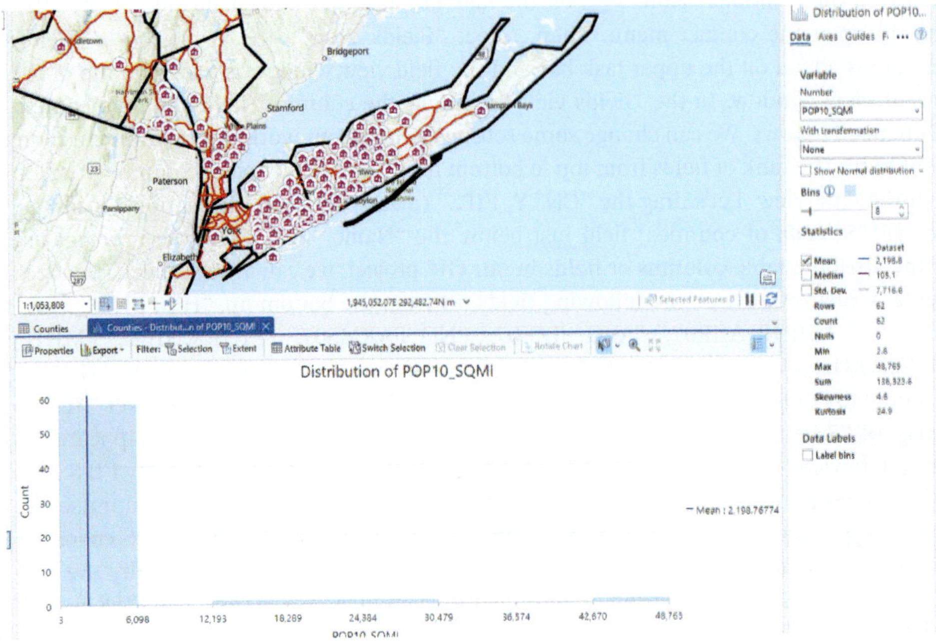


The chart shows the distributions of counties by population density per square mile. Check the "Label bins." We notice that the population densities of 58 out of 62 total counties in the New York State are lower than 6098 persons per square mile. The highest population density by county in New York State is 48,765 persons per square mile. Check the "Show Normal distribution" on the "Chart Properties...." dialog. The normal distribution curve is shown.

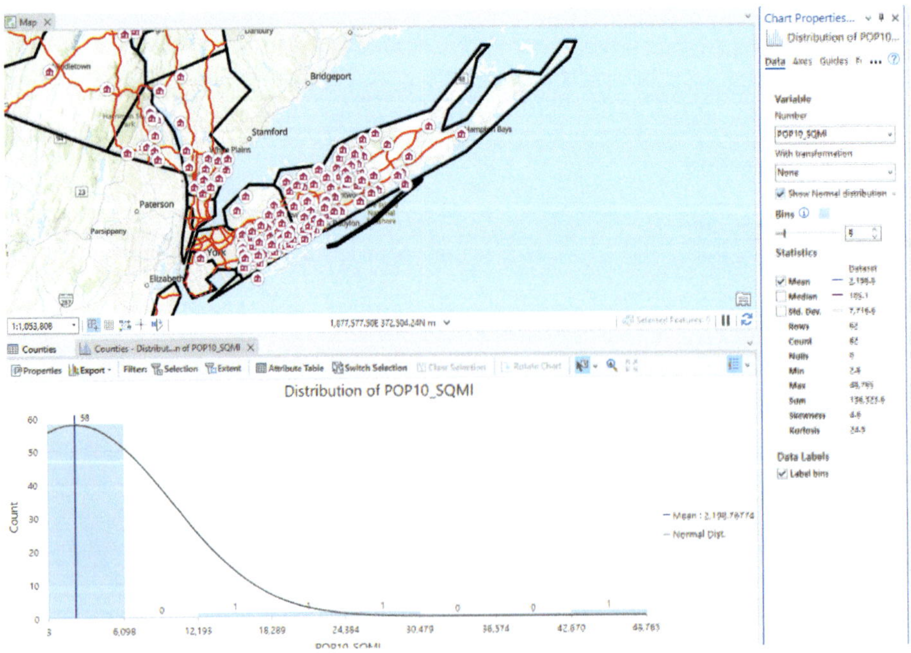

Now, we can close the tab of "Distribution of "POP10_SQMI" chart. Also, close the chart properties dialog. Another simple operation is that we can sort any selected column of field ascending or descending.

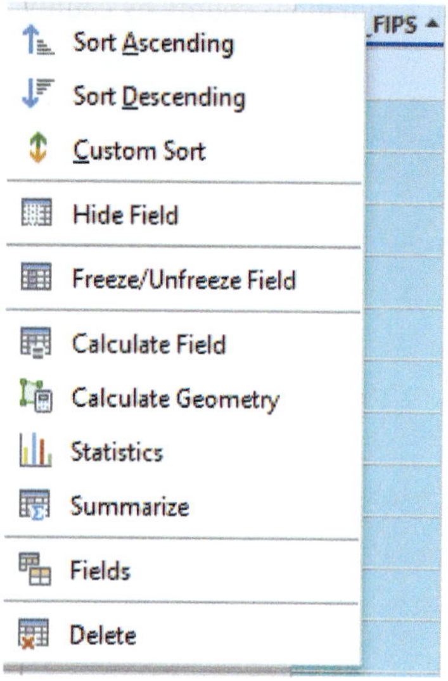


We can also summarize the statistics in a separate stand-alone attribute table for any of the column or field. Highlight and right-click on "CNTY_FIPS" field to select the "Summarize" menu. The summarize dialog opens.

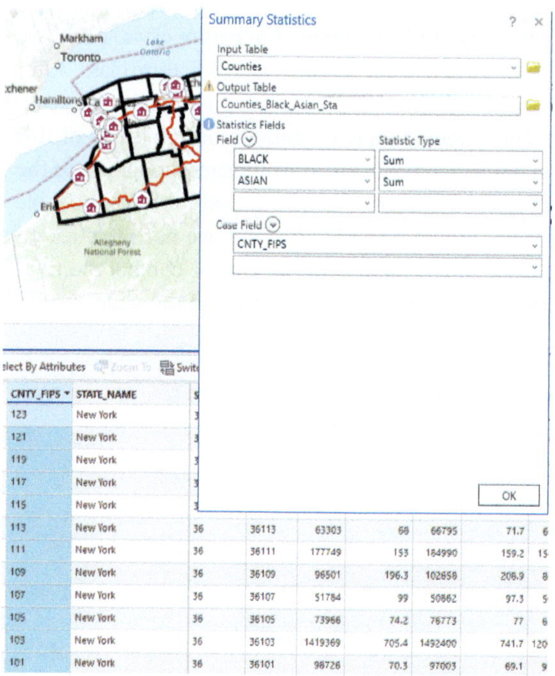

Select "BLACK" as your first statistic field and "SUM" (Summarize) as the statistical method. Then, choose "ASIAN" as the second field and "SUM" again as the statistical method. Click OK to process the new table. You notice the table is added on the Content Pane. Open the table and you will see the following. In this process, we created a stand-alone table in the geodatabase of the GIS project that compares the African American and Asian American populations county by county across the New York State using the county FIPS code.

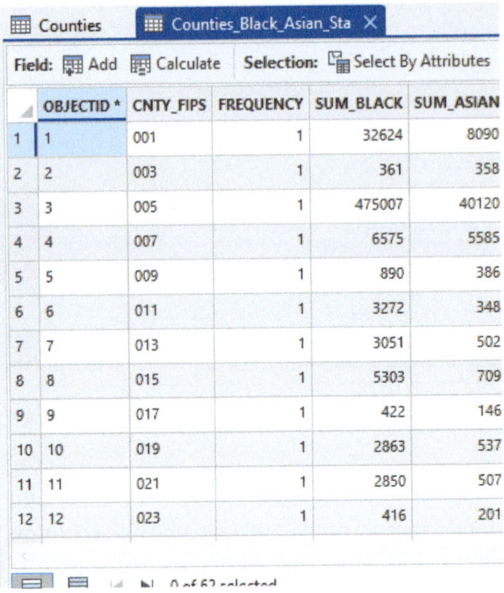


| | OBJECTID * | CNTY_FIPS | FREQUENCY | SUM_BLACK | SUM_ASIAN |
|---|---|---|---|---|---|
| 1 | 1 | 001 | 1 | 32624 | 8090 |
| 2 | 2 | 003 | 1 | 361 | 358 |
| 3 | 3 | 005 | 1 | 475007 | 40120 |
| 4 | 4 | 007 | 1 | 6575 | 5585 |
| 5 | 5 | 009 | 1 | 890 | 386 |
| 6 | 6 | 011 | 1 | 3272 | 348 |
| 7 | 7 | 013 | 1 | 3051 | 502 |
| 8 | 8 | 015 | 1 | 5303 | 709 |
| 9 | 9 | 017 | 1 | 422 | 146 |
| 10 | 10 | 019 | 1 | 2863 | 537 |
| 11 | 11 | 021 | 1 | 2850 | 507 |
| 12 | 12 | 023 | 1 | 416 | 201 |

Close this summary table. Close the attribute table as well for now.

Our next analysis is to add new columns or fields in the attribute table of a feature class. Then, we will calculate the new values or new characteristics for each of the features in this feature class. After we obtain this new characteristic or new field, we can conduct a thematic digital mapping to present the result using a particular characteristic. Our targeted topic to study or our objective of this GIS project is that we would like to calculate the percentage of African American population to the total population in 2010 county by county across New York State. Now, let's open the attribute table of the "Counties" map layer from the Content Pane. In order to add one field, we need to click on the "Options" icon ☰ on the upper-right-hand side of the table window. On the contact menu as follows, select "Fields View." The field view dialog opens on top of the data view of the table.

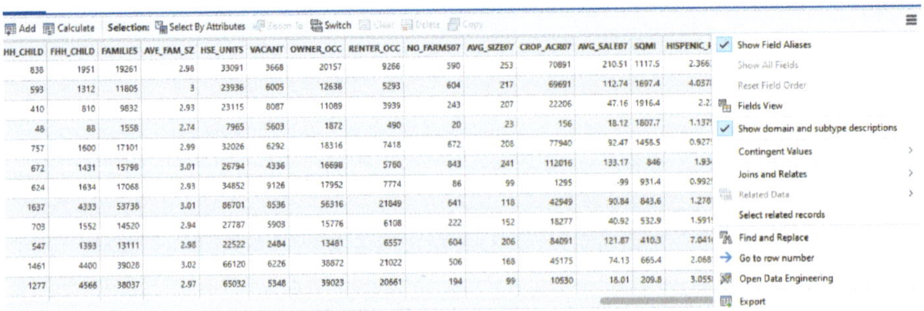


| HH_CHILD | FHH_CHILD | FAMILIES | AVE_FAM_SZ | HSE_UNITS | VACANT | OWNER_OCC | RENTER_OCC | NO_FARMS07 | AVG_SIZE07 | CROP_ACR07 | AVG_SALE07 | SQMI | HISPENIC_I |
|---|---|---|---|---|---|---|---|---|---|---|---|---|---|
| 838 | 1951 | 19261 | 2.98 | 33091 | 3668 | 20157 | 9266 | 590 | 253 | 70891 | 210.51 | 1117.5 | 2.366 |
| 593 | 1312 | 11805 | 3 | 23936 | 6005 | 12638 | 5293 | 604 | 217 | 69691 | 112.74 | 1697.4 | 4.037 |
| 410 | 810 | 9832 | 2.93 | 23115 | 8087 | 11089 | 3939 | 243 | 207 | 22206 | 47.16 | 1916.4 | 2.2 |
| 48 | 88 | 1558 | 2.74 | 7965 | 5603 | 1872 | 490 | 20 | 23 | 156 | 18.12 | 1807.7 | 1.137 |
| 757 | 1600 | 17101 | 2.99 | 32026 | 6292 | 18316 | 7418 | 672 | 208 | 77940 | 92.47 | 1458.5 | 0.927 |
| 672 | 1431 | 15798 | 3.01 | 26794 | 4336 | 16698 | 5760 | 843 | 241 | 112016 | 133.17 | 846 | 1.93 |
| 624 | 1634 | 17068 | 2.93 | 34852 | 9126 | 17952 | 7774 | 86 | 99 | 1295 | -99 | 931.4 | 0.992 |
| 1637 | 4333 | 53738 | 3.01 | 86701 | 8536 | 56316 | 21849 | 641 | 118 | 42949 | 90.84 | 843.6 | 1.278 |
| 703 | 1552 | 14520 | 2.94 | 27787 | 5903 | 15776 | 6108 | 222 | 152 | 18277 | 40.92 | 532.9 | 1.591 |
| 547 | 1393 | 13111 | 2.98 | 22522 | 2484 | 13481 | 6557 | 604 | 206 | 84091 | 121.87 | 410.3 | 7.041 |
| 1461 | 4400 | 39028 | 3.02 | 66120 | 6226 | 38872 | 21022 | 506 | 168 | 45175 | 74.13 | 665.4 | 2.068 |
| 1277 | 4566 | 38037 | 2.97 | 65032 | 5348 | 39023 | 20661 | 194 | 99 | 10530 | 18.01 | 209.8 | 3.055 |

Scroll down to the bottom of the field view of the attribute table. We see the last row by default is named "Field." Or you can click on the bottom row to add a new field. We can type in the new name and select the settings to create a new field. For the "Field Name," we type "Percent_AAmerican." For "Alias," type in "P_Blacks." For "Data Type," select "Double" and keep the default for "Number Format" as "Numerical."

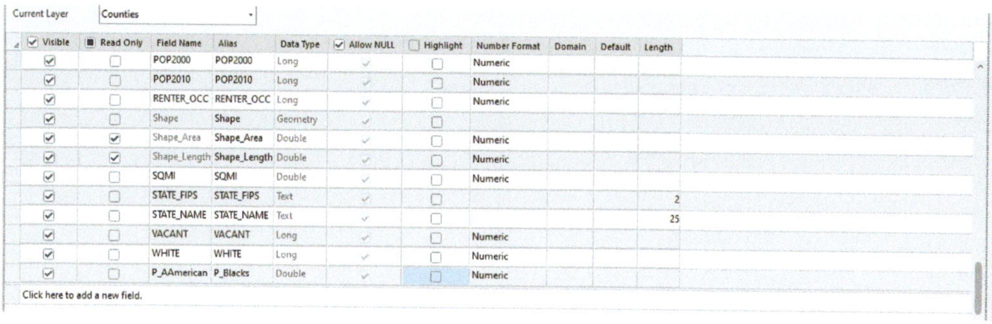


We can close the "Fields View" tab now. We notice that the "P_Blacks" field or Column is added at the far-right end of the attribute table on the data view. Now, we need to conduct the calculations from current fields to populate the values of "P_Blacks" field (Percent of African American population by county). Now, highlight the new field and right-click on it to open the contact menu.

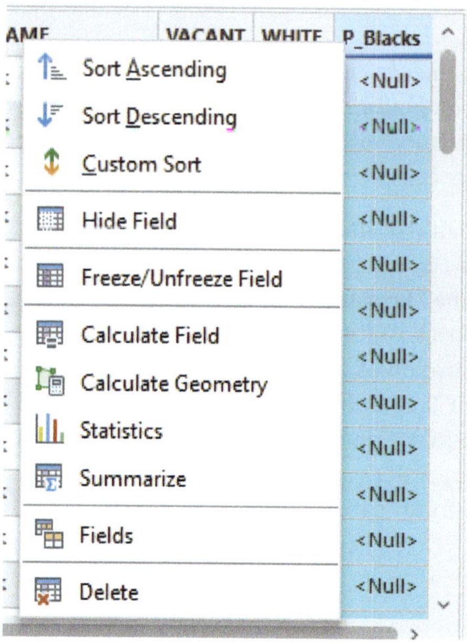


Choose "Calculate Field" (Field Calculator) and the "Calculate Field" dialog opens. It needs to be emphasized that this dialog only deals with the new field "P_Blacks" and there is no need to enter the field name again in this case. Select existing fields and numerical operators to fill "P_Blacks =."

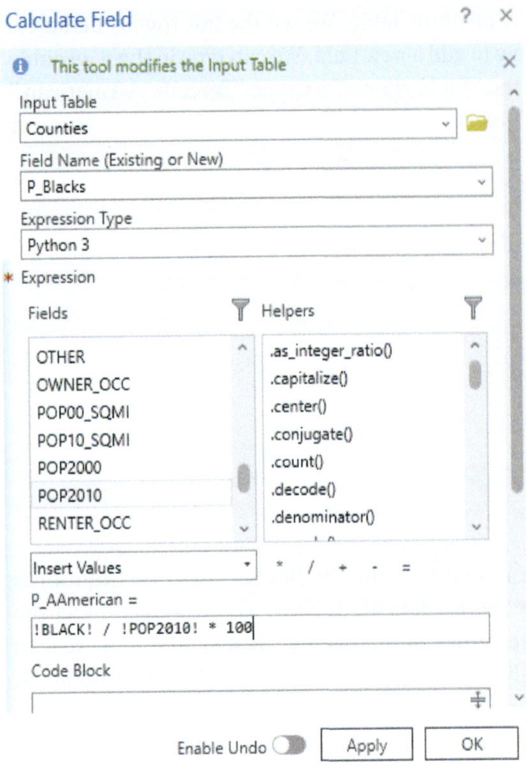


The fill-in should be "!BLACK! / !POP2010! * 100." This means "P_Blacks = Black / POP2010 x 100. Click "Apply" and "OK" to process the data. Now, you notice that the P_Blacks" field is populated. Sort the values by descending, we find that the county has highest percentage of African American population is Kings County, and the percentage is 34.98%.

We can close the attribute table now. Right-click on the title of the "Counties" map layer on the Content Pane. Choose the "Symbology" menu to open the Symbology Pane. It might be displayed on top of the Catalog Pane on the right-hand-side. For the "Primary symbology" drop-down list, select "Graduated Colors." For the "Field," select "P_Blacks." Change the "Classes" to 6. That means we will have a total of 6 categories in this classification of percent of population. Change the classification "Method" to "Equal Interval." Expand the "Color Scheme" and check "Show names." Now, select the color scheme as "Yellow-Orange-Red (Continuous)." Close the Symbology Pane. The thematic map of the percentage of the African American population by counties now is presented in the main mapping window. Checking off the "Cities" map layer, the digital map should look like the following. Save the project.

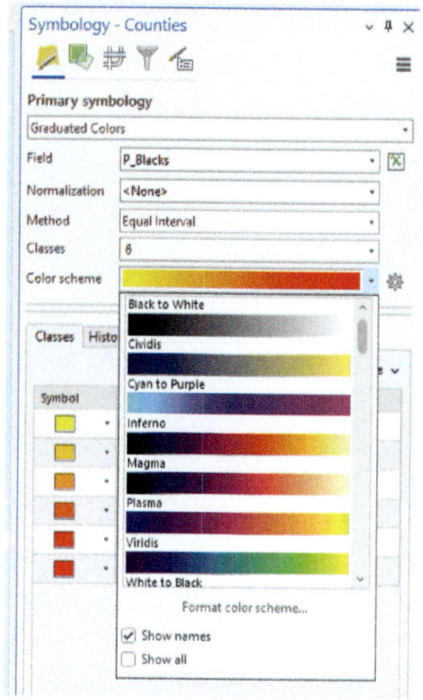


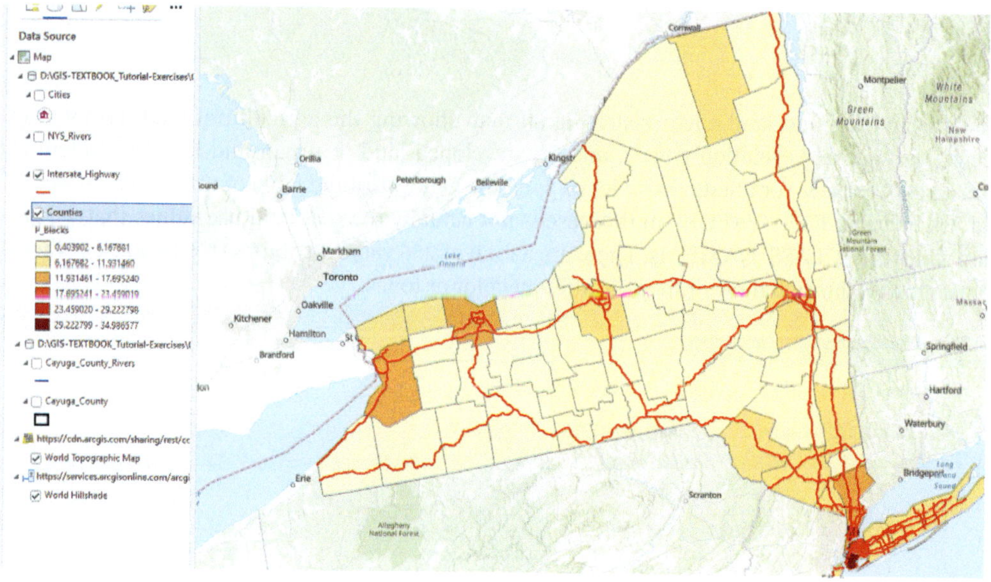


So far, we explored vector GIS database. Let us explore raster database now. Expand the "Folders" category on the Catalog Pane, then the "Chapter02-Tutorial." We see the "New_York_State_Agricultural_Land_Use" folder. Double-click on this folder, it is expanded and the raster data file "Agricultual_Land_Use_2020_36.tif" is shown. Add this raster dataset or map layer into the "Map" window. Although the dataset is added, we still cannot view spatial data on a digital map because it is loaded on the bottom of the Table of Contents on the left side of the software platform. In general, ArcGIS Pro can detect the data format automatically and rank the digital map layers as follows: vector-points, vector-lines, vector-polygon, and raster-grid cells. We can manually rearrange the "Agricultual_Land_Use_2020_36.tif" to put it on the top while the "Drawing Order" is activated on the Table of Contents.

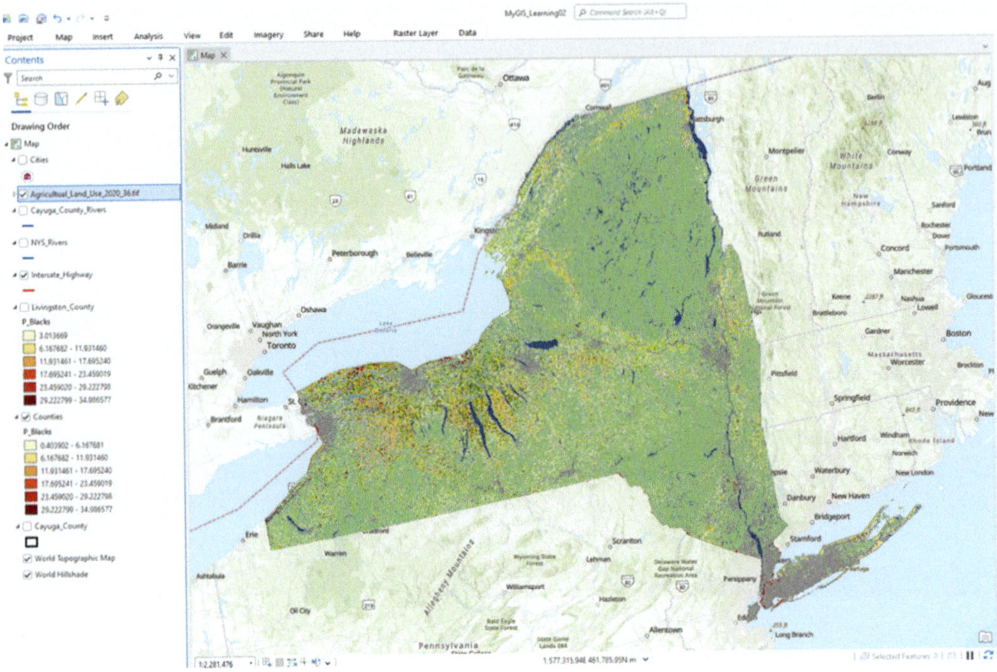


This is a land use and land cover raster digital map showing the agricultural land use by crop species, the natural vegetation covers, and the developed land in urban and human residential areas across the New York State. Now, let us open the value attribute table of this raster dataset to read it. The value field "VALUE" in this case is not actually the real land use values that we are using while the "CLASS_NAME" is. There are a total of 256 value attributes (rows). Each of them is presented on the digital raster map as different color or tones.

Agricultual_Land_Use_2020_36.tif

Field: Add Calculate Selection: Select By Attributes

| | OID | VALUE | RED | GREEN | BLUE | CLASS_NAME | OPACITY |
|---|---|---|---|---|---|---|---|
| 1 | 0 | 0 | 0 | 0 | 0 | Background | 0 |
| 2 | 1 | 1 | 1 | 0.827 | 0 | Corn | 1 |
| 3 | 2 | 2 | 1 | 0.149 | 0.149 | Cotton | 1 |
| 4 | 3 | 3 | 0 | 0.659 | 0.894 | Rice | 1 |
| 5 | 4 | 4 | 1 | 0.62 | 0.043 | Sorghum | 1 |
| 6 | 5 | 5 | 0.149 | 0.439 | 0 | Soybeans | 1 |
| 7 | 6 | 6 | 1 | 1 | 0 | Sunflower | 1 |
| 8 | 7 | 7 | 0 | 0 | 0 | | 1 |
| 9 | 8 | 8 | 0 | 0 | 0 | | 1 |
| 10 | 9 | 9 | 0 | 0 | 0 | | 1 |
| 11 | 10 | 10 | 0.439 | 0.647 | 0 | Peanuts | 1 |
| 12 | 11 | 11 | 0 | 0.686 | 0.294 | Tobacco | 1 |
| 13 | 12 | 12 | 0.867 | 0.647 | 0.043 | Sweet Corn | 1 |
| 14 | 13 | 13 | 0.867 | 0.647 | 0.043 | Pop or Orn Corn | 1 |
| 15 | 14 | 14 | 0.494 | 0.827 | 1 | Mint | 1 |

0 of 256 selected

Close the attribute table. Now, let us enlarge the raster digital map as the Finger Lake Region in the center. Your process can roughly follow the example of following imagery steps.

1. Scale- 1:382,768.

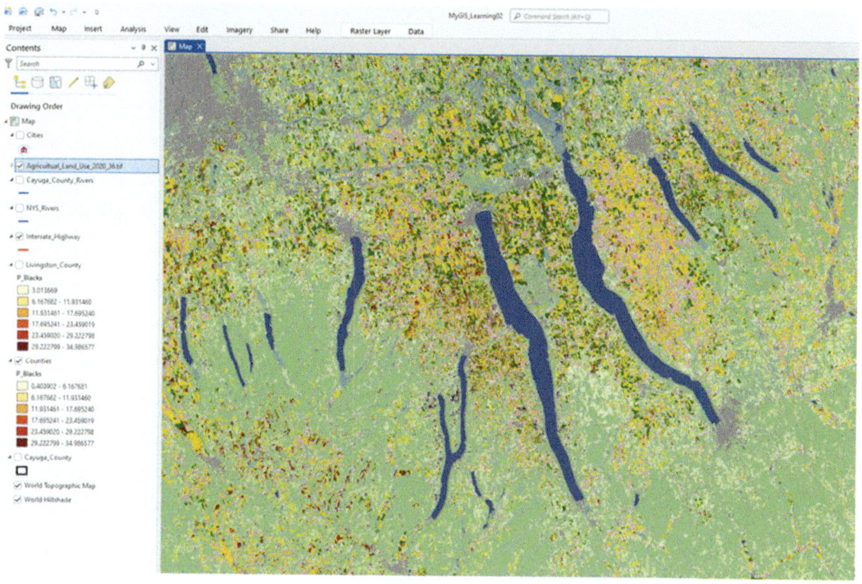


2. Scale- 1:100,340.

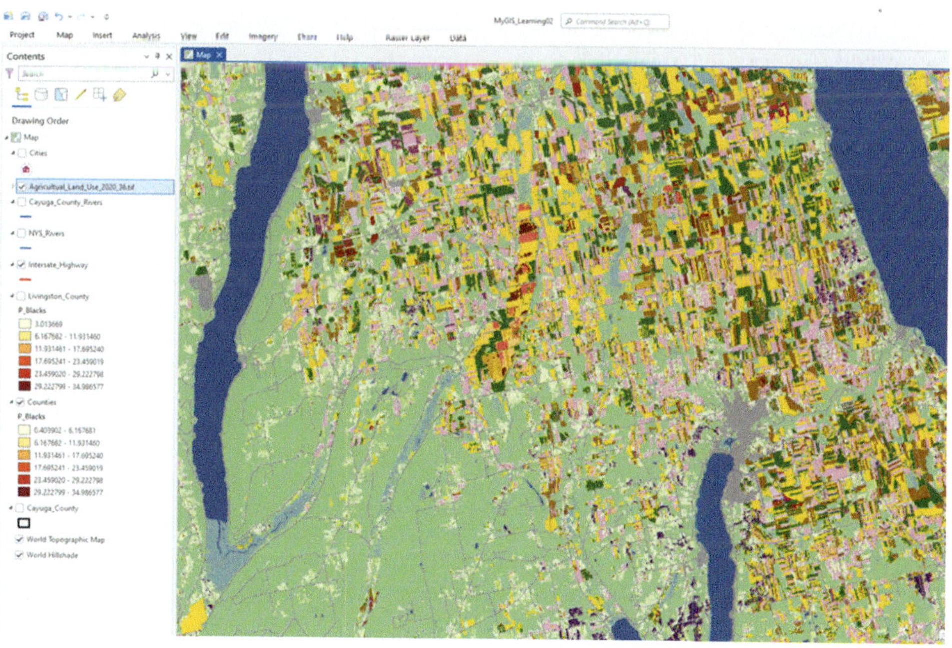

**3.** Scale- 1: 5000.

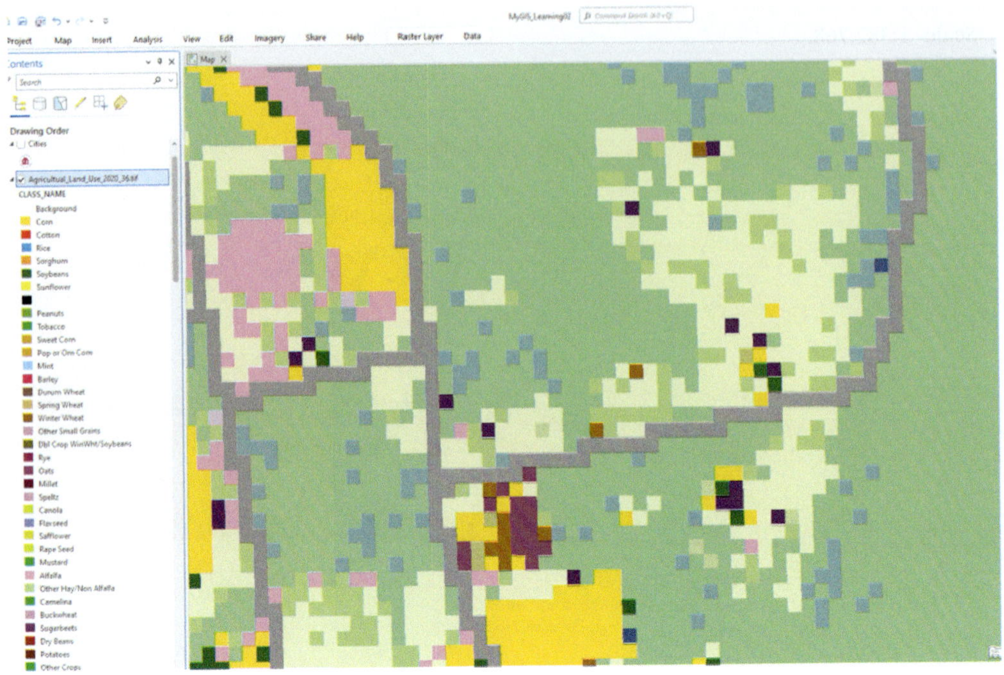


Finally, we can see clearly the grid cells that make this spatial dataset (map layer) on the digital map presentation. Each of the colors of the grid cells is represented in the value attribute table as a row of attributes. Inspect the colors of the grids here and the categories in the Table of the Contents. After the inspections, right-click on the layer on the Table of Contents to "Zoom to the Layer" (show entire map layer).

Congratulations, you accomplished the tutorial.

Save the project file and close the ArcGIS Pro.

# Part II

# Fundamentals of Geographic Information Data Processing

# 3

# Geographic Information Database and Database Management

## Concepts and Theories

Geographic information system (GIS) in essence is a specific type of information system. An information system is defined in computer science as an integrated set of components for collecting, storing, and processing data and for providing information, knowledge, and digital products (Stair and Reynolds 2021). The purpose of building an information system is to solve the real-world problems and for decision making. Information systems use data stored in computer databases to provide the needed information. A database is an organized collection of interrelated data reflecting a major aspect of a firm's activities or an issue of general concerns. Therefore, data or database are only the raw facts, the material for obtaining information, although the two concepts are often inter-exchanged in practices (Stephens 2009). The unique approach of GIS is that it collects, stores, and processes geographic or spatial data for the objects on the Earth's surface.

In computer science, software packages of database management system (DBMS) are developed to handle the storage, retrieval, and updating of data. In designing and managing the DBMS, four different database structures were developed. These are: (i) hierarchical (Figure 3.1), (ii) network (Figure 3.2), (iii) relational (Figure 3.3), and (iv) object-oriented structures (Figure 3.4). A hierarchical database structure is a tree-like structure. Data is stored hierarchically and is represented by using a parent–child relationship. The network database structure allows each child to have multiple parents. This approach addresses the need to model complex relationships. The relational database (RDB) structure is based on normalizing data in rows and columns of the tables and is stored in a fixed structure. The relations of different tables are established by unique identifiers that are contained in both digital tables. RDBs are the most popularly used databases in computer science. In object-oriented databases, data is stored in the form of objects. It defines a database as a collection of objects that stores data members' values, behaviors, and operations.

Fundamentally, the vector GIS database design and implementation in ArcGIS Pro followed both RDB and object-oriented database rules. First, a feature class in a geodatabase is a collection of geographic objects (features) with the same characteristics. Similar to the "person" as the title of the DBMS (Figure 3.4), a geodatabase with a title, such as New York State or the City of Buffalo, is a spatial data collection of the region. Secondly, like the feature classes of "Doctor," "Engineer," "Faculty," or "Student" in our example (Figure 3.4), we have feature classes in a geodatabase as "Streets," "Parcels," "Buildings," etc. In addition, all the attribute tables that are used to illustrate features and objects are designed and implemented with object ID numbers and other IDs to establish the relations of the databases (Figure 3.3). In this chapter, we will discuss the methods on how

*Fundamentals of Geographic Information Systems (GIS) Analysis: Applying ArcGIS Pro*, First Edition. Tao Tang.

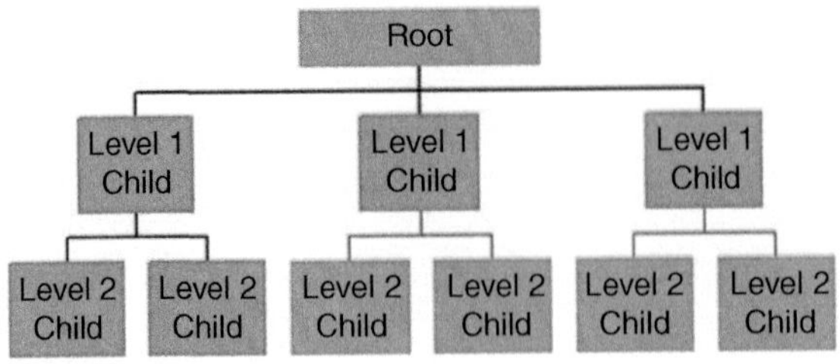


**Figure 3.1** Hierarchical database model in DBMS. *Source:* Adapted from O'Reilly Media, Inc. (2024).

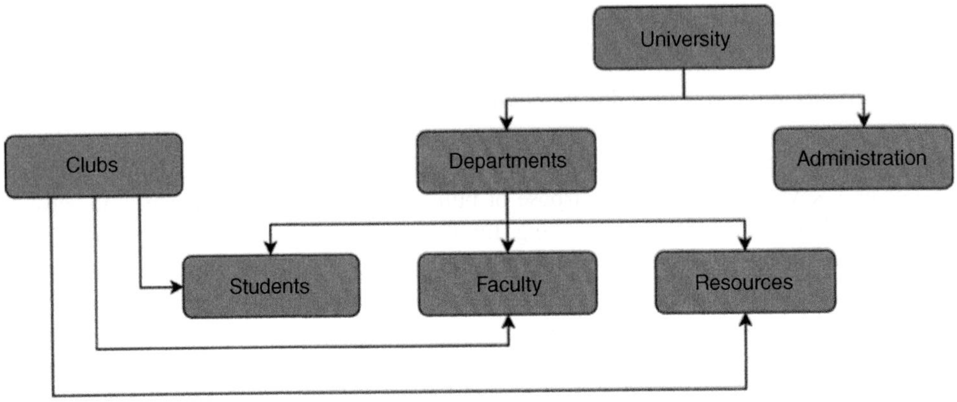


**Figure 3.2** Network model of DBMS using a university campus as an example. *Source:* Adapted from O'Reilly Media, Inc. (2024).

to input geographic objects into GIS databases of both vector type and raster type. We are also going to incorporate nonspatial attribute data into spatial data through the rules of RDB structures.

## Methods of Spatial Data Input in Vector and Raster Data Formats

Data input in GIS is the procedure of encoding both spatial and attribute data into a computer-readable form and writing it into the GIS database. In general, there are a total of six different techniques in the GIS environment for data input. These are: (i) manual attribute input, (ii) digitizing, (iii) heads-up digitizing, (iv) automatic scanning, (v) entry of coordinates and conducting coordinate geometry (COGO), and (vi) conversion of existing digital data from another format.

Manual attribute input is simply typing on the keyboard or importing text or excel tabular data fields into the GIS database. Keyboard entering is an original method for almost all the computerized databases including GIS database. Manual attribute input is a major method for GIS analysts to input attribute data in the GIS database. Attribute data is equally important to that of spatial data presented by digital maps in GIS. Applying database joins and spatial joins based on location information, GIS analysts can obtain comprehensive spatial databases to solve many complicated problems in our society.

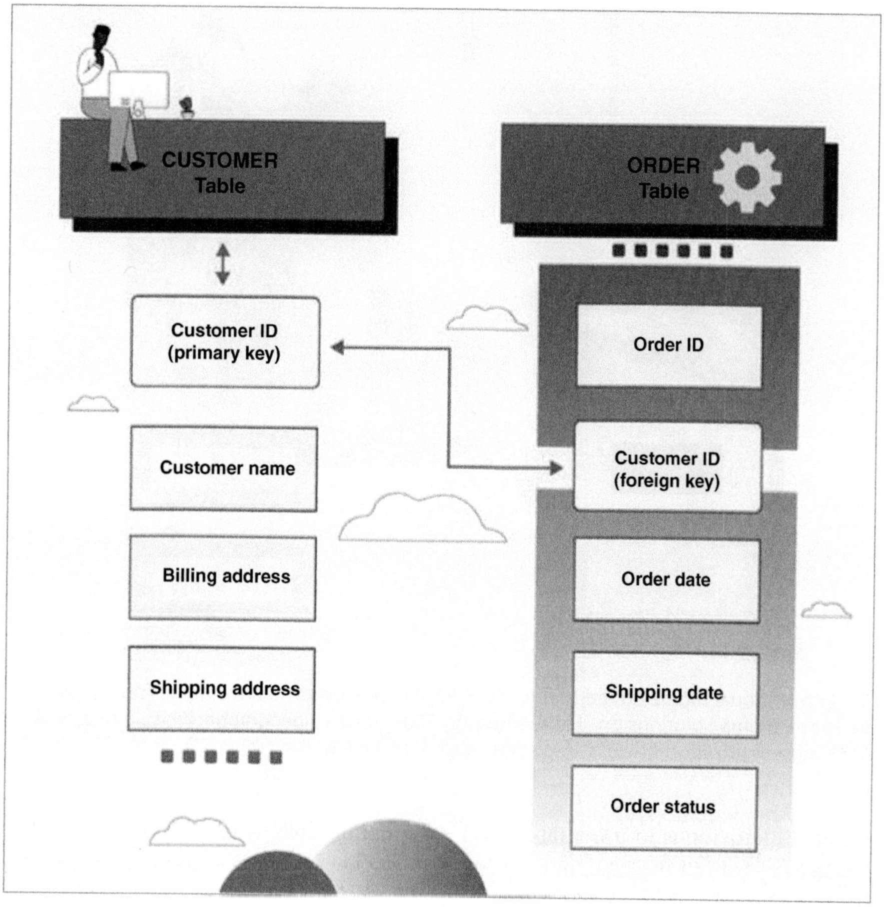


**Figure 3.3** Simple example of an RDB model for a business company managed by DBMS. *Source:* Adapted from O'Reilly Media, Inc. (2024).

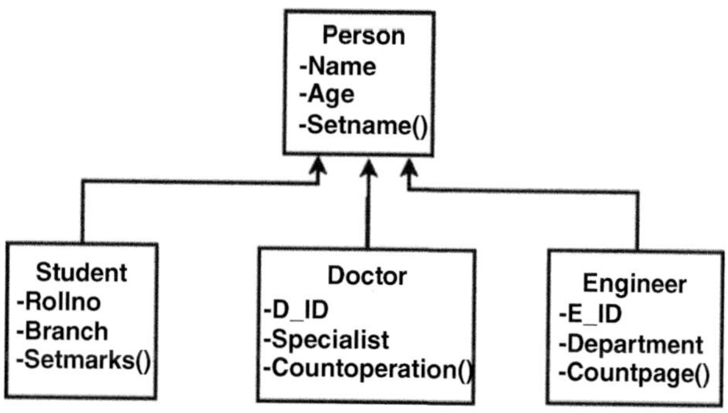


**Figure 3.4** Simple example of an object-oriented database model in DBMS. *Source:* Adapted from O'Reilly Media, Inc. (2024).

**Figure 3.5** Conventional digitizing method to input hard copy maps and drawings using a digitizing table and tablet. *Source:* Sigma Solutions Co., Ltd./Geography Today/https://geographerswasaut.blogspot.com/2015/01/map-digitization-process-in-gis.html/last accessed on November 20, 2024.

Conventional digitizing is to use a tablet and a digitizing table, which is a special table with a grid of fine wiring behind the face, in entering digital codes into the computer to store the geographic objects or features in the GIS database (Figure 3.5). The original purpose for digitizing is to convert hard copy map data into the computerized or digital format. Digitizing technique only creates vector type of GIS data. While raster type of data is mainly collected using various remote sensing methods directly.

Heads-up digitizing combines either digital remote sensing imageries or scanned hard copy maps and drawings as the backdrop on the computer screen. Then, the GIS analyst will be able to trace and record points, lines, and polygons of features on the computer screen. This way, we are heads up working on the computer screens rather than heads down working on a digitizing table. The revolution of heads-up digitizing is that we reduced or eliminated use of digitizing table. Most current digitizing works are accomplished this way now.

In order to capture the data at a fast speed, automatic scanning is used, and the large format and high-resolution scanners for map and engineering drawings are used to convert the spatial data automatically (Figure 3.6). The original results of this type of process are raster data. Then, we can convert it into vector data format using the heads-up digitizing method if it is needed (Figure 3.7). In addition to scanned hard copy maps and engineering drawings, a large quantity of remotely sensed data, such as digital aerial photographs obtained by a human-controlled aircraft or a drone (UAV) and a variety of satellite images, can be used as the digital reference materials for heads-up digitizing, raster-to-vector conversion, and geographic objects or feature extractions.

**Figure 3.6** Large format map and engineering drawing scanner. *Source*: dja65/Adobestock

**Figure 3.7** The heads-up digitizing in an office or outdoors on a mobile device. *Source:* U.S. Bureau of Land Management/Public domain.

Converting tabular data of coordinates using COGO in GIS is to map the location of features and their boundaries measured on the ground and recorded on survey plans, deed descriptions, and other types of physical or electronic land record documents. Survey plans can include references to points with known coordinates that spatially locate or georeferenced objects or features within a particular coordinate system. These points are known as control points, monuments, or cadastral reference points. Therefore, points, or the key points of features can be converted straightforwardly in the digital environment in a GIS database. Using COGO tools in a GIS environment, lines and curves can be constructed using surveyed points with known x, y coordinates or x, y, z coordinates in a 3D digital environment. Polygons, such as parcel areas of land ownership, can be constructed by enclosed COGO lines.

Conversion of existing digital data mainly indicates to translation of other types of spatial data, such as CAD files, to the GIS database format. AutoCAD (DWG/DXF) and Microstation (DGN) files are read-only files in ArcGIS Pro feature datasets. Therefore, GIS analysts need to convert this type of file into the GIS databases using the tools built in the software package. The remote sensing imageries either from space or airborne are by nature aligned with the raster data format in GIS. Therefore, they can be converted to raster GIS data first in the digital environment.

## Incorporating Geographic Features and Events with Their Attribute Data

Almost all of the tabular data used in a GIS system are stored in RDB tables. The design and implementation of the RDB is a major approach in computer science. RDB is a way of structuring information in tables, rows, and columns. Major characteristic of an RDB is its ability to establish links, or relationships, between information that was collected in different field works and in different environments. The main task to achieve in combining information is to join or relate the database tables. This digital process makes the database easy to understand and gain insights about the relationship between various data items.

Practically, in computer science other than in GIS, tabular data are commonly held and manipulated in RDBs such as dBASE, rBase, ACCESS, Oracle, SQLServer, INFORMIX, or other high-end relational database management systems (DBMS). RDBs are the most popular type of databases in the computational science and information technology (IT) industry because they allow users to understand data items easily. These databases are related to each other in one or more tables inside the database and provide relational operations such as to query using structured query language (SQL) and manipulate the data.

Database relationships are associations between tables that are created using join statements to retrieve data in the computational science. All tables in an RDB should have a primary key. A primary key is the unique identifier of the data objects or data items that many other attributes are associated with. The value in the primary column determines which attributes are used to create the primary key. A relationship can be established between two database tables when one table uses a foreign key that references the primary key of another table (Figure 3.8). For example, the Customer ID or ID in the Customers database is the primary key or unique ID numbers of customers in this case (Figure 3.8). When a customer orders a product from this company, the order is registered in the Orders database (Customer Order database). The primary key in this database is Order ID or ID number in the table. Customer ID is also entered in the Orders database as a foreign key. Straightforwardly speaking, the reason it is called foreign key because it is not the primary key of this database and rather it is the primary key of a foreign database or table (Customers database).

The types of relationships in the database can be recognized as following. These relationships are also called cardinality in GIS.

1. One-to-one: Both tables can have only one record on each side of the relationship. Each primary key value relates to none or only one record in the related table (Figure 3.9). For instance, one country has only one capital city in Figure 3.9.

**Customers**

| ID | CustomerName | CustomerAge | CustomerCountry |
|---|---|---|---|
| 1 | Salvador | 23 | Brazil |
| 2 | Lawrence | 60 | China |
| 3 | Ernest | 38 | India |

**CustomerOrders**

| ID | OrderDate | CustomerID | Amount |
|---|---|---|---|
| 1 | 2019-04-29 00:00:00.000 | 1 | 968 |
| 2 | 2019-05-10 00:00:00.000 | 2 | 898 |
| 3 | 2019-10-21 00:00:00.000 | 3 | 47 |

**Figure 3.8** Illustration of customer ID as the primary key in customers database while a foreign key in the customer order database.

2. One-to-many: The primary key table contains only one record that relates to none, one, or many records in the related table (Figure 3.10).
3. Many-to-many: Each record in both tables can relate to none or any number of records in the other table. These relationships require a third table in computational science, called an associate or linking table, because relational systems cannot directly accommodate the relationship (Figure 3.11). In the following case in Figure 3.11, one author can publish many books while one book can be published by several authors.

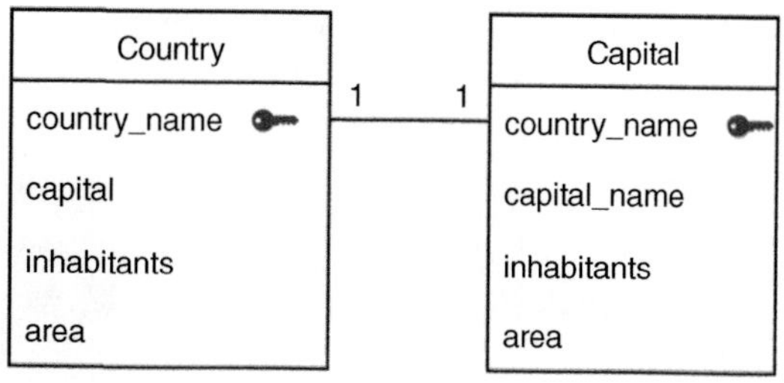


**Figure 3.9** One-to-one relationship.

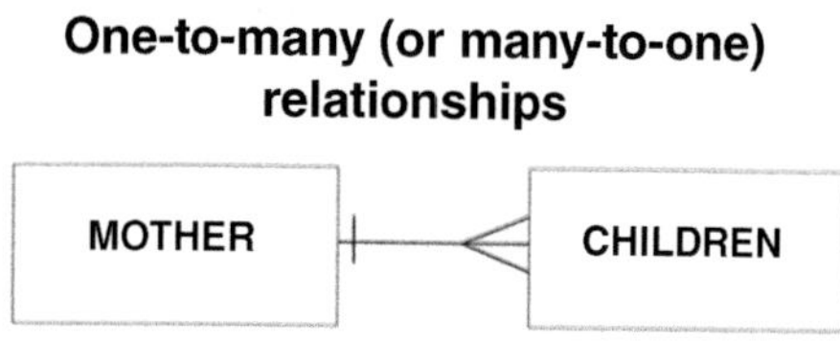


**Figure 3.10** One-to-many relationship.

**Figure 3.11** Many-to-many relationship illustrated by relationships of authors database and published books database.

Based on the perspective of database management, both digitized vector feature class data and input raster data in the GIS database are RDB that can be joined or related to another attribute or spatial database. A table join in GIS appends all the columns from one table into the other table based on the unique ID of the two joining tables. This increases the capabilities for GIS analysts to use GIS to digitally map, visualize, and analyze the data that was collected at different times and for different purposes in order to achieve our current goals in a GIS project.

Unique ID or unique identifier is the common terminology for both primary key and foreign key when we join two or more tables. Joins are most suitable for a one-to-one (1:1) relationship. Even if we can conduct the join operations for one-to-many (1:many) relationships, normally only the first item or feature in the joining table is added to the expanded new table. While join appends attributes to the end of a table, relate creates a temporary table for all the matching records. Relate is ideal for a one-to-many (1:many) relationship. If we do not have any unique field to join, or we have many-to-many relationships, we can try a spatial join in GIS. This process uses location or proximity models to join the attribute fields. In particular, a spatial join can help us to assign attributes to feature classes by the locations of objects or features. This process is unique in the GIS environment only. Spatial join is also considered as a unique process to combine both attribute and spatial data. It will be discussed in Chapter 6.

In contrast to attribute tables of vector data, raster datasets in GIS incorporate attribute tables which are called **value attribute tables**. This is because raster datasets normally contain thousands or millions of grid cell units per digital map layer, so it is not feasible nor practical to link one row of attributes to one grid cell unit (Figure 3.12). In practice, raster datasets or images typically carry grid or pixel values that represent certain classes, groups, categories, or memberships. For example, a land-cover or land use map classified using a satellite image may have some land-cover categories as forest, wetland, crop, and urban. Each of the grid cells is

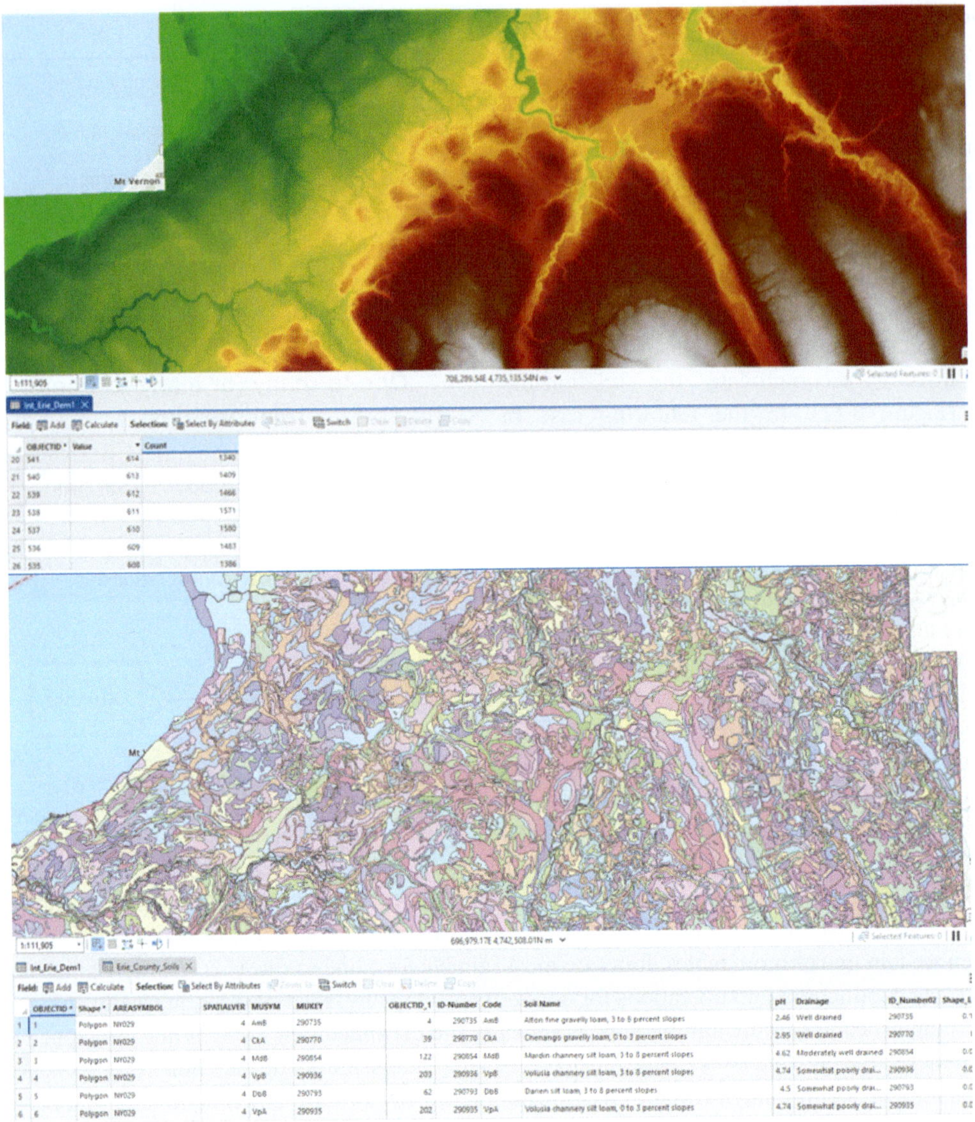


**Figure 3.12** Comparison of raster value attribute table (top) with polygon feature attribute table (bottom) in Erie County, NY.

assigned a number that represents the land-cover class that is contained in the value attribute table as follows.

1. Forest
2. Wetland
3. Crop
4. Urban

In ArcGIS Pro, when a raster attribute table is generated, three default fields are created in the table: OID, VALUE, and COUNT. The content in these fields cannot be edited. The ObjectID (OID) is a unique, system-defined object identifier number for each row in the table. VALUE is a list of each unique cell value in the raster datasets. In the grid cell file format of ArcGIS Pro, only integer data is accommodated. COUNT represents the number of cells in the raster dataset with the cell value in the VALUE column. Cell values represented by NoData are not calculated in the raster attribute table. For instance, the DEM (digital elevation models) of Erie County, NY in raster data format is shown on the top portion of Figure 3.12. The value field (column) in the attribute table is the elevation or altitude by clusters (or class) and the count column is the depiction of a total of how many grid cells on the raster digital map have this specific value. By contrast, the vector polygon map layer of soils in same region showing one polygon is associated with one row (record) in the attribute table with soil names, pH values, etc. Raster attribute tables can be joined or related as well using unique ID of value categories. However, spatial join is not applicable to raster data currently in ArcGIS Pro. Or spatial join is applied in raster by map algebra process now.

## RDB, Spatial, and Structural Querying of Database

SQL is a standardized programming language that contains a set of defined syntax and expressions used for accessing and managing data in databases in general and in the GIS database in particular. Queries are constructed from a command language that lets the users select, insert, update, and locate data. SQL is a standard of both American National Standards Institute and International Standards Organization, although many databases support SQL with proprietary extensions. In summary of SQL statements, it can be identified into following six categories. These are: (i) data definition language (DDL) statements, (ii) data manipulation language (DML) statements, (iii) transaction control statements, (iv) session control statements, (v) system control statement, and (vi) embedded SQL statements (Oracle 2003; Figure 3.13).

DDL statements help us to perform the tasks, such as, creating or altering schema objects; granting and revoking privileges and roles of database management; analyzing information on a table, index, or cluster, and establishing data audit options. DML statements access and manipulate data in existing schema objects. These statements, such as delete or insert, do not implicitly commit the

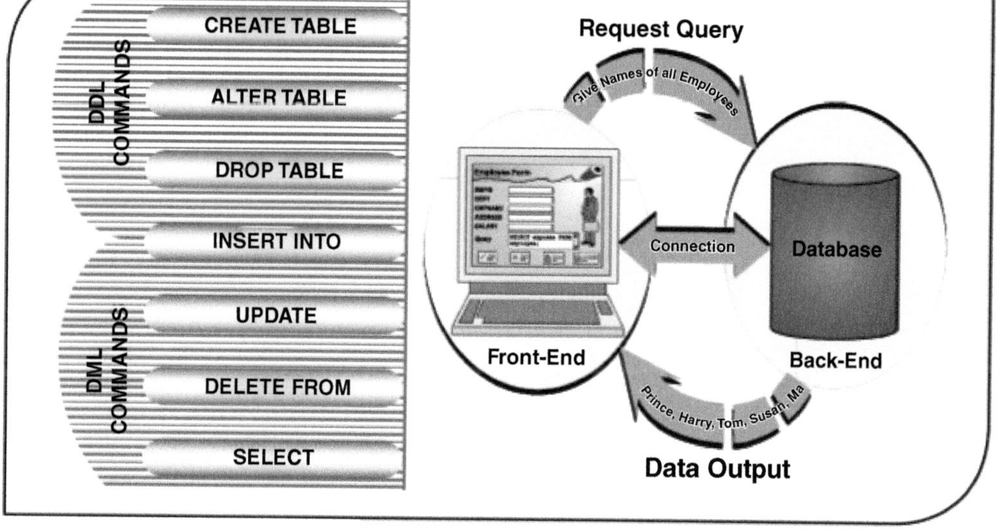


Figure 3.13 SQL in practice. *Source:* Adapted from O'Reilly Online Learning (2024).

current transaction. Transaction control statements, such as commit or rollback, manage changes made by DML statements. Session control statements dynamically manage the properties of a user session. Embedded SQL statements place DDL, DML, and transaction control statements within a procedural language program (Oracle 2003).

Query expressions in ArcGIS Pro adhere to standard SQL expressions. A SQL expression contains a combination of one or more values, operators, and SQL functions that can be used to query or select a subset of features or objects and table records within ArcGIS. Examples of SQL in ArcGIS Pro are (i) STATE_NAME = "New York"; (ii) select * from table where date <'2007-05-29'; or (iii) SUB_REGION = "New England" AND NOT STATE_NAME = "Maine" (Figure 3.14).

Spatial query refers to the process of retrieving a data subset from a digital map layer by working directly with the geographic features or objects in GIS database. In a GIS or spatial database, data are stored in attribute tables and feature-based spatial tables. Although both raster and vector data use Boolean connectors, raster data query uses these connectors to isolate raster cell values from other cell values within one raster map and link the selected cell values to cell values of other raster layers. Spatial queries can only be performed on a single vector file in every process as they can store multiple attributes simultaneously.

The rules of spatial or geographic query are mainly based on the judgment of locations, distances, and proximities of features or objects. Such kinds of rules can be straightforwardly measured and visualized through the embedded network analysis functions. Spatial or geographic query works for both ordinary GIS features and network features that are created in network analysis and modeling in GIS. In ordinary feature class spatial query, the practice in general is presented as a map layer on map layer selection. The result is to produce a new feature class.

By contrast, the spatial query rules are used to append new network features to diagrams based on their location relative to the network features currently represented in the diagram. The Spatial Query rule works from network features that belong to a given source feature class, with those features being optionally filtered out based on specific attribute values. It also appends network features to diagrams that belong to a given source feature class, with those features being optionally filtered out based on specific attribute values.

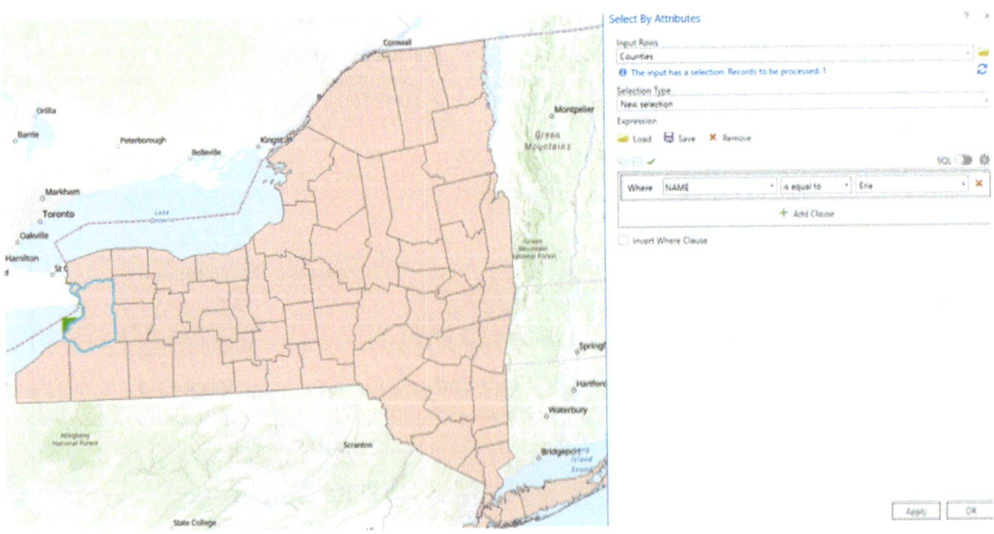


**Figure 3.14** Database query (Select by Attribute-SQL) in ArcGIS Pro.

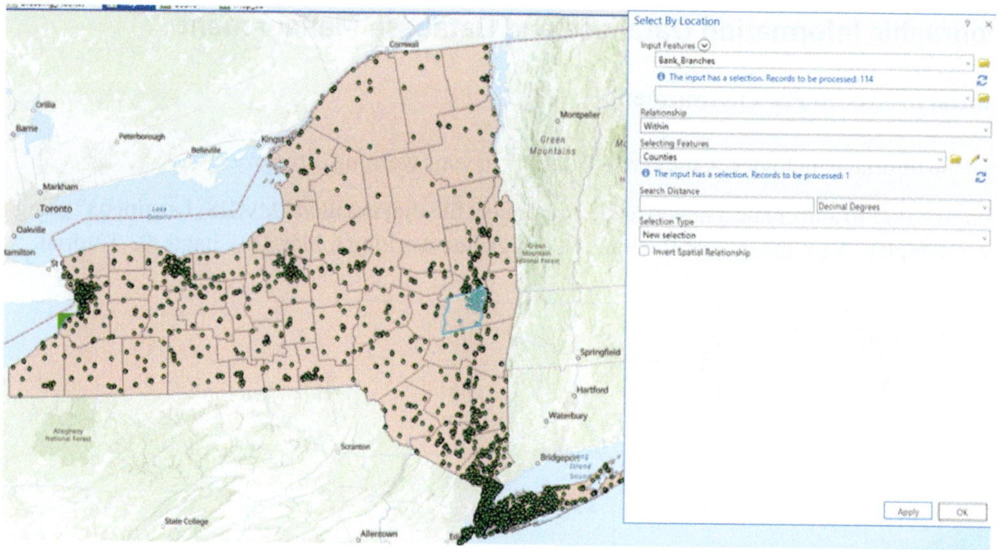


**Figure 3.15** Example of spatial query to yield result of bank branches within the Albany County, NY.

The specified major spatial query rules and statements in the ArcGIS Pro environment are (i) Intersect, (ii) Are within a distance of, (iii) Contains, (iv) Are within, (v) Have a boundary that touches, (vi) Share a line segment with, (vii) Are crossed by the outline of, (viii) Have its center in (Figure 3.15). For instance, Albany County was selected among all the counties across the New York State using a regular query in Figure 3.15. In this GIS project, our objective is to find the bank branches within the county and the list of their names and locations. Therefore, we built the spatial query (Select by Location) to use the selected county (Albany) on the counties feature class to select bank branches in the feature class of Bank_Branches that **are located within** the Albany County polygon.

In summary, GIS is an effective platform that can help us to combine, synthesize, and analyze large datasets from various data sources. This unique capability and characteristic can help us to visualize the situations for answering questions such as "What happened or is happening? At what location?" etc. Therefore, GIS analyses are useful and significant approaches for decision makings on the Earth's surface.

## References

Oracle (2003). Oracle® Database SQL Reference. https://docs.oracle.com/.

O'Reilly Media, Inc. (2024). Introduction to Information Technology. https://www.oreilly.com/library/view/introduction-to-information/9789332525146/xhtml/chapter021.xhtml.

Stair, R. and Reynolds, G. (2021). *Principles of Information Systems*, 4e. p. 438. Independence, KY 41051: Cengage Publisher, ISBN: 978-0357112410

Stephens, R. (2009). *Begin Database Design Solutions*, 2e. p. 515. Hoboken, NJ 07030: John Wiley Publishing Company.

## Geographic Information Database and Database Management

### Tutorial in ArcGIS Pro Environment

Copy the tutorial data folder "Chapter03-Tutorial" to your working directory.

Open ArcGIS Pro and create a new ArcGIS Pro project file that is called "MyGIS_Larning03" using "Map" template. Save the project file into the designated working folder, D:\Chapter03-Tutorial.

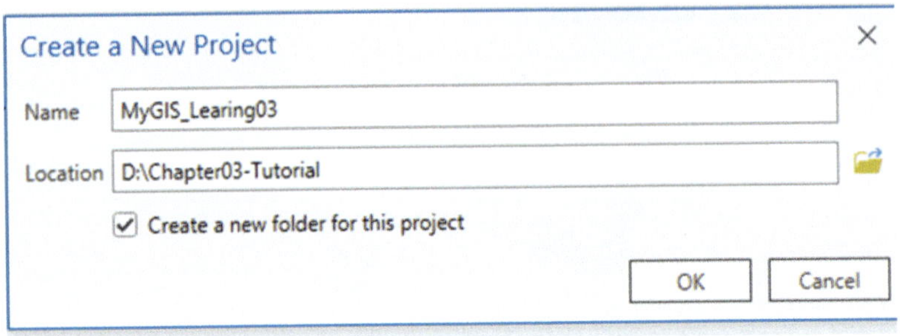


Now, add the entire Chapter03 -Tutorial folder on the Catalog Pane using the contact menu of "Add Folder Connections."

1. Join a standalone data table with the attribute table of a feature class to enlarge the geodatabase and enhance the GIS analysis

Expand the Chapter03 -Tutorial folder then the NYS_Civil_Boundaries.gdb, geodatabase on the Catalog Pane. Add two feature classes, Counties and NYS_Block_Group to the mapping window; also add the standalone table of NYS_Blcokgroup_Population_2020 into the mapping window. Expand the "Maps" category on the Catalog Pane to change the mapping window title from "Map" to "Join_Table."

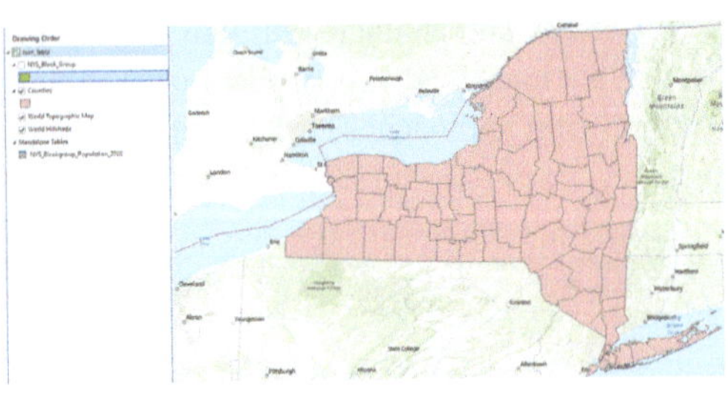

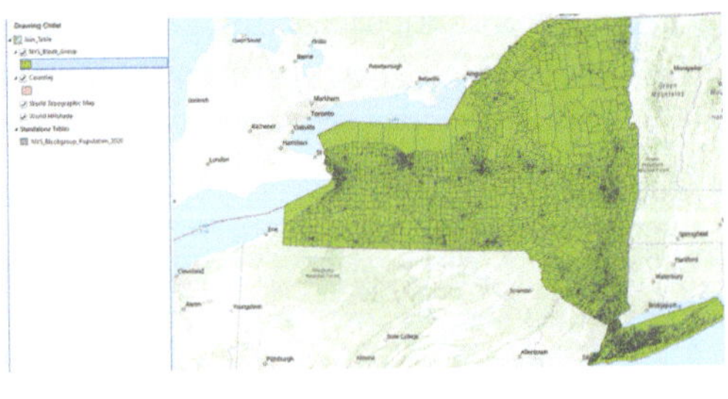

Now open the attribute table of NYS_Block_Group by right-click on the title of the map layer in the Table of Contents. Although we found there are a total of 16,070 block groups geographic areas (regions) that were created by the United States Census Bureau across the entire New York State, there are no population or demographic data fields or columns in this database. Therefore, this digital map has insufficient values now that we can utilize to thematically present the population distributions, such as showing the distribution of gender or ethnic groups across the entire New York State. We need to find and join a tabular dataset with all the thematical information that we need in order to map and analyze the spatial pattern and characteristics of population distribution. In essence, the database items in ArcGIS Pro environment or converted from a different file format to ArcGIS Pro, are relational databases. We can use the unique identification, either a number or a text string to join the two database tables into one in order to enlarge the GIS spatial database. In this case, the unique ID is the "GEOID" field in the attribute table.

Now, close the attribute table and open the standalone table of NYS_Blcokgroup_Population_2020. This is a table that contains the population and demographic information by each of the block-groups across the entire New York State. The dataset is summarized based on 2020 US Census field surveys. This database table was originally downloaded from the US Census Bureau website. The title for each of the fields or columns may be very confusing at this time. It represents a standard "secret" code for a particular demographic data item surveyed by the bureau, such as "Total Population" or "African American Population."

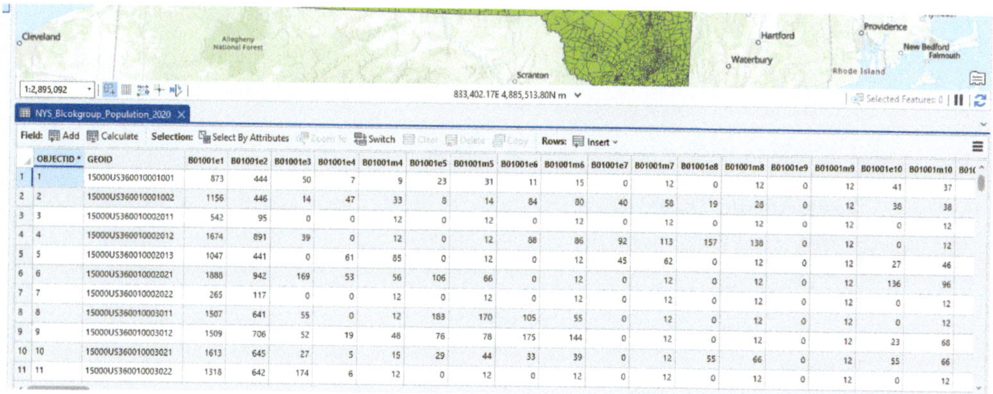


| | OBJECTID * | GEOID | B01001e1 | B01001e2 | B01001e3 | B01001e4 | B01001m4 | B01001e5 | B01001m5 | B01001e6 | B01001m6 | B01001e7 | B01001m7 | B01001e8 | B01001m8 | B01001e9 | B01001m9 | B01001e10 | B01001m10 |
|---|---|---|---|---|---|---|---|---|---|---|---|---|---|---|---|---|---|---|---|
| 1 | 1 | 15000US360010001001 | 873 | 444 | 50 | 7 | 9 | 23 | 31 | 11 | 15 | 0 | 12 | 0 | 12 | 0 | 12 | 41 | 37 |
| 2 | 2 | 15000US360010001002 | 1156 | 446 | 14 | 47 | 33 | 8 | 14 | 84 | 80 | 40 | 58 | 19 | 28 | 0 | 12 | 38 | 38 |
| 3 | 3 | 15000US360010002011 | 542 | 95 | 0 | 0 | 12 | 0 | 12 | 0 | 12 | 0 | 12 | 0 | 12 | 0 | 12 | 0 | 12 |
| 4 | 4 | 15000US360010002012 | 1674 | 891 | 39 | 0 | 12 | 0 | 12 | 88 | 86 | 92 | 113 | 157 | 138 | 0 | 12 | 0 | 12 |
| 5 | 5 | 15000US360010002013 | 1047 | 441 | 0 | 61 | 85 | 0 | 12 | 0 | 12 | 45 | 62 | 0 | 12 | 0 | 12 | 27 | 46 |
| 6 | 6 | 15000US360010002021 | 1888 | 942 | 169 | 53 | 56 | 106 | 66 | 0 | 12 | 0 | 12 | 0 | 12 | 0 | 12 | 136 | 96 |
| 7 | 7 | 15000US360010002022 | 265 | 117 | 0 | 0 | 12 | 0 | 12 | 0 | 12 | 0 | 12 | 0 | 12 | 0 | 12 | 0 | 12 |
| 8 | 8 | 15000US360010003011 | 1507 | 641 | 55 | 0 | 12 | 183 | 170 | 105 | 55 | 0 | 12 | 0 | 12 | 0 | 12 | 0 | 12 |
| 9 | 9 | 15000US360010003012 | 1509 | 706 | 52 | 19 | 48 | 76 | 78 | 175 | 144 | 0 | 12 | 0 | 12 | 0 | 12 | 23 | 68 |
| 10 | 10 | 15000US360010003021 | 1613 | 645 | 27 | 5 | 15 | 29 | 44 | 33 | 39 | 0 | 12 | 55 | 66 | 0 | 12 | 55 | 66 |
| 11 | 11 | 15000US360010003022 | 1318 | 642 | 174 | 6 | 12 | 0 | 12 | 0 | 12 | 0 | 12 | 0 | 12 | 0 | 12 | 0 | 12 |

In order to translate the names or meanings of the fields, we need to read the metadata table that comes with this data table. Toward the right-hand side of this table, we have already translated some fields (columns) for our current GIS project. In this case, the "GEOID2" is the unique ID for matching and joining the information with the attribute table of the NYS_Block_Group digital map layer.

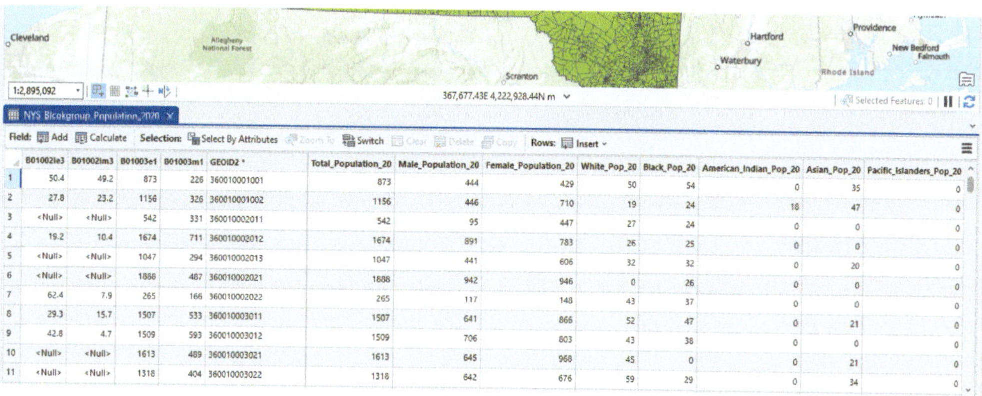


| | B01002le3 | B01002lm3 | B01003e1 | B01003m1 | GEOID2 * | Total_Population_20 | Male_Population_20 | Female_Population_20 | White_Pop_20 | Black_Pop_20 | American_Indian_Pop_20 | Asian_Pop_20 | Pacific_Islanders_Pop_20 |
|---|---|---|---|---|---|---|---|---|---|---|---|---|---|
| 1 | 50.4 | 49.2 | 873 | 226 | 360010001001 | 873 | 444 | 429 | 50 | 54 | 0 | 35 | 0 |
| 2 | 27.8 | 23.2 | 1156 | 326 | 360010001002 | 1156 | 446 | 710 | 19 | 24 | 18 | 47 | 0 |
| 3 | <Null> | <Null> | 542 | 331 | 360010002011 | 542 | 95 | 447 | 27 | 24 | 0 | 0 | 0 |
| 4 | 19.2 | 10.4 | 1674 | 711 | 360010002012 | 1674 | 891 | 783 | 26 | 25 | 0 | 0 | 0 |
| 5 | <Null> | <Null> | 1047 | 294 | 360010002013 | 1047 | 441 | 606 | 32 | 32 | 0 | 20 | 0 |
| 6 | <Null> | <Null> | 1888 | 487 | 360010002021 | 1888 | 942 | 946 | 0 | 26 | 0 | 0 | 0 |
| 7 | 62.4 | 7.9 | 265 | 166 | 360010002022 | 265 | 117 | 148 | 43 | 37 | 0 | 0 | 0 |
| 8 | 29.3 | 15.7 | 1507 | 533 | 360010003011 | 1507 | 641 | 866 | 52 | 47 | 0 | 21 | 0 |
| 9 | 42.8 | 4.7 | 1509 | 593 | 360010003012 | 1509 | 706 | 803 | 43 | 38 | 0 | 0 | 0 |
| 10 | <Null> | <Null> | 1613 | 489 | 360010003021 | 1613 | 645 | 968 | 45 | 0 | 0 | 21 | 0 |
| 11 | <Null> | <Null> | 1318 | 404 | 360010003022 | 1318 | 642 | 676 | 59 | 29 | 0 | 34 | 0 |

After inspecting this table, close it.

Now, highlight and right-click on the title of the NYS_Block_Group feature class (map layer) in the Table of Contents to obtain the contact menu. Highlight and select "Joins and Relates" to open the second contact menu. Then, select "Add Join" to open its dialog.

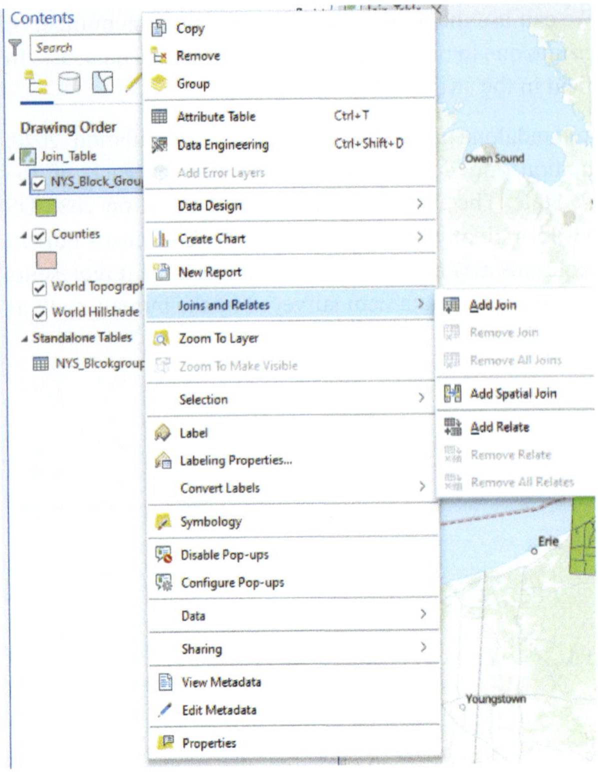


The "Input Table" is automatically selected as the attribute table of NYS_Block_Group. For "Input Join Field," select GEOID. For "Join Table," select NYS_Blcokgroup_Population_2020, the standalone table. For "Join Table Field," change it (select) from GEOID to GEOID2. Check on both "Keep all target features" and "index joined field." The filled result in the dialog should look as follows.

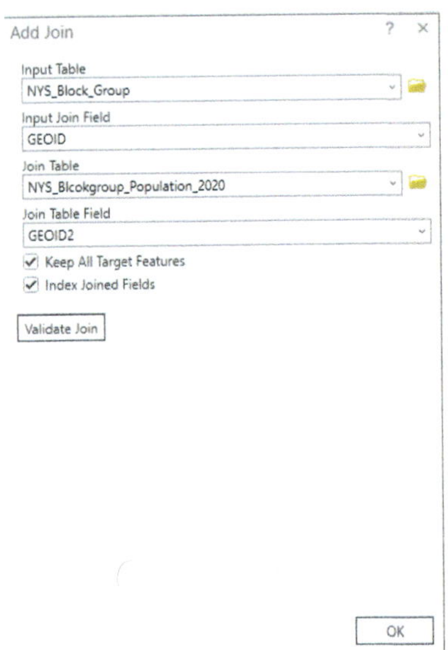

Click "OK" to run the function to process the join. Open the attribute table of NYS_Block_Group feature class (map layer) again, and you may notice the fields or columns of the NYS_Blcokgroup_Population_2020 table have been added on to it and the table is enlarged.

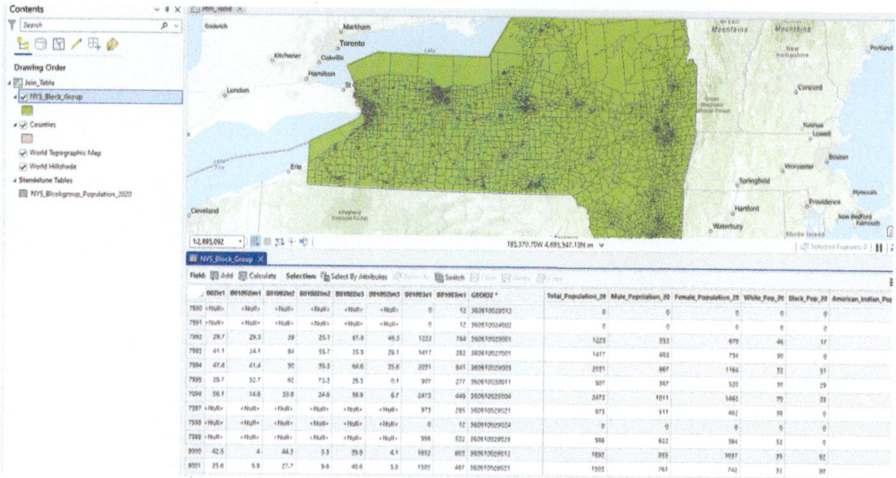


The two tables are joined virtually during this geoprocessing, and population and demographic data are available within this specific project at this time. We can utilize any of these newly added fields to make a thematic map now to visualize or present the data. However, we would like to save a permanent feature class in the geodatabase in order to make the result of this process to be used for other GIS projects as well. This task can be achieved using the export menu. Close the attribute table after examining it.

Right-click on the title of the NYS_Block_Group feature class again in the Table of Contents to get the contact menu. Highlight and click on the "Data" to open the second contact menu. Then, select "Export Features" to open its dialog.

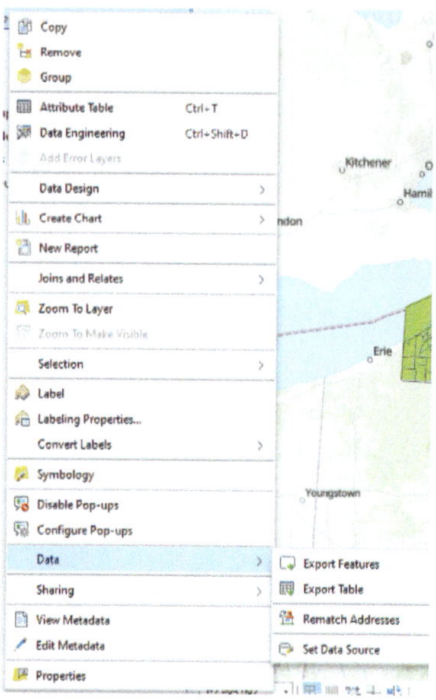

"Input Features" is automatically selected in the NYS_Block_Group feature class. For "Output Feature Class," navigate to your NYS_Civil_Boundaries.gdb geodatabase to save it as "NYS_BG_Population_2020." Keep other parameters as default and process the function.

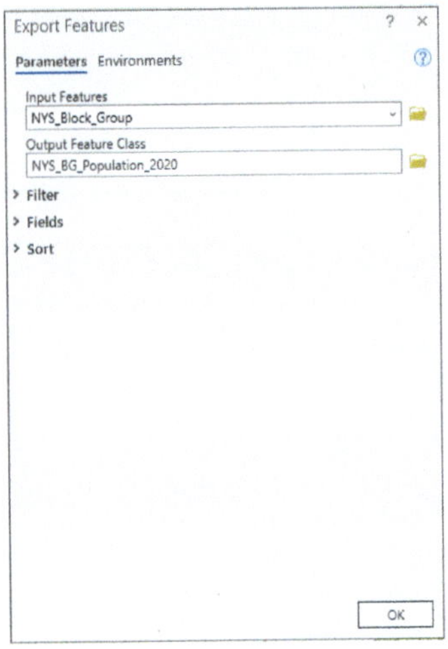


You may notice the newly created feature class is added in the Join_Table mapping window in the same color. We can quickly change its color by clicking on its symbol to open its color palette. Open the attribute table of this feature class and inspect the fields or columns. The difference between this feature class and the NYS_Block_Group feature class is that the added attributes are permanently saved in the attribute table of this feature class.

Now, we can remove the joins of the NYS_Block_Group feature class and remove the entire map layer from the mapping window. Check off NYS_BG_Population_2020 feature class on the Table of Contents for now. Right-click on the title of NYS_Block_Group feature class again to obtain the contact menu. Highlight and click on the "Joins and Relates" again to get the second contact menu. Click on "Remove all Joins." Then, the added fields and columns will disappear.

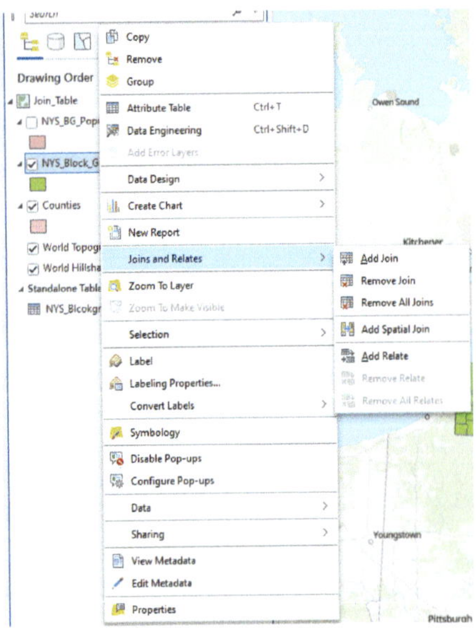

Open the attribute table of the NYS_Block_Group feature class, and you may notice that we came back to the original content again. Now, we can remove this feature class from our mapping window.

Check on the NYS_BG_Population_2020 feature class. Now, we can conduct several thematic mapping processes to analyze the distributions of the total population and population by different gender and ethnic groups across the New York State based on the block-group geographic units and new attribute data. For mapping and data presentation purposes, we can copy and paste our base map, the NYS_BG_Population_2020 feature class in this case, in the mapping window, “Join_Table” multiple times.

Right-click on the title of the map layer in the Table of Contents and click on “Copy” then. Moving one level up at the Table of Contents in the “Join_Table” (mapping window), right-click to get the contact menu, then click on “Paste” to duplicate the base map. Duplicate a total of five (5) copies. We will make five different thematic maps for population distributions.

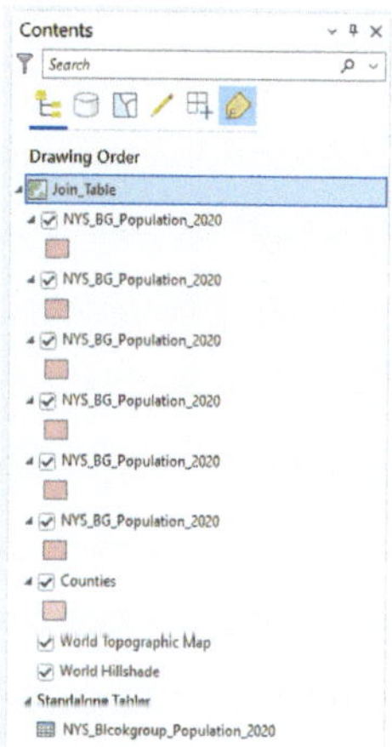


Highlight and right-click on the first base map to obtain the contact menu, then click on the Symbology to open the Symbology Pane.

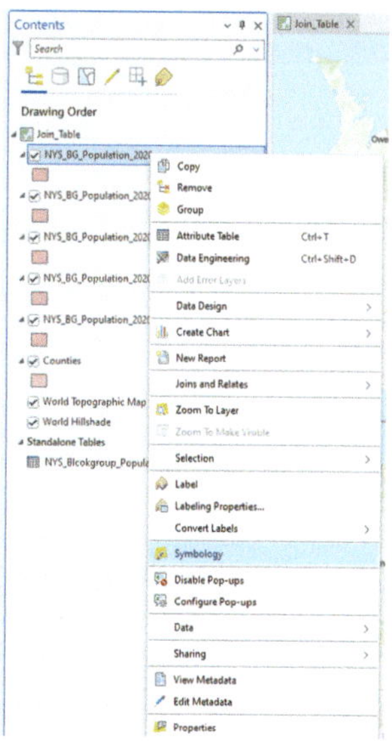

On the Symbology Pane, change the "Primary symbology" to "Graduate Colors." Then, select the field (column) of Total_Population_20 as the "Field." Keep the default "Natural Break (Jenks)" as the method of data classification and change the "Classes" to 6. There are several different methods to classify your data in the attribute table for thematic mapping using the graduate colors method. The Natural Break is a way to keep the range of values at each of the classes the same. However, the values (in this case is the population at the different areas) in practice would vary a lot.

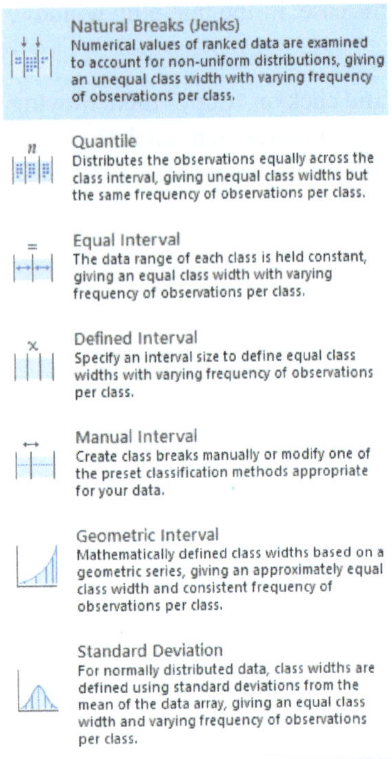


Change the "Color Scheme" to Yellow-Orange-Red (Continuous). This map layer now is showing the total population distribution across the New York State.

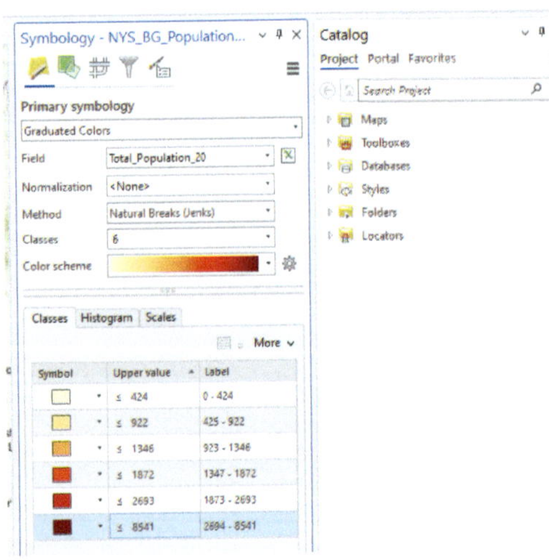

Now, we can click on and change the title (label) of this map layer in the Table of Contents to "NYS Total Population Distribution." The map should look like the following.

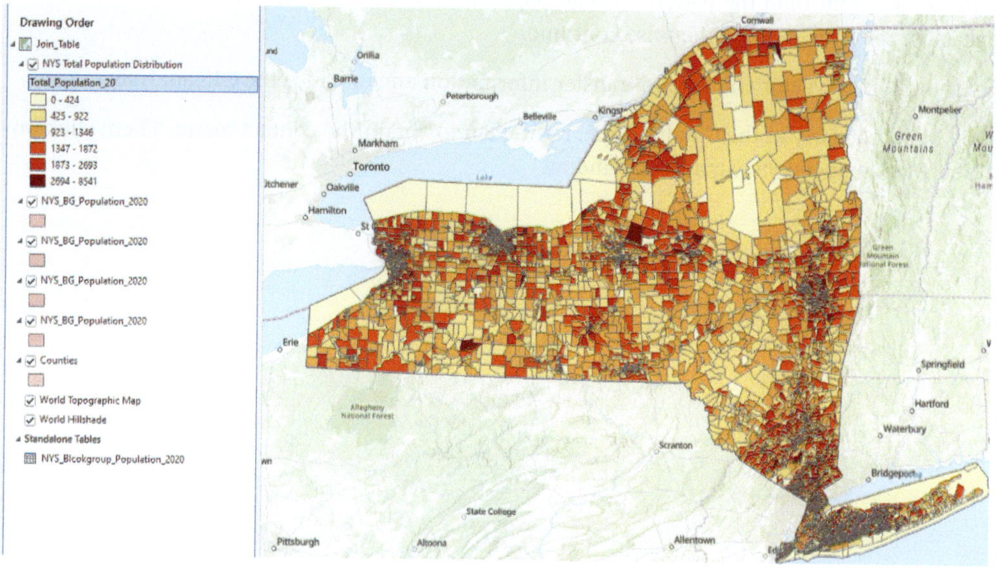


Check off the NYS Total Population Distribution map for now. Using the same steps above to make the next four base maps the following thematic maps:

1. NYS White Population Distribution
2. NYS Black Population Distribution
3. NYS Native Population Distribution
4. NYS Asian Population Distribution

Example of NYS White Population Distribution map is shown as follows.

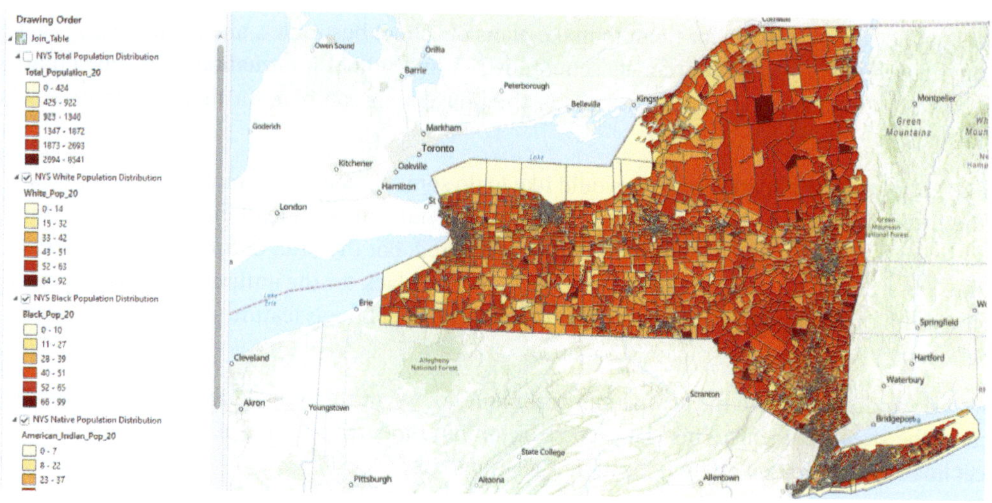

After you accomplished the thematic mapping work, close the Symbology Pane and save the GIS project. In summary, the patterns of population distribution of minority groups, such as native people are different than the pattern of total population distribution. If you would like to take a break, close the ArcGIS Pro, otherwise continue.

**2.** Spatial Join of vector GIS data to transfer information and enhance the GIS analysis

On the Maps category at the Catalog Pane, right-click to obtain the contact menu. Then, click on the New Map to open a new mapping window.

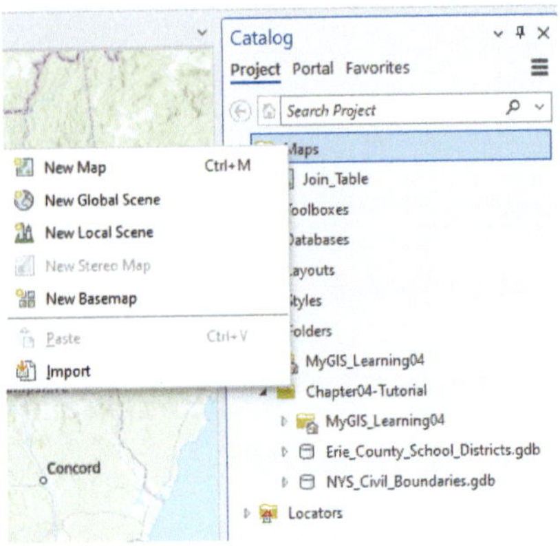


Rename it as "Spatial_Join." Expand the Chapter03-Tutorial folder, then expand Erie_County_School_Districts.gdb geodatabase. Add the following vector feature classes into the "Spatial_Join" window from this geodatabase.

1. Erie_County_Public_Schools
2. Erie_County_School_Districts
3. Streets

The objectives for doing this GIS project are: (i) providing the detailed streets information to each of the public-school districts for them to make plans of school bus routes, and (ii) providing each of the households in the county for its school district and school information from elementary schools to the high schools. So, the parents in the households can plan for their children's K-12 educations.

Before performing the spatial join, we would like to do some thematic mapping to symbolize each of the three map layers with appropriate symbols. For the Erie_County_Public_Schools feature class, select the school symbol, and paint it to the color red. For the Erie_County_School_Districts feature class, select "No Color" (hollow) for the color fill, black for the outline color, and 2 points for the width of the outline school district boundaries. For the Streets feature class, select "Moorea Blue" color.

Now, we would like to use the spatial join technique to assign the school district information to each of the street segments. This is a process of joining the attribute information of polygons to street line segments based on their locational relationships.

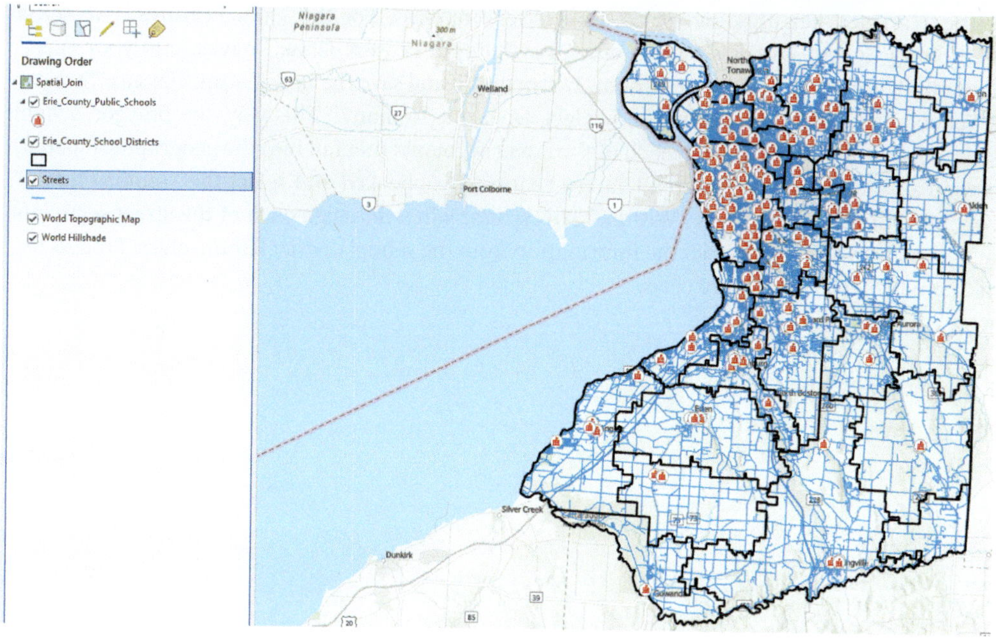


Activate the Analysis ribbon on the main menu of ArcGIS Pro and click on the toolbox icon to open the Geoprocessing Pane. Type in "Spatial" under the Favorite tab to search the spatial join tool. The Spatial Join tool is found and promoted on the top of the list. Click on it to open its dialog.

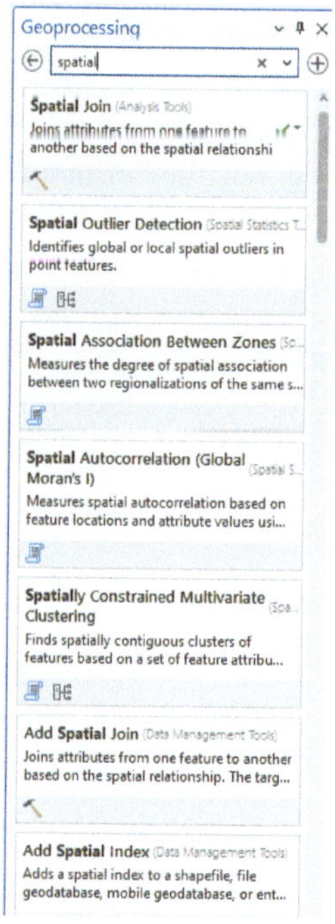

For the "Targeted Features," we select Streets feature class. For the "Join Features," we select Erie_County_School_Districts feature class. For "Output Feature Class," navigate to your \\......\ Chapter03-Tutorial\ Erie_County_School_Districts.gdb and save the new feature class as "Streets_for_School_Bus." For "Join Operation," select "Join one to many." This way, the polygon school district attributes will be joined to each of the street segments that fall into the geographic (spatial) query areas. Make sure the "Keep all Target Features" is checked and select the "Match Option" (geographic or spatial query) as "Within." This spatial join will assign each of the street segments that fall inside of a school district the information from the school district feature class. Process the tool to produce the new feature class.

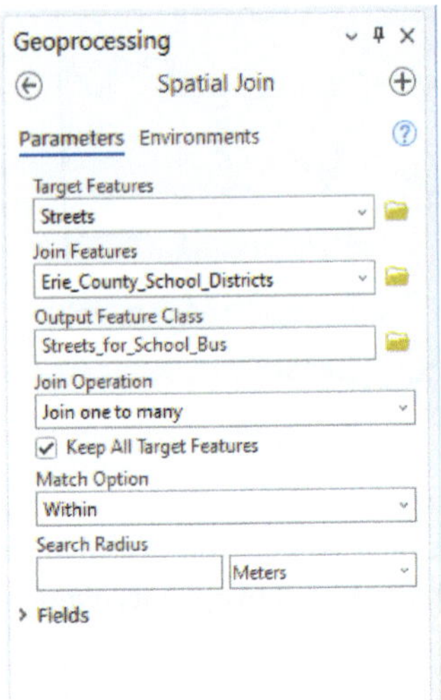


You notice that the new feature class is created and added to the mapping window. Now, we can open the Symbology Pane of the feature class to do some thematic mapping. For the "Primary Symbology," select the Unique Values. For "Field 1," select "SchDist" in the attribute table of the feature class. For the "Color Scheme," select "Enamel."

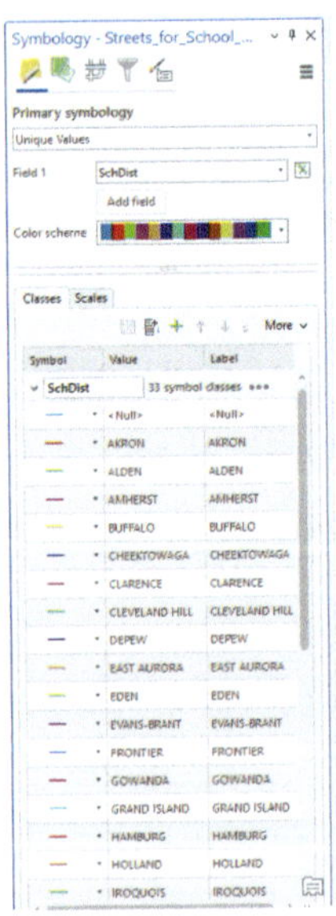

Close the Symbology Pane and geoprocessing Pane. The new map should look like the following.

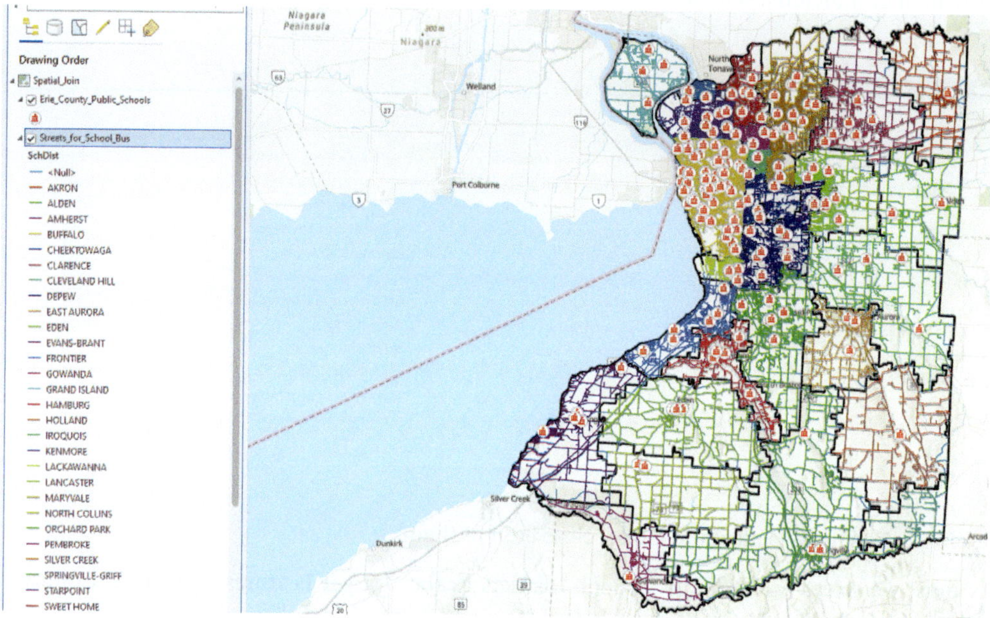


Different colors in the street networks represent the different school districts in Erie County, New York. Then, we can use these street segments to transport the K-12 students to their public schools.

Our second task is to assign attribute information of individual public schools to their school districts. This is so residents in the county can review schools when they browse the school district digital map online. Check off the Streets_for_School_Bus feature class for now. Reload the Geoprocessing Pane from the Analysis ribbon at the main menu. Open the Spatial Join tool again.

Open the attribute table of the Erie_County_School_Districts to inspect it. We do have the brief name of the school district ("SchDist" field), such as "CATTARAUGUS" or "EDEN." However, we do not have any individual school information in the school district. In this case, we have a total of 37 school districts in Erie County, New York.

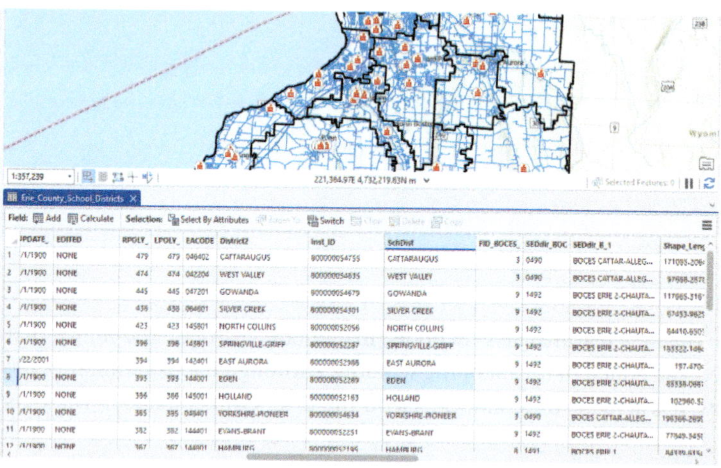

Open the attribute table of Erie_County_Public_Schools feature class, we can find the detail information of individual schools.

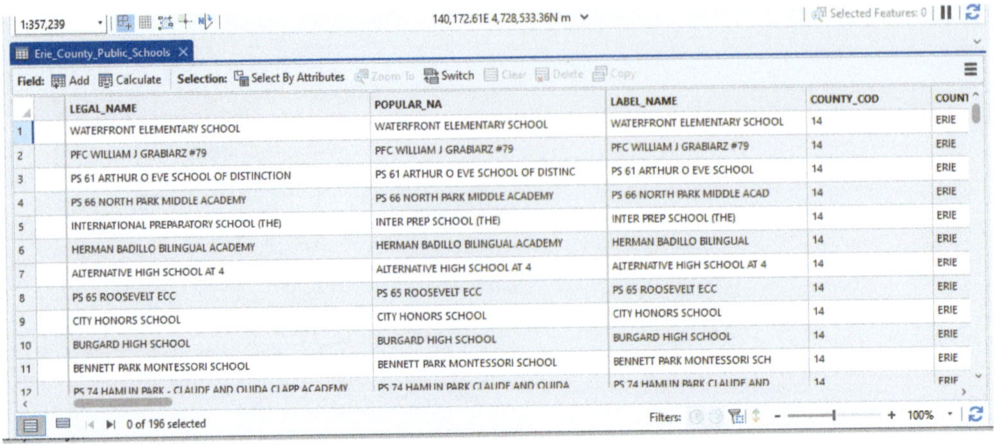


| | LEGAL_NAME | POPULAR_NA | LABEL_NAME | COUNTY_COD | COUNT |
|---|---|---|---|---|---|
| 1 | WATERFRONT ELEMENTARY SCHOOL | WATERFRONT ELEMENTARY SCHOOL | WATERFRONT ELEMENTARY SCHOOL | 14 | ERIE |
| 2 | PFC WILLIAM J GRABIARZ #79 | PFC WILLIAM J GRABIARZ #79 | PFC WILLIAM J GRABIARZ #79 | 14 | ERIE |
| 3 | PS 61 ARTHUR O EVE SCHOOL OF DISTINCTION | PS 61 ARTHUR O EVE SCHOOL OF DISTINC | PS 61 ARTHUR O EVE SCHOOL | 14 | ERIE |
| 4 | PS 66 NORTH PARK MIDDLE ACADEMY | PS 66 NORTH PARK MIDDLE ACADEMY | PS 66 NORTH PARK MIDDLE ACAD | 14 | ERIE |
| 5 | INTERNATIONAL PREPARATORY SCHOOL (THE) | INTER PREP SCHOOL (THE) | INTER PREP SCHOOL (THE) | 14 | ERIE |
| 6 | HERMAN BADILLO BILINGUAL ACADEMY | HERMAN BADILLO BILINGUAL ACADEMY | HERMAN BADILLO BILINGUAL | 14 | ERIE |
| 7 | ALTERNATIVE HIGH SCHOOL AT 4 | ALTERNATIVE HIGH SCHOOL AT 4 | ALTERNATIVE HIGH SCHOOL AT 4 | 14 | ERIE |
| 8 | PS 65 ROOSEVELT ECC | PS 65 ROOSEVELT ECC | PS 65 ROOSEVELT ECC | 14 | ERIE |
| 9 | CITY HONORS SCHOOL | CITY HONORS SCHOOL | CITY HONORS SCHOOL | 14 | ERIE |
| 10 | BURGARD HIGH SCHOOL | BURGARD HIGH SCHOOL | BURGARD HIGH SCHOOL | 14 | ERIE |
| 11 | BENNETT PARK MONTESSORI SCHOOL | BENNETT PARK MONTESSORI SCHOOL | BENNETT PARK MONTESSORI SCH | 14 | ERIE |
| 12 | PS 74 HAMLIN PARK - CLAUDE AND OUIDA CLAPP ACADEMY | PS 74 HAMLIN PARK CLAUDE AND OUIDA | PS 74 HAMLIN PARK CLAUDE AND | 14 | ERIE |

After inspecting the two attribute tables, close both of them.

This time, we select Erie_County_School_Districts as the "Target Features," and Erie_County_Public_Schools as the "Join Features." For "Output Feature Class," navigate to your \\......\Chapter03-Tutorial\ Erie_County_School_Districts.gdb and save the new feature class as "Schools_in_School_District." For "Join Operation," select "Join one to many." Make sure the "Keep all Target Features" is checked and select the "Match Option" (geographic or spatial query) as "Contains." This means the school district polygons contain all the public schools within it.

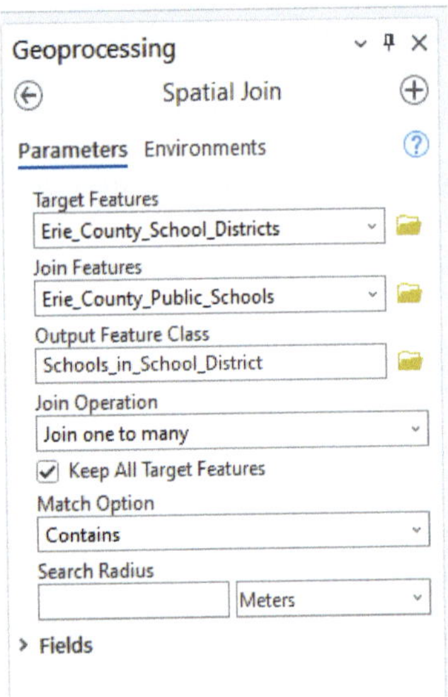


Process the function to create the new feature class "Schools_in_School_District." You notice that the new feature class is created and added in the mapping window. Although it looks similar to the Erie_County_School_Districts, it is actually different. Check off the Erie_County_School_Districts feature class for now. Open the attribute table of Schools_in_School_District. Use the Select by Attributes query on the title bar of this table to build a query Selection: Select By Attributes . SchDist = (is equal to) "EDEN."

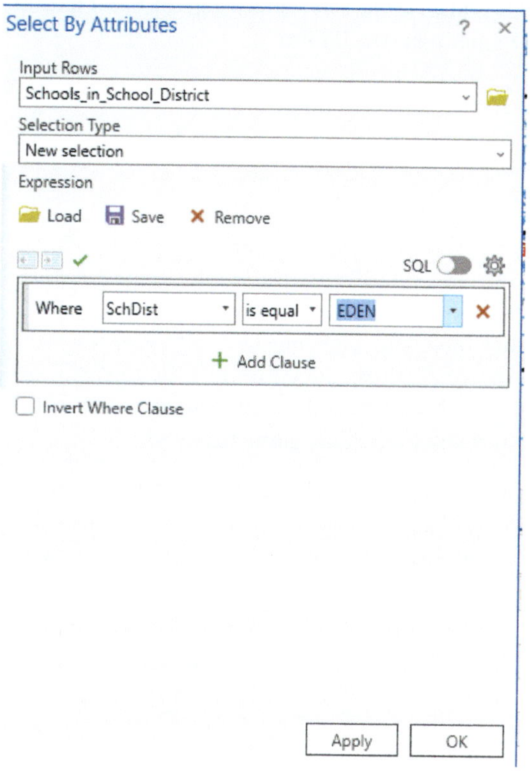


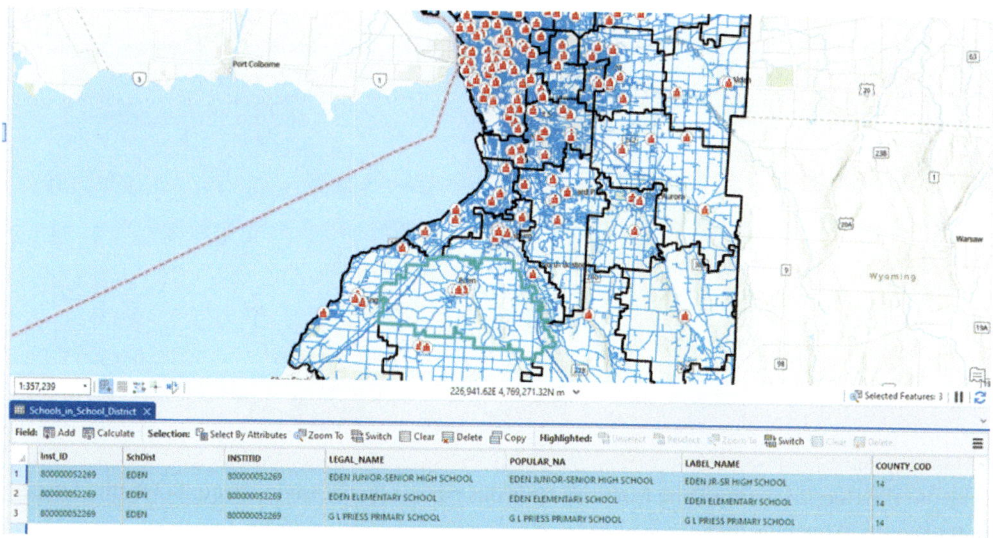

This query selected the Eden School District on the digital map. Meanwhile, three rows that represent three public schools in this school district in the attribute table are selected. Close the Geoprocessing Pane for now. We can reactivate Geoprocessing Pane later.

The spatial join process in this case enlarged the attribute table. Each of the public schools is added to the school district polygons as a particular record. Detail school information is added to this database. After inspections, close the attribute table and clear the selections using the Clear icon at the Selection group on the Map ribbon of the main menu in ArcGIS Pro.

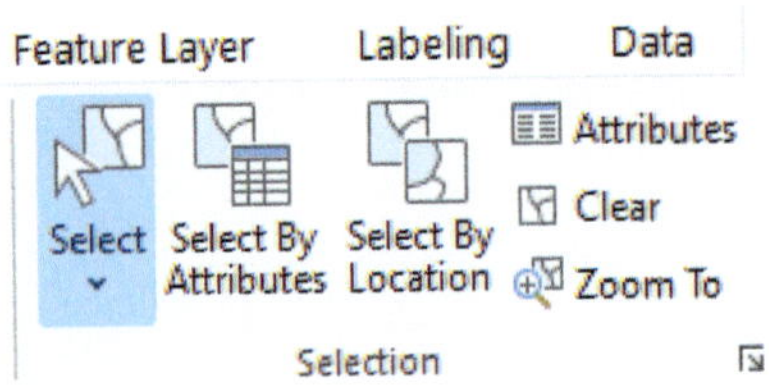


Now, we would like to accomplish the task to geographically (spatially) join the school district information to each of the 351,866 households (parcels) across the entire Erie County, NY. So, each of the families in the county can readily and quickly gain the information about which of the public schools from elementary to high school that their children are going to attend.

Check off the Schools_in_School_District feature class and check on the Erie_County_School_Districts and Parcels feature classes in the Table of Contents. Make sure the Erie_County_School_Districts feature class is symbolized in hollow and move it above the Parcels feature class.

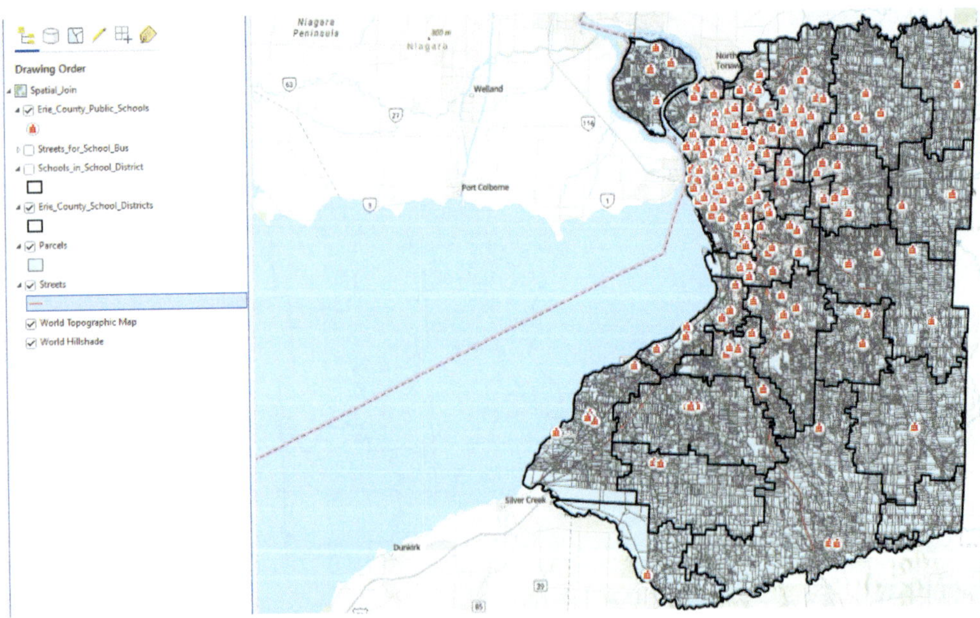


Reactivate the Geoprocessing Pane from the Analysis ribbon on the main menu. Select the Spatial Join tool again to open its dialog.

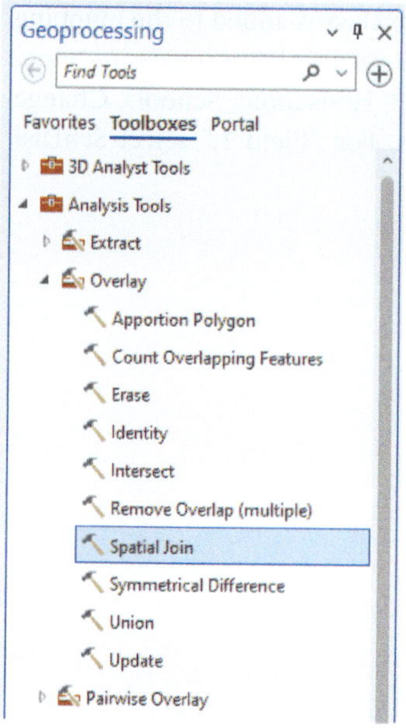


For the "Target Features," select the Parcel feature class. For the "Join Features," select the Erie_County_School_Districts feature class. For "Output Feature Class," navigate to your \\......\ Chapter03-Tutorial\ Erie_County_School_Districts.gdb and save the new feature class as "Parcel_ Household_Schools." For "Join Operation," select join one to many. Make sure "Keep All Target Features" is checked and select the "Match Option" (Geographic or spatial query) as Intersect. In this process, all the parcels that intersect with a school district polygon will be joined to that school district. Process the tool to produce the new feature class at the parcel level. Since ArcGIS Pro needs to process a total of 351,866 parcels or households across the entire Erie County, NY, this process may take a few minutes.

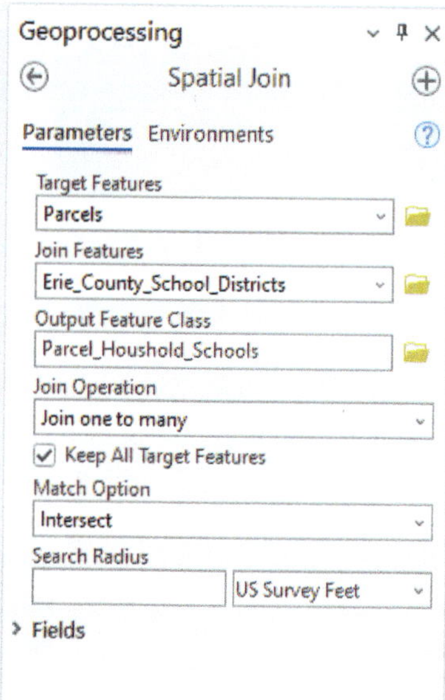

After the process is finished, you notice that the new parcel feature class is added to the mapping window. We can conduct some thematical mapping to make the result of geoprocessing more visible. Open the Symbology Pane of the new feature class, Parcel_Household_Schools. Change the "Primary symbology" from a single symbol to unique values. For "Field 1," select SchDist (School District). For "Color Scheme," select "Paired (12 classes)."

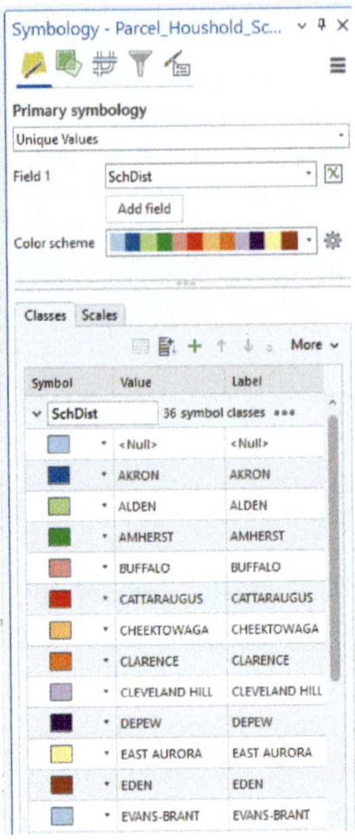


Close the Symbology Pane and the Geoprocessing Pane. The thematical map should look like this.

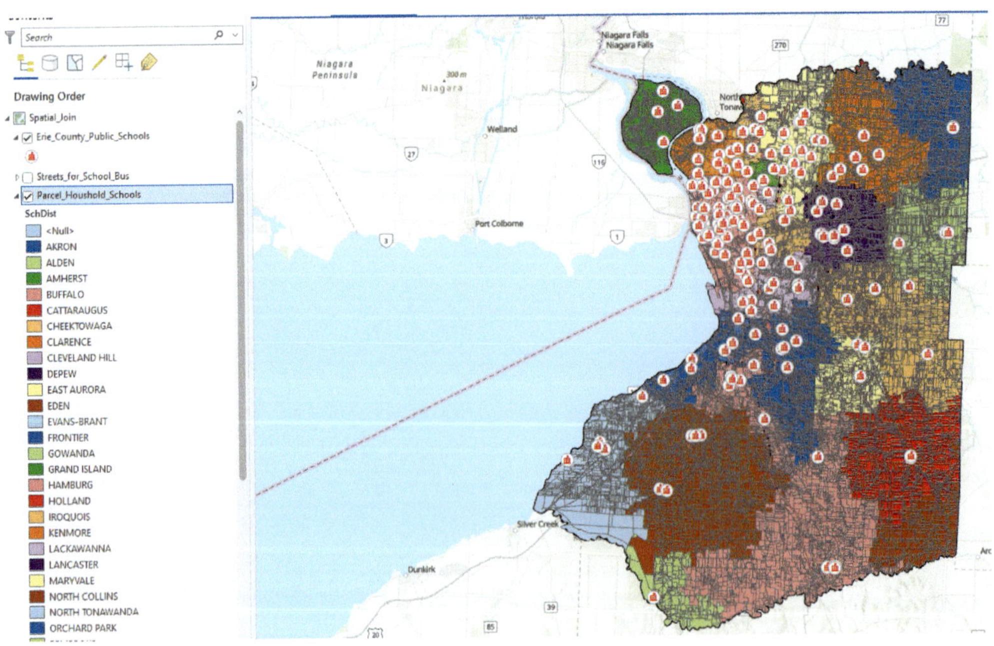

In summary, each of the 351,866 parcels across the Erie County, NY has been added all the detail school district information.

If you would like to take a break, save the project file, and close the ArcGIS Pro. Otherwise, continue.

**3.** Relating the relational database tables to query the geographic (spatial) information

Open ArcGIS Pro with "MyGIS_Learning03" project if it is not open, otherwise continue. In this process, we are going to relate the parcel feature class of Erie County, NY to the United States Congressional Districts. So, the citizens living in this county can find their representative at the US Congress to discuss their concerns and move our society and our country to a better place.

Open another mapping frame (Map) from the Catalog Pane under the Maps category. Rename it as "Relate_Congress." Add Parcels02 and Erie_County_Congressional_Districts feature classes into the mapping window. Parcels02 was created using the Parcels feature class. The attribute data of congressional district polygons were added to the parcel attribute table using spatial join. Since one congressional district contains many parcels or families and households, it is easier and more obvious to relate the attribute tables to find the congressional representatives for each of the individual families living in a particular parcel across the Erie County, NY.

Right-click on the title of Parcels02 in the Table of Contents to open the contact menu. Then, highlight and click on Joins and Relates to open the second contact menu. Select Add Relate, and the dialog opens.

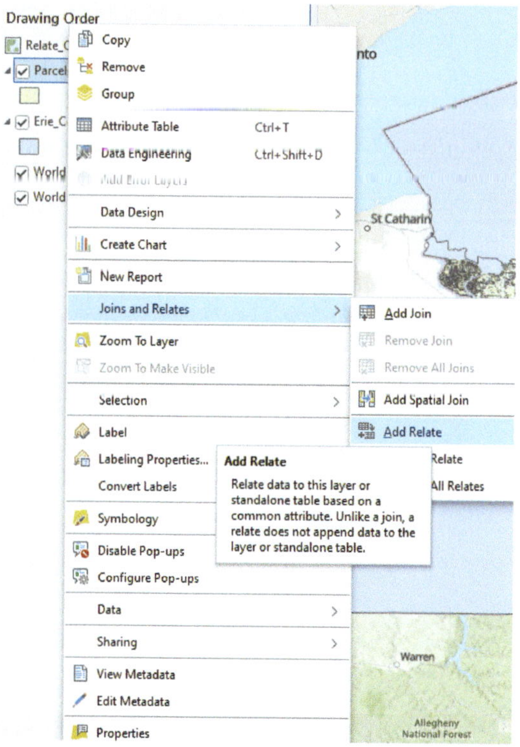


The "Layer Name or Table View" has already been filled by Parcels02. For "Input Relate Field," select NAMELSAD. It is the name of the congressional district. For "Relate Table," select Erie_County_Congressional_Districts feature class. Make sure that "Output Relate Field" is NAMELSAD

as well. Rename the "Relate Name" as "Congressional_Relate." Keep the default "One to Many" for the cardinality. Process the function to build the relation bridge. You notice that the Parcel02 was redrawn.

Now, we would like to conduct one query and relate process to demonstrate the relate bridge exists between the two attribute tables. Open the attribute table of Parcels02 feature class and build the following query using Select by Attributes. This is a new query.

Street = (is equal to) 154 SAUSALITO DR

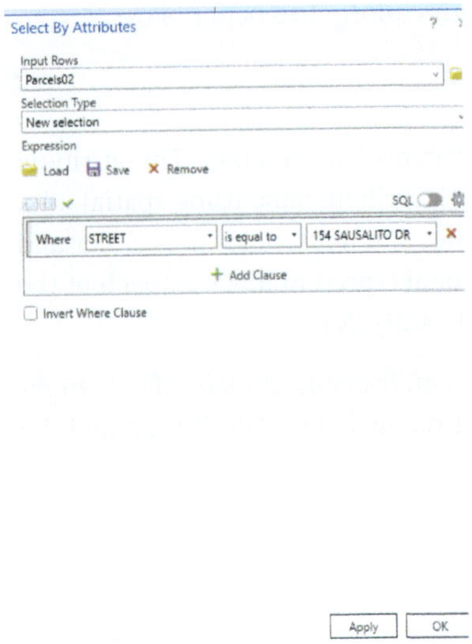


Process the query, you notice one parcel record is selected.

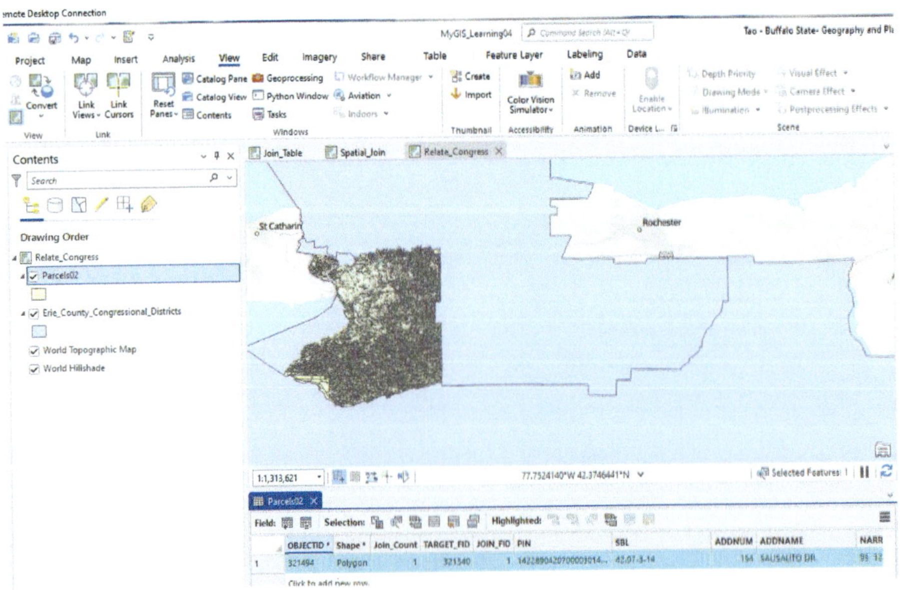

Navigate to Data ribbon now on the main menu of ArcGIS Pro, click on Related Data icon in Relationship group.

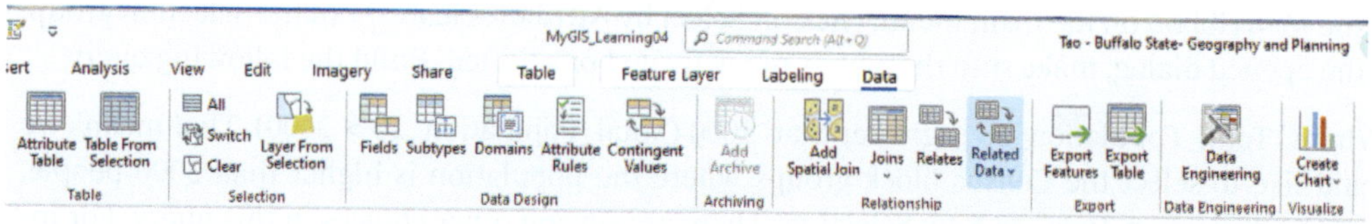


Select Erie_County_Congressional_Districts feature class as the destination table. You notice that the Erie_County_Congressional_Districts attribute table is open, and one record is selected. Pull the lower taskbar to the right, and we find the congressional representative for this particular household is Mr. Chris Jacobs. The live relate only exists in this GIS project until we remove it. To remove the relate, you can use the contact menu at the Table of Contents to open the join and relate menu again to do it. Now, close both tables and clear the selections using the Clear icon at Selection group of Map ribbon on the main menu.

Save the project and close the ArcGIS Pro if you would like to take a break. Otherwise, continue.

**4.** Query geodatabase and produce the geographic (spatial) information needed

In this GIS practice, we would like to analyze and find out how many bank branches are located in the US Census Block-group areas where the population exceeds 3000 persons across New York State. The objective of this data analysis is to review if the most populated census block groups are adequately served by bank facilities for family financial management. In addition, we would like to make a population distribution of the 2020 feature class or map layer for Erie County from the NYS digital population geodatabase.

If the MyGIS_Learning03 project in ArcGIS Pro is open, continue. Otherwise, open the project. Create a new mapping window under the Maps category at the Catalog Pane and rename it as "Query." Add the Bank_Branches and NYS_BG_Population_2020 feature classes from NYS_Civil Boundaries.gdb geodatabase. Conduct some thematic mapping for both map layers using the Symbology Pane.

For the point feature class of Bank_Branches, keep the single symbol format, select the ATM symbol in the symbol gallery, and choose the color as "Dark Navy" and size as 10 points. For the polygon feature class of NYS_BG_Population_2020, choose Graduate Color for the Primary Symbology and Total_Population_20 column as the "Field." Change the "Method" to Quantile and "Class" to 6. Keep or change to the Continuous Yellow-Orange-Red color scheme. The digital map looks like the following.

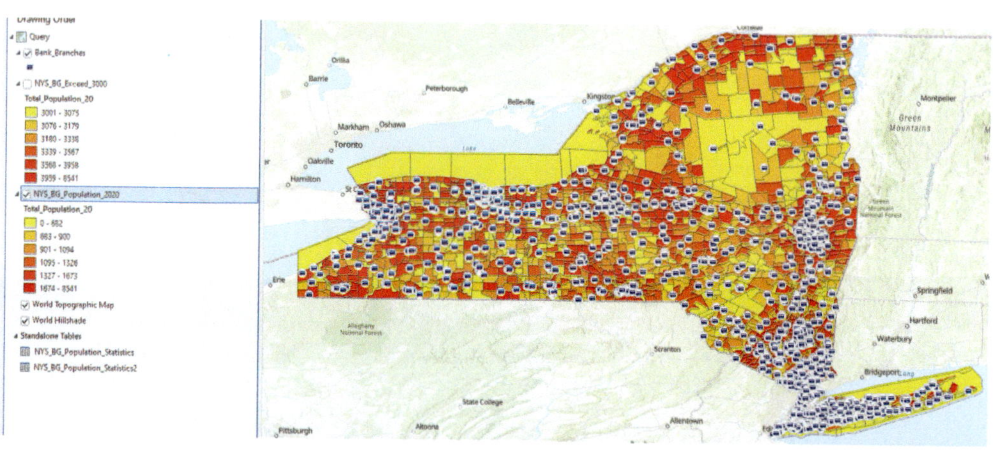

At the Map ribbon on the main menu, click on Select by Attributes icon in the Selection group. In the opened dialog, make sure the SQL is not activated or selected. Build the following query.

"Where" Total_Population_20 is greater than 3000 (Total_Population_20 > 3000). That means we would like to select the Census block groups where the population is higher than 3000 people. Keep the "Selection Type" as New Selection. There are several other choices in this query. For instance, you can add the selection if you already have one selection. Process the query and view the selection result.

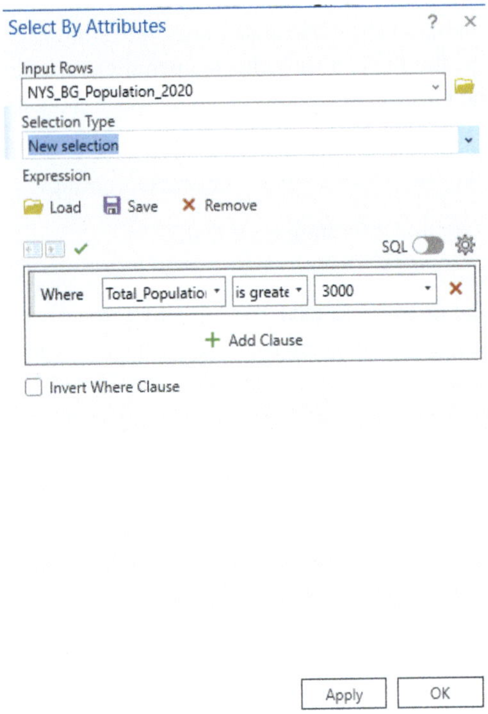


Turn off the Bank_Branches feature class, you notice many block groups across the New York State are selected (the boundaries are in light blue).

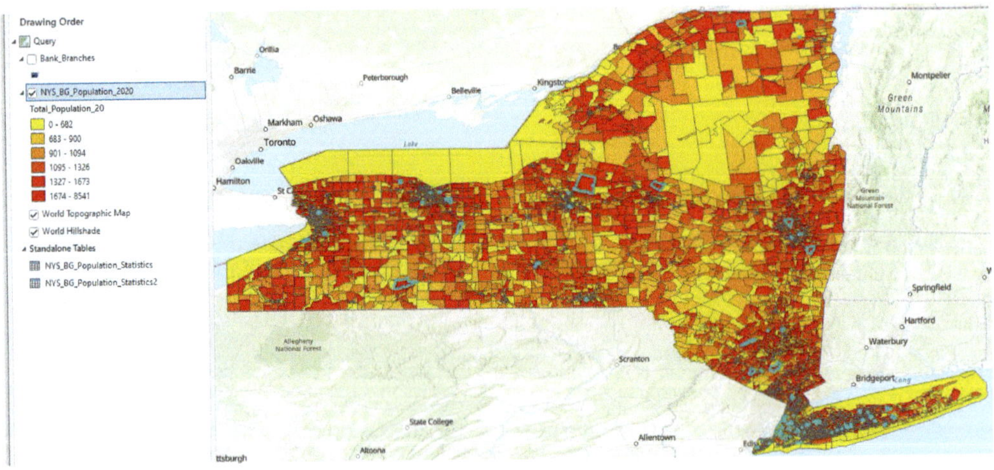

Right-click on the title of the NYS_BG_Population_2020 feature class in the Table of Contents to obtain its contact menu. Highlight the Data menu to get the second contact menu. Select Export Features to export a new feature class.

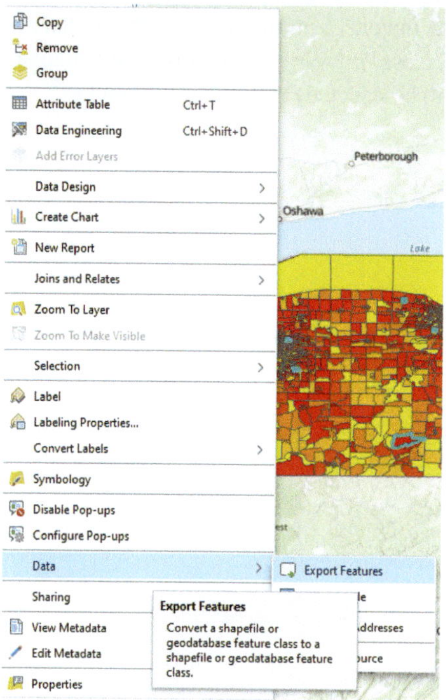


In the dialog of Export Features, the "Input Features" has already been filled with NYS_BG_Population_2020 feature class. For the "Output Feature Class," navigate to your NYS_Civil_Boundaries.gdb geodatabase, save the new feature class as "NYS_BG_Pop_Exceed_3000." Process the data exporting function, you notice the new feature class is produced. Now, we can clear the selection and check off the NYS_BG_Population_2020 feature class on the Table of Contents. The new feature class looks like this.

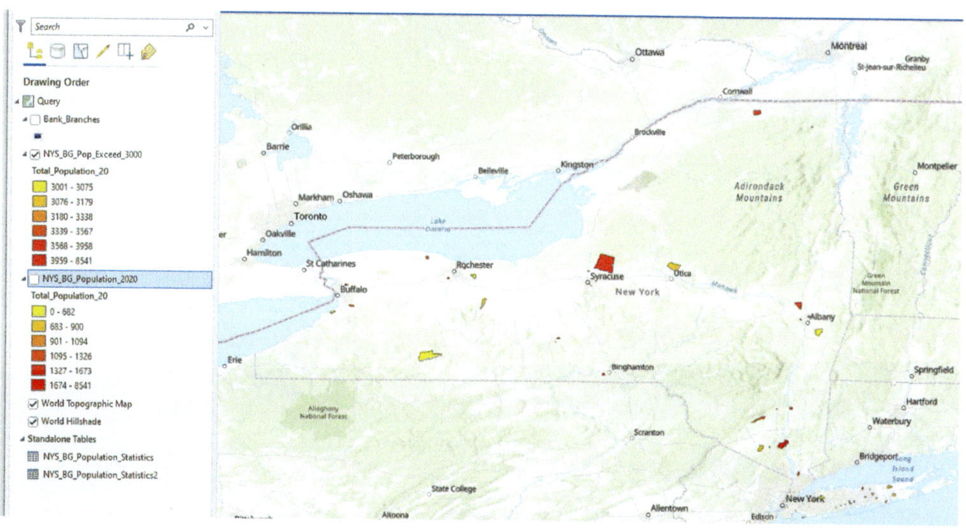

We would like to conduct a spatial (geographic) query now to find out how many bank branches are in these Census block groups of high population density in the New York State. Check on the Bank_Branches feature class. Click on the Select by Location (spatial query) icon in the Selection group at the Map ribbon on the main menu to open its dialog. The design and logic of this query are totally different than that of Selection by Attributes. Here the major selection routes are the spatial or locational relations of features between the targeted feature class and selection feature class. Change the "Input Features" (targeted feature class of the features to be selected) to the Bank_Branches.

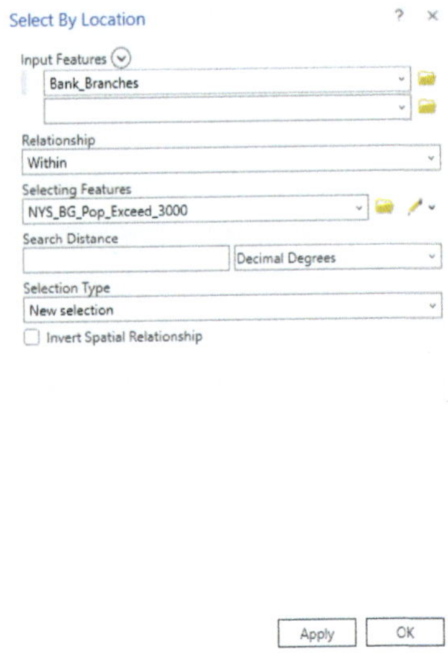


This tool can select features from more than one feature class. In this case, we only need one. For the "Relationship," select Within. For "Selecting Features," choose NYS_BG_Pop_Exceed_3000. The expression of this spatial query is: "Select those bank branches their locations fall within the polygon of US Census Block-groups that the population exceed 3,000 people." Since we are not going to perform the selections based on the distance range of the features, we can leave the rest of the parameters empty. Process the tool to make the selection. Again, this is a new selection. There are several different options for selections in this group that are similar to Select by Attributes. You notice a few bank branches are selected.

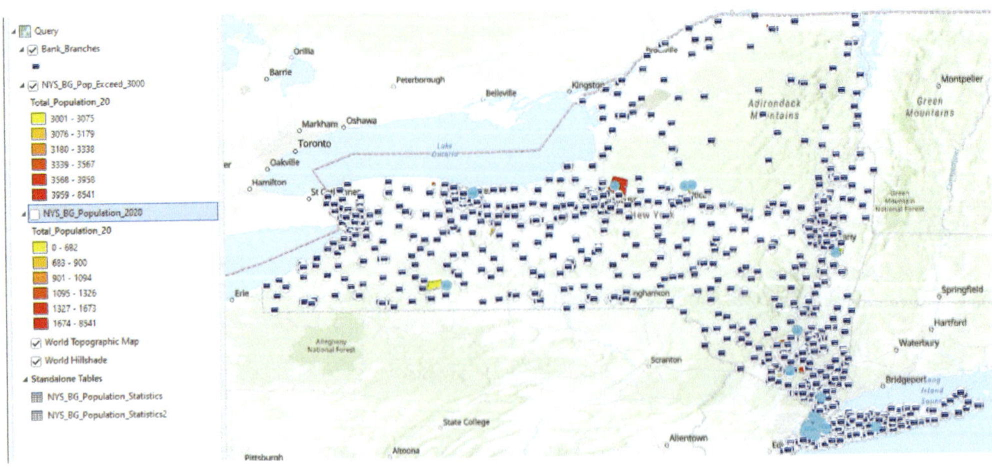

Now, we can conduct the reverse query just for more practice. This step is not necessary but just for practice, as we could just do this spatial query at the beginning of this process. Open the Select by Location (spatial query) again, we change the "Input Features" (target feature class of the features to be selected) to the NYS_BG_Pop_Exceed_3000 this time. For the "Relationship," select Contains. For "Selecting Features," choose Bank_Branches.

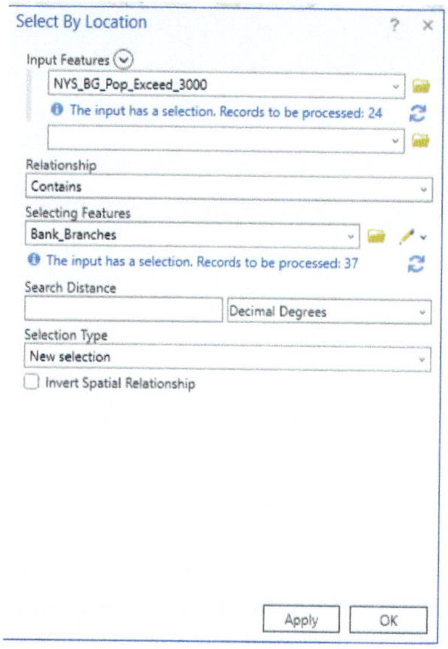


The query statement this time is "Select the US Census block-group features that exceeded the population of 3000, which contains at least one bank branch." Check off the Bank_Branches feature class, and you may notice that some block group polygons are selected. Now, we open the attribute table of NYS_BG_Pop_Exceed_3000 feature class to find out how many of these geographic areas have a bank branch? According to the report of this query from the attribute table, only 24 out of 140 US Census block-group polygons are selected. This means only 24 out of 140 block-group communities where the population is more than 3,000 people across the New York State have one bank branch conveniently located in their block-group community.

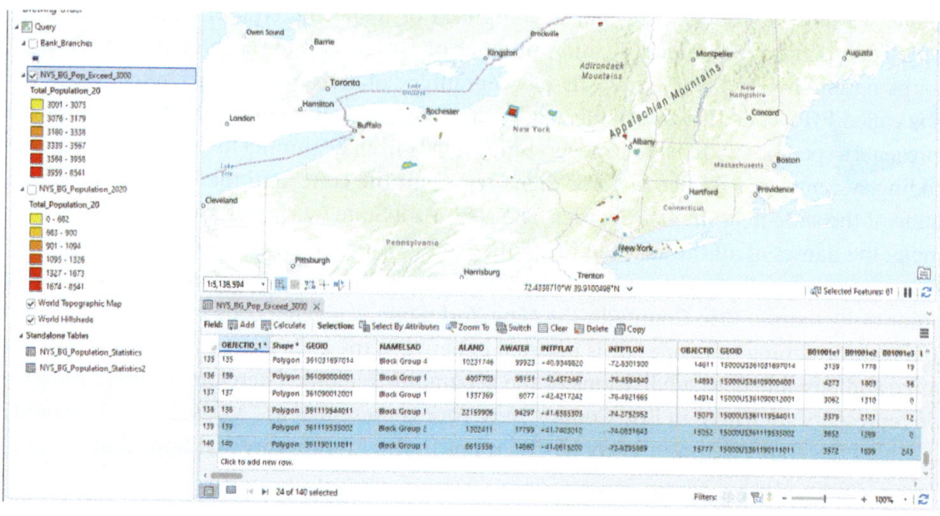

We can click on the Switch icon on the title bar of the attribute table to switch the selection to 116 block-group communities that do not have one bank branch.

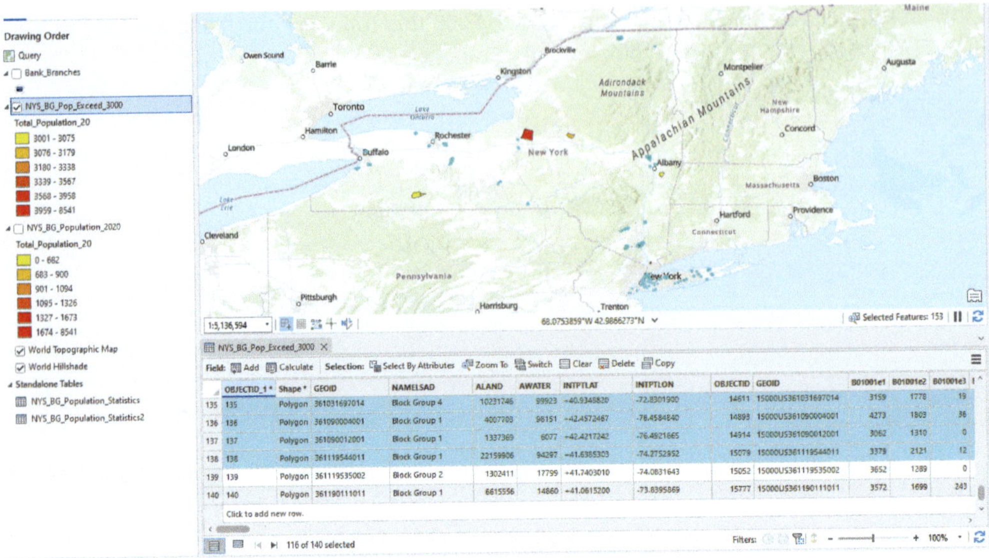


Then, we can export the data for further analysis later. Now, we can close the attribute table and clear the selections. If you would like to take a break, save the project and close ArcGIS Pro. Otherwise, continue to work on the following query project.

Since the field-investigated population survey by the US Census Bureau was published, the Erie County government in New York State would like to study both the new population distribution patterns and the total population change from previous years. In order to conduct this analysis, we are going to do a project to query the NYS GIS database to prepare the new dataset for this task. Check off the feature class of NYS_BG_Pop_Exceed_3000 and check on the feature class of NYS_BG_Population_2020. Open the attribute table of NYS_BG_Population_2020 feature class to inspect its contents.

In this table, we have a GEOID field that is comprised of a specific type of code. In essence, this specific code is the unique identification number or Unique ID for each of the US Census block-groups nationalwide. This Unique ID of geographic subdivisions of the United States of America is called FIPS code. It was established by the US Census Bureau. Each of the groups of digits represents a specific subdivision of geographic regions from beginning to the end of the code. In this coding system, the first two or three digits represent the state, and the number is ranked by the name of the state in alphabetical order. The New York State by its alphabetical order of the name among the names of all the states in the United States is given the code of "36." The county group of the FIPS code follows the same rule and the counties in a state are ranked by the alphabetical order of its names. Erie County in the New York State is given a code of "29." According to the FIPS code designing rule, one "0" is inserted in between the code groups for possible future code change and possible new subdivisions of the geographic regions. Therefore, in this case, the FIPS code represents anywhere in Erie County, New York is presented as "36029.........". We would like to use this PIPS code pattern to select block-groups in the Erie County and hope that we can export the selected data to a new feature class to make our basic database for population studies in the Erie County.

On the technique side, the data format of FIPS code in our attribute table is text. Close the attribute table after inspections. Now, let's open the Select by Attributes dialog from the Selection group at the Map ribbon on the main menu. Making sure the "Input Rows" is NYS_BG_Population_2020 feature class, construct the following query.

"Where the GEOID contains the text of 36029."

Keep the "Selection Type" as its default of New Selection. Process the query. You notice that all the Census block groups in the County of Erie, New York is selected.

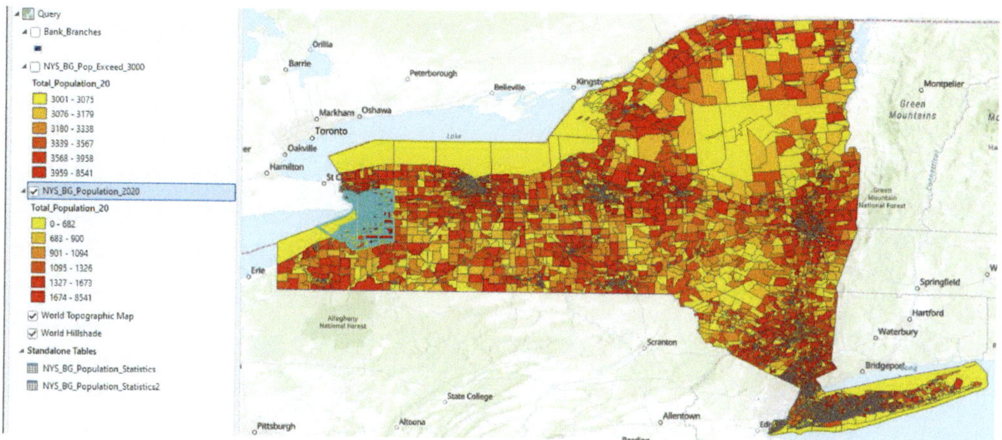


Right-click on the title of this map layer in the Table of Contents to obtain the contact menu. Then, export a new feature class called "Erie_County_BG_Population_2020" to your Erie_County_School_Districts.gdb geodatabase in your working folder. We can clear the selection on NYS_BG_Population_2020 feature class and check off the map layer. You notice the new feature class is produced. Congratulations, you've accomplished this project!

# 4

# Geographic Information Data Gathering and Data Collections

## Concepts and Theories

Geographic or spatial data is important for analyzing how situations in both natural environments and social environments impact our societies. Geographic information systems (GIS) analyses help to make more informative and effective decisions. Therefore, GIS can be utilized for natural science, engineering studies, and the studies of social science. In practice, similar analytical procedures and approaches in GIS can be applied in both natural and social sciences. In the software environment, one set of tools, windows, or geoprocessing can be used to solve various problems that we may encounter and support many different applications. The nature of GIS is not only software application intensive, but also intensive in using large quantities of data. Ultimately, data collection and gathering in GIS is very significant. In many GIS analytical projects, GIS analysts spend a large amount of their total time and effort searching for and gathering the data. Over the years, a great number of GIS data-gathering methods have been developed. We are going to conduct a simple survey of some major approaches used in both science, engineering, and social science to collect data in this chapter.

## Geographic Information Primary (Original) Data Collections

Since humans live on the Earth's surface, we always gather the original or primary data for our concerns on the ground using various scientific or social science methods. These methods have been developed since the beginning of human civilizations, long before GIS was created and developed. These methods are not only embedded in the new inventions and innovations of science and technology but are also incorporated with the governing of human society or government activities. GIS analysts can take advantage to utilize all these methods to collect data. Alternatively, GIS analysts can also take advantage and use the datasets from government agencies and other professional fields (Figure 4.1). Figure 4.1 shows the four different examples of geographic information (GI) collections by their different purposes. The common characteristic of this type of data collection is that it contains street or home addresses as detailed geo-markers of locations for a feature or an event. The street address that we use every day is a relative geo-marker or relative location that coincides with the absolute geographic coordinate system locations, such as latitude and longitude (refer to Chapter 5). A large quantity of data on this type of nature gathered from the ground can be input into a GIS system to analyze spatial (geographic) distribution patterns.

*Fundamentals of Geographic Information Systems (GIS) Analysis: Applying ArcGIS Pro*, First Edition. Tao Tang.

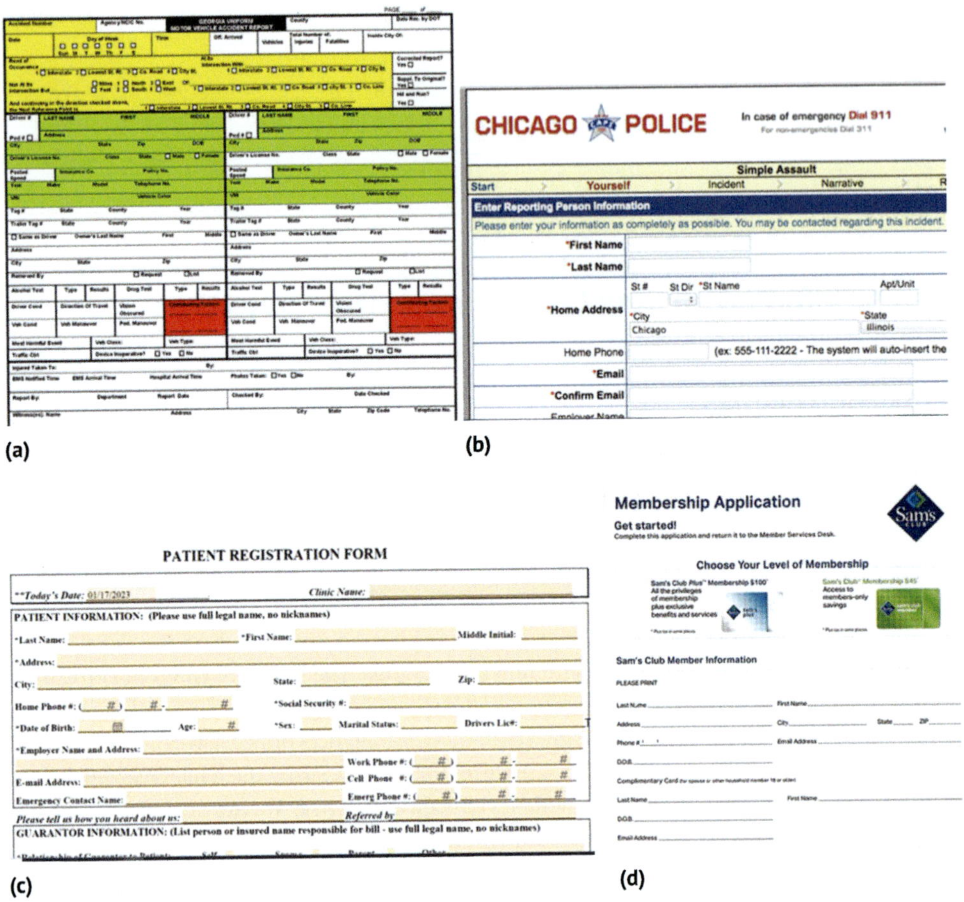


**Figure 4.1** Examples of original GI collection by various public and business sectors: (a) Police accident report form; (b) Police crime report form; (c) Hospital patient registration form; and (d) Sam's Club membership registration.

The so-called primary or original data collections in the discussions of this book mean the developed methods of collecting datasets with or without GIS use originally or at the first place. For instance, the United States Bureau of Census has been collecting human population distribution data for the country since the founding of the nation. Although GIS analysts in many cases do not need to conduct the original or primary data collections themselves, they need to understand the methods. Meanwhile, understanding the original methods of data collections and the history and change of the datasets helps the GIS analysts to utilize the data effectively.

## Geographic Data Collections in Natural Science and Engineering: Survey of Geospatial Data

In science and engineering, surveying on the ground is the original and popularly used method to obtain information on the Earth's surface. Surveying is the science for accurate measurement of natural and artificial features on the Earth. The purpose of surveying is to gather accurate locational and geometric characteristics of features on the Earth's surface. The end products of surveying normally are maps. Conventionally, surveys of the Earth's surface can be classified as geodetic surveys or engineering surveys. A geodetic survey determines the precise location of permanent points or geographic features on the Earth's surface, considering the shape, size, and curvature of the Earth (Figure 4.2). By contrast, the engineering type of survey generally ignores the curvature of the Earth's surface and treats the Earth's surface as a flat surface in the study area (Figure 4.3). Therefore, geodetic surveys are used either to establish fundamental geo-reference systems for features on the Earth's surface or to conduct measurements of high precision and accuracy on an object or a group of objects. Engineering surveys are mainly focused on a small geographic area in order to minimize the inaccuracy introduced when ignoring the curvature of the Earth's surface.

**Figure 4.2** Geodetic survey and measuring the precise and accurate location at a surveying control point. *Source:* From the Geographic Information Authority of Japan.

In practice, since the World War II, surveying and three-dimensional (3D) or topographic map-making in a large area on the Earth's surface were mainly performed by using aerial image interpretations and photogrammetry. Photogrammetry is the art and science of extracting 3D information from photographs or images. The process involves taking overlapping aerial photographs of an object, structure, or space and converting them into 2D or 3D analog or digital models (Figure 4.4). After spending more than one century on its effort to survey and map the entire nation in detail, the United States Geological Survey accomplished 1:24,000 topographic map quadrangles for the entire country in the early 1970s. This is the defining landmark achievement for the original GIS or geospatial data. Meanwhile, the aerial survey and photogrammetric engineering became a profession in the science and engineering even before GIS was fully developed. Although the fundamental geospatial or spatial reference data surveys have already been

**Figure 4.3** Engineering survey. *Source:* Dr. Tao Tang.

**Figure 4.4** Overlapping aerial image taking for aerial photogrammetric engineering of 3D data extractions. *Source:* Dr. Tao Tang.

accomplished in the United States and in most of the developed countries, GIS analysts seldom conduct field survey for large quantity of data collections. It is significant for us to understand that the foundation of GIS database building is from field survey and photogrammetric survey.

Surveying is defined in science and civil engineering as a means of making relatively large-scale, accurate measurements of part of or the entire Earth's surface. Two distinctive purposes of surveying are: (i) to determine the relative horizontal and vertical positions of existing points or features on the Earth's surface in order to be used for the process of mapping or engineering drawings; and (ii) to establish the marks to control constructions or to indicate land partition boundaries. Modern surveying of the Earth's surface can be traced back to the late eighteenth century. Since World War II, many improvements and refinements have been incorporated in all the basic surveying instruments, such as the theodolite. In addition to modification of existing instruments, photogrammetry, or making maps utilizing aerial photographs was introduced in the industry since the 1920s. In 1960s and late 1990s, electronic or digital range finder and laser range finder was introduced in the field survey. Laser range finders were then further developed as light detection and ranging (LiDAR) or laser scanners in the late 1990s. These two improvements of technologies revolutionized the industry of mapping and GIS data collection. Since then, humans can collect large quantities of geographic or spatial data at fast speeds while covering a large area for the purpose of detailed digital mapping. Starting in the late twentieth century, satellite imaging and data collection technology was developed to conduct various surveys, including the geodetic survey, which is a discipline in measuring the shape of the earth.

Generally speaking, surveying can be identified into three broad categories based on the objects to be measured. These are: (i) land surveying – objects are land or on the land; (ii) marine surveying – objects are on the ocean or in the ocean; (iii) astronomical surveying – determine the shape of the entire Earth using objects in the space. In GIS projects, we mainly focus on the land surveying since humans live on the land, and most of our economic development occurs on land.

Land surveying can be further classified as (i) topographical surveys, (ii) cadastral surveys, (iii) city surveys, and (iv) engineering surveys. A topographic survey locates and records all features on Earth's surface in a 3D way. These features can be natural or artificial features. The unique characteristic of a topographic survey is that it provides original z-coordinate values on top of x, y locational values. The conventional result of a topographic survey is a piece or a quadrangle of a topographic map. Maps in general, and topographic maps in particular, are 2D; therefore, the drawing of contour lines and contour intervals were invented to represent the values of the third dimension (z). Topographic surveys and their map products can be further identified as (i) general topographic surveys and maps, or (ii) engineering topographic surveys and maps. Engineering topographic maps tend to show details of features in large map scale with their location, size, height, and any changes in elevation. Topographic surveys and mapping are very important for large construction projects, and they are an essential component for utility management companies, such as the National Grid (electricity transportation) or oil or gas transportation companies.

Cadastral surveying is to establish or reestablish the boundaries of real properties or land ownership. It involves the physical delineation of property boundaries and the determination of the dimensions, locations, areas, and certain rights associated with properties. This is regardless of whether they are on land, water, or defined by natural or artificial features. It is an important component of the legal creation of properties. Cadastral surveying carries both spatial-measurement principles and legal principles. City surveying, by its name, is meant as an extensive coordinated survey of the area within the limits of municipal boundaries for the technical purposes of governing and delivering services to the residents. City survey is valuable during large-scale improvements or extensions of the existing street, water supply, and sewer systems, layout and construction of new roads, and development of the areas in or around a city. In many cases, city surveying carries characteristics of either topographic survey or cadastral survey, or both.

The American Society of Civil Engineers (ASCE 2021) defines engineering surveys as those activities involved in the planning and execution of the surveying process for the planning, design, construction, operation, and maintenance of engineered projects. The practical purpose of engineering surveys is to accomplish human construction projects. Therefore, such kinds of surveys also include locating or laying out alignments, positions, or elevations for the construction of fixed works.

In recent years, a new remote sensing and data collection technology called LiDAR or laser scan was developed. LiDAR technology enables us to collect 3D locations of the objects on the Earth's surface directly. In practice, LiDAR is an equipment that utilizes an invisible laser beam to scan the objects on the Earth's surface or planetary surface to conduct the surveys. LiDAR equipment can calculate 3D locations or coordinates (x, y, z) of points on the surfaces of the objects using the measured range from the device to the target. When the laser beam is reflected from the target, the range is computed by multiplying the return time and the speed of light. The mathematical formula is: $d = c * T/2$; where $c$ is the speed of light, $d$ is the distance between the laser pulse detector and the object, and $T$ is the time spent for the laser light to travel to the object and come back. Since the initial location of the LiDAR equipment or laser pulse generation location (x, y, z) is known, the locations of all the reflection points can be calculated to generate a 3D point cloud. Therefore, the datasets generated by LiDAR are called point clouds.

Since the invention of LiDAR or 3D laser scan, it has been widely used in surveying, geodesy, geomatics, archaeology, geography, geology, and geomorphology for accurate location determinations or making high-resolution and high-accuracy digital maps. LiDAR is the device that can most closely mimic the human eyes; therefore, it is also widely used for self-driving vehicles or target model capture for 3D printing. Based on data collection platform locations, LiDAR can be further classified into two types: airborne LiDAR remote sensing and terrestrial LiDAR remote sensing. Airborne LiDAR remote sensing is when a laser scanner is put onboard of an aircraft or a drone (unmanned aerial vehicle [UAV]) for surveying and data collection (Figure 4.5). Terrestrial LiDAR remote sensing is conducted on the Earth's surface and can be either stationary or mobile (Figure 4.6). Stationary terrestrial scanning is to attach equipment at one location and scan the surrounding environment. Mobile terrestrial scanning is to set up equipment to a moving vehicle

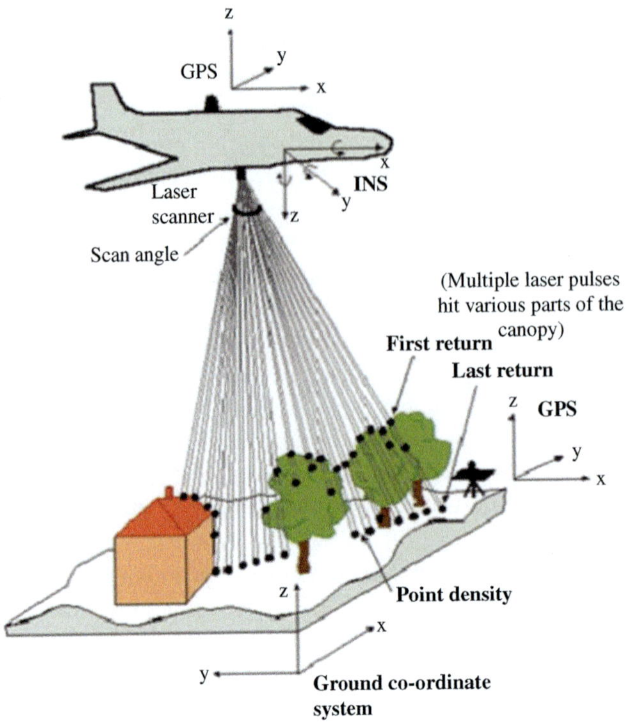


**Figure 4.5** Airborne LiDAR (laser scanner).

(a) (b)

**Figure 4.6** Terrestrial LiDAR: (a) stationary; (b) mobile. *Source:* Dr. Tao Tang.

to collect data along a path (Figure 4.7). Normally, we use more than one LiDAR sensor for accurate data collection. Currently, the most popular application of mobile terrestrial scanning is in automatic-driving or self-driving technology for vehicles. In essence, a multi-sensor LiDAR system in combination with high-resolution digital cameras resembles the functions of the human eyes, and a built-in computer with fast computing to guide the mechanical actions on the vehicle's control system mimics the human brain's judgment and actions.

(a) (b)

**Figure 4.7** Mobile LiDAR data collection (a) along the shoreline of the Lake Erie, and (b) on Interstate highway I-90. *Source:* Dr. Tao Tang.

A case study of GIS data collection is presented here. In 2011, the research team led by Dr. Tang at SUNY Buffalo State University obtained a contract to conduct and accomplish a pilot asset management GIS spatial data collection project using vehicle-based 3D LiDAR experiment equipment for the New York State Thruway Authority. The New York State Thruway Authority would like to utilize this 3D spatial data for highway maintenance needs. The data deliverables were point cloud files in LAS format with real-world x, y, z coordinates and laser beam reflection intensities. These 3D data files will be accompanied with 2D digital side view pictures collected at the same time during the survey for digitally building the surface texture and colors for the features surveyed.

The geographic extent of this pilot project is along the entire extent of the Niagara Thruway (I-190) from its junction with the NYS Thruway (I-90) to Exit 21, at the north end of the North Grand Island Bridge, and the section of NYS Thruway (I-90) from Exit 53 in Buffalo to Exit 57 in Hamburg. The total length of this survey and data collection is roughly 31 miles one way, or 62 miles in both directions. This project involves a vehicle-based research type of laser scanner and gyroscope combination to be loaded on a pickup truck to collect 3D spatial data along a highway (Figure 4.8).

Field data collection was conducted, and both LiDAR point clouds and color imagery data were recorded while driving along the experimental section of the NYS Thruway (Figure 4.9). In order to ensure the high precision and accuracy of the data collection, an RTK GPS field control station was set up for the entire project (Figure 4.10).

Another significant technological advancement for scientific data collection on the Earth's surface is the global positioning system (GPS). GPS was invented by the scientists who worked for the United States Defense Department in the 1980s. The objective of GPS equipment is to accurately measure the locational information, for instance, latitude, longitude, and altitude, of the person who holds the device. The GPS is a satellite-based navigation system that currently consists of at least 24 operational satellites in broadcasting positioning signals (Figure 4.11). Therefore, a hand-held GPS unit can utilize the signals received from at least three of these satellites to compute its location on the Earth's surface (Figure 4.12).

GPS locational equipment has been widely used in combination with various scientific investigations of the Earth's surface, either for in situ measurements or for the sample collections of laboratory analyses. These approaches make the GPS markers readily available with various specialty datasets and very straightforward for GIS visualizations, mapping, and analysis. Since we cannot

**Figure 4.8** Mounting the experiment mobile LiDAR (laser scanner) on the back of the NYS Thruway Authority Pickup Truck for GIS data collection. *Source:* Dr. Tao Tang.

**Figure 4.9** Mobile LiDAR (laser scanner) and imagery data collection along the NYS Thruway. *Source:* Dr. Tao Tang.

**Figure 4.10** RTK GPS control station setting up for the 3D LiDAR and imagery data collections. *Source:* Dr. Tao Tang.

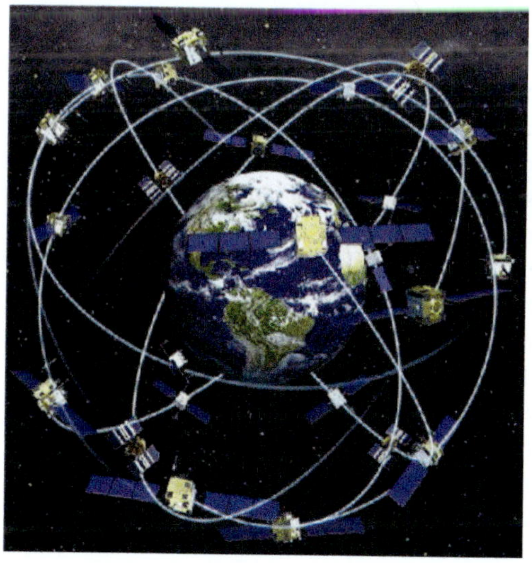

**Figure 4.11** The GPS. *Source:* NASA/Public Domain.

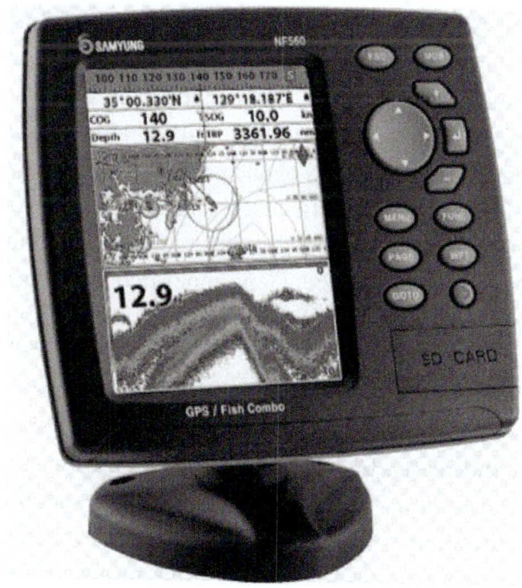


**Figure 4.12** GPS receiver unit.

discuss all the different scientific or social science data that can be readily incorporated into the GIS software environment for analysis, we would like to simplify the discussion and indicate that any scientific datasets associated with locational markers that are obtained in the field with GPS can be incorporated into GIS systems using the coordinate geometry (COGO) process. The product of this process is a new feature class in the geodatabase.

Fundamentally, GPS is an equipment to find and collect locational markers for geographic objects or events. Locational markers are very important for researchers and professionals to understand why some objects or features occupy a certain location or a certain area. Therefore, we can further understand the characteristics of the objects or features. The majority of objects or features that we investigate, study, and explore in different scientific disciplines, such as biology, chemistry, or Earth's science, have geographic locations as one characteristic among many other characteristics. The storage of all these characteristics in the computational databases is becoming increasingly common. Using some GIS methods, we can rediscover and visualize the locational markers of these objects and features and map them on a digital base map. GIS enables us to use spatial locations and distribution patterns to solve real-world problems. For instance, we can plot the COVID-19 infectious disease spreading patterns to help in analyzing the situation and prevent its further infection of people living in a region.

## Geographic Data Collections in Social Sciences

Generally speaking, social scientists help society to understand how a particular region looks or how we as humans change its appearance and function. It is also in social science that we often engage in partnerships that effectively work across disciplines, geographic communities, and organizations. Toward this end, GIS is an effective tool to create and evaluate alternative designs or geo-designs to make better decisions interactively and iteratively. For example, we can utilize GIS and land cover data to conduct land-use planning, green infrastructure planning, and urban planning. GIS has become a powerful utility for social scientists to conduct their work and research. Employing GIS capabilities, social science experts can capture and geocode data; visualize and disseminate data; analyze behaviors and values in the datasets that can be used to address long-standing issues of social and environmental concerns.

Similar to scientific research data, data in social sciences cover a broad range of materials, from structured numerical datasets to interviews, field notes, and documents collected for ethnographic studies, for instance. Data generated in social sciences can be both quantitative and qualitative. Notably, within the field of social sciences, you often work with data originating from human participants. This means that we are often handling personal data with specific sensitive concerns. Personal datasets are defined as any information relating to an identified natural person known as "a data subject." An identifiable natural person is someone who can be identified, either directly or indirectly, by reference to an identifier such as a name, an identification number, locational information, or to one or more factors specific to the physical, physiological, genetic, economic, cultural, or social identity of that person. Personal data can include a variety of information, such as names, home addresses, phone numbers, IP addresses, etc.

Certain personal datasets are considered particularly sensitive and thus require specific protection. Examples of sensitive personal data include data revealing religious affiliation, sexual orientation, racial or ethnic origin, genetic or health-related data, etc. When handling this type of dataset, GIS analysts must provide extra care to the security of the data. We can technically convert the sensitive data to nonsensitive data, such as stripping the field of owner or occupant names of a dataset with home addresses when we do the geocoding processes and publish the digital map online.

By nature, large quantities of social science datasets are collected in qualitative formats, such as interviews, town hall meetings, summary of focus group discussions, or visual observations. Qualitative data tends to be nonnumerical and descriptive. In contrast, quantitative data is numerical and can be readily used to conduct statistical analysis. Both qualitative and quantitative data carries spatial or locational characteristics. Or, in other words, it can be converted into spatial or geographically enabled data to be visualized and analyzed on the digital maps in GIS.

Interviews in social science are a method of data collection that involves two or more people exchanging information through a series of questions and answers. The questions are designed by a researcher to elicit information from interview participants on a specific topic or a set of topics. The original results of interviews are qualitative and can be converted to quantitative later on. The purpose of town hall meetings is for local and regional government officials to hear the community's views on public issues to gather information or data for decision-making or police initiatives. There have been no specific rules or guidelines for holding a town hall meeting. The format of the meetings can vary. Town hall meetings are commonly held in a range of venues, including schools, libraries, municipal buildings, and churches. Recently, a few government officials have also experimented with digital formats for town hall meetings. In this type of meeting, attendees can voice their opinions and question elected officials or political candidates. Town hall meetings are the standard practices for data gathering and decision-making in urban planning processes. The original results of this type of meeting are qualitative.

A focus group is a small group of demographically diverse people by age, gender, profession, or ethnicity who are invited to participate in market research of a company or in response to a political campaign or a social issue. Focus groups are a type of method to gather qualitative data. Observations of the group's dynamic, their answers to questions, and even their body language can guide the analyses of consumer decisions, products, and services. Focus groups are often used in marketing and business studies, library science, as well as other social science disciplines. They can provide more natural feedback than individual interviews and are easier to organize than experiments or large-scale surveys.

One of the most popularly used data sources in social sciences is the United States Census Bureau demographic data (Figure 4.13). Demographic data is comprised of an array of socioeconomic information, including the breakdown of a population by gender, age, ethnicity, income, employment status, education, home ownership, and even internet access. Demographic data also provides a generalization of the population in a particular geographic region, based on a sampling of people in that region. Furthermore, demographic data captures a snapshot in time, such as 2020. Each of the demographic datasets has its imprint of time, for instance, the total population of Erie County, New York, in the year 2010. Comparing demographic data at different points in time, we can analyze historical changes, such as whether the population in a county is aging or new people are moving to a community. Demographic data is directly marked with locations and geographic regions and is normally organized by specific structures. Worldwide, governments of many countries collect demographic data, and the United Nations is the agency to gather the demographic data for all the nations. In the United States, the Census Bureau under the US Department of Interior is the federal government agency that is responsible for conducting surveys and gathering demographic data for the entire country.

**Figure 4.13** US Census Bureau collects demographic data every 10 years by mail order. *Source:* Dr. Tao Tang.

How does the US Census Bureau obtain all this type of demographic data? According to Article I, Section 2 of the United States Constitution, the Census Bureau conducts an on-ground household survey every 10 years. The survey is designed to count every resident in the United States. The survey and data collection are also called the decennial census. The data collected by the decennial census determines the number of seats each state has in the US House of Representatives and is also used to distribute hundreds of billions of dollars in federal funds to local communities and local governments.

During the survey, the decennial census asks questions of people in homes and group living situations, including how many people live or stay in each home and the sex, age, and race of each person. The goal of the on-ground survey is to count everyone once, only once, and in the right place. In between the two surveying years, such as 2010 and 2020, the Census Bureau publishes estimated population and demographic components of change, such as births, deaths, and migration, every year based on the original survey conducted previously. This data can be sorted by characteristics such as age, sex, and race, as well as by national, state, and county. The Census Bureau also created geographic subdivisions below the county level for data gathering, statistics, and dissemination. These are called tracts, block groups, and blocks.

US Census demographic data, therefore, is considered the original and authoritative data in our country. As the Census Bureau illustrated, this demographic data "tells us who we are and where we are going as a nation, and helps our communities determine where to build everything from schools to supermarkets, and from homes to hospitals. It helps the government decide how to distribute funds and assistance to states and localities" (US Census 2022). The advantage of US Census demographic data to GIS analysis is that every dataset is either collected by geographic locations or regions or carries the locational markers that can be readily incorporated into GIS systems for mapping and analysis. In the following chapters, we will discuss the US Census demographic data in detail.

## Remote Sensing as the Method of GI Data Collections

In addition to point-based on-ground surveys and data collections, area-based surveys of photogrammetry and remote sensing methods have been developed to enhance data collection of large geographic areas at fast speeds. Remote sensing (RS) is defined in science as the measurement or acquisition of data and properties of objects or phenomena on the Earth or planet surface by an equipment without physical contact with the objects (Jensen 2007). Remote sensing is widely used as a technology to collect data in numerous fields, including GIS, geography, land surveying, and most earth science disciplines, such as hydrology, ecology, meteorology, oceanography, glaciology, and geology. It is also a valuable data collection practice in military, intelligence, commercial, economic, planning, humanitarian fields, etc.

According to the platforms of data collection equipment, remote sensing can be identified as: (i) orbital or space remote sensing; (ii) airborne remote sensing; and (iii) ground remote sensing (Figure 4.14). In orbital or space remote sensing, sensors or equipment are mounted on-board a spacecraft; either a satellite, space shuttle or space station that is orbiting the Earth. The advantages of space remote sensing are: (i) it can cover everywhere on the Earth's surface without disputing territories and ownership of airspaces; (ii) it can collect data in a large geographic area in a relatively more efficient way; (iii) it has a relatively low cost to data users and data consumers and

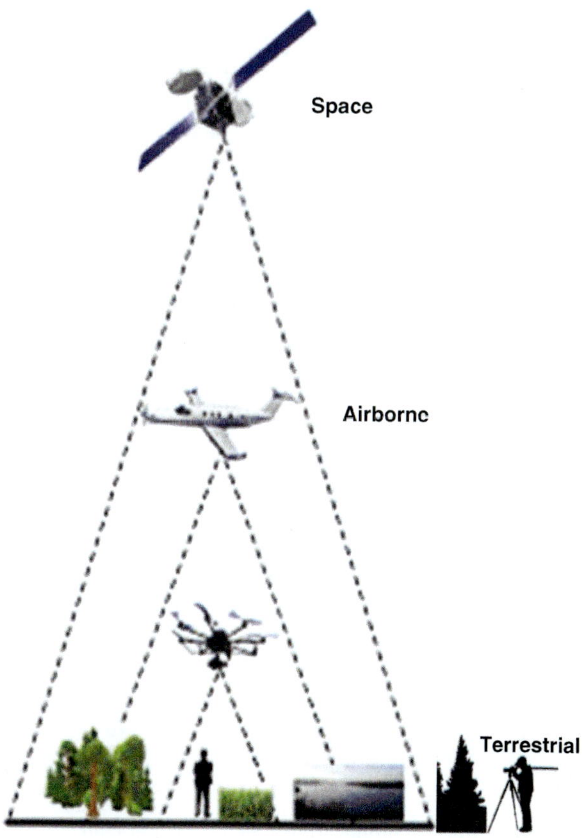


**Figure 4.14** Platforms of remote sensing for GIS data collections. *Source:* Dr. Tao Tang.

is easy to collect data multiple times in a predefined survey period. Earth observation satellites can be launched and operated by either a country's government or a private company. Therefore, the cost in obtaining satellite imagery data ranges from a free download to a couple of hundred dollars per square kilometer.

The sensors or equipment of airborne remote sensing for data collections are mounted on human-operated airplanes, or more recently on UAV or drones. By the nature of altitude of remote sensing platforms, an advantage of airborne remote sensing is the capability of offering very high spatial resolution images and datasets, simply because the sensor is closer to the Earth's surface. However, recent technology development of satellite imaging sensors drastically reduced the gap of spatial resolution. Satellite imagery sensers are now capable of obtaining images with half (0.5) meter ground spatial resolution. Space remote sensing in essence is a part of "rocket science" as the saying goes. Therefore, it does cost a lot of money to initiate and to establish the system in collections and disseminations of the data. As soon as the space remote sensing system, such as satellite systems, is established, the data resources can be shared by various users, even worldwide by all the countries. Therefore, the cost to individual data users can be drastically reduced or even free. In contrast, although the total cost of airborne remote sensing by human-operated airplanes is much lower than the satellite-based space remote sensing, the data collected often support a particular purpose and/or in a particular region. The end users of the data products must take on the entire cost of the data collection. In practice, airborne remote sensing missions are often carried out as one-time operations, whereas Earth observation satellites offer the possibility of continuous monitoring of the Earth over a long period.

Since the 1990s, large area surveys and photogrammetric engineering have been mainly conducted or expanded based on the images that were collected from human-operated airplanes, the high spatial resolution satellite images from the space, or aerial images from UAV or drones (Figure 4.15). Among the two airborne remote sensing technologies that are popularly used currently, namely human-operated aircraft and UAVs or drones, the UAV or drone airborne remote sensing is much more cost-effective than the human-operated aircraft ones. Fundamentally, the advantage of airborne remote sensing is that the data users can deploy the data collection events at any time and anywhere on the Earth's surface in order to achieve their goals (Figure 4.16). For instance, temporal resolutions with the desired frequency of observations can be easily achieved using UAVs (Tang and Zhang 2018; Tang, Jiang, and Perrelli 2020a, b; Tang et al. 2021).

Ground-based platforms, as their name suggests, are those remote sensors or devices taking data on the ground. These devices can be handheld cameras (film or digital), cranes, ground vehicles,

(a) (b)

**Figure 4.15** Two types of popularly used aerial image collection drones (UAVs). *Source:* Dr. Tao Tang.

**Figure 4.16** Airborne drone (UAV) GIS data collection. *Source:* Dr. Tao Tang.

etc. The most popularly used data are the Google Street View images online. Figure 4.17 shows the Google remote sensing vehicle that is used for street view data collection.

In summary, GIS is a comprehensive way of analyzing the geographic or spatial characteristics of data. On one hand, GI technology can be widely used by both scientists and social scientists. On the other hand, all the diversified data collection methods and collected data in both science and social science disciplines can be input into the GIS systems for visualization, analysis, and modeling. Several scientific methods and engineering technologies, such as GPS, remote sensing, and photogrammetry, have been developed worldwide for GIS data collections.

**Figure 4.17** Google remote sensing vehicle for data collection. *Source:* Dr. Tao Tang.

## References

ASCE (American Society of Civil Engineers). (2021). Policy statement 333 – Engineering surveying definition. https://www.asce.org/advocacy/policy-statements/.

Jensen, J.R. (2007). *Remote Sensing of the Environment – An Earth Resource Perspective*, 2e. p. 592, Hoboken, New Jersey: Prentice Hall.

Tang, T., Radomski, M., Stefan, M., et al. (2021). UAV-based high spatial and temporal resolution monitoring and mapping of surface moisture status in a vineyard. *Journal of Papers in Applied Geography* 6 (4): 402–415. https://doi.org/10.1080/23754931.2020.1798808.

Tang, T., Jiang, L., and Perrelli, M. (2020a). Data collection and analyses applying unmanned helicopter (UAV) remote sensing to survey water chestnut invasive species. *International Journal of Data Analytics* 1 (1): 38–51.

Tang, T., Jiang, L., and Perrelli' M. (2020b). Low-altitude unmanned helicopter (UAV) remote sensing for spreading control of water chestnut invasive species. In: The Erie Canal System, New York. *Middle States Geographer*. In press.

Tang, T. and Zhang, J. (2018). Time scale sensitive sensor applications in collecting and analyzing geographic event data. *Annals of Geographic Information Science (GIS)* 24 (4): 241–253. https://doi.org/10.1080/19475683.2018.1520299.

US Census Bureau. (2022). Decennial Census. https://www.census.gov/programs-surveys/censuses.html.

## Tutorial in ArcGIS Pro Environment

1. Digitizing the road center lines and house footprint polygons for a subdivision in the Town of Amherst, and tree location points for Delaware Park, Erie County, New York

Copy or download the "Chapter04-Tutorial" folder from the textbook data site to your working location (either C:\ or D:\ drive of your computer, or your portable drive or share of a network drive).

Open ArcGIS Pro and create a new project called "MyGIS_Learning04" with Map template. Save the project file in the Chapter04-Tutorial folder.

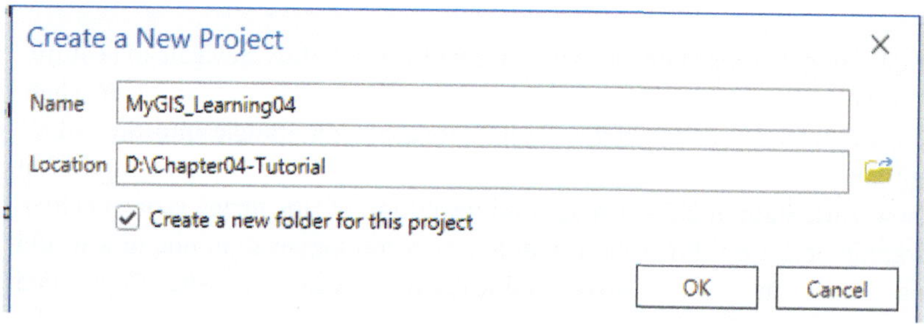


Now, under the category of "Folders" on the Catalog Pane, add the entire folder of "Chapter04-Tutorial." So, we can see all the folders and files.

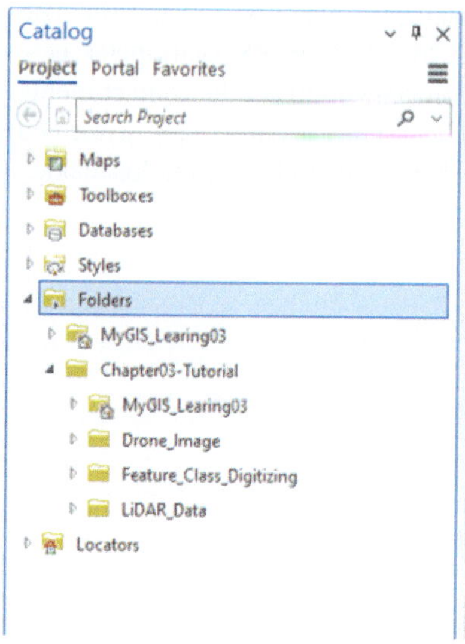


Digitizing as well as all other modifying works in ArcGIS Pro database or geodatabase are classified as the tasks (category) of editing in the ArcGIS Pro software environment. The workflow is that we need to create a file geodatabase first. If we have a file geodatabase that we would like to store the feature classes in already, we can digitize feature classes at any time we need. Secondly, we need to create empty feature classes to hold the digitized features later. Feature classes can be either created directly in a geodatabase or can be created in a feature dataset that serves as a database organizer in a geodatabase. In practice, there are two major ways to create a feature class: (i) using the "Create Feature Class" wizard (dialog) and (ii) using the "Create Feature Class" tool in the Geoprocessing Pane or the Toolbox. Finally, we can bring each of the empty feature classes from the geodatabase to the "Map" window and use the "Edit" ribbon on the main menu to digitize (create) features. When we create empty feature classes in a geodatabase, we need to fulfill two major initiation criteria. First, we need to determine the geometric type of the feature class, namely point, line, polygon, or others. As soon as the geometric type of features is decided and the feature class is made, we cannot change it. The only way is to delete it if that geometric type is not what we want. Second, we need to determine the coordinate system or georeference system that this feature class will use. If the coordinate system of the feature class that we input does not match the backdrop remotely sensed images or scanned map images, the digitized features cannot be saved in the newly created feature class.

Now, let us input the first remotely sensed image to prepare for the digitizing. Navigate to Feature_Class_Digitizing -> Tree-Point-Feature-Digitizing folder in your Chapter04-Tutorial folder. Then, drag and drop the "w_10711068_12_19200_4bd_2021.jp2" imagery quadrangle into the "Map" window. This is a high spatial resolution aerial remote sensing image (aerial photograph) that was collected by the New York State in 2021. The ground resolution of this image is one (1) foot. That means any geographic object on the ground with a dimension bigger than one foot would be detected in this image. The image is also orthographically re-projected to ensure the unified scale-down representations across the entire region. The geographic area of this image is the Delaware Park and old cemetery field in Buffalo, New York. Our task in this GIS project is to create the "Delaware_Park_Trees" point feature class and digitize the point locations of trees in the southeast portion of this aerial image.

Since we have already created the "MyGIS_Learning04" geodatabase when we created this GIS project, we can elect to create and save the new point feature class into the "MyGIS_Learning04" geodatabase, MyGIS_Learning04.gdb. Right-click on MyGIS_Learning04.gdb to get the contact menu. Highlight the "New" menu and then click on "Feature Class." The "Create Feature Class" wizard on the Geoprocessing Pane is overlaid on top of the Catalog Pane. Type in the following on this dialog.

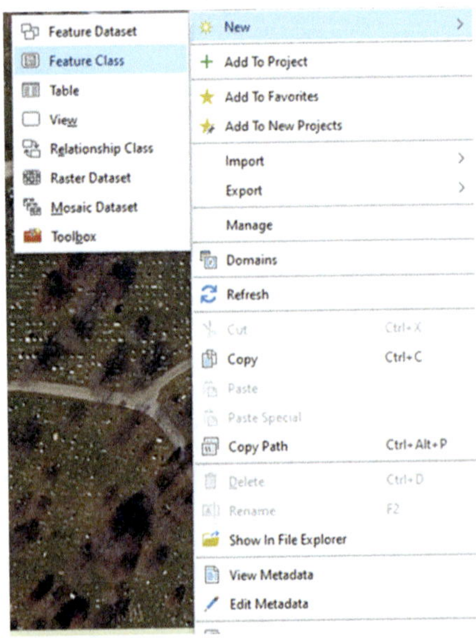

Name: Delaware_Park_Trees

Feature Class Type:

Select "Point" for "Type of features stored in the feature class." Leave other parameters as default, click on "Next."

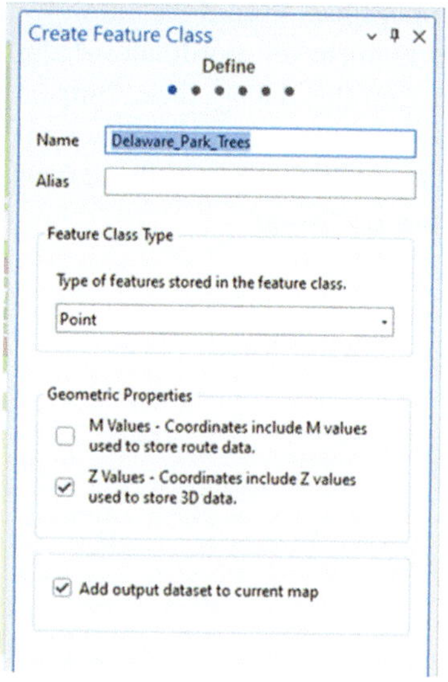


The second page is a domain or attribute table editing dialog for us to fill in the fields or columns in the attribute table of this new feature class. Each row we add here will be a column in the attribute table of this new feature class. We will practice entering two rows as follows. The data type on this page means the data format of the field or column in the attribute table.

| **Field name** | **Data type** |
|---|---|
| Tree_Species_ID | Short Integer |
| Tree_Species | Text |

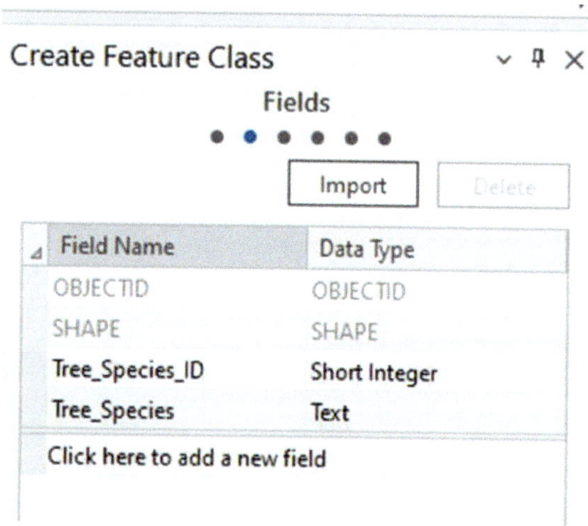

Keep the other default attributes. Click on "Next."

This page is the "Spatial Reference" or, more precisely, the coordinate system editing dialog. It is very important for us to get the same coordinate system (geographic or projected) that "w_10711068_12_19200_4bd_2021.jp2" image has for this feature class. So, we can overlay this feature class on top of the aerial image to digitize and save the digitized features. At this time, we do not know specifically which coordinate system the aerial image has. Even if we know, we may not remember the specific name. The easiest way is to import whichever coordinate system the image has. We click on the "Add Coordinate System" icon on this dialog, then, "Import Coordinate System..." menu. Navigate to the imagery file location to load its coordinate system. After loading the coordinate system, if no new coordinate system is added, that means the default coordinate system in the dialog is the same as that to be added. Click on "Next."

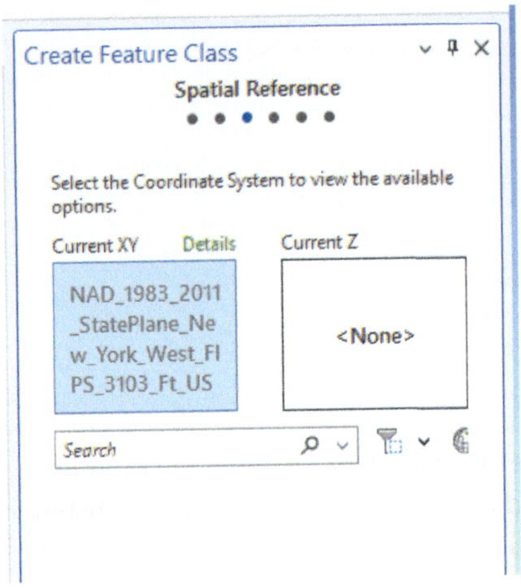


This page is a dialog that controls the "Tolerance." Or in other words, this page is to set up the precision and accuracy of your new feature class. Keep the default and click on "Next."

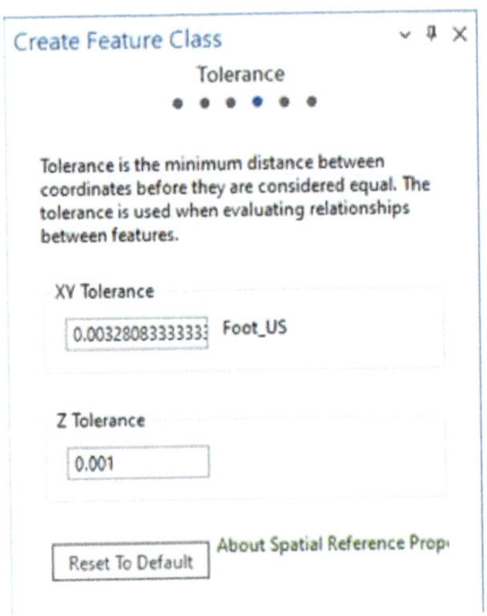

On this page, accept the default resolution and click on the "Next."

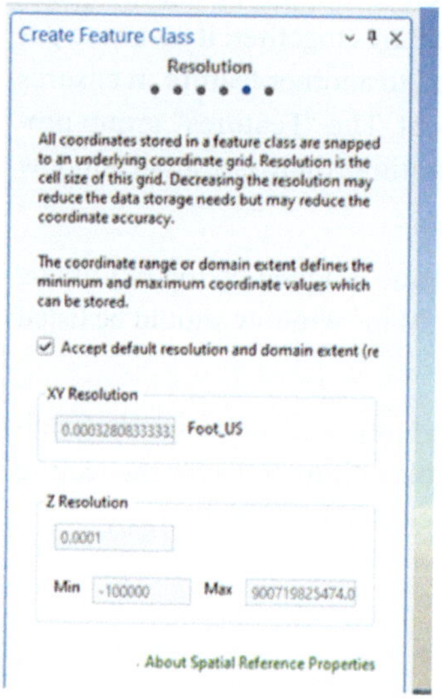


Then, keep the "Storage Configuration" as the default and "Finish" the feature class creation. Now, you notice that the new empty point feature class is created in the geodatabase. Drag the "Delaware_Park_Trees" point feature class to the "Map" window. Symbolize the points with "Circle 4" and font size 15.

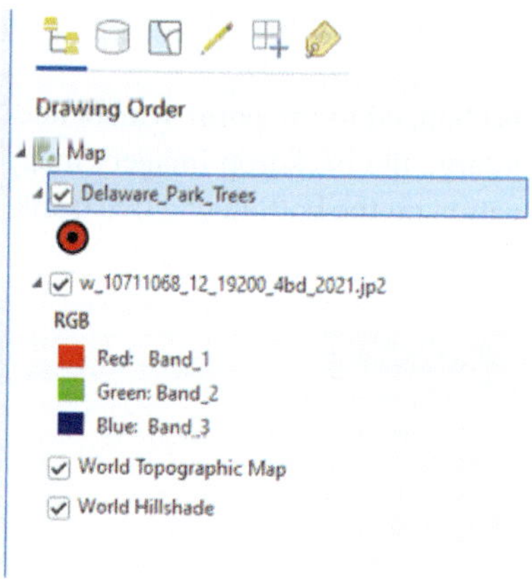


Now, let's zoom into the southeast quarter of the image and digitize the tree points. Click on the "Edits" at the main menu on the top. You notice that the ribbon is changed to the following.

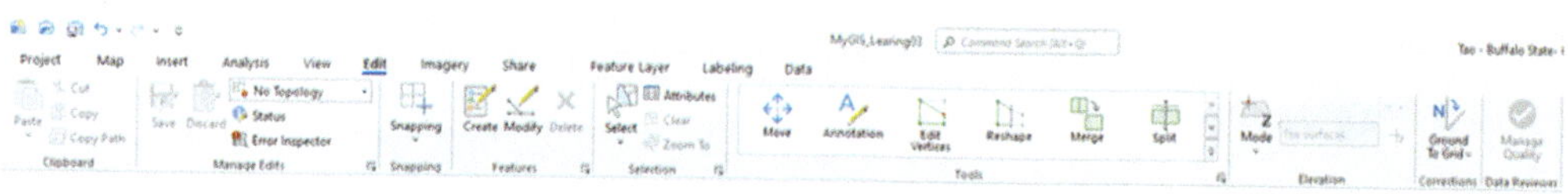

The icons on this ribbon are grouped by their functions. For instance, "Manage Edits" can help to save the edited features, and "Snapping" controls in which editing environment the digitizing features will snap on each other. Snapping does not mean to glue the features together; it means to put the feature or component of a feature at the exact location in relation to another feature. It ensures the topological integrity of connectivity, adjacency, and containment. The "Features" group provides us with the tools that can create or modify features. In the meantime, there is a small window on this ribbon from which the users can grab a specific tool directly.

Click on the Create icon in the Features group, Create Feature Pane is added on top of Catalog Pane. All the editable feature classes in the Table of Content of the "Map" window should be listed here. In our case, we only have one map layer at this time.

Click on the map layer Title "Delaware_Park_Trees" to expand the template icon to see the available tools for digitizing. ArcGIS Pro automatically selects the suitable editing tools for the feature class that you prepare to edit. The first tool for digitizing points is selected by default.

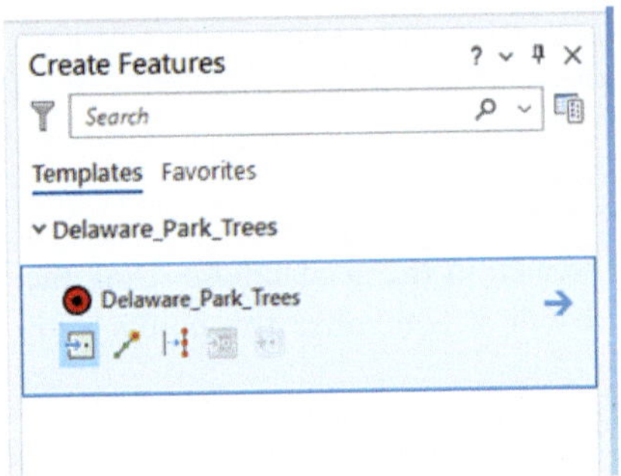


Moving the cursor to the "Map" window, you may notice it has changed to the point symbol and is ready for you to digitize. You can use the middle roller to enlarge the backdrop imagery map. Then, you can click the left button on the mouse at the exact location on the bottom of a tree trunk like the following.

In ArcGIS Pro editing mode, the last digitized feature is always put in the selection mode (light blue). If you are not satisfied with the last digitized feature, you can always push the delete key on the keyboard to delete it and try again. If you are not satisfied with several previously digitized features, you can always use the select tool to select them and delete them before saving the feature class file. The digitized features will not be saved in the feature class until you click on the "Save" icon 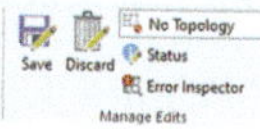

. Clicking the "Disregard" icon will erase everything that you digitized up to that moment. Users should periodically save their editing work. You can also click on the arrow to flip the page to "Active Template" to add attribute data for the digitized feature one by one.

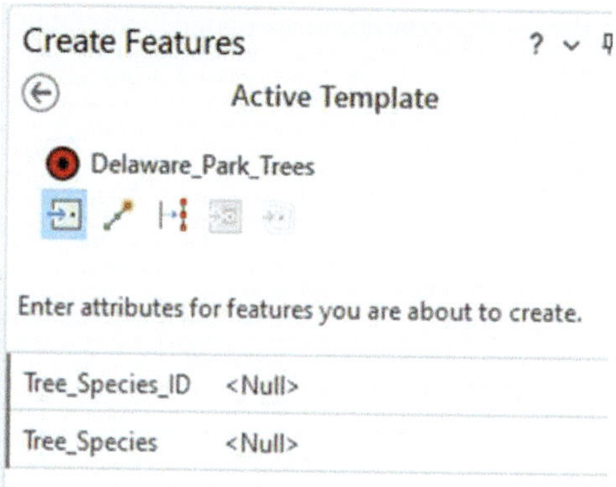


If you would like to move or pan the mapping area, you can click on the "Finish" (F2) icon 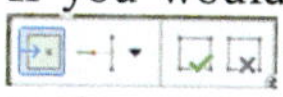 to temporarily stop the editing. Clicking on the point entering tool will lead you back to the editing mode. You are expected to digitize all the trees in the southeast corner of this image in Delaware Park as we did in the following. After saving the edits, you can close the editing template. You can elect to open the attribute table to enter the attribute then. In real-world practice, the tree species need to be entered with field-recorded ground truth.

Close the editing template and save the project. Now, we accomplished the Delaware Park Tree Inspection Project. We are going to move on to digitizing some features for the Town of Amherst, Erie County, New York.

Under the "Maps" category on the Catalog Pane, right-click on "Maps" to open a new "Map" window, which is called "Map 1" now. Change the name from "Map 1" to "Amherst_Digitizing."

Now, we would like to expand the "Folders" category and navigate to the "Feature_Class_Digitizing" folder. We would like to create a new file geodatabase for the Town of Amherst to digitize a couple of feature classes. On the Catalog Pane, right-click on the "Feature_Class_Digitizing" folder to get the contact menu, then highlight the new menu to open the second contact menu. Then, click on "File Geodatabase," and the new geodatabase is created. Name it as "Town_Amherst_Data.gdb."

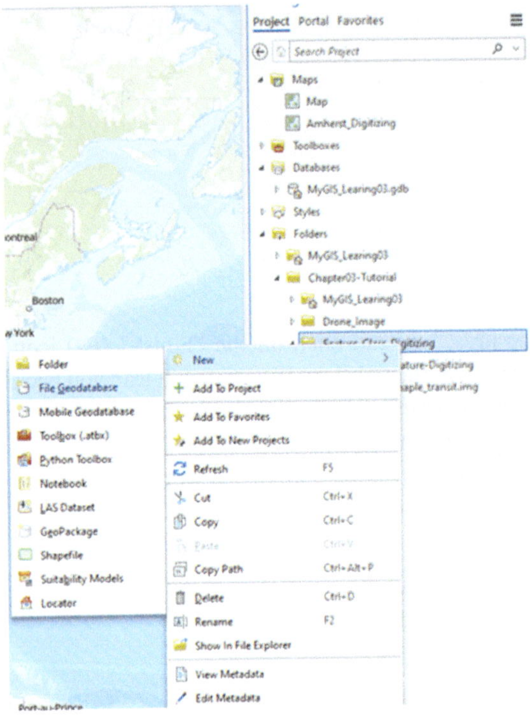


We continue this process to create new feature datasets and feature classes for the town. It is always good to consider some database designs and organization. Since the GIS geodatabase is used by various functioning departments of the town's government, such as the Police Department, Engineering Department, Tax Assessor's Office, etc., it is very important that the digital map layers and feature classes are organized by their functions. We will use the feature datasets to classify different data groups.

Right-click on the "Town_Amherst_Data.gdb" on the Catalog Pane to obtain the contact menu. Then, highlight the "New" menu to get the second contact menu. Click on "Feature Dataset" to create a feature dataset.

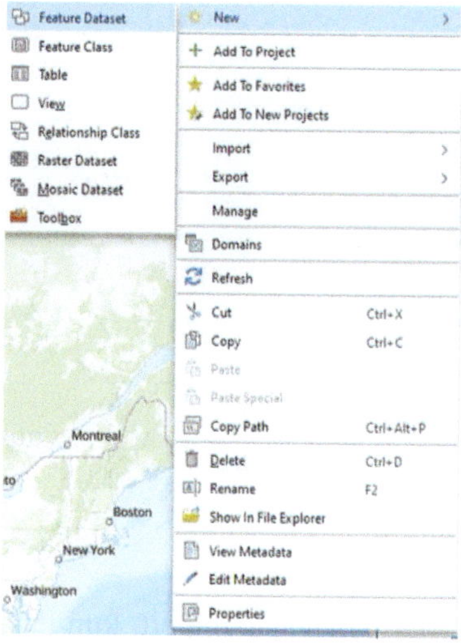


Call it "Transportation." The Geoprocessing Pane with "Create Feature Dataset" is overlaid on top of the Catalog Pane. For the Feature Dataset Name fill in "Transportation." For the Coordinate System import the coordinate system file from the aerial image of "amherst_at_maple_transit.img." Click on the small globe icon to open the Coordinate System dialog. Then, click on the small globe icon here to get the "Import Coordinate System..." menu.

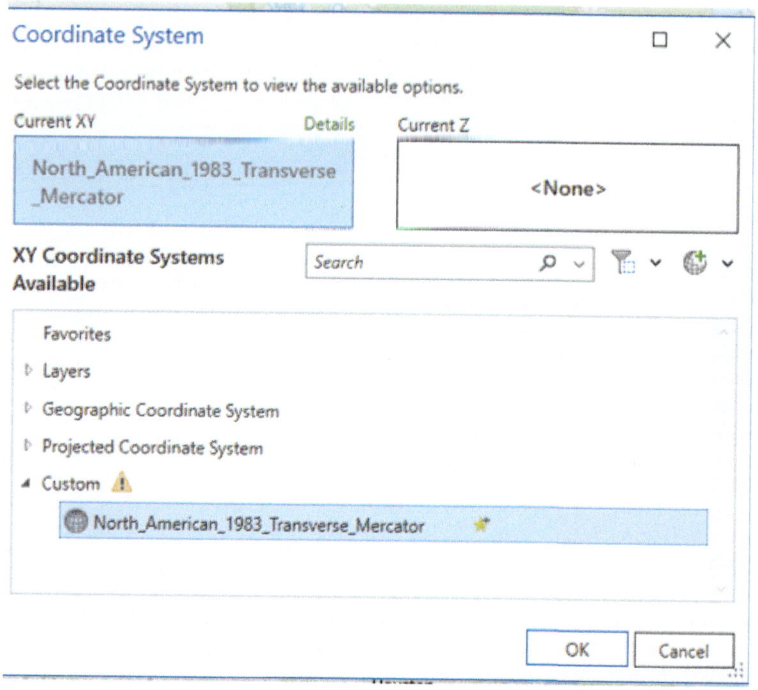

Import the coordinate system of amherst_at_maple_transit.img.

Highlight the file and click “Ok,” then “Ok” on the coordinate system dialog to import. Run the geoprocessing tool, and the feature dataset is created.

Now, we would like to create the Streets (street center lines) feature class within the Transportation feature dataset. In the future, we can put a bicycle route feature class, railroad feature class, airport polygon feature class, etc. into this feature dataset. Right-click on the Transportation feature dataset to obtain the contact menu. Then, click on the “New” menu to get the second contact menu and click on Feature Class to open the “Create Feature Class” wizard (pane).

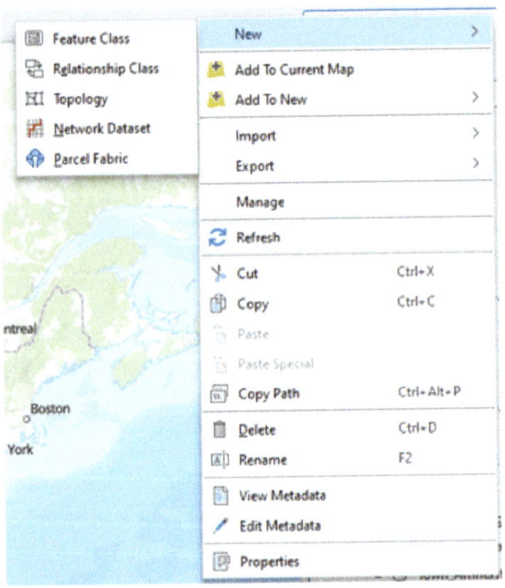


Fill in the Name as “Streets” and select the type of features stored in the feature class as “Line.” “M Value” is a coordinate value related to time and sequence. We do not need to check it. “Z Value” is the Z coordinate value. Since we only make a 2D feature class in this case, we can check it off. Click “Next.”

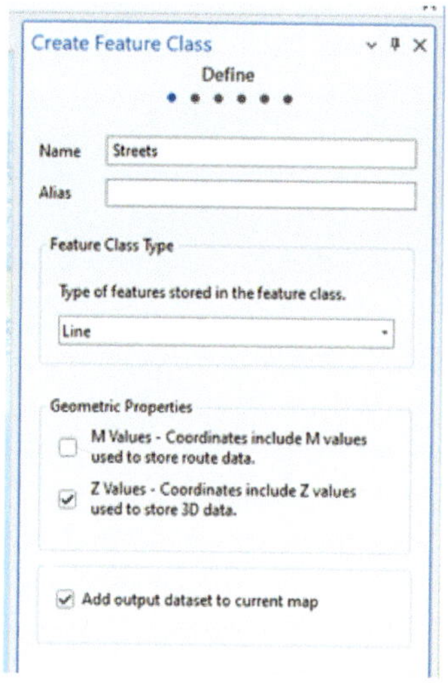


This is a dialog to initiate the fields in the attribute table of the feature class, called the domain editor. Enter the following on top of the default fields. If you accidentally added one more row here, you can right-click on it and cut it off.

| **Field name** | **Data type** |
|---|---|
| Street_Name | Text |
| Number_of_ lanes (each direction) | Short Integer |
| Full_Name (including street type) | Text |

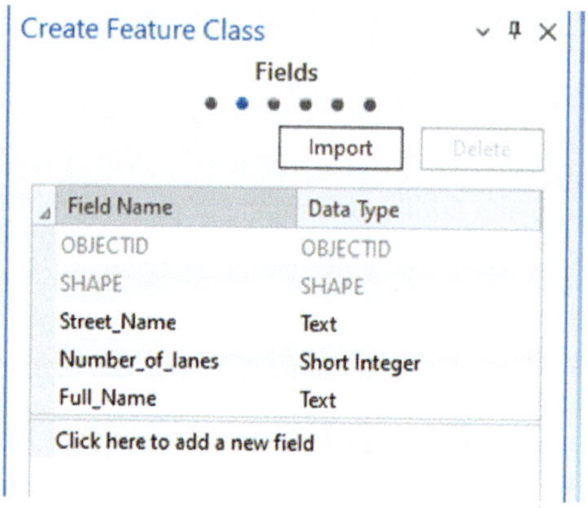


Then, click on "Next."

This is the page of the coordinate system dialog; we can continue to the next page since we have already set it up for the entire geodatabase dataset.

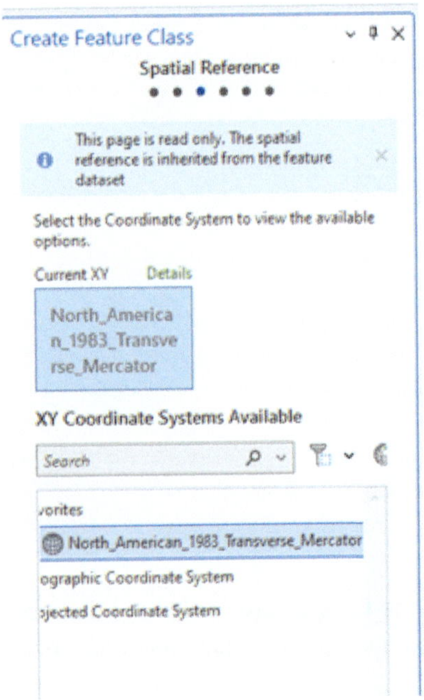


Keep the defaults on all the rest of the pages and finish the geoprocessing to create the feature class. Now, you see the background base map is redrawn, and the new empty feature class is automatically loaded in the Table of Contents.

Following the same procedure and steps above to create the "Land_Use" feature dataset in the "Town_Amherst_Data.gdb." Make sure that you copied (imported) the coordinate system from the aerial image of "amherst_at_maple_transit.img."

Now, let's make two polygon feature classes called "Building" and "Parcel" in the "Land_Use" feature dataset. Building is a polygon we can digitize according to the footprint (roof print) of the houses and buildings on the aerial image. Parcel of land means one piece of real property as identified by the county tax assessor's parcel number (APN) that is one contiguous parcel of real property. A parcel of land is normally owned by a person or a family, or a public or private entity. A parcel of land is also a property that is defined for tax purposes. We will also practice parcel polygon digitizing in this project.

In this case, we are going to use the "Create Feature Class" tool in Geoprocessing Pane or the Toolbox to create "Parcel" polygon feature class.

Click the "Analysis" tab on the main menu to activate its ribbon.

Then, click on the Tools icon in the Geoprocessing group Tools . Then, the Geoprocessing Pane opens on top of the Catalog Pane. In the Geoprocessing pane, type in "Create feature Class" to search the tool. We find "Create Feature Class" on the top of the list, and it is indicated that the tool is in the "Data Management Tools" box.

Click on the tool panel to open it. You notice it is different than the tool that we used previously. The default geodatabase was set as "MyGIS_Learing03.gdb." That is not where we would like to save the data.

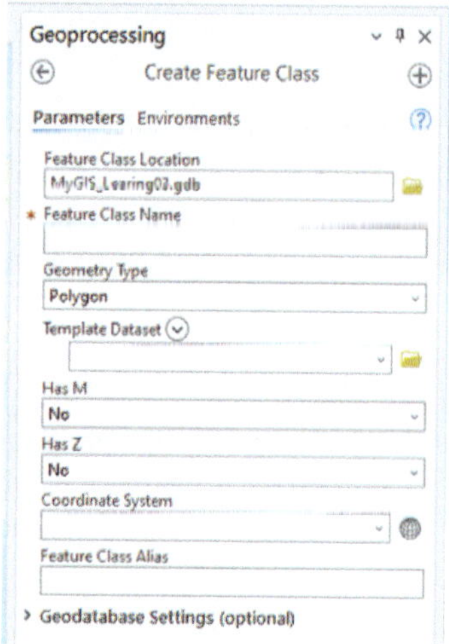


Use the folder icon to change the "Feature Class Location" to Land_Use feature dataset in the Town_Amherst_Data.gdb (D:\Chapter03-Tutorial\Feature_Class_Digitizing\Town_Amherst_Data.gdb).

Type in "Parcel" for feature class name. Change (or keep) the Geometry Type to "Polygon." Keep defaults for other parameters and run the geoprocessing. Turn on the tab of Catalog Pane, and you may notice that the empty feature class of Parcel is created in the Land_Use feature dataset.

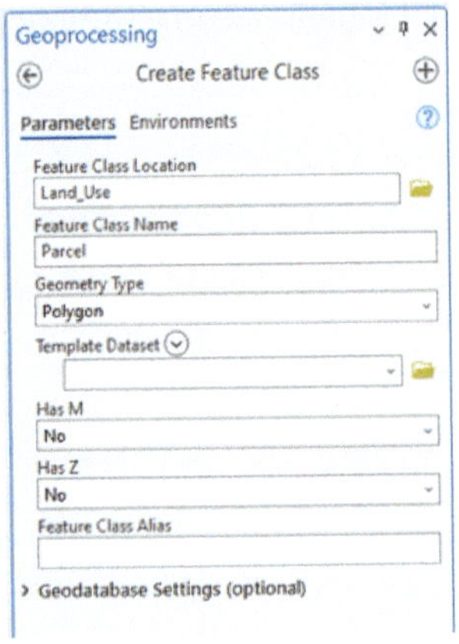


Now, we turn the Geoprocessing Pane on again and follow the same steps above to create another feature class that is called "Building" in the Land_Use feature dataset. After this process is accomplished, we should have all the empty feature classes that we need for digitizing geographic objects of features. Close the Geoprocessing Pane.

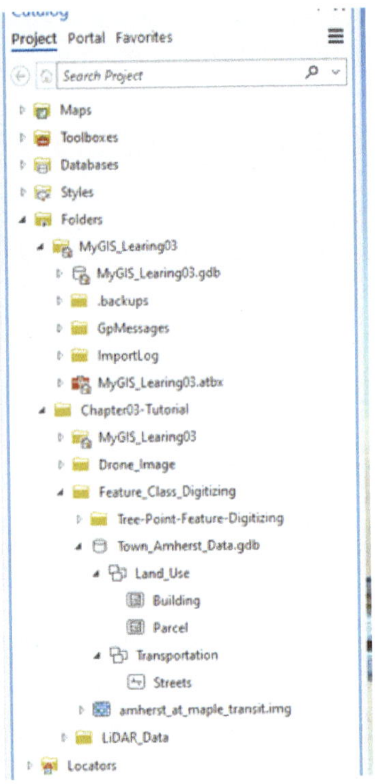


Now, we are ready to digitize the features and populate the feature classes. Drag and drop the digital aerial image, "amherst_at_maple_transit.img" from the Catalog Pane to the "Amherst_Digitizing" window. This is an orthographic aerial image around the intersection of Maple Road and Transit Road in the Town of Amherst, Erie County, NY. The image was a mosaic of four downloaded imagery quadrangles from the New York GIS Clearinghouse website. The ground resolution of this image is one (1) foot, and the image has four color bands.

Activate the "Edit" ribbon by clicking the title on the main menu. Click on the Create icon in the Features group to open the digitizing template. The three editable map layers are all loaded on this template.

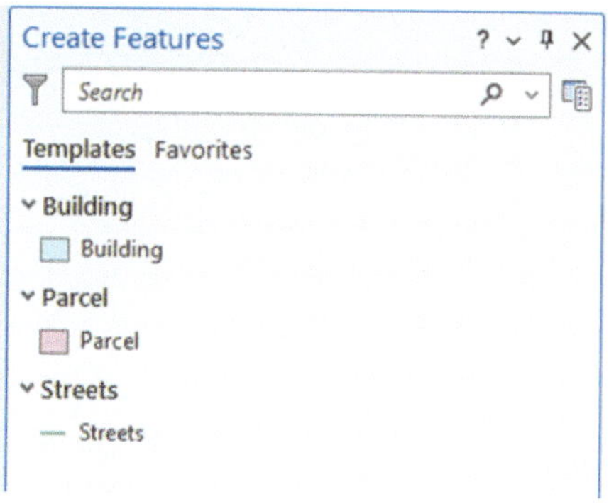


First, we are going to digitize the street center lines for the Streets feature class. We need to enlarge the image map to get details for digitizing the features. This way will ensure the precision and accuracy when we digitize the features. You can choose a strategy of separating four parts or six parts of the entire region to digitize from one corner to another. The following shows the detailed view at the southeast corner of the image.

Expand the template of Streets, you see the line drawing tool is selected by default. When you select the line drawing tool, the following panel is also available in the template. During the digitizing, clicking on the Finish icon will deactivate the editing mode. Then, we can navigate to different parts of or zoom in or out on the imagery map for digitizing the features. Click on a drawing tool, such as the line drawing tool in the template will lead us back to the digitizing mode. During the time when you are digitizing a feature, if you are not satisfied with the outcome, you can click on the Cancel icon to cancel the process right away and start over again.

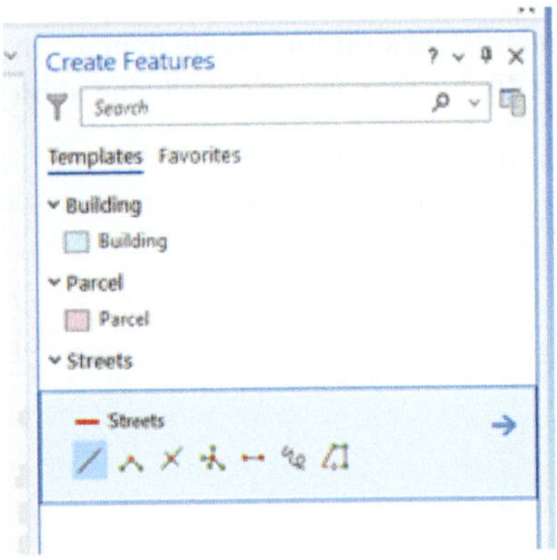


On the Table of Contents, there is one tab or category for listing the map layers by "Editing" in addition to listing the layers by "Drawing Order," "Data Source," and several others. When this tab is selected, the user can view which map layer can be edited and which one cannot as follows.

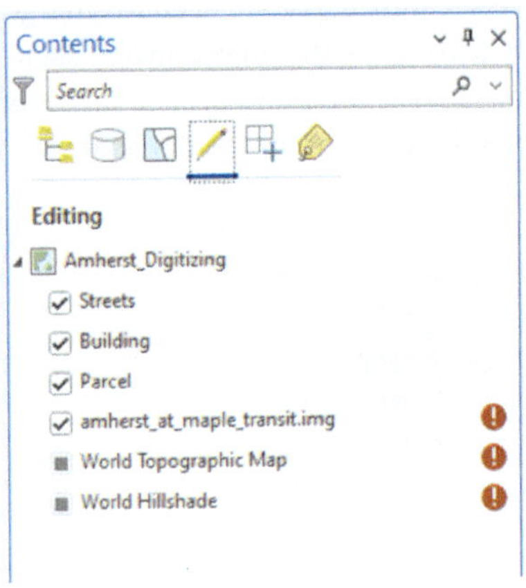


Before digitizing features, we need to check and set up the snapping modes. Open the "Snapping" dialog, we will see the following. The general slide on the top of the dialog controls the snapping on or off. There is a total of seven snapping mode that you can choose. You can also choose them one by one. If a particular snapping mode is selected, it turns to light blue. Snapping is a significant function to be set up in the beginning of your digitizing project in order to ensure the topological correctness of connectivity, adjacency, and containment of features in a feature class.

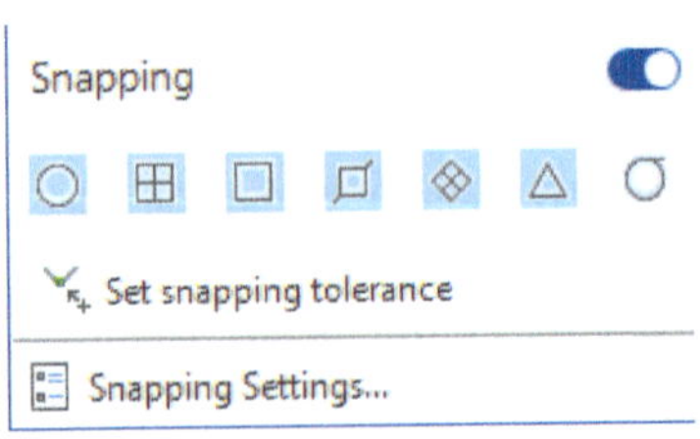


Let's digitize the street centerlines in the Streets feature class now. Single-click on the map is to start, and double-click is to finish the digitizing of a line segment. After each feature is digitized, we can flip to the domain or attribute editor to enter the attribute right away. We can also elect to enter all the attributes later in a table mode. When a feature is digitized, it is automatically put into the selected mode (light blue). So, we can use the delete key on the keyboard to delete it if not satisfied. When we digitize the next, the previous one will be automatically dropped from the selection.

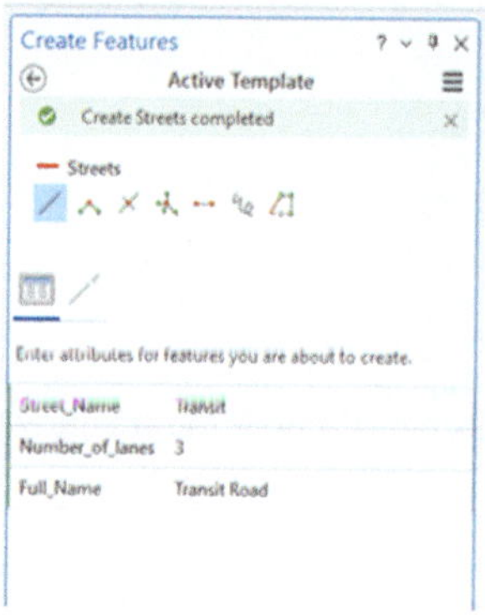


We should accomplish all the digitizing work for the Street feature class like the following.

Click on the "Save" icon in the Manage Edits group on the Edit ribbon to save all the features digitized. If you click on the "Discard" icon, everything digitized will be erased. If you stop the editing process, answer "Yes" for the following dialog in order to save all the work. Again, if the answer is "No," everything will be deleted.

Now, let's check off the Streets feature class and digitize the "Building" polygons. Since there are many buildings in the study area, we will focus on the immediate area where Maple Road and Transit Road intersect to digitize this part.

Make sure "Snapping" is on and the following snapping modes are selected. Enlarge the imagery map and activate the polygon drawing tool .

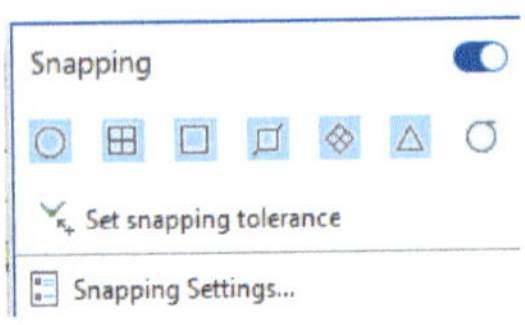


You notice that when we digitize a polygon, we actually drag a three-edge rubber net. Simply put, a polygon needs to have at least three edges. We can click on the corners of the building footprint (roof print) to digitize each of the building polygons. Again, the double click is to finish the digitizing of the feature. After the visible toolbar in the "Map" window is dismissed, you can always right-click in the window to get the contact menu for the same type of tools on the menu. We accomplish the digitizing of house footprints for the "Building" feature class as follows. Save the edits by clicking on the "Save" icon . We can come back later to complete the rest of the area.

Now, we can check off the "Building" map layer on the Table of Contents and digitize the parcels into the Parcel feature class. In this case, we need to use the auto-complete polygon tool in order to enforce the adjacency rule of topology.

Expand the Parcel template symbol to get the digitizing tool bar. The first icon from the left side is the polygon drawing tool. The second is the autocomplete polygon tool that helps us ensure the legal and topological rule of adjacency of land parcel ownerships in this case.

We can focus on the southeast corner of the community where we digitized the Building feature class to practice parcel digitizing.

Since there are no land markers for the boundaries of land parcel ownership on the ground, we will use our judgment to guess the boundaries. In real-world practice, we need to use construction surveys or auto-CAD drawings to measure the dimensions for each of the parcels that we digitize. Now, make sure the snapping function is on like the following, and digitize the first parcel using the polygon drawing tool.

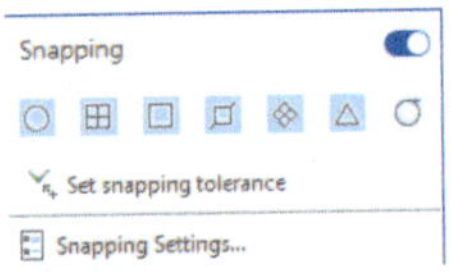

After digitizing the first parcel polygon, the rest of the polygons next to each other need to be digitized using the autocomplete polygon tool. Activate the autocomplete polygon tool and move the cursor into the digitizing site. Click either inside or on the corner of the adjacent polygon to start. Drag the rubber band to trace the boundaries of the digitizing parcel. In the end, you need to complete inside one of the adjacent polygons and double-click to complete the new polygon.

In this practice, digitize two rows of parcels of land polygons in this community. In many cases, the owner-built fences around the properties can help the GIS analysts locate the boundaries of parcels.

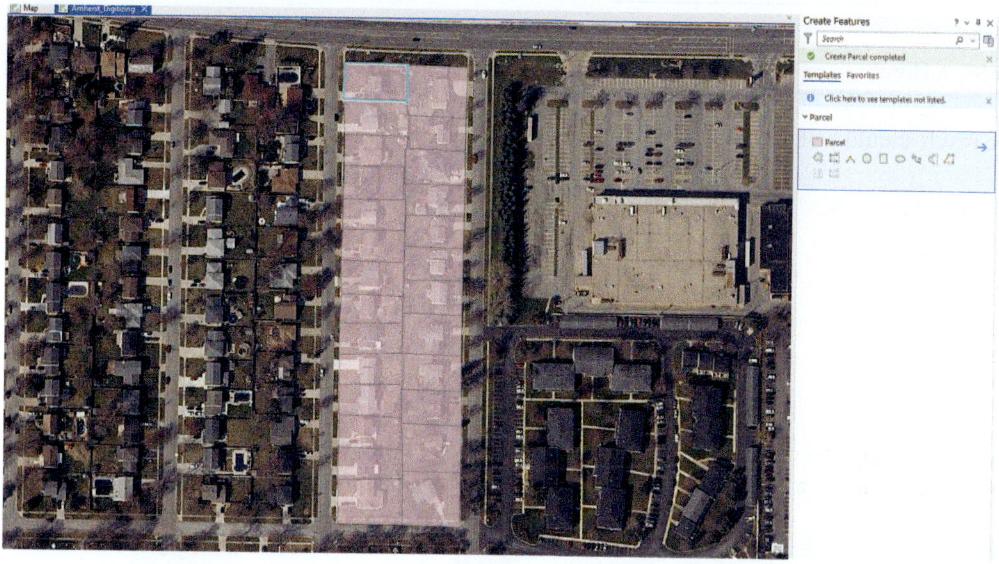


Now, save the edits using the "Save" icon 

In the editing mode, we can also open the attribute table to add attribute fields. In this case, we would like to add two fields. One is the "Owner_Names" and the other is the "Street_Address."

Now let us open the attribute table of Parcel feature class from the Table of Contents. Then, we navigate in the table window to the "Fields View."

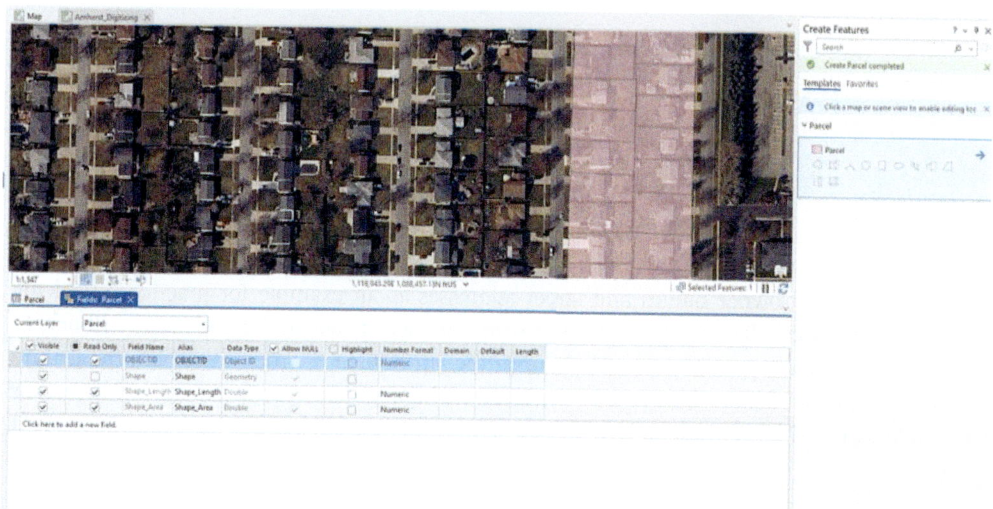

Add the two following fields.

| Field name | Data type |
|---|---|
| Owner_Names | Text |
| Street_Address | Text |

The result should look like this.

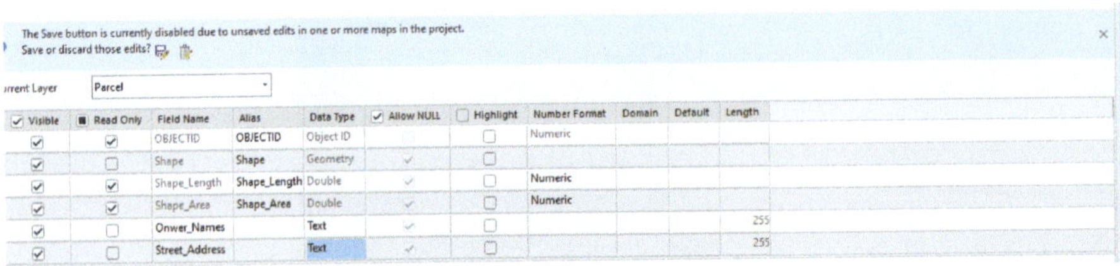

Save the changes and close the "Field View." We have two new attribute fields now. But we need to fill the values for each of the rows in the table.

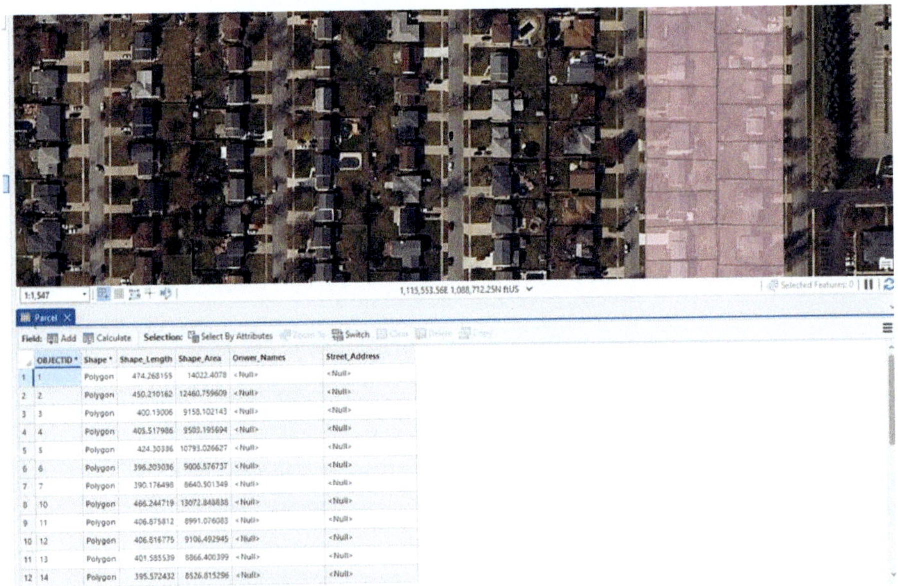

Let's enter two parcel owner names and street addresses as an example for practice. Select the southeast corner parcel that we digitized first using the select tool under the "Map" ribbon in the selection group on the main menu. Then, open the attribute table. You notice both the feature on the digital map and the attribute row in the table are highlighted. Type in the following just for practice in this case. This information is not based on the ground truth but was just made for student practice.

| Owner_Names | Street_Address |
|---|---|
| Derek Goodchild and Jennifer Goodchild | 689 Palmdale Dr. |

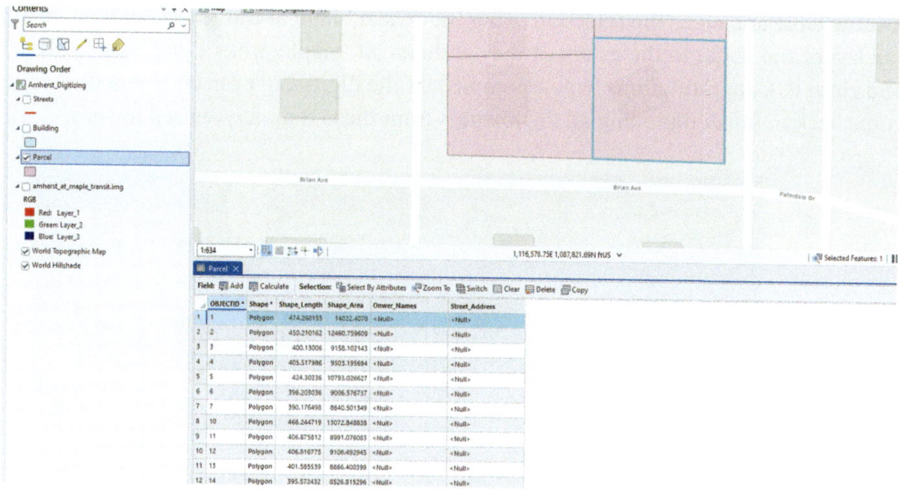


Then, go to the Edit ribbon and save edits using the "Save" icon 

Now let's go back to the Map ribbon and select the next parcel that faces to the other street. Enter the following and save the database from the Edit ribbon.

| **Owner_Names** | **Street_Address** |
|---|---|
| Berry Smith and Elizabeth Smith | 635 Patrice Terrace |

The result should look like this.

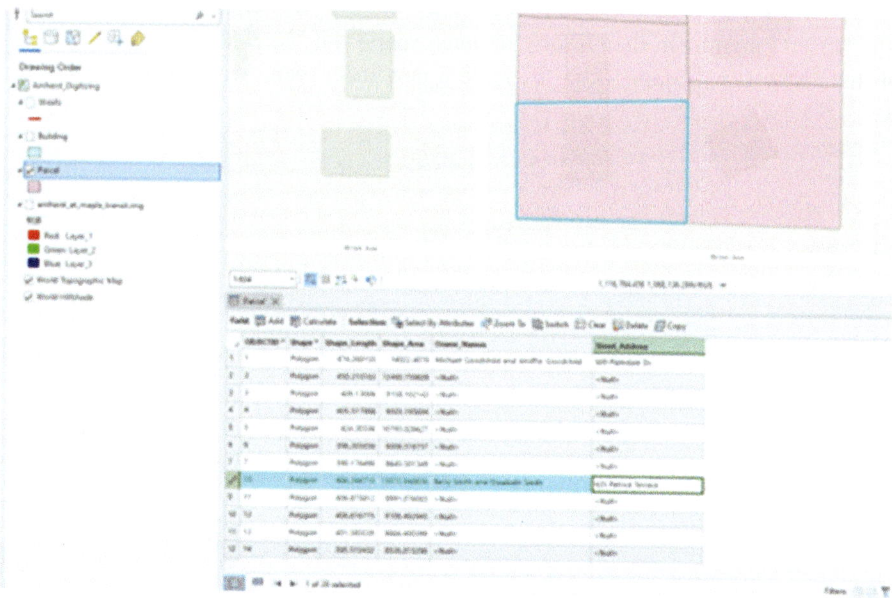

Close the attribute table and clear the selection using the "Clear" icon at the "Map" ribbon. Turn on all the map layers and zoom to the extent of the "amherst_at_maple_transit.img" image. Save the project and close it. Congratulations, you accomplished the digitizing project! If you like, you can always come back to finish digitizing all the features from these remotely sensed images.

Save the project and close the ArcGIS Pro.

2. <u>Converting a quadrangle of LAS (LiDAR or laser scan) file to the digital elevation models (DEM)</u>

Open ArcGIS Pro and the MyGIS_Learning04 project.

Navigate to the "Insert" menu. On the "Insert" ribbon, expand the "New Map" icon on the left end. Click on the "New Local Scene" icon to create a new local scene window. Rename it to "Amherst_3D."

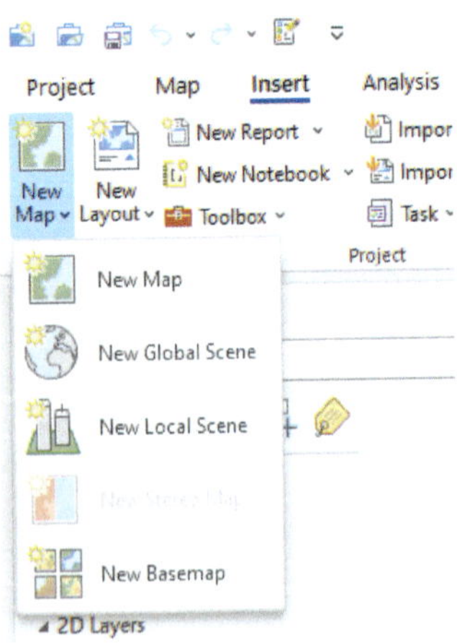

Now, we are ready to convert the downloaded LiDAR 3D point cloud data to the ArcGIS format. Expand the LiDAR_Data folder in the Chapter04-Tutorial folder on the Catalog Pane. We see the LAS file, e1395n2347_2019.las.

In order for this dataset to be presented properly and to be processed to some other type of 3D data, we need to use Create LAS Dataset tool in the Data Management Tools box to convert it to ArcGIS Pro compatible format. Navigate to "Analysis" menu and click on the Tools box or geoprocessing icon 

. The Geoprocessing Pane is open. Type in the searching window "Create LAS...," "Create LAS Dataset" tool occurs on the top of the tool list.

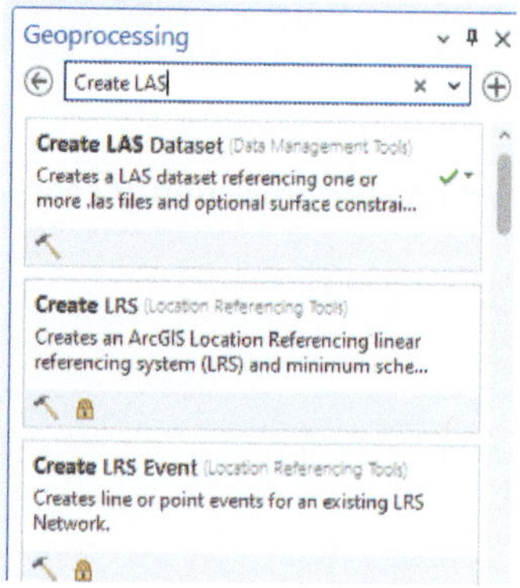


Click on this panel to open the tool.

For "Input Data," navigate to D:\Chapter04-Tutorial\LiDAR_Data\ and load e1395n2347_2019.las file. For "Output LAS Dataset," navigate to D:\Chapter04-Tutorial\LiDAR_Data\LAS_Conversion\ folder, type in "East_Amherst_LiDAR" (.lasd)

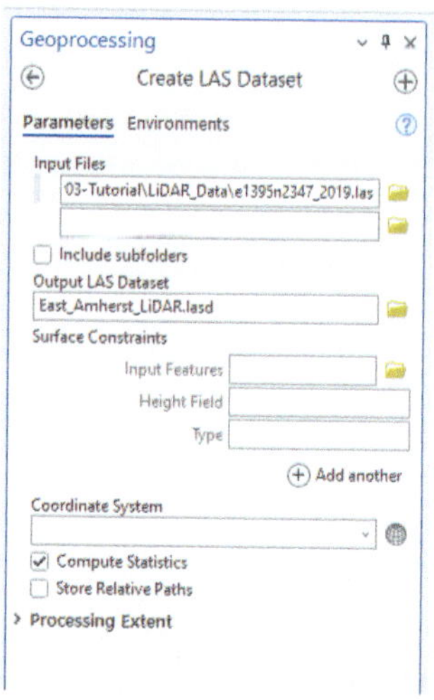


For the Coordinate System setting, go to the "Environments" setting tab (page). For the "Output Coordinate System," click on the small globe icon to open map projection (Coordinate System) dialog.

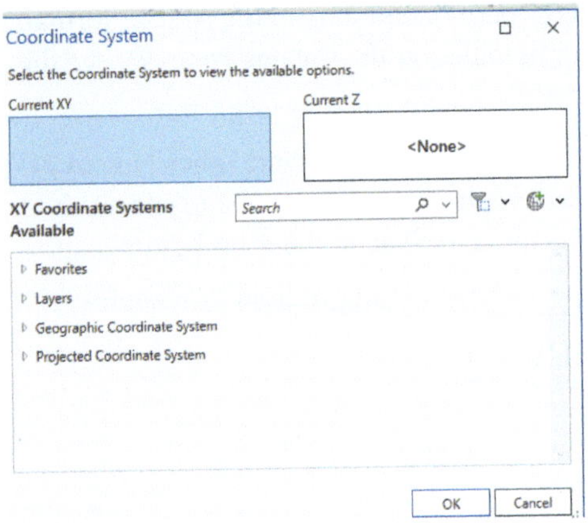


Select Projected Coordinate System -> Continent -> North America, and set the projection or coordinate system to “NAD_1983_Contiguous_USA_Albers.”

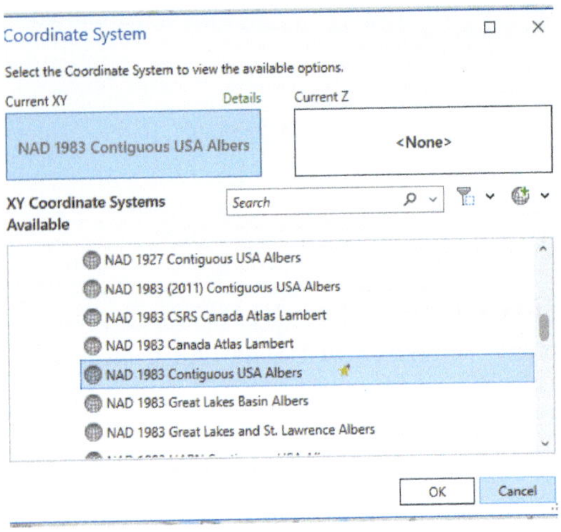


Click “OK” to install it.

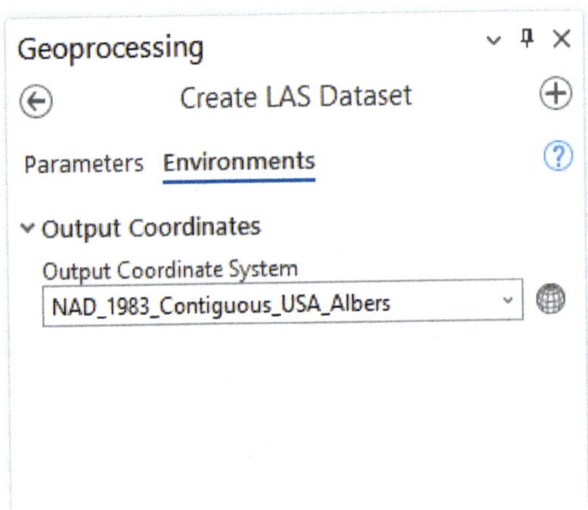

Flip to the parameter tab (page), you will also see selected type is installed in the "Coordinate System" window. Click "Run" to process the tool. The new 3D LiDAR point cloud is created and presented in the local scene window "Amherst_3D."

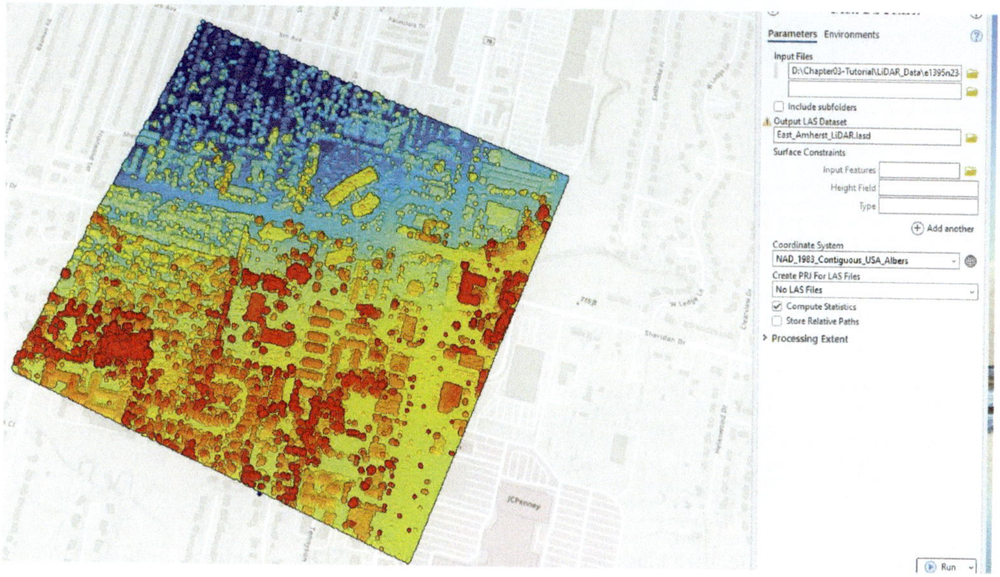


The color of points now is arranged by the elevations. Close the Geoprocessing Pane.

We accomplished the importing of a LiDAR survey data file in ArcGIS Pro. Save the project and close the ArcGIS Pro.

3. UAV (drone) aerial image-derived vector data transforming and georeferencing

Popularly used commercial UAVs, such as DJI drones, can be used as an equipment to conduct near-surface remote sensing and imagery data collection. While the imagery data in either JEPAG, Tiff, or any other file format can readily be inputted into ArcGIS Pro, the original imagery datasets do not have a georeferenced coordinate system attached to them. In this tutorial, we introduce a straightforward method for the GIS analysts in the field to conduct fast transformation of a vector feature layer after it is digitized from a drone image. The other way for drone imagery data processing in GIS is to use the digital image processing capability of a GIS or remote sensing software to rectify and georeferencing the aerial images directly.

Let's start to explore a UAV image now. Open ArcGIS Pro and the MyGIS_Learning04 project. Add a "Map" window and rename the title of the mapping window to "UAV_image_transfer." Expand the Drone_Image folder in the Chapter04-Tutorial folder. Then, add the image raster file "DJI_0020.JPG" to the mapping window. This is an image that was collected in the field using a DJI Marvic Pro 2 drone. Because the drone image does not have a geographic coordinate system, it is not compatible with the ArcGIS Pro basemap layers. In order to view the aerial image, we need to right-click on the title of the image in the Table of Content on the left-hand side of the window to get the contact menu. Then, select "Zoom To Layer." The aerial image should look like the following.

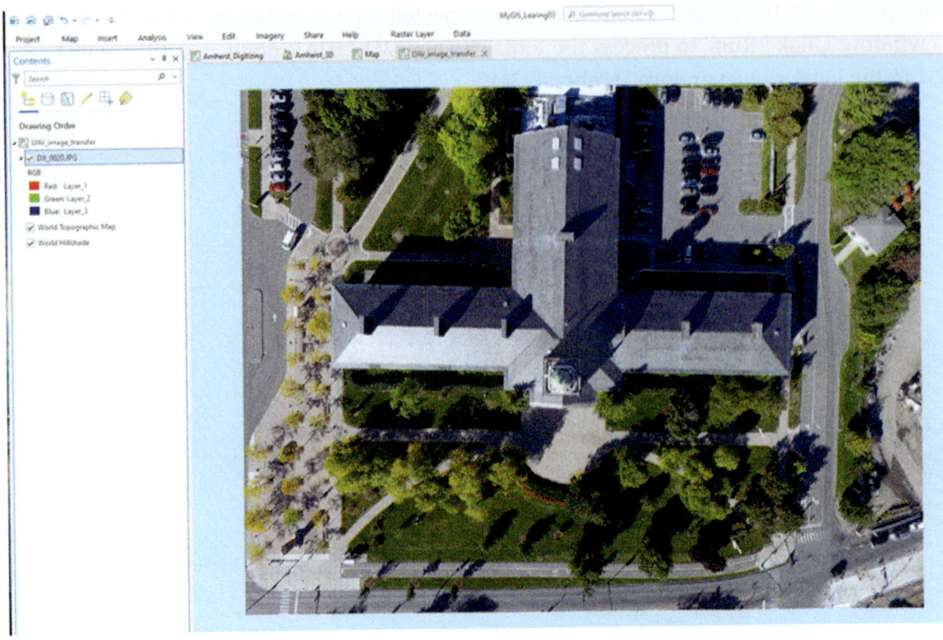


Create an empty polygon feature class and name it as "Buffalo_State_Building" in the premade UAV_Geodatabase.gdb in the Drone_Image folder. Import the drone image (DJI_0020.JPG) coordinate system (without georeferenced coordinate system) for this polygon vector feature class. You will be given a warning as follows. But this is fine and we can just finish the creation of the feature class.

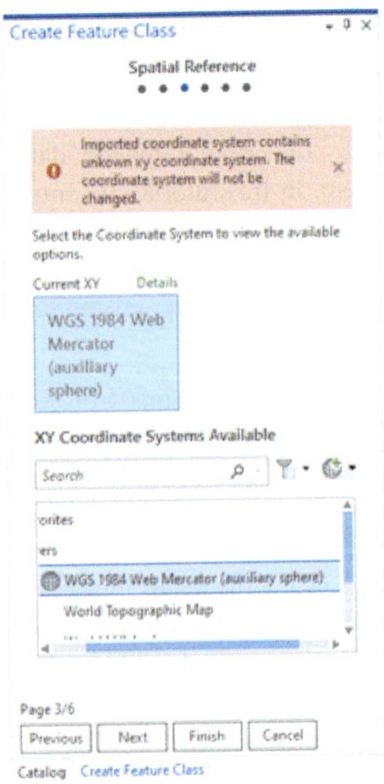

Bring the empty "Buffalo_State_Building" feature class into ArcGIS Pro - "UAV_image_transfer." Digitize the building as one vector polygon in this feature class using the tools under the Edit ribbon. Save the feature class in the UAV_Geodatabase.gdb.

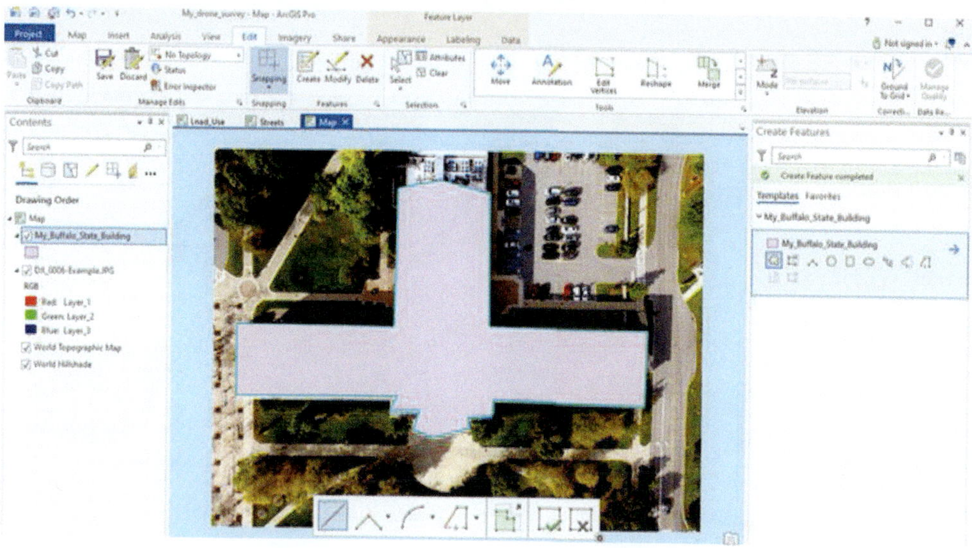


Save the edits and close the "Create Features" tool.

In ArcGIS Pro, we can spatially adjust the vector feature class, "Buffalo_State_Building" in this case, to a georeferenced coordinate system using the "Edit" menu and "Transformation" tool.

Expand the Done_Image folder at the Catalog Pane. From NYS_orthographic_Image folder add w 10681068_12_15100_4bd_2014.jp2 to the "UAV_image_transfer" mapping window. Zoom to the layer in order to see it. You notice this georeferenced aerial image is in the right location on the Earth's surface.

If necessary, add your "Buffalo_State_Building" feature class (one polygon).

In this case, the aerial image is georeferenced with the coordinate system.

But "Buffalo_State_Building" feature class (one polygon) does not have a georeference and real-world coordinate. They are not showing in the same geographic areas in the window. We can only show either one by clicking on the "Zoom To Layer" on its contact menu in the table of contents.

We are going to use the "Transformation" tool in the "Edit" menu to conduct spatial adjustment to bring "Buffalo_State_Building" feature class to the standard coordinate system of w_10681068_12_15100_4bd_2014.jp2 aerial image.

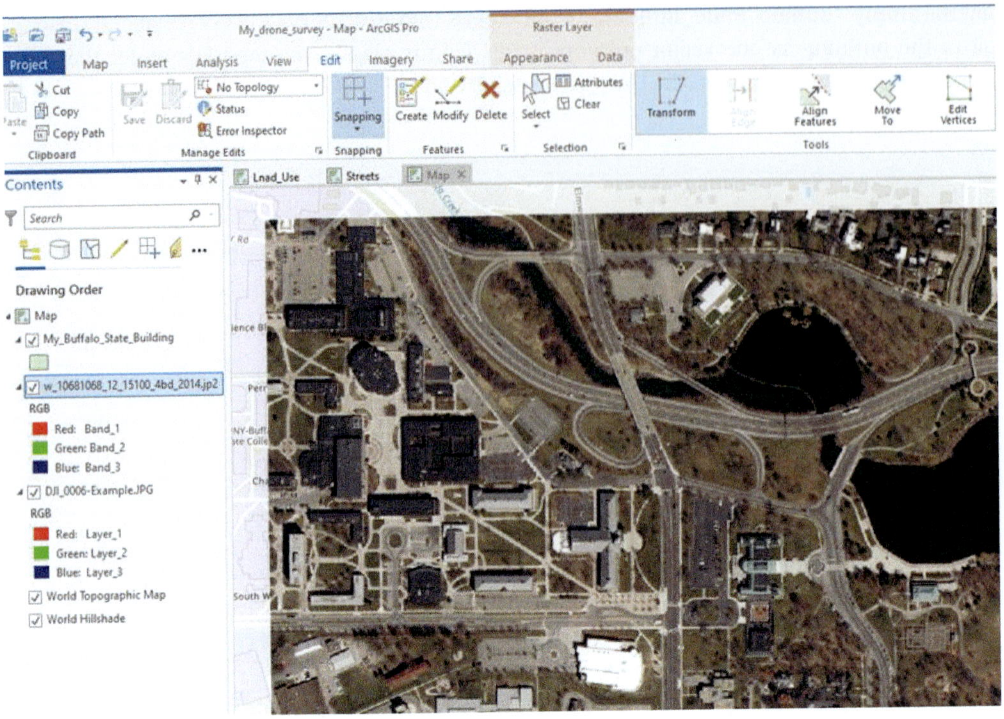


Click on the "Transformation" tool  in the "Tools" window under the "Edit" menu.



The "Modify Features" tool opens on top of the Catalog pane.

Right-click on the DJI_0020.JPG image in the table of contents, then choose "Zoom To Layer" menu, the mapping window is switched to the drone image and digitized building side. Click on "Add New Links" icon in the tool dialog. Click on one corner of the building polygon (Feature Class) to start the rubber band.

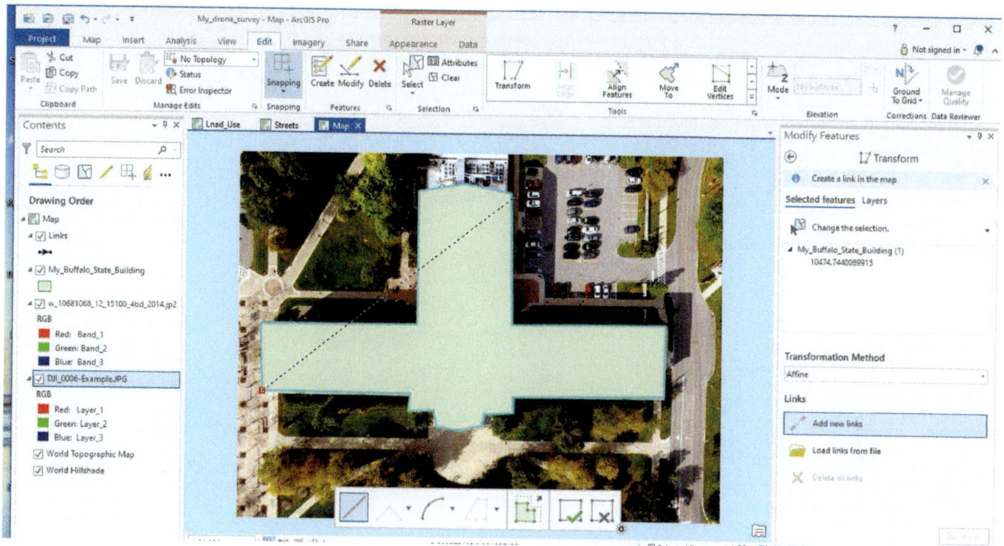


Now, we can make four (4) adjustment links to process the data and transform the coordinate system alignment.

We need to flip the two map layers like I did using the "Zoom To Layer" menu to make the links for at least four exact same points as follows. The following are the map-flipping and link-making sequences. Please be aware that the virtual rubber band connects from one map layer to the other. You must ensure all four right point connections in order to make a right transformation. You need to flip the maps from "Buffalo_State_Building" to the aerial image while the rubber band is alive.

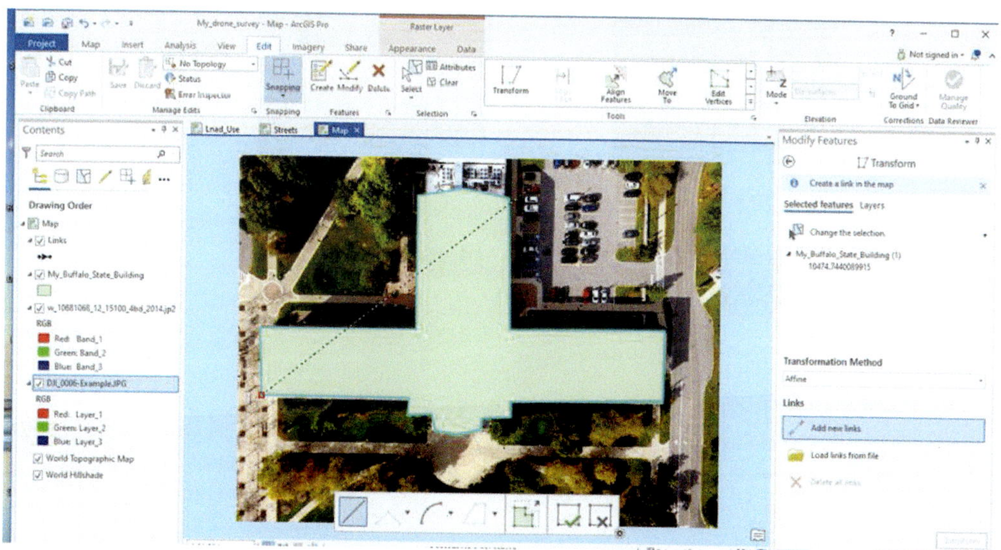

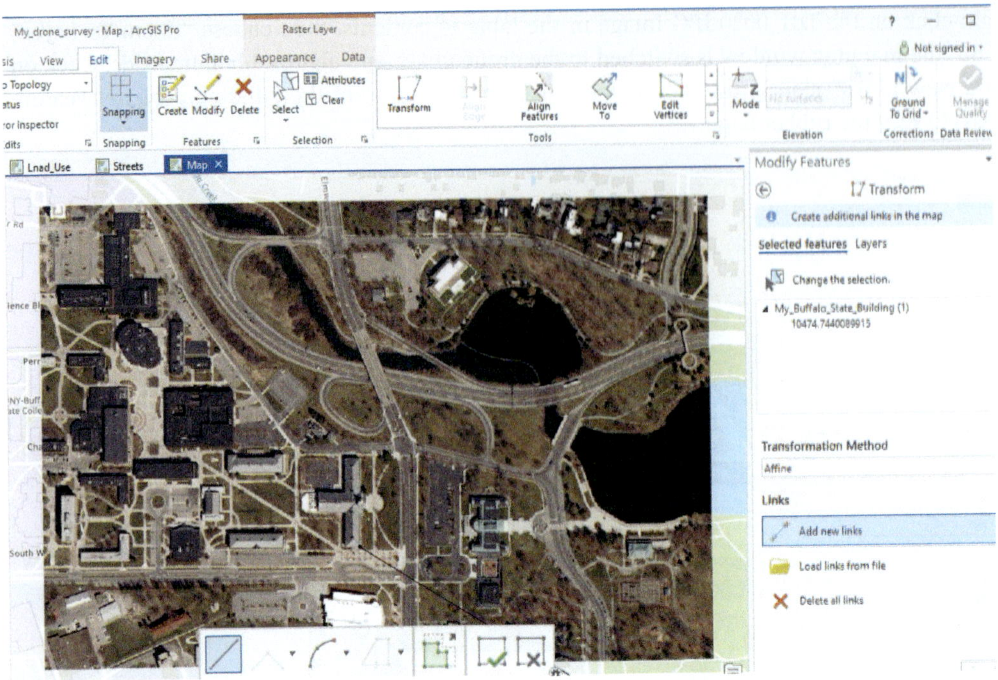


The final stage before spatial adjustment should look like the following on the two map layers, namely "Buffalo_State_Building" and NYS orthographic image–w_10681068_12_15100_4bd_2014.jp2.

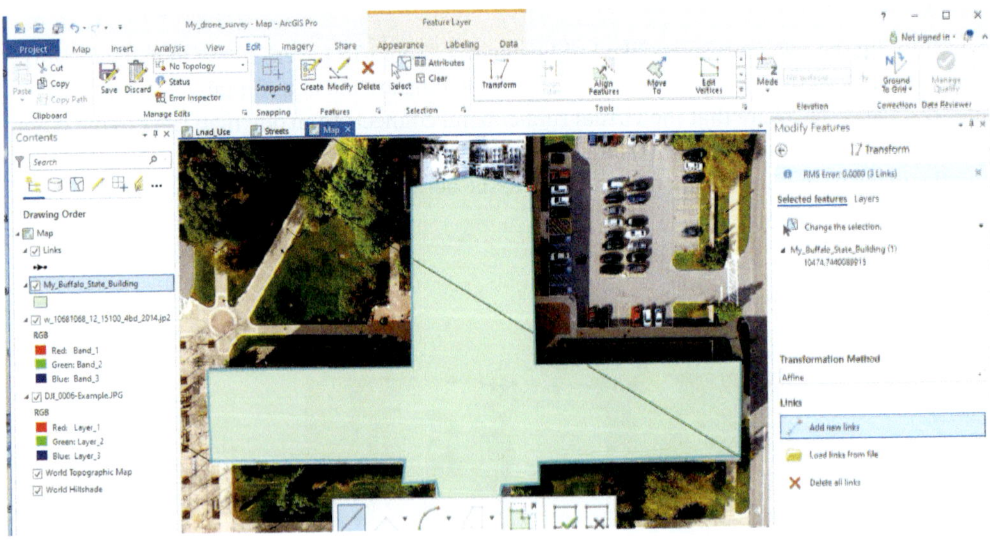

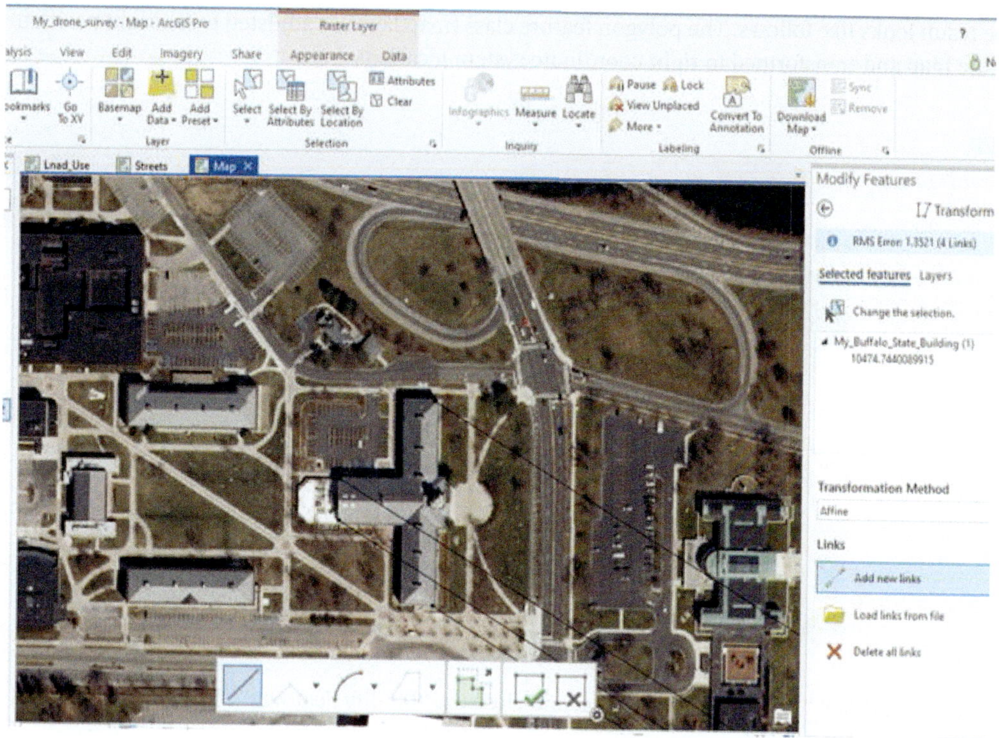


Now, we can click on the "Transform" on the tool as shown in the following to make the spatial adjustment and alignment.

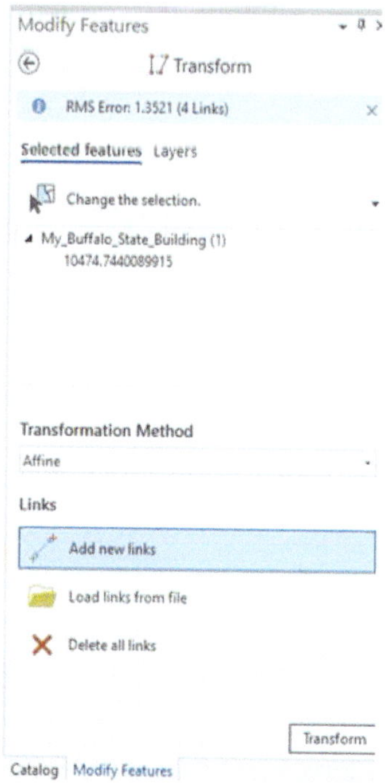

The result looks like follows. The polygon feature class (map layer) is adjusted to the right position on the map and transformed to right coordinate system locations.

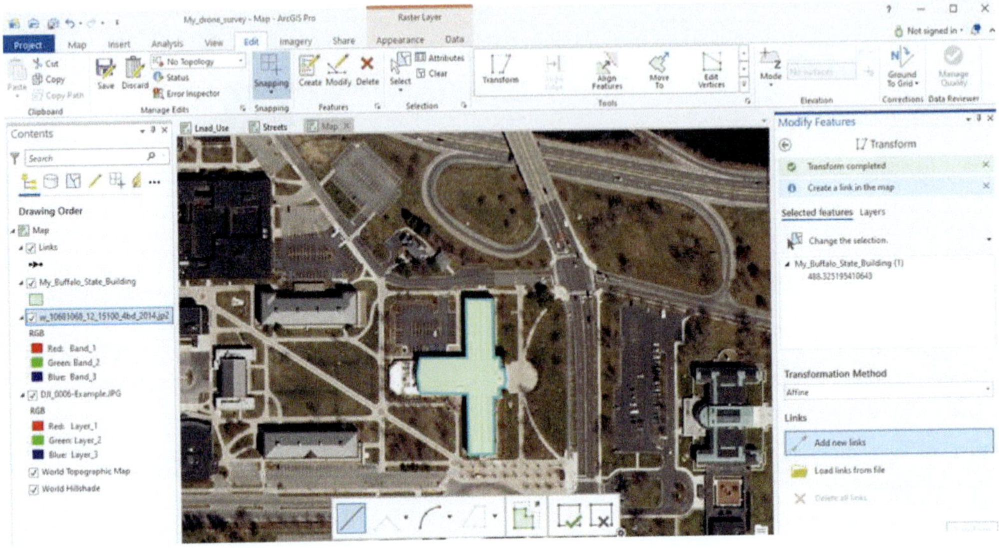


Save edits and click “Yes” for the following, and close the “Modify Features” (Transformation) tool to stop editing.

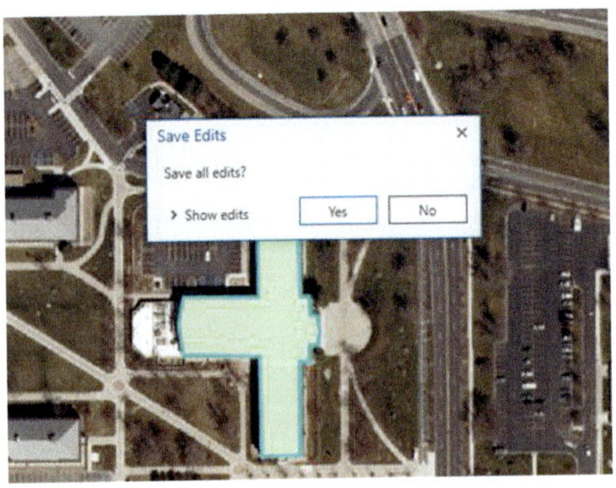


Congratulations! You accomplished the vector feature class data transformation and georeferencing.

# 5

# Coordinate Systems, Map Projection, and Geo-referencing in GIS

## Concepts and Theories

The core of geographic information systems (GIS) and GIS analysis is to find locations of objects or features and their distribution patterns on the Earth's surface. As indicated by ESRI, the company that created the software of ArcGIS and ArcGIS Pro, GIS in essence is a locational science (ESRI 2018). In order to measure and determine the locations of the objects and features on the Earth's surface, we need first to measure and describe the shape of the Earth. Secondly, we need to determine the reference sphere or standard starting point for measuring or computing the Earth's surface. Thirdly, we need to establish and maintain some specific coordinate systems on the Earth's surface in which our measuring units and digitally mapped features can be referred to and stored in computational databases.

## Modeling of the Earth

The scientific discipline that studies the methods of accurately measuring or surveying the Earth's geometry, shape, and size is called geodesy. Geodesy is rooted in mathematics and astronomy. Stars and planets outside of the Earth have been used as reference objects. Mathematics has been used as a tool to calculate the results. According to geodetic modeling and calculations, scientists recognized that the Earth is an irregularly shaped ellipsoid. An ellipsoid is a 3D geometric figure that has an elliptical shape. Earth's ellipsoid shape is constantly measured and updated periodically. Each of these established ellipsoid models is also used to define a datum. A datum can be comprehended as a specific reference system established by a 3D earth model for the precise pinpointing of locations and accurate mapping. Some of the ellipsoid models are shown in Table 5.1 (GIS-Geography 2022).

The geoid is a model of global mean sea level that is used to measure precise surface elevations (NOAA 2021). In other words, the geoid is a 3D spherical surface along which the gravity potential everywhere is equal and to which the direction of gravity is always perpendicular to the modeling object, the surface of the Earth.

*Fundamentals of Geographic Information Systems (GIS) Analysis: Applying ArcGIS Pro*, First Edition. Tao Tang.

**Table 5.1** Ellipsoid models and datums popularly used worldwide.

| Name | Year | Semimajor axis (equator radius) | Semiminor axis (polar radius) | Users |
|---|---|---|---|---|
| Clarke | 1866 | 6,378,206.4 m | 6,356,583.8 m | North America |
| International (Hayford) Ellipsoid | 1924 | 6,378,388.0 m | 6,356,911.9 m | Most of the world |
| WGS72 | 1972 | 6,378,135.0 m | 6,356,750.5 m | NASA |
| GRS80 | 1980 | 6,378,137.0 m | 6,356,752.3 m | Worldwide |
| WGS84 | 1984 | 6,378,137.0 m | 6,356,752.3 m | Current worldwide |

## Horizontal and Vertical Datums for Measuring and Representing GIS Data

A datum is defined as a plane, a straight line, or a point that is used as a reference for measuring the dimensions of a target. In GIS, our target is the Earth, and we differentiate the vertical datum and horizontal datum. A vertical datum is a surface of zero elevation or altitude to which heights of various points on the Earth's surface are referenced. The theoretical base for establishing the vertical datum is the geoid model and long-term ocean surface altitude surveys. A vertical datum is also called a zero-elevation surface or zero-level reference. The currently used vertical datum in North America is the North American Vertical Datum of 1988 (NAVD 88) (NOAA 2021).

According to NOAA (2018), a horizontal datum is a specified coordinate system for a collection of positions on the surface of the Earth. Traditionally, horizontal datums were established by classical geodetic surveying methods, which model the entire Earth in an elliptical shape in 3D. Nowadays, the details of the shape of the entire Earth are mainly measured by the global positioning system (GPS). The current horizontal datum for the United States and its territories is effectively part of a geometric datum, and it is called the North American Datum of 1983 (NAD 1983). One popularly used horizontal datum worldwide is the World Geodetic System (WGS 1984), which was defined by the US Department of Defense and is commonly used in civilian GPS for positioning (Figure 5.1). Another popularly used horizontal datum in the United States is the North American Datum of 1927 (NAD 1927). The NAD 27 was based on the model of the Clarke Ellipsoid of 1866, and the NAD 83 is based on the model of the Geodetic Reference System of 1980. In ArcGIS Pro, each of the spatial datasets, either a feature class or feature dataset in vector data format, or a grid dataset in raster format is referenced by a particular datum, namely NAD 83, NAD 27, or WGS 84.

## Geographic Coordinate System and Projected Coordinate Systems for GIS Data Referencing

A coordinate system is defined in mathematics as a set of either two-dimensional (2D) or three-dimensional (3D) perpendicular number lines used to describe the numerical values of any point in the space. These perpendicular lines intersect each other at zero, and this point is called the origin. These mathematical coordinate systems are also called Cartesian coordinate systems, named after the French mathematician who established them. In 2D space, the geometry is defined by two

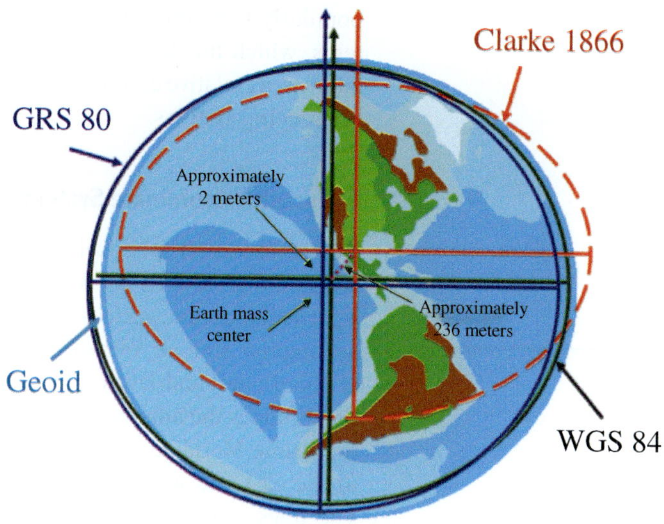


**Figure 5.1** Models of referenced ellipsoid for different datums. *Source:* Adopted from US National Oceanic and Atmospheric Administration – NOAA.

perpendicular numbered lines: the horizontal x-axis and the vertical y-axis. Any point or location can be described using an ordered pair of numbers referenced to these axes. In Figure 5.2, the point has coordinates (−3,2).

In 3D space, the Cartesian coordinate system is illustrated on three mutually perpendicular coordinate axes: the x-axis, the y-axis, and the z-axis (Figure 5.3).

Many GIS software, such as ArcGIS Pro can accommodate a Cartesian coordinate system when we digitize the objects or features in the database. The conversion or referencing to a real-world coordinate system can be conducted after the data objects have already been input into the GIS database.

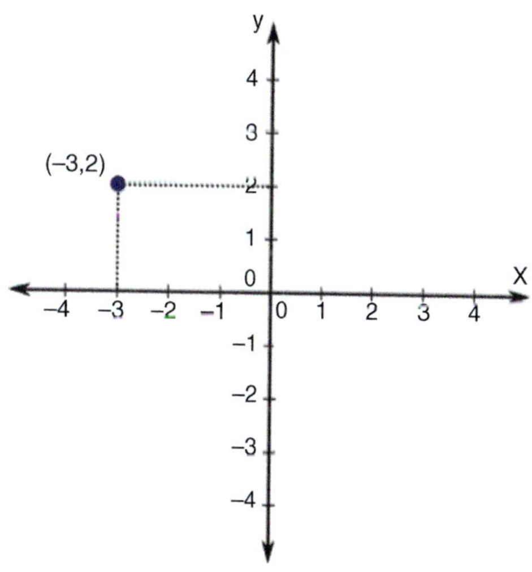


**Figure 5.2** Cartesian coordinate system in 2D space.

In Geography or GIS analysis, we need to relate or reference each of the x, y, or z coordinates and their values directly to the real-world locations of objects or features on the Earth's surface. Therefore, geographers or scientists studying the Earth's surface developed the Geographic Coordinate System (GCS) and a few projected coordinate systems for mapping and spatial or geographic data survey and management purposes. In practice, objects and features that are stored and analyzed in GIS are more often interpreted and referenced by relative locations, such as street address, postal code, etc. In this textbook, we identify the GCS and the projected coordinate systems, such as UTM or SPCS, as absolute coordinate systems. These coordinate systems can precisely pinpoint the accurate locations of geographic objects or features on the Earth's surface. By contrast, we call those more

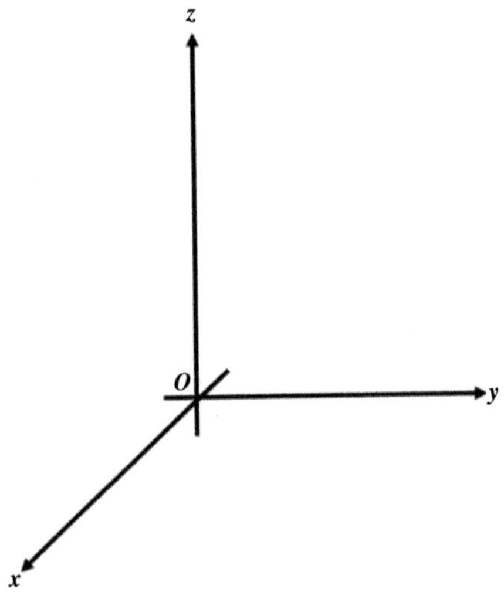


**Figure 5.3** Cartesian coordinate system in 3D space.

popularly used relative locational reference systems, which impact our daily life, such as street address, as relative coordinate systems, or relative referencing systems.

### Geographic Coordinate System

GCS utilizes a 3D spherical surface to define locations on the Earth in 2D. A GCS is established on a particular datum, such as NAD 83. GCS includes an angular unit of measurements of longitude (x coordinate) and latitude (y coordinate) values. Longitude is the angular distance accounted from the Prime Meridian (0°) at Greenwich, England, to the east and west up to a maximum of 180°. The lines of half circle representing the vertical or north-south curved lines on the Earth's surface are called Meridians. The Prime Meridian (0°), the starting line, was chosen in 1884 at an international conference and by international treaty. Latitude is the angular distance accounted from the Equator (0°) to the north and south up to a maximum of 90°. The equator is the great circle halfway in between the North Pole and South Pole (Figure 5.4).

Latitude and longitude values are traditionally measured in degrees, minutes, and seconds (DMS). One degree (1°) equals 60 minutes (60′), and one minute (1′) equals 60 seconds (60″). Latitude values in the Southern Hemisphere or south of the equator are designated as "S" and those in the Northern Hemisphere or north of the equator are designated as "N." Longitude values in the Western Hemisphere are designated as "W" and those in the Eastern Hemisphere are designated as "E." For example, the GCS coordinates of the Empire State Building in New York City, NY, are latitude: 40°44′54″N, and longitude: 73°59′04″W.

Since the computer does not read and accommodate the mixture of numbers and letters as designations in its database, we must use decimal degrees and plus (+) or minus (−) signs to represent latitude and longitude in the spatial databases in a GIS system. In this conversion process, latitude of the northern hemisphere is designated as positive values, and that of the southern hemisphere is designated as negative values. The longitude of the eastern hemisphere is designated as positive values, and that of the western hemisphere is designated as negative values. The decimal degree format of the above location at the Empire State Building, which we need to convert before inputting into GIS, is latitude: 40.748817°, and longitude: −73.985428°.

### Map Projections

Since either hard copies or digital maps that we use are 2D, and the Earth's surface and the objects on that surface are 3D, we need to conduct the map projections. We need to project the curved Earth's surface on to a flat surface in order to accurately represent real-world features on a map. Therefore, map projections must be conducted both in traditional cartography and in GIS today. Map projection is a process to systematically transform the 3D or curved surface of the Earth that might be presented on a globe onto a plane or flat surface. In a map projection process, cartographers or GIS analysts can either transform the entire Earth's surface or only part of it.

Visualizing the map projection using a globe model is like casting a shadow of the globe onto a projection screen or a projection surface using a light bulb. The fundamental factors to make a map projection are (i) projection surface, (ii) the light source and location for the projection, and (iii) the orientation of the globe (Figure 5.5). According to the projection surface of a map, map projections can be classified genetically as (i) planar or azimuthal projection, (ii) cylindrical projection, or (iii) conical (conic) projection (Figure 5.6). Planar map projections are to project the spherical surface onto a flat paper or plane. Conical map projections are to project the spherical surface onto a cone placed over the globe. Then, the map maker can cut the surface from the base to the apex of the cone to unravel it into a flat surface. Cylindrical projections are to project the spherical surface onto a cylinder wrapped around the globe. The cylinder is then cut lengthwise and flattened to form a map. Both cylindrical and conical surfaces are called developable surfaces since these surfaces need to be cut and flattened in order to make a map.

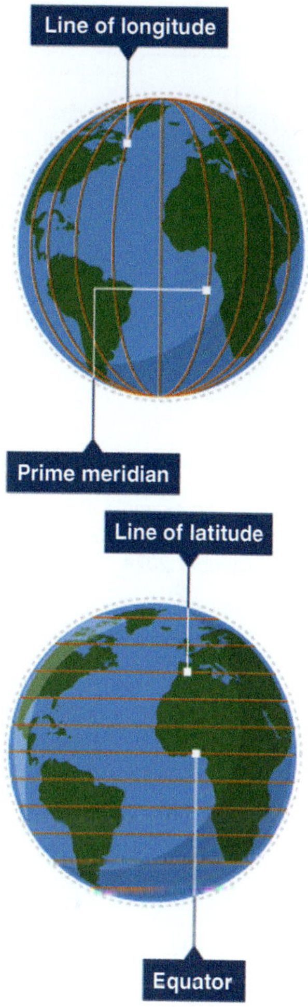


**Figure 5.4** Latitude and longitude coordinates were established on a globe as the Geographic Coordinate System. *Source:* DrillingFormulas.Com, 2016.

According to the projection light location, map projections can be classified as (i) gnomonic map projections, (ii) stereographic map projections, and (iii) orthographic map projections (Figure 5.7). The light source location is at the center of the globe in a gnomonic projection. The light source of a stereographic projection is at the opposite end of where the map projection plane touches the reference globe. The light source of an orthographic projection is placed at an infinite distance to the projection surface. Therefore, the light beams of the orthographic projection are considered parallel to one another. The objects at different locations in the map projection area are considered to have the same scale for size reductions.

According to the orientation of the globe in relation to the projection surface, the map projection can be identified as (i) normal, (ii) transverse, or (iii) oblique (Figure 5.8). Using cylindrical projections as an example, a normal cylindrical projection means the imaginary line connecting the North Pole and South Pole in the globe is parallel to the wall of the cylinder. Transverse cylindrical projection means the line is perpendicular to the wall or both North Pole and South Pole touch the wall of the cylinder. Oblique cylindrical projection indicates that the imaginary line connecting the North Pole and South Pole is neither parallel to nor perpendicular to the wall of the cylinder.

Transforming an object, such as a continent, from the surface of a globe in 3D to a flat surface on paper or on a computer screen will distort either the shape or the size (scaled-down distances) of features (Figure 5.9). In cartography and GIS, we recognize two major quality properties of map projection: conformality and equivalence. The original meaning of conformality is the retention of correct angles of latitude and longitude lines during the map projection process. This is because

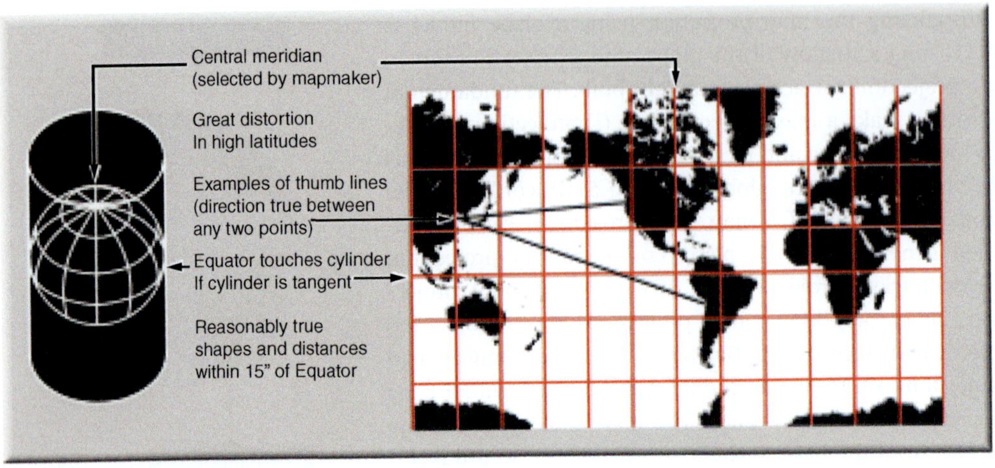


**Figure 5.5** An example of cylindrical map projection surface and globe orientation in the process of projecting a world map. *Source:* Adapted from US Geological Survey: http://www.usgs.gov/, last accessed on 23 November 2024.

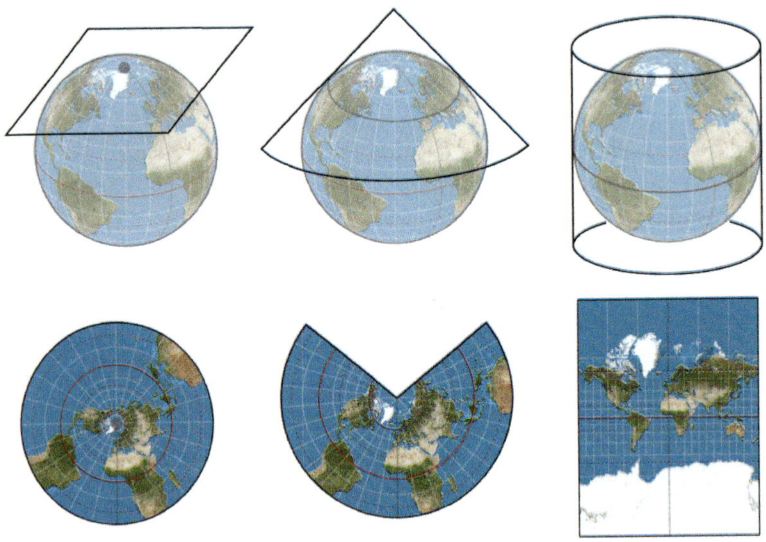

**Figure 5.6** Different map projection surfaces. *Source:* UCGIS – University Consortium for GIS, 2024/ University Consortium for Geographic Information Science.

if the angles of latitude and longitude and direction lines are copied correctly in the transformation, the shapes of local objects that occupy small areas would be correct. In practice, we refer to a conformal map projection as a projection that preserves the correct shapes of features at the level of small areas.

Equivalence is the condition in which a unit area drawn anywhere on the projected map always represents the same number of square units on the surface of the globe (Campbell 2001). In practice, if a projected map can be used to measure areas, it is called equal-area map projection. If a

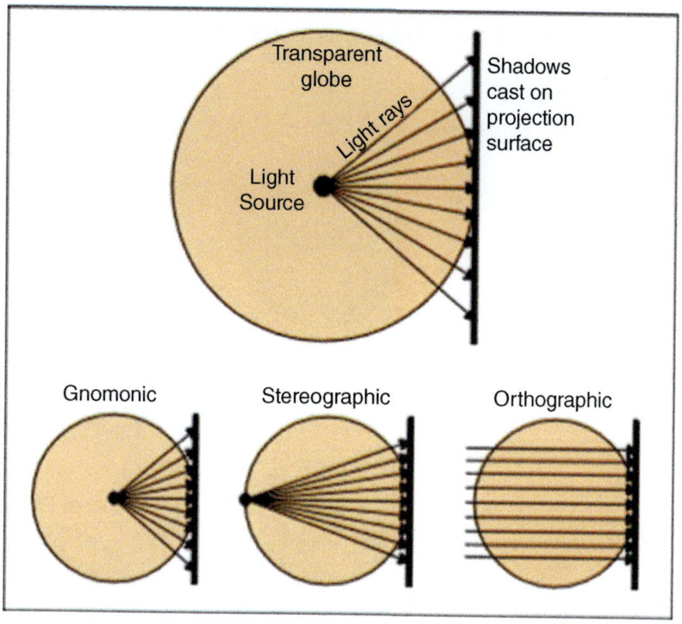


**Figure 5.7** Different light source locations for map projections. *Source:* Adapted from Heitzler et al. (2019).

projected map can be used to measure the distances, in particular along the great circle lines such as the equator or meridian circles, it is referred to as an equidistant map projection. Conformal map projections can be used to measure correct directions from one place to another on the Earth's surface. This is a significant function that has been used in guiding world travelers and for ocean navigations since the great exploration era in the seventeenth and eighteenth centuries (Figure 5.10).

In GIS and mapping industries, the "best map projection" does not exist. GIS analysts or cartographers need to select different types of map projections based on the application and purpose. Distortions of map projections of distance, direction, area, or shape are also closely related to the area of the study or mapping region of GIS projects and the map scale. Large-scale digital maps portray small geographic areas, and small-scale digital maps portray large geographic areas on the Earth's surface. In essence, no matter which type of map projection we select, the errors or distortions are minimal when we conduct a GIS project in a local region or study a small geographic area, such as the Town of Amherst or County of Erie, New York. By contrast, if we are analyzing or presenting features across the entire New York State, we do need to choose a map projection with a clear purpose in order to avoid large accumulations and unwanted distortions.

Based on the above discussions of map projections, mapmakers or cartographers, can create many different types of projections by combining the three fundamental elements: projection surface, orientation of the projection surface with the globe, and light source locations. Since the 1500s, many geographers and mapmakers studied these map projection concepts and created a variety of map projections. Most of these projections are named after the inventor, for instance, the Mercator Projection (named after Gerardus Mercator) and the Lambert Conformal Conic Projection (named after Johann Heinrich Lambert). In summary, the reason that GIS analysts need to deal with map projections is simply because field-collected datasets in science or social science

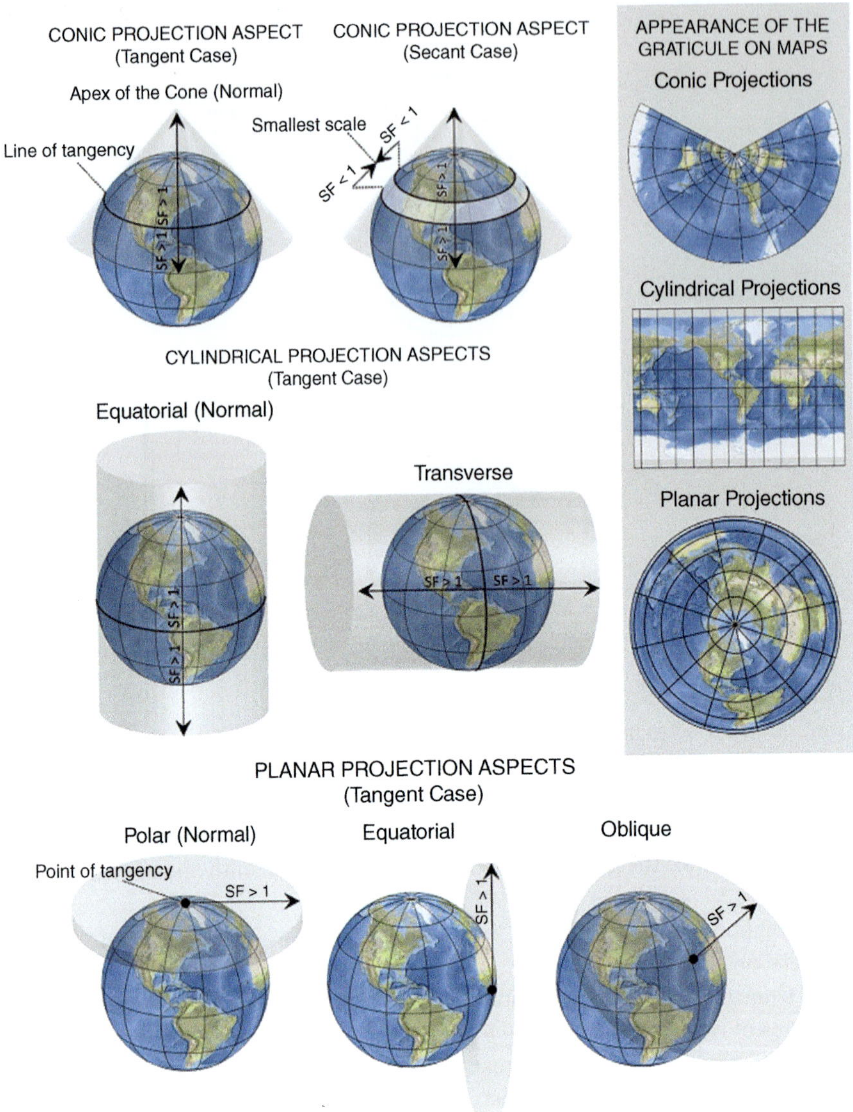


**Figure 5.8** Orientation of the globe with map projection surfaces. *Source:* Kutztown University of Pennsylvania/https://faculty.kutztown.edu/courtney/blackboard/Physical/05Project/project.html, last accessed on 23 November 2024.

that are stored in GIS databases can only be presented, visualized, and analyzed by using a digital map on the computer screen that is referenced by a real world related coordinate system (CS). Generally speaking, cylindrical map projections are mainly used to make and present worldwide digital maps. Conic map projections are more suitable to project a nation in mid-latitude regions, such as the United States of America. Planar map projections are more suitable to create Arctic area or Antarctic area maps. In order to ensure that different feature classes, vector map layers, or raster datasets in the same study area in a GIS mapping window can be aligned to each other, the same datum, coordinate system, and map projection need to be applied to all of them.

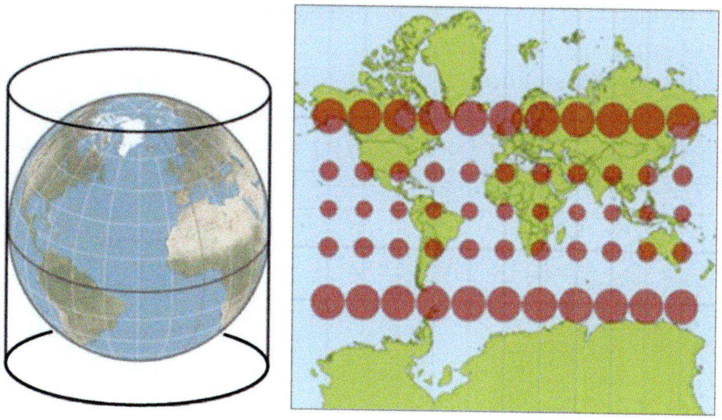

**Figure 5.9** Example of distortion distributions on a world map made by cylindrical projection.

## Projected Coordinate Systems

GCS was developed on a globe that represents the 3D Earth's surface. GCS utilizes angular distance units of degrees, minutes, and seconds as the original measuring units. In GIS systems, GCS coordinates of latitude and longitude are converted to decimal degrees with a large number of decimal digits in locating points or features on the Earth's surface. Meanwhile, positive and negative values need to be designated in order to differentiate the locations in different hemispheres. Therefore, there is a need to establish a detail-oriented and locally focused coordinate system to accommodate the accurate and precise measurement of location on the ground to facilitate local GIS analysis, mapping, and survey purposes. This is the reason that the Universal Transverse Mercator (UTM) coordinate system was created and is widely used in the world.

### UTM Coordinate System

UTM coordinate system is a projected coordinate system. That means the coordinate system, or more precisely, a group of coordinate systems, was established on the projected surfaces using one type of map projection. Reading its name, "Universal" stands for worldwide application. That means this coordinate system applies to the entire world. The map projection is called "Transverse Mercator." Mercator map projections are named after Gerardus Mercator, a sixteenth-century geographer and cartographer in Europe who lived during the years of 1512–1594. Gerardus Mercator was the first person to invent the method of cylindrical map projection in making world navigation maps. Therefore, the cylindrical map projection is also called the Mercator map projection. The world map normally shown between the latitudes of 85°N to 85°S made by cylindrical projection is also called the Mercator World

**Figure 5.10** Mercator conformal cylindrical map projection that yields correct directions while distorting the size and shape of countries (Routley 2021/Visual Capitalist.).

Map (Figure 5.11). Owing to his pioneer works in mapmaking, Mercator is also considered to be the funding father of cartography and mapmaking.

The UTM coordinate system utilizes a transverse Mercator map projection or transverse cylindrical map projection to make 60 coordinate system zones worldwide (Figure 5.12). That means the UTM system was constructed through 60 separate transverse Mercator projections by repeatedly rotating the cylinder at 6° longitude intervals. In the projection process, the line or

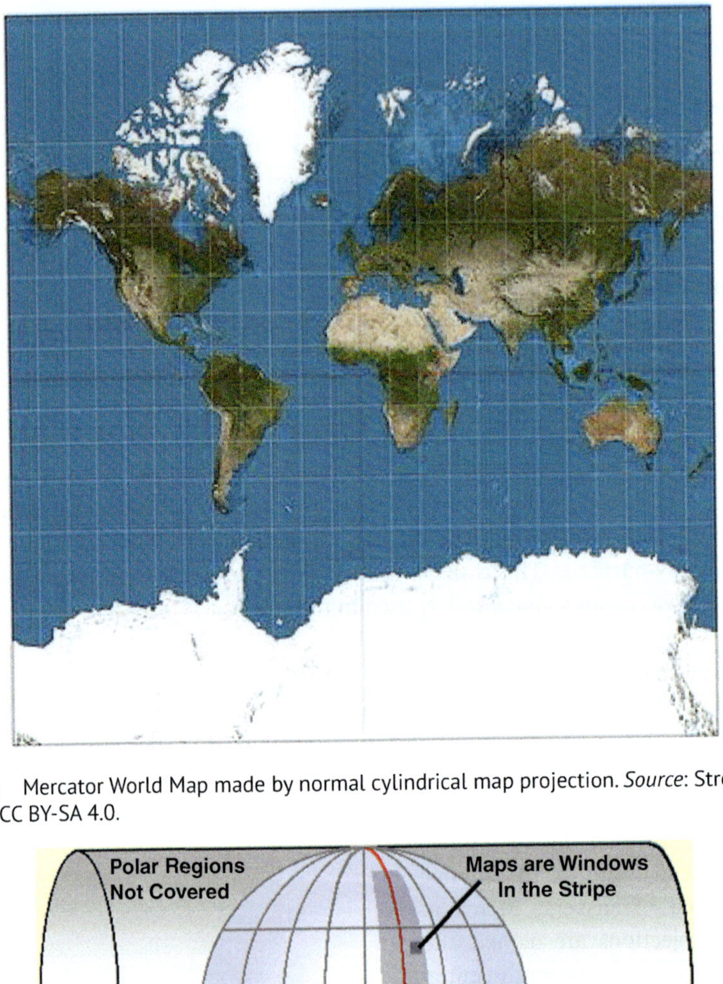

**Figure 5.11** Mercator World Map made by normal cylindrical map projection. *Source*: Strebe/Wikimedia/CC BY-SA 4.0.

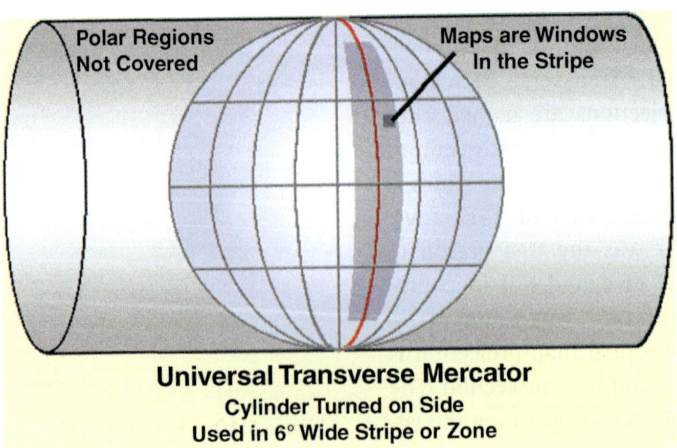


**Figure 5.12** Multiple transverse Mercator map projections in the making of the UTM coordinate system. *Source:* DrillingFormulas.Com, 2016/Drilling Formulas.

circle where the globe touches the cylinder wall is called the tangent line or tangent circle. In the process of making UTM zones, the two meridians (half circles) representing tangent circles are also referred to as central meridians because the tangent circle is the center of the projected zones.

Since there is no distortion (0 error) on the tangent circle or central meridian lines at each of the map projections, the further away from the central meridian line, the more distortions that are introduced. Therefore, we cut off the projected map at the 3° angular distance line from the central meridian on each side, east or west, to form an elongated longitudinal zone with 6° (Figure 5.13). This multiple map projection process slices and flattens the Earth's surface like peeling an orange.

In order to cover the entire range of 360° around the Earth, the UTM system consists of 60 zones, each zone being 6° of longitude in width. The UTM zones are numbered 1 through 60, beginning at 180° longitude and increasing from the west to the east (Figure 5.14). In the continental United States, the mapping territory expands from zone 10 to zone 19 (Figure 5.15).

The specific zone number is difficult to memorize. The mathematical approach to using longitude to find your UTM zone number is as follows:

1. Take your longitude coordinate in decimal degrees and add 180. Most often, people will choose a coordinate in the northwest corner of their area of study or GIS project and assign it this zone number even if the data straddles two zones.
2. Then divide the value by 6 and round up to the next highest whole number. This would be your UTM zone number.

For example, the UTM zone number for 39° W would be found through these steps:

$$-39 + 180 = 141$$

$$141/6 = 23.5$$

Round up to 24

Therefore, 39° W is in UTM zone number 24. If the study area is north of the Equator, then it would be in zone 24 North; and if it is south of the Equator, it would be in zone 24 South.

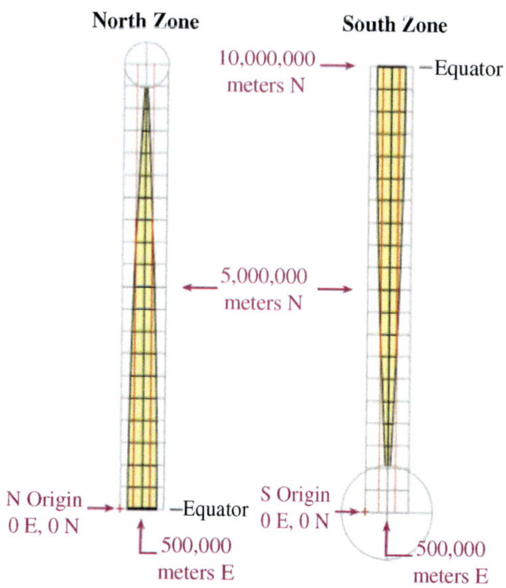


**Figure 5.13** X (Eastings) and Y (Northings) coordinate set up in a single UTM zone. *Source:* From the Pennsylvania State University.

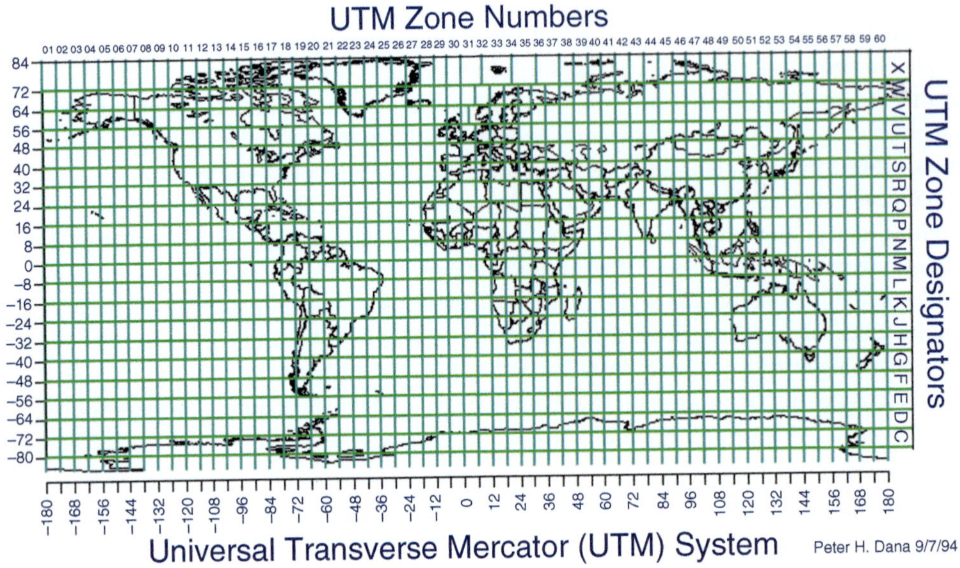


**Figure 5.14** Distribution map of UTM zones worldwide. *Source:* Adapted from Dana (1994).

On each of the projected and flattened zones, a coordinate system like the Cartesian coordinate system is established. Two fundamental rules were set up in this case. In order to avoid using negative values and numbers, we only use Quadrant I of the Cartesian coordinate system to establish UTM coordinates (Figure 5.16). Secondly, within each zone, coordinates are measured as northings and eastings in meters. Eastings are x-coordinates, and northings are y-coordinates in a mathematical sense.

A northing value in the northern hemisphere specifies the number in meters that a point is located north of the Equator (0 m). The northing of a point south of the Equator is equal to 10,000,000 m minus its distance from the Equator. In both the northern and southern hemispheres, northing values increase from south to north. A point south of the Equator with a northing of 7,587,834 mN is 10,000,000 − 7,587,834 = 2,412,166 m south of the Equator. The easting coordinate of a point is measured from the false origin 500,000 m to the west of the central meridian of the UTM zone. Within a zone, easting values increase from west toward east. A point lying 10 m east of the central meridian has an easting of 500,000 + 10 = 500,010 mE. The easting of a point 300 m west of the central meridian would be 500,000 − 300 = 499,700 mE. The east–west distance between two points is obtained by the difference of their easting values. The distance between the above points is 500,010 – 499,700 = 310 m.

The UTM coordinate system provides a constant distance relationship anywhere on the mapped coordinate zone in the original unit of meters. This gives GIS analysts a map to work with that is always in meters. The results of measurement and analysis on this type of digital maps can be easily translated to other distance measuring units. For example, the UTM coordinate of the Empire State Building in New York City, NY is 585,650.84 mE and 4,511,369.00 mN, Zone 18. The most popularly used high-technology equipment of field survey and location finding, the GPS can measure both UTM and the latitude and longitude coordinates at a particular location on the Earth's surface. This process is achieved through internal computations in the GPS unit. The UTM coordinate system is a more accurate georeferencing system in GIS for mapping and survey use.

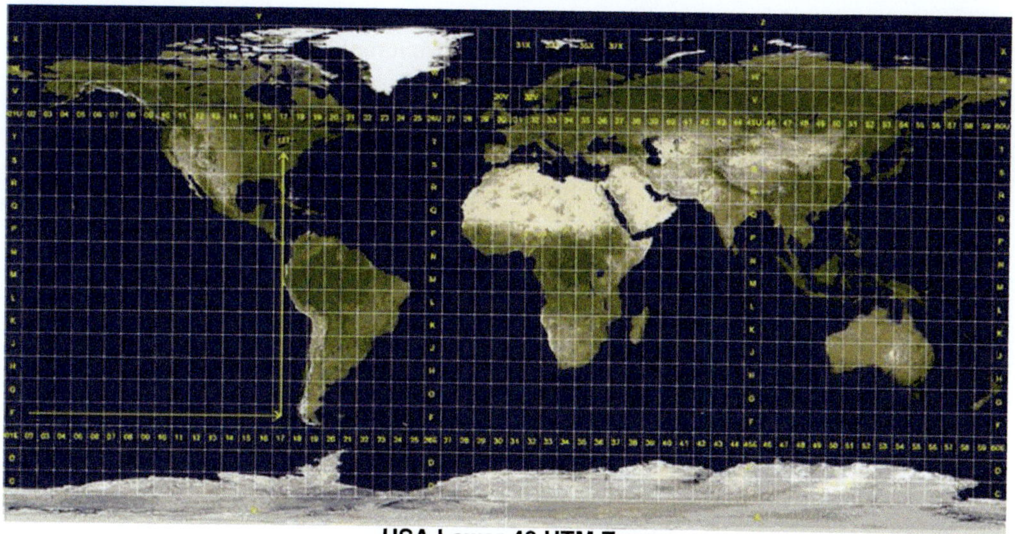

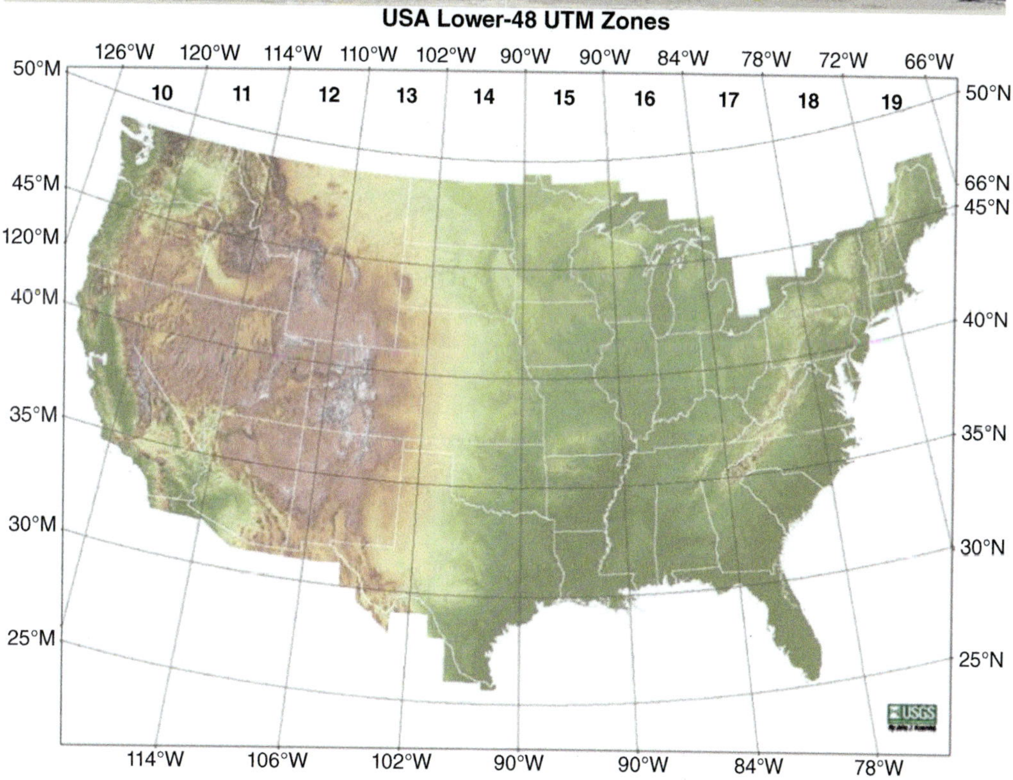


**Figure 5.15** UTM zones in the lower 48 states of the United States. *Source:* U.S Geological Survey (USGS)/ Public Domain.

### Sate Plane Coordinate System

The state plane coordinate system (SPCS) is a projected or plane coordinate system that was originally created in the 1930s to support surveying, engineering, and mapping activities throughout the United States and its territories. SPCS is only used in the United States, and it projects each

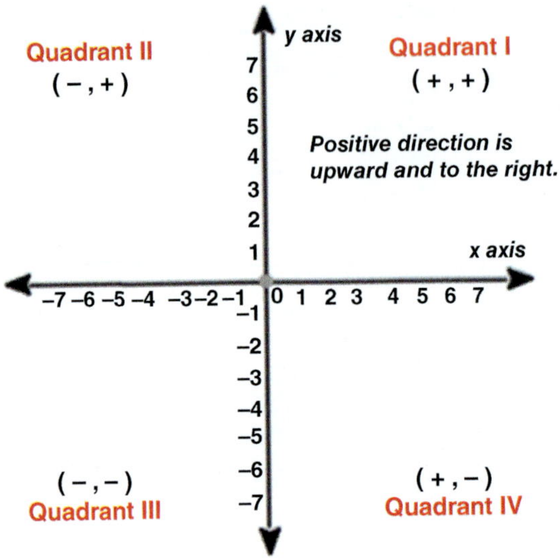


**Figure 5.16** Quadrants of the Cartesian coordinate system.

individual state to somewhere from one to six coordinate zones to facilitate detailed local survey, mapping, and construction purposes. In practice, people refer to SPCS directly as the state-owned coordinate system. For example, "Wisconsin State Coordinate System" means SPCS in Wisconsin, and "New York State Coordinate System" means SPCS in New York State.

The purpose of establishing the SPCS is similar to that of the UTM coordinate system. In each state of the United States, local land surveys, the construction of buildings and towns, and engineering projects need to be conducted in a local flat coordinate system in which the real-world distance unit can be measured and presented on the maps. The entire United States is projected and flattened onto roughly 120 different coordinate zones (Figure 5.17) in the SPCS system, such as "New York State West Zone (3103)." A Mercator map projection is used for north-south elongated zones, and a Lambert conformal conic map projection is used for east-west orientation zones. In either case, the tangent circle or central meridian of the map projection generally runs down the approximate center of the zone.

A Cartesian coordinate system is created for each zone by establishing an origin some distance, usually 2,000,000 feet, to the west of the central meridian of the zone and some distance to the south edge of the zone (USGS 2021). This ensures that all coordinates fall into the Cartesian Quadrant I with positive values. The x-axis of this origin runs through the zone from west to east, and the y-axis runs from north to south. Distances from the origin are generally measured in feet but sometimes are in meters. X distance values are typically called "eastings" because they are measured to the east from the origin, and Y distance values are typically called "northings" because they are measured to the north from the origin.

## Relative Geographic Coordinate (Referencing) Systems

The absolute location of a point or a feature on the Earth's surface in GIS is mainly marked and referenced by one of the three coordinate systems, namely GCS, UTM, or SPCS. However, relative location information of features and objects on the Earth's surface is much more popularly used in

**Figure 5.17** Projected coordinate zones of the SPCS in the United States. *Source:* Adopted from https://www.e-education.psu.edu/.

daily life practices. For instance, people ask the questions, "Where do you live?" or "What is your home address?" People seldom ask each other, "What is your location in latitude and longitude?" Real-world practices of geographic or spatial thinking every day demonstrate that the referencing or coordinate systems of relative locations are equally important to those of absolute locations (Figure 5.18). In GIS, we conduct many queries and data analyses based on relative georeferencing (coordinate) systems. In our discussions in this textbook, the term coordinate is borrowed from the concept of absolute coordinate systems because this type of georeferencing system, such as street address or postal code, can also help users pinpoint locations on the Earth's surface.

Three major characteristics of a relative geographic coordinate (referencing) system are:

1. One system only applies to one country or a certain geographic domain (countries). For instance, in the postal or zip code system established in the United States, an example code such as "14222" only applies in the United States and its territories. Crossing the border to a

**Figure 5.18** An example of the unique location of a residence marked by street address. *Source:* Dr. Tao Tang

different country, the postal code system would be totally different. The postal code at King's Circle in Toronto, Canada is "M5S 1A1."

2. Relative geographic referencing systems are generally organized hierarchically according to the geographic extent of marked regions or for end-locations.
3. Locational information items of relative geographic referencing systems are logically connected and topologically related to one another on digital maps. Location of features or objects can be traced or geocoded. The geocoded features and their locations can form a new digital map layer or a feature class in the GIS environment.

## Important Relative Geographic Referencing Systems Used in GIS

### Street or Postal Address and Postal Code System

Postal address and the postal code system contain two levels of comprehension in its geographic concepts: location and region. The original purpose of establishing the postal address and postal code system in a country was mainly for mail and parcel deliveries. In everyday practice, a country's postal services or international shipping companies, such as the United Parcel Service (UPS), Inc. or Federal Express (FedEx), Inc., can utilize the postal address and postal code to pinpoint a location on the ground for their deliveries. From a geographic perspective, a postal code is a locational marker or geo-marker of a small region. A postal address or street address is a geo-marker of a specific location. Technically speaking, a postal code is a series of letters or digits, or both, sometimes including spaces or punctuation, for the purpose of sorting mail for delivery. Postal addresses originated from street addresses and are largely similar to street addresses. Street address, or more precisely, the relative geo-marker of the human living environment, was developed simultaneously with human civilization in different countries and territories, as well as in various languages. Street address is intertwined with human settlement.

First, postal address and postal code systems, in particular the postal code subsystems, are country- or territory-specific. Different countries have different postal address and postal code formats.

Postal code is also known locally in various English-speaking countries worldwide as a postcode, postal code, or ZIP code. In the United States, postal code is also called ZIP code. According to the United Nations (Universal Postal Union 2021), 160 countries require the use of a postal code now for the shipping and delivery industries. Postal code and postal address systems are hierarchical by the designations of geographic regions. In the United States, postal or ZIP codes are typically made by five (5) digits, for instance, 14222 (Figure 5.19). By contrast, the postal codes in Canada are written in a different format, such as M5S 1A1. In English language, the full postal address is written in a backward order by its rank of location or regions, except the postal code. For instance, 1300 Elmwood Ave., Buffalo, New York, 14222, USA. In this case, 1300 Elmwood Ave. is the street address location or geo-marker of the object. Buffalo is the city of the location, and New York is the state of the location. 14222 is the postal or ZIP code of the area of the location. USA is the abbreviation of the county name of the location.

In the GIS environment, we can query and trace the targeted objects or features in the database using the postal or street address and present these features on a digital base map of the study area. Meanwhile, we can also create a new digital map layer or geographic feature class in GIS using a data table with postal or street addresses of features. This process is called geocoding. A comprehensive software engine of geocoding is developed in ArcGIS Pro. Geocoding processes can be

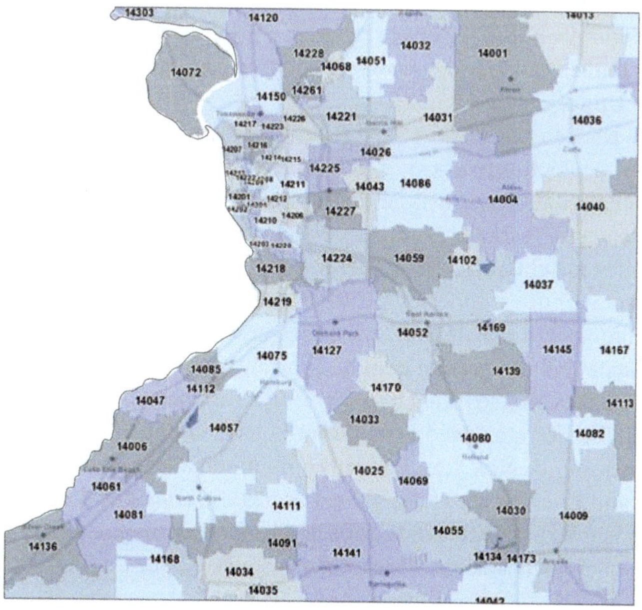


**Figure 5.19** An example of postal or ZIP code geographic areas in Western New York State. *Source:* US Postal Services.

used for tabular data that contains a street address, a postal or ZIP code field, a name of a place or region, or even a location that is identified by a code. With more help from artificial intelligence (AI) and machine learning (ML), we can develop more sophisticated and comprehensive geocoding modules and tools in GIS for analyzing both quantitative and qualitative datasets.

### US Census Geographic Areas or Regions

In order to survey and manage distributions of human population and population change in the United States, the Census Bureau in the United States Department of Interior developed the US Census Geographic Area System. According to a published statement of the US Census Bureau (US Census Bureau 2022), geography is central to the work of the Census Bureau. Geography provides the framework for survey design, sample selection, data collection, tabulation, and dissemination. Geography provides meaning and context to Census statistical data. "Given the diversity of our population, our economic activities, and our geographic areas, use of the latest and best geographic methodologies is critical to the Census Bureau's ability to serve as the leading provider of statistical and geospatial data. Our geographic area concepts, information, and statistical data must keep pace with the needs of the researchers and analysts who work to understand the changing distribution and characteristics of our people, places, and economy" (US Census Bureau 2022).

The US Census Geographic Area System is hierarchical, and the top-level units are the same as the country's government administration regions and subdivisions, namely states and counties. Owing to the geographic discontinuity of the existence of cities, towns, and rural communities, these geographic entities are not incorporated as the direct ground survey and data collection units like the states and counties. Below the county level, the Census Bureau created statistical areas of Census Tract, Census Block Group, and Census Block. A county is subdivided into several to several tens of these Census Tracts. Census Tracts are further subdivided into Census Block Groups, and then the Census Blocks. A Census Tract generally contains a population size between 1200 and 8000 people, with an average of 4000 people. The geographic size of a Census Tract changes largely, although continuity of the regions is ensured. Currently, the total number of counties and county-equivalent administrative subdivisions in the United States is 3234. The total number of Census tracts in the United States is 85,528.

Embedded in the US Census Geographic Area System is the Federal Information Processing Standards (FIPS) code, which serves as the unique identifier, or unique ID, of hierarchical geographic regions in the United States. FIPS codes are numeric codes assigned by the National Institute of Standards and Technology (NIST). FIPS codes are assigned alphabetically by geographic name for states and counties. FIPS codes for smaller geographic entities are usually unique within larger geographic entities. For example, FIPS state codes are unique within nation and FIPS county codes are unique within state. Since counties nest within states, a full county FIPS code identifies both the state and the nesting county. Typically, US states are identified by a two-digit number, while US counties are identified by a three-digit number. To make the county FIPS codes unique, the state FIPS codes are added to the front of each county. For example, a FIPS code of 36029 represents Erie County, New York State in computational or GIS databases. New York State is designated as 36 and Erie County is designated as 029. FIPS code is a mandatory standard for the protection of sensitive or valuable data collected by US federal government as well as by the state and county level of governments. Please refer to Table 5.2 for the FIPS code of the states in the United States.

Below the county level, Census tracts are coded with six (6) digits. Therefore, the FIPS code, or unique ID, of a Census tract is an 11-digit numerical code. For instance, in the Census tract, the code is designated as 000110 in Erie County (029), New York State (36), and the FIPS code is

**Table 5.2** List of FIPS code of states in the United States.

| State abbreviation | FIPS code | State name | State abbreviation | FIPS code | State name |
|---|---|---|---|---|---|
| AK | 02 | Alaska | MS | 28 | Mississippi |
| AL | 01 | Alabama | MT | 30 | Montana |
| AR | 05 | Arkansas | NC | 37 | North Carolina |
| AS | 60 | American Samoa | ND | 38 | North Dakota |
| AZ | 04 | Arizona | NE | 31 | Nebraska |
| CA | 06 | California | NH | 33 | New Hampshire |
| CO | 08 | Colorado | NJ | 34 | New Jersey |
| CT | 09 | Connecticut | NM | 35 | New Mexico |
| DC | 11 | District of Columbia | NV | 32 | Nevada |
| DE | 10 | Delaware | NY | 36 | New York |
| FL | 12 | Florida | OH | 39 | Ohio |
| GA | 13 | Georgia | OK | 40 | Oklahoma |
| GU | 66 | Guam | OR | 41 | Oregon |
| HI | 15 | Hawaii | PA | 42 | Pennsylvania |
| IA | 19 | Iowa | PR | 72 | Puerto Rico |
| ID | 16 | Idaho | RI | 44 | Rhode Island |
| IL | 17 | Illinois | SC | 45 | South Carolina |
| IN | 18 | Indiana | SD | 46 | South Dakota |
| KS | 20 | Kansas | TN | 47 | Tennessee |
| KY | 21 | Kentucky | TX | 48 | Texas |
| LA | 22 | Louisiana | UT | 49 | Utah |
| MA | 25 | Massachusetts | VA | 51 | Virginia |
| MD | 24 | Maryland | VI | 78 | Virgin Islands |
| ME | 23 | Maine | VT | 50 | Vermont |
| MI | 26 | Michigan | WA | 53 | Washington |
| MN | 27 | Minnesota | WI | 55 | Wisconsin |
| MO | 29 | Missouri | WV | 54 | West Virginia |
| | | | WY | 56 | WYOMING |

36029000110 (Figure 5.20). A Tract is subdivided into Census Block Groups by adding 1-digit to form a 12-digit FIPS code. Every Census Block Group has a unique 12-digit FIPS code.

The Block Group's unique identifier is the 12th digit of the FIPS Code. This number determines the first digit of all the census blocks contained within a block group. For instance, census Block Group 1 includes any block numbered 1000–1999. The FIPS code of a Census Block is a 15-digit identifier. This code contains the two-digit state FIPS code, the three-digit county FIPS code, the six-digit census tract code, and the four-digit tabulation block code. Continuing our previous example, we have Census Block code as 1005. Therefore, the FIPS code of the Census Black Group is 360290001101, and that of the Census Block is 360290001101005 (Figure 5.21).

**Figure 5.20** US Census 11-digit numerical coded tract geographic areas in the County of Erie, NY.

**Figure 5.21** US Census 15-digit numerical coded blocks of geographic areas in the Metro Buffalo region, County of Erie, NY.

US Census demographic data collected and statistically summarized by Census tracts or blocks is a major data source for the studies of all the social science disciplines. It is also one of the major databases to be used for various analyses in GIS. Similar to the postal address and postal code systems, the Census blocks or tracts can also be used to trace features, geocoding, and identify the characteristics of features. More importantly, the attribute-rich data the US Census Bureau collected, including population, household income, and the population of various ethnic groups, can be used for GIS and spatial analysis in social sciences.

## Telephone Coding System

The telephone coding system is another hierarchical relative locational reference system. Telephone number prefixes of country codes are a component of the international telephone numbering plan and are necessary when dialing a telephone number to establish a call to another country. Country codes are dialed before the national telephone number, and they are unique IDs of the countries. The map in Figure 5.22 shows the country codes for the phone calls worldwide (Figure 5.22).

Within a country, regional or area codes are established for locating individual ground line telephones conventionally. In the United States, an area code is a three-digit number that comes before the seven digits specific local number. It is also called a Numbering Plan Area (NPA) (Figure 5.23). For instance, the area code for the municipality of Buffalo region in the New York State is 716. Within the seven-digit local code, three are for the prefix and four are for the suffix. While the prefix of the local number gives an idea of the more specific geographic area, such as a town or neighborhood, the area code denotes the larger region, either an entire district of a city or even an entire county in a state. With the recent development of telecommunications and cellphone use, although cellphone objects carried by individual humans can travel across the nation or even internationally, the original three-digit area code is still registered for each of the devices. The US Telephone Coding System provides the telephone area code boundaries in the United States that can be used directly for telecommunication and GIS analyses. The following is a map of the US telephone area codes. The telephone coding system can also be used for querying and tracing objects and features and to conduct geocoding in GIS database.

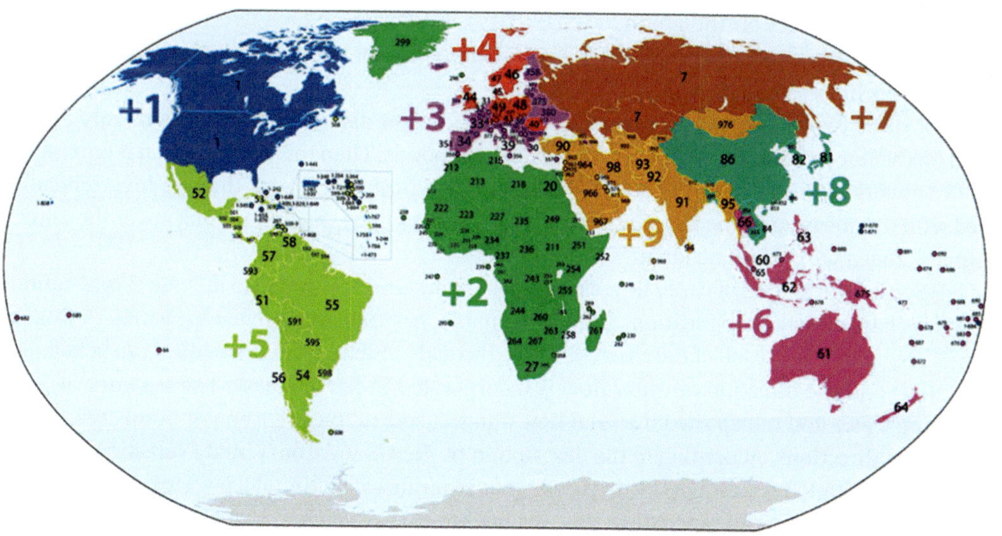


**Figure 5.22** Worldwide digital map showing the telephone codes by country.

Figure 5.23 NPA of the United States of America.

**The National Hydrography Database and Watershed Boundary Database Coding System**

Up to now, most relative georeferencing or coordinate systems have been developed either for surveys or for management or engineering issues in human society. Readers may wonder if there is any relative georeferencing system for natural features or objects in the United States. The National Hydrography Database (NHD) and Watershed Boundary Database (WBD) Coding System created by the US Geological Survey and US Environmental Protection Agency (EPA) during the last 10 years (since 2012) is one of this type of georeferencing system. NHD is a digital vector dataset used by GIS to define the spatial locations of surface waters, and it represents the water drainage network of the United States with features such as rivers, streams, canals, lakes, ponds, coastlines, dams, and stream gages on the ground (USGS 2019). WBD is used broadly in applications from scientific research to regulatory work. It is a companion dataset to the NHD and a component of the NHD-Plus high-resolution dataset.

Both of these two subsystems are established in GIS vector data format. NHD is mainly represented on the line feature GIS database except lakes and ponds. Then this reorientation is enhanced by GIS network system models. WBD fundamentally are nested polygons that are hierarchically coded with unique geographic IDs. Each of the drainage basins or watersheds that cover the entire country is coded with a unique identifier (unique ID).

In NHD, the information of water flow channel linear features is encoded to include classification and other characteristics, delineation, geographic name, position, and related measures. A Reach Code (RC) is assigned to each of the river channels, through which other information can be related to the NHD and the direction of water flow is incorporated in the database. The network of RCs delineating water and transported material flow allows users to trace movement in upstream and downstream directions. According to the description of the US Environmental Protection Agency (US EPA 2017), an RC is a unique 14-digit code that identifies a continuous piece of surface water with similar hydrologic characteristics. It is assigned to each receiving water body by the United

States Geological Survey's (USGS) NHD. Since the year 2014, the EPA Toxics Release Inventory (TRI) Program began asking facilities to optionally report RCs on the Form R to indicate where they were discharging chemicals into streams or bodies of water. The reporting of RCs enables the EPA and other researchers to accurately map discharges to the correct stream when conducting GIS analyses.

In the WBD GIS database, hydrologic units (HUs) are defined and delineated (USGS 2019). HUs represent the area of the landscape that drains to a portion of the stream network. More specifically, a HU defines the areal extent of surface water drainage to an outlet point on a dendritic or tree-like stream network. Multiple outlet points can take place in a HU where the stream network is not dendritic. A HU may represent all or only part of the total drainage area to an outlet point, so multiple HUs may be required to define the entire drainage region at a given outlet. HU boundaries in the WBD are determined based on topographic, hydrologic, and other relevant landscape characteristics without regard for administrative, political, or jurisdictional boundaries. The WBD seamlessly represents HUs at six required and two optional hierarchical levels. These six required units are named as hydrologic region, hydrologic subregion, basin, subbasin, watershed, and sub-watershed. Two additional optional levels of units can be further divided below the sub-watershed level.

The HU in the WBD form a standardized system for organizing, collecting, managing, and reporting hydrologic information for the United States. The HUs in the WBD are arranged in a

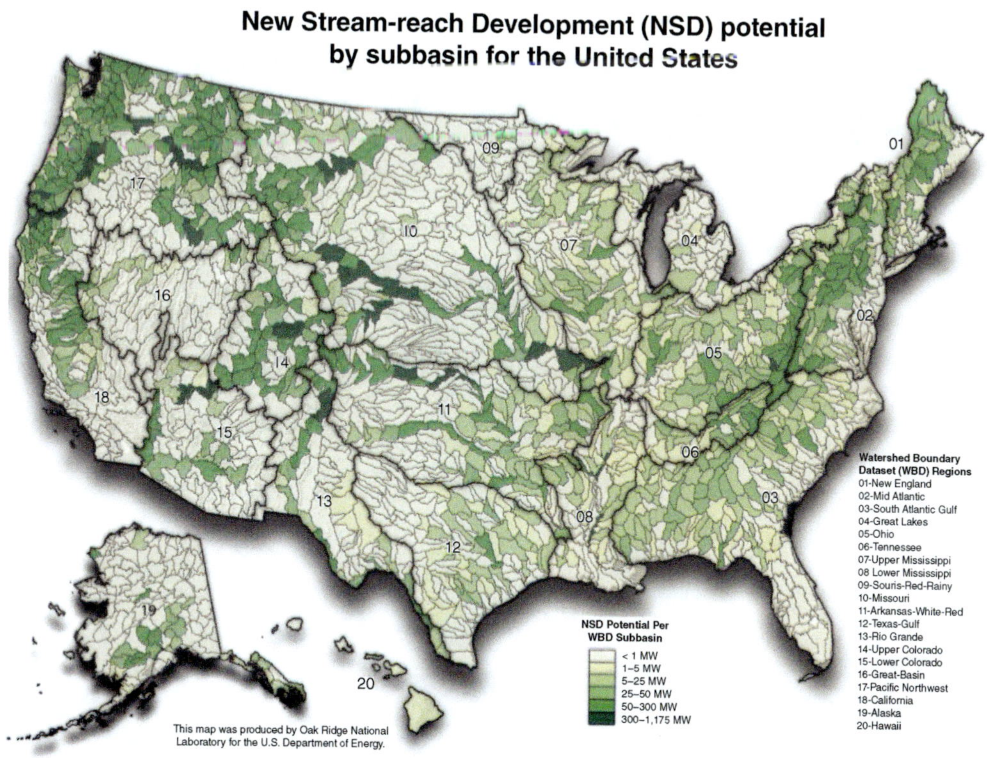


**Figure 5.24** Subdrainage basin level digital map of NHD and WBD for the entire United States.

nested, hierarchical system with each HU in the system identified using a unique code. Hydrologic unit codes (HUC) are developed using a progressive two-digit system, where each successively smaller areal unit is identified by adding two digits to the identifying code of the smaller unit within which it is nested. WBD contains eight levels of progressive HUs identified by unique 2- to 16-digit codes. The dataset is complete for the United States to the 12-digit HU. The 14- and 16-digit HUs are optional and are not complete for the entire nation.

The format of a 14-digit RC in NHD is in two parts. The first eight digits are the HUC code for the subbasin in which the reach exists, and the last six digits are a sequential number that is assigned when reach codes are allocated in the subbasin. Each RC occurs only once throughout the entire nation, but a RC may be on a number of contiguous features if they all belong to the same reach. Once assigned, RC is permanently associated with its reach. If the reach is deleted from the GIS database, its RC is retired. Both WBD and NHD GIS datasets for the entire United States or part of it can be downloaded and used in a software environment for analysis and mapping purposes. The following are a high-level WBD dataset map of the entire United States (Figure 5.24) and a sample data map of NHD (Figure 5.25).

**Internet Protocol Address System**

IP address stands for Internet protocol address. An IP address is a unique string of characters that identifies each computer using the Internet Protocol to communicate over a network. When

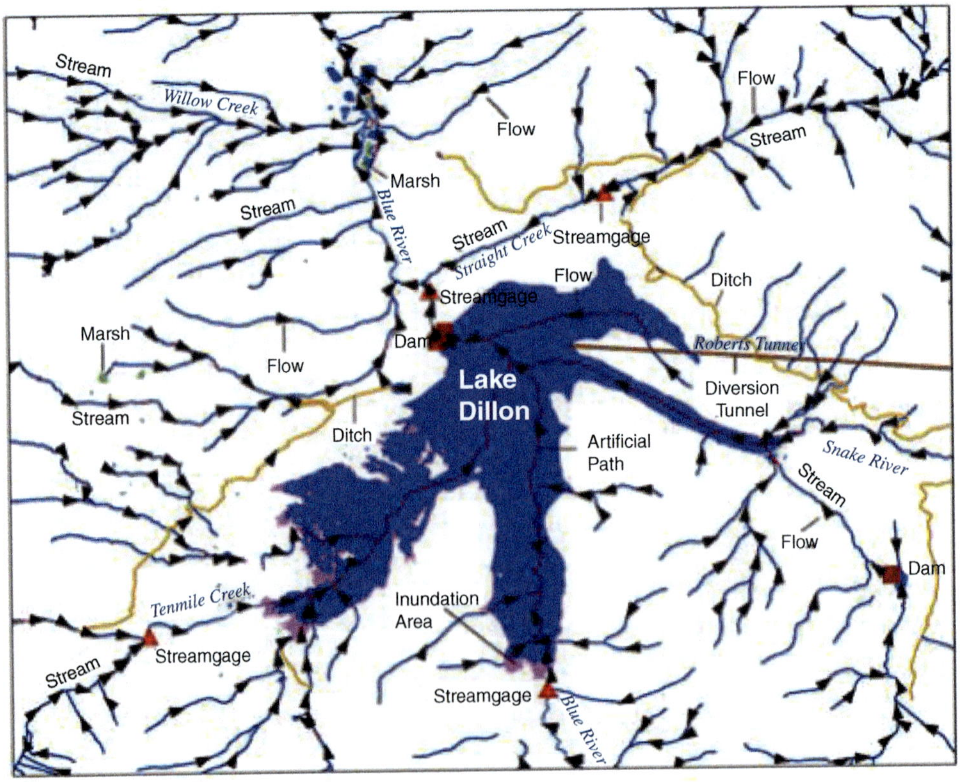


**Figure 5.25** Example of NHD dataset.

connected to the internet, the IP address allows the computers to send and receive information. Therefore, an IP address can be used to trace and track down the location of that particular computer. An IP address serves two principal functions in a computer communication network, host identification and location addressing. IP address can also be used to conduct tracing and geocoding in GIS environment. Figure 5.26 is an example of IP address of a computer in Buffalo, New York (Figure 5.26).

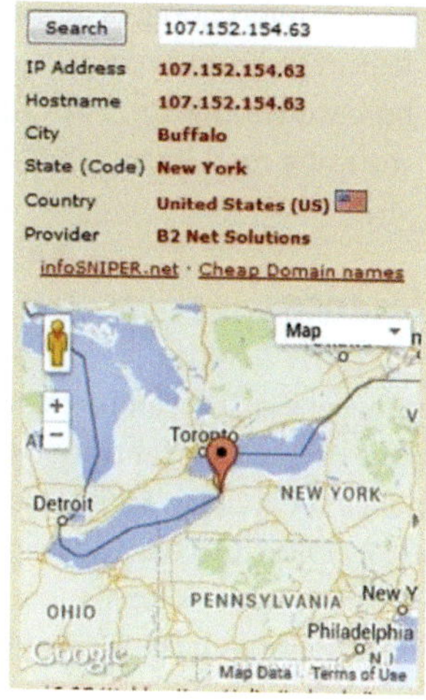


**Figure 5.26** IP address of a specific computer in Buffalo, NY.

## Linear Referencing System

Real-world objects such as roads, recreational trails, railroads, water or oil pipelines, and rivers are represented as linear features in the GIS database in 2D space. Since linear features are 1D themselves, the locations of these features can be referenced by a single value to a standard point. Linear referencing is the method to store and geographically locate objects on a line feature or a group of connected line features using relative values. In other words, linear referencing is essential and a very significant part of network analysis and maintenance. Linear referencing can be applied to highways, streets, liquid material transportation lines, electricity transportation lines, etc. (Figure 5.27).

The significance of using linear referencing is:

1. Many locations are recorded as events along linear features. For example, locations of traffic accidents are recorded using the distance to a mile-marker of a state highway. In this case, linear referencing using the mile marker from a standard starting point is necessary.

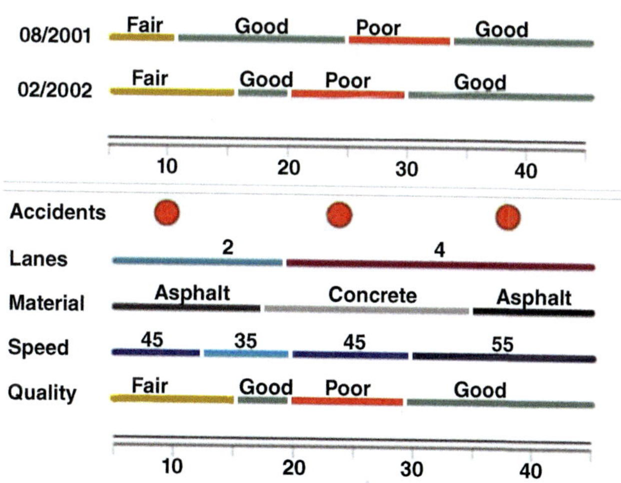


**Figure 5.27** Example of linear referencing system on highways and streets networks. *Source:* Adapted from ESRI – ArcGIS Users' Guide.

2. Linear referencing is also used to associate multiple sets of attributes with portions of linear features without requiring that the underlying lines be segmented (split) each time the attribute values change. For example, the pavement changes and road maintenance groups are on the same road line.

## References


Christopherson, R.W. and Birkeland, G. (2019). *Elemental Geosystems*, 9e, p. 539.

Prentice Hall Higher Education Publisher (PEARSON).

Campbell, J.C. (2001). *Map Use and Analysis*, p. 372. McGraw-Hill Higher Education.

Dana, P. (1994). Electronic Navigation, Precise Positioning, and Geographic Information Systems Research and Development. http://www.pdana.com/PHDWWW.html.

GIS-Geography. (2022). Ellipsoid/Spheroid – Our Oblate Spheroid Planet Earth. https://gisgeography.com/.

Heitzler, M., Hans-Rudolf, B., Schenkel, R., and Hurni, L. (2019). The light source metaphor revisited—bringing an old concept for teaching map projections to the modern web. *ISPRS International Journal of Geo-Information* 8 (4): 162. https://doi.org/10.3390/ijgi8040162.

NOAA (National Oceanic and Atmospheric Administration). (2018). National geodetic Survey. https://geodesy.noaa.gov/datums.

NOAA (National Oceanic and Atmospheric Administration). (2021). What is the geoid? https://oceanservice.noaa.gov/.

Routley, N. (2021). The Problem with Our Maps. https://www.visualcapitalist.com/problem-with-our-maps/.

US-EPA. (2017). Envirofacts. https://enviro.epa.gov/enviro/.

US Geological Survey. (2019). National Hydrography Dataset (ver. USGS National Hydrography Dataset Best Resolution (NHD) for Hydrologic Unit (HU) 4-200. https://www.usgs.gov/national-hydrography/access-national-hydrography-products (accessed 23 October 2019).

USGS. (2021). What is the State Plane Coordinate System? Can GPS provide coordinates in these values? https://www.usgs.gov/.

Universal Postal Union. (2021). "Addressing knowledge center," General Addressing Issues (PDF). https://www.upu.int/en.

US Census Bureau. (2022). Information about the Geography Program.

## Tutorial in ArcGIS Pro Environment

1. Coordinate geometry (COGO) process to build new geodatabase feature classes with field surveyed locations and data

Copy the tutorial data folder "Chapter05-Tutorial" to your working folder.

Open ArcGIS Pro and create a new ArcGIS Pro project file called "MyGIS_Larning05" using "Map" template. Save the project file and other files come with it into the designated working folder, C:\ or D:\Chapter05-Tutorial.

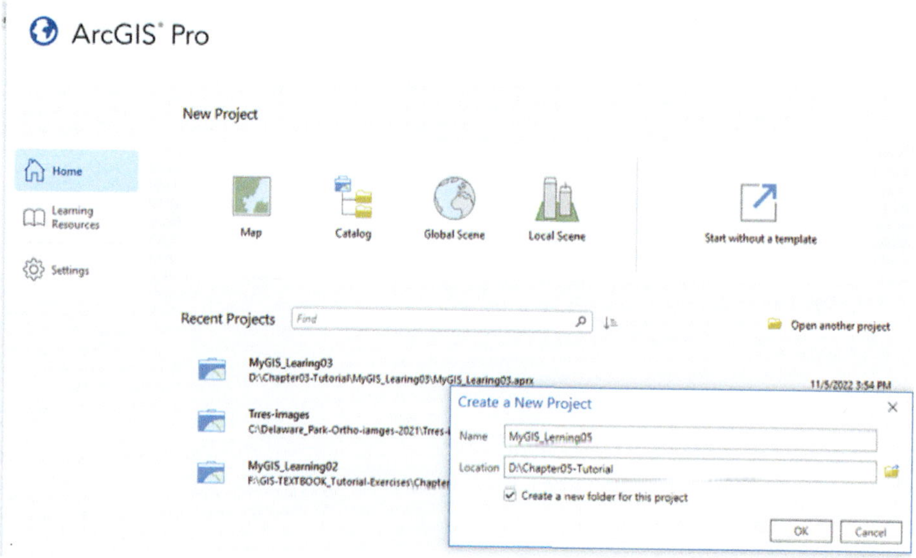


Now, add the entire folder of "Chapter05-Tutorial" on the Catalog Pane of your project window.

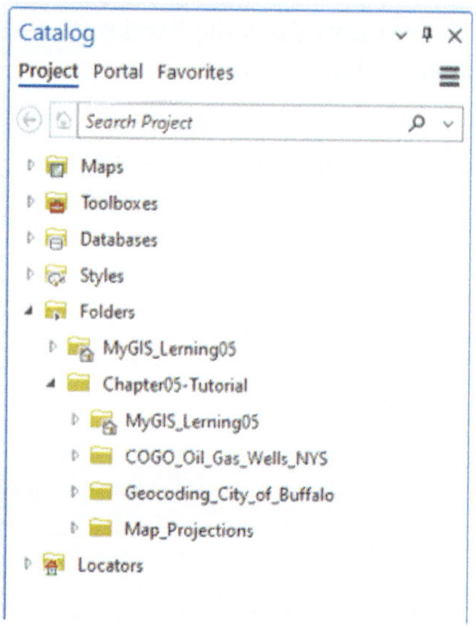

We are going to conduct a coordinate geometry (COGO) process to produce a new point feature class from a standalone excel data table that was downloaded from the Department of Environment Conservations (NYS-DEC) of New York State. First, we need to change the name of the "Map" window to be task-specific. Rename it as "COGO" at the Catalog Pane.

Let's expand the "COGO_Oil_Gas_Wells_NYS" folder. Generally speaking, we need to bring in our own base map in order to make the newly created feature class align with other datasets. Expand geodatabase: NYS_Civil_Boundaries.gdb. Add the "Counties" feature class to the "COGO(Map)" window.

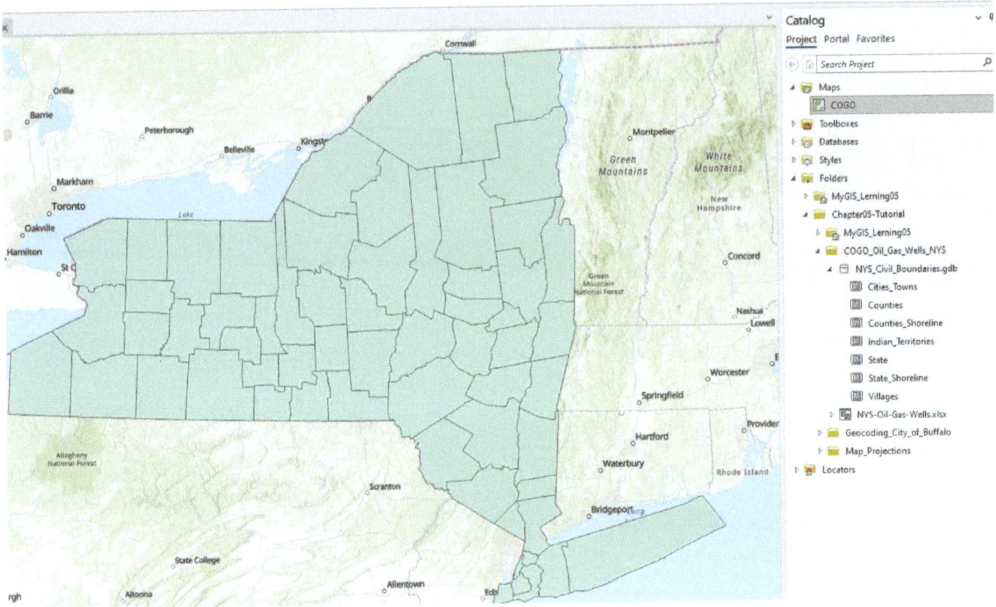


Although ArcGIS Pro can read the Excel file in varieties, the right route is to import an Excel file to ArcGIS Pro as a standalone stable in the ArcGIS geodatabase format. We are going to import the database. ArcGIS Pro works well with CSV or text format. Therefore, we open the Excel software and save the "NYS-Oil-Gas-Wells" spreadsheet to a CSV (Comma delimited) copy.

Now, we can close the Excel and go back to the ArcGIS Pro window to refresh the "COGO_Oil_Gas_Wells_NYS" folder (right-click on it to get the contact menu). The newly created NYS-Oil-Gas-Wells.csv should be here. Right-click on the NYS_Civil_Boundaries.gdb to import the table into the geodatabase.

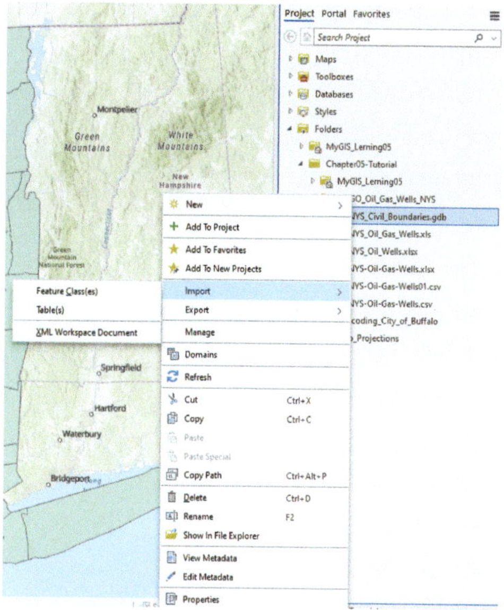


Table to Geodatabase geoprocessing tool opens. Navigate to your working folder "COGO_Oil_Gas_Wells_NYS" to fill the "Input Table" with NYS-Oil-Gas-Wells.csv. Run the tool to convert the table.

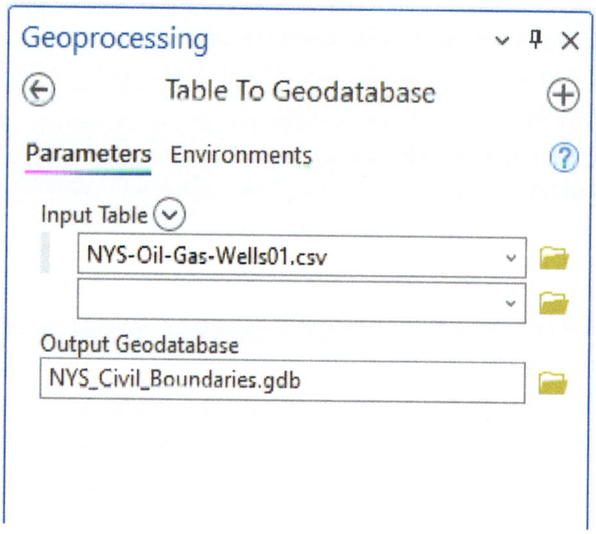


Add the standalone NYS_Oil_Gas_Wells table to the "COGO" (Map) window, and it appears on the Table of Contents. At the Table of Contents, right-click on the title of the table to obtain the contact menu. Then, select "Display XY Data" (COGO).

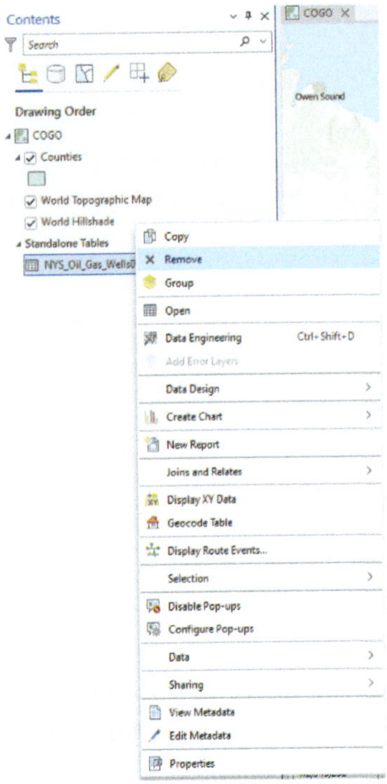


The following dialog opens. Keep the "Input Table" as default. Click on the small window of "Output Feature Class" to change the file name (last part of the link) to "Oil_Gas_Wells." Make sure the "X Field" is Surface_Longitude and the "Y Field" is Surface_Latitude.

Make sure the "Coordinate System" is GCS_WAS_1984. This is very significant. If you change to other coordinate systems, the new feature class could be shifted and land in Venezuela, South America (spoken from my experience). Since we only make a 2D geodatabase feature class now, we can leave the "Z Field" empty. Click "OK" to process the tool to produce your Oil_Gas_Wells feature class.

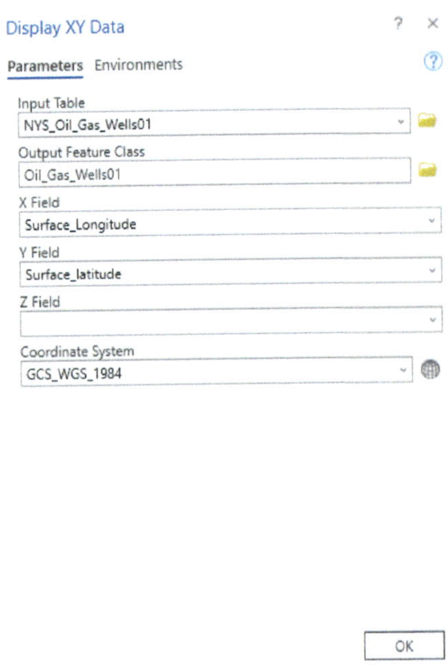

Now, we have our oil and gas wells in the New York State feature class. Next, we would like to thematically map the wells and present this digital map to the New York State Governor and Legislature to discuss the regulations and environmental protection law initiatives.

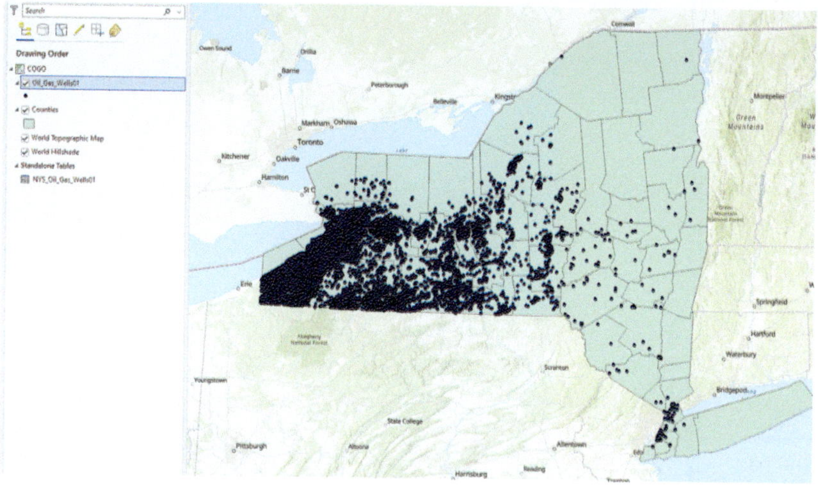


Right-click on the title of the Oil_Gas_Wells map layer in the Table of Contents to get the contact menu. Then, select "Symbology" to open the Symbology Pane. On the drop-down list of symbolizing methods, select "Symbolize your layer by category."

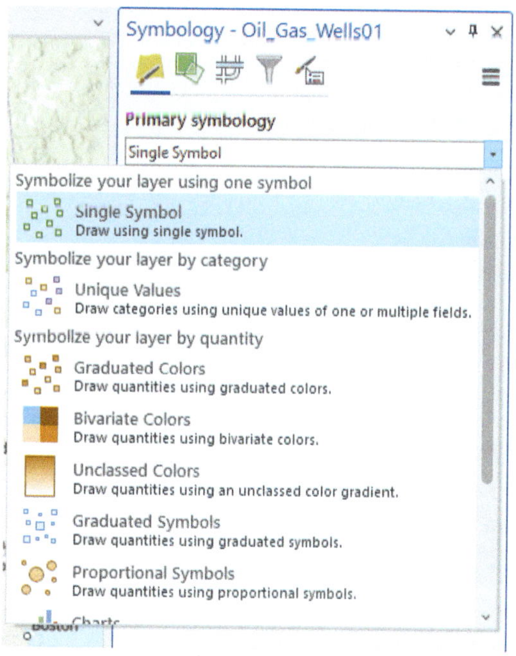


For "Field 1," select "GeneralWellType" field in the attribute table. Choose the "Color scheme" as Basic random. Then, close the Symbology Pane.

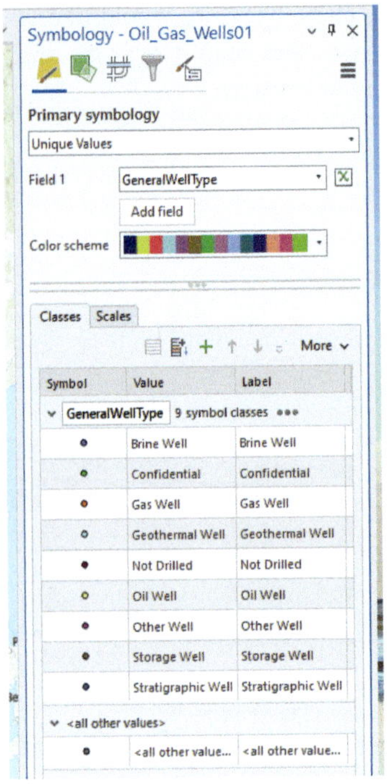


Now, we have the result of digital map to present. The project is accomplished. Save the project and close the ArcGIS Pro.

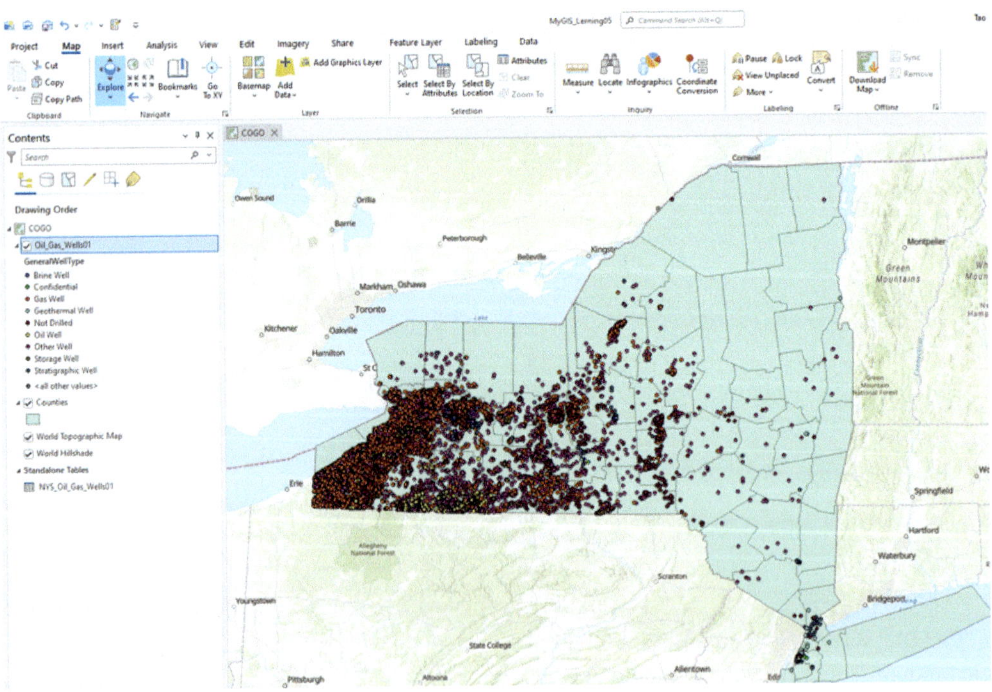

**2.** Geocoding with relative geo-reference systems to produce new geodatabase feature classes

In this exercise, we are going to geocode the data tables for creating one digital map of business company and store locations and one homicide digital map from the crime database for the City of Buffalo, New York. This is a GIS project to serve the city mayor and the city council for decision-making. The original datasets were downloaded from the internet website portal of the open data project in the City of Buffalo, NY.

Open ArcGIS Pro and "MyGIS_Larning05" project if necessary.

Create a new mapping window in the Catalog Pane under the "Maps" category. Change the name to "Geocoding." Now, collapse the "COGO_Oil_Gas_Wells_NYS" folder and expand the "Geocoding_City_of_Buffalo" folder on the Catalog Pane. In this folder, we have City_of_Buffalo.gdb geodatabase, Buffalo_Business_License_Locations.xlsx table, and five other Excel tables from the Police Department of the city. This is the police officer field-recorded data in 2022. The department keeps records of data on a yearly basis. Our exercise is to use the existing feature class "Buffalo_Streets" in the City_of_Buffalo.gdb as the base map and the Excel table of Buffalo_Business_License_Locations to make a new feature class of all the business locations in the territory of the city. Meanwhile, we also need to create another new feature class or map layer from the "Crime-Murders" table.

Like we did previously. We are going to lower down the ArcGIS Pro window and re-save the Excel file "Buffalo_Business_License_Locations" and "Crime-Murders" to CSV format in Excel first. Now, we refresh the Chapter05-Tutorial folder, you notice the newly saved CSV tables are added on the list.

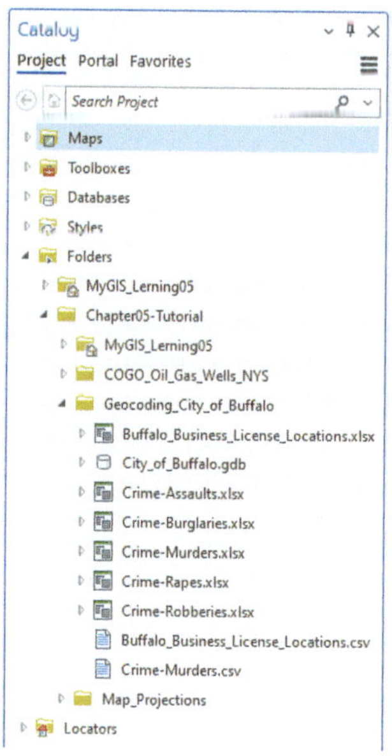


Now, let's import the standalone tables Buffalo_Business_License_Locations.csv and Crime-Murders.csv into the City_of_Buffalo geodatabase.

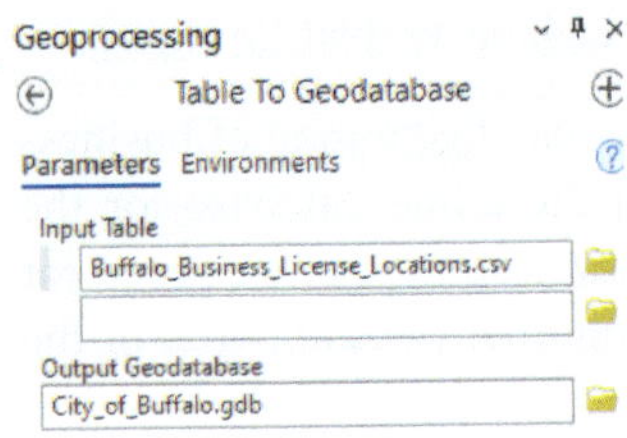


From the City_of_Buffalo geodatabase, add the Buffalo_Streets feature class and the two newly imported tables into the "Geocoding" (Map) window.

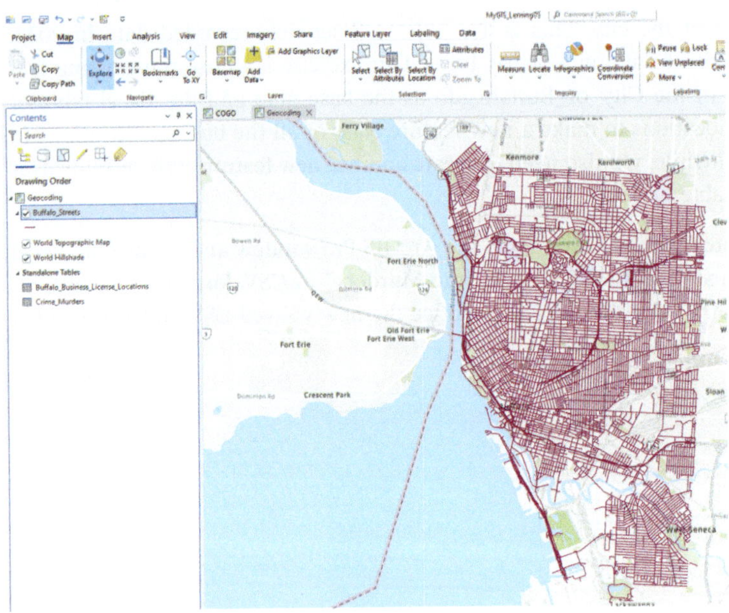


Before geocoding the tables to create feature classes, we would like to inspect the attribute table of the base map Buffalo_Streets feature class and the two standalone tables. First, we open the attribute table of the Buffalo_Streets feature class. The fields (columns) of "leftFromAddress," "LeftToAddress," "RightFromAddress," "RightToAddress," and "CompleteStreetName" are very important columns for street centerline feature class to be used for geocoding processes. Without these fields, the geocoding processes cannot be accomplished using a street center line as the base map. During the geocoding process, the street addresses will be matched against these ranges of "leftFromAddress," "LeftToAddress,"...

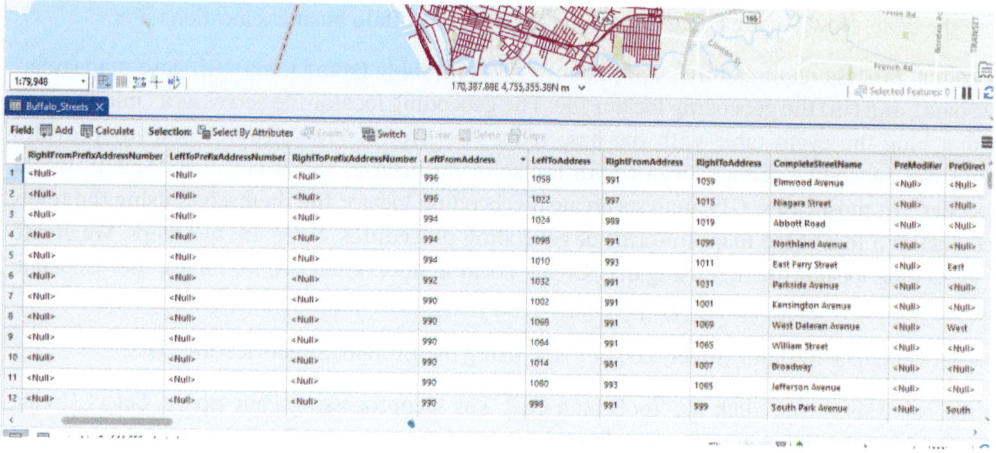


Now, close this attribute table and open the two standalone tables one by one. The "Address" field in these tables will be used to match the relative reference locations provided by a base map to produce a new point feature class.

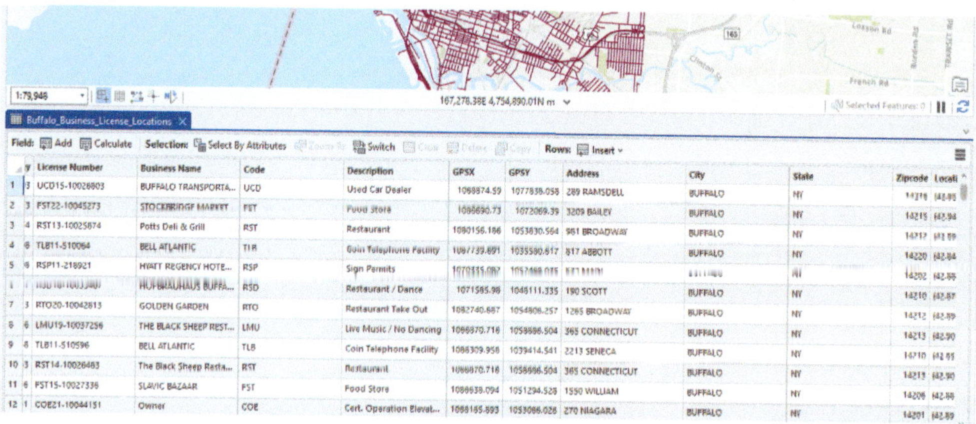


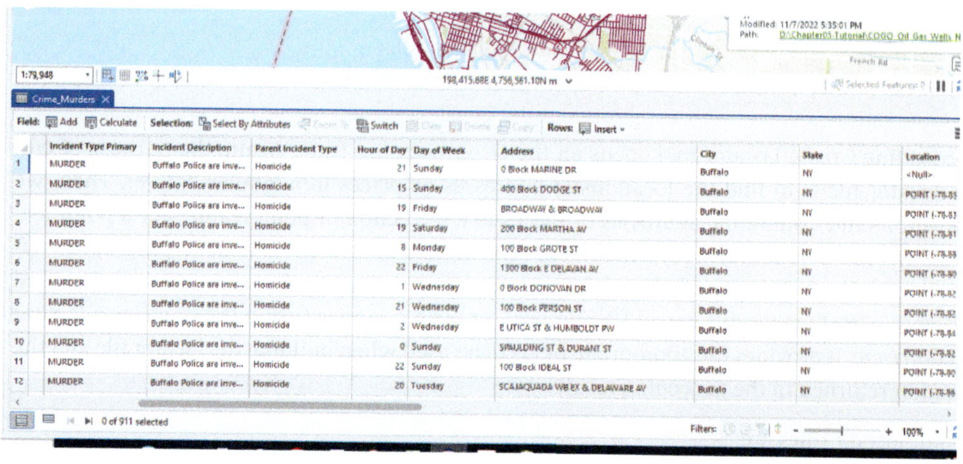

Close the table. Now, we are going to geocode the data of Buffalo business locations first.

Geocoding process requires three components: (i) event table (standalone), (ii) base map (reference data), and (iii) the geocoding locator file. The geocoding locator file serves as a "middleman" for matching the event table with the base map. The ArcGIS Pro software company provides a universal geocoding service locator with the software license, and it is called "ArcGIS World Geocoder." In most cases, GIS analysts create a geocoding locator file themselves using the reference data or a local base map. In terms of geocoding procedures, there are two steps. We create the geocoding locator file first using the "Create Locator" tool. Then, we would use the "Geocode Address" tool to geocode the event table to produce a digital map layer or a feature class.

Now, let's create a "Buffalo_Street_Locator" file using the Buffalo_Streets feature class.

On the Analysis menu, click the Tools icon. The Geoprocessing Pane opens. Select Create Locator tool in the Geocoding Tools toolbox under the Geoprocessing Toolboxes tab.

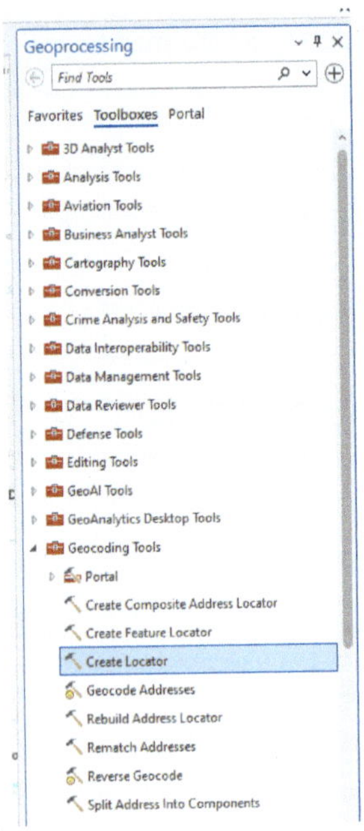


The Geocoding Create Locator tool opens on the Geoprocessing Pane. Again, the purpose of creating a locator file is to find the locations of addresses or places in a tabular dataset. Then, we can convert a table containing addresses or places to a collection of point features or a point feature class in GIS. The locator role defines the type of reference data or relative geographic reference system that is being used. This reference feature location can be street centerline, postal code region, parcel area (with parcel ID number), or even counties (using the name as a locator connection), etc. It provides the appropriate fields to be used when building the locator file and the information returned in the geocoding results.

The major locator roles are:

1. **PointAddress**: The point address locator role allows us to create locator files for a unique address at one reference feature location for the reference feature class. This locator role uses

feature classes with polygon or point geometry as the primary reference data. Each feature in the primary reference data corresponds to a single address, for instance, building or house polygon data with street address.

2. **Parcel:** Targeted to parcel polygons or points with street addresses or parcel land ownership ID numbers.
3. **StreetAddress:** A street address based on street segments where the house number is interpolated from a range of address numbers, including street name and street type extension. The reference feature class should be street lines.
4. **Point of Interest:** Names or IDs of administrative divisions, placenames, businesses, landmarks, and geographic feature names that are connected to their locations.
5. **DistanceMarker**: A point marker that represents the linear distance along a transportation route. It is typically quantified by kilometers or miles from a designated origin location, for instance, the interstate highway mile markers.
6. **DistanceRange:** This locator role allows us to create a locator file for street segments with distance marker ranges.
7. **Postal:** Postal code polygon or point as the reference of locations to create the locator file.
8. **Localities:** Administrative zones or regions, such as cities or states, are used to locate new features.

Only one primary table can be used per role when building a locator with Create Locator tool. On this tool dialog, fill in Buffalo_Streets for "Primary table(s)." For "Role" select StreetAddress, and you notice some changes under the "Field Mapping." Now, we need to fill in the following. Because this is a complicated process, we only need to fill in the required criteria.

| Field name | Alias name |
|---|---|
| Feature ID | <None> |
| Street Join ID | <None> |
| *Left House Number From | LeftFromAddress (from attribute table) |
| *Left House Number To | LeftToAddress |
| *Right House Number From | RightFromAddress |
| *Right House Number To | RightToAddress |
| *Street Name | StreetName |

Leave other fields in default (<None >). Name the locator file as "Buffalo_Street_Locator" and save it to D(C):\......Chapter05-Tutorial\Geocoding_City_of_Buffalo folder. For the Language Code, select English.

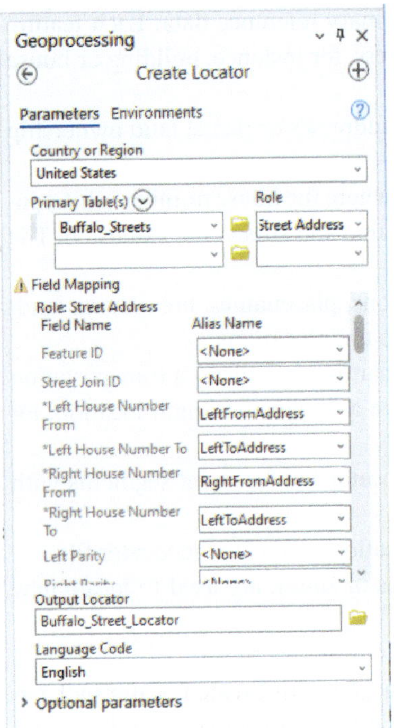


Run the tool to create the locator file. Close the tool and Geoprocessing Pane. The locator file, Buffalo_Street_Locator.loc is now created in the Geocoding_City_of_Buffalo folder.

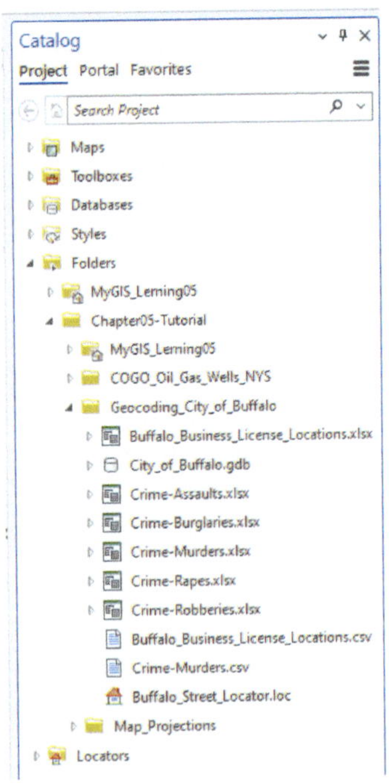

To make sure the geocoding criteria are met, we open and check the properties of the locator file that was just created. Right-click on the "Buffalo_Street_Locator.loc" locator file to open the Locator Properties dialog. Highlight the "Geocoding Options" and expand the "Match Options." Make sure "Match with no zones" is a "Yes." If it is a "No," change it from "No" to "Yes."

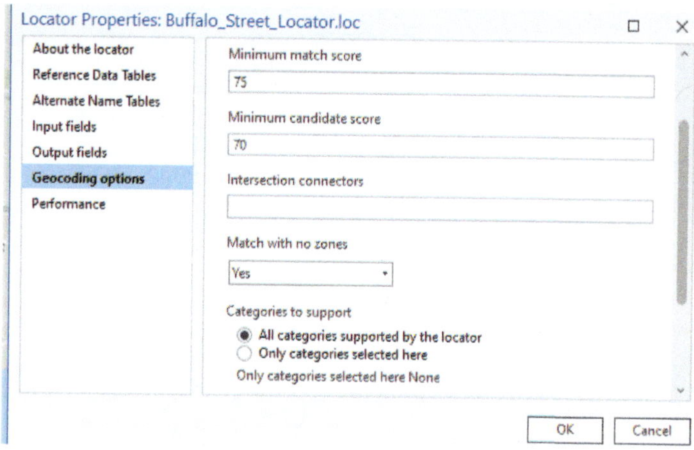


Click "OK" to close the dialog.

If the Geoprocessing Pane is not open, on the Analysis tab of the main menu, click the Tools icon to open the Geoprocessing Pane. On the Geoprocessing Pane, type "Geocode addresses" in the search box. Click the Geocode Addresses tool to open the tool's dialog on the Geoprocessing Pane. For the Input Table box, either drag and drop the Buffalo_Business_License_Locations table from the Table of Contents or navigate to the City_of_Buffalo.gdb to load it. For the Input Address Locator box, navigate and load Buffalo_Street_Locator.loc locator file. For Input Address Fields, select "Single Field" and Alias Name as "Address." For Output Feature Class, navigate to City_of_Buffalo.gdb geodatabase and name it as "Buffalo_Business_Locations."

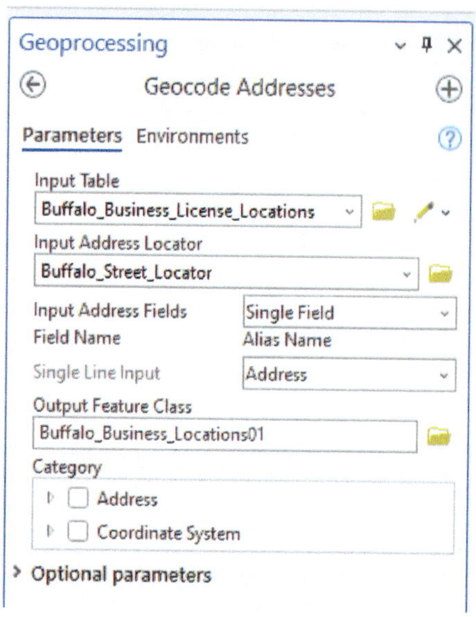

Click Run to process the data. You may notice the geocoded Buffalo Business Locations feature class is loaded in the "Geocoding" window.

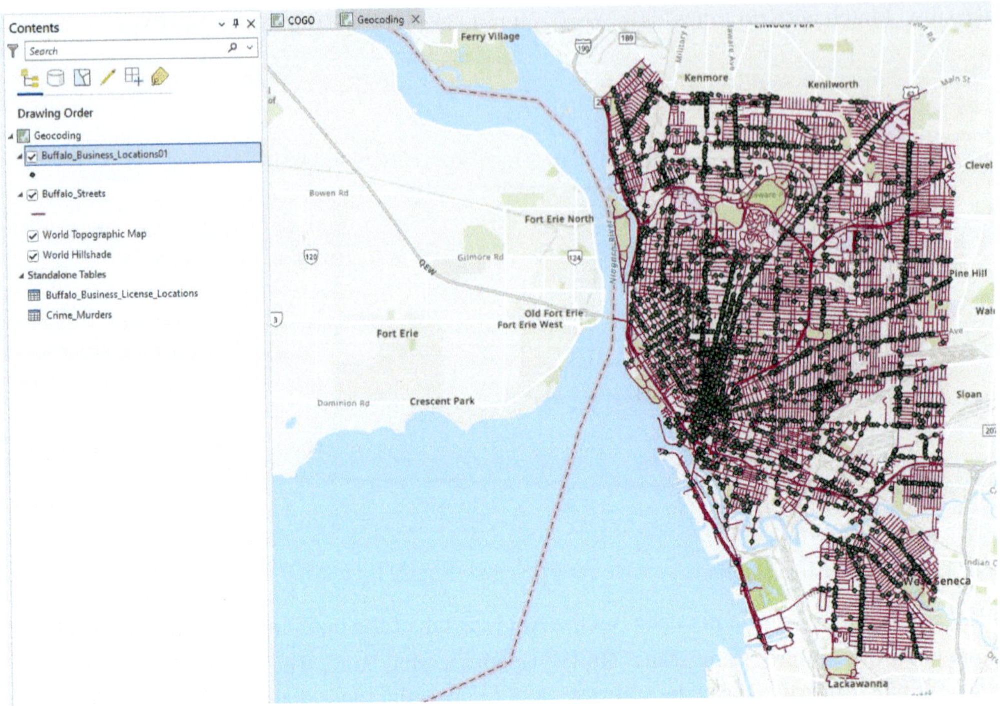


We can quickly symbolize this using a single symbol method. You can pick a specific symbol in the point symbol collection like I did for choosing the "Industrial Complex" symbol.

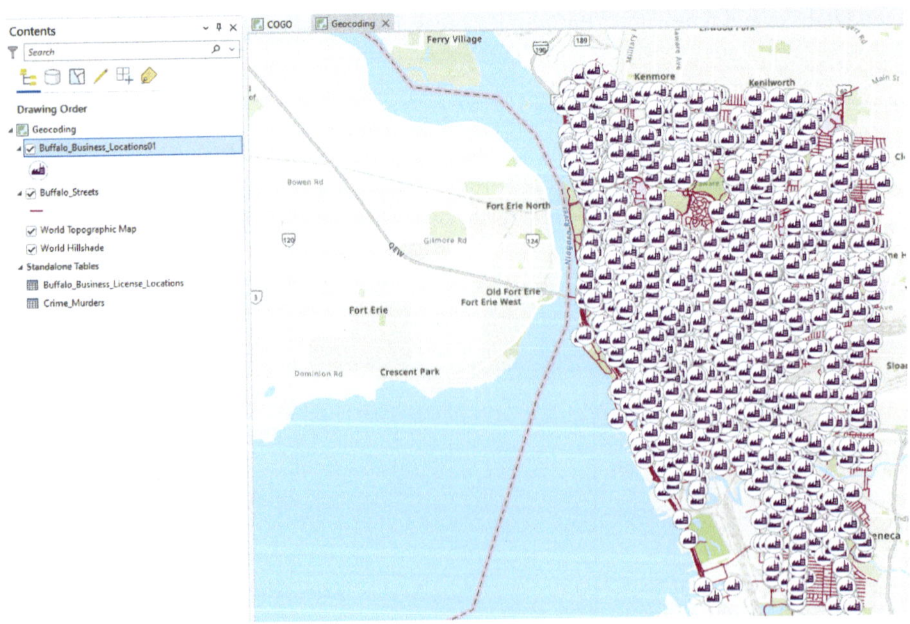

Now check off the Buffalo_Business_Locations feature class in the Table of Contents. Let's conduct another geocoding process using the Crime_Murders table and Buffalo_Street_Locator.loc locator file. If the Geocode Addresses tool in Geoprocessing Pane is not open, open it. Similar to previous steps for geocoding Buffalo_Business_Locations to a feature class, fill in the fields on the dialog of Geocode Addresses tool as follows. Save the Output Feature Class as "Buffalo_Crime_Homicides" in the City_of_Buffalo.gdb geodatabase.

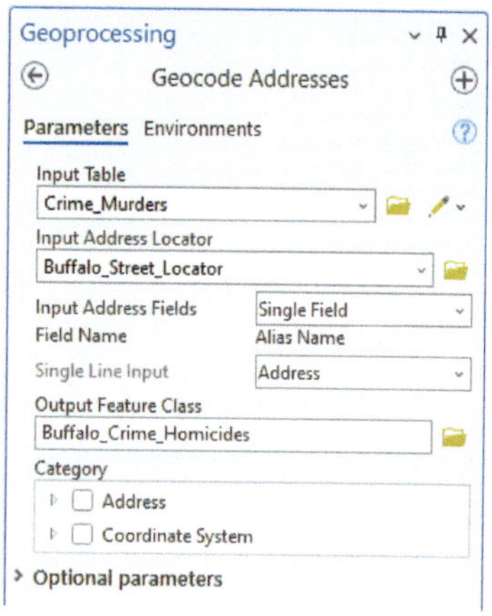


Process the tool, we will see the following map layer is added in the "Geocoding" window.

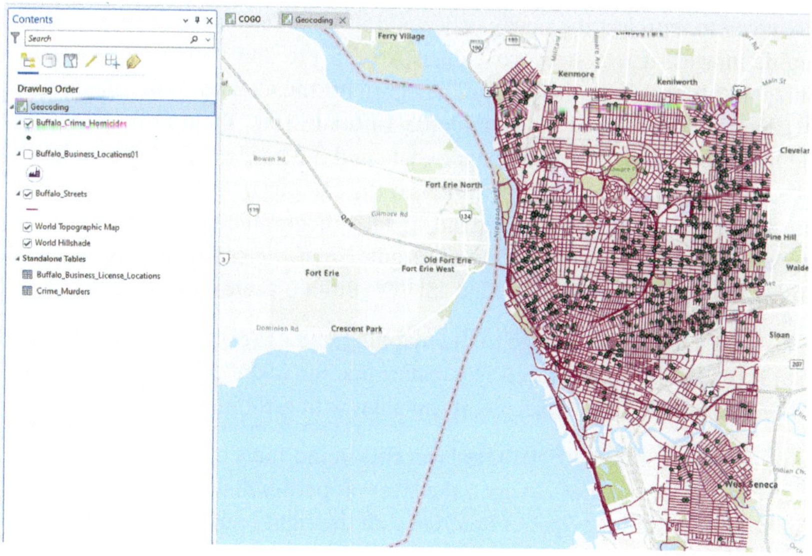

A simple thematic mapping can be done using the “Police Station” symbol.

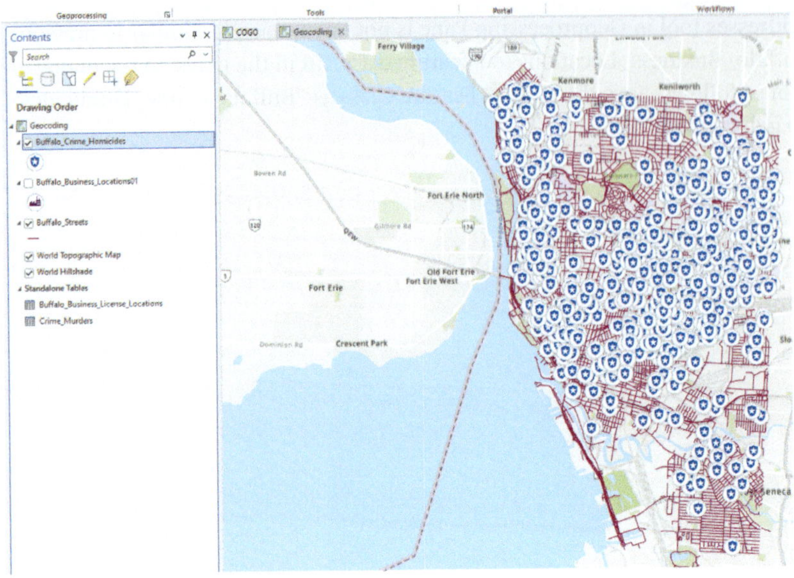


Save the project and close the ArcGIS Pro.

**3.** Projection of feature class coordinate system from one to another

In this exercise, we are going to reproject the newly collected LiDAR or laser scan dataset to align with digital orthographic images. If it is necessary, open ArcGIS Pro and “MyGIS_Larning05” project. Add a new mapping frame from the “Maps” category on the Catalog Pane. Rename it as “Projection_Point_Cloud.” Browse your ......\Chapter05-Tutorial\ Map_Projections folder. In this project, we would like to use both 2D aerial imagery data and the newly collected 3D LiDAR point cloud downloaded to study the geographic features on the campus of SUNY- Buffalo State. Previously, we downloaded the datasets and converted the digital orthographic images in the ArcGIS grid format in the geodatabase called Digital_orthoimages.gdb. Then, we conducted a mosaic of four imagery quadrangles into one image that is called “SUNY_Buffalo_State_Ortho_2021.”

Add the SUNY_Buffalo_State_Ortho_2021 image to the mapping frame, Projection_Point_Cloud. Add the LiDAR data, e1381n2340_2019.las as well. We realize that the LiDAR data does not fall in the right location. We need to project the dataset. So, it can align with orthoimages.

Highlight and right-click on the e1381n2340_2019.las layer title in the Table of Contents to obtain the contact menu. Then click on “Properties” to open the layer properties dialog. On the left-hand side, highlight the Source tab. Expand the Spatial Reference on the right-hand side of this dialog, we see the statement “Unknown Coordinate System.” That means this dataset does not have a coordinate system.

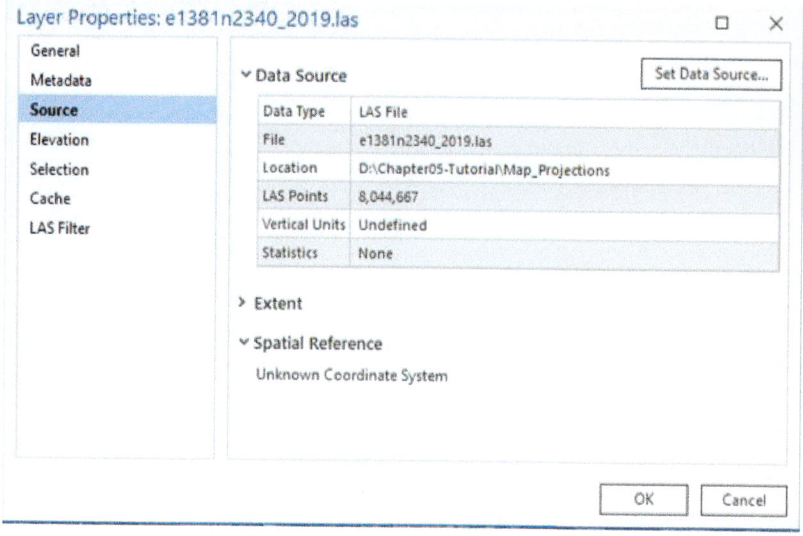


Now, switch to Metadata on the left-hand side. We see no metadata recorded in the form on the right-hand side in this case. However, the LiDAR dataset does carry an original map projection, USA Contiguous Albers Equal Area Conic. We found this information from the original metadata document at the New York State GIS Clearing House website. This means that, although this dataset says it does not have a coordinate system, the original coordinate system and map projection were set to USA Contiguous Albers Equal Area Conic.

Close the Layer Properties dialog.

ArcGIS Pro can automatically detect the coordinate system and defined map projection of a dataset. As long as a dataset carries its own map projection definition file, ArcGIS Pro can automatically match this dataset to other datasets of different coordinate systems in the mapping window. Now, we need to define this LiDAR dataset to its own map projection. Then, the ArcGIS Pro will be able to match it to the right location with the digital orthographic image.

Click on the Analysis tab on the top to open the Geoprocessing Pane. Open the Define Projection tool on the Geoprocessing Pane. You can search for tools or browse the Data Management Tools category.

For Input Dataset or Feature Class, select the e1381n2340_2019.las.

For the Coordinate System, click the small globe icon to change the coordinate system. Navigate to Projected Coordinate System -> Continental -> North America. Choose USA Contiguous Albers Equal Area Conic map projection.

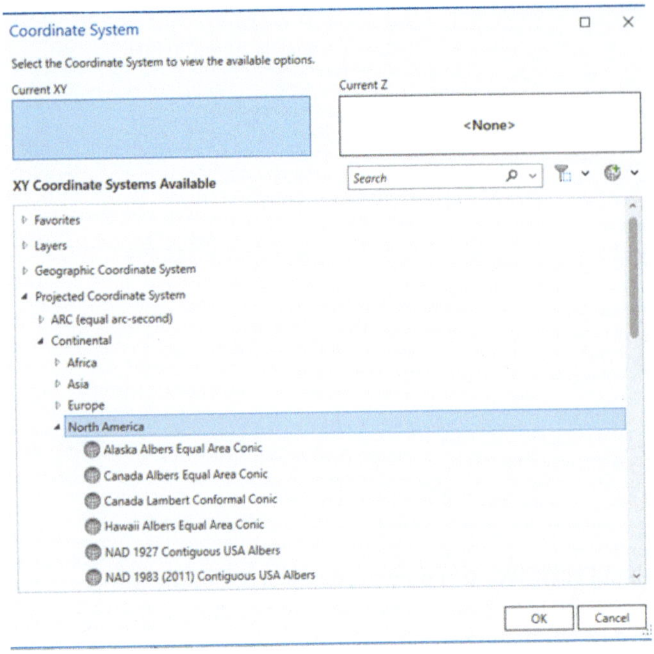


Right-click on the file name to get the contact menu. Select "Copy and Modify..." On the opened dialog, change the Latitude of Origin to 23 and click Save. Click Ok to fill in the new coordinate system into the Define Projection tool.

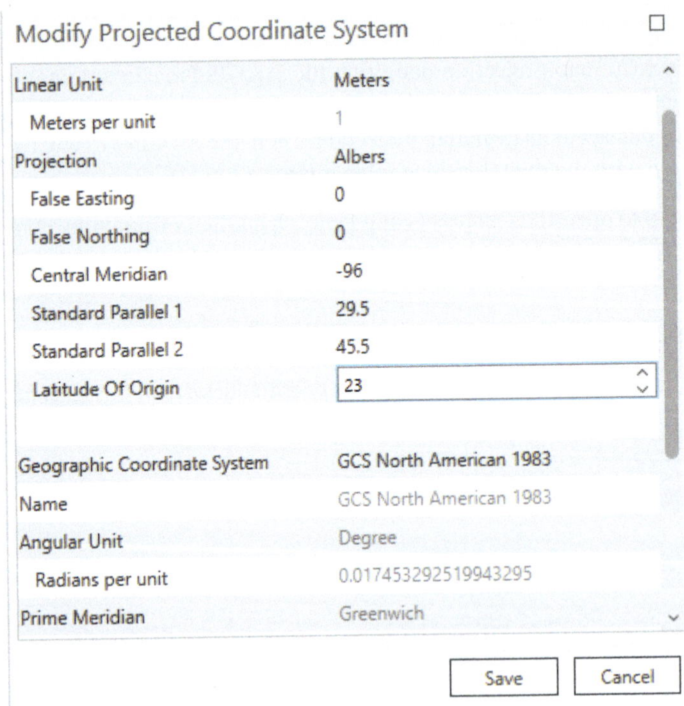

Run the tool to make changes.

You notice that the LiDAR data flies to the right location and aligns with the digital orthoimage.

The LiDAR point cloud dataset consume a lot of computer processing time and storage space. So, the background setting in ArcGIS Pro is adjusted by default so that the point cloud only shows when we zoom into a large scale or read in detail.

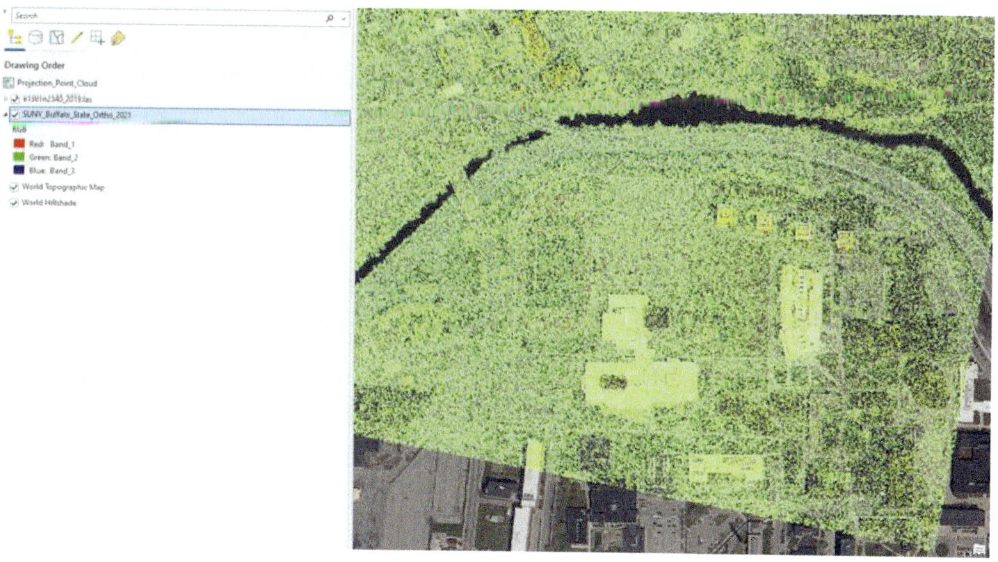

Now we can convert this map composition to a local scene to analyze the datasets. Highlight the map frame name (Projection_Point_Cloud) in the Table of Contents and activate the View menu on the top. Expand the Convert drop down list in the View group on this ribbon. Then, select "To Local Scene." 

Then, a new local scene with the same name is created and a 3D digital model is shown.

The Scene is a 3D visualization window that we will discuss in the later chapters.

Close the Geoprocessing Pane. Save the project file and close the ArcGIS Pro.

# Part III

# Fundamentals of Geographic Information Analysis

# 6

# Basic Spatial (Geographic) Analyses in Vector Data Environment

## Concepts and Theories

The focus of geographic information systems (GIS) is to analyze spatial or geographic data. Spatial analysis is a type of digital geographic analysis that seeks to explain patterns of human behaviors or natural events based on their locations. Fundamental operations of spatial analysis in both vector and raster data environments can be organized into the following six broad categories:

1. Distance calculation, extract spatial data and create new datasets
2. Overlay spatial data and create new datasets
3. Add and calculate attribute fields in spatial datasets
4. Summarize, calculate statistics, and aggregate attributes or features for spatial datasets
5. Model relationships of spatial objects
6. Discover geographic patterns of spatial objects and features

In this chapter and Chapter 7, we are going to discuss the analytical methods and practical operations based on the above categories. In ArcGIS Pro, geoprocessing is a specific term that is used interchangeably with operations of spatial analysis. According to Tulane University (2022), geoprocessing is a combination of software processes for analysis, modeling, and automated GIS tasks. In essence, geoprocessing is a framework and set of tools in GIS for processing geographic and related data. Typically, geoprocessing tools conduct operations on datasets such as a vector feature class, raster data, or tabular data and create one or more output datasets as the result.

## Distance Calculation, Extract Spatial Data, and Create New Datasets

Fundamentally, geography, or geographical analysis, is rooted in measuring location, distance, area, as well as proximity of objects on the Earth's surface. Simple spatial computations in GIS started from measuring, calculating, and presenting distances and areas on the Earth's surface. These operations are performed in both vector and raster datasets or map layers, but mainly started in vector feature distance calculations. Distance is defined as the magnitude or size of displacement between two locations (two points) on the Earth's surface. Five common measurements of distance are identified in GIS: (i) Euclidean distance, (ii) Manhattan distance, (iii) cost distance, (iv) geodesic distance, and (v) network distance.

*Fundamentals of Geographic Information Systems (GIS) Analysis: Applying ArcGIS Pro*, First Edition. Tao Tang.

Euclidean distance is the shortest distance between two locations or points. Euclidean distance can be measured either in 2D environment or 3D environment. Euclidean distance is the straight-line distance between two points. The definition is as follows:

$$d(\mathbf{p}, \mathbf{q}) = \sqrt{\sum_{i=1}^{n} (q_i - p_i)^2}$$

In this equation, $d$ is the distance, and **p** and **q** are two points in $n$ space referenced by a coordinate system. In the real world, only flying objects above the Earth's surface can follow Euclidean distance. Therefore, it is also called flying distance. The Euclidean distance in 2D environment is referred to as follows. A variety of GIS computations apply Euclidean distance. Figure 6.1 is an example of calculating medical helicopter flying distance from the closest hospitals to each of the grid cell locations.

Since many natural or human-made environmental factors impact the real-world distance to travel or to reach the destinations, several other distance measuring methods were developed in GIS. In the case of planned cities like Manhattan, NY, the streets often meet each other at right angles (90°). This layout divides the city into a collection of blocks and forces a traveler to travel only in east-west or north-south directions (Figure 6.2). In such a case, the shortest distance between the two points is shown in red in Figure 6.2. This kind of distance measurement is called the Manhattan distance (M), or the taxicab distance.

Cost distance in practice takes one or more environmental variables into consideration in formulating the measurements of distance of the two locations in GIS. Cost distance does not just account

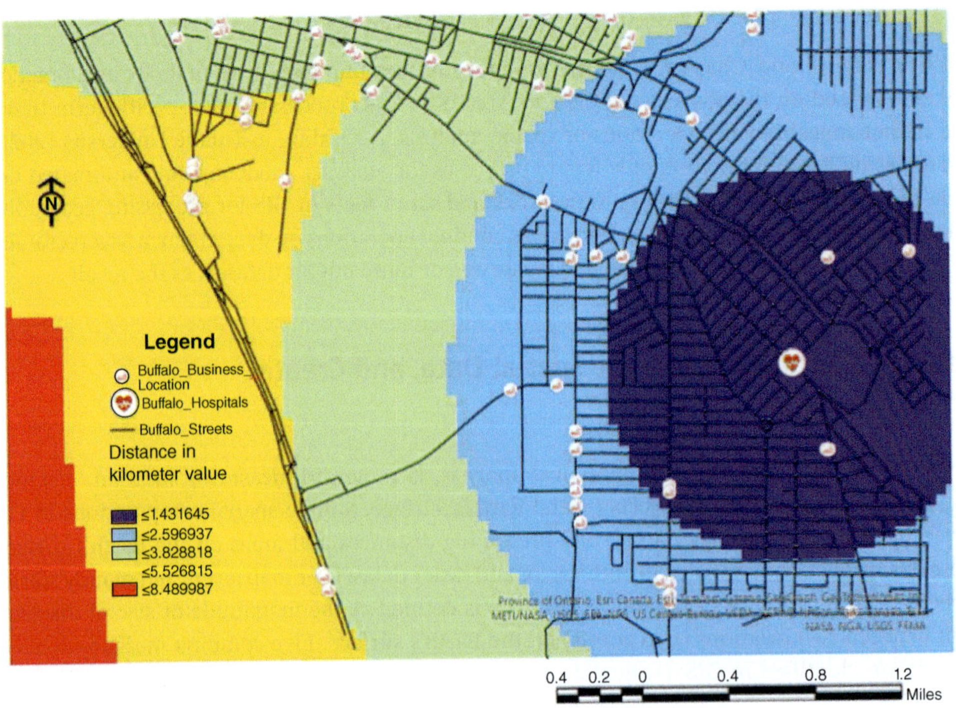


**Figure 6.1** Euclidean distance computation for closest hospital in the study region.

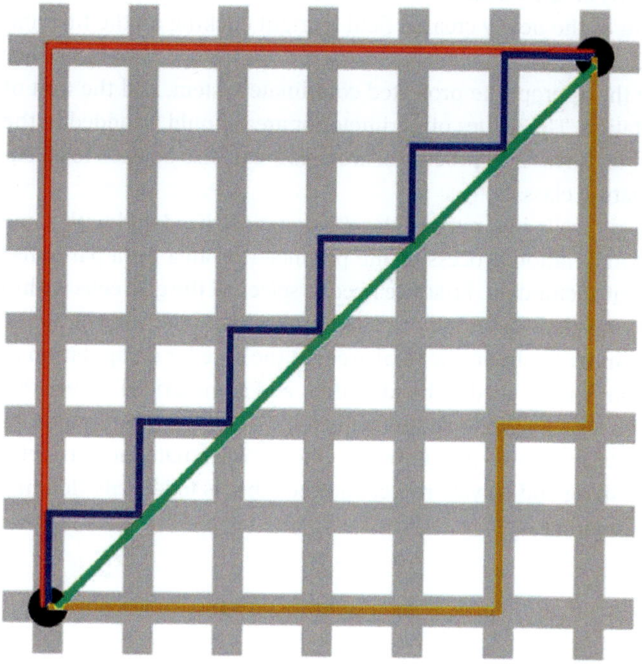

**Figure 6.2** Comparison of Euclidean distance and Manhattan distance. *Source:* Psychonaut/Wikipidia/ CC BY 4.0.

for the slopes and topography for travelling, although the majority of cost distance measurements are made by using raster digital elevation models (DEMs). By its name, many factors can affect the real-world distance measurement, such as driving speed limitations, traffic conditions, pay tolls or fees for facility use, pavement, or natural hazards on the routes. In order to determine the minimum cost distance in a certain environment, cost surfaces or cost factor map layers are often constructed to facilitate the process.

Euclidean and Manhattan distances are suitable for measurement of distance on the projected coordinate systems or flattened mapping surfaces. Geodetic distance is the distance measurement that considers the curvature of the Earth's surface. This measurement is more precise and accurate for the long-distance computations. On the Earth's surface, many natural or human-made network systems exist. For example, there are human-built highway networks across the entire nation in the United States, street networks exist in the urban regions, and the natural river systems form a connected network to discharge the rainfall water back to the ocean. Network distance, as the name suggests, is the distance (or cost, time, etc.) involved in traversing a network. In most instances, network distance refers to the shortest path measured between pairs or a list of points in considering the travel weight, such as travel distance or travel time on the network edges and nodes. The network systems and the GIS analysis on the network systems will be discussed in detail in Chapter 10.

Perimeter is the total distance of boundaries around a polygon or the edges around a shape. The area of a polygon is defined as the total space taken up by a 2D (flat) surface or shape of the object. Both perimeters and areas in a polygon feature class are considered as the geometric attributes. These can be computed using the "Calculate Geometry" menu in the attribute table of the polygon feature class. First, create or add a new attribute field (column) to hold the area or perimeter values. Then, name the newly created field with a meaningful title, either for the perimeter or area. Obtain the contact menu of the newly created field by right clicking on the title and then selecting "Calculate Geometry." When the "Calculate Geometry" dialog opens, select either perimeter or area as desired, choose the appropriate projected coordinate system, and the unit of length or area to process the computation. The values of perimeter or area should be added to the field in the attribute table of polygon features. This process or workflow can also be used to calculate the length of line features in a feature class.

Since the edge length and area of grid cells in a raster dataset are standardized and with same values across the entire space, the computation process of the perimeter or area for a particular polygon feature is different from that of vector data. First, we need to select all the grid cells within the targeted polygon feature. Then, sum the total number of the grid cells. Finally, multiply the standard grid cell area by the total number to obtain the total area. In the case for computation of perimeter, outside cells of the polygon feature need to be identified and summarized. Then, this number is multiplied with the grid cell edge length to obtain the result. Examples of computations for perimeters and areas in GIS projects are shown in Figure 6.3. One is to calculate the perimeter of Lake Placid, NY, to build a paved hiking and bicycle route, and the other is to identify flooding areas in the region after the disaster (Figure 6.4).

Figure 6.3 Planning for a paved hiking and bicycle route around the Lake Placid, NY.

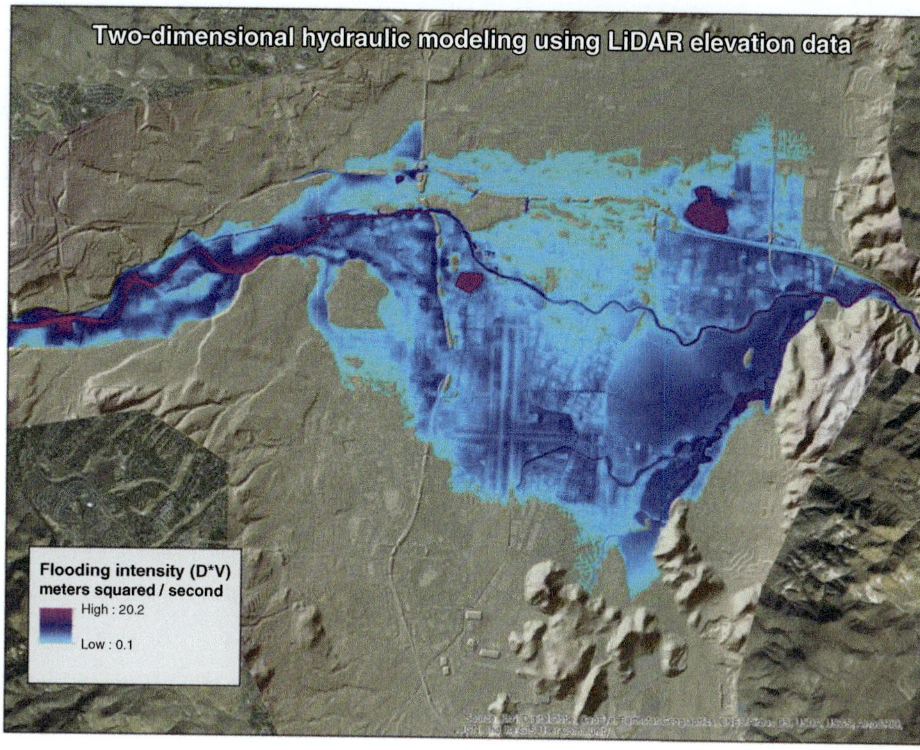


**Figure 6.4** Federal Emergency Management Agency (FEMA) flooding area map. *Source:* Dr. Tao Tang.

Data extraction, in essence, is to select data from existing datasets with some specific criteria and form new datasets to find answers to questions and to explore issues for decision making. Querying databases is a basic operation for data extraction in GIS. In the vector data GIS environment, we can conduct either attribute queries or spatial or geographic queries. GIS analysts can choose visual selections using the specific shape of selectors, such as a circle, a rectangle, or a polygon, or criteria-based SQL selections in the database. This type of operation normally occurs on one map layer or one feature class in an ArcMap or ArcScene window. The result of this type of operation can be directly exported to create a new feature class or dataset. For instance, setting up the SQL "Landcover type = 'Forest'" can help us to extract those land areas that are covered by forest to create a new map layer and save it in the database.

Spatial query refers to the process of retrieving or extracting a data subset from a digital map layer by working directly with the feature locations or geometry. The important portion of this type of query is the criteria of relative locations and geometric proximity of the targeted features with the selection features. This type of operation is normally conducted between two or more map layers or feature classes. One of the map layers with either selected or non-selected features that serves as a selector. People more often consider this type of operation or geoprocessing as "feature-to-feature" selection. For instance, we can identify a point-on-point selection in which both the selector map layer and the target map layer are point feature classes. A point-in-polygon selection is that the target map layer is a point feature class, and the selector is a polygon feature class or map layer. A line-in-polygon selection is such that the target map layer is a line

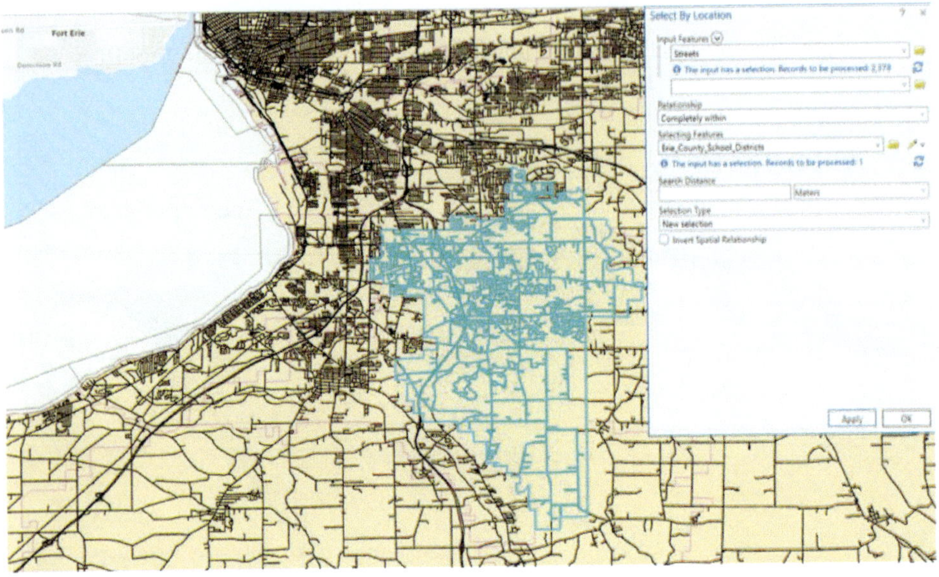


**Figure 6.5** Example of a line-in-polygon spatial query to select street lines in the Orchard Park School District, Erie County, NY, for planning of school bus routing.

feature class, and the selector is a polygon feature class. Other operations include line-on-line and polygon-in-polygon selections (Figure 6.5). The spatial query rules or criteria are mainly of the following seven categories (ESRI, 2020):

1. Intersect
2. Are within a distance of
3. Contains
4. Are within
5. Have a boundary that touches
6. Share a line segment with
7. Are crossed by the outline of

## Buffering Operations

A buffering operation is to create buffer polygons around input features to a specified distance or a set of distances under one variable. The objective of buffering analysis is to extract or differentiate features by a specific geographic region or a set of geographic regions. More precisely, a buffering operation classifies the features on a digital map layer or in a feature class into two types: (i) within the buffer or (ii) out of the buffer. Buffering can be done on all three types of vector data: point, line, and polygon (area), but the result is always a new map layer of one polygon or a set of polygon features. Combining the operations of buffering, layer-on-layer spatial query, and exporting data, a new feature class or map layer of a subset of the original targeted features or objects can be created (Figure 6.6).

A buffering method based on different distances is called a variable buffering. For example, the noise level in the area surrounding a street network may be based on the traffic volumes. Therefore, a variable buffering can be created to illustrate the noise level by using a larger distance

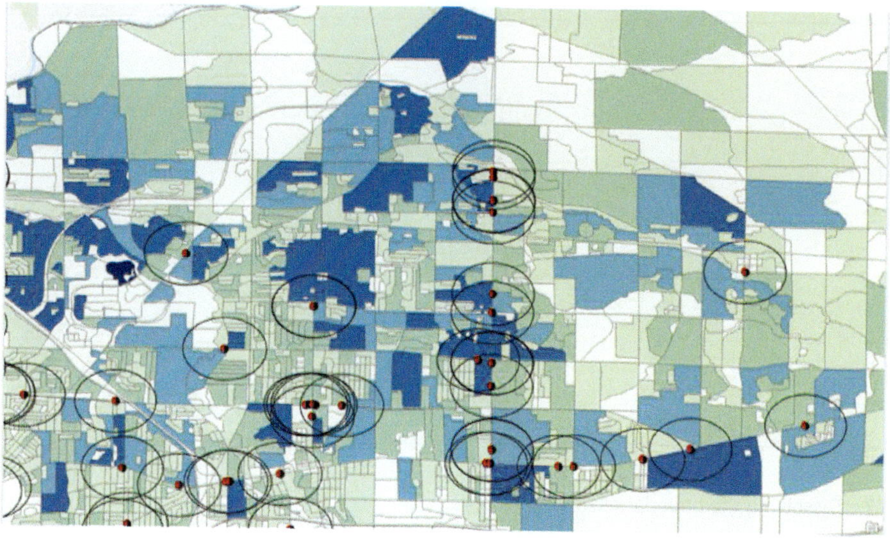

**Figure 6.6** Ring buffering of bank branches in the Town of Amherst, NY, to evaluate their serves in the proximity of distance of 0.5 mile.

for high-traffic roads or multiple-lane roads and a shorter distance for single lane and quieter roads.

Buffers that are created around point features typically look like circles, and this type of buffer is also called a ring buffer. Buffers are elongated around linear features, and they are more often called buffer zones. In application, multiple buffers surrounding the same feature, or a group of features, can be created for analyzing the environment or situations in neighborhood of one feature or a set of features. For polygon features that are buffered, there are two types of buffers in practice. Bidirectional buffers are polygons that are buffered from the boundary outwards as well as inwards. Setback or negative buffers are polygons that are only buffered from the boundary inward.

Buffering operations can help GIS analysts to access the geographic environment surrounding the features for effective decision-making and management. Examples of applying buffering around point features include cellphone communication or broadcasting towers to evaluate cellphone signal quality and in fire hydrants to evaluate response time to fires on buildings and houses surrounding them. Recently, ring buffers around a point feature and infection density mapping have been widely used for the analysis of infectious diseases, such as analyzing distributions of the COVID-19 virus. Line buffers along a river can be used to judge the price of flooding insurance for houses in a region. In practice, boundaries of buffer polygons created around all the different features can be dissolved into one polygon or polygons with certain criteria. Figure 6.7 shows the polygon dissolving process.

Buffering operations are unique in GIS. It is one of the core geoprocessing operations in computational environment. The operation itself looks very simple and straightforward, but it impacts graphic information conveyed to users and decision makers. Buffering in geoprocessing is rooted in the geographic concepts of proximity, closer distance versus farther distance, adjacency, and the close environment of the objects and features that we study. Geoprocessing of buffers surrounding features and objects is one of the earliest analytical operations developed in the GIS software environment. The operation of buffering normally is conducted on one map layer or one feature class and constitutes part of the basic analysis in GIS.

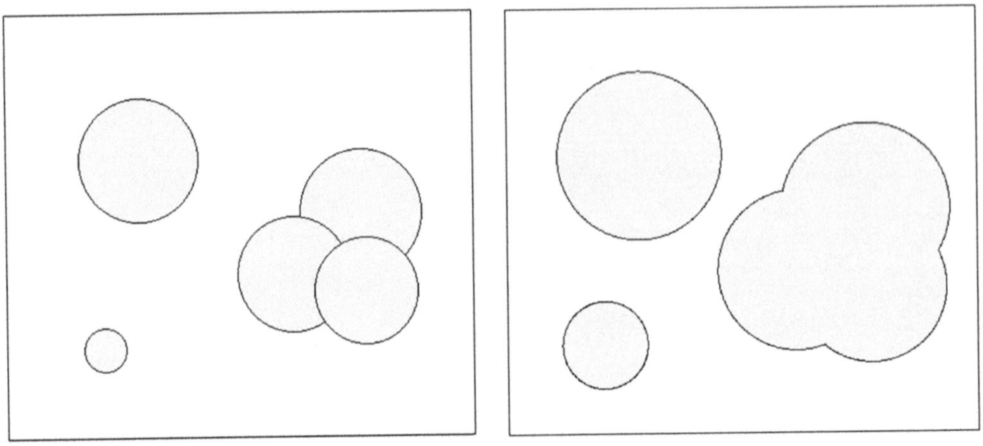

**Figure 6.7** Dissolving the boundaries of buffers.

## Overlay Spatial Data and Create New Datasets

Geographic or spatial data overlay is another fundamental GIS operation. In many cases, spatial data overlay is also called map overlay. However, more recently, we found that spatial data does not have to be presented only by map in the digital or computational world, although mapping is still the most popular method. Recently, we invented local Scene or global Scene in ArcGIS Pro that are not maps by specific definition. Meanwhile, we can use 3D (three-dimensional) voxels to present spatial data as well. Map overlay may not be a valid term in the future, while spatial overlay will continue. Spatial data overlay is an operation that superimposes multiple data sets, which represent different themes, together for the purpose of identifying relationships between them. An overlay creates a composite spatial dataset or a digital map by combining the geometry and attributes of the input datasets on certain rules. The rules of spatial data overlay are established in considering the expected results of output spatial data. Tools are developed in most GIS software following these fundamental rules. Spatial data overlay can be performed in both vector and raster datasets, and it is a multi-map layer process that requires an input of two or more map layers.

Broadly speaking, the spatial data overlay process in GIS is a process of information comprehension. For instance, if we would like to do a GIS project "finding the soils in my home backyard," we overlay the soil distribution digital map layer and land parcel ownership digital map layer to make a new spatial database called "soils in parcels," a new comprehensive database. The operation can combine spatial datasets from various data sources and with various themes or contents. In vector-based GIS, spatial data overlay also follows the topological rules of connectivity, adjacency, and containment in between the two input map layers or feature classes or among various input map layers. Therefore, spatial data overlay is also called topological overlay in vector-based GIS processing.

Operations of spatial data overlay in vector data can be classified either according to the topological relations of feature classes or map layers, or according to the operation rules and results. The ultimate objective or purpose of an overlay operation is either to combine the attributes (characteristics) or to combine geometric shapes and boundaries, or both, for two or more feature classes. Since the fundamental rule of GIS vector databases is that one feature class can only contain one specific type of geometric shapes, namely, point, line, or polygon, GIS analysts again need to differentiate the target input feature class or combination feature classes in this classification. Similar to the spatial query classifications discussed previously, the following are five different types of spatial data overlay operations. Although the names are similar to the spatial queries, the operation focus here is not on extracting, selecting, or sub-setting spatial data. The focus is rather on combining attributes or geometric shapes.

1. Point-in-polygon overlay
2. Line-in-polygon overlay
3. Polygon-on-polygon overlay
4. Point-on-line overlay
5. Line-on-line overlay

In practice, many of these operations are achieved by the geoprocessing operation in ArcGIS Pro that is called "spatial join." Fundamentally, operations of spatial join are a kind of spatial database overlay without combining or changing the geometric boundaries or shapes of the features. In this paired geometry of operations, either feature class ideally can be used as the target to combine data. However, GIS analysts need to consider both geographic cardinal relations of the two feature classes, and the meaning of the results when they conduct the operation. For instance, if we select the point feature class as the target feature class to combine data, if one polygon contains many points, the operation is simple and straightforward. The attributes of the polygon feature would be added to each of the points within the polygon. In the attribute table of the resulting feature class, we will see value repetitions of the attributes originated from polygon feature class. This result may very well be meaningful in real-world practice. For instance, if polygons are school districts that carry the total student population of the district and points are postal or street address points of residential families, the operation assigns the attribute of total student population of the school district to each of the household families. This is very important information for each of the families. By contrast, if polygon feature class is used as the target, the GIS analyst needs to select an operational statistical method in the tool process. The GIS analyst may have to select one from the following: "first value" (means picking the first point encountered only), "last value," mean or average value, sum value, etc. In our example above, if each of the household points carries the total number of people that live in that family, the selection of either "first value" or "last value" does not yield a meaningful result. The selection of either average value or sum value will yield the result that presents different meanings.

According to the operation rules and results, spatial data overlay can be classified into many categories. In most cases, these operation names are directly used as the names of tools in the GIS software environment. For instance, various tools for spatial data overlay geoprocessing are collected in the "Overlay" sub-toolbox (Tool Set) under the "Analysis Tools" toolbox in ArcGIS Pro software.

**Union:** The operation computes a geometric union of the input features. All features and their attributes will be written to the output feature class. All of the input feature classes must be polygons. Point and line features are not allowed as the input feature class. Output feature class is also of polygons combining all the boundaries from all the input polygon feature classes (Figure 6.8).

**Intersect (intersection):** The operation calculates the geometric intersection of any number of feature classes or map layers. The common geographic region of a feature class, or a portion of feature class, that are among all input feature classes will be written to the output feature class

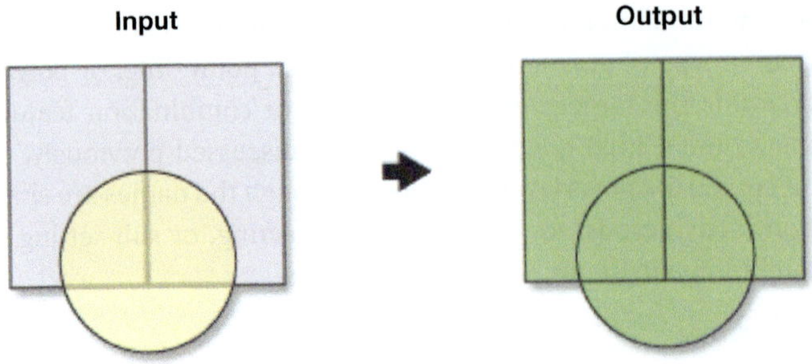


**Figure 6.8** Union operation of map overlay. *Source:* Adapted from Environmental Systems Research Institute, Inc.

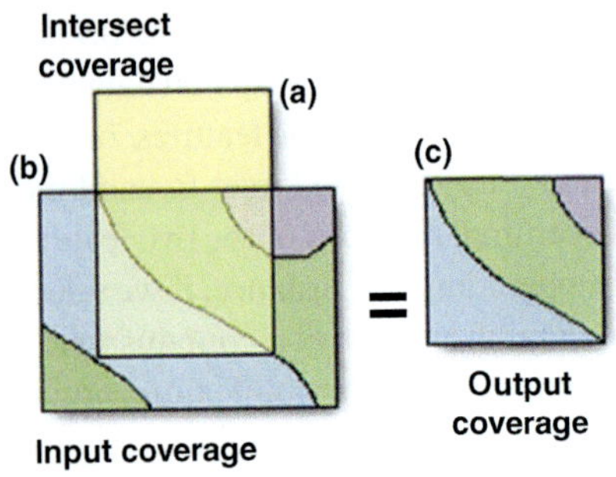


**Figure 6.9** Intersect operation of map overlay (adapted from ESRI, Inc.). (a) Parcel feature class before intersect (b) soils feature class before intersect. (c) Output feature class and attribute table after intersecting.

(Figure 6.9). The meaning of intersect is to combine features and their attributes of all the input datasets in the intersected geographic area (Figure 6.10).

In ArcGIS Pro, the inputs of the intersect operation can be any combination of geometric types, namely point, multipoint, line, or polygon. The output geometric type can be set up as either the same as the inputs, which will return the lowest geometry dimension among the inputs, or point or line feature classes (point = 0-dimension, line = 1-dimension, and polygon = 2-dimension). This means that the feature class among the inputs with the lowest dimension determines the geometric type of the output feature class. The output of a point or line feature class can be selected as the choice of the GIS analyst.

**Identity:** The operation computes a geometric intersection of the input features and identity features. The input features, or portions thereof, that overlapping identity features will obtain the attributes of those identity features. This operation is different to intersect that differentiate input feature classes and identity feature classes on the input side. The output feature class copies the common geographic areas of the input feature classes. The identity feature class or map layer must be the polygon feature class or the same geometric type as the input feature class (Figure 6.11).

**Spatial join:** This operation matches rows of attribute tables from the join layer to the target layer based on a spatial relationship of the objects or features on these feature classes or layers. The output feature class is modified from that of the target feature class. When a match is found during the process, a field (column) is added to the output feature class containing the shape and attributes from the target map layer, and the matching attributes from the join map layer. The spatial relationship is determined by the geometric types of the input layers and the chosen matching rules. Table 6.1 presents the spatial relationships (ESRI 2020).

**Symmetrical difference:** This operation computes a geometric intersection of the input and update feature classes, returning the input features and update features that do not overlap. Features or portions of features in the input and update feature classes that do not overlap will be written to the output feature class. The input and update feature classes or map layers must be in

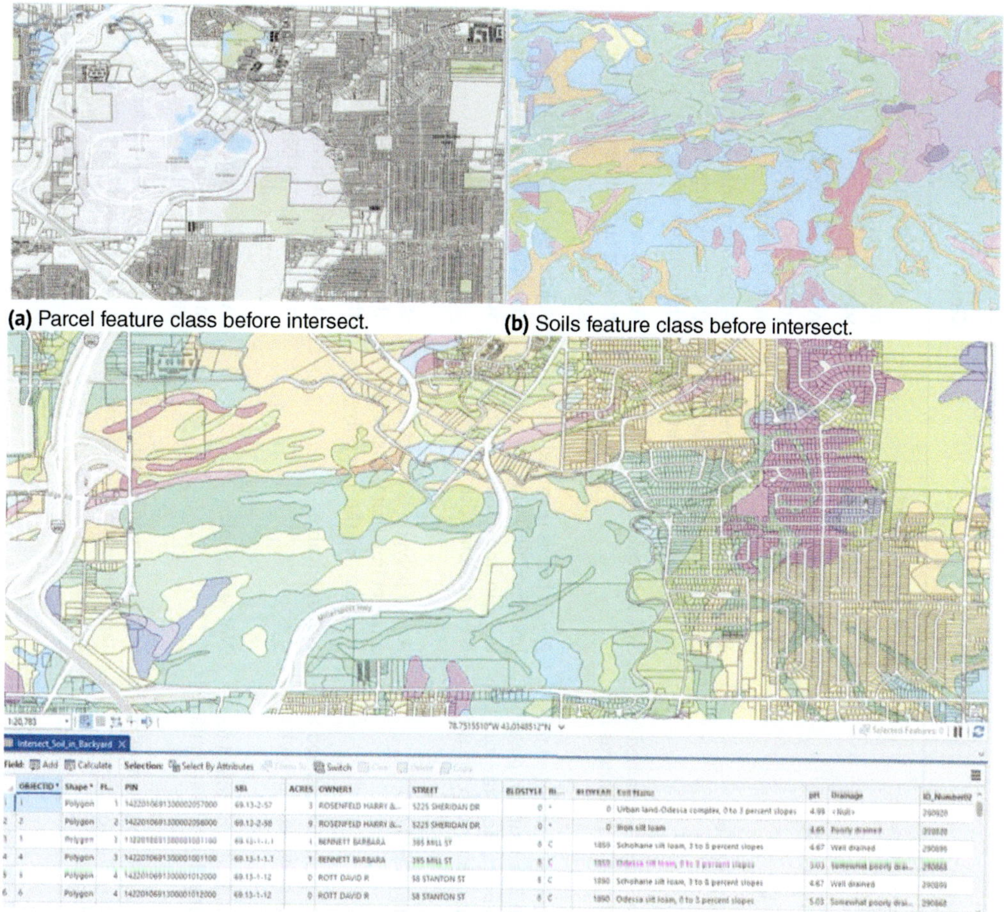

**(a)** Parcel feature class before intersect. **(b)** Soils feature class before intersect.

**(c)** Output feature class and attribute table after intersecting.

**Figure 6.10** Example of map overlay by intersect operation to find the soil types in the backyard of the households in the Town of Amherst, Erie County, NY.

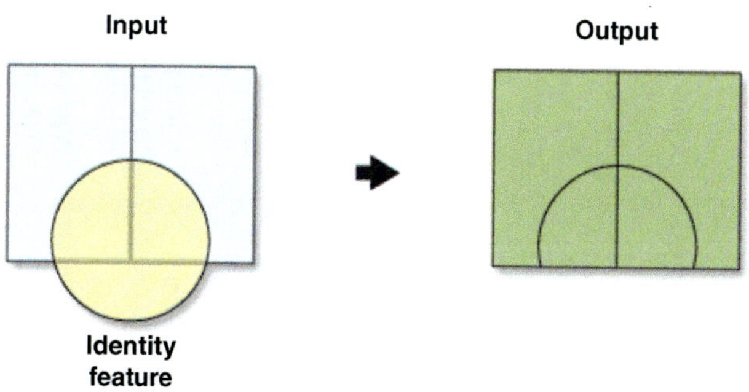


**Figure 6.11** Identity operation of map overlay. *Source:* From Environmental Systems Research Institute, Inc.

**Table 6.1** Classifications of map overlay by spatial relationship of features.

| Target feature | Join feature | Intersect | Contains | Within | Closet |
|---|---|---|---|---|---|
| Point | Point | A join point is matched to a target point at the same location. | Same as intersect | Same as intersect | A join point is matched to the nearest target point within the search radius. |
| Point | Line | A join line is matched to a target point that it intersects. | Not applicable | A join line is matched to a target point contained within the line. | A join line is matched to the nearest target point within the search radius. |
| Point | Polygon | A join polygon is matched to a target point that is inside or on the boundary of the polygon. | Not applicable | A join polygon is matched to a target point that is inside the boundary of the polygon. | A join polygon is matched to the nearest target point within the search radius. |
| Line | Point | A join point is matched to a target line that it intersects. | A join point is matched to a target line that contains it. | Not applicable | A join point is matched to the nearest target line within the search radius. |
| Line | Line | A join line is matched to a target line that it intersects. This includes lines that cross or are coincident. | A join line is matched to a target line that completely contains it. Here the join line is coincident with the target line. | A join line is matched to a target line that falls completely within the join line. Here the target line is coincident with the join line. | A join line is matched to the nearest target line within the search radius. |
| Line | Polygon | A join polygon is matched to a target line that it intersects. | Not applicable | A join polygon is matched to a target line that is inside the boundary of the join polygon. | A join polygon is matched to the nearest target line within the search radius. |
| Polygon | Point | A join point is matched to a target polygon that it intersects or is inside of. | A join point is matched to a target polygon that contains it. | Not applicable | A join point is matched to the nearest target polygon within the search radius. |
| Polygon | Line | A join line is matched to a target polygon that it intersects or is inside of. | A join line is matched to a target polygon that contains it. | Not applicable | A join line is matched to the nearest target polygon within the search radius. |
| Polygon | Polygon | A join polygon is matched to a target polygon that it intersects or is inside of. | A join polygon is matched to a target polygon that contains it. | A join polygon is matched to a target polygon that is inside the boundary of the join polygon. | A join polygon is matched to the nearest target polygon within the search radius. |

*Source:* From Environmental Systems Research Institute, Inc.

the same geometric type, namely, point, line, or polygon (Figure 6.12).

**Erase:** This operation creates a feature class by overlaying the input feature class with the erase feature class. Only those portions of the input features falling outside the erase features are copied to the output feature class. The geometry of output feature class copies that of the input feature class. The "Erase" operation does not combine two feature classes and it only transforms the input feature class (Figure 6.13).

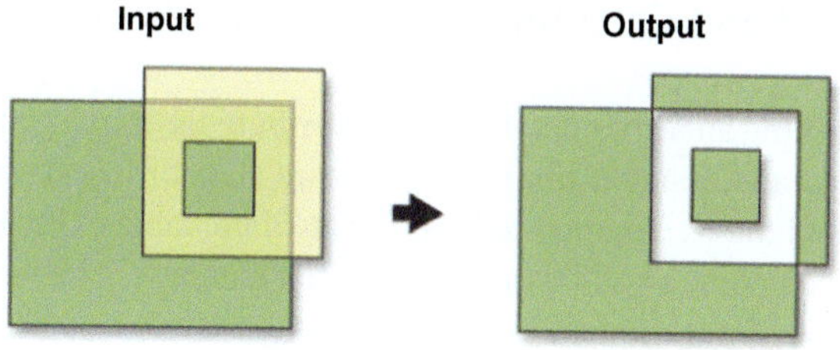


**Figure 6.12** Symmetrical Difference operation of map overlay. *Source:* From Environmental Systems Research Institute, Inc.

**Clip:** This operation cuts out a piece of input dataset or a group of datasets using one feature class in another dataset as a cookie cutter. This is particularly useful for creating a new dataset. The output is also referred to as a study area or area of interest (AOI) that contains a geographic subset of the features from the input dataset. Spatially, the clip operation is one type of dataset overlay. This operation is most often used for data extraction. Therefore, the tool is collected in the "Extract" toolbox in ArcGIS Pro (Figure 6.14).

**Splits:** The operation splits an input feature class with an overlaying feature class to create a subset of a few feature classes as output. The overlaying feature class for splitting the input must be a polygon feature class with one unique attribute value that can be used for splitting. The "Split" operation is also collected in the "Extract" toolbox in ArcGIS Pro (Figure 6.15).

**Figure 6.13** Erase operation of map overlay. *Source:* From Environmental Systems Research Institute, Inc.

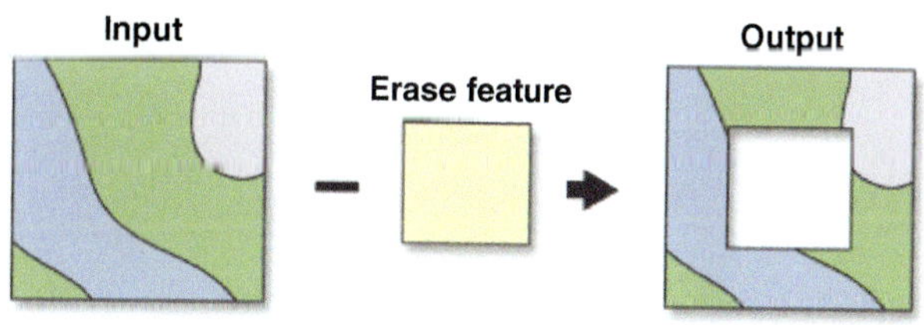


**Figure 6.14** Clip operation of map overlay. *Source:* From Environmental Systems Research Institute, Inc.

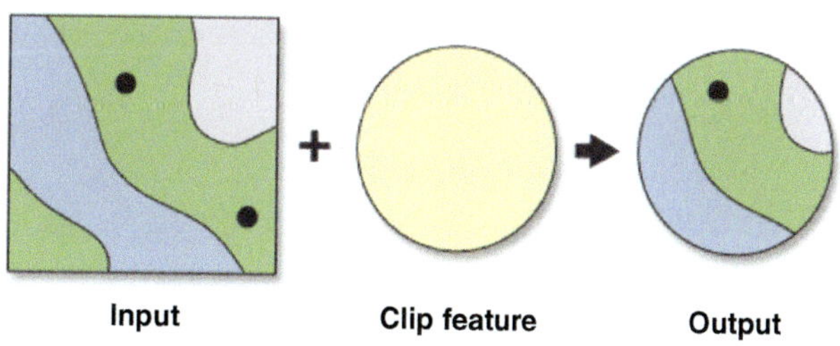


**Figure 6.15** Split operation of map overlay. *Source:* From Environmental Systems Research Institute, Inc.

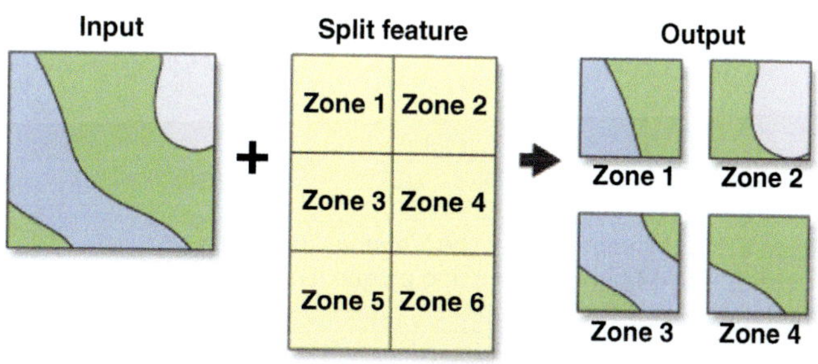

## Add and Calculate Attribute Fields in Spatial Datasets

Joining an attribute table to the attribute table of a geographic feature class allows GIS analysts to add additional attributes to a feature class or map layer based on a unique identifier (unique ID) field that is common to both the attribute tables. In GIS, this is one of the most popular processes to combine or add data in the spatial database. The input table is updated or added to contain the fields from the join table. Optionally, we can also select which fields from the join table will be added to the input table as well (Figure 6.16).

In GIS, the join operation follows rules of relational database, and the joins are temporary. If we would like to update the feature class with the new table contents permanently, we need to export the feature class or map layer while the join of the tables. Joining the contents of a table to another table based on a common attribute field is a popular practice in database management systems. Another join attribute operation in GIS is the spatial join. Spatial join operations can be conducted in map or spatial data overlays as illustrated above. It can also be achieved by adding spatial join tools in the data management toolbox in ArcGIS Pro.

During the geoprocessing of adding a spatial join, the attributes of the joined feature class will be added into the attribute table of the target feature class. In this process, merging rules allow GIS analysts to specify how values from two or more input fields are merged or combined into a single output value in the attribute table.

There are several merge rules that we can select to determine how the output field will be populated with values.

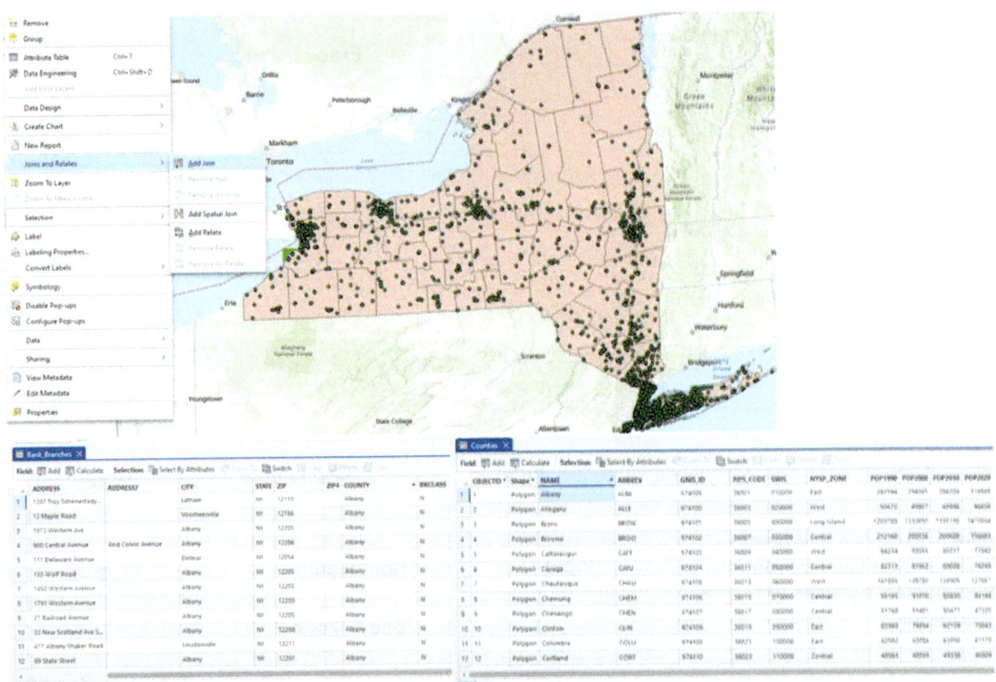


**Figure 6.16** Example of join the "Counties" polygon attributes with population information to the "Bank_ Branches" point feature class using unique ID of county names in GIS to evaluate the bank service to the population across the NY State.

1. First—use the input fields' first value.
2. Last—use the input fields' last value.
3. Join—concatenate (join) the input field values.
4. Sum—calculate the total of the input field values.
5. Mean—calculate the mean (average) of the input field values.
6. Median—calculate the median (middle) of the input field values.
7. Mode—use the value with the highest frequency.
8. Min—use the minimum value of all the input field values.
9. Max—use the maximum value of all the input field values.
10. Standard deviation—use the standard deviation classification method on all the input field values.
11. Count—find the number of records included in the calculation.

In the GIS environment, we can use field calculations to create an expression to perform simple or complex calculations on existing or newly created field values in a table. The GIS analysts may want to perform a mathematical calculation on a set of field values for a single record or for all records in the attribute table. Field calculations work on string, number, and date fields and automatically calculate selected records if the feature class, map layer, or table has a selection set presently.

In practice, we use the "Field Calculator" or "Calculate Field" to accomplish the calculations. When a table view is open in an ArcGIS Pro project, the "Calculate Field" icon or the contact menu by right-clicking on the field (column) title will open the Calculate Field geoprocessing tool, so you will always build your expressions as part of a geoprocessing task. Specify the parameters to build an expression using the provided Fields list and functions. Field calculations of attributes of geographic feature classes in GIS create new data and add the capabilities of analysis.

## Summarize, Calculate Statistics, and Aggregate Attributes or Features for Spatial Datasets

Spatial data of each of the feature classes can be summarized or aggregated to form new information and help with decision making. These processes can be very simple, such as summarizing a particular attribute for a feature class, conducting simple feature class statistics, or plotting a chart from the spatial objects collected in a feature class for visualization of attribute data. Meanwhile, many new aggregate functions and geoprocessing have been created to handle large datasets, or the "big data." Here, we would like to discuss some simple functions first. Then, we will briefly illustrate some of the advanced summarizing functions for large datasets.

The summarize statistic function in the GIS software environment creates a summary table grouped by the values in the selected attribute field or fields. This process produces simple statistical results of the selected fields (columns), such as sum, mean, median, standard deviation, maximum, minimum, range, etc. The function of "creating field statistics" charts and groups values in the attribute table of features to visualize distributions; groups' data and time values into time bins to visualize change over time; or groups text string categories into bar graphs to compare data counts.

Recently, many summarizing data functions were developed to compute large quantities of spatial data. These functions contain tools that calculate total counts, lengths, areas, and basic descriptive statistics of features and their attributes within areas or near other features. We discuss a few of these approaches here in order to show the significant value of summarizing spatial datasets. All these functions and geoprocessing exceed the range of basic vector GIS analysis.

1. **Aggregate points:** This operation aggregates points into polygon features or bins. A polygon is returned with a count of points as well as optional statistics at all locations where points exist.
2. **Describe dataset:** This operation summarizes features into calculated field statistics, sample features, and extent boundaries.
3. **Join features:** Joins attributes from one layer to another based on spatial, temporal, or attribute relationships, or a combination of those relationships.
4. **Summarize attributes:** This function calculates summary statistics for fields in a feature class.
5. **Summarize within:** This operation overlays a polygon feature class or map layer with another map layer to summarize the number of points, length of the lines, or area of the polygons within each polygon and calculates attribute field statistics about those features within the polygons.

## Model Relationships of Spatial Objects and Features

The term "spatial relationships" refers to the way objects or features are arranged in relation to one another in the geographic space presented in either a 2D or 3D environment. For example, we can describe them as adjacency, contiguity, overlap, proximity, etc. When we run operations on the datasets or geoprocessing using a tool like spatial join, we are essentially using their spatial relationship with other features in geographic space. All the established concepts of topology are, in part, the explanations of spatial relationships.

In addition to analyzing spatial patterns, GIS analysis can be used to examine or quantify relationships among features and feature classes. Many modeling spatial relationship tools and functions in GIS have been developed to construct spatial weights, matrices or model spatial relationships using various analysis techniques, including regression, forest-based approaches, maximum entropy methods, etc. Some of the modeling operations are generally basic in the GIS environment. Others are advanced analytical functions that involve separate spatial analytical theories, such as spatial autocorrelation, network analysis, and geographically weighted regression (GWR), which we will discuss later in following chapters.

Currently, a total of twelve spatial relationship modeling approaches have been developed (ESRI, 2020). These are: (i) colocation analysis, (ii) generalized linear regression (GLR), (iii) ordinary least square (OLS) modeling, (iv) exploratory regression, (v) forest-based classification and regression, (vi) generate network spatial weighted modeling, (vii) generate spatial weights matrix, (viii) GWR, (ix) local bivariate relationship analysis, (x) multiscale geographically weighted regression (MGWR), (xi) presence-only prediction modeling, and (xii) spatial association by zones. Among these modeling approaches, the first three are fundamental methods that we are going to discuss in this chapter. Other approaches will be discussed in later chapters.

### Colocation Analysis

This modeling process measures local patterns of spatial association, or co-location, between two types of point features or two feature classes of points using the colocation quotient statistics.

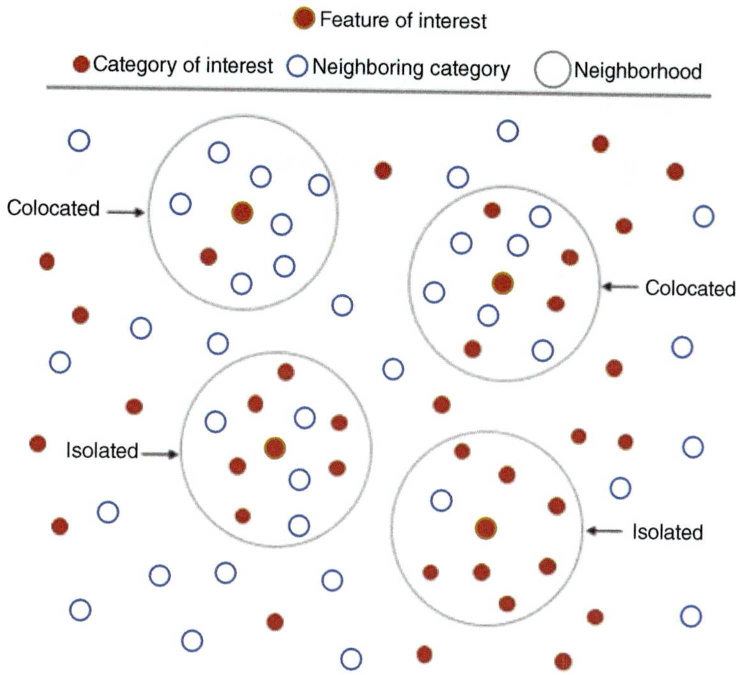


**Figure 6.17** Colocation analysis in GIS. *Source:* From Environmental Systems Research Institute, Inc.

Only point feature classes are accepted as input in this process. GIS Analysts specify the Input Features of Interest either on the same feature class or on a separate feature class with Neighboring Categories of points. The output feature class contains all the Input Features of Interest with fields containing the results of local colocation quotient, symbology bin, and p-values (Figure 6.17).

## Generalized Linear Regression

This operation performs GLR to generate predictions or to model a dependent variable in terms of its relationship to a set of explanatory variables of a feature class or map layer (Figure 6.18). This function or tool can be used to fit continuous (OLS), binary (logistic), and count (Poisson) models. The major difference of this approach to conventional statistical linear regression is that the input in the GIS environment is a feature class or a map layer. The input feature class contains the dependent and independent variables in its attribute table. The output new feature class or map layer will contain the dependent variable estimates and residuals for

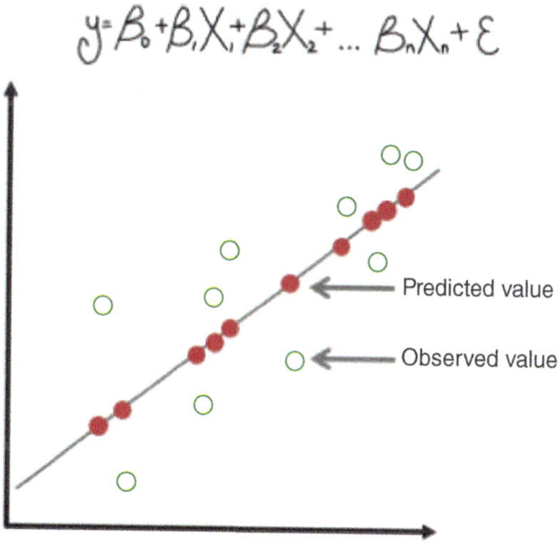


**Figure 6.18** Model of Generalized Linear Regression (GLR). *Source:* From Environmental Systems Research Institute, Inc.

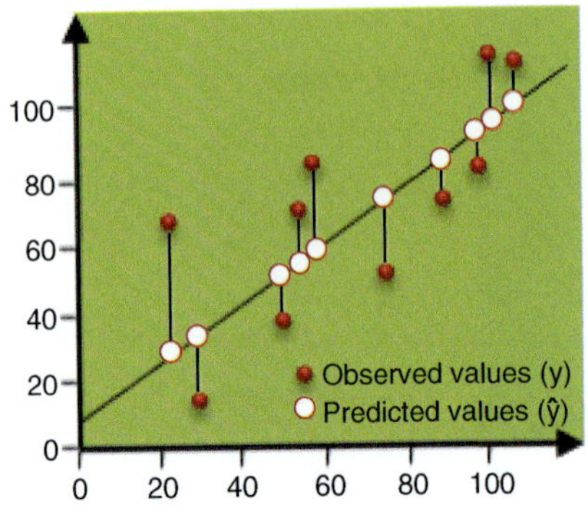


**Figure 6.19** Model of global ordinary least square (OLS) linear regression. *Source:* From Environmental Systems Research Institute, Inc.

each of the feature locations. In this process, the GIS Analyst can select one or more independent explanatory variables in the regression model.

### Ordinary Least Square Modeling

This operation conducts global Ordinary Least Square (OLS) linear regression to generate predictions or a model for a dependent variable in terms of its relationships to a set of explanatory variables (Figure 6.19). The input is a feature class that contains the dependent and independent variables in the attribute table for analysis. The output feature class will copy the geographic locations and feature geometry. The output will receive dependent variable estimates and residuals in the attribute table.

GIS analysts can also build the geoprocessing models to streamline the prediction and analysis of spatial relationships of objects and features in their projects (Figure 6.16). In the GIS environment, models are workflows that string together sequences of geoprocessing tools, feeding the output of one tool into another tool as input (Jensen and Jensen, 2013). ModelBuilder is used to create, edit, and manage geoprocessing models that automate individual operations or tools. ModelBuilder can also be considered as a visual programming language for building workflows in GIS to solve the problems.

A model is represented as a diagram, flowchart, or workflow that chains together sequences of processes and geoprocessing tools, using the output of one process as the input to another process. It provides advanced methods for extending GIS functionality by allowing GIS analysts to create, save, and share the models as methods to solve problems and conduct projects (Figure 6.20). Figure 6.16 is a diagram of a model created in the model builder window in ArcGIS Pro.

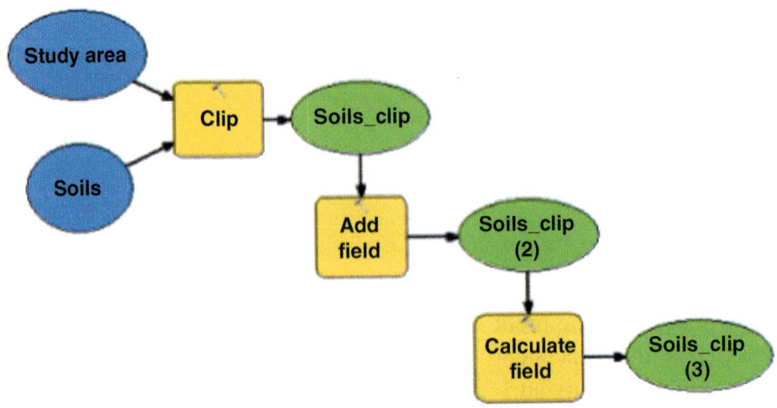


**Figure 6.20** Example of a geoprocessing model in ArcGIS Pro Model Builder

## Discover Geographic Patterns of Spatial Objects and Features

Identifying geographic patterns is important for understanding how geographic phenomena or objects behave. Although we can get a sense of the overall pattern of features or objects and their associated values by mapping them, quantifying and analyzing distribution patterns of spatial objects is our ultimate goal in GIS. This process makes it easier to compare patterns for different distributions or different distributions during different time periods. Often the functions for analyzing patterns are considered as more in-depth analyses. In practice, GIS Analysts often ask these questions when they conduct projects: "Are the objects or features in the dataset, or the values associated with the features in the dataset, spatially clustered?"; "How big are the areas of the clusters?"; and "Is the clustering becoming more or less intense over time?" (ESRI, 2020).

The following is a list of major functions that have been developed and are available in GIS for basic spatial pattern analysis: (i) calculating feature density, (ii) calculating average nearest neighbor, (iii) finding hot spots among features, (iv) finding point clusters, (v) feature class or map layer-based regressions, and (vi) spatial autocorrelation. Among these six different geoprocessing or analyses, the first four of them are basic approaches that we are going to discuss briefly here. The last two approaches will be discussed in later chapters separately.

### Calculating Feature Density

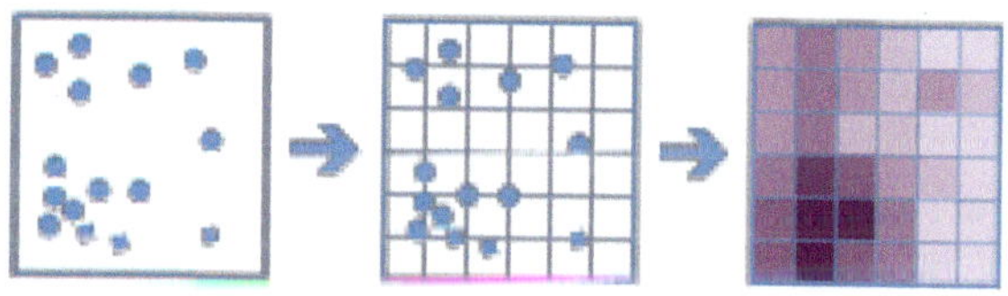

Figure 6.21 Method of calculating feature density in GIS. *Source:* adapted from ESRI, Inc.

This geoprocessing calculates a magnitude-per-unit area from the input of either a point feature class or a line feature class that fall within the study region by each of standard grid cells (Figure 6.21). The standard output of this analysis is a raster grid cell dataset that can be converted back to a vector feature class. Two types of density analysis can be further identified by weighting options to calculate density: uniform density or density and kernel density. The uniform option sums all the values within the neighborhood and divides them by the area. The kernel option weights values in the neighborhood by distance from the feature of interest and applies a kernel function to fit a smooth tapered surface to each of the points.

### Calculating Average Nearest Neighbor

This analysis calculates a nearest neighbor index based on the average distance from each feature to its nearest neighboring feature in a feature class (map layer). This is a basic method to quantify the distribution of geographic objects or features in GIS. This process does not result in a new feature class or map layer. It returns five values in the feature attribute table: observed mean distance, expected mean distance, nearest neighbor index, z-score, and p-value (Behrens et al., 2018). The z-score and p-value results are measures of statistical significance that inform us the confidence of predictions of Nearest Neighbor Index and help the GIS analysts to make a decision of whether or not to reject the null hypothesis (Figure 6.22).

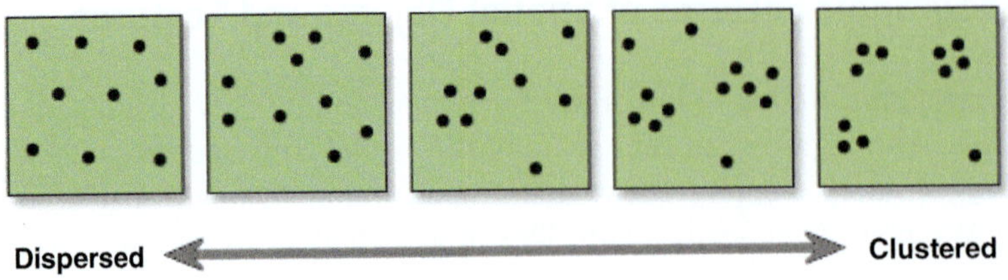


**Figure 6.22** Nearest Neighbor Index calculations for feature distributions. *Source:* From Environmental Systems Research Institute, Inc.

### Finding Hot Spots among Features

This analysis uses a set of features or a feature class to identify statistically significant hot spots and cold spots using the Getis-Ord Gi* statistic model (Figure 6.23). The function calculates the Getis-Ord Gi* statistic (pronounced G-i-star) for each of the features in a dataset or feature class. Z-scores and p-values in the resultant tell us where features with either high or low values cluster spatially. This process works by looking at each feature within the context of neighboring features. A feature with a high value alone may not be a statistically significant hot spot. The quantification of a hot spot with statistically significant features will have a high value and be surrounded by other features with high values as well (Figure 6.24). The local sum for a feature and its neighbors is compared proportionally to the sum of all features. When the local sum is very different from the expected local sum, and when that difference is too large to be the result of random chance, the process would generate a statistically significant z-scores as references.

The Getis-Ord local statistic is given as:

$$G_i^* = \frac{\sum_{j=1}^{n} w_{i,j} x_j - \bar{X} \sum_{j=1}^{n} w_{i,j}}{S \sqrt{\frac{\left[ n \sum_{j=1}^{n} w_{i,j}^2 - \left( \sum_{j=1}^{n} w_{i,j} \right)^2 \right]}{n-1}}} \qquad (1)$$

where $x_j$ is the attribute value for feature $j$, $w_{i,j}$ is the spatial weight between feature $i$ and $j$, $n$ is equal to the total number of features and:

$$\bar{X} = \frac{\sum_{j=1}^{n} x_j}{n} \qquad (2)$$

$$S = \sqrt{\frac{\sum_{j=1}^{n} x_j^2}{n} - (\bar{X})^2} \qquad (3)$$

The $G_i^*$ statistic is a $z$-score so no further calculations are required.

**Figure 6.23** The Getis-Ord $G_i^*$statistic model.

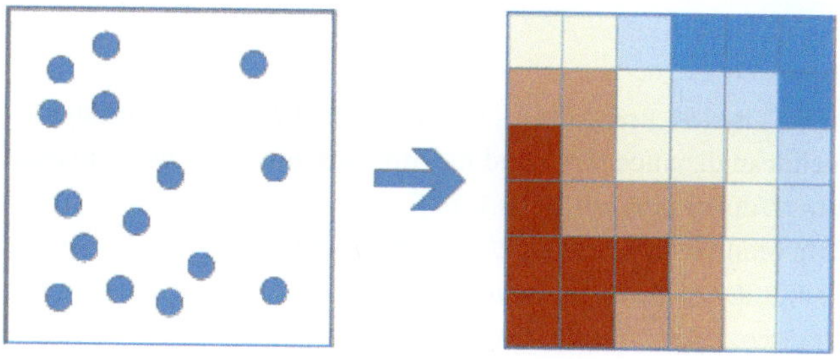

**Figure 6.24** Identifying the hotspot areas by analyzing the feature distribution patterns. *Source:* From Environmental Systems Research Institute, Inc.

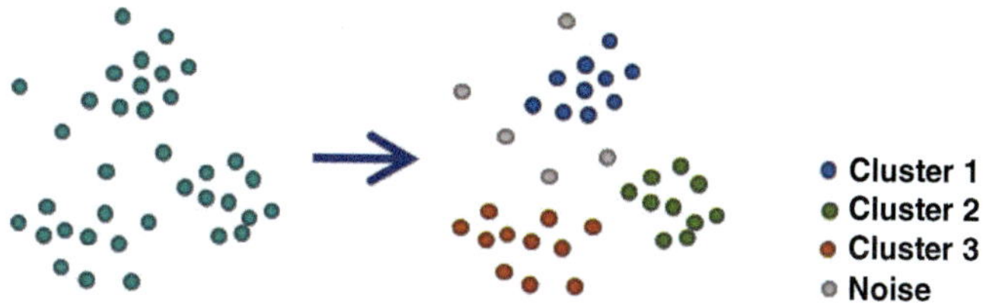


**Figure 6.25** Feature distribution density-based cluster analysis. *Source:* From Environmental Systems Research Institute, Inc.

In summary, this analysis identifies statistically significant spatial clusters of many features (hot spots) and few features (cold spots).

## Finding Point Clusters

This analysis identifies clusters of point features in surrounding noisy environment based on their spatial distribution or both spatial and temporal distributions. In essence, this is a process based on density of features (Figure 6.25). The density-based clustering methods can be further identified as one of three options:

1. Defined distance (DBSCAN)—utilizes a specified distance to separate dense clusters from sparser noise. This method is the fastest in computation. However, it is only appropriate when we know a very clear search distance to use, and this distance needs to work for all potential clusters. This method also allows the users to use the time criterion, which we can find clusters of points both in space and/or by time periods.
2. Self-adjusting (HDBSCAN)—treats the distances for separating clusters as a variable with a range of values from sparser noise. This is a clustering method which is data-driven and applies basic AI (artificial intelligence) functions.
3. Multi-scale (OPTICS)—uses the distance between neighboring features to create a reachability plot, which is then used to separate clusters of varying densities from noise. This method also allows the users to use the time criterium so that we can find clusters of points both in space and/or by time periods.

## References

Behrens, T., Schmidt, K., Viscarra Rossel, R.A., Gries, P., Scholten, T., MacMillan. R.A. (2018). Spatial modelling with Euclidean distance fields and machine learning. *European Journal of Soil Science* 69 (5): 757–770, https://doi.org/10.1111/ejss.12687.

ESRI, Inc., (2020). Help Manual Online. https://pro.arcgis.com/

Jensen, J.R. and Jensen, R.R. (2013). *Introductory Geographic Information Systems*, p. 400. Boston, New York, San Francisco, Upper Saddle River, NJ: Pearson Publisher.

Tulane University. (2022). Geographic Information Systems Library Guide. https://libguides.tulane.edu/geographicinformationsystems.

# Tutorial in ArcGIS Pro Environment

## Erie County Flood Prevention and Flood Zone Insurance Payment Calculation Project

In this project, we are going to explore the conventional analytical methods in vector datasets.

Copy the tutorial data folder "Chapter06-Tutorial" to your working directory.

Open ArcGIS Pro and create a new ArcGIS Pro project file called "MyGIS_Larning06" using "Map" template. Save the project file into the designated working folder, C:\ or D:\......Chapter06-Tutorial.

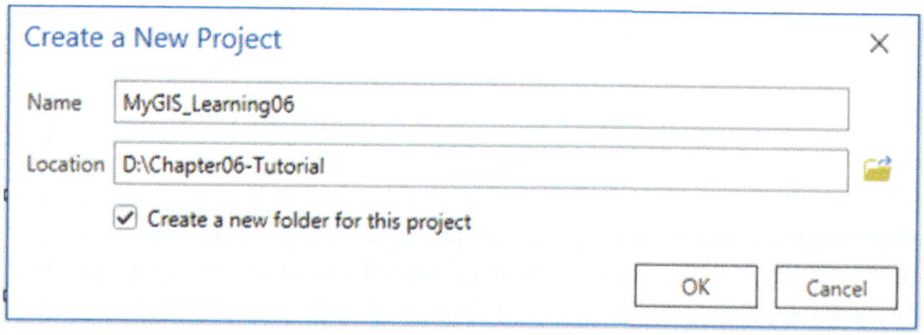


Now, add the entire folder of "Chapter06-Tutorial" on the Catalog Pane in your project window.

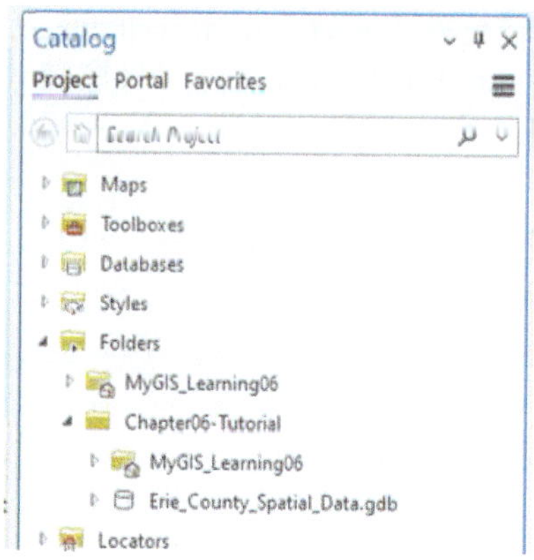


You may notice the geodatabase, Erie_County_Spatial_Data.gdb is in this folder. This is our collected spatial database for this project.

First, change the "Map" title to "Map_Overlay_Buffering" on the Catalog Pane. You may notice that the title of the mapping window was also changed. Then, we can add the feature class "Watersheds_Erie_Niagara" into the Map_Overlay_Buffering window. This map layer is a clip from the national drainage basin and hydrography database that was downloaded from the US Environmental Protection Agency (US-EPA). The national hydrography digital database was developed by both US-EPA and US Geological Survey. Opening the attribute table, we have two nationally recognized drainage basins in this case: the Niagara River Drainage Basin and the Buffalo-Eighteen Mile Rivers Drainage Basin. Now, we can close the attribute table.

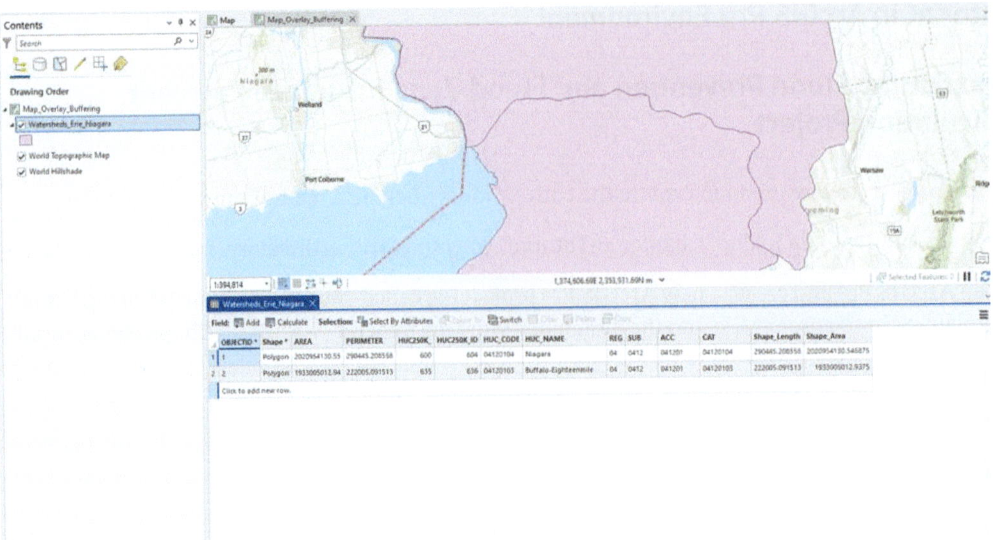


In the flooding hazard prevention project, Erie County government would like to educate its residents to help them understand which drainage basin or watershed that their houses, properties, and lands stand in. This is so people may be fully aware if there is a flood event, which river system will flood their land parcels and properties. The GIS Analyst is asked to overlay the watersheds feature class to the parcel land ownership feature class to assign the watershed attributes to each of the parcels.

Now, let's add the "Parcels" feature class in the Map_Overlay_Buffering window as well. Since the map layer of the two drainage basins does not cover the entire Erie County, we would like to exclude those parcels or land areas that are outside of these two drainage basins first. We would like to use the "Clip" tool to do this.

On the Analysis menu, click on the toolbox icon to open the Geoprocessing Pane. Activate the Toolboxes tab and expand the Analysis Tools box, then Extract tool set to activate the Clip tool.

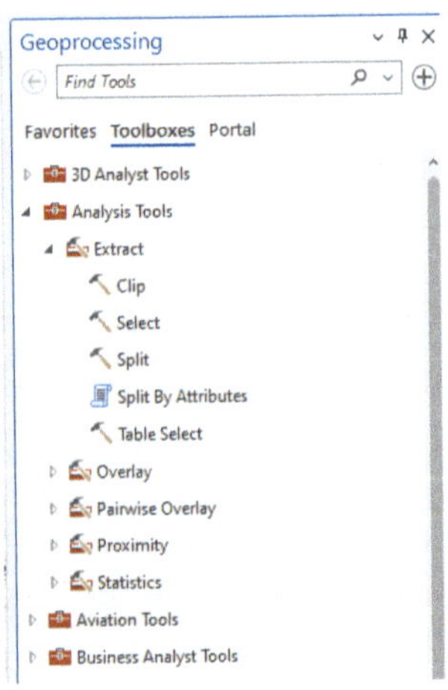

On the "Clip" tool dialog, fill in Parcels for the "Input features or Dataset." For the "Clip Features," fill in the Watersheds_Erie_Niagara feature class. For the "Output Features or Dataset," click on the folder icon and navigate to the geodatabase, Erie_County_Spatial_Data.gdb in your Chapter06-Tutorial folder. Name the new feature class as "Parcels_Clip."

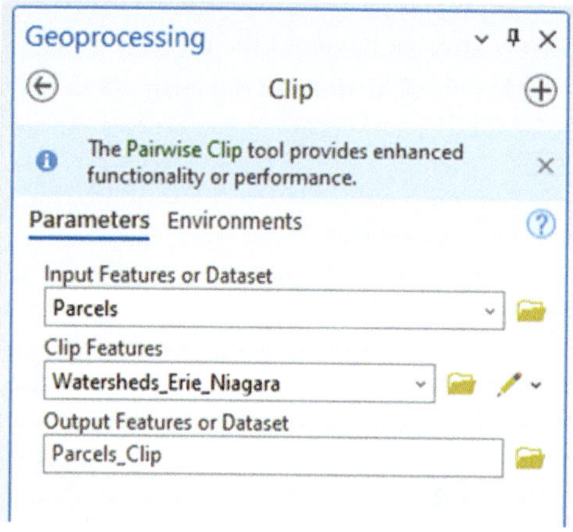


Click "Run" to process the tool. You may notice that the new map layer is added into the mapping window. The new feature class is clipped exactly on the boundaries of the watersheds.

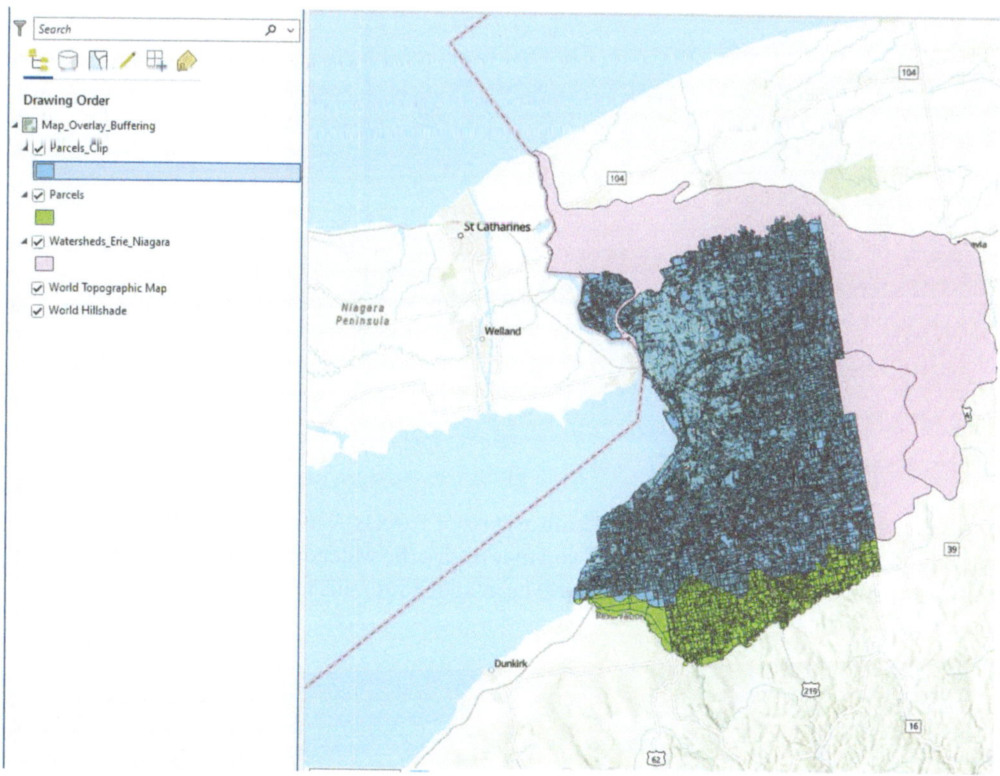

Close the Geoprocessing Pane for now. Let's turn off the original Parcels feature class. The next task is to assign the watershed name (HUC_Name) to each of the thousands of parcels. Thereafter, the owners of the parcel of land can check which drainage basin or watershed their property is located in.

Open the Geoprocessing Pane again from the Analysis menu by clicking on the toolbox icon .

While the Toolboxes tab is activated, collapse the Extract toolset and expand the Overlay toolset. In this case, we are going to use the "Identity" tool to process the data and add the attribute fields. Open Identity tool and fill in the following.

| | |
|---|---|
| Input Features: | **Parcels_Clip** |
| Identity Features: | **Watersheds_Erie_Niagara** |
| Output Feature Class: | **Parcels_Watershed** |

Navigate to save the Output Feature Class to your Erie_County_Spatial_Data.gdb. Keep the Attributes to Join as the default "All attributes."

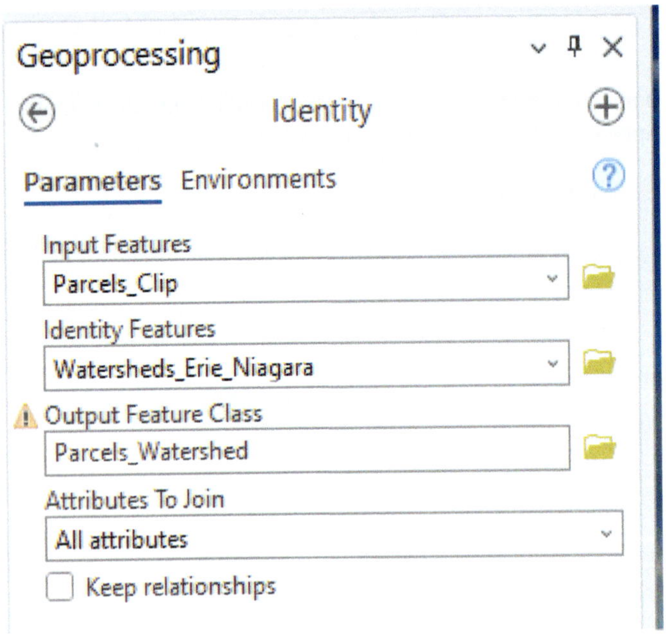


Process the tool. You may notice the new feature class of Parcels_Watershed is added in the mapping window. Open the attribute table, and you may notice the watershed name (HUC_Name) and other fields from the identity feature class have been added into this table.

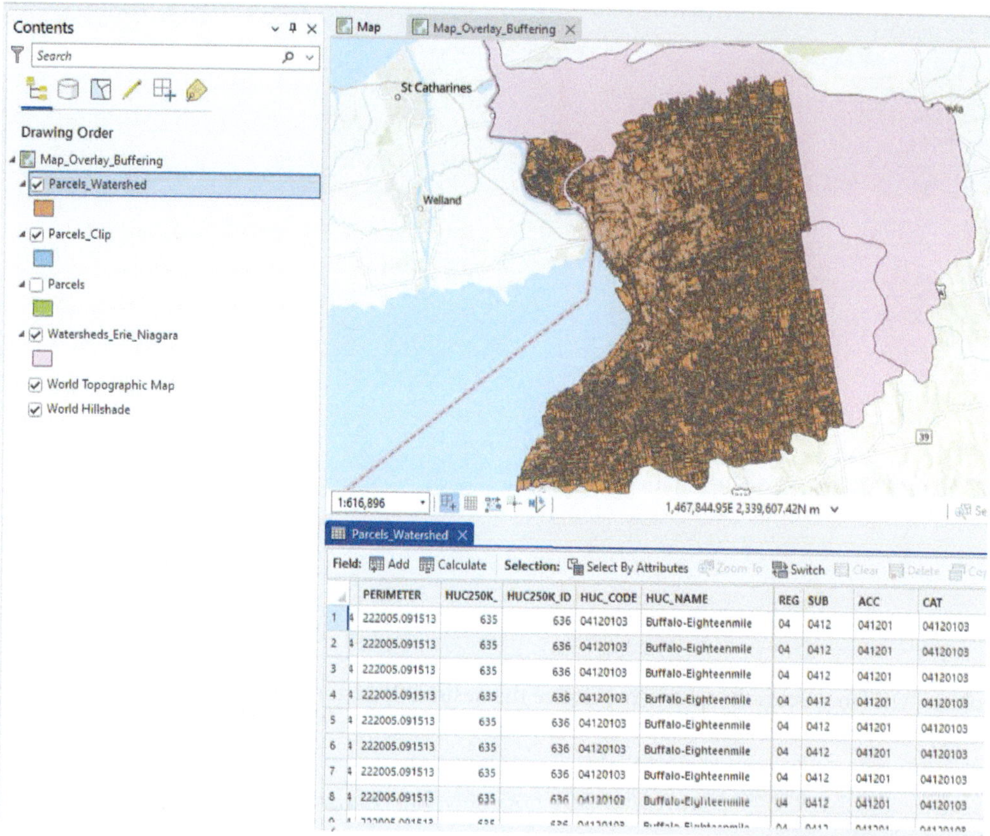


Now, let's check off all the local map layers except the base map layers. Our next task is to select the AE and AH flood zones on the FEMA (the Federal Emergency Management Agency) flood zone map in Erie County, NY. FEMA is the US federal government agency to manage and mitigate all kinds of natural or human made disasters in the territory of the United States of America. Then, we will overlay the newly created AE and AH flood zone feature class on the parcel feature class in Erie County to obtain the residents who live in the flood zones for flooding prevention and insurance purposes. According to FEMA published online documents, Special Flood Hazard Areas (SFHA - Zone) are identified on the Flood Insurance Rate Map. The 1-percent (1%) annual chance flood is also referred to as the base flood or 100-year flood. The areas of this flooding frequency are designated Zone A. The Zone A can be further classified as variety of types using the added letter label. Zone AE is the base floodplain where base flood elevations are provided. That means as soon as the river channel cannot hold all the water flow, the Zone AE will be flooded. Zone AH is the areas with a 1% annual chance of shallow flooding, usually in the form of a pond, with an average depth ranging from 1 to 3 feet. Both of AE and AH zones have a 26% chance of flooding over the life of a 30-year house mortgage.

Add the "FEMA_Flood_Zones" feature class into the Map_Overlay_Buffering window. Activate the Map ribbon on the main menu. Click on the Select by Attributes icon (Select By Attributes) in the Selection group. The Selection by Attributes dialog opens. Make sure the "Input Rows" is FEMA_Flood_Zones and keep the "Selection Type" as the default of "New selection." Formulate this query: "Zone = (is equal to) AE or Zone = (is equal to) AH."

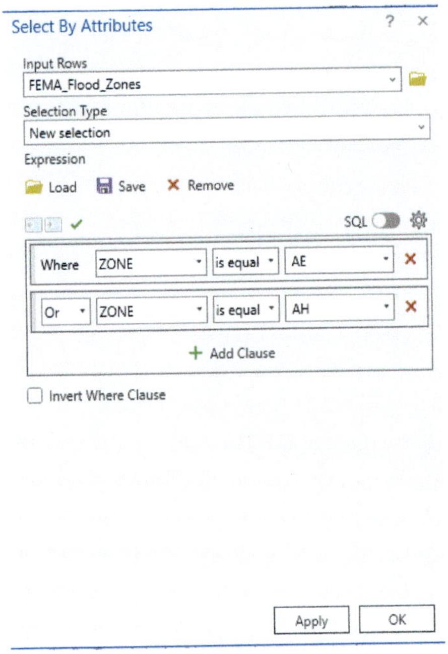


Click on "OK" to process the query. You notice the desired flood zones are selected.

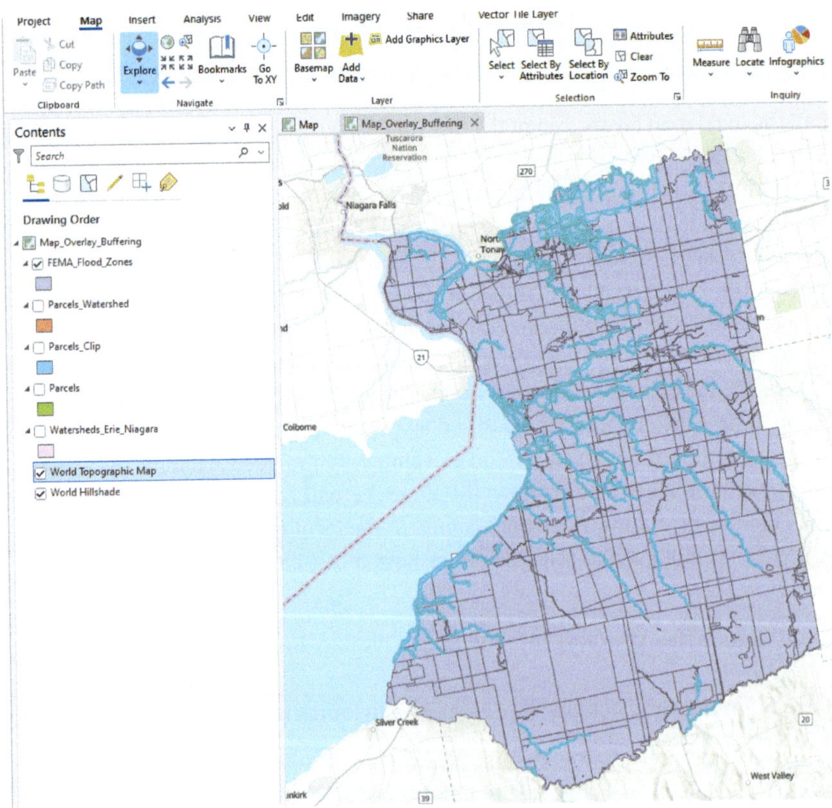

Now, we export the selected data to a new feature class. Highlight and right-click on title of FEMA_Flood_Zones map layer in the Table of Contents to open the contact menu. Click on the "Data" to get the second contact menu. Then, click on "Export Features."

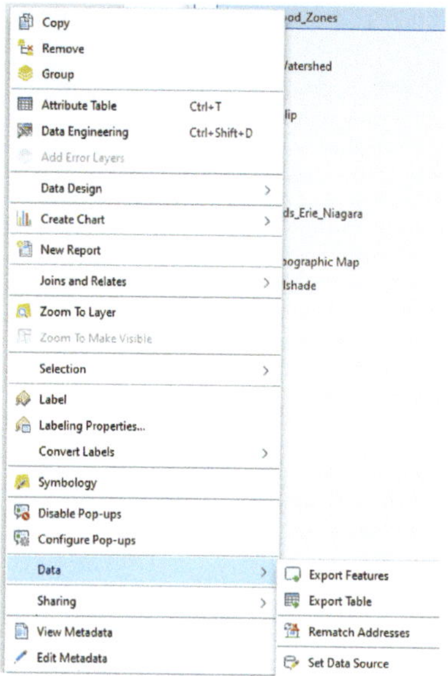


The dialog opens. For the "Output Feature Class" navigate to your Erie_County_Spatial_Data.gdb using the folder icon . Rename the new feature class as "Flood_Zone_AE_AH." Click "Ok" to process the export. You may notice the new feature class is added to the mapping window.

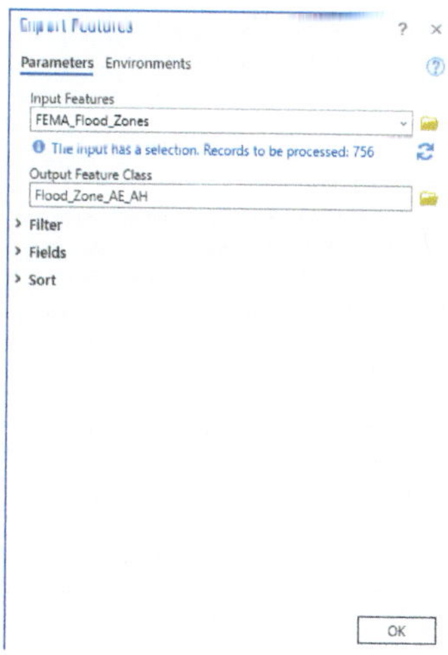


In the ArcGIS Pro environment, the export process is used for both converting ArcGIS file format to a different format and converting from one feature class to another. In order to make the flood hazard zone coverage exclusively for protecting people's properties, we will need to extend the flooding zones 200 feet from the edges and boundaries. Therefore, we will make a 200 feet buffer around the "Flood_Zone_AE_AH" polygon feature class.

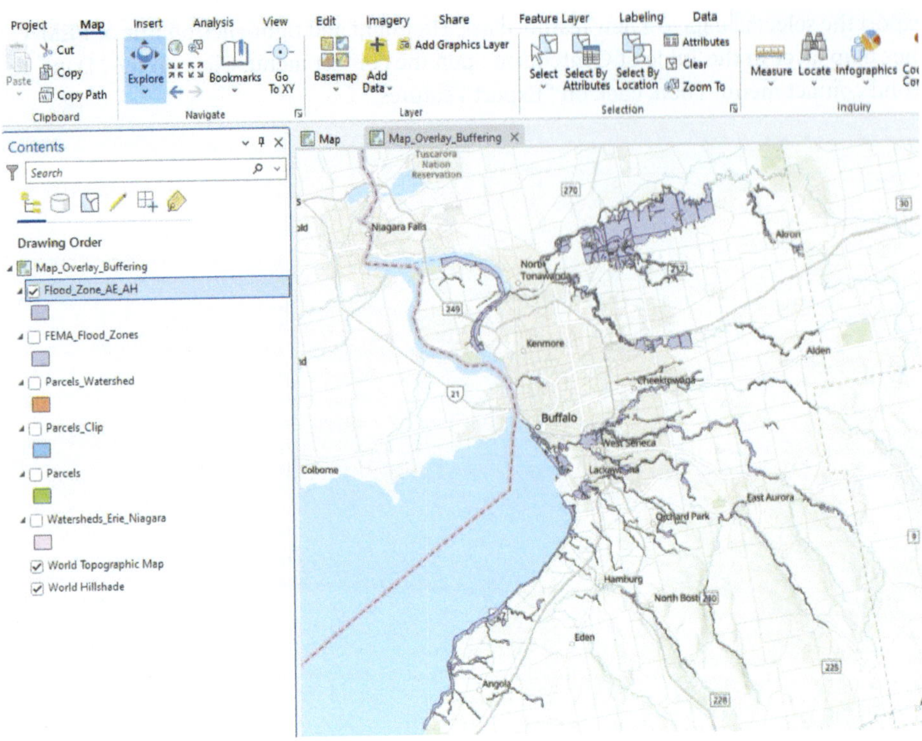


On the Analysis ribbon of the main menu, click the Tools icon to open the Geoprocessing Pane. Activate the Toolboxes tab and expand the Analysis Tools box, then Proximity toolset.

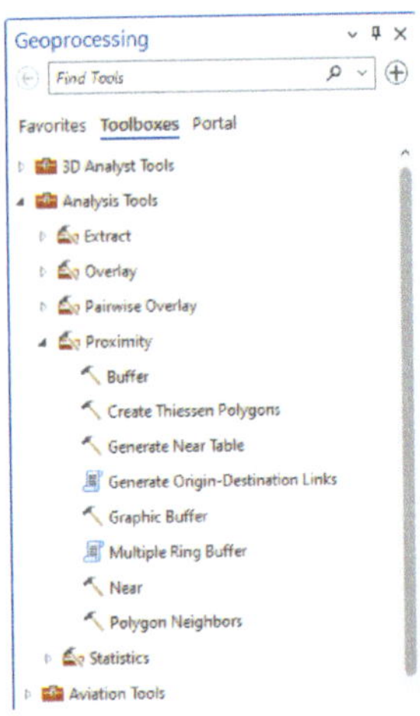


Click on Buffer tool to open the dialog. For "Input Features," select Flood_Zone_AE_AH. For "Output Feature Class," navigate to your Erie_County_Spatial_Data.gdb and name the new feature class to "Flood_Zone_AE_AH_Buffer_200." For the "Distance (Value or Field)." Keep the default of "Linear Unit." Select the unit as "US Survey Feet" and type in 200. Keep the default for other parameters.

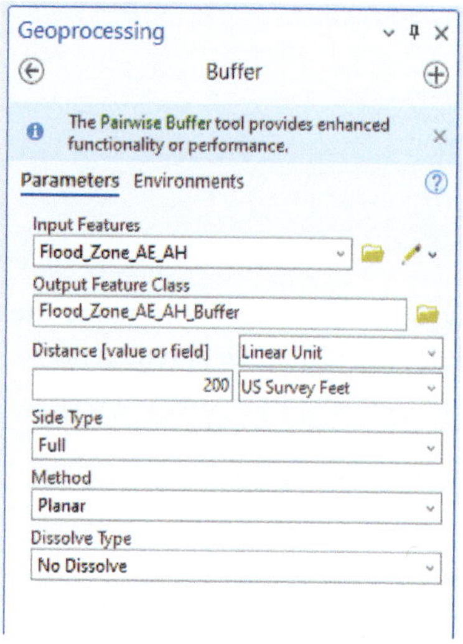


Process the tool to create buffers of 200 feet from the flood zone polygons.

You may notice that the feature class of Flood_Zone_AE_AH_Buffer_200 is added to the mapping window. We can close the Geoprocessing Pane now and inspect the differences of map layers of Flood_Zone_AE_AH and Flood_Zone_AE_AH_Buffer_200.

Now, we would like to conduct a map overlay of intersect to combine parcel land ownership data with the flood hazard zones. Go to the Analysis ribbon on the main menu and select Tools icon to open the Geoprocessing Pane. Activate the Toolboxes tab and expand Analysis Tools box, then the Overlay toolset. Highlight and click on the Intersect tool to open its dialog.

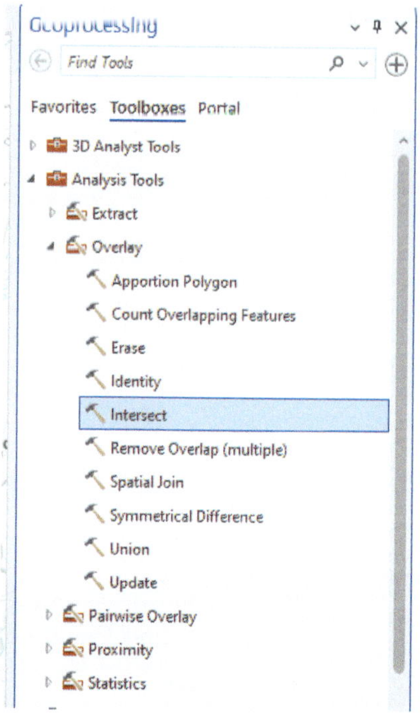

For the "Input Features," fill in the following.

| Input Features | Ranks |
|---|---|
| Parcels | Keep blank |
| Flood_Zone_AE_AH_Buffer_200 | Keep blank |

For the "Output Feature Class," navigate to your Erie_County_Spatial_Data.gdb and name the new feature class as "Parcels_in_Flood_Zone." Keep the default for the rest of the parameters.

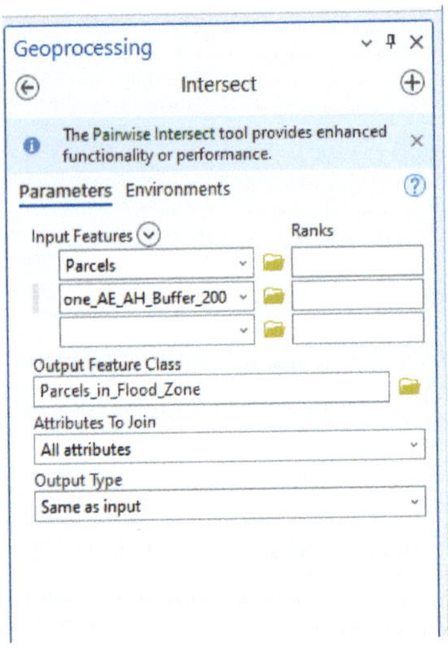


Process the tool to produce the Parcels_in_Flood_Zone feature class.

Open the attribute table of this Parcels_in_Flood_Zone feature class or map layer, and you will see the flood hazard zone is added in the table for each of the parcels as well.

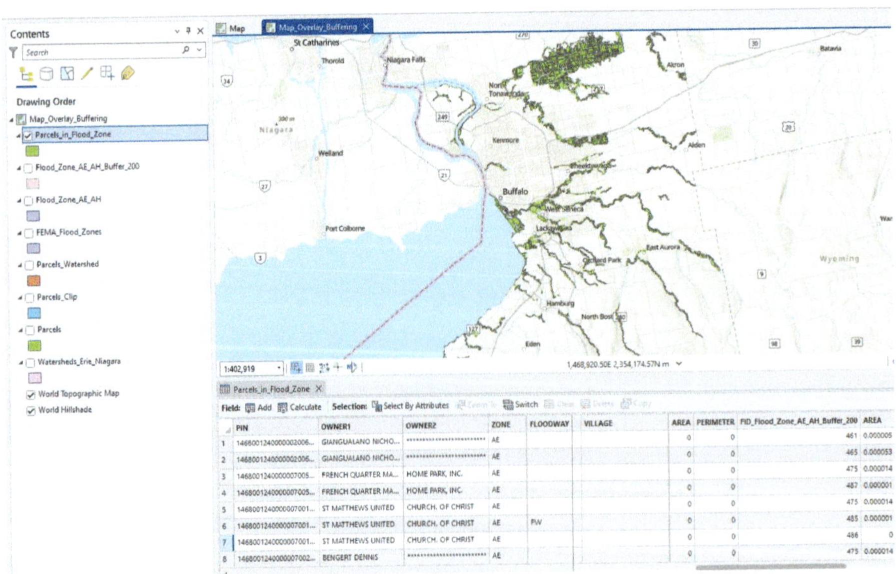

Now, the government officials of Erie County, NY can inform these residents to prepare for possible flooding events and purchasing the flood insurance in case the property is damaged by flooding. Next, the county government needs to plan for the evacuation routes for the flooding hazard zone residents. We need to create a streets and roads map layer in the communities that fall into the flood zones.

Add the Streets feature class into the "Map_Overlay_Buffering" window.

In this case, we are going to explore the tools in Pairwise Overlay toolset and use the Pairwise Clip tool to process the data. Tools in Pairwise Overlay toolset are mainly designed for handle large quantity of data. These tools utilize computational parallel processing method to attempt to obtain the result fast and more accurately.

Open the Geoprocessing Pane from the Analysis ribbon on the top. Expand the Analysis Tools box while the Toolboxes tab is activated. Expand the Pairwise Overlay toolset and collapse all other toolsets in this toolbox.

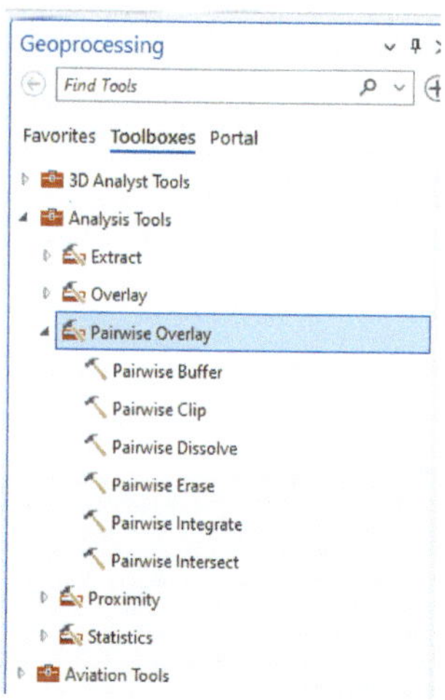

Highlight and click on Pairwise Clip tool to open its dialog. For the "Input Features," select Streets. For the "Clip Features," select Parcels_in_Flood_Zone. For the "Output Feature Class," navigate to your Erie_County_Spatial_Data.gdb and name it as "Streets_Flood_Evacu." Run the Pairwise Clip tool. The Streets_Flood_Evacu feature class is added on the mapping window. Check off the Streets map layer; we can see that the newly created Streets_Flood_Evacu feature class matches the parcels in the flood zones. Close the Geoprocessing Pane.

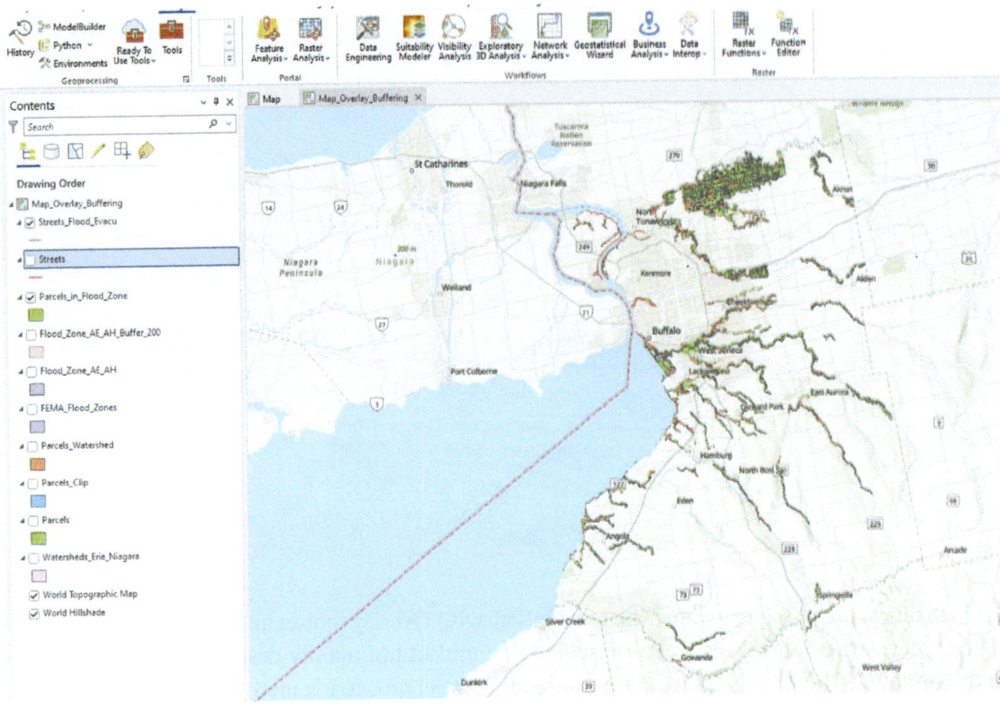


If you would like to take a break, save the project and close the ArcGIS Pro. Otherwise, continue.

Our next task in this project is to assign the AE and AH flood zones attributes to the Streams line feature class. This is so we can identify among river segments which ones actually create flooding hazards to human settlements and society in the County of Erie, New York. We can also present digital maps of the result to the public for people's awareness. Now, let's add the Streams feature class from the Catalog Pane to the mapping window (Map_Overlay_Buffering). Check the Flood_Zone_AE_AH feature class and make it visible in the mapping window. Activate the Analysis ribbon at the main menu and click on the Tools icon  in the Geoprocessing group. The Geoprocessing Pane opens. Activate the Tools tab and expand the Analysis Tools box, then, the Overlay toolset. Highlight and click the Identity tool to open it.

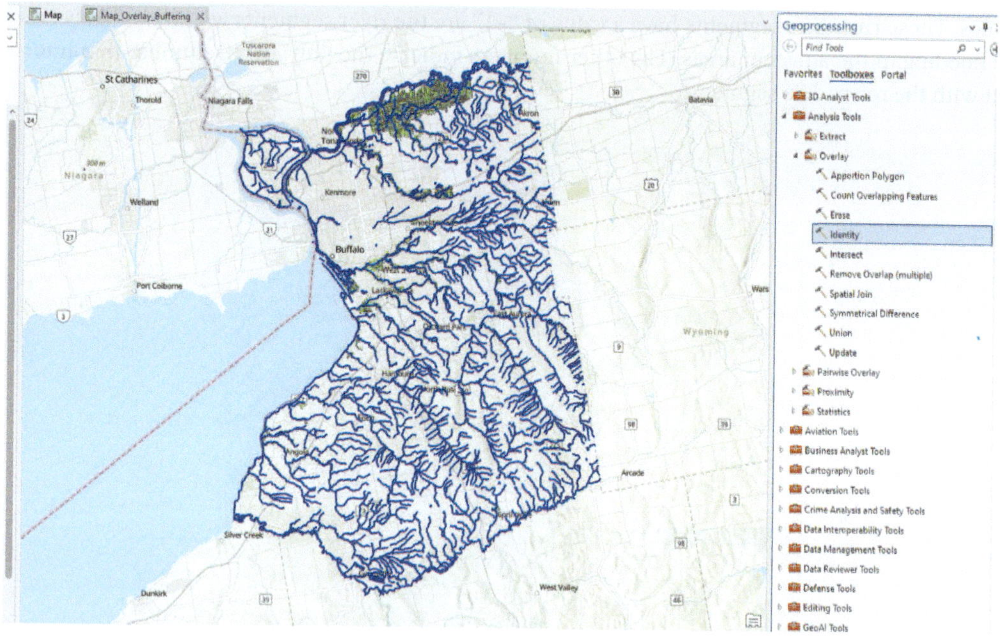


On the dialog of Identity tool, fill in the Streams for the "Input Features." For "Identity Features," fill in the Flood_Zone_AE_AH feature class. For "Output Feature Class," navigate to your Erie_County_Spatial_Data.gdb and name it as "Streams_Flood_Identities."

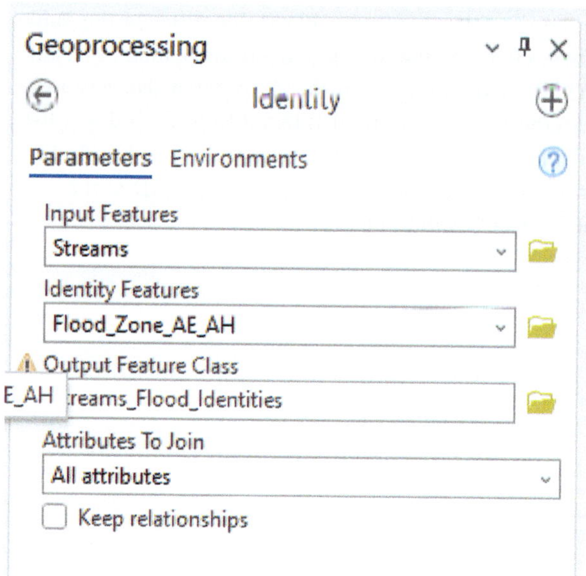


Process the tool to produce the Streams_Flood_Identities line feature class. Although this feature class looks exactly same to the original Streams feature class, we can open the attribute table to find the identities of flooding hazard in the river line segments the Streams_Flood_Identities feature class. "FID_Flood_Zone_AE_AH" field (column) in the following image shows the identity values. Those river line segments have a value of "–1" are the river segments would not likely create flooding in the adjacent areas. The other numbers indicate the polygon ID number that intersect with the reiver line segment.

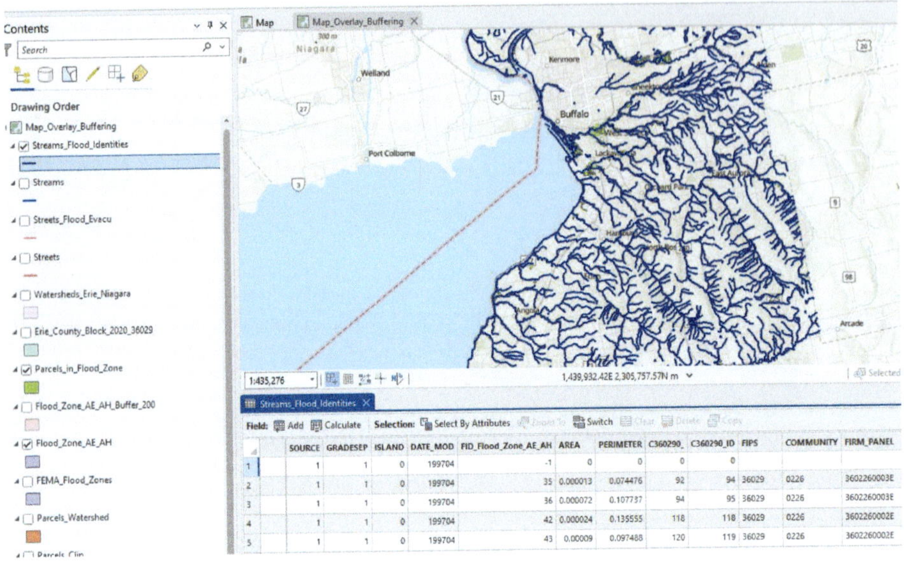

We can go further to produce a thematic digital map to show the rivers might produce flooding to humans living environment in the Erie County, NY as well as with those "good behaved" rivers. Close the attribute table. Right-click on the title of Streams_Flood_Identities in the Table of Contents to obtain the contact menu. Then, click on Symbology to open the Symbology Pane. For the "Primary Symbology," select the Graduated Colors from the drop-down list. Keep the "Normalization" blank. For the coloring "Method," select Manual Interval from drop-down list. For the "Classes," select two (2). Then, adjust the "Upper Value" of legend to "<= 0" for the first class (row), and the second class will be automatically adjusted to "<= 756" (largest ID number of flood zone polygons. For the "Color Scheme," select "Cyan to Purple."

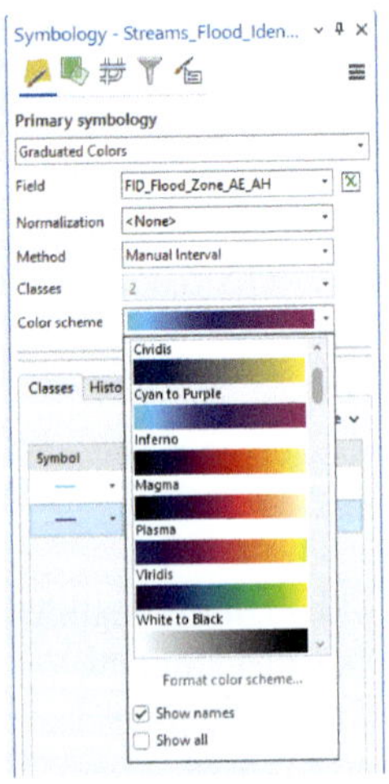

The thematic map shows the flooding river channels and non-flooding rivers as follows.

The purple river channel segments are the flooding hazard making rivers, while light blue colored river channels are "well behaved" rivers. Close the Symbology Pane.

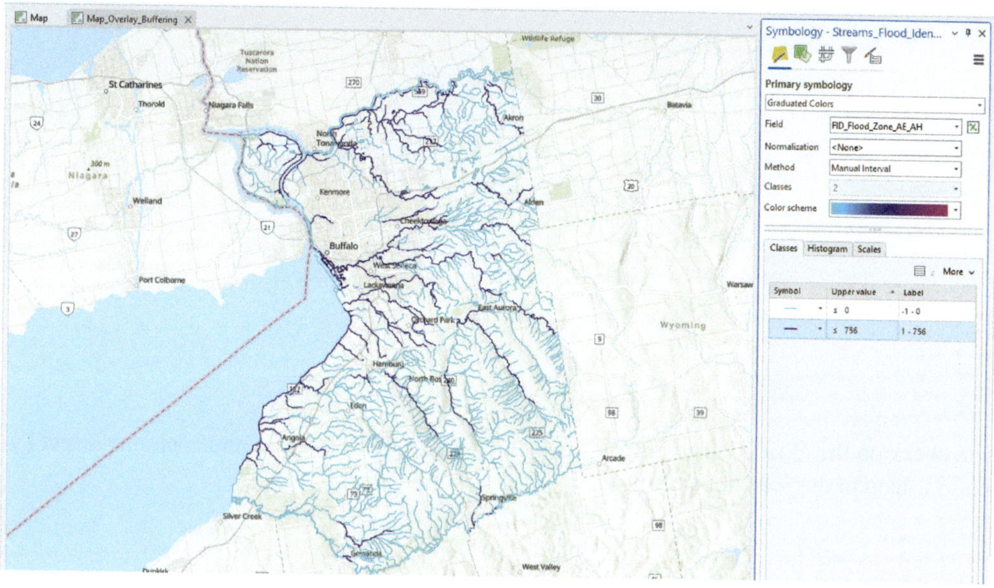


If you would like to take a break, save the project and close the ArcGIS Pro. Otherwise, save the project and continue.

For those of you who came back from a break, open ArcGIS Pro and the project file. Our final task in this project is to count the total population who live in the Buffalo-Eighteenmile watershed (drainage basin) areas based on the US Census Bureau year of 2020 field survey. This process will enable us to know in detail the maximum number of people who would be impacted when we encounter a flooding disaster event in that drainage basin. Add the "Erie_County_Block_2020_36029" feature class to the mapping window (Map_Overlay_Buffering). Moving the Watersheds_Erie_Niagara feature class just above the "Erie_County_Block_2020_36029" feature class in the Table of Contents. Check on the Watersheds_Erie_Niagara feature class if it is necessary. While Watersheds_Erie_Niagara feature class is highlighted in the Table of Contents, adjust the Transparency at the Layer ribbon on the main menu to 50% in order to create the graphic effect that we can view the map layer underneath it. Now, we are going to use geographic query (Select by Location) to conduct feature class to feature class selections.

First, check off the "Erie_County_Block_2020_36029" feature class for now. Activate the Map ribbon. Click on "Select" tool in the Selection group. Then, click on anywhere in the Buffalo-Eighteenmile watershed to select it on the Watersheds_Erie_Niagara.

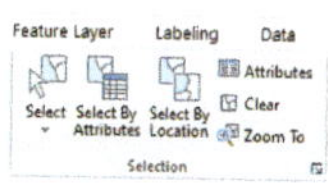

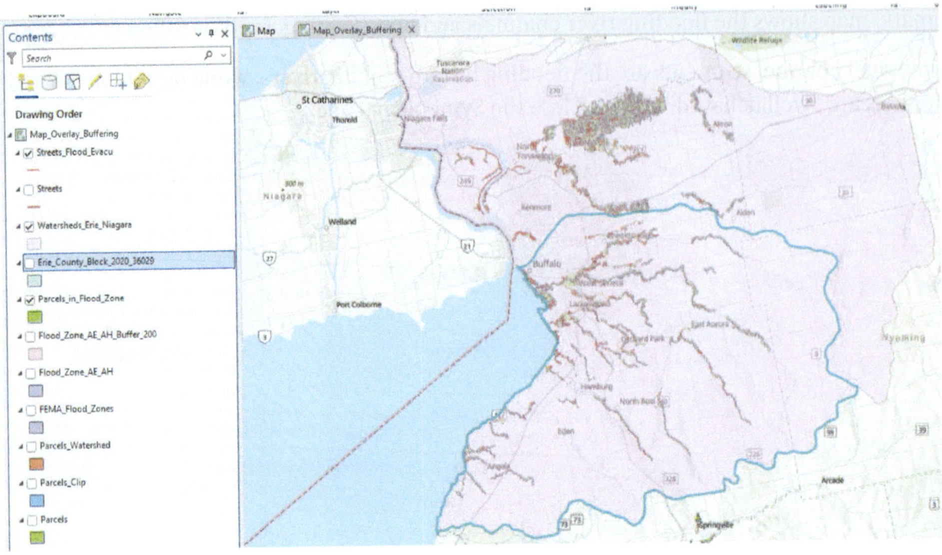


Then, check on the "Erie_County_Block_2020_36029" feature class again and click on "Select by Location" icon in the Selection group.

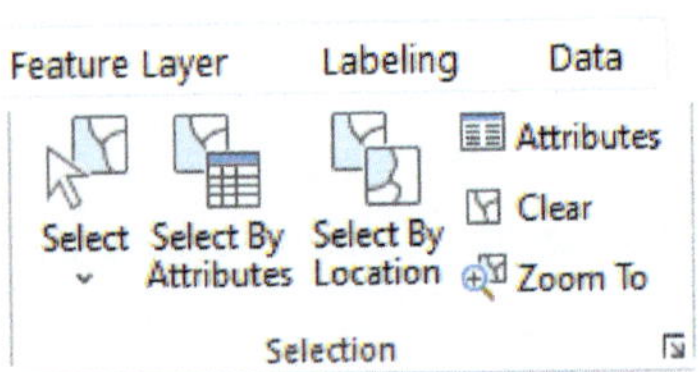


The Select by Location dialog opens. For the "Input Features," choose Erie_County_Block_2020_36029. For the "Relationship," browse through the entire list of querying methods and select "Have their center in" on the bottom of the list. For the "Selecting Features," choose Watersheds_Erie_Niagara feature class. Keep other parameters as the default.

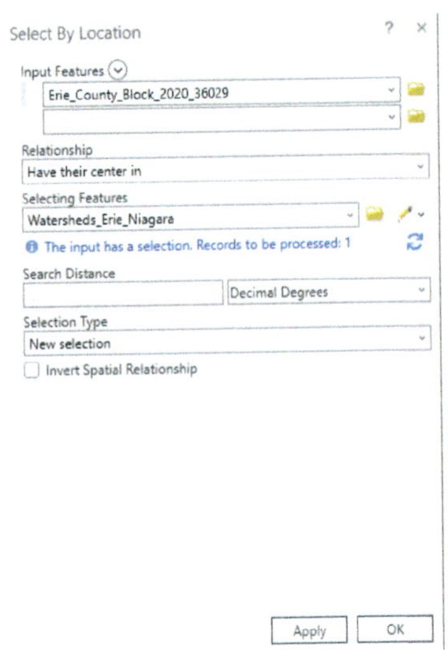

Click on "Ok" to accomplish the process. You may notice that the block polygons that their centers fall into the Buffalo-Eighteenmile Watershed are all selected.

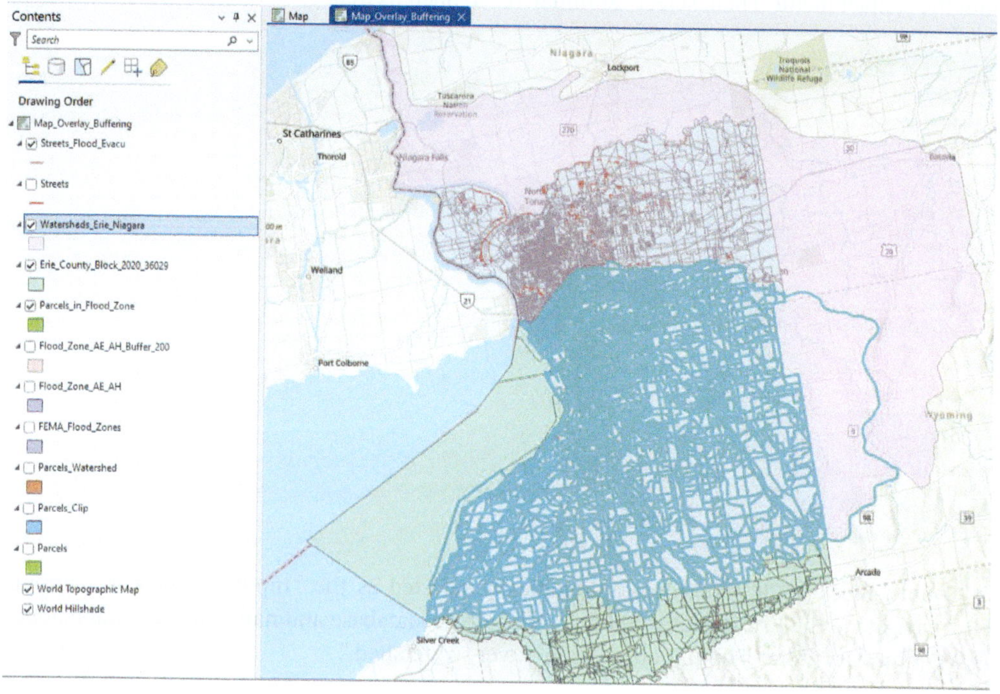


Check off the Watersheds_Erie_Niagara feature class. Open the attribute table of Erie_County_Block_2020_36029 feature class.

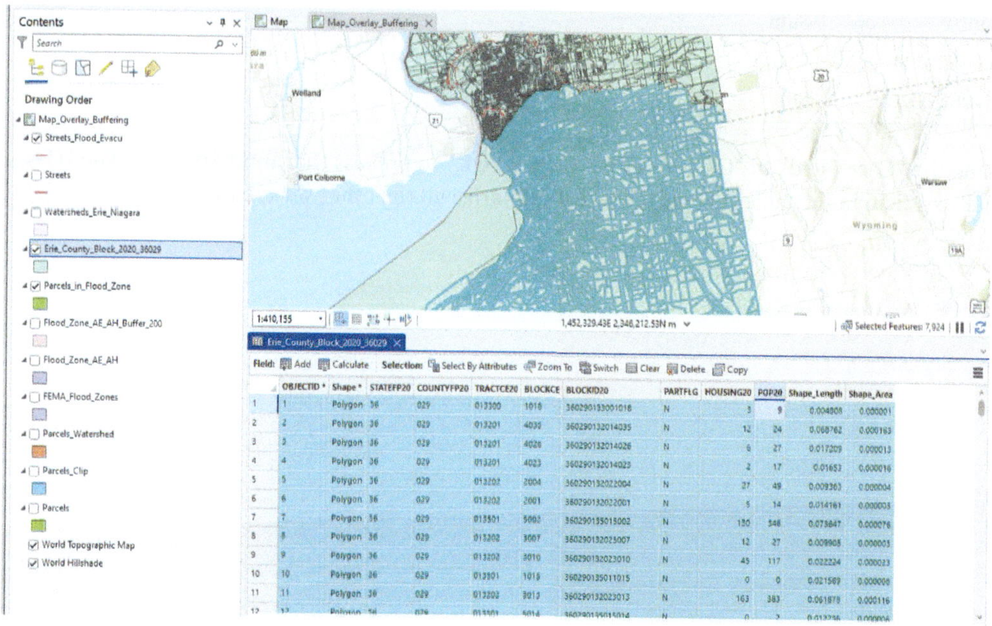

Column (field) "POP20" stands for the total population in the block. "HOUSING20" stands for the total households or families living in the block. Right-click on the title of "POP20" column to obtain the contact menu. Then, select "Summarize" to open its dialog.

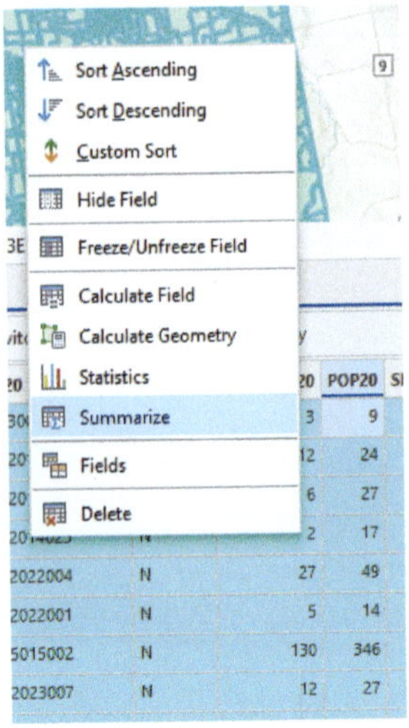


Erie_County_Block_2020_36029 was automatically selected as the "Input Table." For "Output Table," navigate to the Erie_County_Spatial_Data.gdb geodatabase and name the new standalone table as "Summary_Total_Population_Buffalo_River_Watershed."

For the "Statistics Fields," fill in as follows.

| Field | Statistic Type |
|---|---|
| POP20 | Sum |
| BLOCKID20 | Count |
| HOUSING20 | Sum |

Chang the "Case Field" to "COUNTYFP20," click "OK" to process the summary table. The "Case Field" is a field in the table that we use it to summarize all the other selected values.

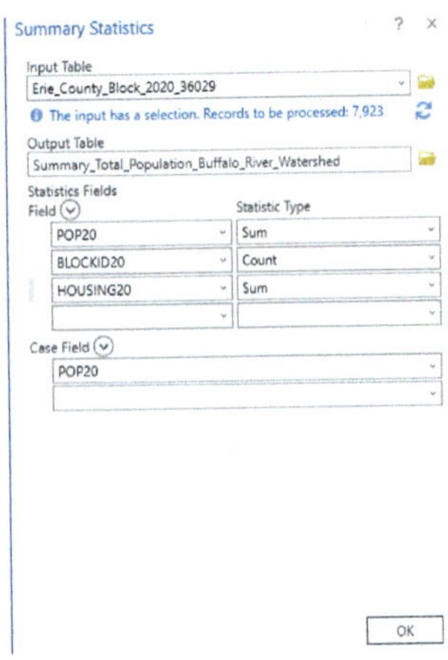

Close the attribute table of the Erie_County_Block_2020_36029 feature class. Open the stand-alone table of "Summary_Total_Population_Buffalo_River_Watershed" from the Table of Contents that was just created. We found that there is a total of 492,380 people are living in the Buffalo-Eighteenmile Watershed, and these account for a total of 228,259 families. The total community blocks in the watershed are 7923.

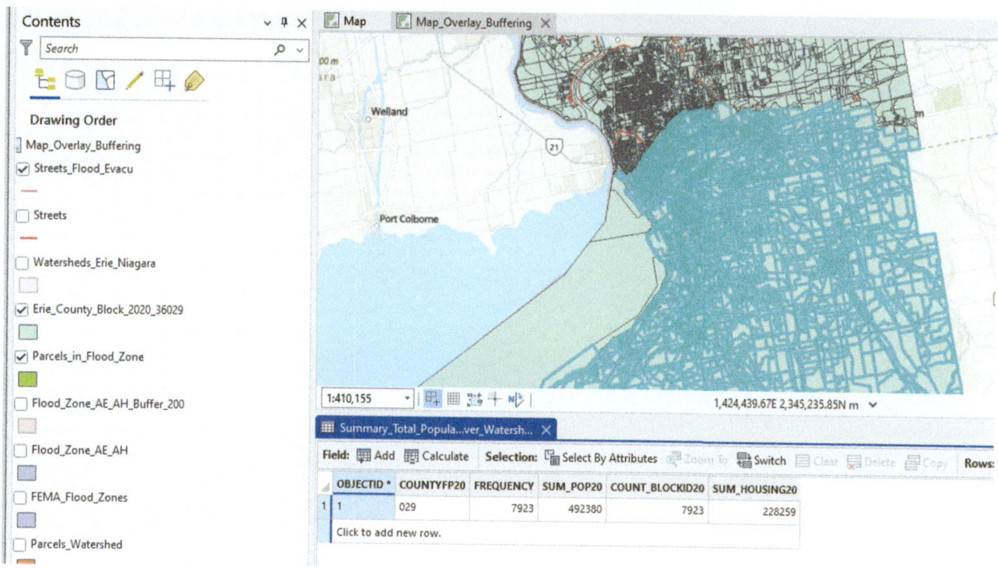


Close the summary table. Activate the Map ribbon on the main menu if it is not activated and click on "Clear" (Clear all the selections) icon to clear all the selections on multiple map layers.

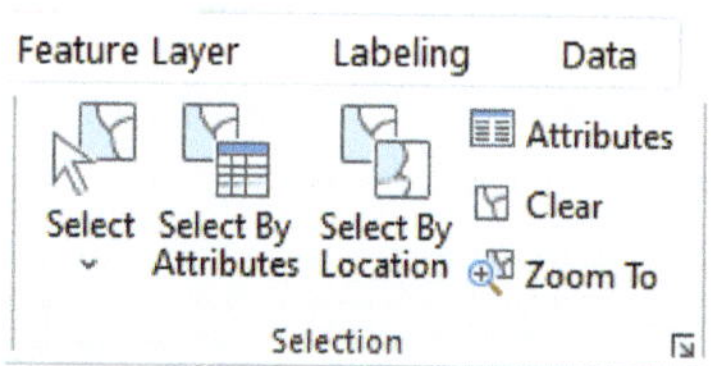


Now, we have completed the Erie County Flood Prevention GIS project for the county government using the vector GIS database. Congratulations!

# 7

# Basic Spatial Analyses in Raster Data Environment

## Concepts and Theories

Raster data analysis is focused on grid cells or raster units of datasets. Raster data analysis can be performed at the level of individual grid cells, groups of grid cells, or grid cells within an entire raster dataset or map layer. Some operations of raster data use a single raster map layer; others use two or more raster map layers. Raster analysis is similar in many ways to vector analysis. However, there are some major differences. The major difference between the analysis and modeling of raster and vector datasets comes from the fundamental nature of the data structure themselves. Vector geographic information system (GIS) analysis is focused on the geographic objects and features. While raster GIS analysis is focused on the grid cells containing geographic objects or features. Therefore, the root of reasoning and analysis is different. More emphasis in raster analysis is put on the computation of value changes at microgeographic units – grid cells. Some geoprocessing or operations in vector data, such as buffering, may not be viewed as significant in raster data analysis. Meanwhile, many unique operations have been developed and applied to raster data only are considered significant methods in spatial analysis.

Since both vector and raster databases utilize the common coordinate systems as their reference frameworks, all the analytical methods used in vector data can be applied to raster data analysis as well. However, raster GIS analysis carries its own unique approaches. Raster GIS analysis enforces spatial relationships solely on the locations of the grid cells. Raster GIS operations performed on multiple input raster datasets generally output values of grid cells that are the result of mathematical computations on a cell-by-cell basis. The value of the output for one cell is normally independent of the value or location of other input or output grid cells.

Raster data are more suitable for continuous than discrete geographic objects, such as chemical pollution concentrations, relative humidity in the air, slope of terrain, elevation or altitude, or slope aspect. Continuous geographic phenomena or objects change smoothly across a landscape or Earth's surface, making them far better suited to the structure of raster data than to that of vector data (Figure 7.1). As more and more GIS spatial analysis methods have been developed and implemented in the software environment, there is a general trend to integrate geoprocessing operations for both vector and raster datasets. Many of the advanced operations in the GIS software environment take the feature classes of vector data as input, while these operations generate raster grid cells as the output. In this discussion, we treat this type of operation as raster data analysis as well since the output results are in raster data format.

In practice, ArcGIS software historically was mainly developed in the structure of a vector database first. Therefore, the majority of fundamental, analytical, and geoprocessing functions for

*Fundamentals of Geographic Information Systems (GIS) Analysis: Applying ArcGIS Pro*, First Edition. Tao Tang.

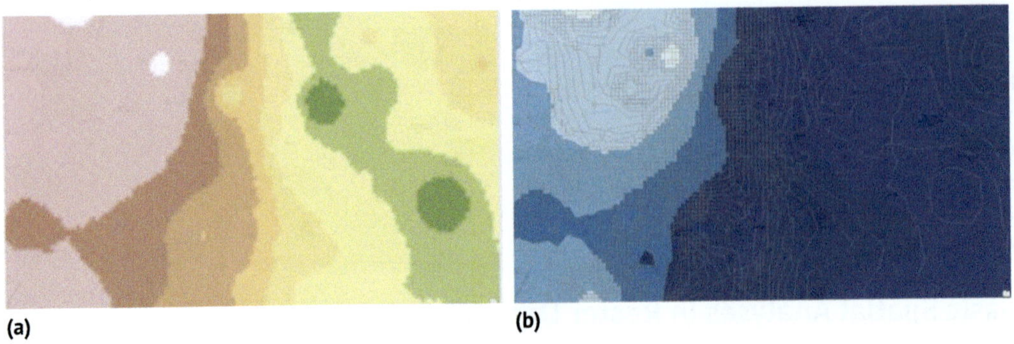


**Figure 7.1** Annual precipitation distribution illustrated by (a) raster data and (b) vector data at the same region.

raster data, which were developed relatively more recently, are collected under the umbrella of one of the software extensions called Spatial Analyst. Secondly, although the majority of raster datasets were collected in the field by the science of remote sensing and these data originally come in imagery format, imagery data is only one type of raster dataset. All the rules in raster operations are applicable to satellite or airborne (aerial) image data. However, the raster datasets and operational concepts for raster data cannot be simply substituted as that for the imagery data. According to the geographic or numerical relationships of the grid cells involved into the geoprocessing operations, the raster GIS analysis can be classified into six types. These are (i) distance operations, (ii) local operations, (iii) neighborhood or focal operations, (iv) zonal operations, (v) global operations, and (vi) spatial data (map) algebra.

Raster data is unique in presenting locations with grid cell structure. A dataset or a map layer is only georeferenced by some control points or the four corners of the digital map. Each of the grid cells occupies the same area and dimensions according to the spatial resolution of the raster data. The relative locations among the grid cells are referenced to the control points or grid cells. This nature makes the computation for distances or distance relationships of geographic objects, or location coincidence in raster data, much more straightforward than that in vector data.

## Distance Operations

Distance analysis is a fundamental concept in geography and one of the most popular applications in GIS. In its simplest form, distance is a measure of how far away one object or feature is from another. A straight line is the shortest possible measure of the distance between two locations. This type of distance measure is called Euclidean distance (Figure 7.2). This type of computational work is applied in both vector data analysis, such as buffering, and raster analysis. In real-world practice, there are other factors that need to be considered. For instance, if there is a barrier in the way, travelers need to detour around it, which increases the total distance to get from one location to another. Another major factor is the texture of the Earth's surface, which can increase the actual distance traveled if the Earth's surface changes from flat to rough terrain. Many other factors can also influence the effort required to travel from a starting point to an end point. Some of these factors include the steepness of the slopes on the terrain, whether the object or feature will be moving with the wind or against it, the mode of transportation, and the types of land cover being traversed. All these environmental factors can be added in to computation of real-world distance in GIS. There is also a difference in distance measures if the calculations are done using a planar (2D) or a geodesic (3D) method. Taking all these elements of impacting factors into consideration, the methods of distance analysis or distance factor-embedded analysis developed for the raster data structure are more effective in GIS. In essence, distance operations or any processes in searching distance relations resulting in ArcGIS Pro are mainly developed in raster data analytical modules.

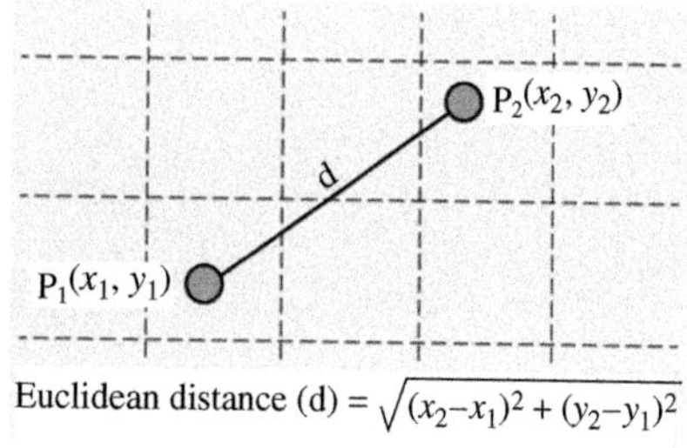


**Figure 7.2** Calculation of a Euclidean distance.

In ArcGIS Pro, the Spatial Analyst extension provides functions or tools that enable the GIS Analysts to calculate distance for a range of scenarios. The tools have several parameters that we can use to apply various modifiers for the different operations. As a result, more accurate analytical outputs are generated for making better decisions. Distance analysis can be further divided into two main tasks: (i) calculating the distance for the closest or lowest cost route; and (ii) connecting a series of locations with optimal (lowest cost) route of paths over a distance surface (ESRI 2020).

In practice, two aspects need to be considered when calculating distance: (i) determine how far away the target object is; and (ii) define how that determined distance is encountered by a traveling object, or the environmental factors of traveling. In GIS, the calculation of distance (how far away) is fundamentally achieved by Euclidean distance or straight-line distance. It is the shortest distance between two points. It is also called the flying distance because it is the distance an airplane would travel between the two points when flying above the surface. Euclidean distance surface maps in raster format can be generated among features in this computation. The input data can be either a vector feature class or a raster dataset.

However, there are other factors that can alter the straight-line distance calculation, such as barriers and surface distance. In a real-world situation, a barrier can be a river, lake, highway, or a building that prevents the traveling object from traveling directly from one location to another. In this case, the GIS analyst needs to know the shortest possible distance between the two locations after accounting for the additional distance necessary to detour around the barrier. In order to overcome the barriers, either natural or human-made, straight-line distance needs to be adjusted by cost factors or cost surfaces. One cost surface is made by one cost factor, such as the elevation or slope of terrain. The least-cost distance between the locations can be computed using these cost surfaces in GIS. In order to analyze the least-cost distance, the conditional raster cost surface dataset is a necessary input. The most popular cost surface raster dataset with straightforward visual understanding is the digital elevation model (DEM) and its product of slope surface of the study area.

How distance is encountered by a traveling object is often the interaction that is of significant interest in geography. The traveling object can be a living entity, such as a person or an animal, or a nonliving object, such as liquid or gas objects in a pipeline or vehicles on a road. There are multiple factors that influence how a traveling object encounters the distance. Each affects the rate at which the distances are consumed. The following two factors are considered to control the rate of a traveling object to cover the distance: (i) cost surfaces (factors), and (ii) characteristics of the traveling objects. Cost factors can be any natural or human-made factors that serve as barriers in the traveling environment. The characteristics of the traveling object can modify how distance is encountered in the real world. The major characteristics are: (i) mode or capability of travel, (ii) initiation cost, (iii) maximum capacity, and (iv) travel direction.

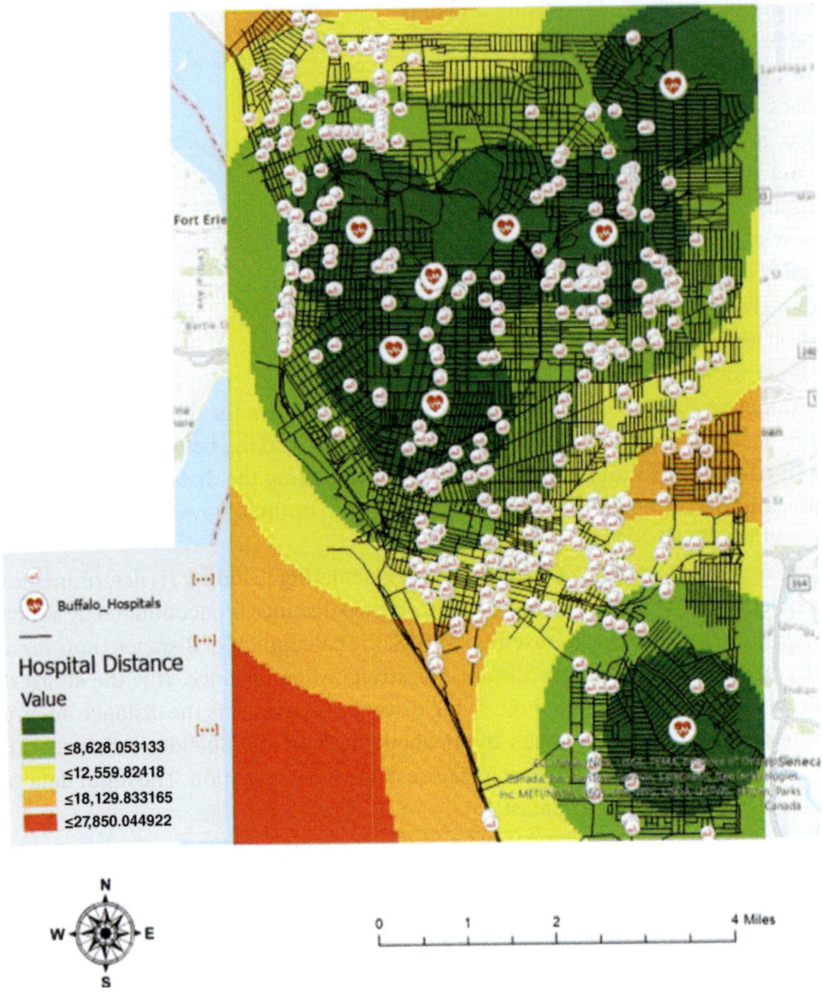


**Figure 7.3** Euclidean distance computations showing the distance to the closest hospital.

In GIS practice, the distance can vary depending on if a planar or geodesic spatial reference system is used in the calculations. The distance analytical functions and tools are mainly collected in the Spatial Analyst Tools box and Distance sub-toolbox in ArcGIS Pro. For instance, the Euclidean distance tool calculates, for each grid cell, the Euclidean distance to the closest source (input) feature (Figure 7.3). The Euclidean Allocation tool calculates, for each grid cell, the nearest source (input) feature with Euclidean distance (assigning the source feature identity; Figure 7.4). The Cost Distance tool calculates the least accumulated cost distance for each grid cell from or to the least-cost source (input) feature over a cost surface (second input).

## Raster Buffering

Buffering in GIS is a procedure to conduct zoning on a specified distance (radius or width) to geographic objects or features. In raster data, objects or features are represented by the grid cells.

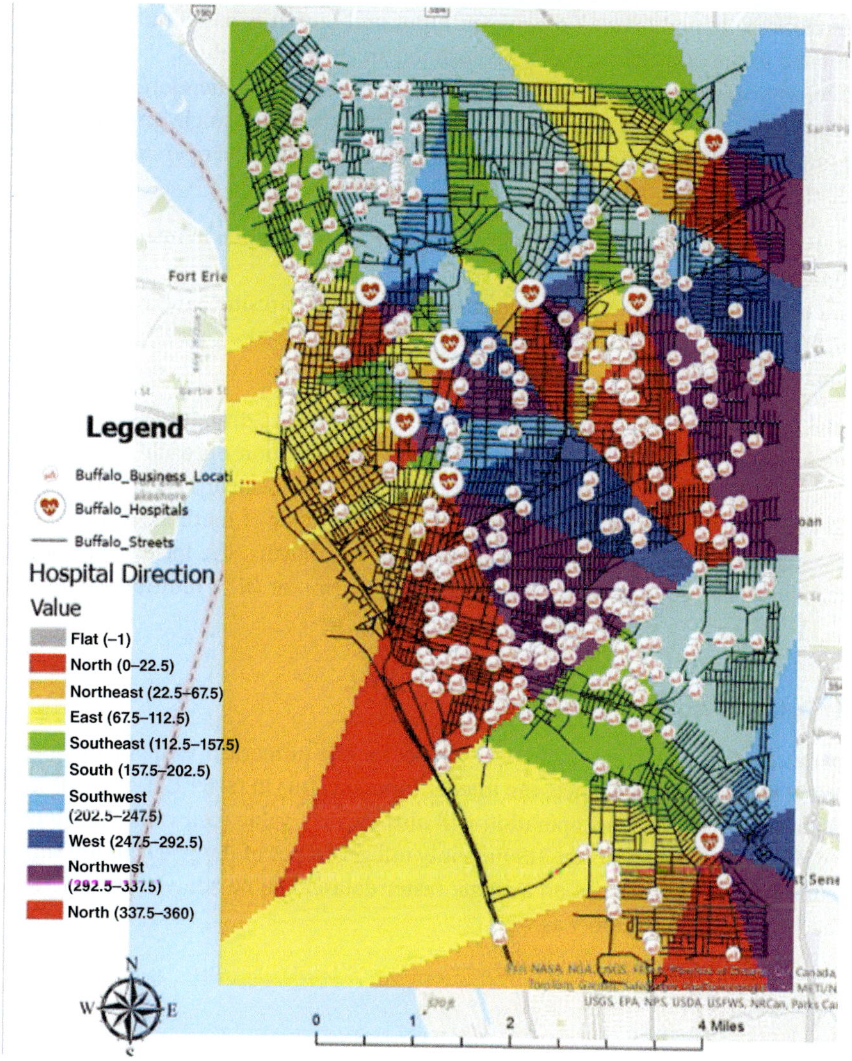


**Figure 7.4** Euclidean direction to the closest hospital at each of the grid cells.

Therefore, raster buffering is, in essence, the classification of grid cells according to whether the cells lie inside or outside the buffer region. Output grid cells are either assigned an in value or an out value. In value cells are those where the cell-to-cell distance is less than the specified buffer distance, while grid cells of out-value are further than the buffer distance. The distance is computed by Euclidean distance, and the result shows the classification of grid cells.

## Local Raster Operations

Local raster operations are also called single-cell operations or cell-by-cell operations. The operation computes a raster output dataset, where the output value at each cell location is a direct function of the value associated with the grid cell of that location on one or more input raster datasets. The relationship and operation between input and output is based on single cells. The result is not related to the grid cells in the neighborhood area or global values.

Raster grid cells normally carry a single numerical value in each unit. Therefore, it is very simple and straightforward to conduct mathematical computations on either single grid cell dataset or map layer or multiple grid cell datasets or map layers. The mathematical operations on grid cells can be further classified into following:

1. Arithmetic operations, such as +, −, ×, absolute value, decimal value to integer conversions
2. Logarithmic operations
3. Trigonometric operations, such as sine, cosine, tangent, arcsine, arctangent, arccosine (Figure 7.5)
4. Power function operations, such as square or square root

Although the results for some of these operations in practice are difficult to be used to illustrate their meaning in the real world, such as tangent computation on each of the grid cells on a raster soil pH value map layer (What does this raster map mean?), various operations based on grid cell values do provide GIS analysts a great number of methods to generate new spatial values. According to the input raster datasets or map layers, the local raster operation can be further identified as (i) a single raster dataset operation or (ii) a multiple raster dataset operation.

## Single Raster Dataset Mathematical Operation

These are operations that utilize mathematical means for the numerical value of the grid cells, one raster dataset at a time. For instance, the integer function (Int) in raster GIS converts grid cell values to integer by truncation. This operation will also create a value (cell value) attribute table for the raster dataset (Figure 7.6). In ArcGIS Pro, only integer values of the grid cells are supported currently. Meanwhile, grid cell values on a single raster dataset can be added, subtracted, multiplied, or divided by a constant number as well.

input grid

| | | | | |
|---|---|---|---|---|
| 12 | 49 | 20 | 46 | 26 |
| 6 | 37 | 20 | 46 | 28 |
| 49 | 45 | 38 | 15 | 6 |
| 23 | 5 | 35 | 17 | 17 |
| 22 | 48 | 3 | 45 | 23 |
| 42 | 14 | 23 | 32 | 42 |
| 31 | 23 | 41 | 2 | 40 |

output grid

| | | | | |
|---|---|---|---|---|
| −0.537 | −0.954 | 0.913 | 0.902 | 0.763 |
| −0.279 | −0.644 | 0.913 | 0.902 | 0.271 |
| −0.954 | 0.851 | 0.296 | 0.65 | −0.279 |
| −0.846 | −0.959 | −0.428 | −0.961 | −0.961 |
| −0.009 | −0.768 | 0.141 | 0.851 | −0.846 |
| −0.917 | 0.991 | −0.846 | 0.551 | −0.917 |
| −0.404 | −0.846 | −0.159 | 0.909 | 0.745 |

**Figure 7.5** Sine operation on local cells. *Source:* Adapted from ESRI – ArcGIS Users' Guide.

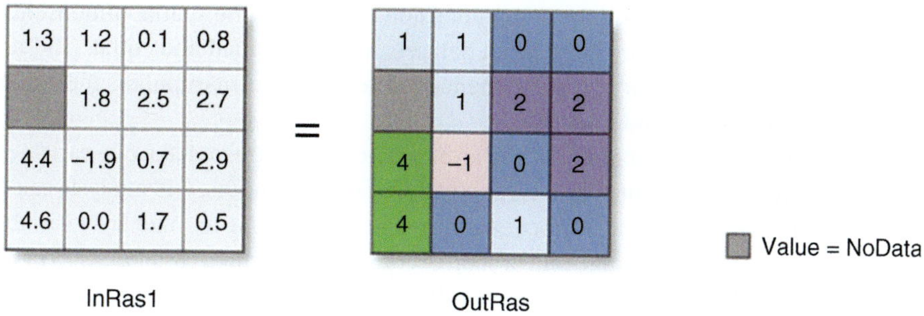


**Figure 7.6** Processing of Int operation. *Source:* Adapted from ESRI, Inc.

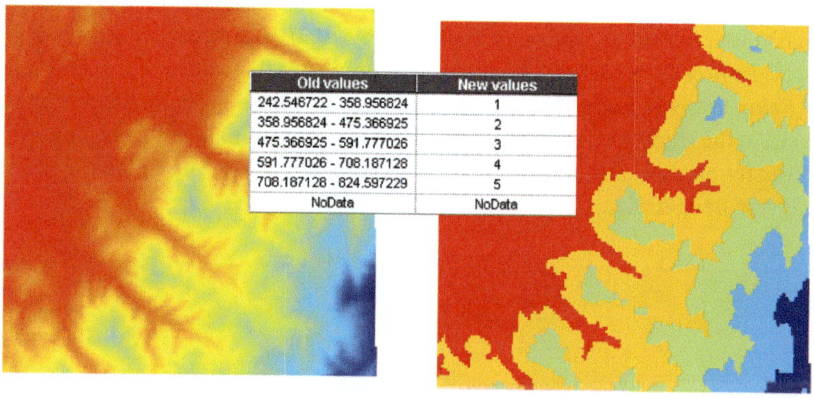


**Figure 7.7** Example of reclassification of DEM. *Source:* David Harbor, Washington and Lee University.

### Reclassification of Single Raster Dataset

This is an operation to reclassify or recode the grid cell values of a single raster dataset using a specific set of criteria or a particular algorithm. For instance, it can reclassify DEMs into (i) elevation <500 m as 1 or low elevation, (ii) elevation 500–1000 m as 2 or medium elevation, (iii) elevation 1000–3000 m as 3 or high elevation, and (iv) elevation > 3000 m as 4 or very high elevation. Reclassification, or the process of recoding, is one of the most important and popularly used raster GIS operations (Figure 7.7). The following is a set of general criteria for rainfall storm classification. Light rain—when the precipitation rate is <2.5 mm (0.098 in) per hour. Moderate rain—when the precipitation rate is between 2.5 mm (0.098 in) – 7.6 mm (0.30 in) or 10 mm (0.39 in) per hour. Heavy rain—when the precipitation rate is >7.6 mm (0.30 in) per hour. In practice, a GIS analyst can reclassify the rainfall distributions on a raster dataset or grid cell map based on these criteria.

### Multiple Raster Dataset Local Operations

This type of operation, in essence, is a kind of raster spatial data or map overlay. In general, it is similar to the vector data or map overlay process that needs at least two datasets as the input. The objective of this type of process is the same to that in vector data environment, and it is to comprehend the various raster datasets. However, two fundamental differences exist: (i) Raster operations are based on grid cells, as the basic unit for comprehension. Therefore, the spatial dimensions of the grid cells, or spatial resolution, for all the input datasets must be unified and equal. (ii) Raster operations theoretically can be a variety of mathematical methods to manipulate the original values of the input raster datasets and grid cells.

In the raster GIS environment, a dataset or a map layer can be directly treated like one variable for computations. In the process of combing and comprehending raster datasets, GIS analysts can formulate sophisticated mathematical models or equations with each of the raster map layer variables. This type of complex mathematical spatial data or map overlays is often called raster algebra or map algebra. In ArcGIS Pro, a "Map Algebra" tool has been developed in handling a large variety of multiple raster data analysis needs. We will only discuss simple multiple raster dataset local operations here and dedicate one section later in particular to discuss the map or raster algebra.

Simple mathematical or statistical functions can be easily applied to multiple raster dataset overlays (Figure 7.8). For example, we can produce the resulting raster dataset or map layer to comprehend the maximum, minimum, mean, or standard deviation of the grid cell values of a group of raster datasets (Figures 7.9 and 7.10). We can also summarize the values of the grid cells on a variety of raster map layers to judge the overall value or quality for a particular factor or event at different locations in the study area or region, or account the value of majority cells, minority cells, or the value diversity on different input raster datasets (Figure 7.11). The following are several examples of this type of operation.

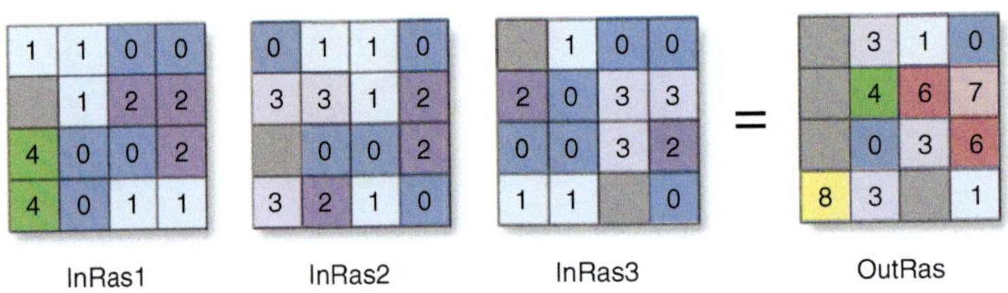


**Figure 7.8** Summary of the input grid cell values.

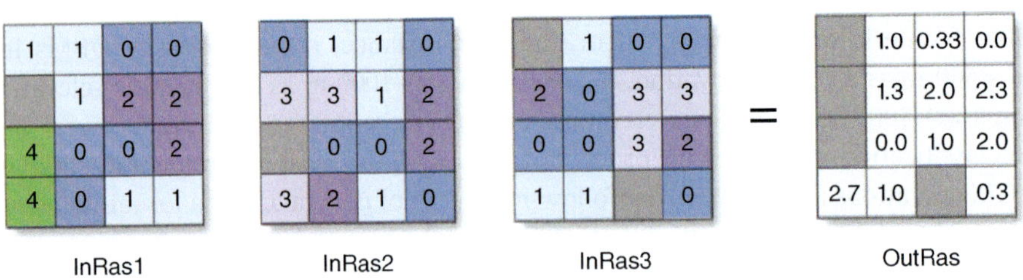


**Figure 7.9** Mean of the input grid cell values.

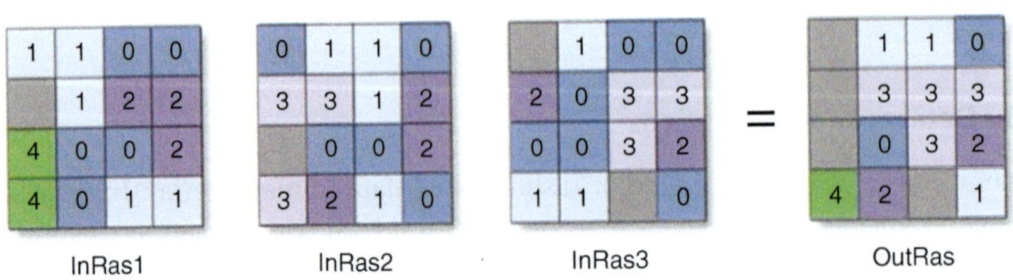


**Figure 7.10** Maximum of the input grid cell values.

**Figure 7.11** Majority of the input grid cell values. *Source:* Adopted from ESRI, Inc.

## Neighborhood or Focal Operations

Grid cell operations can also be conducted according to cell values in the neighborhoods. In this type of operation, the result of a particular grid cell in the output raster dataset is the result of a calculation performed on a group of cells determined by a window of cells (a kernel) surrounding the target cell of computation. This type of operation originated in the digital imagery processing works of remote sensing science for sharpening or classifying the image datasets. Therefore, an operation of this type is called a neighborhood or focal operation.

In digital image processing, the neighborhood analytical methods are also called spatial convolution filtering (Jensen 2015). The most popularly used moving windows or moving operators for assigning the output cell values are either square or rectangular in shape. Other shapes of the moving windows can be circles, annulus, or wedges, according to Jensen (2015). The functioning window can be set up as 3 × 3 cells, or 5 × 5 cells for example (Figure 7.12). A variety of methods of the neighborhood or focal analysis, such as mean or majority, can be computed. According to the purpose of reclassifying the remote sensing imagery pixels or grid cells, these methods are referred to as low-pass filtering or high-pass filtering in digital image processing. Any operation resulting in the smoothing of the color or tone of the image is called low-pass filtering. This type of process smooths the edges of objects on raster image and remove the high-frequency fluctuations of cell values. By contrast, any operation that intensifies the edges and variation of the fluctuation values of the grid cells is called high-pass filtering. A neighborhood algorithm calculates the mean value of a 3 × 3 cell kernel and places the output value in the location of the central (targeted) grid cell as shown below.

input grid

| | | | | |
|---|---|---|---|---|
| 24 | 50 | 7 | 41 | 32 |
| 30 | 27 | 8 | 22 | 39 |
| 14 | 16 | 21 | 16 | 41 |
| 38 | 6 | 44 | 8 | 7 |
| 36 | 7 | 32 | 30 | 32 |
| 38 | 20 | 28 | 28 | 21 |
| 32 | 2 | 13 | 35 | 49 |

output grid

| | | | | |
|---|---|---|---|---|
| 32.8 | 24.3 | 25.8 | 24.8 | 33.5 |
| 26.8 | 21.9 | 23.1 | 25.2 | 31.8 |
| 21.8 | 22.7 | 18.7 | 22.9 | 22.2 |
| 19.5 | 23.8 | 20.0 | 25.7 | 22.3 |
| 24.2 | 27.7 | 22.6 | 25.6 | 21.0 |
| 22.5 | 23.1 | 21.7 | 29.8 | 32.5 |
| 23.0 | 22.2 | 21.0 | 29.0 | 33.3 |

**Figure 7.12** Value assignment with a 3 × 3 moving window in the neighborhood of the targeted grid cell. *Source:* Jensen (2015)/Prentice Hall Press.

## Zonal Operations

Operations of this type utilize one input raster dataset or map layer as the zoning standard to perform functions based on a group of cells in one of the inputs within a common zone (value). All the grid cell values from various input raster datasets are changed according to the computational method selected for the operation (Figure 7.13). In raster GIS analysis, a zone means all the grid cells that have the same unique value inside. In other words, each zone must have a unique identity. Practically, any integer or string (text) field of unique values in the zone input can be specified to define the zones (Figure 7.14). This analytical method has high potential to be utilized by urban and regional planners for zoning analyses. Zoning refers to municipal or local laws or regulations

input grid

| | | | | |
|---|---|---|---|---|
| 53 | 57 | 33 | 10 | 14 |
| 78 | 31 | 12 | 22 | 55 |
| 32 | 9 | 9 | 85 | 26 |
| 6 | 54 | 33 | 85 | 94 |
| 75 | 25 | 76 | 49 | 27 |
| 48 | 16 | 67 | 23 | 89 |
| 36 | 46 | 82 | 97 | 74 |
| 45 | 86 | 44 | 42 | 35 |

zone grid

| | | | | |
|---|---|---|---|---|
| 1 | 1 | 1 | 2 | 2 |
| 1 | 1 | 1 | 2 | 2 |
| 1 | 1 | 1 | 2 | 2 |
| 5 | 5 | 1 | 2 | 2 |
| 5 | 5 | 1 | 2 | 2 |
| 5 | 5 | 10 | 91 | 91 |
| 5 | 5 | 10 | 91 | 91 |
| 5 | 5 | 10 | 10 | 10 |

output (zonal sum) grid

| | | | | |
|---|---|---|---|---|
| 423 | 423 | 423 | 467 | 467 |
| 423 | 423 | 423 | 467 | 467 |
| 423 | 423 | 423 | 467 | 467 |
| 437 | 437 | 423 | 467 | 467 |
| 437 | 437 | 423 | 467 | 467 |
| 437 | 437 | 270 | 283 | 283 |
| 437 | 437 | 270 | 283 | 283 |
| 437 | 437 | 270 | 270 | 270 |

**Figure 7.13** Example of summary of the grid cell values in each of the specific zones.

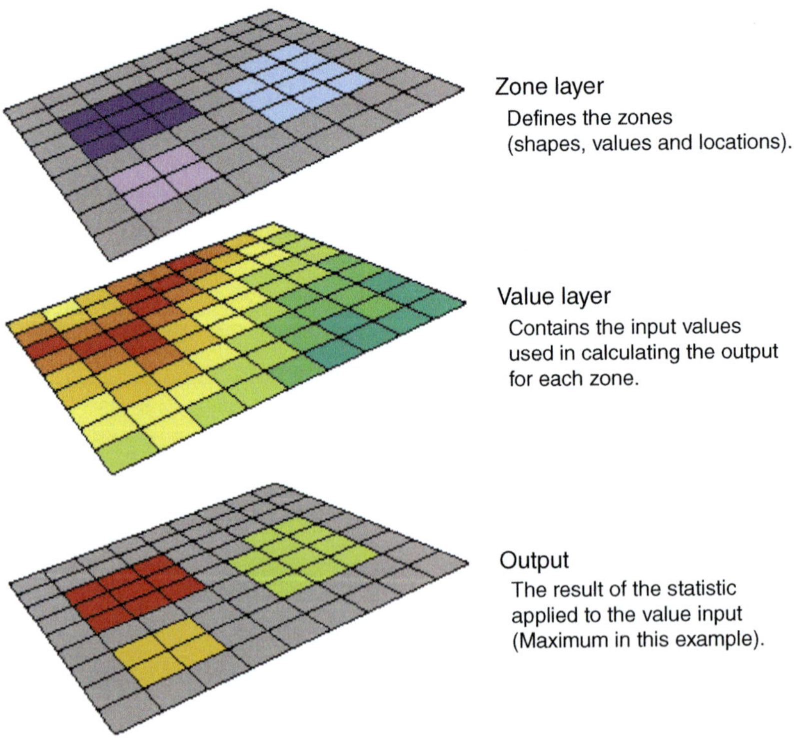


**Figure 7.14** Example for calculating maximum for the value map layer with input zones. *Source:* Erdem et al. (2021)/with permission of Elsevier.

that the local governments enforce for how real property can and cannot be used in certain geographic areas. Figures 7.13 and 7.14 are examples to summarize the values for all the input grid cells in a zone to assign this value to every grid cell in the zone.

## Global Operations

Global operations perform analysis based on the input of the entire grid cell dataset. This type of operation is also called per-raster dataset analysis. Computations of Euclidean distance or weighted distance for all the grid cells to the closest source or target feature or target cell location are global operations. However, calculations for Euclidean distance or weighted distance between two features or two grid cell locations are not a global operation. The results of global operations that are normally presented on a digital map or digital model can also be used to illustrate specific geographic relationships, such as distance, elevation difference, or slope between the features or grid cells.

Grid cell-based modeling operations in GIS are typical global analyses. Basic calculations of all the grid cells in one raster dataset, such as elevation difference to the neighboring grid cell or slope to the neighboring grid cell can be accomplished in the first step of modeling process. Then, flow accumulation or hydrologic processes can be simulated to create a stream network and delineate a group of nested watersheds (DeMers 2005; Chang 2008).

In order to model water or any liquid material flow accumulation, we need to determine the flow direction first. The initial method to compute the flow directions on digital raster elevation models was proposed in 1974 (Mark 1974a,b; Mark 1975). Since then, a few methods were developed (Greenlee 1987; Tarboton 1997; Qin et al. 2007). In this case, we will only illustrate the simplest one, D8 computational method of flow directions. D8 flow method models flow direction from each grid cell to its steepest downslope adjacent cell. Starting from the DEM from the region of interest, the operation calculates the slopes of a total of eight (8) adjacent cells to the targeted grid cell using the elevation differences (Figure 7.15). Then, it creates the flow direction value of the grid cell that follows the steepest slope.

The function of flow accumulation then calculates the flow along with the flow direction from the source to accumulate the value or weight of the grid cells passed through. If no specific weight or weight raster dataset is provided, a weight of 1 is applied to each cell, and the value of cells in the output raster is the number of cells that flow into each cell. In the diagram below (Figure 7.16), the left image shows the direction of flow traveling from each cell, and the right image shows the number of cells that flow into each cell. In practice, we can change the values for each of the grid cells in the flow accumulation operation according to a real-world situation.

Flow accumulation and watershed delineation simulations and modeling in GIS for river systems are very important for scientists to combine large field-measured datasets as the input to answer question, such as, "What will happen if we have this....?". A great number of applications exist for us to explore.

The example of raster analysis application using flow direction and accumulation modeling functions is shown in Figure 7.17. The overland flow directions are in different colors presented by grid cells. The sub-watersheds of the river system are generated or delineated by the raster grid cells of flow directions and flow accumulations. The drainage points

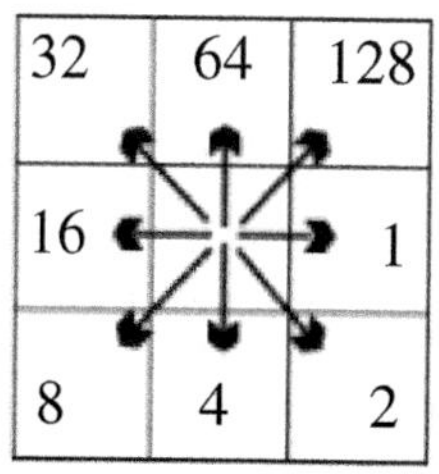


**Figure 7.15** Slope computational model of a total of eight (8) adjacent cells to identify the water flow direction (steepest slope); (DeMers 2005).

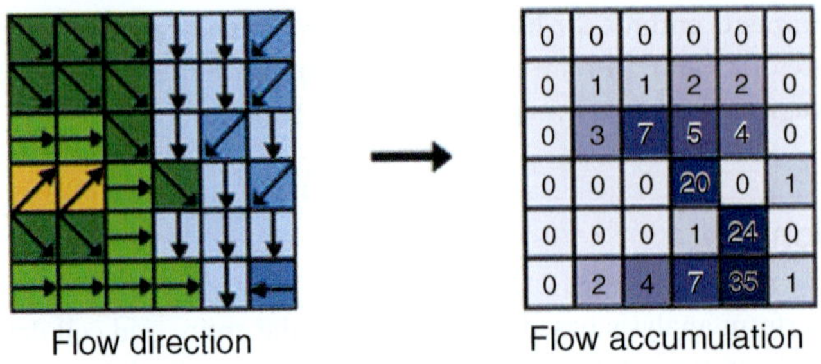


**Figure 7.16** Raster model simulation of water flow and flow accumulations. *Source:* DeMers (2005)/ John Wiley & Sons.

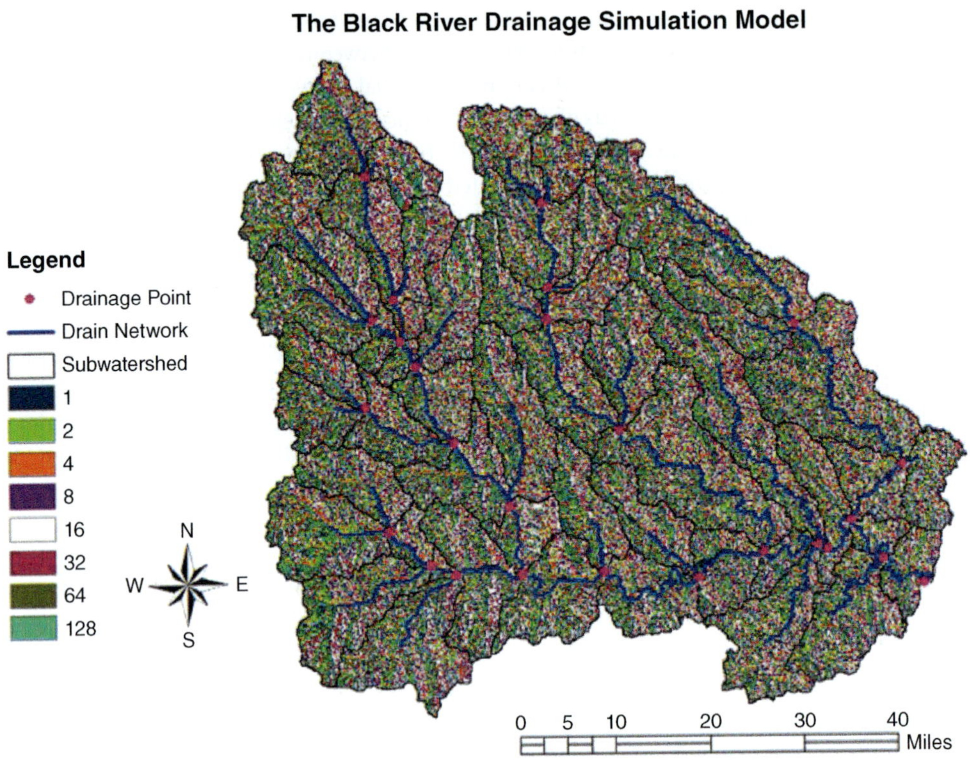


**Figure 7.17** Flood simulation modeling using raster dataset in the Black River, Beijing, China.

of the sub-watersheds of the river system are also delineated. Figure 7.17, on one hand, presents a digital map. On the other hand, it is a digital spatial simulation model.

## Spatial Data (Map) Algebra

Map algebra, or more precisely speaking, what should be called spatial data algebra, is basically to do mathematical computations with maps or spatial data. It is often the case that each of the map layers or spatial datasets in map algebra is treated as a variable in a mathematical function or a model. Map algebra only applies to raster data. This is because first raster data is field-based data, and each field or grid carries a simple numerical value that can be directly manipulated by mathematical methods. Secondly, the value attribute table of grid cells carries only one unique attribute per map layer, which makes mathematical computations relatively simple processes. Finally, in raster datasets, cell locations have been defined, and cell location matching between different datasets or map layers is not problematic. Therefore, sophisticated mathematical models can be built using raster datasets as inputs or variables to accomplish the tasks.

In practice, most people call the raster grid cell-based mathematical or statistical computations in the spatial database as map algebra. In terms of the computational process, it is actually doing the mathematical or statistical computations by grid cell locations. The locational relationships of grid cell are defined by a specific spatial model, such as searching the lowest elevation grid cell in an eight-neighborhood grid cells or using a nine-grid-cell moving window to calculate the mean value within the window. This type of operation should be more precisely called as spatial data algebra other than map algebra.

In the ArcGIS Pro software environment, the toolbox of "Map Algebra" exists in the Spatial Analyst Tools box, and a comprehensive tool, "Raster Calculator," was developed in the toolbox. In essence, map algebra function or tool can be used to perform any of the local, neighborhood, zonal, or global analysis. The reason that we discuss the map or spatial data algebra as a separate section here is because the operations can go beyond the basic analysis of the raster datasets.

One of the examples for map algebra applications is to assess the overall natural hazard risks in a geographic region (Figure 7.18). Each of the individual natural hazards can be inputted as a raster map layer. The values can be computed in GIS, and each of the disasters happening in the region can also be weighted according to its occurrence or degrees of severeness.

Another successful application of map algebra is to calculate soil erosion in a farming area, applying the Universal Soil Loss Equation by the US Department of Agriculture (USDA). Soil erosion is the wearing away of the topsoil on the surface of the earth. Soil erosion is mainly caused by floods, rainwater impacts, winds, ice, human and animal activities, etc. As urbanization is increasing and agricultural practices are intensifying, soil erosion has become more and more severe over the years.

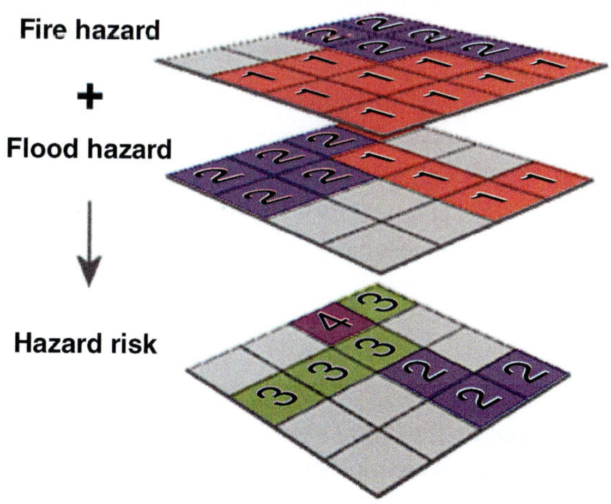


**Figure 7.18** Computation of overall natural hazard risks in a geographic region. *Source:* Adapted from ESRI – ArcGIS Users' Guide.

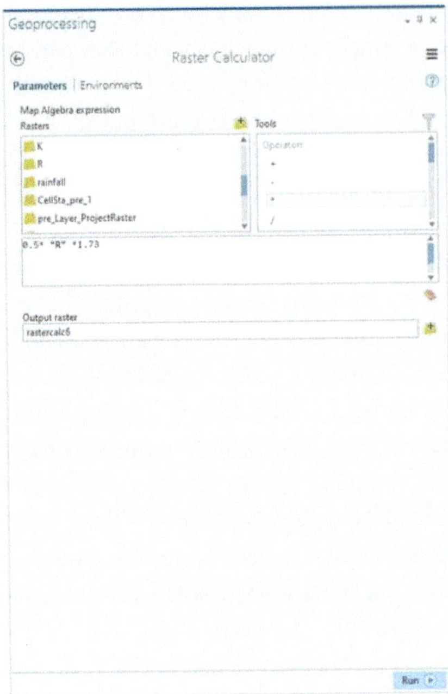


**Figure 7.19** Raster calculator for map algebra in GIS environment.

The Universal Soil Loss Equation (USLE) as a complete model was first published by the USDA in 1965 (USDA 1965). It is an empirical model that has been applied around the world to estimate soil erosion by raindrop impact and surface runoff. Since the development of GIS software and raster databases in the 1970s and 1980s, USLE computations have been fully transferred to be conducted in the GIS environment. Each of the original factors in the USLE model can be obtained from the field or converted from other types of datasets to one of the raster datasets or map layers with grid cells. These are: (i) R – rainfall and runoff; (ii) K – soil erodibility; (iii) L – slope length; (iv) S – slope steepness; (v) C – land cover or vegetation cover and management; and (vi) P – erosion control practice factor. The model is as follows:

$$A = R*K*L*S*C*P$$

While A is the prediction of average annual soil loss (ton/ha/yr).

In the raster data or map algebra, each of these factors is an input of the raster map in the "Raster Calculator" function. In the "Raster Calculator" window, each of them is treated as a variable for computations using the above model (Figure 7.19). The result of this geoprocessing is a raster map of the prediction of average annual soil loss (A- ton/ha/yr).

## References

Chang, K.T. (2008). *Introduction to Geographic Information Systems (GIS)*, 4e. p. 450. Boston, Burr Ridge, IL, New York, San Francisco, Bangkok, Taipe: McGraw Hill Publication, Inc.

DeMers, M.N. (2005). *Fundamentals of Geographic Information Systems (GIS)*, 3e. p. 468. New York, Chichester, Weinheim, Brisbane, Singapore, Toronto: John Wiley & Sons, Inc.

Erdem, F., Atun, R., Avdan, Z.Y. et al. (2021). Drought analysis of Van Lake Basin with remote sensing and GIS technologies. *The Egyptian Journal of Remote Sensing and Space Sciences (EJRS)* 24 (3): 1093–1102.

Greenlee, D.D. (1987). Raster and vector processing for scanned linework. *Photogrammetric Engineering and Remote Sensing* 53 (10): 1383–1387.

Jensen, J.R. (2015). *Introductory Digital Image Processing, A Remote Sensing Perspective*, p. 623. Upper Saddle River, NJ: Pearson, Inc.

Mark, D.M. (1974a). Line intersection method for estimating drainage density. *Geology* 2: 235–236.

Mark, D.M. (1974b). On the interpretation of till fabrics. *Geology* 2: 101–104.

Mark, D.M. (1975). Computer analysis of topography: A comparison of terrain storage methods. *Geografiska Annaler* 57A: 179–188.

Qin, C., Zhu, A.X., Pei, T. et al. (2007). An adaptive approach to selecting a flow partition exponent for a multiple flow direction algorithm. *International Journal of Geographical Information Science* 21 (4): 443–458.

Tarboton, D.G. (1997). A new method for the determination of flow directions and upslope areas in grid digital elevation models. *Water Resources Research* 33 (2): 309–319.

USDA. (1965). Rainfall erosion losses from Cropland East of Rocky Mountains – guide for selection of practices for soil and water conservation. *Agriculture Handbook* 282: 48.

## Tutorial in ArcGIS Pro Environment

GIS project for identifying breeding season habitats of the red-winged blackbirds for the natural environment conservations in Erie County, New York.

The red-winged blackbird is a migrating bird species that is spread throughout the eastern part of North America. The birds spend winter seasons in the Southern United States and Central America, reaching as far as the northern part of Costa Rica and south of Mexico. Every spring and summer season, the birds fly to local Western New York areas for breeding. The Erie County government collaborates with the New York State Department of Environmental Conservation (New York DEC) to identify the breeding habitat areas and study the natural conditions for the protection of this species (Figure 7.1).

As a GIS analyst of the county, you are going to utilize the raster datasets to conduct geoprocessing and analyses. Our task is to identify the areas in the county with the most favorable habitat conditions for this bird species to breeding.

Copy the tutorial data folder "Chapter07-Tutorial" to your working directory.

Open ArcGIS Pro and create a new ArcGIS Pro project file that is called "MyGIS_Learning07" using the "Map" template. Save the project file into the designated working folder, c:\ or D:\...... Chapter07-Tutorial.

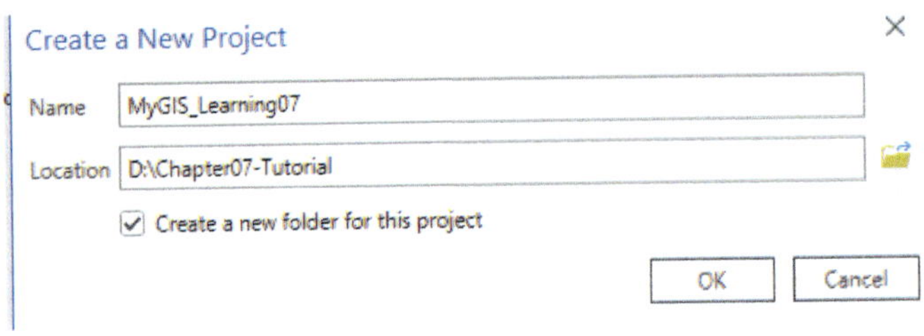

**Figure 7.1** Red-winged blackbird. *Source*: Dr. Tao Tang.

Now, add the entire Chapter07-Tutorial folder on the Catalog Pane using the contact menu of "Add Folder Connections."

According to the report of a local bird-watching group, migrating red-winged blackbirds are spotted in the Spring or Summer seasons year by year in Western New York. However, the number of reported discoveries in the natural field is trending downward. These birds come to the territory of Erie County to breed the next generation. Erie County government and the New York DEC would like to analyze the breeding habitat environment of the red-winged blackbirds.

Field investigations and studies have led scientists to summarize the following favorable habitat conditions for red-winged blackbirds to breed. (i) This bird species is likely to lay eggs in woody wetland areas close to swamps. However, the majority of these birds would avoid the water-rich areas in the wetland regions. (ii) The breeding habitats that are generally in the lowland regions within the elevations (altitudes) from 200 feet to 250 feet are most likely elevations. Meanwhile, the breeding areas need to be flat or gentle in slope, normally with the slope not exceeding 4° (degrees). We are going to utilize raster datasets and raster analytical methods to identify the breeding habitat areas of red-winged blackbirds in Erie County, NY in this project.

Let's add the three original datasets that we collected from various sources to the mapping window from the Catalog Pane. We downloaded these datasets from either the New York State GIS Clearing House website (https://gis.ny.gov/), from the Cornell University Geospatial Information Repository (CUGIR) website, or from the New York State DEC. These are: (i) Erie_County_Crop_Lands in raster dataset, (ii) Erie_Dem (Digital Elevation Models) in raster dataset, and (iii) Erie_County_Soils in vector polygon feature class.

In preparation to conduct this GIS project, we merged a total of 30 downloaded digital elevation model (DEM) quadrangles to produce the Erie_Dem raster dataset. However, the geographic extent of this dataset is too big and beyond the county boundary, so we need to trim it first. At the Analysis ribbon on the main (top) menu, click on the Tools icon to open Geoprocessing Pane. Type in "clip raster" to search the tool. You may notice that the "Clip Raster" tool is promoted to the top of the list.

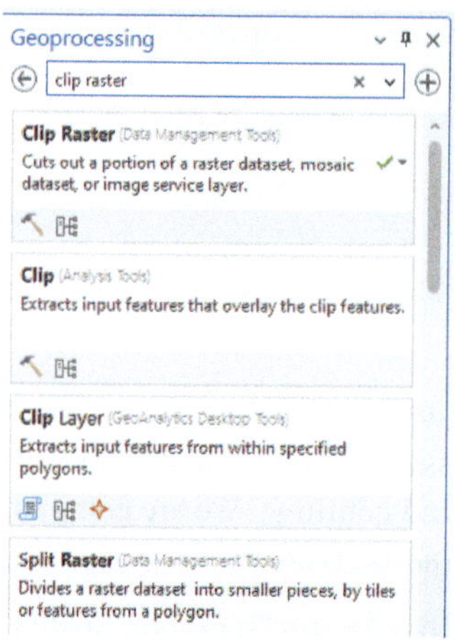


Click on the panel to activate the tool. For "Input Raster," select Erie_Dem. For "Output Extent," select Erie_County_Crop_Lands. For "Output Raster," navigate to your Erie_Redwinged_Blackbirds.gdb and name it as "Erie_County_DEMs." Flip to the "Environments" tab and set up the cell size in raster analysis "is similar to Erie_Dem." Process the tool, and you may notice the new dataset is added to the mapping window.

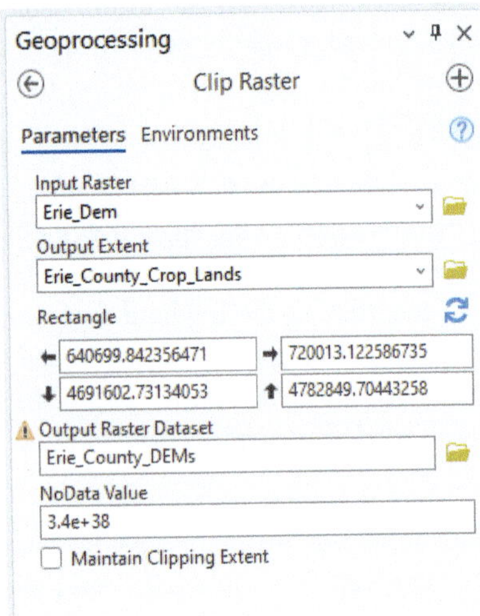


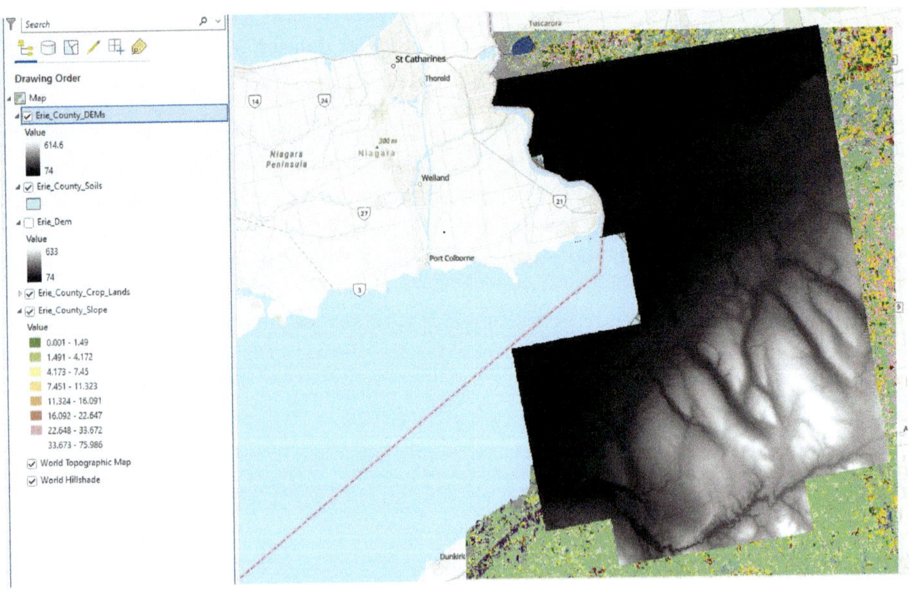


Now, we can remove the Erie_Dem raster dataset from the mapping window. At the Catalog Pane, rename the "Map" window to "Breeding_Habitat." If the Geoprocessing Pane is closed, open it from the Analysis ribbon on the main menu; otherwise, activate it and continue. We are going to create a raster slope map layer using the elevation models of Erie_County_DEMs.

In this exercise, we are going to use the functions and tools in the Spatial Analyst Tools box. Spatial Analyst is one of the extensions of ArcGIS Pro that is separately licensed. We need to check and make sure that we have adequate software licenses. Click on the project at the main menu on the top of the software window. Then, highlight the "Licensing." We have the following list and statement.

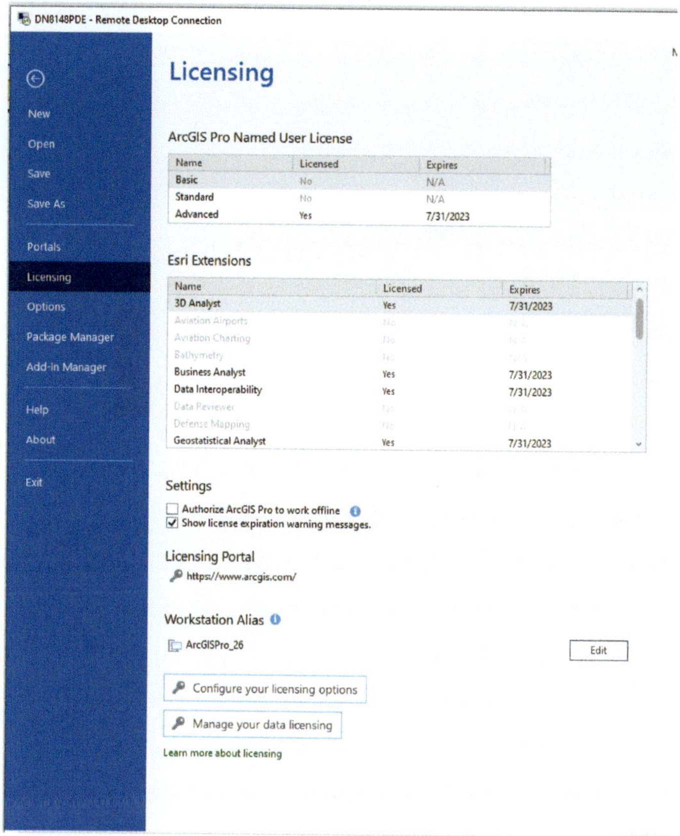


Browse down the list, and we see that we do have the Spatial Analyst extension license. This process of license checking is very important. If we do not have the license, even if we have all the tools available in the toolbox in the Geoprocessing Pane, we will not be able to use them.

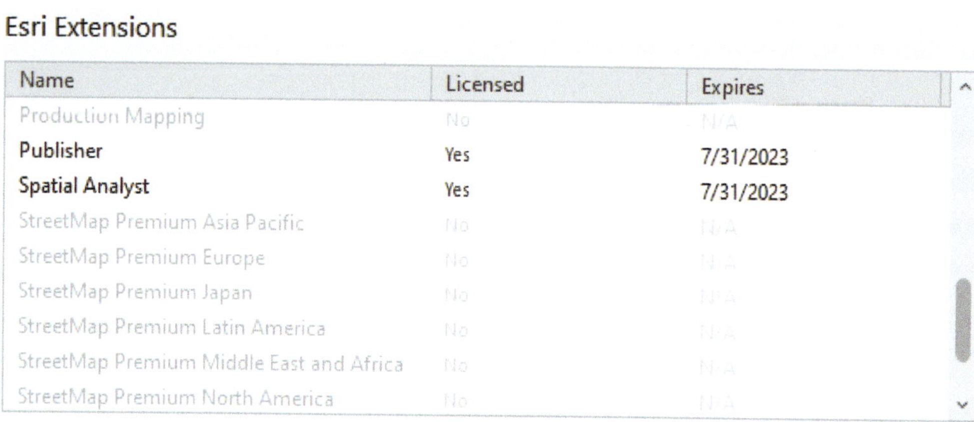


Now, let's go back to the mapping window, Breeding_Habitat (clicking the back arrow). On the Geoprocessing Pane, activate the Toolboxes tab, expand the Spatial Analyst Tools box. This is a collection of tools using the Spatial Analyst extension license. In this case, we expand the Surface toolset.

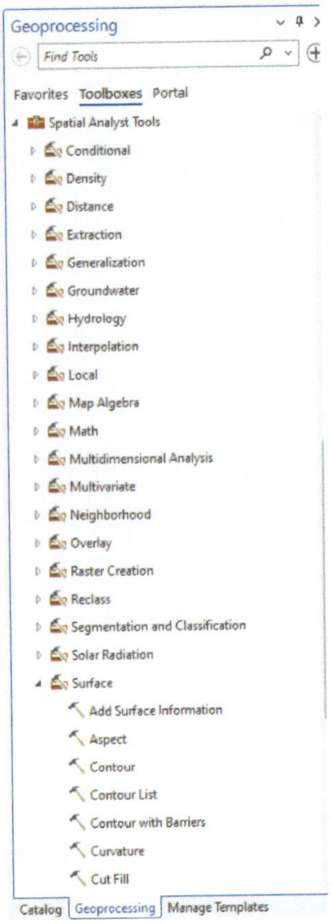


Select the Slope tool on the bottom of the list, and the dialog opens. For the "Input raster," choose Erie_County_DEMs. For the "Output raster," navigate to your Erie_Redwinged_Blackbirds.gdb and name it as "Erie_County_Slope." Keep the "Output measurement" as Degree and other parameters as default. Process the tool, and you may notice that the new Erie_County_Slope raster map layer is created.

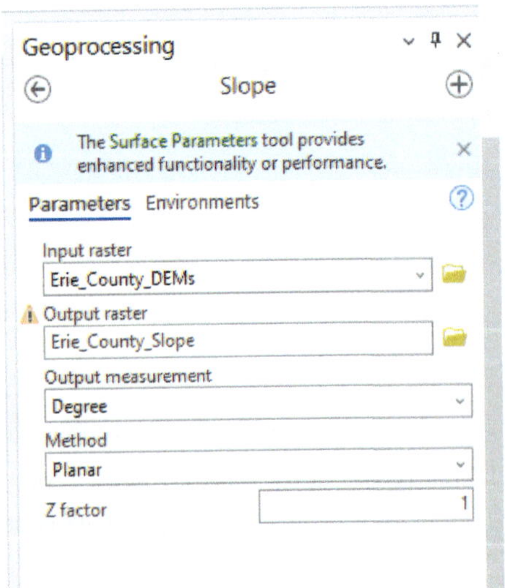

Use the Symbology Pane to classify the slope degrees in eight categories and select the "Color Scheme" as the Slope that is listed on the list. You need to check the "Show names" to see the names. Then, close the Symbology Pane.

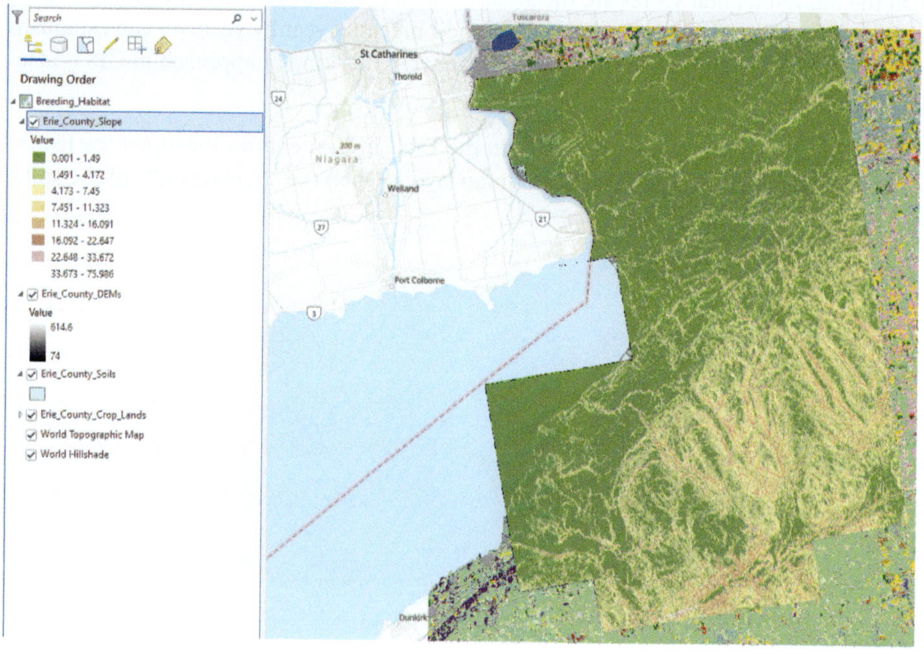


According to discussions in a meeting with local experts in ecology, four major environmental factors heavily impact or contribute to the breeding habitat of red-winged blackbirds.

1. Vegetation or land cover type should be woody wetlands without any agricultural cultivations or farming. The area should not be in the core regions of water-filled wetlands or wetlands with ponds.
2. Elevation or altitude above sea level should be in the range of 200–250 feet. The areas should have a very gentle slope, and the slope should be lower than 4°.
3. The soil types should be wetland soils with some kind of muck material accumulations.

So far, we have collected and processed a total of four geographic or spatial datasets for this study area, the Erie County. These are three raster datasets, namely, Erie_County_Crop_Lands, Erie_County_DEMs, and Erie_County_Slope; and one vector feature class that is called "Erie_County_Soils." In order to achieve our objective to find the most favorable habitat areas across the county, we need to conduct three steps. First, we need to select the suitable soils that the ecological experts identified from the vector soils feature class (SSURGO). Then, we will convert soil polygon features to a raster dataset. SSURGO stands for the Soil Survey Geographic Database that was developed by the Natural Resources Conservation Service at the US Department of Agriculture (USDA). It is the detail level of soil classification using the comprehensive soil taxonomy. Each soil type at this level has a unique combination of physical, chemical, nutrient, and moisture properties. In the second step, we will identify the most favorable habitat conditions for each of the four parameters, namely vegetation or land cover, soils, elevation or altitude, and slopes. Then, we will reclassify each of the categories according to one unified raster scoring range that we choose. This range is

1–10, where 1 is the lowest or most unfavorable score for that category of criteria, and 10 is the most favorable score for the category of the criteria. Finally, we will use the Map Algebra tool to add all the scores of the four factors together with a distributed weight to each of them. The areas that obtained the highest habitat scores are the areas of the most favorable habitat conditions and likely where the red-winged blackbirds will conduct their breeding year by year. We can establish these areas as protected regions within the county. In general, this kind of approach in GIS is also called suitability analysis.

Bring the vector feature class, Erie_County_Soils, in the Table of Contents on the top of the list in drawing order. If necessary, check off some raster map layers. Open the attribute table of the soils map layer.

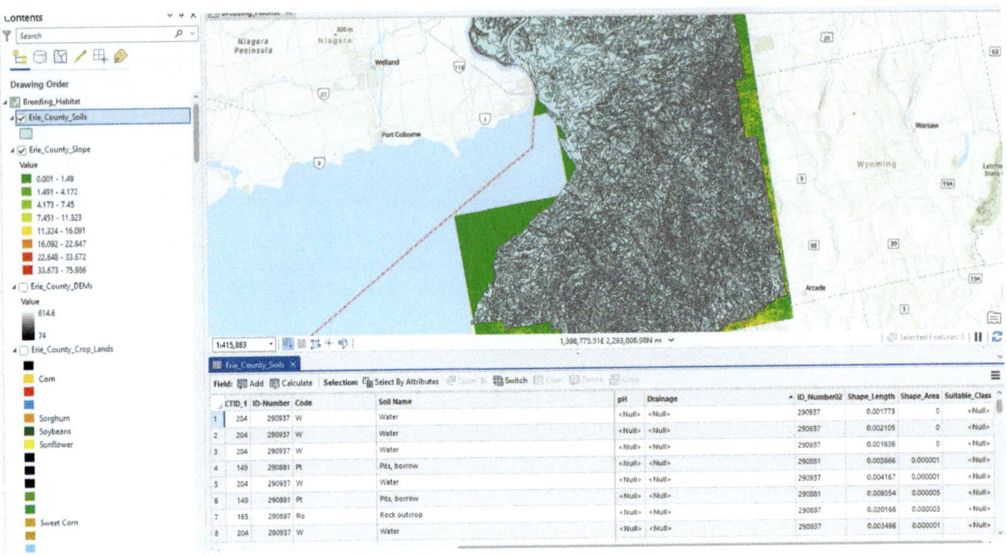


We would like to add another field to classify the soil suitability for the breeding habitat of the red-winged blackbirds. Now, we click on the contact menu icon on the upper right-hand side of the attribute table ≡ . The following contact menu opens. Select "Fields View" to open its tab.

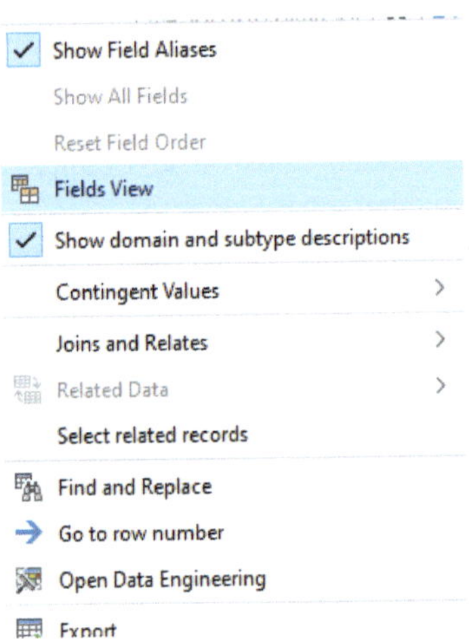

On the bottom of this dialog, add a new field that is called “Suitable_Class” and select the data type as “Short” (means short integer). Save the new update using the “Save” icon in the “Fields” ribbon.

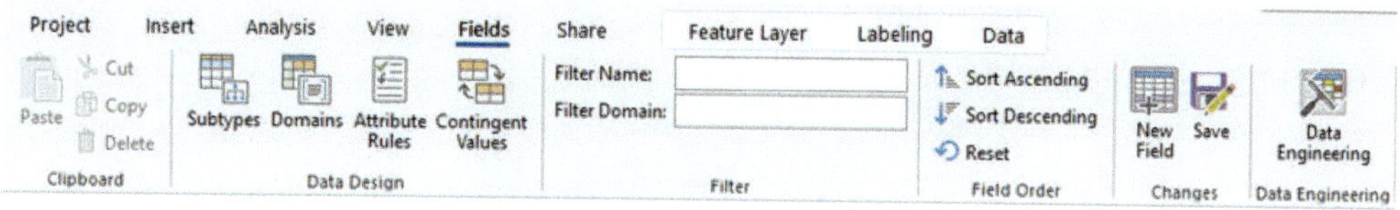


Close the Fields View tab. You may notice that the new field (column) is added to the attribute table on the far right-hand side.

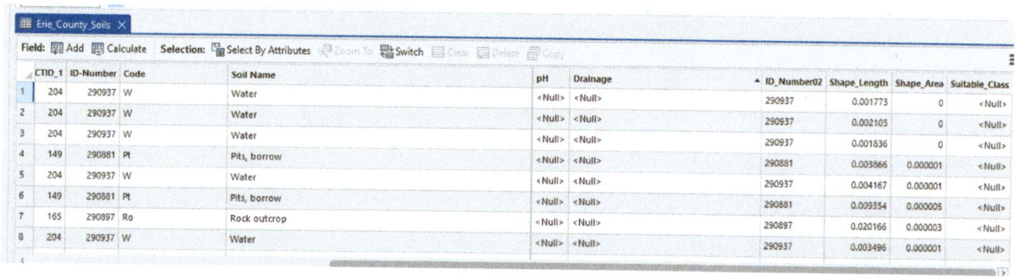

| | CTID_1 | ID-Number | Code | Soil Name | pH | Drainage | ID_Number02 | Shape_Length | Shape_Area | Suitable_Class |
|---|---|---|---|---|---|---|---|---|---|---|
| 1 | 204 | 290937 | W | Water | <Null> | <Null> | 290937 | 0.001773 | 0 | <Null> |
| 2 | 204 | 290937 | W | Water | <Null> | <Null> | 290937 | 0.002105 | 0 | <Null> |
| 3 | 204 | 290937 | W | Water | <Null> | <Null> | 290937 | 0.001836 | 0 | <Null> |
| 4 | 149 | 290881 | Pt | Pits, borrow | <Null> | <Null> | 290881 | 0.003866 | 0.000001 | <Null> |
| 5 | 204 | 290937 | W | Water | <Null> | <Null> | 290937 | 0.004167 | 0.000001 | <Null> |
| 6 | 149 | 290881 | Pt | Pits, borrow | <Null> | <Null> | 290881 | 0.009354 | 0.000005 | <Null> |
| 7 | 165 | 290897 | Ro | Rock outcrop | <Null> | <Null> | 290897 | 0.020166 | 0.000003 | <Null> |
| 8 | 204 | 290937 | W | Water | <Null> | <Null> | 290937 | 0.003496 | 0.000001 | <Null> |

Close the attribute table. Activate the Map ribbon on the main menu. Click on Select by Attributes icon in the Selection group of the Map ribbon  . The dialog opens. Fill the query as MUSYM = “Cd” or MUSYM = “Lb” or MUSYM = “Ld” like the following. Keep the default of “New selection.” Click “OK” to process the query to select the features. The code “Cd” is Canandaigua mucky silt loam soil; the code “Lb” is Lakemont mucky silt loam soil; and the code “Ld” is Lamson mucky very fine sandy loam soil.

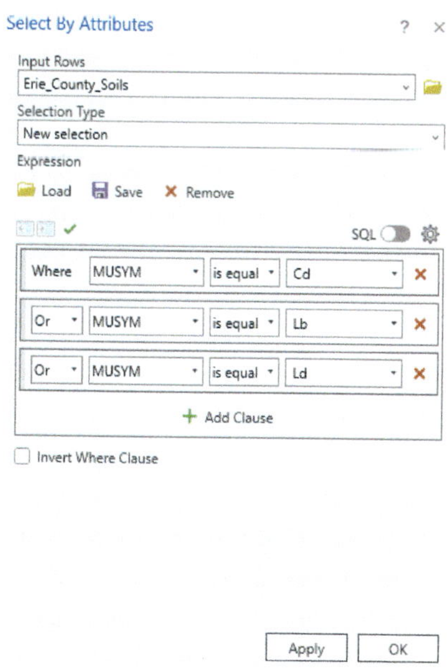

Now, open the attribute table and activate the "Show selected records' icon. Right-click on the title of the "Suitable_Class" field (column) to obtain the contact menu. Then, select Calculate Field. The dialog opens.

Keep everything else as the default, and type in "10" for "Suitable_Class =." Process this calculation. You may notice that "10" is added to all the selected rows for Suitable_Class.

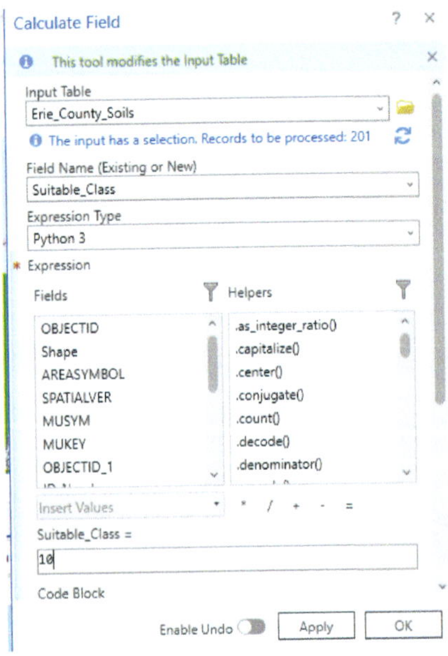

Close the attribute table and clear all the selections at the Selection group  .

Now, we repeat the above steps (from the bottom of page 7) to select polygon features using the attribute query of MUSYM = "Ed" or MUSYM = "Pa" and assign the "Suitable_Class =" to "9." The code "Ed" is Edwards muck, and the code "Pa" means Palms muck.

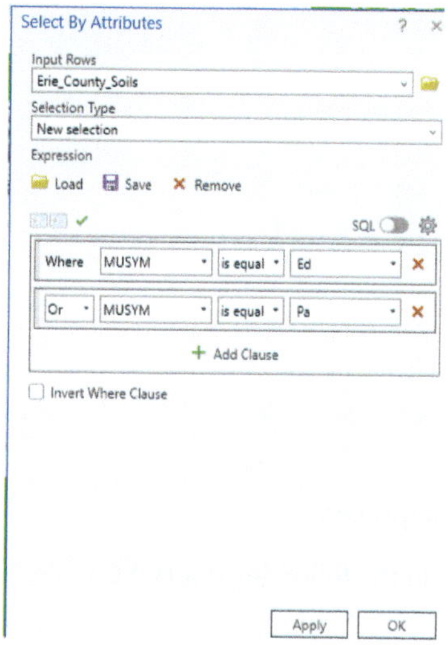


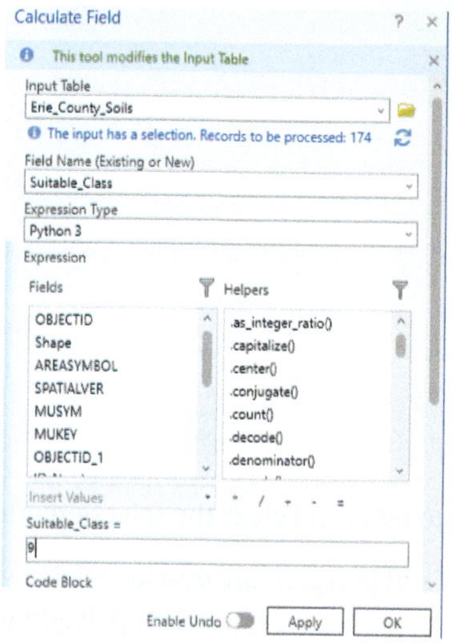


Repeat these steps again with the attribute query MUSYM = "Ha" and assign the "Suitable_Class =" to "8." "Ha" means Halsey silt loam.

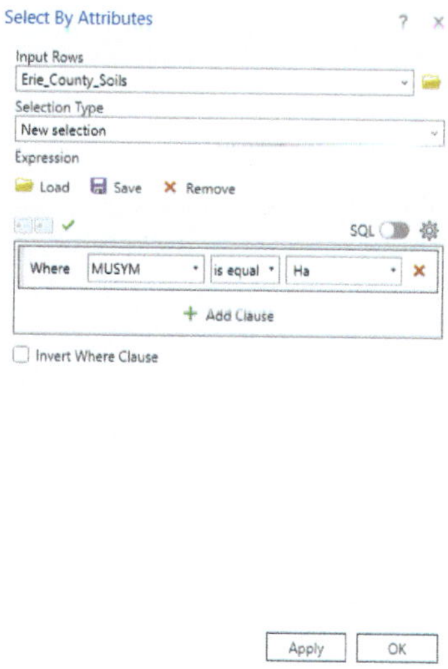


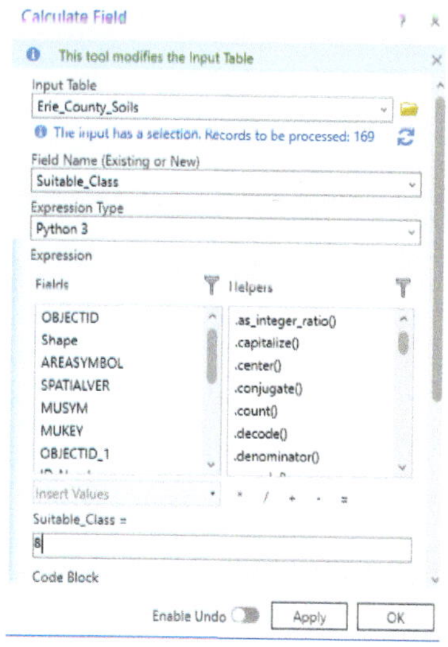

Each time, close the attribute table and clear the selection after the process.

Make sure all the features of soil polygons are deselected. Now, build the following query, Suitable_Class >= 0 as following. Process the query to select polygon features of soils.

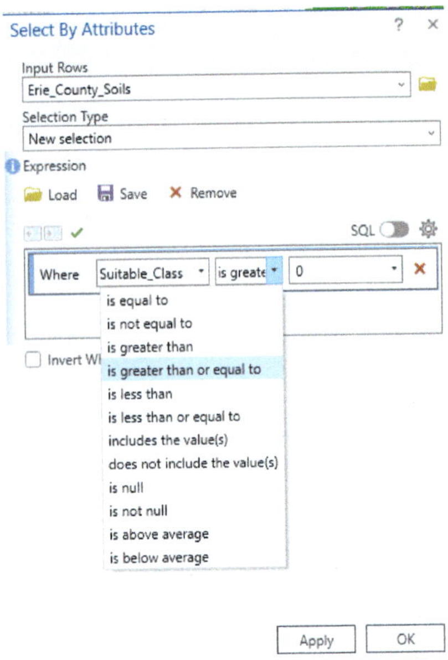


Open the attribute table, and you may notice that all the features that have a suitable class are selected.

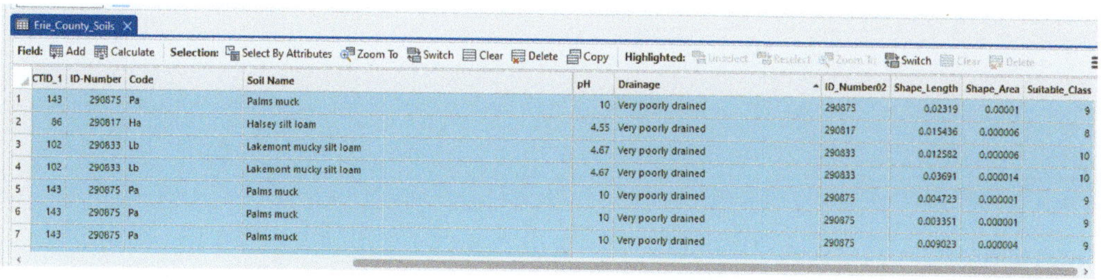

| | CTID_1 | ID-Number | Code | Soil Name | pH | Drainage | ID_Number02 | Shape_Length | Shape_Area | Suitable_Class |
|---|---|---|---|---|---|---|---|---|---|---|
| 1 | 143 | 290875 | Pa | Palms muck | 10 | Very poorly drained | 290875 | 0.02319 | 0.00001 | 9 |
| 2 | 86 | 290817 | Ha | Halsey silt loam | 4.55 | Very poorly drained | 290817 | 0.015436 | 0.000006 | 8 |
| 3 | 102 | 290833 | Lb | Lakemont mucky silt loam | 4.67 | Very poorly drained | 290833 | 0.012582 | 0.000006 | 10 |
| 4 | 102 | 290833 | Lb | Lakemont mucky silt loam | 4.67 | Very poorly drained | 290833 | 0.03691 | 0.000014 | 10 |
| 5 | 143 | 290875 | Pa | Palms muck | 10 | Very poorly drained | 290875 | 0.004723 | 0.000001 | 9 |
| 6 | 143 | 290875 | Pa | Palms muck | 10 | Very poorly drained | 290875 | 0.003351 | 0.000001 | 9 |
| 7 | 143 | 290875 | Pa | Palms muck | 10 | Very poorly drained | 290875 | 0.009023 | 0.000004 | 9 |

Click the Switch icon on the title bar of the attribute table Switch . You may notice, the selection was switched to <Null> values in the column of Suitable_Class. Now, we use the Calculate Field dialog to fill these soil polygon features as "3." This means that the soil type is not suitable for the breeding habitat of the red-winged blackbirds. Click "OK" to process the value assignment.

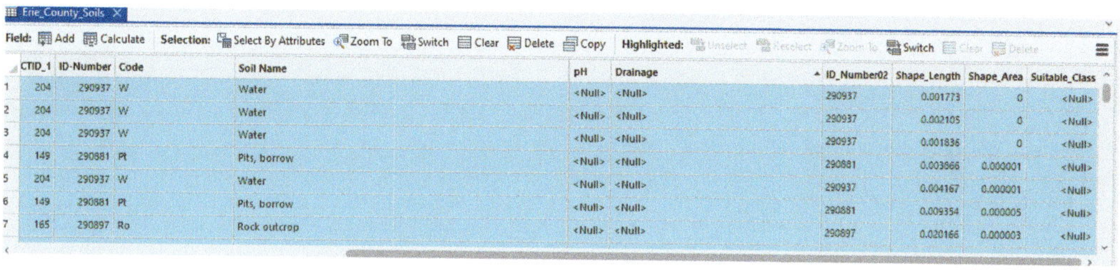

| | CTID_1 | ID-Number | Code | Soil Name | pH | Drainage | ID_Number02 | Shape_Length | Shape_Area | Suitable_Class |
|---|---|---|---|---|---|---|---|---|---|---|
| 1 | 204 | 290937 | W | Water | <Null> | <Null> | 290937 | 0.001773 | 0 | <Null> |
| 2 | 204 | 290937 | W | Water | <Null> | <Null> | 290937 | 0.002105 | 0 | <Null> |
| 3 | 204 | 290937 | W | Water | <Null> | <Null> | 290937 | 0.001836 | 0 | <Null> |
| 4 | 149 | 290881 | Pt | Pits, borrow | <Null> | <Null> | 290881 | 0.003866 | 0.000001 | <Null> |
| 5 | 204 | 290937 | W | Water | <Null> | <Null> | 290937 | 0.004167 | 0.000001 | <Null> |
| 6 | 149 | 290881 | Pt | Pits, borrow | <Null> | <Null> | 290881 | 0.009354 | 0.000005 | <Null> |
| 7 | 165 | 290897 | Ro | Rock outcrop | <Null> | <Null> | 290897 | 0.020166 | 0.000003 | <Null> |

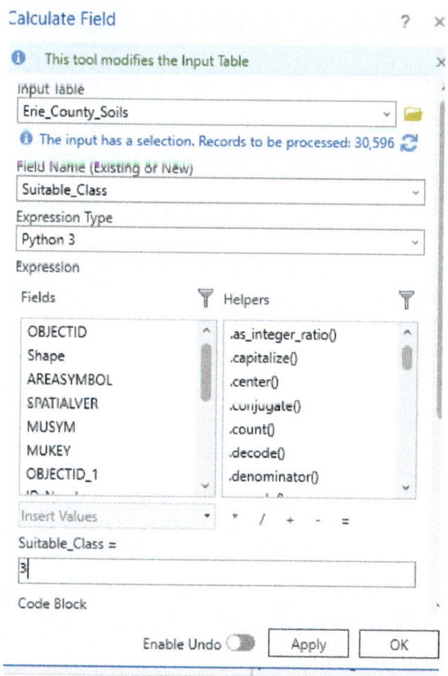


Close the table and clear the selection.

Now, we convert this soil polygon feature class to a raster dataset using the suitability class. On the Map ribbon at the main menu, click on the Tools icon to open the Geoprocessing Pane. Activate the Toolboxes tab. Expand the Conversion Tools box on the list, then expand the To Raster toolset. Select and click on the Polygon to Raster tool to open its dialog.

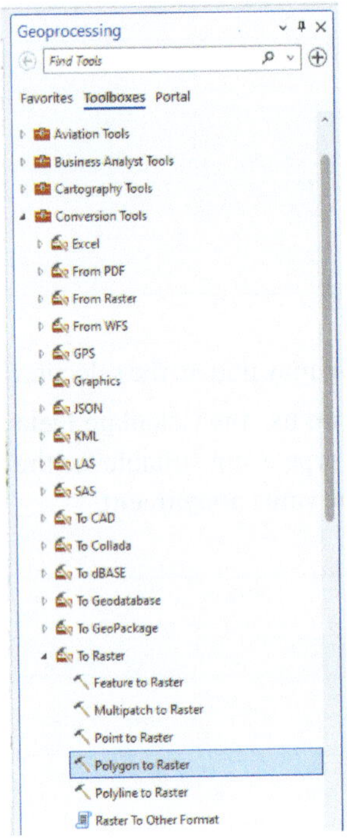


For "Input Features," select Erie_County_Soils. For "Value Field," select Suitable_Class field in the attribute table. For "Output Raster Dataset," navigate to your Erie_Redwinged_Blackbirds.gdb geodatabase.

Save the new dataset as "Soil_Suitability." Keep other parameters as default. Run the function to create the raster dataset.

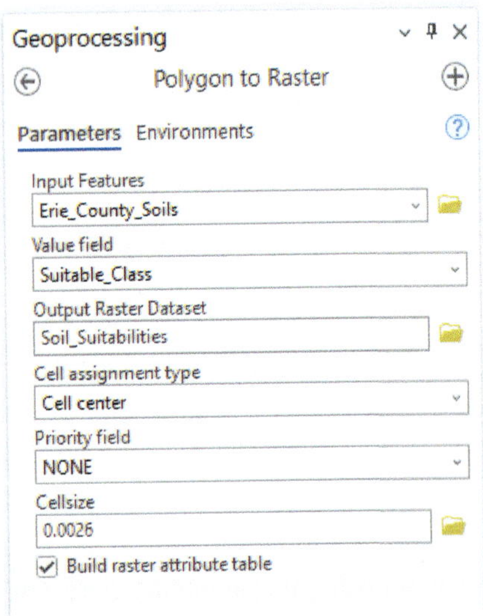

Move the newly created raster map layer to the top and inspect it. Close the Geoprocessing Pane.

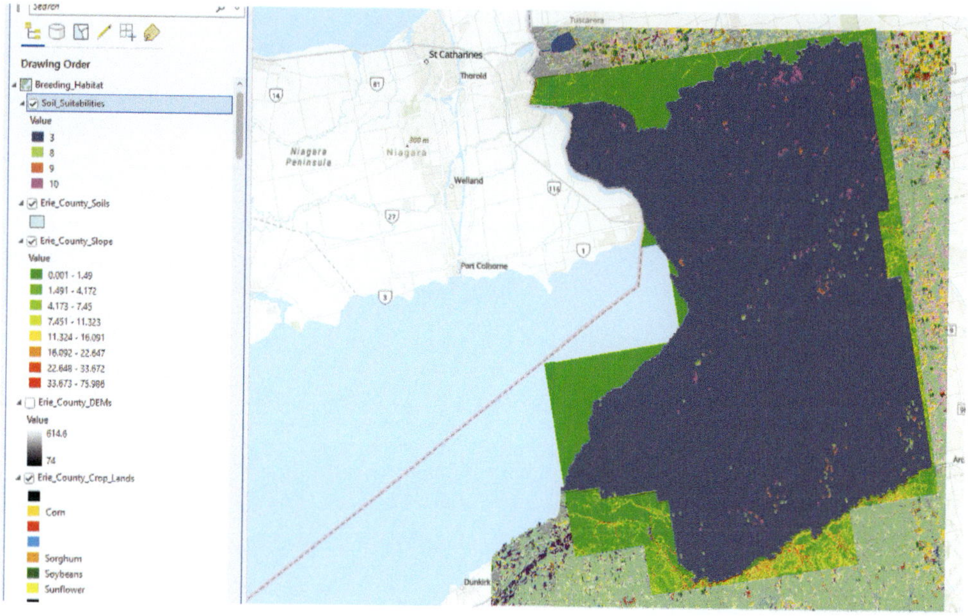


Now, we need to reclassify each of the other raster datasets according to the suitability for breeding habitat conditions for the red-winged blackbirds. First, let's reclassify the elevations or altitudes using the Erie_County_DEMs raster dataset. Check off other map layers and bring the Erie_County_DEMs map layer on the top of the drawing list. Open Geoprocessing Pane from Analysis ribbon at the main menu. Activate the Toolboxes tab. Expand the Spatial Analyst Tools toolbox, then the Reclass toolset. Highlight and click on the Reclassify tool to open its dialog.

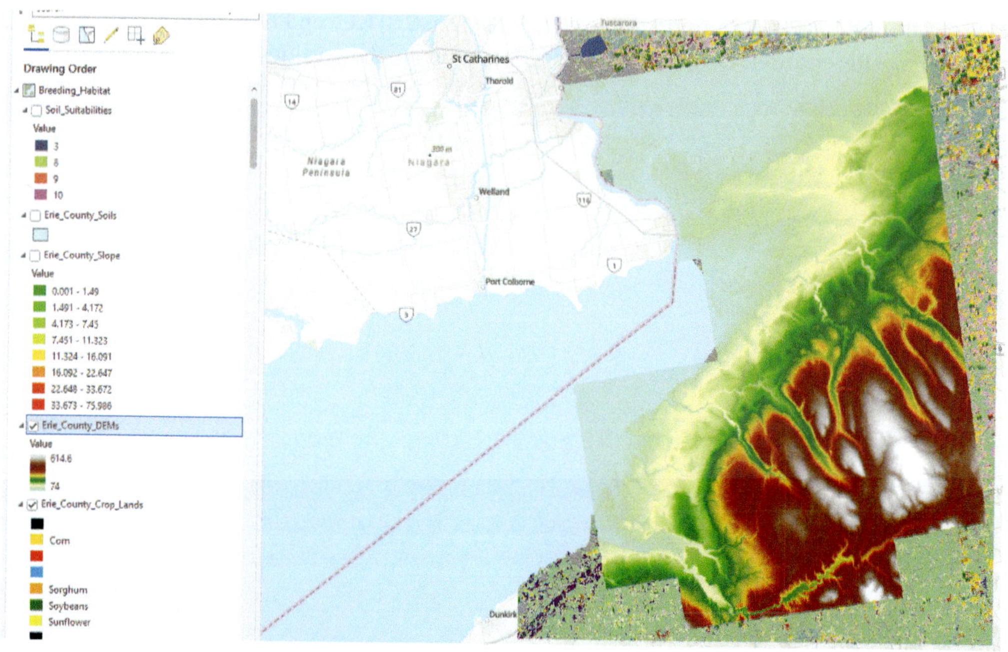

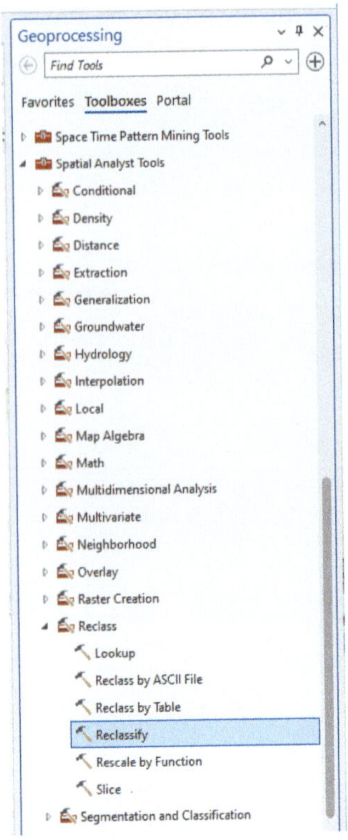


Select Erie_County_DEMs as the "Input Raster." For "Reclass Field," select VALUE. Click on the Classify tab to modify the classification ranges and categories. The second dialog of Classify opens. Choose three (3) categories. According to discussions with ecological experts, only the elevation range of 200–250 feet is suitable for red-winged blackbirds to breed. Therefore, we set up the upper value of category 1 at 200 feet, upper value of category 2 as 250 feet, and the category 3 follows the upper value of the entire elevation range. Click OK to promote these values to the Reclassify dialog.

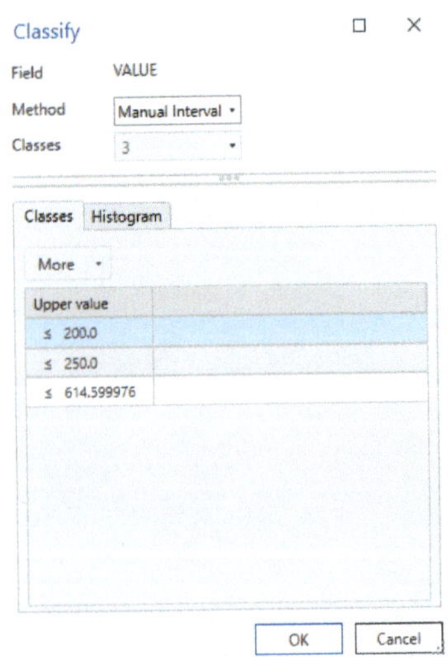

In the Reclassify dialog, type in the new values, category 1 = 3, category 2 = 10, category 3 =2, and NODATA = 0. In this case, we need to press the enter key on the keyboard to register the new value. For the "Output raster," navigate to your Erie_Redwinged_Blackbirds.gdb geodatabase and save the new raster dataset as "Elevation_Reclass."

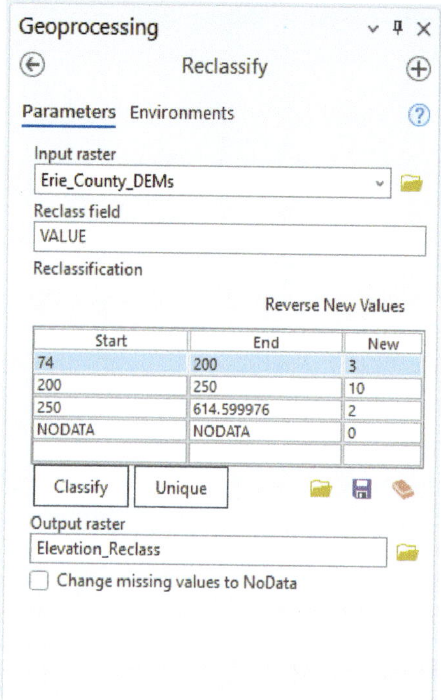


Click the Environment tab on the tool and set the "Mask" as Erie_County_Soils. Process the function to obtain the new raster dataset. You may notice that the new classification of elevations is added in the mapping window.

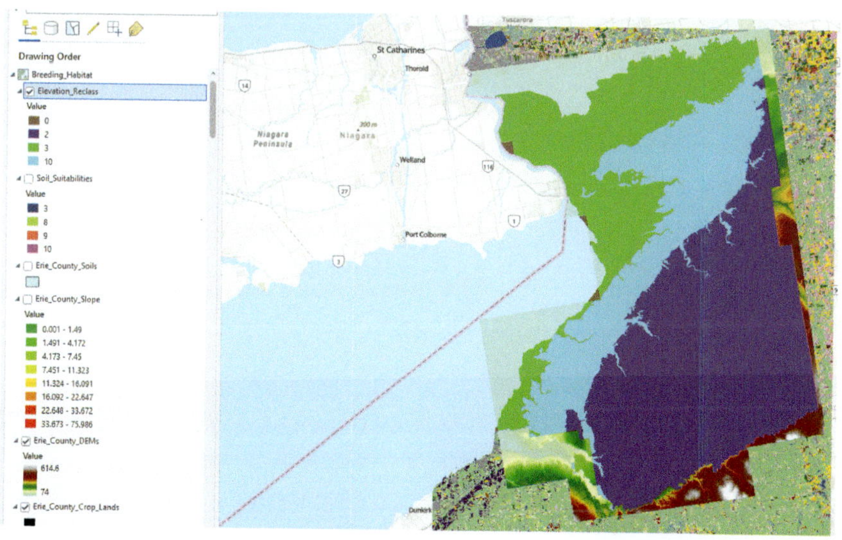

Secondly, we will reclassify the Erie_County_Crop_Lands raster dataset for the suitability to habitat conditions. Move the Erie_County_Crop_Lands map layer to the top of the drawing list in the Table of Contents. Open the value attribute table of the Erie_County_Crop_Lands raster dataset. Use the Field View tab in the attribute table window to create a new field (column) called "Suitable_Class."

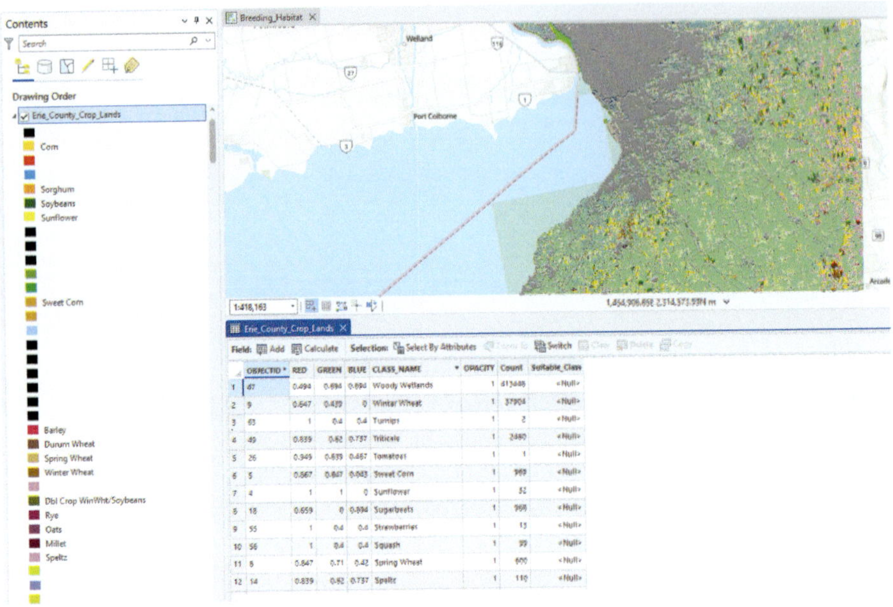


Now, close the attribute table. Activate the Map ribbon on the main menu. Click on the Select by Attribute icon in the selection group  to construct a query. According to the discussions with ecological experts, the woody wetlands are the suitable vegetation or land cover for the breeding habitat. Therefore, we construct the following query, CLASS_NAME = Woody Wetlands for the Erie_County_Crop_Lands raster dataset.

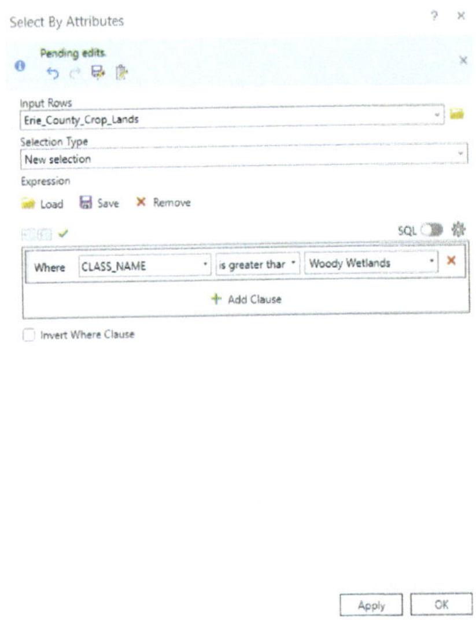

Click on "OK" to process the query. You may notice that the grid cells of woody wetlands are selected. Open the attribute table again to assign the value of "Suitable_Class" as "10" to the selected grid cells.

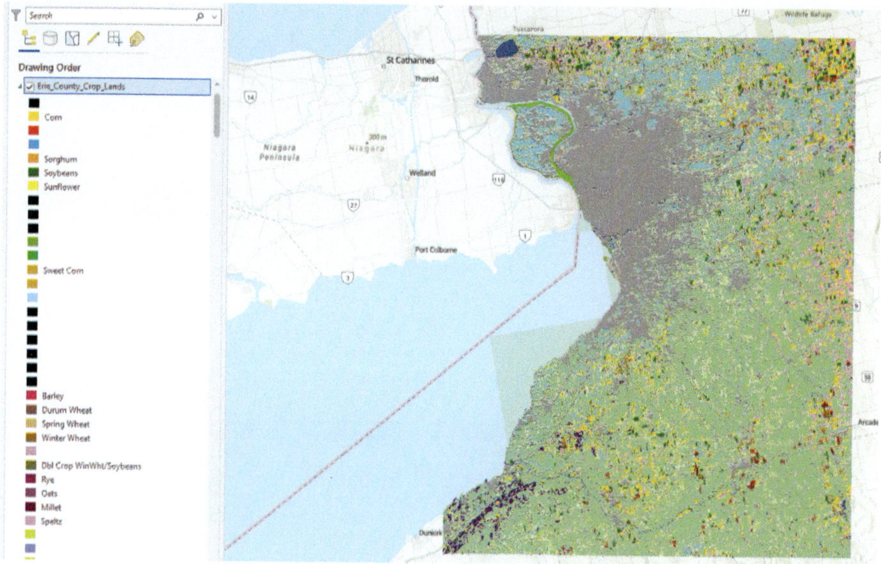


Clear the selection after the value is assigned and close the attribute table. Now, we follow the same steps above to select the Deciduous Forest, CLASS_NAME = Deciduous Forest. Open attribute table to assign this category the value of "Suitable_Class" as "5."

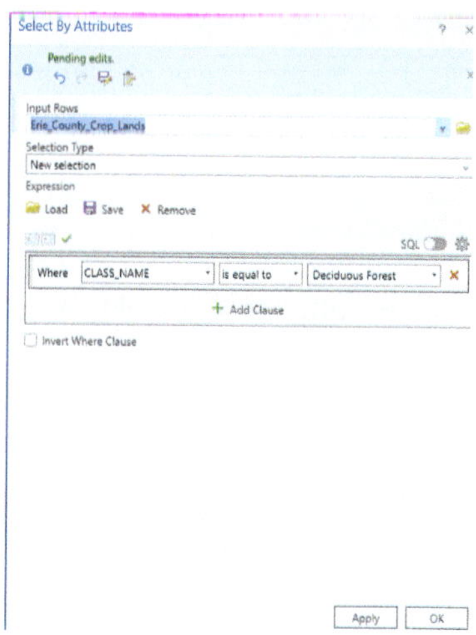


After assigning the value for the selected grid cells, clear the selection and close the attribute table. Now, we are going to assign the value of suitable class for all the other land use or land cover types including the crop vegetation and human urban development as "3." Open the Select by Attribute dialog again, and we build the following query and select the grid cells first, Suitable_Class >= (is greater than or equal to) 0.

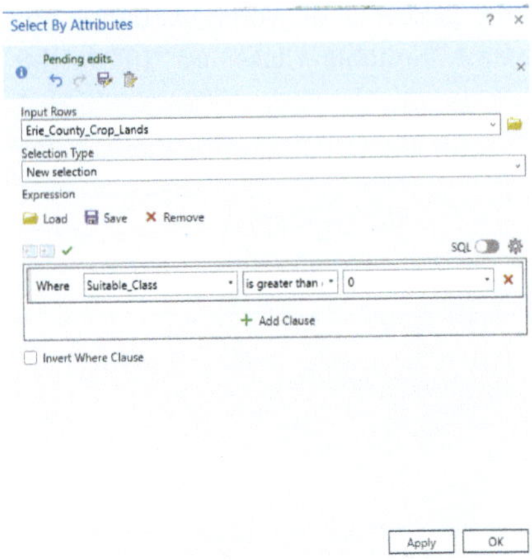


The selected map layer looks like this.

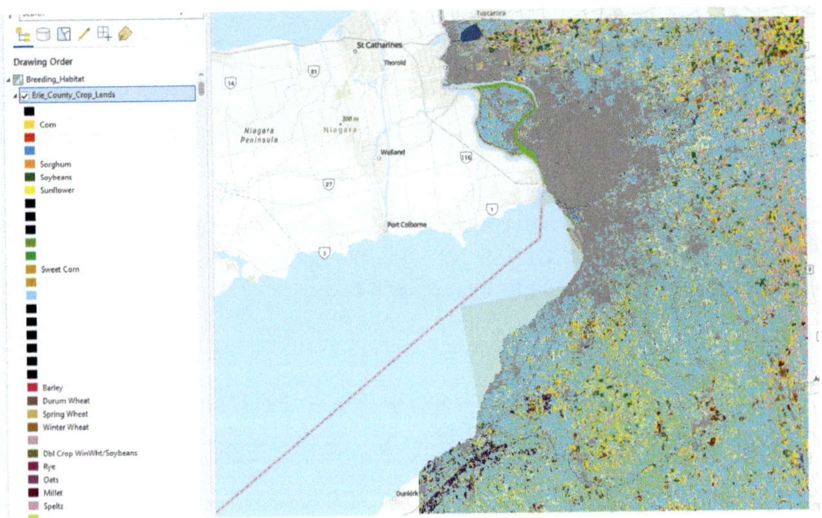


Now, open the attribute table and click on the Switch icon  on the title bar of the table. You may notice the selection is switched.

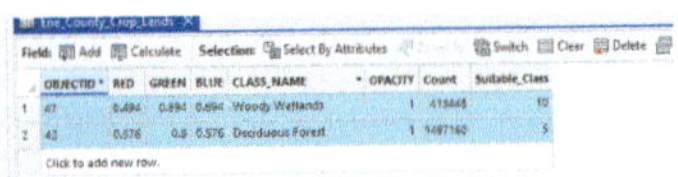


Selection before clicking on the Switch

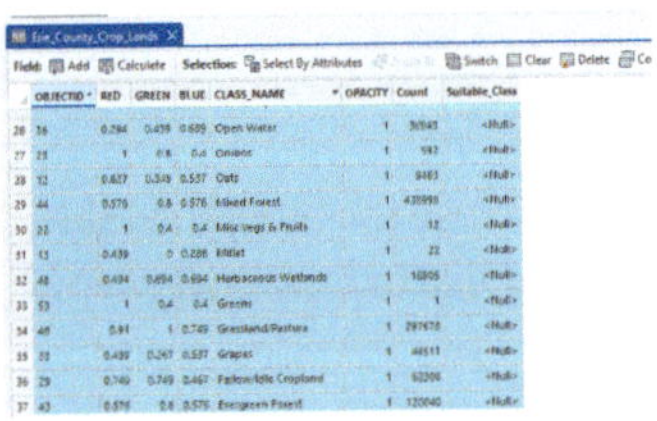


Selection after clicking on the Switch

Right-click on the title area of the Suitable_Class field to obtain the contact menu. Then, click on Calculate Field to open its dialog.

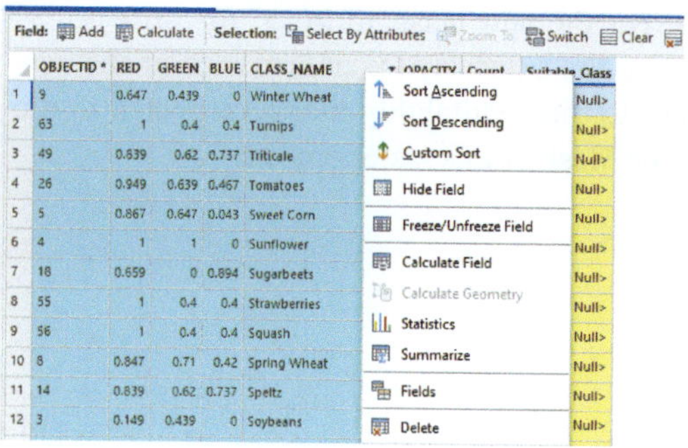


Type 3 to form the equation as Suitable_Class = 3.

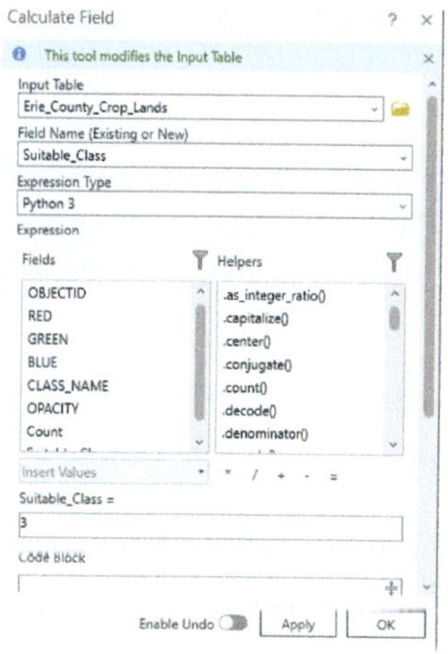


Process the function to fill the value in the Suitable_Class field. Clear the selection at the "Map" ribbon on the main menu and close the attribute table. Now, we are going to use the Reclassify tool in the Reclass toolset at the Spatial Analyst Tools box to create a vegetation suitability raster dataset. Open the Reclassify tool dialog on the Geoprocessing Pane. For "Input raster," select Erie_County_Crop_Lands. For the "Reclass field," choose Suitable_Class. It is automatically provided in three categories.

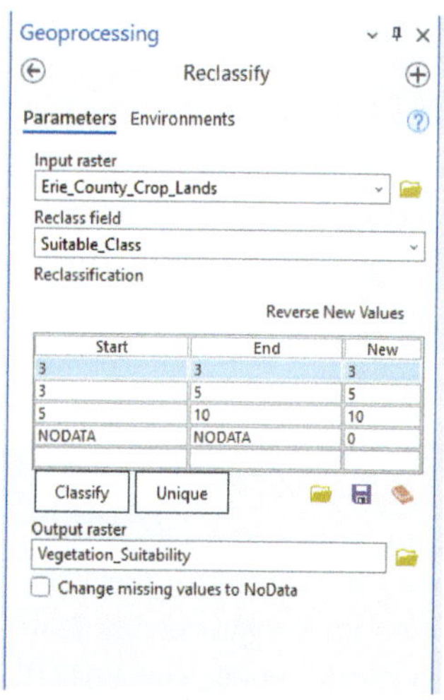

Click on the Classify to verify the original classes. For the new class fill in "3" for value 3, "5" for value 5, and "10" for value 10. You do need to use the enter key to confirm each of the new values. This process will change the grid cell values to these suitability numbers. Switch to the Environments tab and select Erie_County_Soils as the Mask of this raster process.

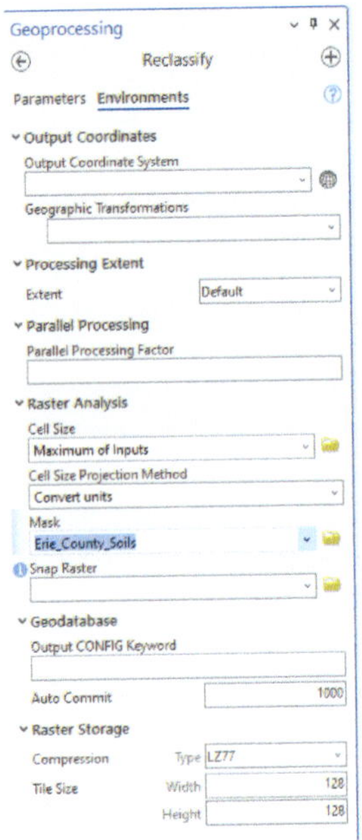


For "Output raster" navigate to your Erie_Redwinged_Blackbirds.gdb and save it as "Vegetation_ Suitablity." Run the function to obtain the result.

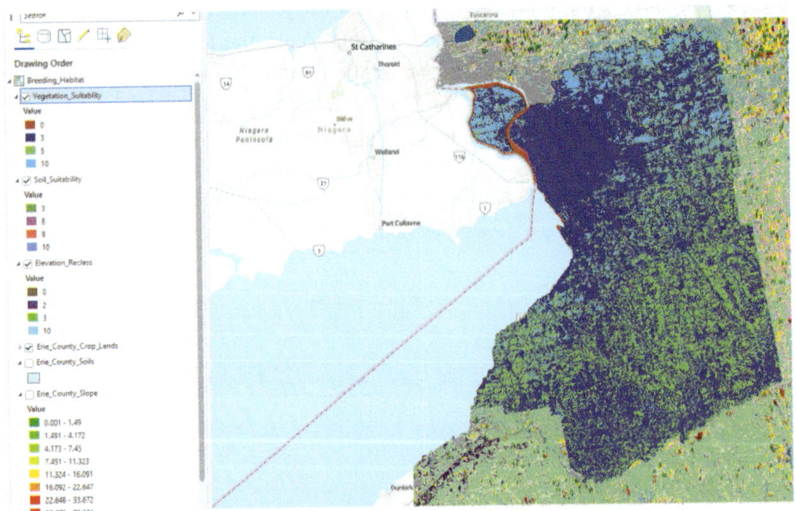


Finally, we are going to create the slope suitability raster map layer from the original Erie_ County_Slope raster dataset to unify the evaluation criteria and grid cell size. Move the Erie_ County_Slope layer on the top of the list in the Table of Contents. Open the Geoprocessing Pane and select Reclassify tool again. For the "Input raster," select Erie_County_Slope. The VALUE filed is automatically promoted as the classification field. Click on the Classify to open its dialog in justifying the classes and value ranges. Use the Natural Break (Jenks) method to change the classification to three categories. Then, manually type in 4 (4 degrees) for the first class, 10 (10 degrees) for the second category, and leave the third category to the highest value. Click OK to promote this classification of original slope degrees to the Reclassify dialog.

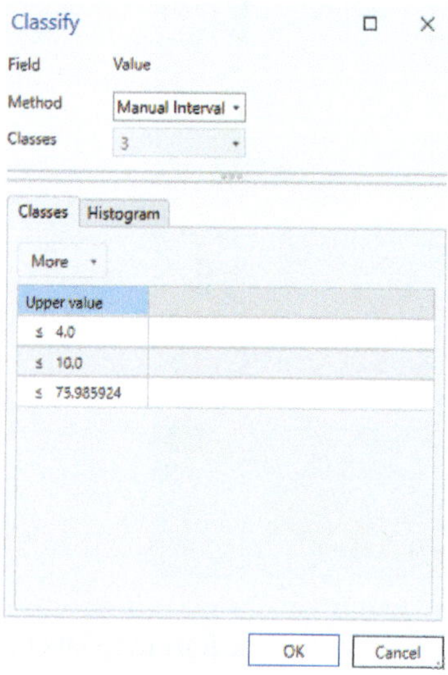


For the “New” values, manually type in “10” for the first class that the slope is lower than or equal to 4 degrees. Type “3” for the second category and “1” for the third one. For the no data, type in “0.” Navigate to your Erie_Redwinged_Blackbirds.gdb and save it as “Slope_Suitability.”

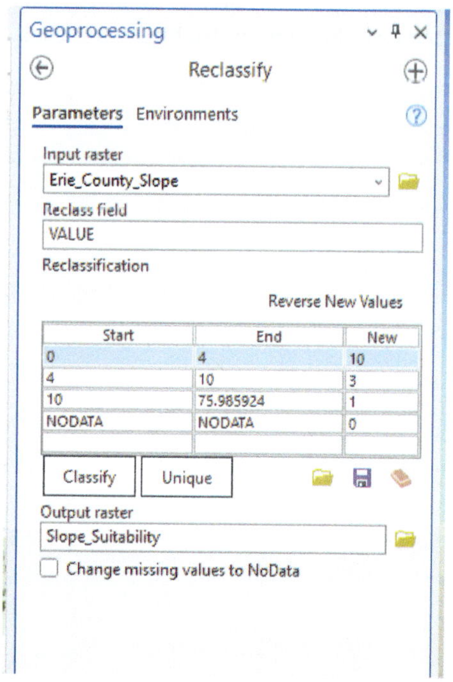

Meanwhile, switch to the Environments tab and select Erie_County_Soils as the Mask of this raster process. Process the tool or function to get the result.

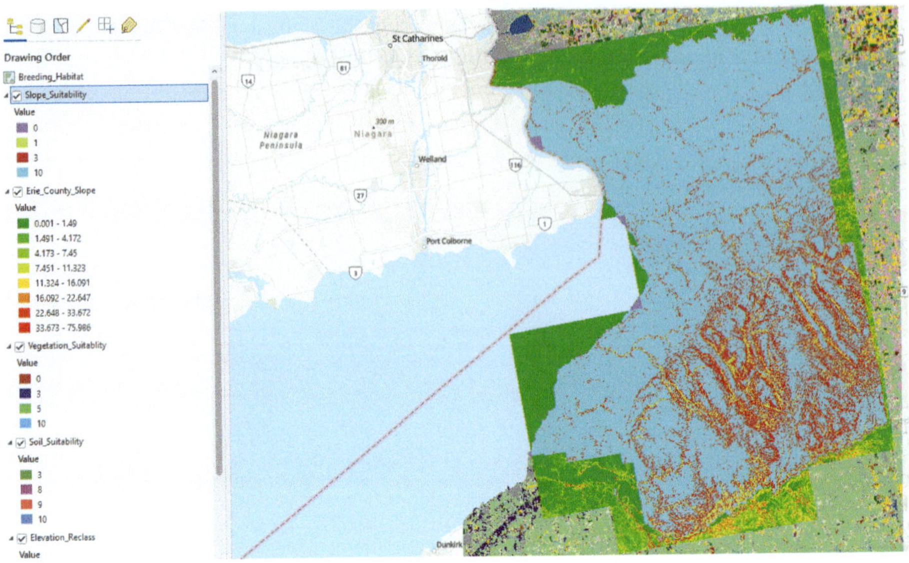


Now, we are going to use the raster map algebra tool (function) to overlay all these four map layers of habitat factors with differentiated weight parameters to evaluate the most suitable breeding habitat regions for the red-winged blackbirds. Vegetation cover and elevation factors are more important than the other two factors. Therefore, we use a weight of 0.3 for vegetation/land cover and elevation. The other two factors are assigned to weight of 0.2.

Open the Geoprocessing Pane if it is necessary, otherwise continue. Activate the Toolboxes tab. Expand the Spatial Analyst Tools box, then expand the Map Algebra toolset. Highlight and click on the Raster Calculator tool to open its dialog.

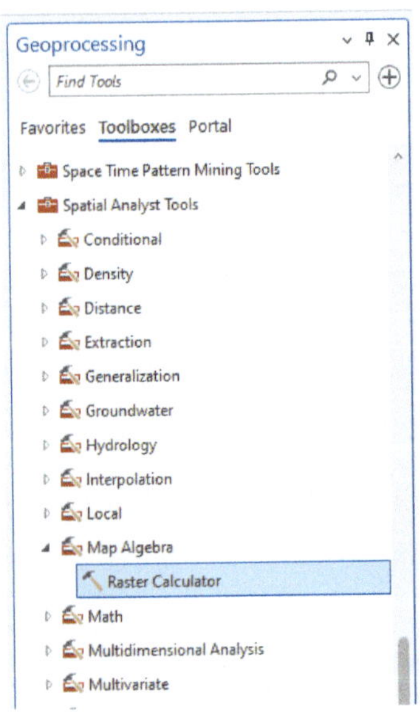

Using the keyboard to type in the number of weights, clicking on the raster datasets on the list of left-hand side, and clicking on the Operators of the Tools for multiply (*) or plus (+) operators, ext. to construct the following map algebra equation in the Output raster window.

**0.3 x "Vegetation_Suitability" + 0.3 x "Elevation_Reclass" + 0.2 x "Soil_Suitability" + 0.2 x "Slope_Suitability"**

(0.3 * "Vegetation_Suitablity" + 0.3 * "Elevation_Reclass" + 0.2 * "Soil_Suitability" + 0.2 * "Slope_Suitability")

For "Output raster," navigate to your Erie_Redwinged_Blackbirds.gdb and save it as "Red_Winged_Black_Bird_Breeding_H" (D:\Chapter07-Tutorial\Erie_Redwinged_Blackbirds.gdb\Red_Winged_Black_Bird_Breeding_H).

Run the Raster Calculator tool, you notice the new raster dataset is created.

This is a raster red-wringed blackbird breeding suitability map that shows the suitable habitat areas. Close the Geoprocessing Pane.

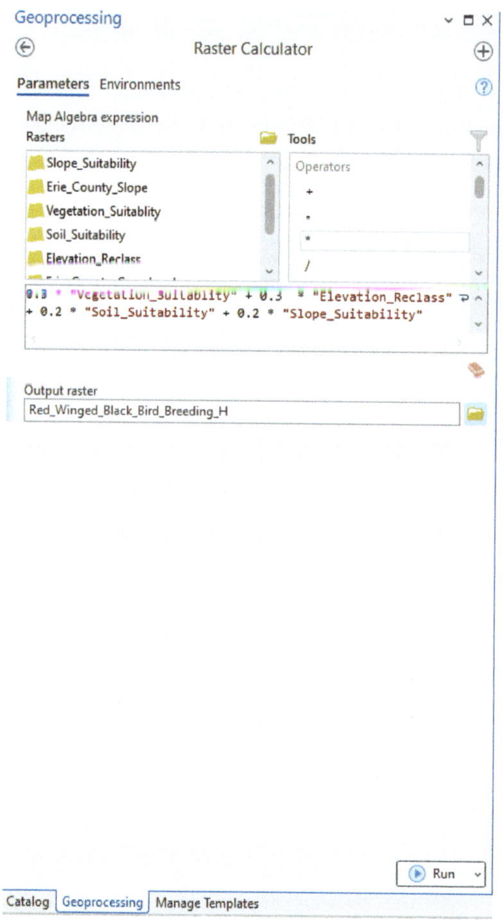

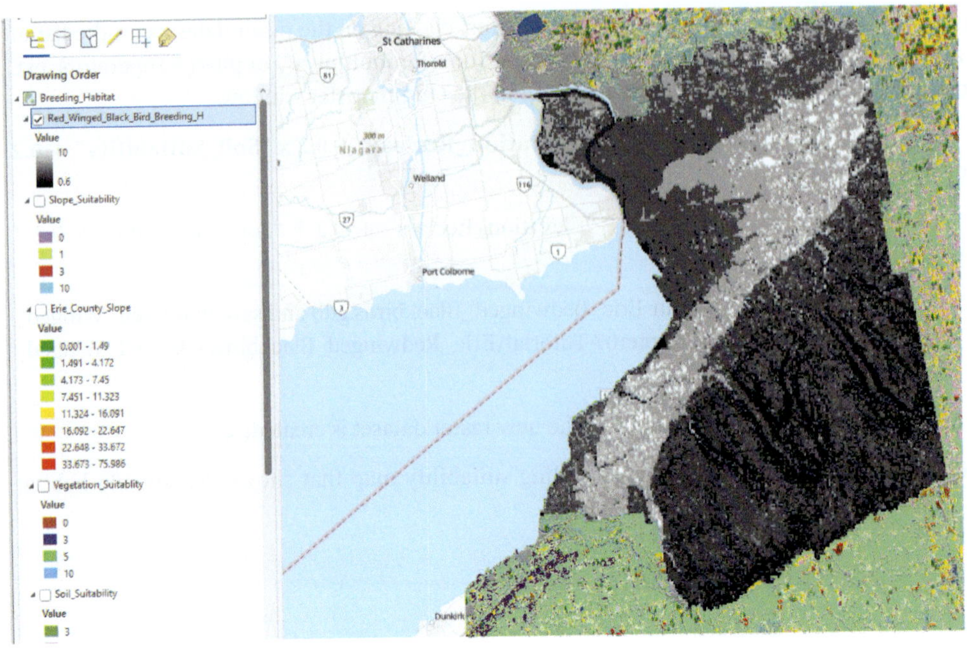


We can open the Symbology Pane for Red_Winged_Black_Bird_Breeding_H raster map layer to do a little bit of thematic mapping. Select Condition Number color scheme and check the Invert. This way makes green color represents a higher score of suitability. The color scheme of the map is changed.

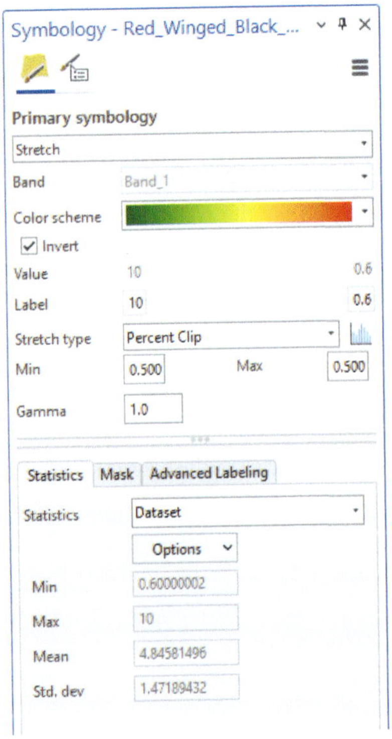

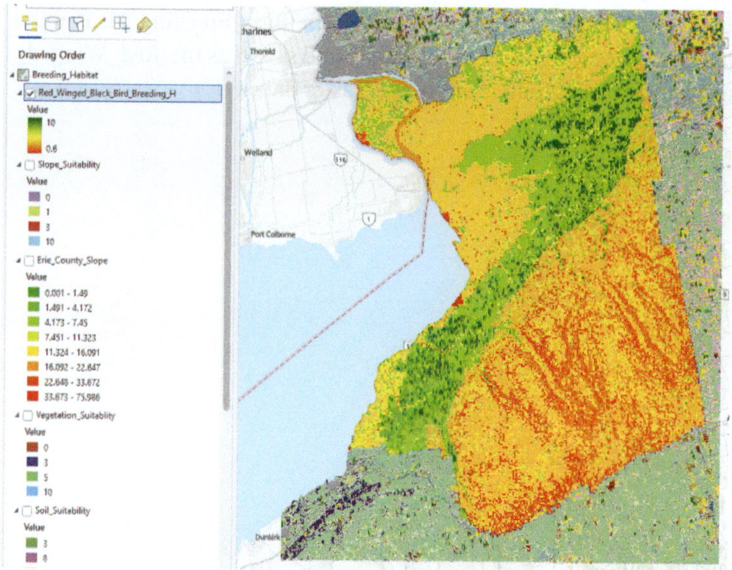


Close the Symbology Pane.

Since the patches of grid cells on this raster map look more fragmented, we would like to use raster neighborhood smoothing techniques to consolidate the breeding habitat areas. Before we can do that process, we need to convert the Red_Winged_Black_Bird_Breeding_H raster dataset to an integer format raster dataset.

Open the Geoprocessing Pane from the Analysis menu. Activate the Toolboxes tab and expand the Spatial Analyst Tools box. Then, expand the Math toolset. Highlight the Int (Integer) tool and open it.

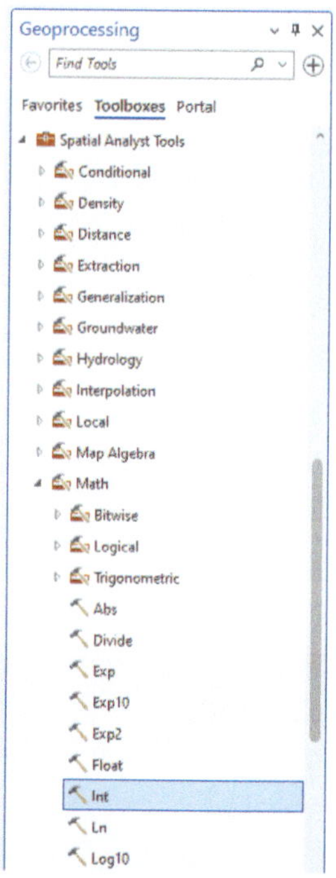

For the "Input raster or constant value," select Red_Winged_Black_Bird_Breeding_H. For the "Output raster," navigate to your Erie_Redwinged_Blackbirds.gdb and save it as Int_Red_Winged_Black_Bird_Breeding_H. Process the tool to produce an integer raster dataset.

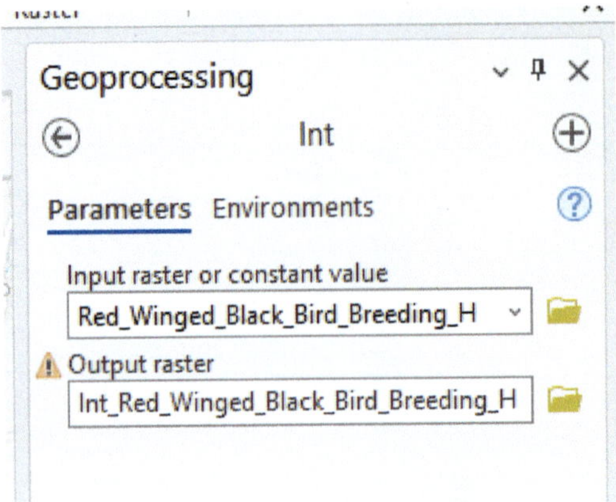


Now, we go back to the Toolboxes tab on the Geoprocessing Pane. Collapse the Math toolset and expand the Neighborhood toolset. Highlight and click on the Focal Statistics tool to open its dialog.

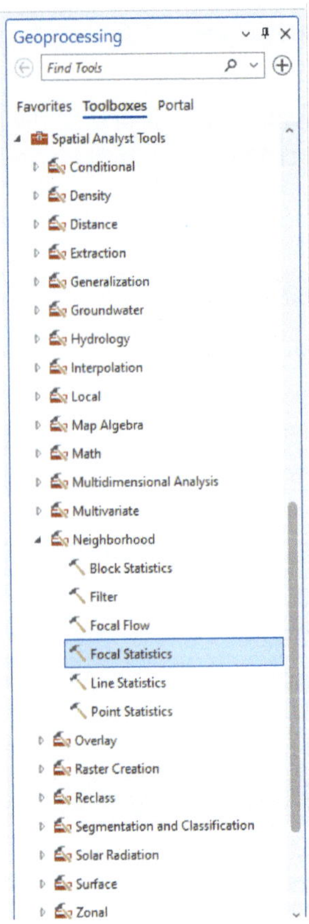

For the "Input raster," select Int_Red_Winged_Black_Bird_Breeding_H. For the "Output raster," navigate to your Erie_Redwinged_Blackbirds.gdb and save it as Breeding_Habitat_Final. Keeping the neighborhood as rectangle format, change the width and height of the cells from 3 to 4. This means that we are going to use a 4 by 4 window of grid cells to smooth the cell values in a local area. For "Statistics Type," select the Majority. That means when calculating a value of the cell, it will use majority values in the moving window of 4 × 4. Process the tool, and the new raster dataset is produced.

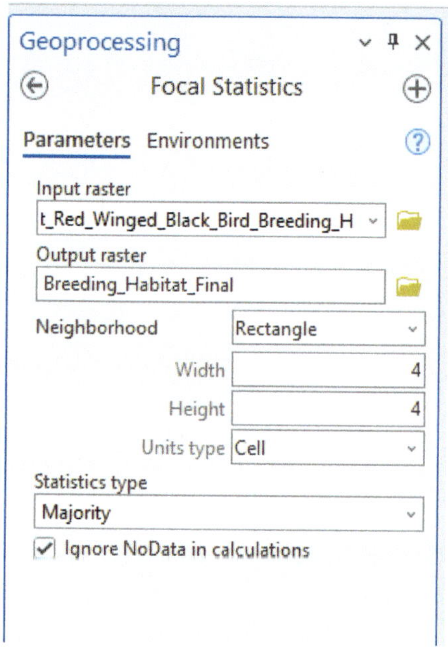


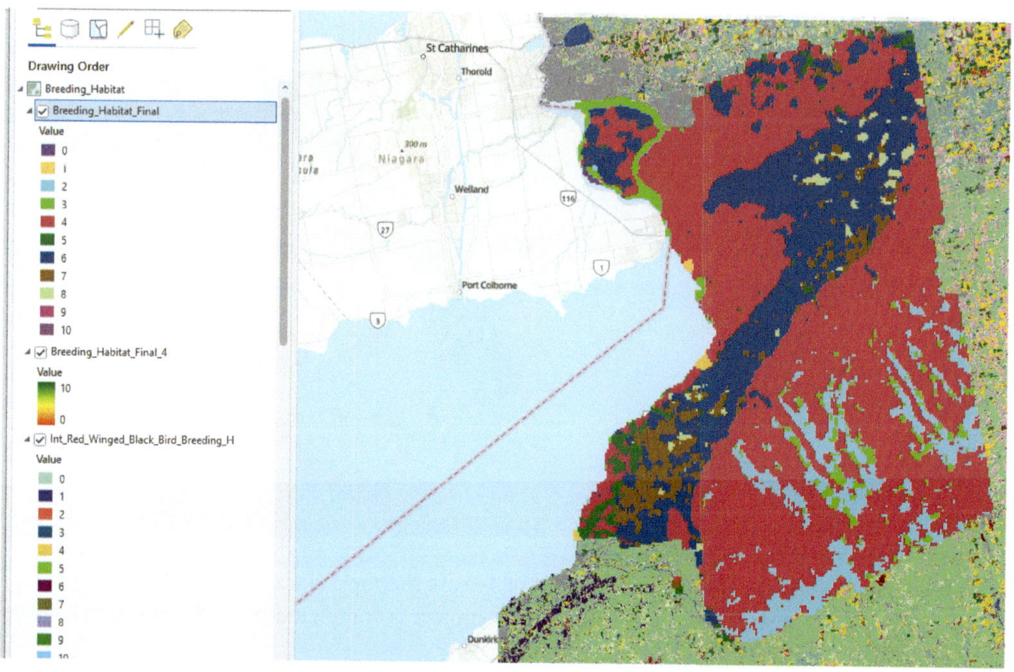

We would like to make this final raster map of red-winged black bird breeding habitats more readable. To do this, we will conduct some thematic mapping procedures. Close the Geoprocessing Pane and open the Symbology Pane. Change the “Primary Symbology” to Stretch. For the “Color Scheme,” select the Condition Number and check the “Invert.” The map should present like follows. Close the Symbology Pane.

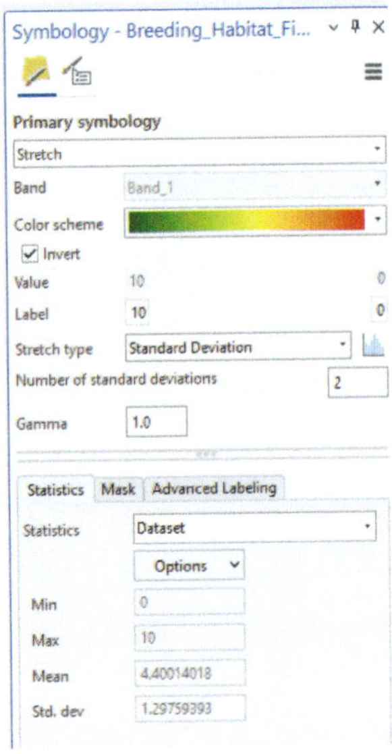


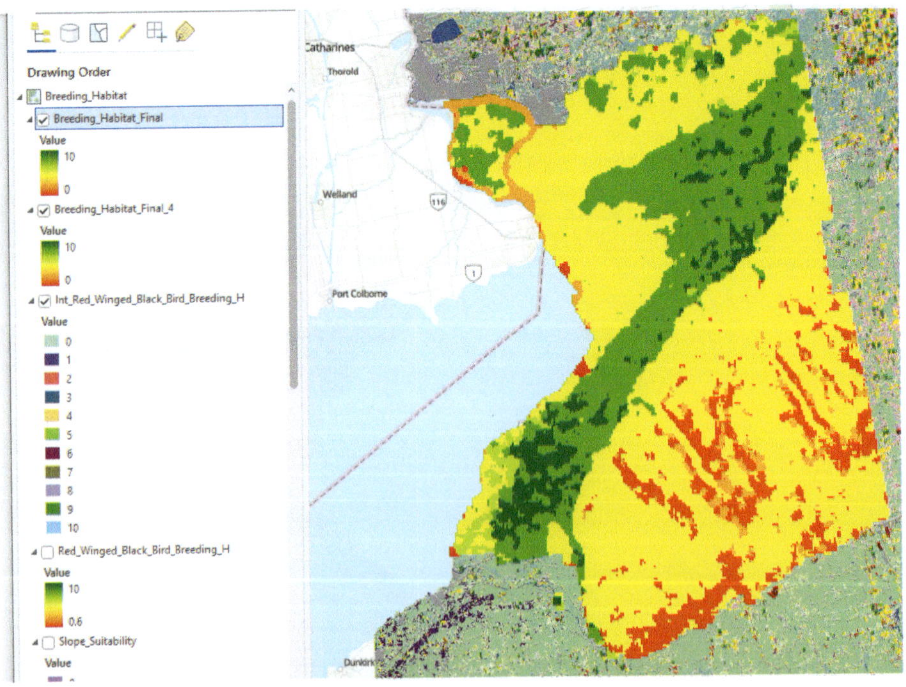


The dark green regions (with the highest score of 10) are the most suitable areas for the breeding of red-winged black birds in the Erie County. This project is completed! Save the project file and close ArcGIS Pro.

# 8

# Three-dimensional (3D) Spatial Analyses

## Concepts and Theories

In essence, the Earth itself and all the geographic objects on the Earth's surface are generally in three-dimensional (3D) shapes. Owing to the limitations of human development of technologies, geographers or cartographers conventionally can only use either flat paper maps or 3D physical models, such as a globe to visualize, analyze, and comprehend spatial objects. This was the original objective of map projections in cartographic science in flattening the curved Earth's surface itself from the real world and in facilitating the drawing of 3D real-world objects on a two-dimensional (2D) paper map (Robinson et al. 1995; McMaster et al. 2005). Based on the paradigm theory of applying 2D geographic space to represent 3D geographic space, this entire set of frameworks was developed. With new developments in computational science and engineering, technologies of digital 3D visualization, presentation, and virtual reality enabled us to directly survey, input, analyze, model, and output spatial data in the 3D environment now. In the future, digital 3D maps, scaled models, and virtual reality in portable computational and communicational devices will largely replace the 2D digital maps.

Similar to 2D maps, 3D geographic information systems (GIS) datasets illustrate spatial objects in greater detail by adding another dimension or coordinate (z) on top of x, y coordinate settings. The 3D illustrative method in GIS enables the explanatory representation of scaled-down real-world objects. 3D digital models, which are beyond a 2D map, assist in visualizing and surveying a large number of different features. For instance, 3D digital models or scenes can show the height or elevation of a hotel or a mountain and not just its location but accurate elevation. Currently, 3D software functions have been used along with 2D GIS and then merged in a 3D data processing setting in many cases.

In the last few decades, GIS has made significant contributions to our society by using digital maps and spatial analysis as an essential tool to solve problems. Conventionally, GIS information was based on a 2D environment on digital maps, which apparently limited its usage in many applications. Incorporating 3D data presentation and analytical methods in GIS advances the entire experience, making GIS visualization more personal and enabling detailed information to be obtained. The following are some important 3D GIS applications.

*Fundamentals of Geographic Information Systems (GIS) Analysis: Applying ArcGIS Pro*, First Edition. Tao Tang.

## Urban and Regional Planning

Owing to the problem of improper natural or society resource allocations, many cities worldwide face shortages of basic amenities like water, electricity, green space, or parks. Incorporating 3D GIS technology and 3D simulations can help local government agencies, architects, and engineers envision, evaluate, and analyze as to how certain changes in the study region will look like, and how these changes can meet the needs in the future to help these government agencies to answer "what if" questions. Nowadays, 3D digital models imbedded in GIS are widely called "BIM – Building Information Models" that consist of building information, satellite imagery and street traffic data that can be used by urban planners to find alternative solutions effectively and address emergency situations efficiently.

### Building Information Modeling

In the last few years, building information modeling (BIM) as a new subdiscipline of Urban Planning, Architecture, and Civil Engineering has been developed. BIM is an important scaled modeling technology that illustrates a real-world environment in 3D in a computerized environment (Kensek and Noble 2014). The combination of BIM and GIS provides the necessary components of a virtual robust model for knowing how to build. The development of 3D GIS and BIM can help produce error-free and more effective building management plans that would eventually allow us to analyze detailed data for building maintenance and renewal.

### Civil Engineering Design and Coastal Development

Current construction design in building constructions, mining operations, and ocean transportation port development and maintenance are mainly based on 2D drawings and digital maps. 3D GIS can efficiently and effectively plan and utilize the resources for designing and implementation purposes. 3D GIS design and implementation can also provide some level of understanding of the economic activities and environmental conditions at the construction and building sites.

### Disaster Response

In natural disaster prevention and management, 3D GIS can help people and local governments conduct virtual reality visualizations or simulations. 3D GIS is an essential tool to analyze disaster results and predict possible hazard events. This would also include providing necessary information such as (i) coordinating the point or area of action and the best route for rescuing; (ii) detailing terrain visualizations of the rescuing or mitigation areas; and (iii) finding proper and safe routes for evacuations.

Similar to 2D GIS analysis, we need to create, manage, and analyze the 3D GIS databases in a 3D environment. 3D GIS database means that every geographic object on the Earth's surface is referenced by coordinates of x, y, and z location values. GIS analysis has changed the dynamics of Geography and Earth Science. With the development of digital medium, modern GIS interface lets its users and GIS practitioners not only envisage and evaluate datasets, but also manage geographical facts and realities. In order to manage and display the objects or features in 3D, ArcGIS Pro developed the Arc-Scene windows that are parallel to the "Maps" window for handling 2D digital map layers. Two types of scene windows or data visualization modules were developed. These are the Local Scene and Global Scene. The major difference between Global Scene and Local

Scene is that the Global Scene takes into consideration the Earth's surface curvature. Users can display a large area, or entire continent or the entire Earth in the global scene. Local Scenes typically deal with small areas and start-ups in the building of 3D objects and features based on a flat surface and projected coordinate system.

The users of the global scene in ArcGIS Pro can zoom in and out to visualize the entire Earth in 3D. Global scenes support World Geodetic System 1984 (WGS 84) or Chinese geodetic reference system for geospatial information and BeiDou/COMPASS 2000 (CGCS 2000) datums in the geographic coordinate systems. Using a local scene to display your spatial data is recommended if the spatial reference of the data or models is not in WGS 84 or CGCS 2000, to avoid the heavy distortions the global scene may cause. All the 2D digital map layers by either vector datasets or raster datasets can be directly opened and viewed in local scene mode. In the local scene, data transferred from the 2D display can remain in the same projected coordinate system without conducting a reprojection. Editing, analysis, or measurement functions in a local scene are similar to that in a 2D mapping window ("Maps"). The window of a local scene is mainly used to visualize data in small geographic areas such as cities or construction sites. Local scenes support geographic, projected, and custom coordinate systems.

Both vector and raster GIS data structures can be used to store and present 3D spatial objects or features. In essence, 3D GIS data incorporates z-value into its location or definition (x, y, z). In practice, there is no restriction or rule enforcement that the z-value must be a real-world altitude, elevation, geological depth, or feature height measurement. GIS analysts or practitioners often take advantage of the graphic visualization capability in a 3D environment to use the z-value to present numerical values in the attribute table of geographic features, such as chemical concentrations, relative humidity, or even the suitability of a location. In this regard, the z-value in 3D can be identified as physical z-values that are the representations of feature altitude or elevation, and attribute z-values that are the representations of quantity of a feature attributes. According to ESRI (2022), 3D GIS data can be fundamentally identified as one of two types: 3D feature or object-based data and 3D surface-based data.

### 3D Feature Data

3D feature or object-based data represents discrete objects similar to that in a 2D environment, and the 3D information for each object is stored in the feature's geometry. These are points, lines, and polygons. It might be very useful to establish cubes as another basic 3D GIS feature type to facilitate applications and analysis in the future. 3D feature data can support potentially many different z-value feature locations for each x, y location. For example, on a vertical line in 3D, a point feature can occupy one upper vertex or one lower vertex, and these two point features have the same 2D coordinates, but each has different z-values. Another example of 3D feature data would be a representation of a multiple-floor building in urban areas, whose roof, interior floors, and foundation would all contain different z-values for the same 2D coordinate locations that they share. Other examples of 3D feature data, such as an aircraft's flying route or a walking trail up on a mountain, would only have a single z-value for each x, y location.

3D object-based data still represents discrete objects like in a 2D environment. Point features might represent mountain peaks, treetops, corners of a skyscraper in a city, or bottom locations of a water well. Lines might represent roads, high-power electricity transportation lines, stream center lines, or ridgelines of a mountain. Polygons might represent floors of a building, lakes, or floors of an underground subway station or a mining site. The established and coordinated location registered cubes in 3D GIS can represent in partial a unified 3D space, such as part of a living room in an apartment in an urban building, or a cube of water body in a specific part of the Lake Erie (one of the Great Lakes in North America).

Recent breakthroughs in laser scan or light detection and ranging system, normally using the laser beam (LiDAR) field survey technology in engineering enabled GIS Analysts to use invisible laser beams to scan the 3D geographic space on the Earth's surface. Laser scan survey or LiDAR remote sensing collects 3D points that represent the real-world features and feature locations in x, y, and z. LiDAR field survey or remote sensing bridges the gap between 3D data collections and inputs from the real world to digital GIS systems. The original results of the LiDAR survey or data collection are called point-cloud files. Field-collected point-cloud datasets either from an airborne remote sensing platform or a ground-based platform are genetically in ASCII format. A binary format called LAS was developed to manage and standardize the point-cloud datasets for storage, analysis, and dissemination. A LAS dataset can also store references to feature classes that contain surface constraints. The so-called surface constraints are break-lines, water surface polygons, area boundaries, or any other type of surface feature that is to be enforced in the LAS datasets. Ideally, we can construct 3D feature classes of points, lines, polygons, and possible cubes in a GIS environment by extracting the original point cloud files collected by LiDAR remote sensing. This type of function and workflow is still in the process of being developed today in a GIS software environment.

Two operations are very popularly performed in 3D feature data for GIS analysis and visualizations. These are Z-value extrusion and feature class draping. Z-value extrusion means to extrude the altitude or height values of features of a feature class in the 3D viewing window, such as in local scene in ArcGIS pro. If we would like to show building features in 3D, we can extrude these features using an attribute, such as building height or number of stories. We can also extrude features using an arbitrary value in the attribute table as the z-value, such as air pollution by counties or cities. If a 3D surface model for the study area such as digital elevation model (DEM) is loaded in the window of a local scene or global scene, we can use the values in the model as the base z-values for the feature class, no matter how it is an extruded 3D feature class or a 2D feature class. This is called the draping of feature classes. This process in essence digitally fits human-built or natural features, such as buildings, river lines, and lake polygons on top of terrain models.

### 3D Surface-based Data

3D surface-based data represents altitude or height values over an area. The 3D data values for each location within that area can be either stored in a grid cell or deduced from a set of 3D faces using an irregular triangulated network (TIN). The altitude or height value in each of the grid cells is considered uniform. TINs can only be generated from a point feature class. Many GIS Analysts consider 3D surface data as 2.5D data owing to it only supports one single z-value for each x, y location. For instance, the altitude above the sea level for the 3D surface of Earth will only return a single value. Meanwhile, 3D surface-based data cannot be used to illustrate multiple-floor buildings in urban areas.

A 3D-surface data model is a digital representation of geographic objects or features, either physically existing or hypothetically projected in 3D space (ESRI 2020). In other words, 3D-surface models in a GIS software environment, particularly in ArcGIS Pro can be applied to represent real-world elevations of terrain, which is called digital terrain model (DTM). Alternatively, the z value in a 3D-surface model can be utilized to present other variables as well. Some examples of 3D surfaces are a digital terrain model, average household income in an urban area by accounting units such as block groups, petroleum deposits under the Earth's surface, and projected percentage of a specific religion in a geographic region. 3D surfaces could be field measurement derived or imaginary. An example of a field measurement-derived surface is the air pollution or contamination quantities predicted by the point datasets of field measurements. When we conduct computer simulations in GIS or play video games in a digital or computational environment, we use imaginary 3D surfaces.

**Figure 8.1** Digital elevation model (DEM).

3D surface data can either be illustrated by raster grid cells or vector TIN. 3D raster data are field-based or continuous data. Since this type of raster grid cell carries the elevation or altitude values, we refer to it as DEM (Figure 8.1). Similarly, we also call some elevation and height values carrying raster datasets digital surface models (DSMs). Sometimes the datasets are named DTM. The major difference between DEM and DSM is that DEM only references the altitude or elevation of the natural Earth's surface. While DSM by its name includes both human-built buildings and houses in the urban areas and the trees or bushes on top of the bare Earth's surface.

In a GIS analytical environment, 3D surface data can also be illustrated in a 2D environment or vice versa. Toward this end, many of the analytical methods and functions in ArcGIS Pro are shared among the modules of "Spatial Analyst," which mainly deals with 2D raster datasets, "3D Analyst," which mainly deals with 3D GIS data, and the "Spatial Statistical Analyst" which mainly conduct spatial and statistical interpolations using field measured data (Figure 8.2). The common and simple reason for this approach is that the results of all these analytical methods generate the raster grid cell surfaces that can be presented in either 2D or 3D. Among many interpolation methods, the most popularly used ones are: IDW (Inverse Distance Weighted), Spline, Kriging, and Nearest Neighbor. We will discuss each of these in Chapter 9 in detail.

TIN data is a form of vector GIS data. TINs are constructed by triangulating a set of points (vertices or notes). The points are connected by a series of edges to form a network of irregular triangles (Figure 8.3). Delaunay triangulation is one of several triangulation methods used that allow the minimum degrees of interior angle of all triangles to be maximized. The edges of TINs form contiguous, nonoverlapping triangular facets and can be used to capture the position of linear features on the surface representing the terrain that plays an important role, such as mountain ridgelines or stream or river lines (Franklin 1973). The graphics below show the nodes and edges of a TIN (top) and the nodes, edges, and facets of a TIN model (bottom).

Because nodes can be placed irregularly over a surface that is represented by triangles, TIN models tend to present higher resolutions in areas where the surface is highly rouged or where

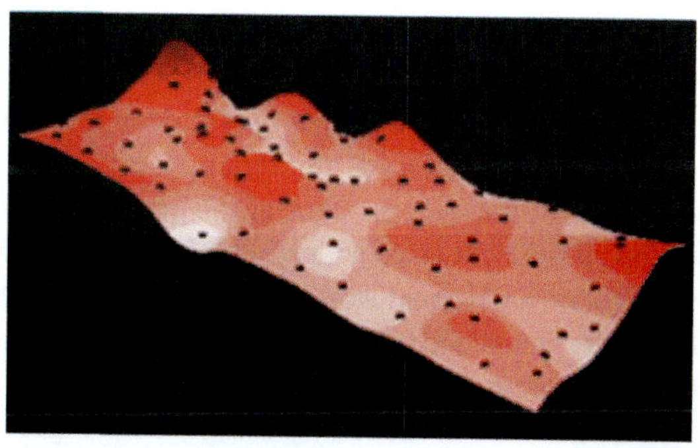

**Figure 8.2** Example of interpolation – Chemical pollution concentration in a study area.

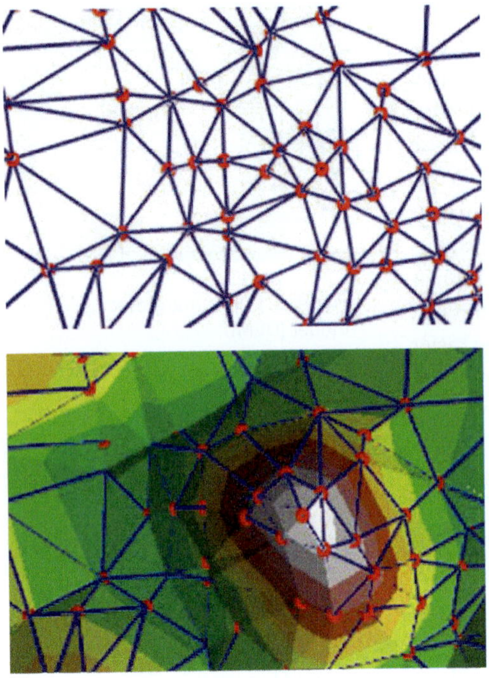

**Figure 8.3** TIN data model in GIS.

more details are desired while presenting relatively lower resolutions in areas that are flat. The input field sample point data that are used to create a TIN model remains in the same position as the nodes in the TIN. This preserves the precision of the input data. TIN models are typically used for high-precision 3D modeling of smaller areas, such as to calculate surface areas or the volume of rock and sediment to be removed in civil engineering applications.

3D GIS data analysis can provide many solutions to the real-world studies of engineering, urban planning, and environmental condition assessment. Here, we would like to discuss two fundamental approaches using 3D GIS data. These are computations of slope and aspect using DEM data and the computations of visibility or viewsheds in a geographic region. The slope is a numerical value that describes the steepness of the terrain surface along a line and is determined by calculating the ratio of the vertical distance to the horizontal distance (rise over run) between the two notes of the line. In practice, the slope for a particular location is computed as the maximum rate of change of elevation between that location and its surrounding locations. Aspect is the horizontal direction of the steepest slope, measured clockwise in degrees from 0° to 360°, where 0° is north-facing, 90° is east-facing, 180° is south-facing, 270° is west-facing, and returning to the north-facing position by 360° (Figure 8.4). In a 3D GIS environment, slopes, and aspects at each of the grid cells can be computed by DEM. The resulting raster dataset or map layer of slopes can be in slope degrees from 0° to 90°, or in slope percentage, of which 100% slope equals to 45°. Higher than 45° slopes are more than 100%. The aspect map layer shows the horizontal direction of the face of the steepest slope at the grid cell location. This geoprocessing is very important for setting up solar panels in collecting solar energy for electricity.

The following is a real-world example of using DEMs to study the slope degrees and slope aspects to identify suitable locations for solar energy farming in Erie County, New York. Figure 8.5 shows the combinations of the DEMs for the entire county. Figure 8.6 is the processed result of slopes and slope distribution in the county using DEMs. Figure 8.7 is the computational result of aspects and

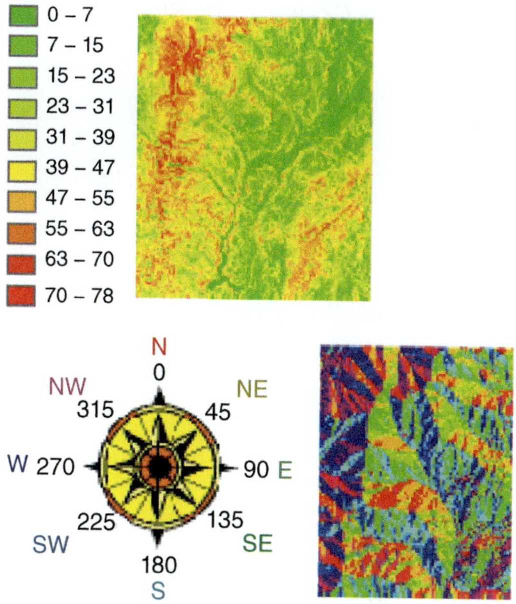


**Figure 8.4** Slope aspect on the digital map.

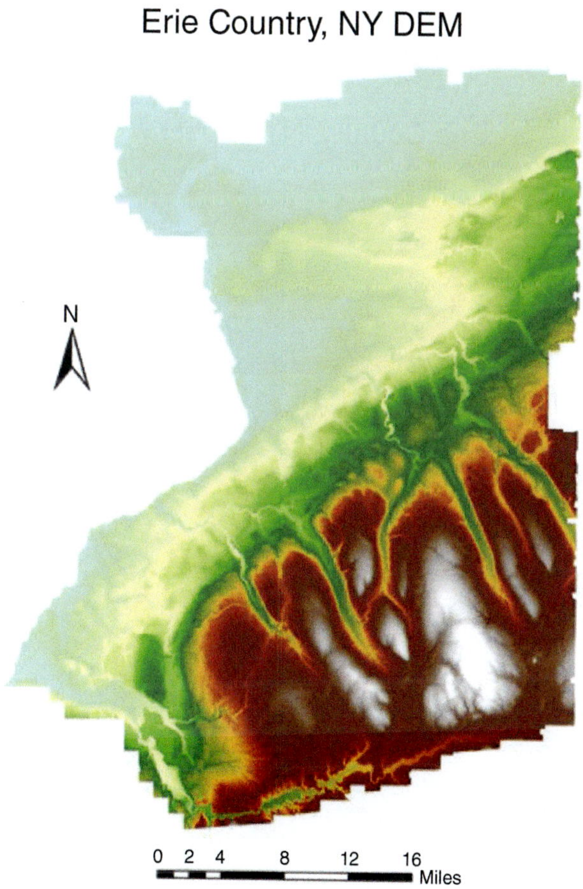


**Figure 8.5** DEM in Erie County, New York. *Source:* Dr. Tao Tang

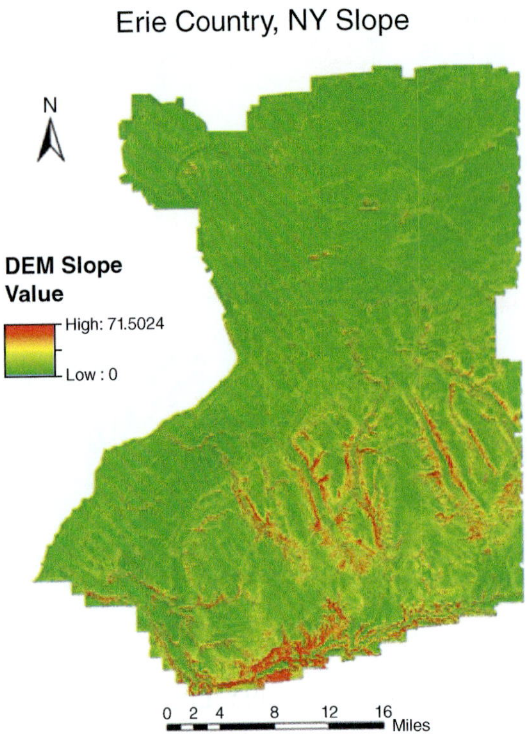


**Figure 8.6** Slopes derived from DEM in Erie County, New York.

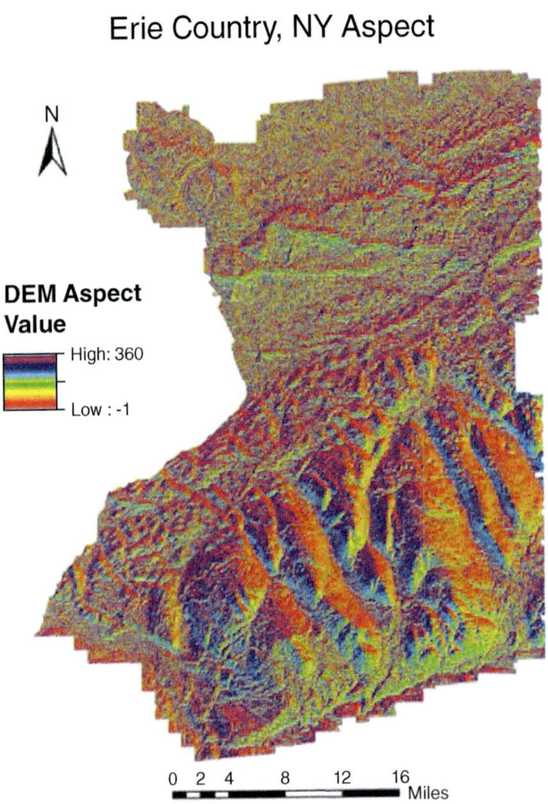


**Figure 8.7** Aspects derived from DEM in Erie County, New York.

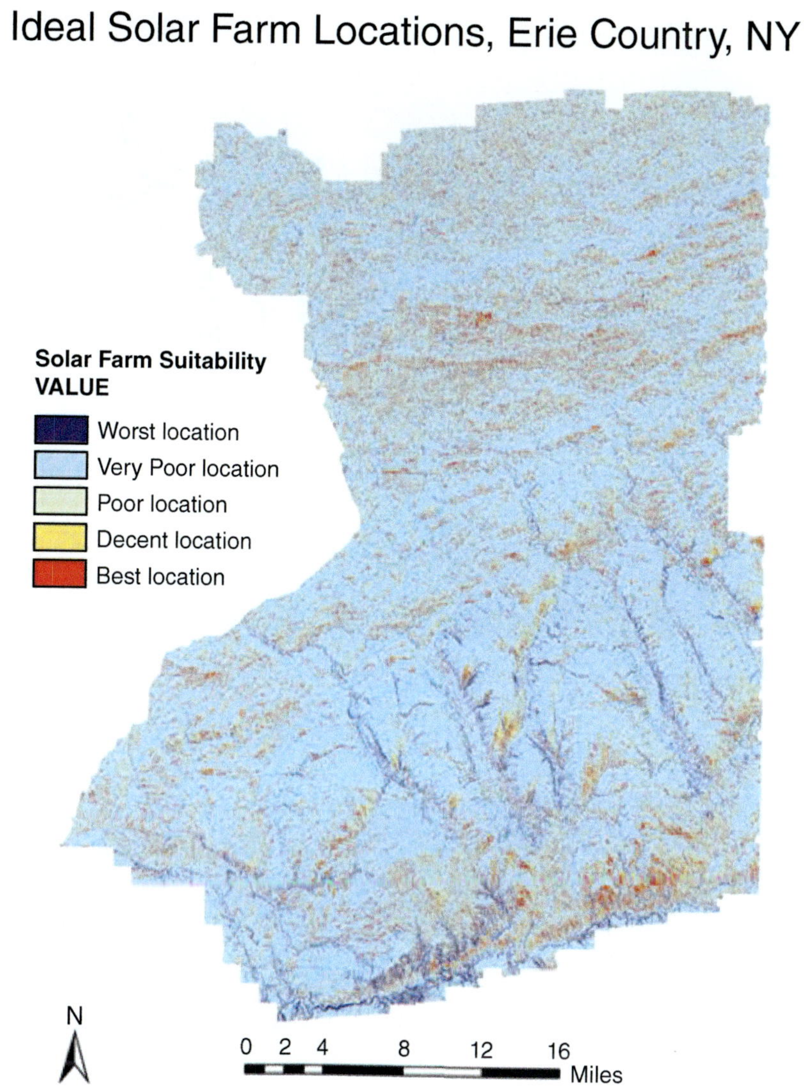


**Figure 8.8** Ideal solar farm locations identified and suggested to the Erie County, New York. *Source:* Dr. Tao Tang.

aspect distributions in the county. Based on the analyses of slope degrees and slope-facing directions of terrain in the study area, the ideal solar farm locations across Erie County, New York were identified on the raster digital map (Figure 8.8).

Visibility analysis using 3D surface data or DEM takes into consideration what can be seen by an observer from a given location (Figures 8.9 and 8.10). The result presents the location which can be seen or cannot be seen. The areas that can be seen are called within a viewshed, and the straight line in the 3D space connects and marks the observer to the object is called line of the sight. The analysis utilizes the elevation value of each grid cell of the DEM to determine visibility to or from a particular observation point or grid cell. Visibility analysis is not only significant in evaluating the key location in the mountainous region to set up forest fire observation towers, but also commonly used to locate TV or cellphone signal communication towers.

**Figure 8.9** Visibility analysis in rural areas. *Source:* Adopted from ESRI (2022).

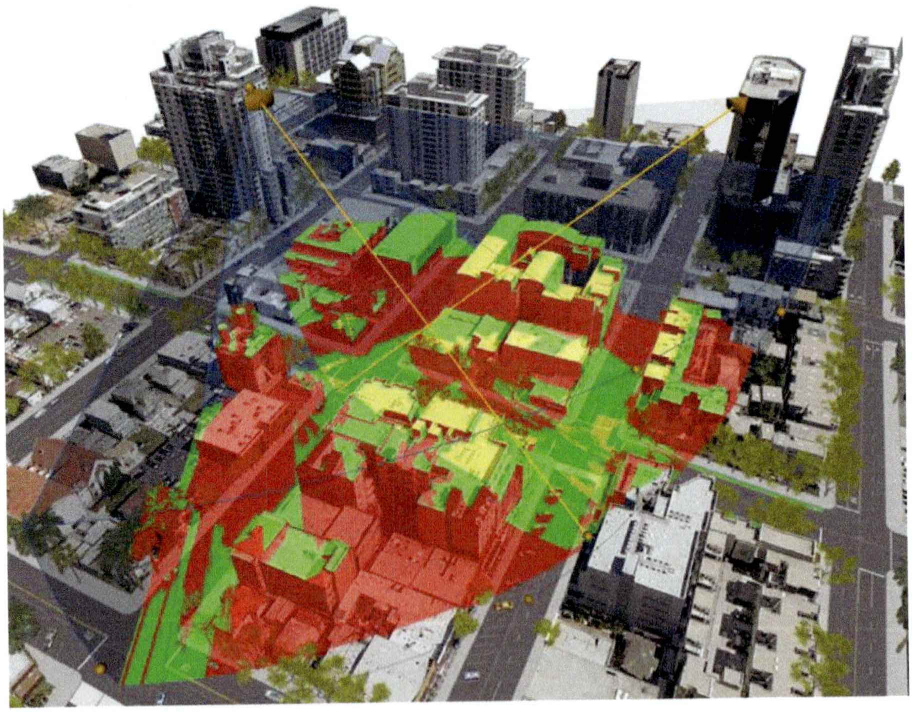

**Figure 8.10** Visibility analysis in urban areas. *Source:* Adopted from ESRI (2022).

## References

ESRI. (2022). *ArcGIS Pro Help Manual.* https://pro.arcgis.com/en/pro-app/latest/help/main/welcome-to-the-arcgis-pro-app-help.htm

Franklin, W.R. (1973). Triangulated irregular network program.

Kensek, K. and Noble, D. (2014). *Building Information Modeling: BIM in Current and Future Practice.* Wiley. ISBN 978-1-118-76630-9

McMaster, R.B., Slocum, T.A., and Kessler, F.C. (2005). *Thematic Cartography and Geographic Visualization*, 2e, 518 pages. Upper Saddle River, NJ: Prentice Hall.

Robinson, A., Morrison, J., Muehrke, P., Kimmerling, A., and Guptill, S. (1995). *Elements of Cartography*, 6e, 674 pages. New York: Wiley.

## Tutorial in ArcGIS Pro Environment

Copy the tutorial data folder “Chapter08-Tutorial” to your working directory.

Open ArcGIS Pro and create a new ArcGIS Pro project file that is called “MyGIS_Larning08” with the “Map” template. Save the project file into the designated working folder, c:\ or D:\...... Chapter08-Tutorial.

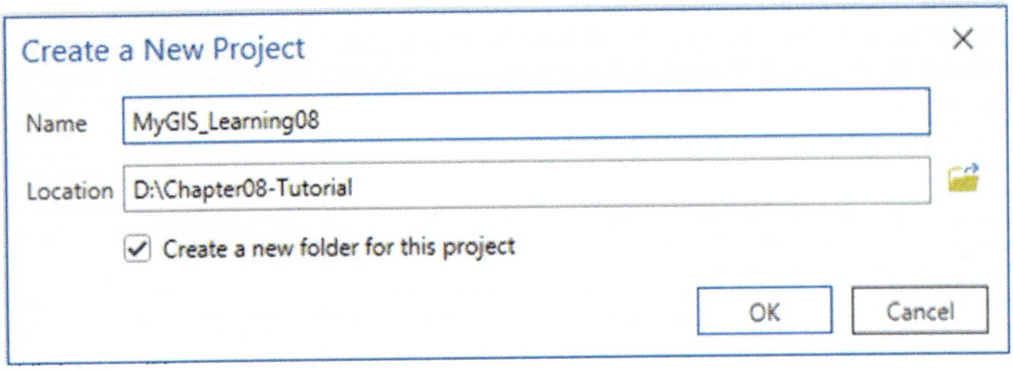


Now, highlight and right click on the Folders on the Catalog Pane to get the contact menu. Then, click on “Add Folder Connections” to add the entire Chapter08-Tutorial folder on your working drive.

1. SUNY-Buffalo State University Project

Expand the Chapter08-Tutorial folder and SUNY_BSC_Data.gdb geodatabase. This geodatabase contains four (4) digital elevation models (DEM) and six (6) digital aerial images that cover the entire area of the SUNY-Buffalo State University campus. Add the six aerial images to take a look. We would like to mosaic these six images together. So, we can use them as one dataset. Remove all the six individual images from the mapping frame (Map).

Activate the Analysis ribbon in the main menu and click on the Tools icon to open the Geoprocessing Pane. While Toolboxes tab is activated on the Geoprocessing Pane, expand Data Management Tools box and “Raster” toolset.

Select Mosaic Dataset sub-toolset -> "Create Mosaic Dataset" tool. This is a tool to create an empty raster dataset that can hold all the different individual raster files.

For "Output Location," navigate and select the SUNY_BSC_Data.gdb geodatabase.

For "Mosaic Dataset Name," type in "Buffalo_State_Aerial_Image."

For "Coordinate System," clicking on the small globe icon next to it to navigate to the SUNY_BSC_Data.gdb geodatabase to import the coordinate system of w_10651066_12_19200_4bd_2021 image.

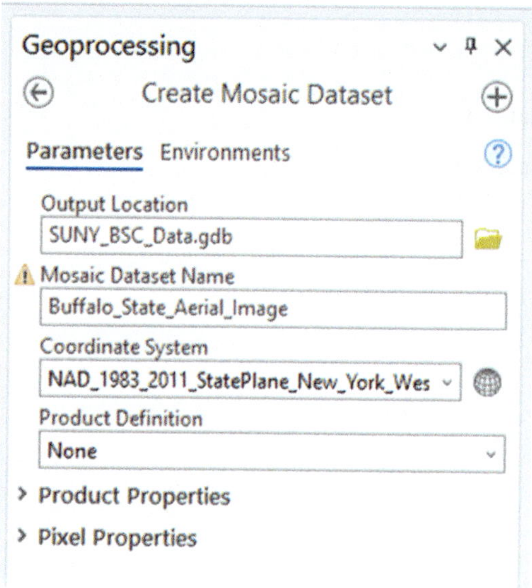


Run the tool to create an empty mosaic raster dataset holder. Now, go back to the previous page on the Geoprocessing Pane to activate the "Add Rasters To Mosaic Dataset" tool.

For the "Mosaic Dataset," choose the newly created "Buffalo_State_Aerial_Image." For "Raster Type," select "Raster Dataset." For the "Input Data," select "Dataset" and add all six raster aerial images starting from w_10651066_12_19200_4bd_202.

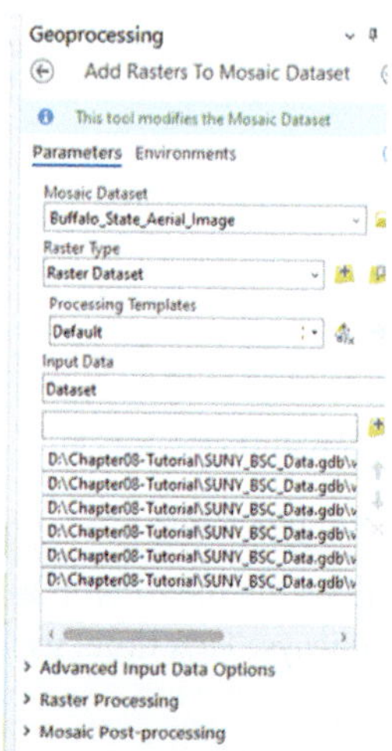

Run this tool to produce the imagery mosaic of "Buffalo_State_Aerial_Image."

Use the "Mosaic" tool in the "Raster Dataset" tool set to mosaic the four DEM into one piece. Open the "Mosaic" tool. The major difference for this process is that we need to designate one of the original DEM files (raster dataset) as the output dataset. In the process, the original dataset would be replaced as the new file that includes all the content of the four files.

For the "Input Rasters," navigate and select the SUNY_BSC_Data.gdb geodatabase and load each of the four DEM files, such as "DEM_e1381n2339_2019." For the "Target Raster," select "DEM_e1381n2339_2019."

Keep the rest as default.

Process the tool to create the mosaic DEM raster dataset.

Now, "DEM_e1381n2339_2019" file become the mosaic file.

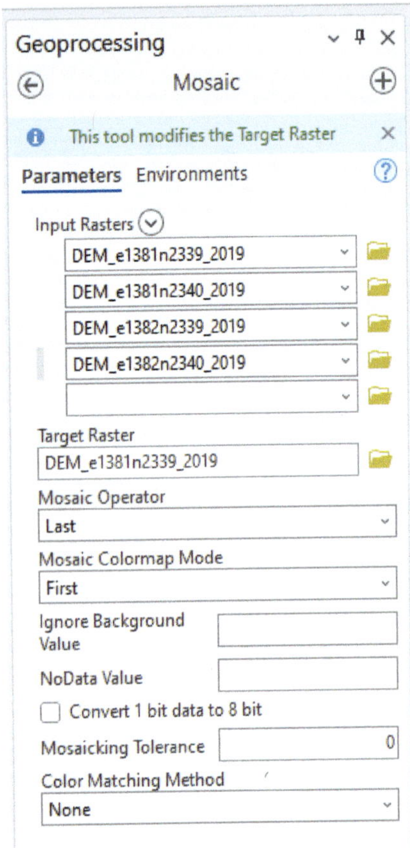


We can now remove the mosaic raster from this mapping window and close the Geoprocessing Pane.

Rename the "DEM_e1381n2339_2019" file as "DEM_Buffalo_State" on the Catalog Pane.

Now, we would like to open a "Local Scene" window to visualize and analyze the datasets in a three-dimensional (3D) way. We can add a "Local Scene" window in several ways. This time, we are going to use the "Insert" menu on the top in the main menu.

Activate the "Insert" menu on the main menu. There are two icons "New Map" and "New Layout" 

in the "Project" group of this ribbon. The "New Map" is a collection of icons to open a new mapping or a 3D scene window. The "New Layout" is an icon to open a new layout window with different paper size settings to share a static copy or print or plot a hard copy map. Highlight the "New Map" icon and click on the down arrow, we get the following menu.

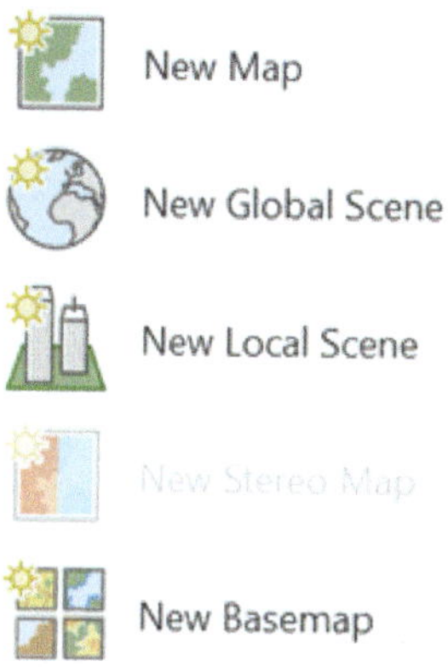


Click on "New Local Scene" icon to open a new local scene window. Rename this new local scene window as "Buffalo_State_3D." You may notice that the title of this window has changed too. Activate the "Map" window and highlight and right click on the map layer of "Buffalo_State_Aerial_Image" to obtain the contact menu. Copy the imagery layer.

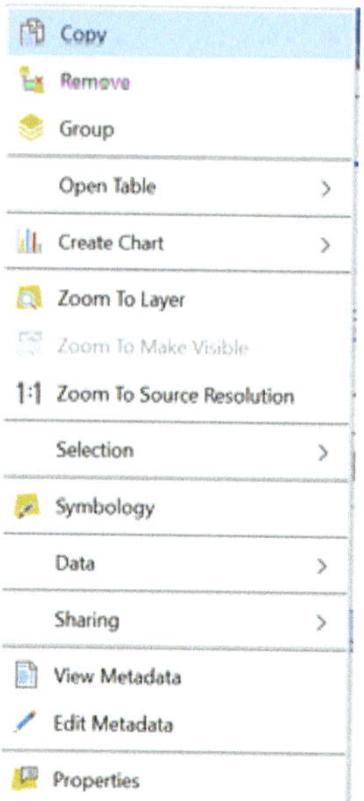


Then, activate the "Buffalo_State_3D" scene window, highlight the scene name and right click on it to get the contact menu. Click on the "Paste," you may notice that the digital image layer is loaded in this scene as the following.

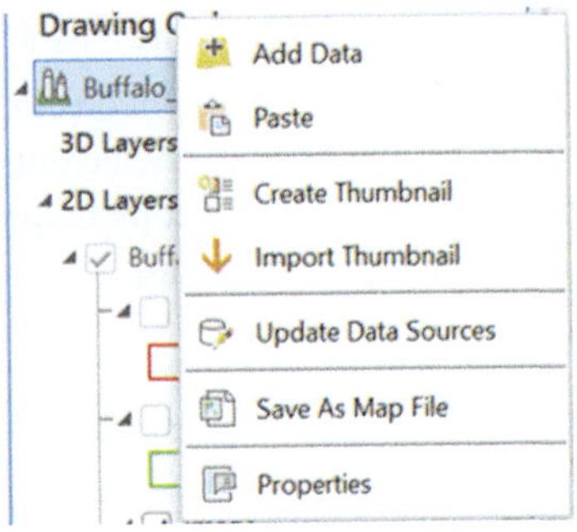


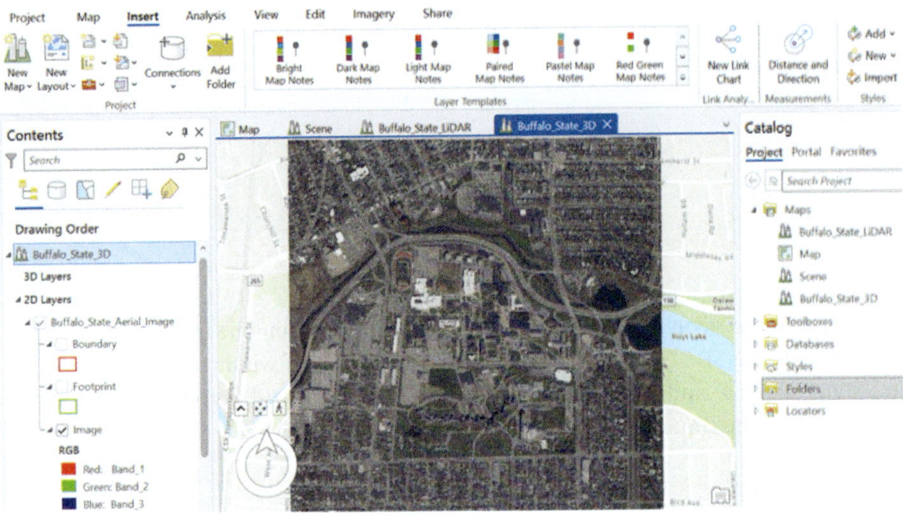


In the Table of Contents in a scene (ArcScene), 3D and 2D map layers are recognized and separately listed. Now, add the "BSC_Building_Footprint" feature class from the "SUNY_BSC_Data.gdb" geodatabase to the "Buffalo_State_3D" scene window. Right now, the "BSC_Building_Footprint" map layer is listed as a 2D layer. Using the navigational compass to title the scene layers like following.

Let's conduct an extrusion for the "BSC_Building_Footprint" map layer to make it in a 3D layer. Extrusion is only available in scenes in ArcGIS Pro. Highlight the "BSC_Building_Footprint" map layer in the Table of Contents, three main menus namely "Feature Layer," "Labeling," and "Data" appear. Activate the "Feature Layer" menu, you may see the following ribbon.

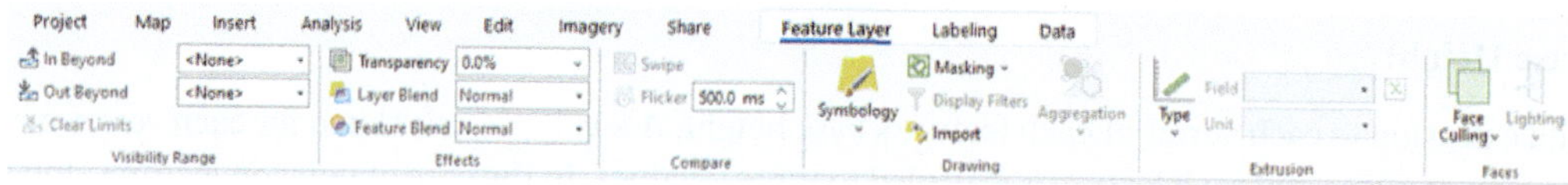


Click on the "Type" drop down arrow and select "Min Height (Minimum Height)." Fill in the "Field" as "Building_H" and "Unit" as "Meters." You may notice that the buildings on this map layer are extruded to the height of "Building_H" field in its attribute table. Meanwhile, the "BSC_Building_Footprint" map layer is moved to the 3D Layers list.

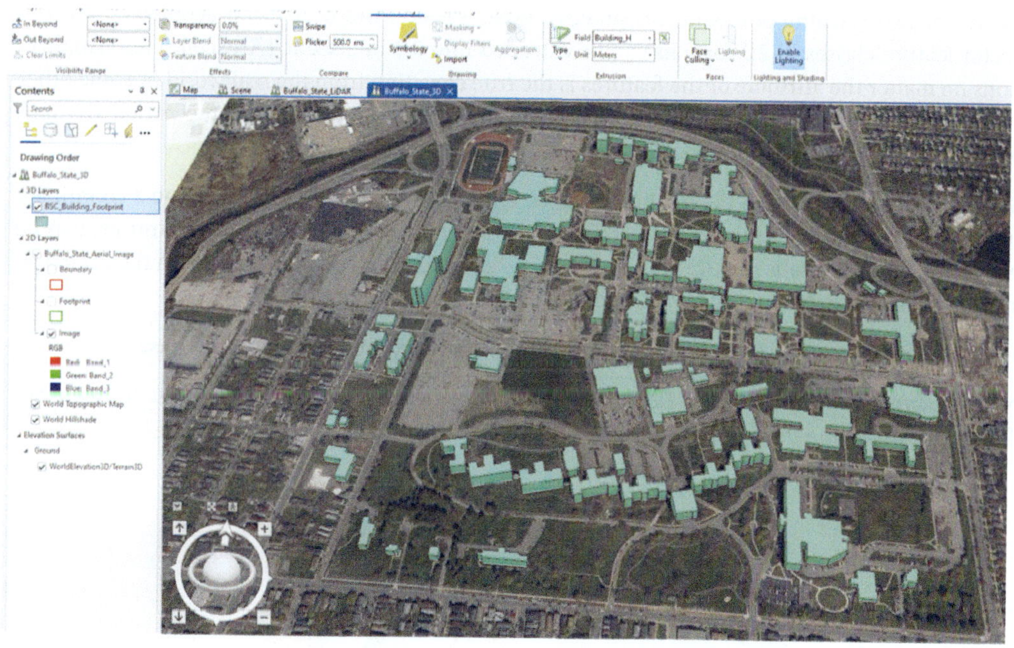

The following is the explanation of a feature extrusion type from the ArcGIS Pro help menu (ESRI 2020).

**None**

Features are not extruded. This is the default.

**Min Height**

Add extrusion to each feature's minimum height. A z-value is calculated by adding the extrusion height to the minimum z-value of each feature. They are extruded to a flat top at that value. This option is not available for point features.

### Max Height

Add extrusion to each feature's maximum height. A z-value is calculated by adding the extrusion height to the maximum z-value of each feature. They are extruded to a flat top at that value. This option is not available for point features.

### Base Height

Add extrusion to each vertex of each feature's base height. A z-value is calculated for each vertex of the feature's base and each feature is extruded to those z-values. In the case of polygon features, top and base faces are created to connect to the extruded walls to construct a closed shape.

### Absolute Height

Each feature is extruded to the specified z-value from a field or expression as a flat top, regardless of the z-values of each feature.

There are two learning points in this process. First, all the third-dimension geometry values of vector feature classes in 2D formats can be extruded into 3D in a scene (ArcScene) for visualizations no matter the attribute of the features is the true elevation from the real world or some other variables, such as time, quantities of money, humidity, etc. In essence, some of the virtual 3D maps with real-world locations showing can be considered as the virtual cartograms.

Secondly, point features on the 2D maps can be extruded to vertical lines along the z coordinate in a 3D scene. Line features on the 2D maps, such as border lines of the two countries, can be extruded to vertical walls in a 3D scene with one-grid cell width. Polygon features on the 2D maps, such as building footprint, can be extruded to blocks in a 3D scene.

In our practice, we can either select minimum height or maximum height in this case.

Now, we can add the "DEM_Buffalo_State" into the scene. Change the color scheme from gray to "Elevation #1" using the Symbology Pane. It looks like this.

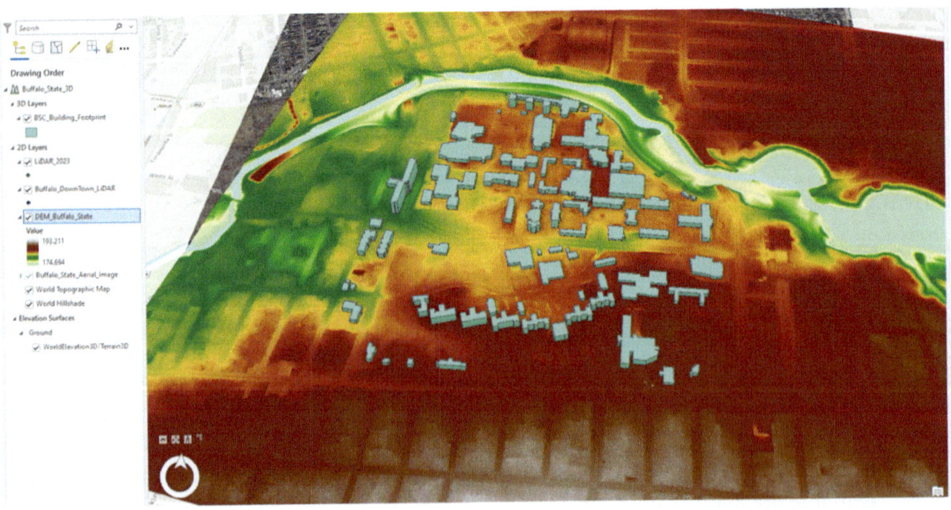


We would like to make the ground in a 3D digital model as well. Highlight the "Ground" in the Table of Content, "Elevation Surface Layer" menu is available on the top now.

While, "Elevation Surface Layer" menu is highlighted, check "Shade Relative to Light Position" and "Navigate Underground" (this allows you to view the underground features and extrusions, such as oil wells).

Now, increase the "Vertical Exaggeration" from 1.00 to 10.00. After rearrange the scene, the image should look like this.

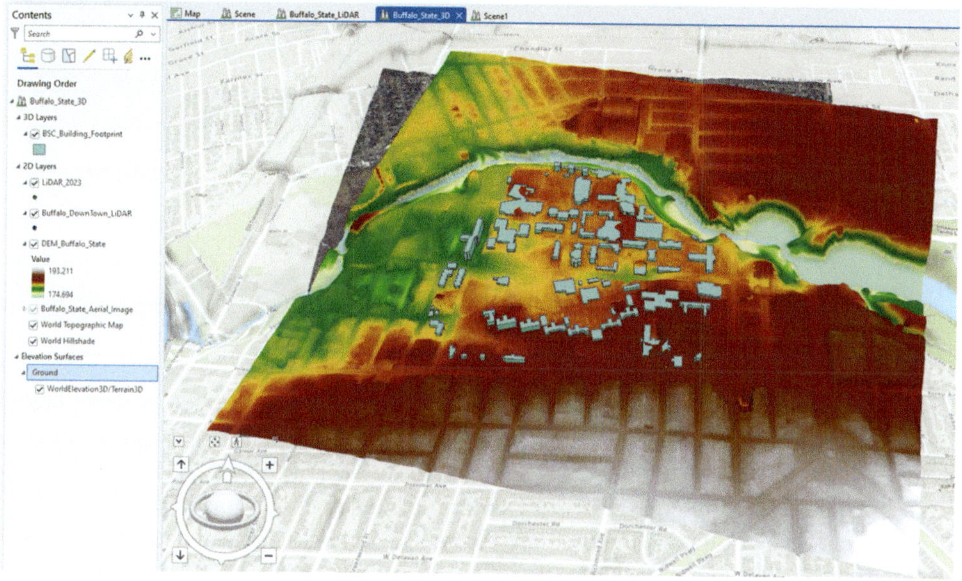


We can rearrange the map layers in the Table of Content to put the Buffalo State University aerial image on top of the digital terrain. The compass at the lower left corner of the scene window can help to navigate in this digital scene. The inner ring is a control of the tilting angle of the scene. The outer ring controls the direction rotations with the Northern direction is marked as the big arrow. Now, it looks more 3D and close to the real environment on the ground. We can also zoom in to view the bluff of the Scajaquada Creek next to the university campus.

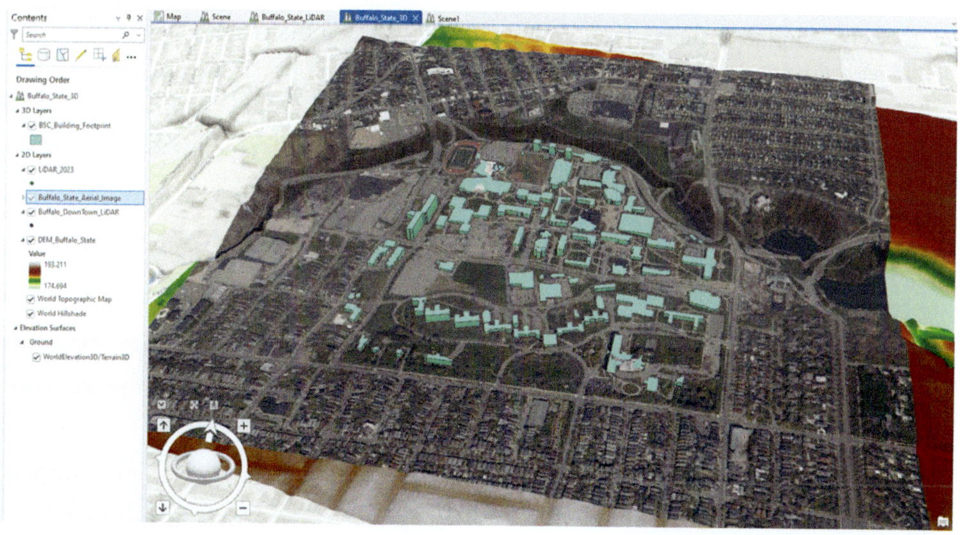

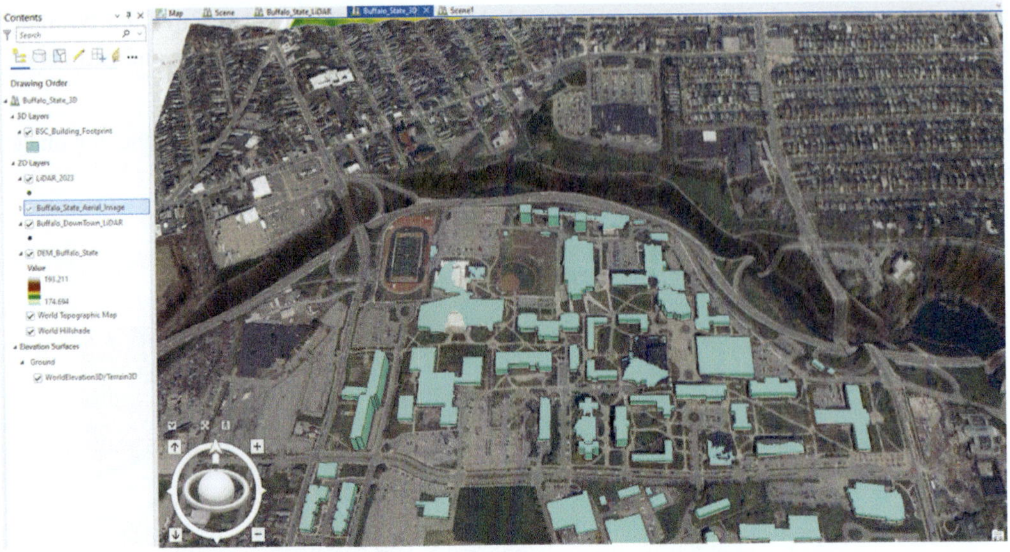


LiDAR Stands for Light Detection and Ranging. It is a remote sensing technology using invisible laser beam to scan the objects on the Earth's surface to create 3D digital models of the real-world in the GIS databases. LiDAR data is originally collected as a "point cloud" of individual points reflected from everything on the Earth's surface, including human made structures, vegetation, and terrain. LiDAR data is the first type of high-resolution 3D digital data that we ever collected for the GIS and computations.

Now, let's explore the LiDAR point cloud data. The widely used point cloud files are LAS file. The LAS is a file format designed for the interchange and archiving of LiDAR point cloud, which support the x, y, z coordinate values. Create a new scene window in the "MyGIS_Larning08" project in ArcGIS Pro. Rename the new scene window to "Buffalo_State_LiDAR."

Four LAS point cloud data quadrangles that cover the SUNY – Buffalo State University were download from the New York State GIS Clearing House website in the "LiDAR-2019" folder. We can use the "Create LAS Dataset" tool in the "Data Management Tools" toolbox and "LAS Dataset" toolset to mosaic or merge these four quadrangles. Open the Geoprocessing Pane and the tool dialog.

For "Input Files," navigate to ".......\Chapter08-Tutorial\LiDAR-2019" folder to add the LAS files, such as "e1381n2339_2019.las," one by one. For "Output LAS Dataset, type in "Buffalo_State_LAS." Leave "Coordinate System" empty and other criteria as default.

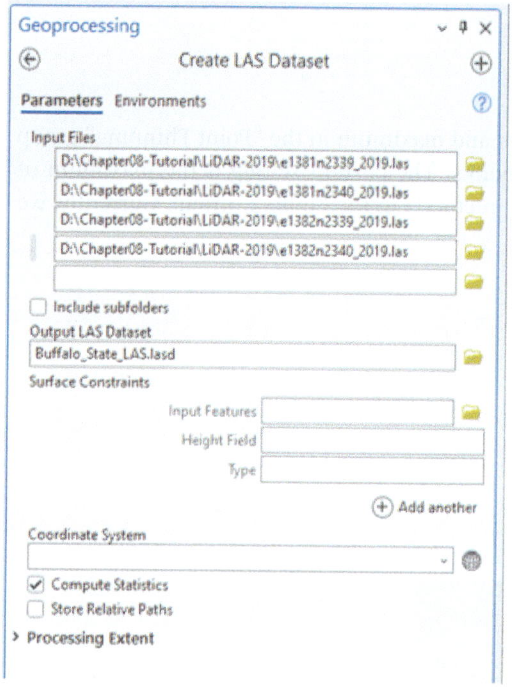


Run the Geoprocessing to create the point cloud data mosaic. The mosaic point cloud layer is added to the scene.

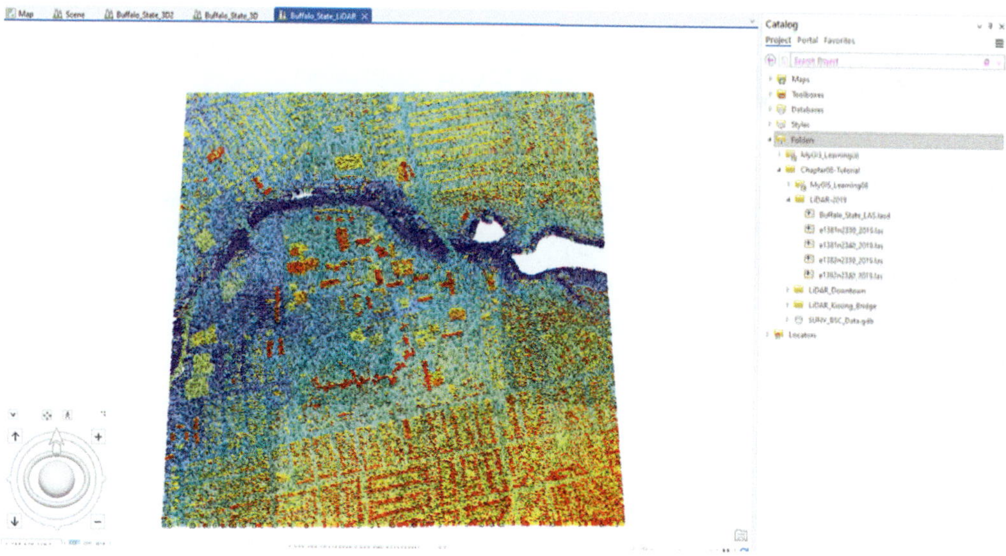


Highlight the Buffalo_State_LAS.lasd layer in the Table of the Contents, we can get three menus or ribbons on the top of ArcGIS Pro window. These are "LAS Dataset Layer," "Data," and "Classification."

Activate the "LAS Dataset Layer" ribbon on the main menu. This is a ribbon of the functions for basic point cloud layer management.

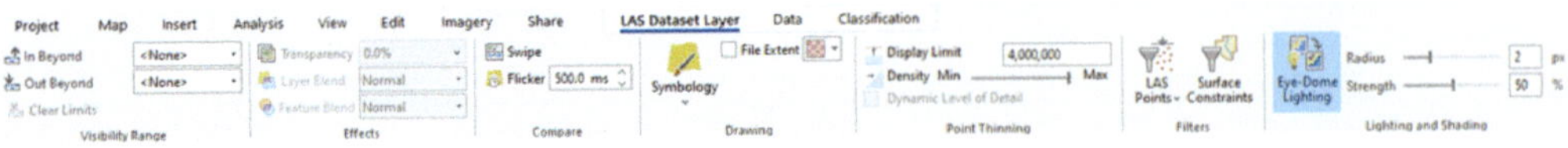


Change the point density in between the minimum and maximum in the "Point Thinning" group can help the users to view different number of points. The current setting is the maximum of 4,000,000 points. We can try the different filtering methods in the "Filters" group. Currently, we have all the LiDAR points.

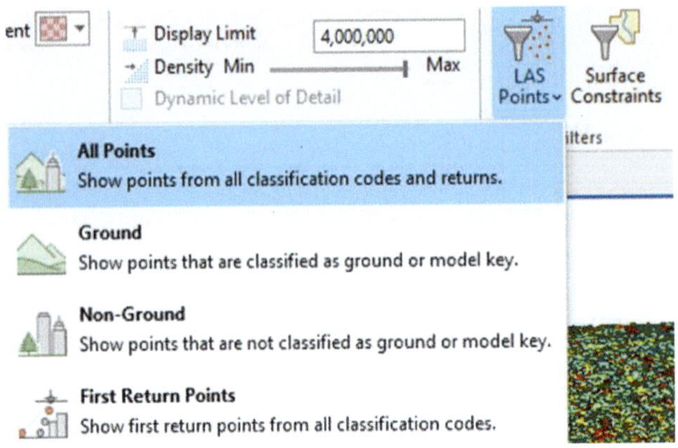


Let's select the "Ground" on this drop-down menu. You noticed that all the buildings on campus are cut off.

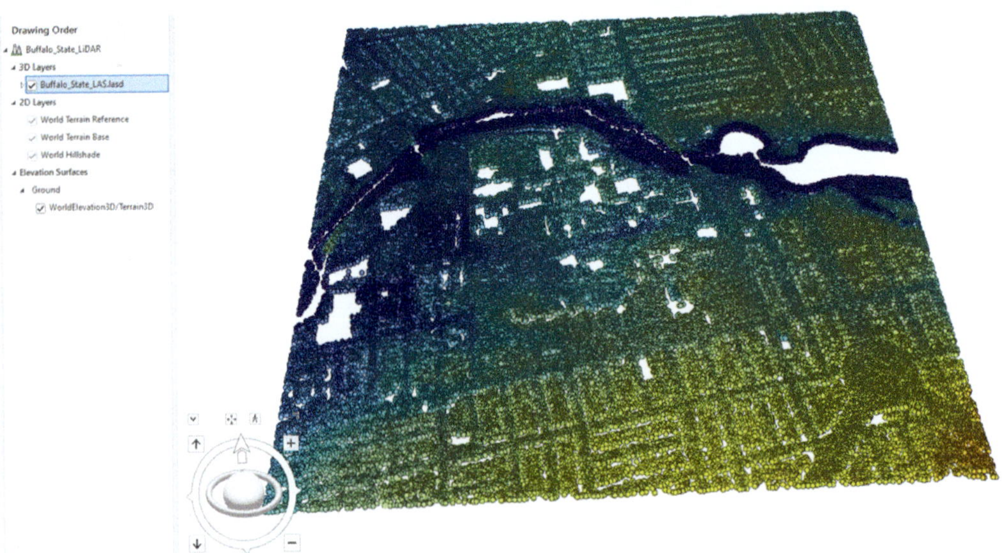

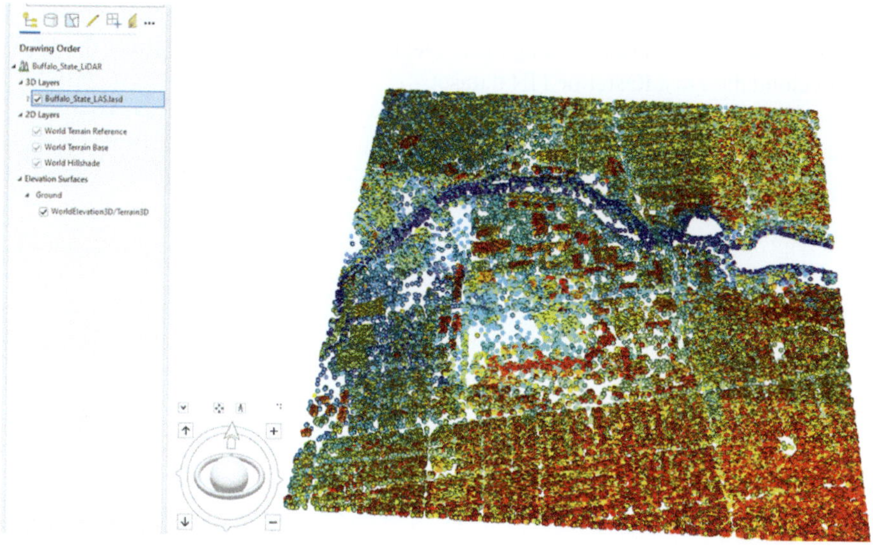


Now, let's try the "Non-Ground" option. The data view turns to the following.

This filtering result supposed to only keep the laser scan reflection points above the ground. However, we still get some returning points from the ground in this case. After these evaluations, we can return to the "All Points." Since invisible laser beams can penetrate tree canopies, we can technically identify the first return, second return, or third return of one laser beam reflections. The first return is the reflection by highest object surface at the place and normally the strongest return. In our case of this dataset, we do not have second or third returns. Therefore, "First Return Points" are similar to "All Points" in this case.

Let's activate the "Data" ribbon now. This is a ribbon contains some basic tools to analyze the point cloud data.

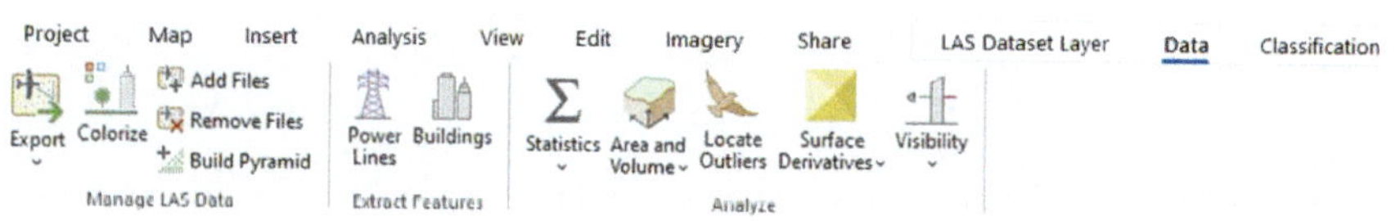


For instance, the "Surface Derivatives" icon in "Analyze" group allows the user to expand and call the Geoprocessing tools to make aspect, contour lines, or slope layer.

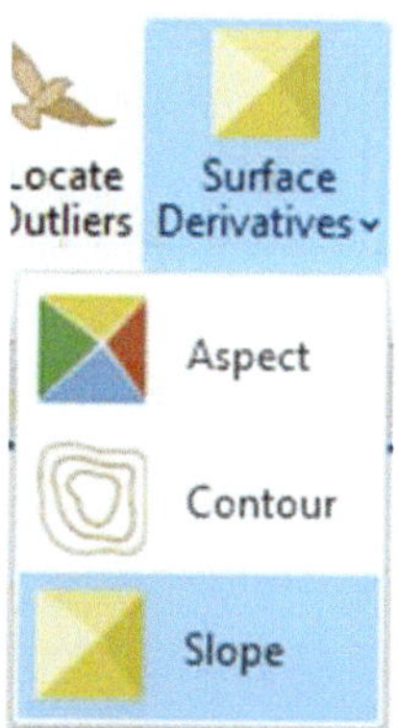

"Statistics" icon allows the user to calculate point cloud statistics, such as mean height or elevation. The "Export" drop-down list in the "Manage LAS Data" group enable us to extract the LAS data or Export point cloud data to a Raster or TIN dataset.

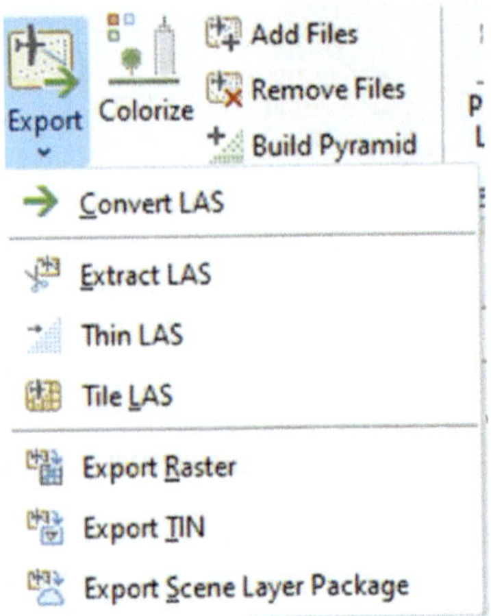


Let's select the "Ground" filtering again at the "LAS Dataset Layer" ribbon again. Then, switch back to "Export" -> "Export Raster" on the "Data" ribbon. The "LAS Dataset to Raster" tool in the Geoprocessing Pane opens. For "Output Raster," navigate to your SUNY_BSC_Data.gdb and name it (type in) "My_BSU_DEM." Keep the other criteria as the default like follows.

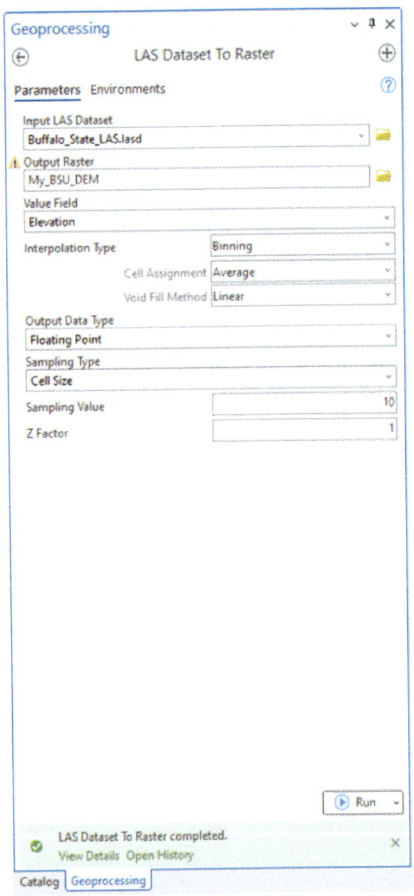

Run the tool to create the high-resolution DEM from the LiDAR point cloud data. In essence, most high spatial resolution and detail oriented DEMs that we use to conduct 3D analyses are made this way. LiDAR field 3D data collection technology helped us to revolutionize the 3D analysis and mapping method in the GIS environment.

Although we created the "My_BSU_DEM" raster layer, we cannot view it in this scene simply because the coordinate system of it does not match with the layers here. We can remove it from the Table of Contents. If we would like to view the "My_BSU_DEM" raster data layer, we can activate a 2D "MAP" frame and add it.

Now, let's explore the "Classification" ribbon. On this ribbon, we can find the newly developed point cloud classification tools.

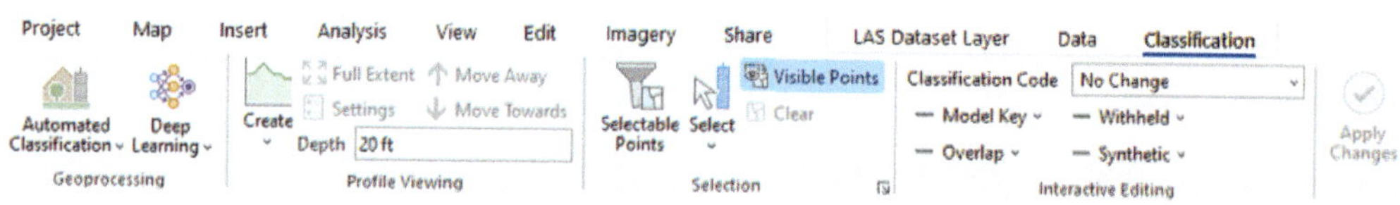


Up to today, classification and extraction of 3D point cloud for different features on the Earth's surface are still the bottleneck for 3D GIS analysis. On one hand, we have the capability to collect millions and millions of laser beam reflection points at different localities on the objects. On the other hand, we do not have enough good methods to classify the points by their x, y, z coordinate locations and extract features in 3D. This is the reason why deep learning and artificial intelligence (AI) are heavily used in this frontier area.

In this case, two group of classification tools have been developed. One type is a group of "Deep Learning" (AI) tools. This type of approach involves preparing a subset of point cloud for data analysis training and using deep-learning models. We are not going to illustrate in detail here. People are interested for this approach should read more specific reference materials and conduct experiments using LAS datasets.

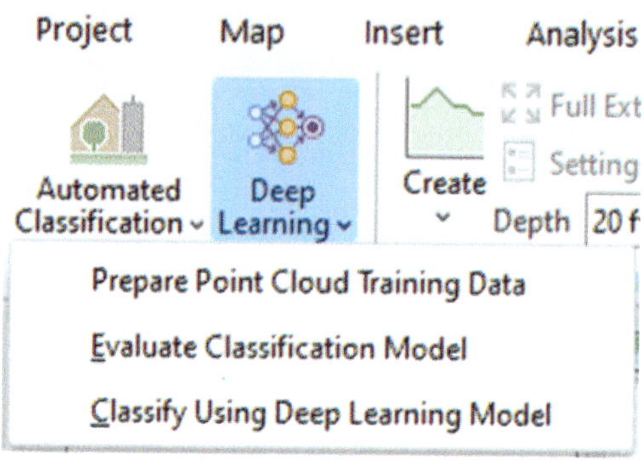

The second group of tools are organized for simple and automated classifications.

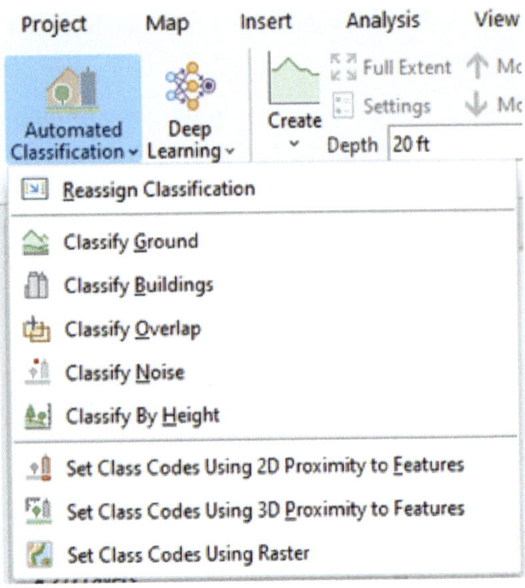


In current stage, we can attempt to classify the ground (terrain), the buildings, the noise or errors in the dataset, or classify features by their heights. The standard of the classification is as following.

ArcGIS uses the LAS classification scheme defined by the American Society for Photogrammetry and Remote Sensing (ASPRS).

| Classification value | Classification type |
|---|---|
| 0 | Never classified |
| 1 | Unassigned |
| 2 | Ground |
| 3 | Low vegetation |
| 4 | Medium vegetation |
| 5 | High vegetation |
| 6 | Building |
| 7 | Low noise |
| 8 | Model key/Reserved |
| 9 | Water |
| 10 | Rail |
| 11 | Road surface |
| 12 | Overlap/Reserved |
| 13 | Wire – Guard |
| 14 | Wire – Conductor |
| 15 | Transmission tower |
| 16 | Wire – Connector |

| Classification value | Classification type |
|---|---|
| 17 | Bridge deck |
| 18 | High noise |
| 19–63 | Reserved for ASPRS definition (LAS 1.1 to 1.3 support up to class code 31) |
| 32–255 | User definable (only supported in LAS 1.0 and certain versions of 1.4) |

Now, let's explore the "Classify Ground" geoprocessing. Click on the icon, the "Classify LAS Ground" tool opens on the Geoprocessing Pane. Fill in DEM resolution as one (1) US Survey Feet and use "Standard Classification" for the ground Detection Method in this time.

Run the tool to classify the laser beam reflection points on the ground.

This tool requires the input LAS dataset to have a projected coordinate system. Only the last return of LAS points with class code values of 0, 1, and 2 would be evaluated as ground points. Standard Classification method has a tolerance for slope variation that allows it to capture gradual undulations for topographical changes on the ground but do not capture the type of sharp reliefs that would be captured by the aggressive option, such as a wall of a building.

Close the Geoprocessing pane.

Now, let's classify the features with different heights. Click on the "Classify by Height" icon to open the "Classify LAS by Height" tool in the Geoprocessing Pane. By default, the class code 3, 4, 5 have been inputted with the height values.

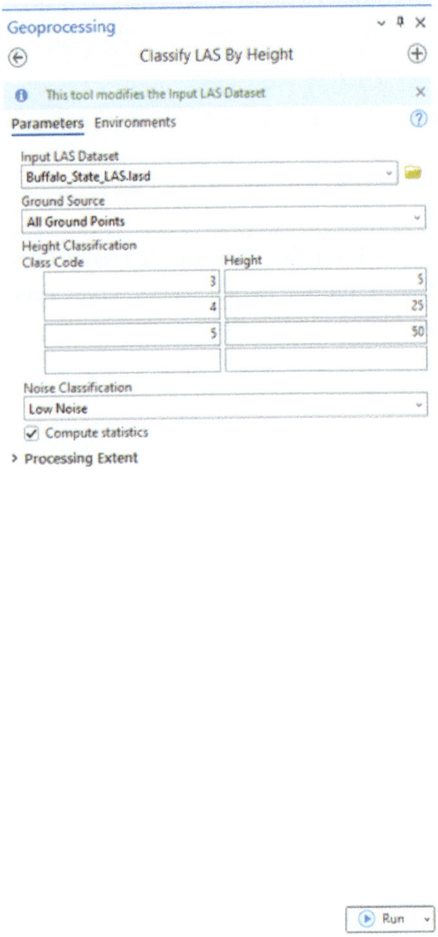


Referring the above table of the standard code established by US-ASPRS, code 3 is low vegetation, code 4 is medium vegetation, and code 5 is high vegetation. Do you think that high vegetation type of trees in Buffalo can reach 50 feet? Do you think the low vegetation such as bushes are typically taller than 5 feet? This is the reason that we need AI and machine learning for doing a good job. Meanwhile, although humans, such as trained GIs analyst can produce very accurate results using visual interpretation, it is very labor intensive when deal with large quantity of data points in a large area.

Change the heights of code 3, 4, and 5 to "3," "10," and "15." Change the "Noise Classification" to "Low Noise." Run the tool to classify the point cloud data.

Close the Geoprocessing Pane.

Now, let's classify the buildings. Click the icon "Classify Buildings" to open the "Classify LAS Building" tool. Change the "Minimum Rooftop Height" to 10 US Survey Feet, and the "Minimum Area" to 30 square US Survey Yards. Use the standard method.

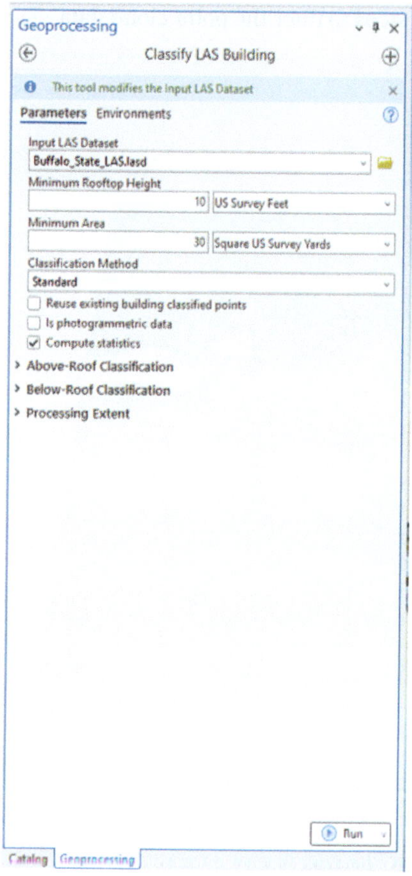


Run the tool to classify buildings. Close the Geoprocessing Pane after the function is finished.

Now, we can click on "Selectable Points" icon  in the "Selection" group on the "Classification" ribbon to show different classes using LAS filter. The "Layer Properties" dialog opens.

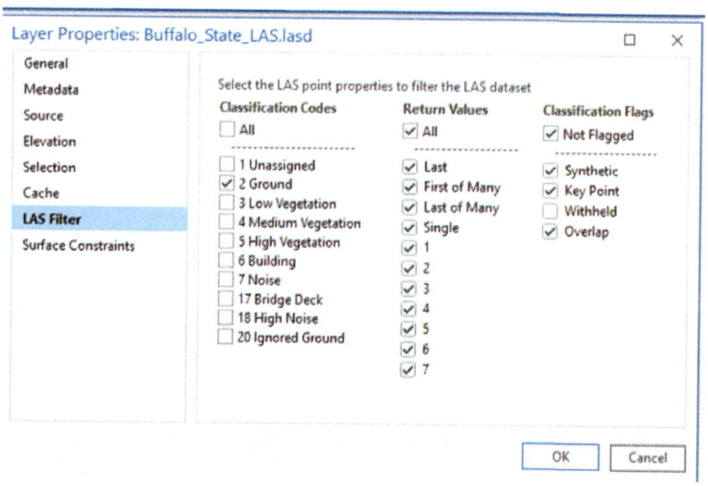

We can use the classification codes that we created in above steps to filter the point cloud data.

The point cloud of ground is like the following.

The point cloud of buildings looks like this.

The low, medium, and high vegetation covers look like to following.

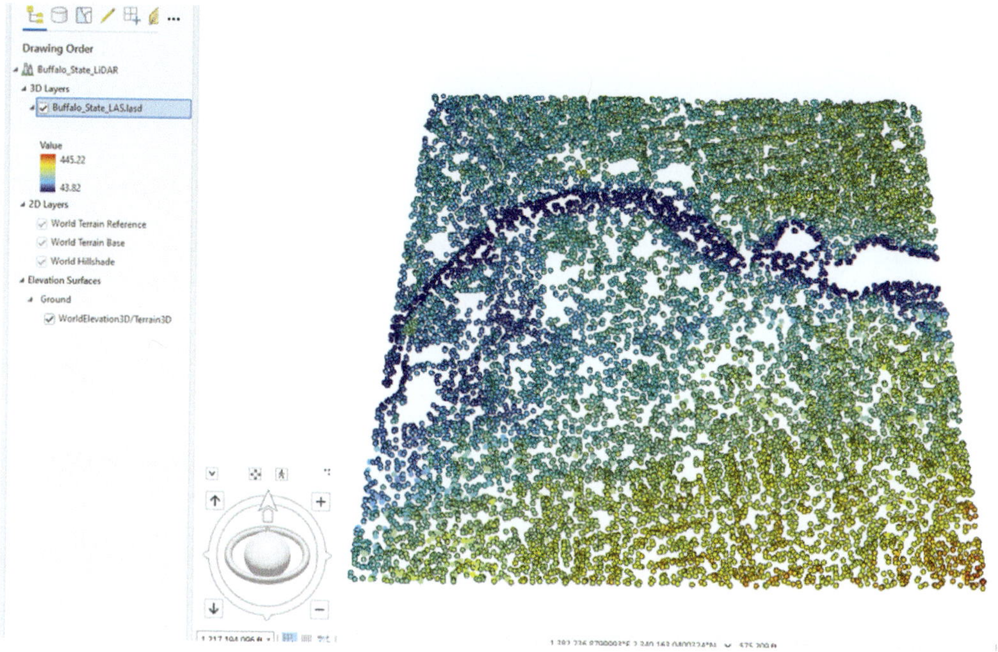


1. Low vegetation

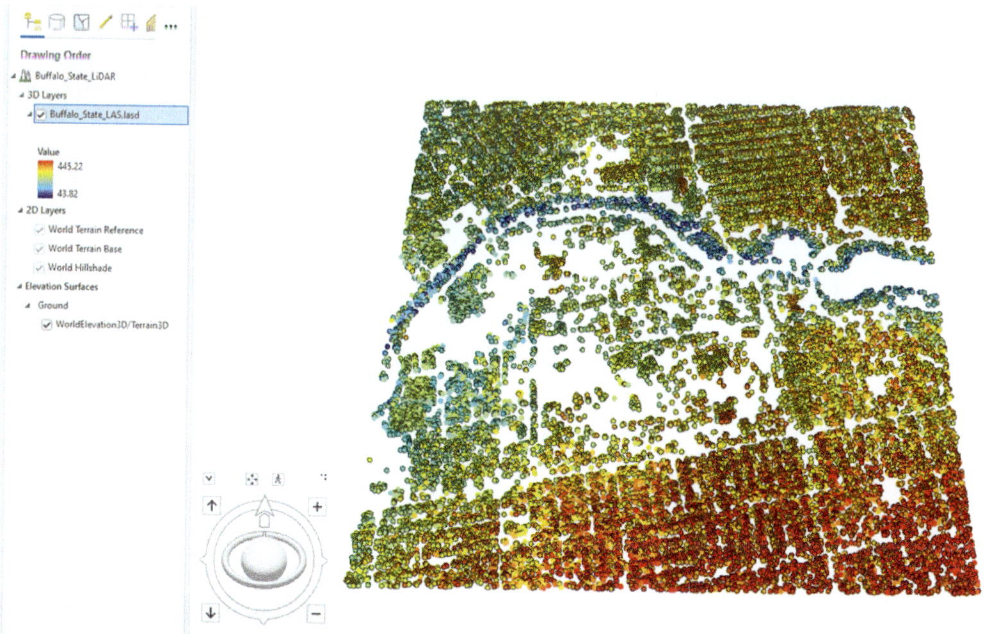

2. Medium vegetation

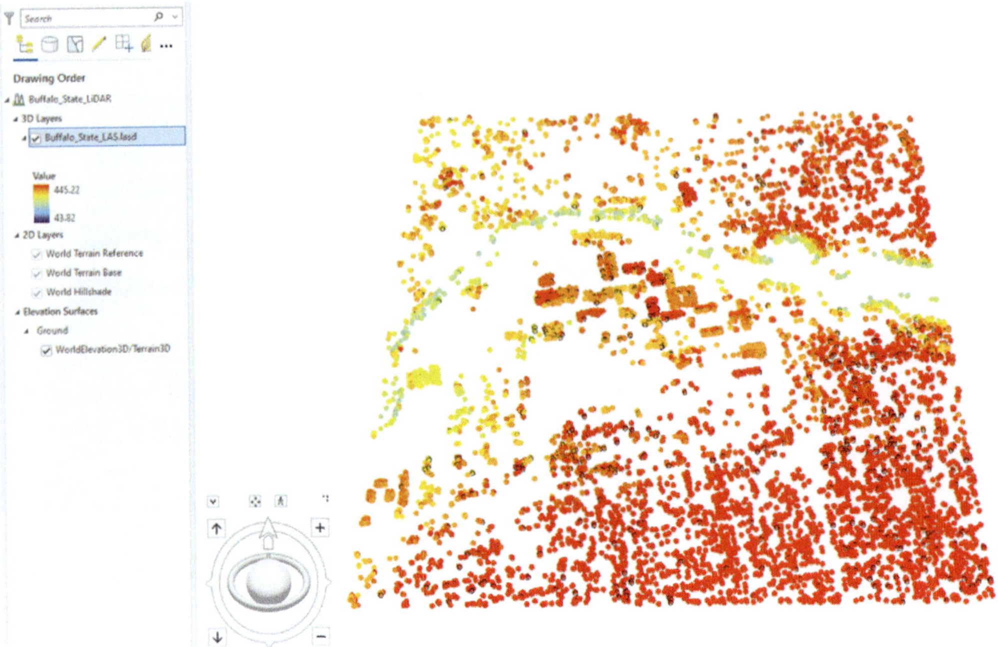


3. High vegetation

Now, let's export one of the filtered classes, "Ground" as a separate point cloud dataset.

Make sure the "Ground" filter is the following.

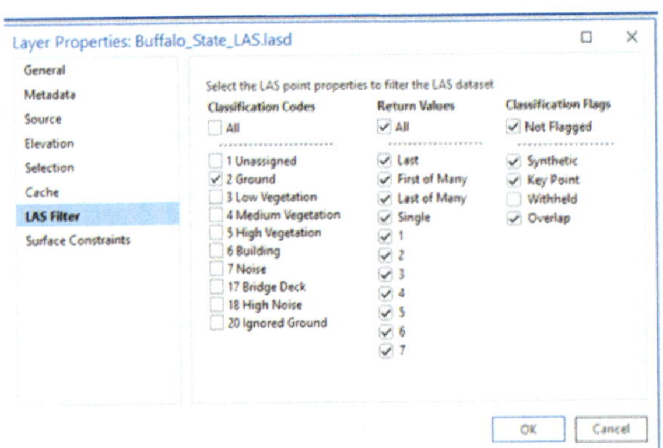


Right-click on "Buffalo_State_LAS.lasd" layer on the Table of Contents to open the contact menu, then click on the "Data" menu to get the second contact menu like the following.

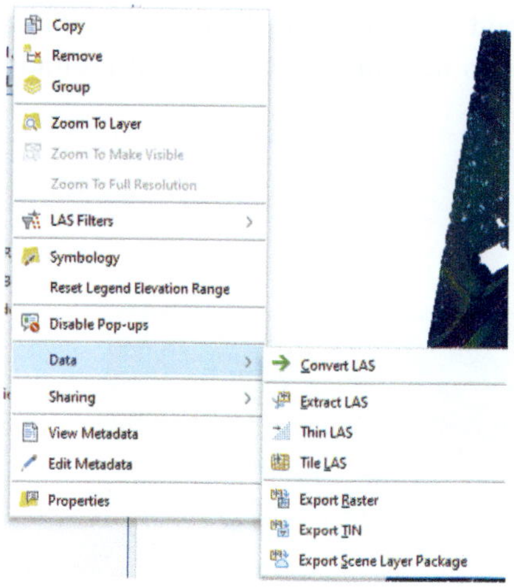


Select "Extract LAS" to open the "Extract LAS" tool on the Geoprocessing Pane.

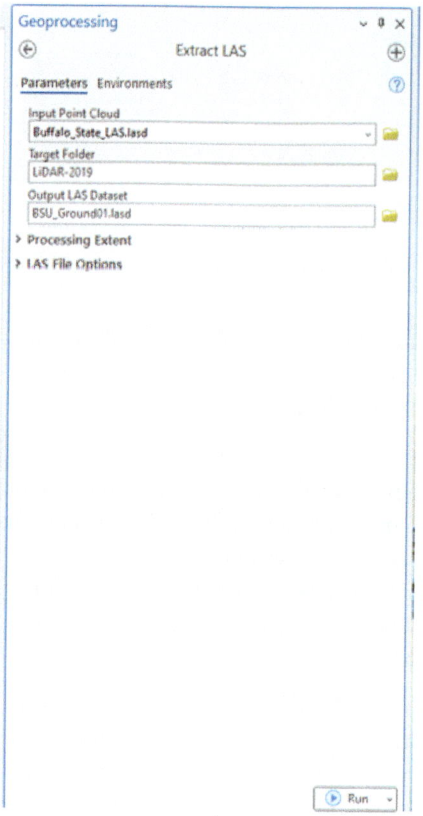


For the "Target Folder," navigate and select your "......LiDAR-2019." For "Output LAS Dataset," type in "BSU_Ground." The output will be saved in your original exercise folder unless you direct to a specific one.

Run the tool to create the new "BSU_Ground" point cloud dataset. Close the Geoprocessing Pane. This is a permanent dataset that we can use later to determine the details of ground elevations.

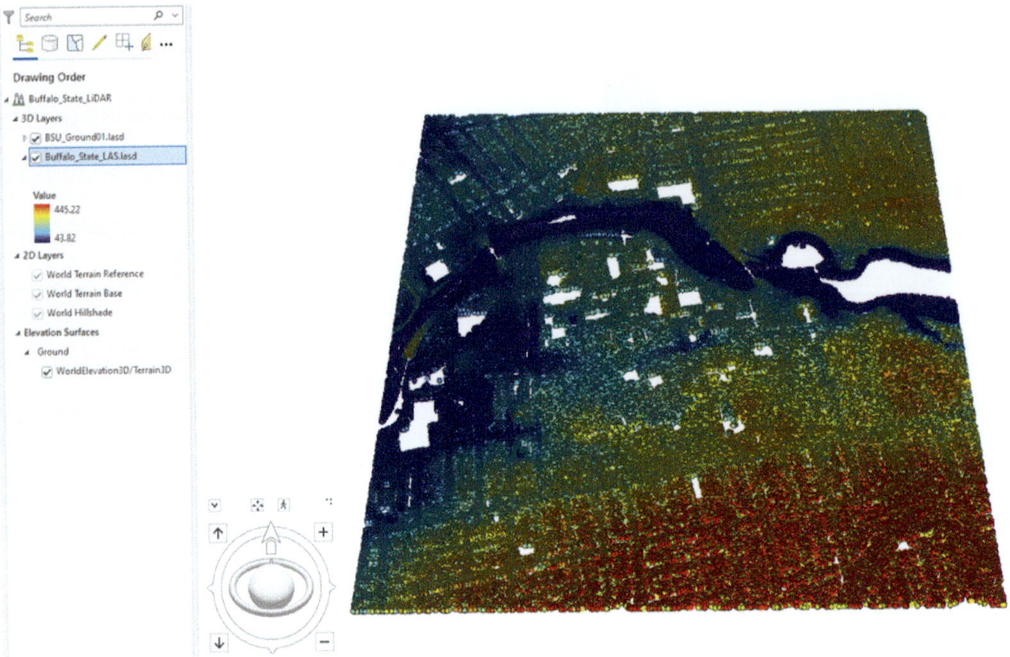


**2.** Kissing Bridge Ski Resort Project

In this project, we will use field collected LiDAR point cloud data to create a triangulated irregular network (TIN) 3D model in the study area. Then, we will convert the TIN model to raster DEM. Finally, we will use the DEM models in the study area and the selected observation sites to produce the viewshed area map layer.

The viewshed analysis in 3D identifies the cells in an input raster that can be seen from one or more observation locations. Viewshed analysis has many important applications in our modern society. The following are a few examples for the questions to be answered. (i) Which areas can be seen from a fire lookout tower set up on a mountain top? (ii) How frequently can a land-fill or solid-disposal site be seen from an near-by highway? (iii) Where should the next communications or cellphone signal broadcasting tower in a series be located in a region?

First, we would like to merge the four quadrangles of point cloud that we downloaded from the New York State GIS Clearinghouse website. Open the "Create LAS Dataset" tool in the Geoprocessing Pane. For the "Input File," navigate to your "LiDAR_Kissing_Bridge" folder to load the four quadrangles, such as e1406n2306_2019.las, one by one. For "Output LAS Dataset," navigate to your "LiDAR_Kissing_Bridge" folder again and type in "Kissing_Bridge_LAS" as the new file name (Chapter08-Tutorial \LiDAR_Kissing_Bridge\Kissing_Bridge_LAS.lasd). Run the tool to process the data. You notice that the merged file was added to the scene window.

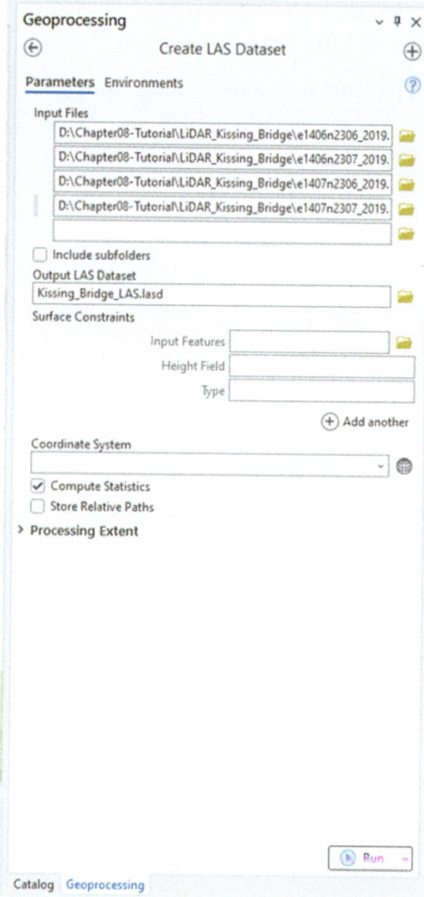
Geoprocessing
Create LAS Dataset
Parameters Environments
Input Files
D:\Chapter08-Tutorial\LiDAR_Kissing_Bridge\e1406n2306_2019.
D:\Chapter08-Tutorial\LiDAR_Kissing_Bridge\e1406n2307_2019.
D:\Chapter08-Tutorial\LiDAR_Kissing_Bridge\e1407n2306_2019.
D:\Chapter08-Tutorial\LiDAR_Kissing_Bridge\e1407n2307_2019.
Include subfolders
Output LAS Dataset
Kissing_Bridge_LAS.lasd
Surface Constraints
Input Features
Height Field
Type
Add another
Coordinate System
Compute Statistics
Store Relative Paths
Processing Extent
Run
Catalog Geoprocessing

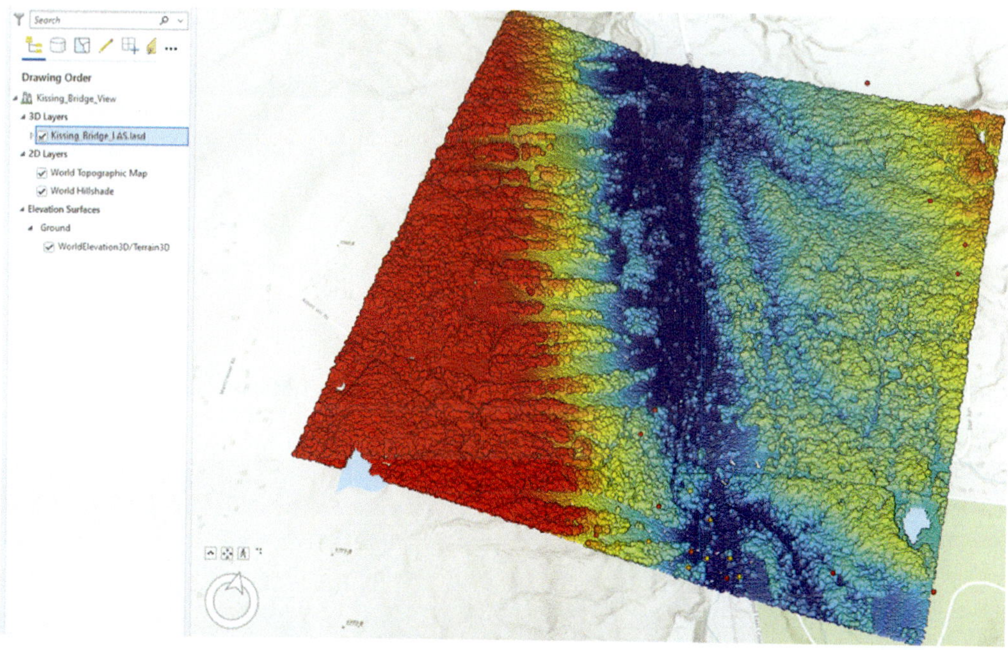
Search
Drawing Order
Kissing_Bridge_View
3D Layers
Kissing_Bridge_LAS.lasd
2D Layers
World Topographic Map
World Hillshade
Elevation Surfaces
Ground
WorldElevation3D/Terrain3D

Close the Geoprocessing Pane for now. While the "Kissing_Bridge_LAS.lasd" layer is activated on the Table of Contents, click the "Data" menu on the top. Expand the drop-down list of "Export" icon . Select "Export TIN" to open the "LAS dataset to TIN" tool in Geoprocessing Pane.

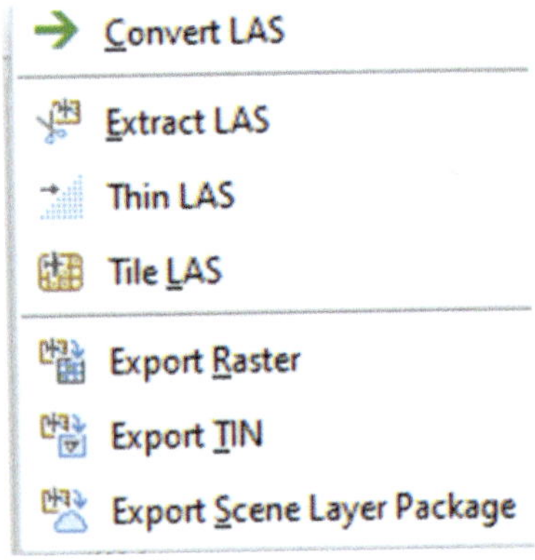


For the "Output TIN," navigate to your LiDAR_Kissing_Bridge folder and type in "KB_TIN" as the file name (\Chapter08-Tutorial\LiDAR_Kissing_Bridge\KB_TIN). For the "Thinning Type," select "Window Size" (using a moving window). For the "Thinning Method," select "Close to Mean Z" (mean Z values).

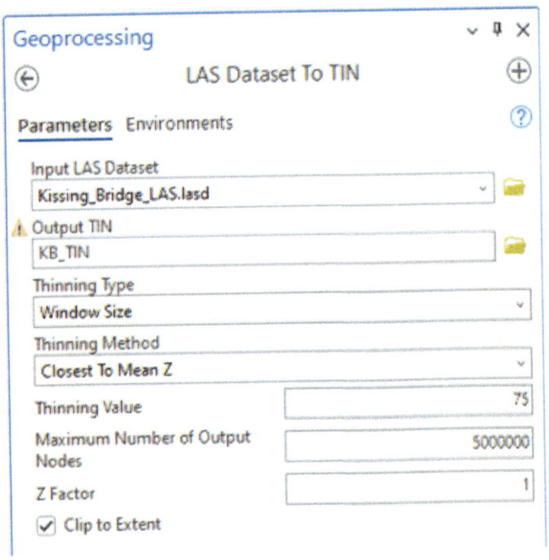


Run the tool to produce TIN model. Check off the "Kissing_Bridge_LAS.lasd" layer, the KB_TIN layer is now visible.

In order to conduct different scenarios of the viewsheds to facilitate a cellphone signal communication tower in this area, we need convert this TIN dataset to a raster dataset.

We go back to "Toolboxes" on the Geoprocessing Pane and search "TIN to Raster." The tool is found on the top of the list.

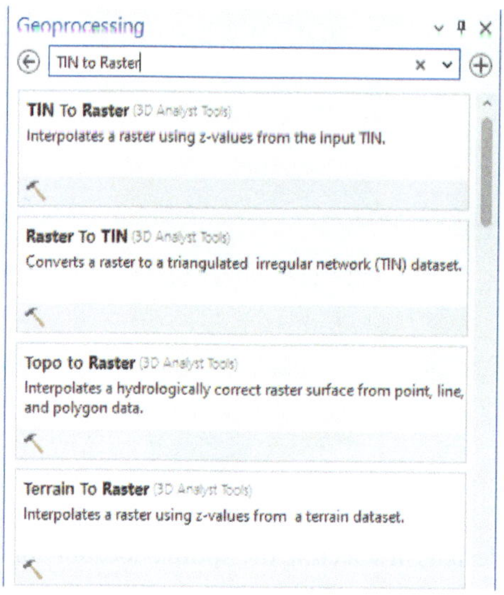


Activate the tool. For "Input TIN," select "KB_TIN." For "Output Raster," navigate to Kissing_Bridge.gdb (geodatabase) and type in "KB_DEM" as the raster dataset name. Change the "Sampling Distance" to "Cell Size" and keep other criteria as the default.

Run the process to create the raster dataset.

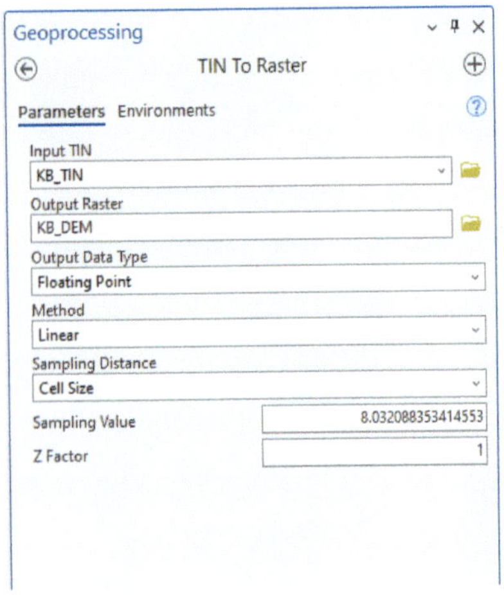


If the newly produced DEM raster file is not added in the scene window, add it. Check off all the other layers except KB_DEM. Symbolize this layer using the "Elevation #8" color ramp.

Now, let's conduct live scenarios for different viewshed designs in selecting the optimal location for a cellphone communication tower in this region.

We will use interactive viewshed creation methods to analyze the data to get the result. The exploratory analysis viewshed can be created in a variety of ways based on your workflow of following.

1. Interactive placement: place one or more viewsheds in the scene interactively when a user clicks to specify the watch out locations.

2. Interactive orientation: the user can click to place the observer (origin) location and change its view directions when placing one or more viewsheds within scene.
3. Observer from camera: assuming there is a camera position that the user can choose. Then, the user to "Apply" the camera position to obtain the viewshed.
4. Along a line: using a line feature class as an input, the user can find the visible areas in the study region along that line.
5. From layer: using a feature layer of points that serve as observation points in the study area, users can create a comprehensive raster map of viewsheds to show the places that can be seen at these points.

Now, let's activate the Analysis ribbon on the main menu.

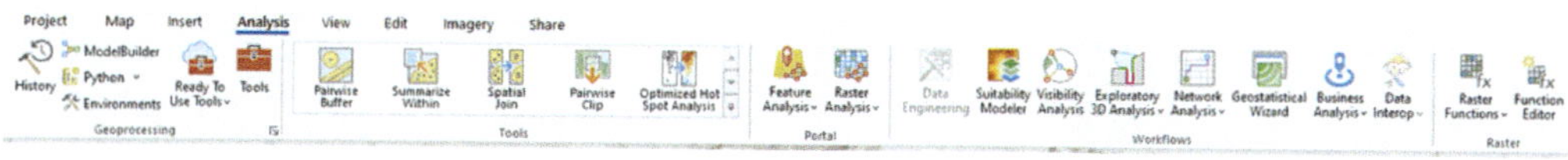


Expand the "Exploratory 3D Analysis" drop-down list in the "Workflows" group.

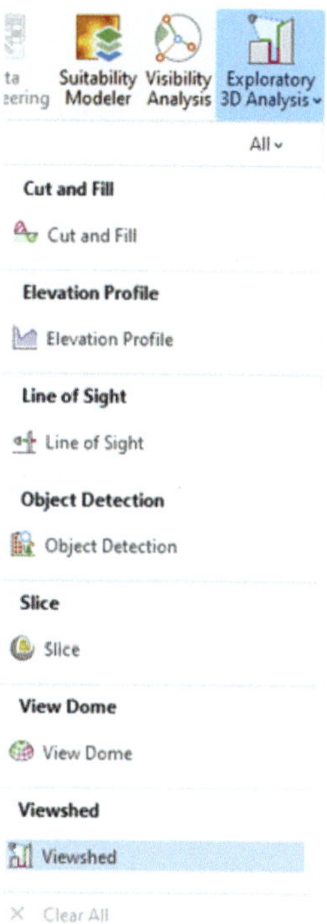

Select "Viewshed," the "Exploratory Analysis – Viewshed" dialog opens on the Geoprocessing Pane.

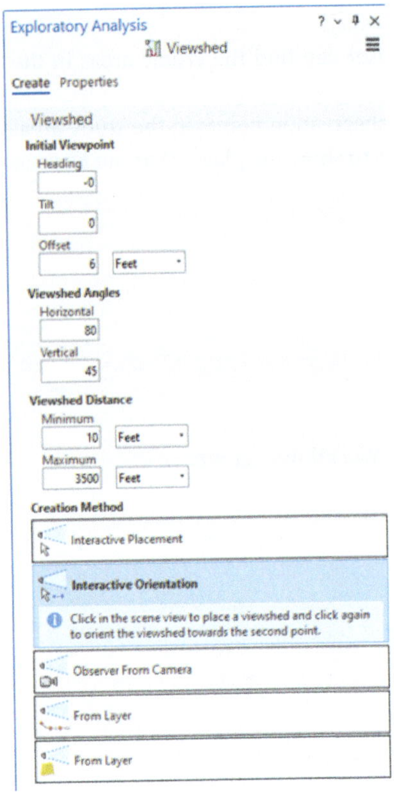


In this case, we are going to explore the "Interaction Orientation" method. Change to "Horizontal" angle of the Viewshed to 80°. Change the "Maximum" Viewshed Distance to 3500 feet. Activate the "Interaction Orientation" category and move the cursor to the scene window. Click on the location where you would like to set up as the observation point. You notice one viewshed model on the map is formed. The red areas mean the places that the people cannot see directly at the observation point. In this option, the user can turn the entire viewshed around to analyze the visible regions in different directions.

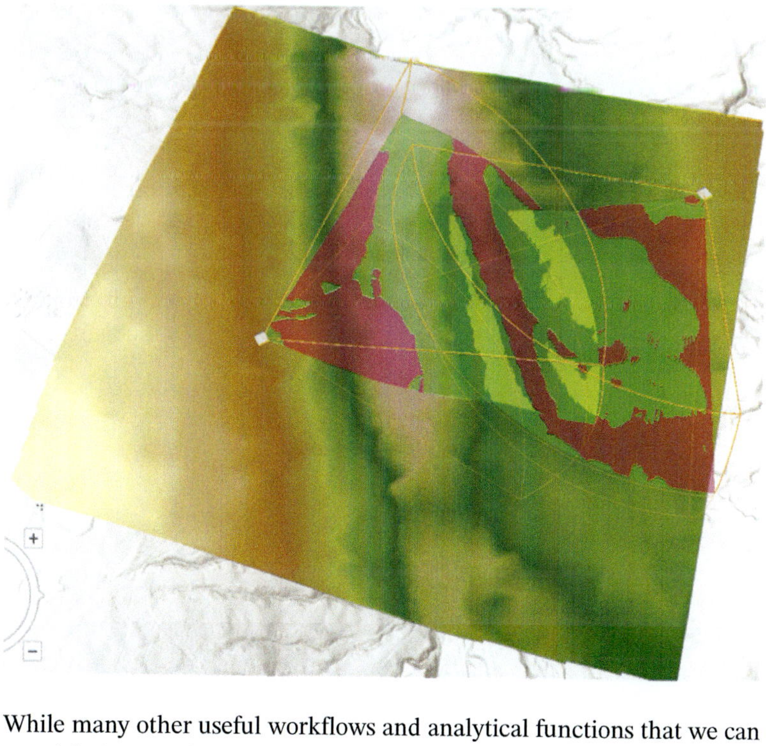


While many other useful workflows and analytical functions that we can use, we accomplish the tutorial of 3D analysis now. Save the project and close the ArcGIS Pro.

# Part IV

# Advanced Concepts and Applications of Geographic Information Analysis

# 9

# Spatial Interpolation and Spatial Statistical Analysis

## Concepts and Theories

Spatial statistical analysis is to analyze data that has spatial or locational characteristics using the methods of statistics. Spatial statistical analysis looks for patterns or correlation in recorded field observations that are generally called events, objects, or features across geographic space. It can also predict or interpolate the values and value distribution of a particular attribute of object based on measured values at sampled locations. According to the analytical objectives and the number of variables involved, spatial statistical analysis can be classified into three categories: (i) simple spatial statistical analysis, (ii) single variable spatial statistical analysis, and (iii) multiple variable spatial statistical analysis.

## Spatial Descriptive Statistics

Spatial descriptive statistics is the intersection of spatial analysis and descriptive statistics; these methods are used for a variety of purposes in geographic analysis, particularly in quantitative data analyses in Geographic Information Systems (GIS). In essence, this type of approach in GIS combines basic statistical functions either with methods of digital map presentations or with analysis of feature locations or distribution patterns.

**Mode** is the value that appears most often in a set of data. Mode computation can be used for land use and land cover analysis. Figure 9.1 is the remote sensed data can be used to compute frequency of land use/land cover in the region.

**Median** is the numerical value separating the higher half of a data sample or a population of data from the lower half. Median equals to the mean only for the data of symmetric distribution (Figure 9.2). Figure 9.3 map produced by US Census Bureau is the Median Household income of the United States by County.

**Mean** is obtained by dividing the sum of observed values by the number of observations, $n$. When mean equals to median, it is referred the data have normal distribution (Figures 9.4 and 9.5).

$$\text{Mean} = \frac{\sum_{i=1}^{n} X}{n}$$

**Standard deviation** is the amount of variation or dispersion from the mean value (Figures 9.6 and 9.7).

*Fundamentals of Geographic Information Systems (GIS) Analysis: Applying ArcGIS Pro*, First Edition. Tao Tang.

**Figure 9.1** Example of a remotely sensed image showing that most of land use in the area is for transportation. *Source:* Dr. Tao Tang.

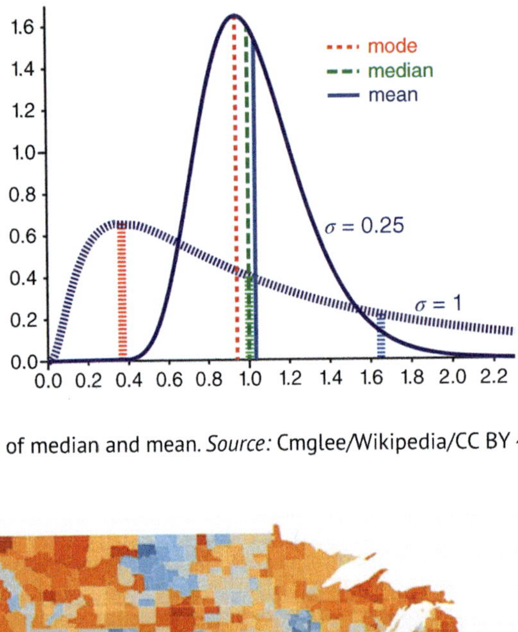


**Figure 9.2** Difference of median and mean. *Source:* Cmglee/Wikipedia/CC BY 4.0.

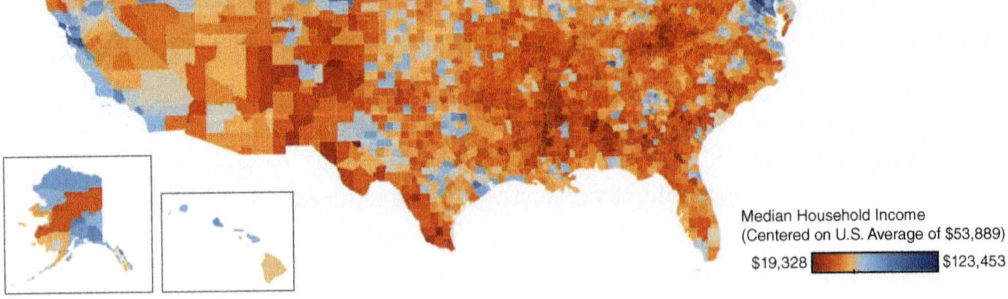


**Figure 9.3** Median household income map of the United States. *Source:* U.S. Census Bureau/Public Domain.

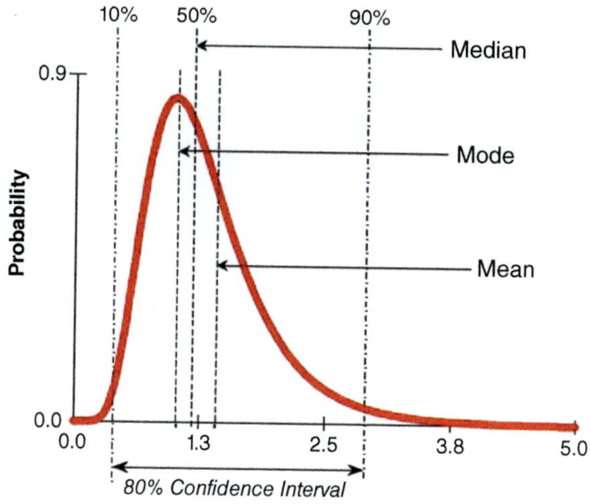


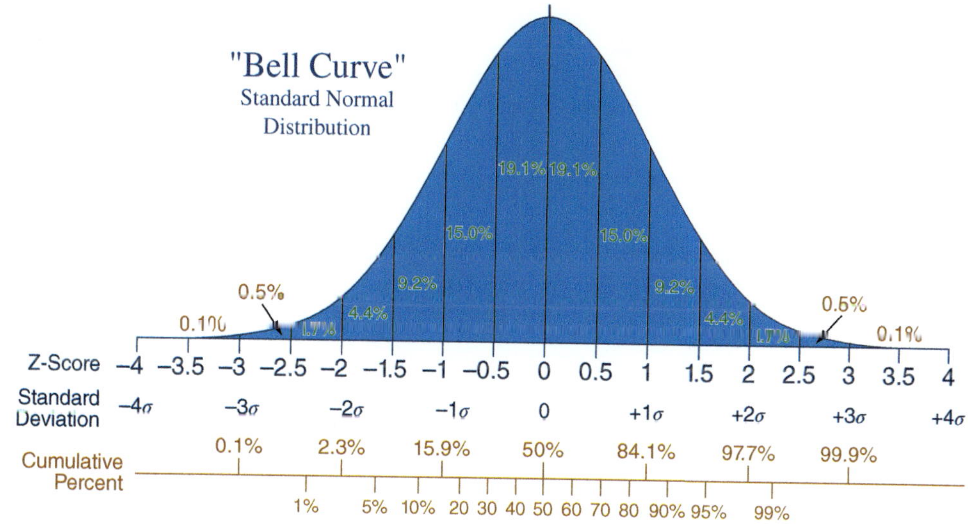


**Figure 9.4** Mean location difference between the normal and abnormal distributions. *Source:* Adapted from Normal distribution, Wikipedia: https://en.wikipedia.org/wiki/Normal_distribution, last accessed on 29 November 2024.

$$\sigma = \sqrt{\frac{1}{N}\sum_{i=1}^{N}(x_i - \mu)^2}$$

where $\sigma^2$ is the variance.

**Skewness** measures asymmetry of the distribution of data values (Figures 9.8 and 9.9).

$$\text{skew} = \frac{1}{n} \quad \sum\left(\frac{x_i - \mu}{\sigma}\right)^3$$

**Kurtosis** is the measurement of "peakedness" of the probability distribution of data values (Figure 9.10).

$$\sum\frac{(X - \mu)^4}{\sigma^4} - 3$$

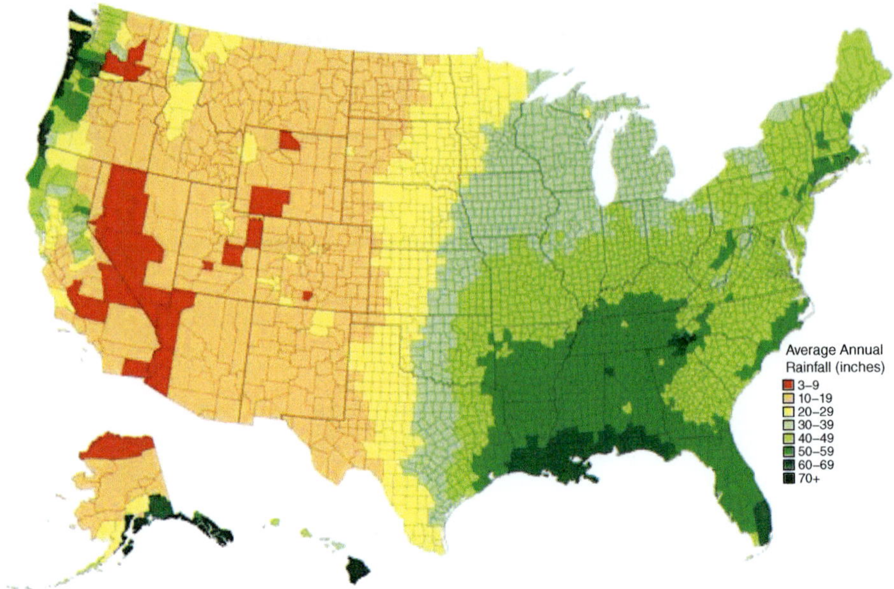


**Figure 9.5** Mean annual precipitation (rainfall) by county in the United States.

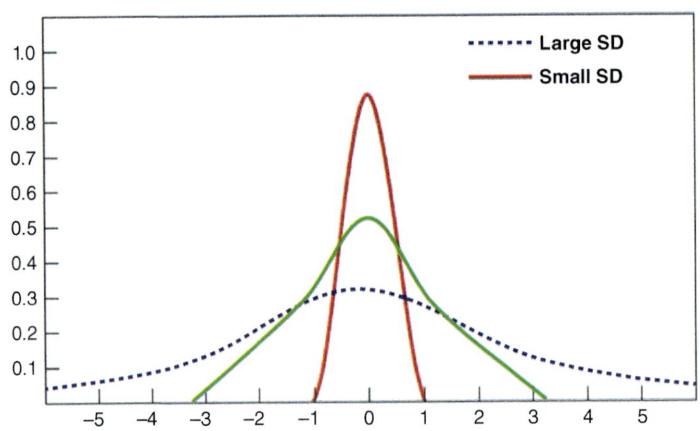


**Figure 9.6** Comparison of standard deviations. *Source:* Adapted from Wikipedia.

## Analysis of Geographic Mean Center

The geographic mean center is the average coordinate location of all the features in the study area, accounting x-, y- in two-dimensional (2D) space and x-, y-, z- in 3D space. The purpose of computing the mean center is to measure geographic distributions of objects or a feature class. Geographic mean center is also called the center of concentration for a set of features. It can be calculated for point, line, or polygon features. In 2D environment, the x coordinate and y coordinate of the center point can be calculated directly. A new point feature class will be generated in this process. The mean of the x-coordinate and the y-coordinate of all features is as follows:

$$\overline{X} = \frac{\sum_{i=1}^{n} x_i}{n}, \quad \overline{Y} = \frac{\sum_{i=1}^{n} y_i}{n}$$

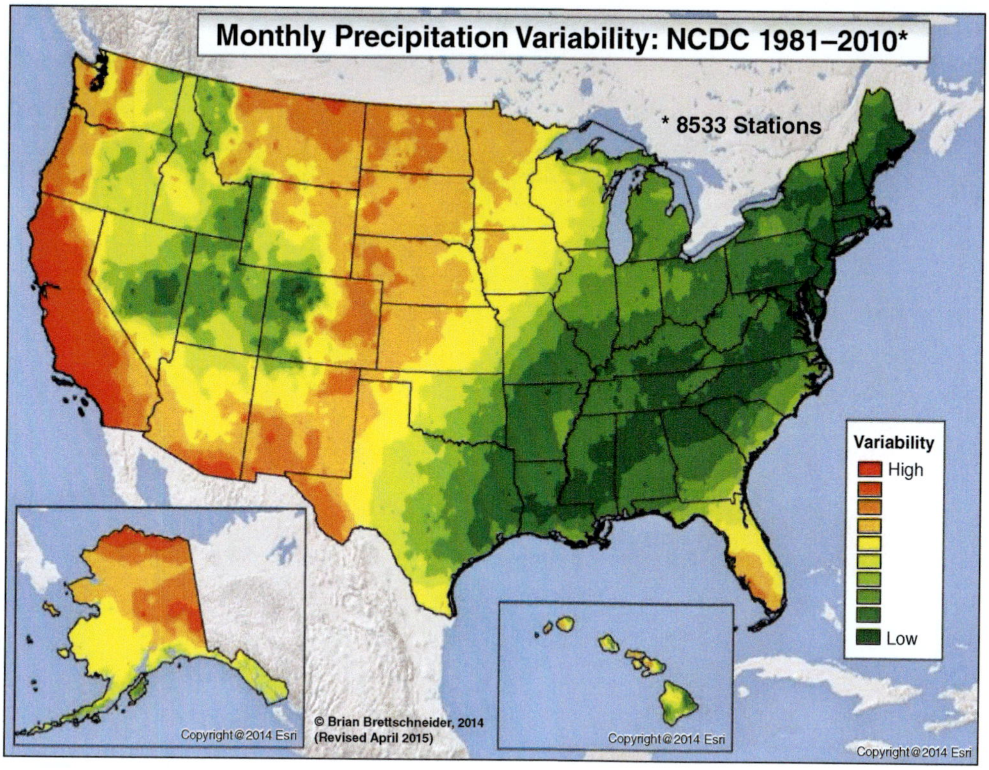


**Figure 9.7** Map of standard deviation of monthly precipitation (rainfall) in US, 1981–2010.

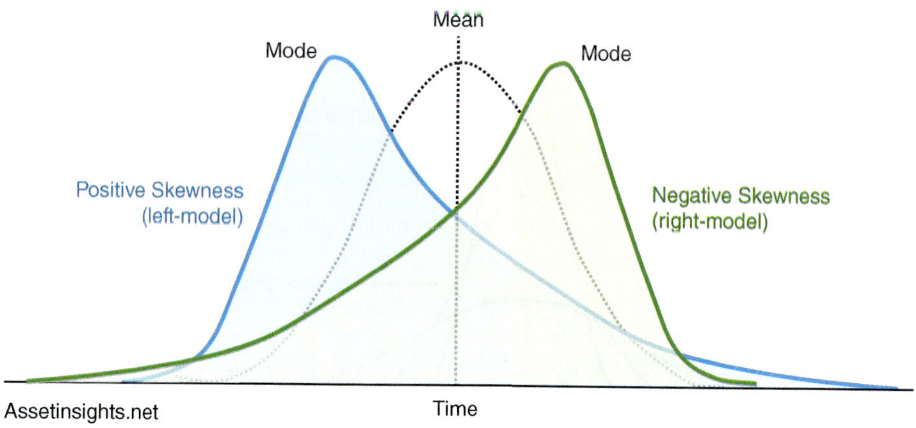


**Figure 9.8** Positive skewness and negative skewness. *Source*: Adapted from Skewness, Wikipedia: https://en.wikipedia.org/wiki/Skewness, last accessed on 29 November 2024.

If the weighted field of the mean center is set, the position of the center point needs to consider the weight value, and the calculation model is as follows:

$$\overline{X}_w = \frac{\sum_{i=1}^{n} w_i x_i}{\sum_{i=1}^{n} w_i}, \quad \overline{Y}_w = \frac{\sum_{i=1}^{n} w_i y_i}{\sum_{i=1}^{n} w_i}$$

$w_i$ is the weight of the feature $i$

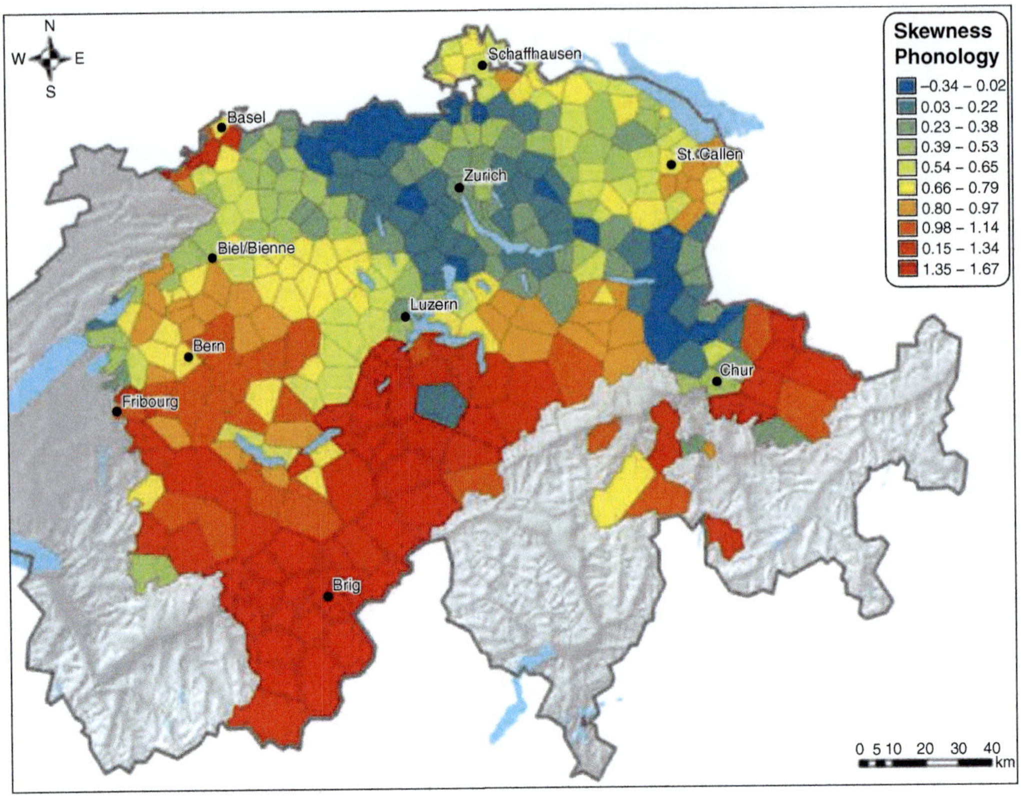


**Figure 9.9** Skewness map of German ethnic group in various social subdivisions in Switzerland.

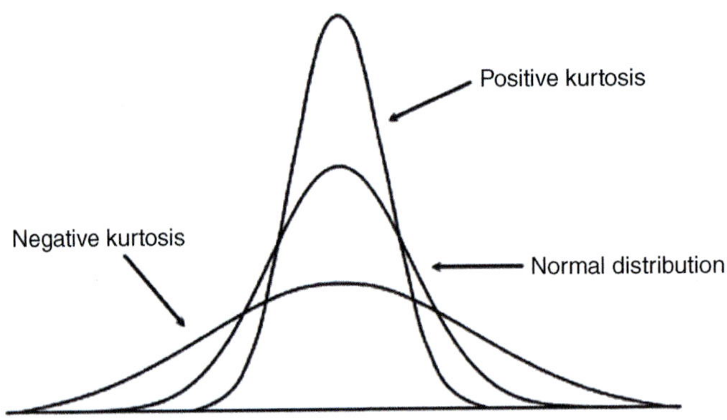


**Figure 9.10** Comparison of peaks of Kurtosis. *Source:* U.S. Census Bureau.

There are a great number of potential applications of GIS mean center analysis. The following are some of these examples. A crime GIS analyst would be able to see if the mean center for burglaries shifts when evaluating daytime versus nighttime incidents. This can help police departments better allocate resources. A wildlife biologist can calculate the mean center of elk observations within a park over several years to see where elk congregate in summer and winter to provide better information to park visitors. A GIS analyst can assess the level of service by

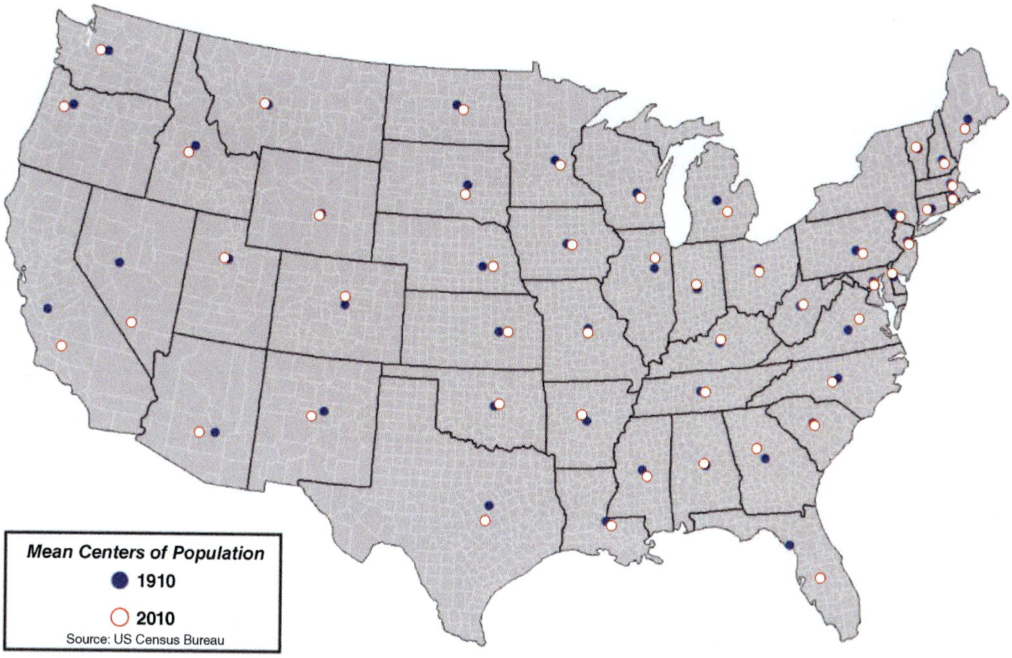


**Figure 9.11** Mean center change by states in the United States.

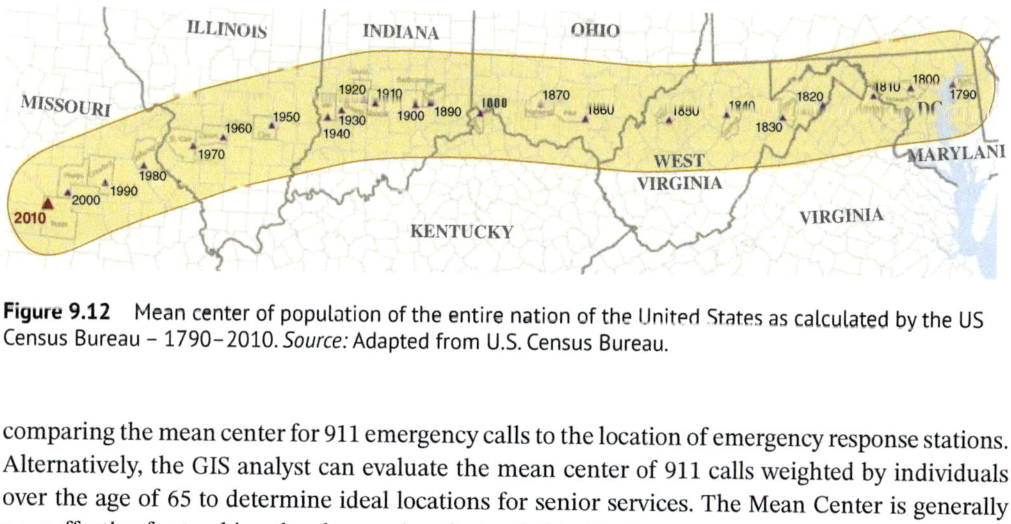


**Figure 9.12** Mean center of population of the entire nation of the United States as calculated by the US Census Bureau – 1790–2010. *Source:* Adapted from U.S. Census Bureau.

comparing the mean center for 911 emergency calls to the location of emergency response stations. Alternatively, the GIS analyst can evaluate the mean center of 911 calls weighted by individuals over the age of 65 to determine ideal locations for senior services. The Mean Center is generally very effective for tracking the changes in a feature's distribution over time and can also be used to compare the distribution of multiple features (Figures 9.11 and 9.12).

## Standard Distance Analysis of a Set of Features

Standard distance is an average value representing the distance in units from the mean center to each of the original features in determining the mean center location. The standard distance is a radius or distance value. This distance value is usually plotted on a map as a circle for a visual indication of the dispersion of features. Big circle means features are more dispersed, and small

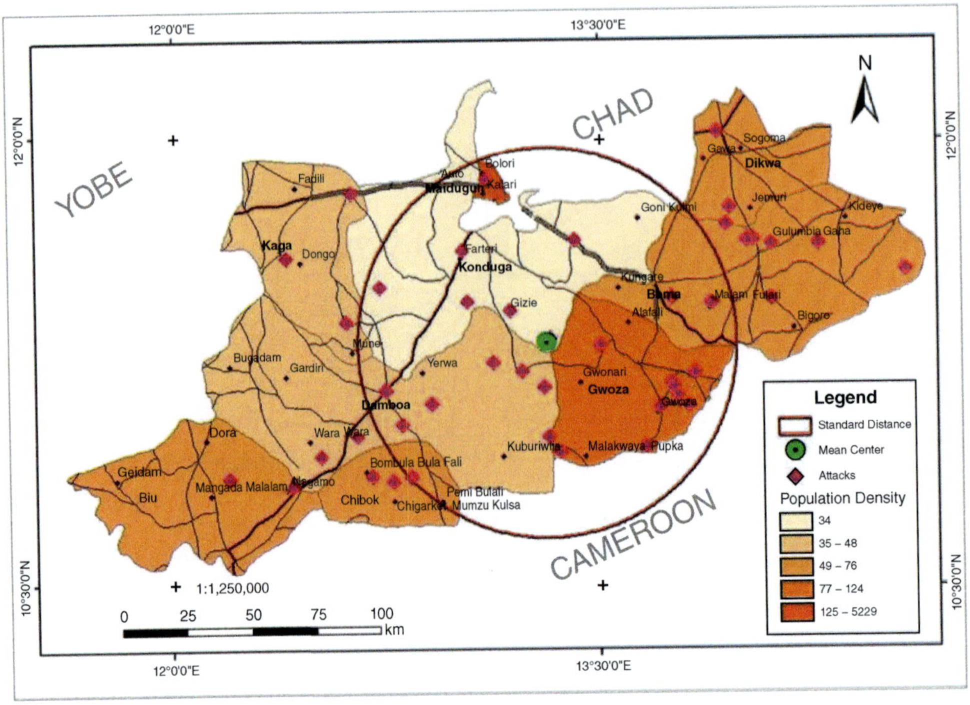


**Figure 9.13** Mean center and standard distance of terrorist attacks in Borno State, Nigeria. *Source:* Kappo et al. 2016/IJSRCSEIT/CC BY 4.0.

circle means features are more concentrated in the study area. Standard distance provides a single summary measure of feature distribution around their mean center. The following is the model to calculate the standard distance. Figure 9.13 is an example of standard distance measuring for terrorist attacks in Borno State, Nigeria (Kappo et al. 2016).

$$\mathrm{SD} = \sqrt{\frac{\sum_{i=1}^{n}(x_i - \overline{X})^2}{n} + \frac{\sum_{i=1}^{n}(y_i - \overline{Y})^2}{n}} \tag{1}$$

where SD is the standard distance, $x_i$ and $y_i$ are the coordinates for feature $i$, $\{\overline{X}, \overline{Y}\}$ represents the mean center for the features, and $n$ is equal to the total number of features.

## Nearest Neighbor Analysis

The nearest neighbor analysis measures the distance between each point feature or feature centroid of a polygon feature and its nearest neighbor's point or centroid locations. Then, the average of all these nearest neighbor distances is computed. If the average distance is less than the average for a hypothetical random distribution (<1), the distribution of the features being analyzed is considered clustered. If the average distance is greater than a hypothetical random distribution (≥1), the features are considered dispersed and randomly distributed (<2) or dispersed and regularly dispersed (>2). Nearest Neighbor analysis provides a numerical value that describes the extent to which a set of points or polygon centroids are clustered or uniformly spaced. Distribution patterns of complete spatial randomness (CSR) are generated by the means of two original hypotheses: (i) all locations or places are equally likely to be the recipient of a geographic object (event) and (ii) all the geographic objects are located independently of one another (Figure 9.14).

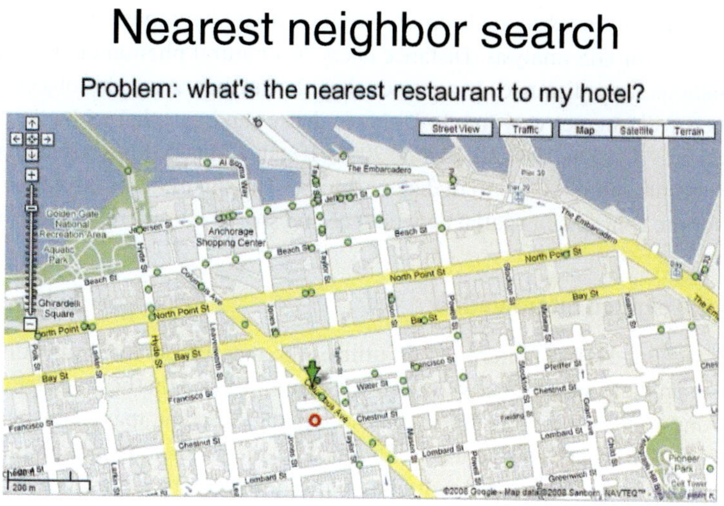


**Figure 9.14** Searching for the nearest restaurant to me. *Source:* From Google Maps.

## Spatial Inferential Statistics

While spatial descriptive statistics summarize the characteristics of a dataset and help us to illustrate the dataset on a digital map, spatial inferential statistics allow us to make predictions based on a dataset and present the predictions on digital maps. Most of the time, we can only acquire sample data from the entire population of the dataset, because it is too difficult or expensive to collect the whole dataset from the Earth's surface or real-world human society. In essence, the term inferential here means interpretation or generalization using the sampling data (Rao 2018; Salas-Parra et al. 2023). In GIS, we call this type of computational processes as interpolations as well. The fundamental principle of Inferential Statistics is to use random and unbiased sampling methods to collect data. This will ensure that the sampling dataset we use is a representative of the entire population. "Spatial statistics are used for a variety of analysis types, including pattern analysis, shape analysis, surface modeling and surface prediction, spatial regression, statistical comparisons of spatial datasets, statistical modeling, and prediction of spatial interaction" (ESRI 2022).

### Single Variable Spatial Statistical (Geostatistical) Analysis

In this textbook, we identify statistical analyses of single variables or multiple variables in spatial or GIS analysis. The so-called single variable means that we only take one characteristic or attribute of the geographic objects or features into consideration in GIS operations. In many cases, the geographic locations are treated as the independent variable (the second variable) in the single-variable spatial statistical analysis. The point is that geographic locations and distances are unique. When it is used as a variable in the spatial statistical analysis, it is fundamentally spatial or geographic. Geographic locations and distances can be sampled but are not statistical in nature. In practice, many professionals and researchers call the single-variable spatial statistical analysis as geo-statistics or geostatistical analysis.

Theory and the approach of traditional statistics are based on the assumption of independent observations. Independence of observations or variables is utilized as a suitable benchmark to identify statistically significant non-independent phenomena in the real world (Getis 1999). In contrast, geographic studies analyze relationships of objects on the Earth's surface to their locations and distances to each other. Distance decay hypothesis is one of the original theories in geographic or GIS analysis. Distance decay is a natural phenomenon observed among locations, populations, and distance. It assumes that the further apart the objects are, the less likely these objects will interact very much. Distance decay can be explained by the "First Law of Geography" (Tobler 1970, 1979, 1989; Goodchild 2004). It states, "Everything is related to everything else, but near things are more related than distant things." The First Law of Geography established the foundation for the concepts of spatial dependence and spatial autocorrelation (SAC).

### Inverse Distance Weighted Interpolation

Inverse distance weighted (IDW) interpolation utilizes the measured values surrounding the prediction location to predict a value for any unsampled location, based on the assumption that things that are close to one another are more alike than those that are farther apart. The measured values closest to the prediction location have more influence on the predicted value than those farther away. IDW assumes that each measured point has a local influence that diminishes with distance. IDW gives greater weights to points closest to the prediction location, and the weights diminish as a function of distance. This is the reason it is called IDW.

Practically, IDW interpolation is considered one of the deterministic operations in GIS simply because its function is mainly made by mathematical models. IDW works only on one set of objects or one feature class at a time. The result of IDW operation is normally a raster surface of grid cells that contain the prediction values. Since weights are proportional to the inverse of the distance between the observation point and the prediction location, the power function value p was developed in the computational process. The higher the p-value, the faster the weights decrease as the distance increases. High p-value creates more local impact on the immediate surrounding prediction locations. It will produce more variations on the prediction surface.

Since objects or features that are close to one another are more alike than those that are farther away, as the locations get farther away, the measured values will have little relationship to the value of the prediction location. In order to speed up the computation, we can exclude the more distant points that will have little influence on the prediction or interpolation. As a result, we can define a search neighborhood with a limited number of measured values. The search neighborhood restricts how far and where to look for the measured values to be used in the prediction. In Figure 9.15, five measured points (neighbors) will be used when predicting a value for the location without a measurement, the yellow point (ESRI 2020).

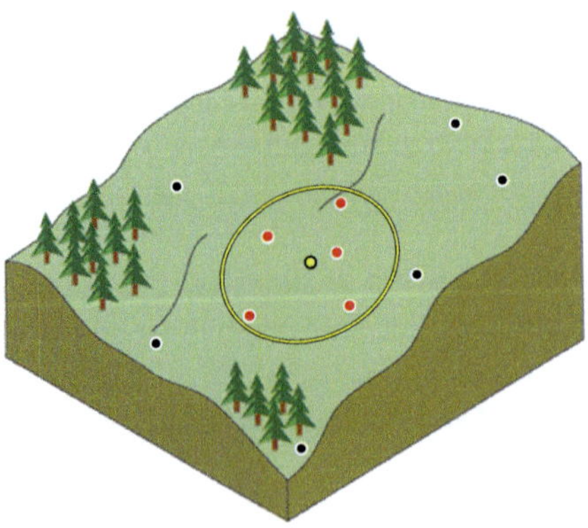

**Figure 9.15** Inverse Distance Weighted (IDW) interpolation. *Source*: From Environmental Systems Research Institute, Inc.

## Spatial Autocorrelation

Autocorrelation is a measure of correlation (similarity) between nearby observation values in statistics. Autocorrelation can be applied to predict or understand spatial distributions of geographic features or objects that is called SAC. Autocorrelation can also be applied on time scale to analyze the time dependence (Griffith 2005). In that case, it is called temporal autocorrelation. SAC measures correlation between feature locations and feature values simultaneously. In essence, it is an operation to predict or interpolate values of a feature class using the sampled values and a space dependence model (Griffith 2005).

SAC is the term used to describe the presence of systematic spatial variation in a variable or a feature class, and positive SAC, which is most often encountered in practical situations, is the tendency for areas or sites that are close together to have similar values. There are several statistical techniques for detecting the presence of SAC. Moran's I Index is one of the major estimators of SAC. Given a set of features and an associated attribute, it evaluates whether the pattern expressed is clustered, dispersed, or random. The following is the model to calculate the Moran's I Index.

$$I = \frac{N}{W} \frac{\sum_i \sum_j w_{ij}(x_i - \bar{x})(x_j - \bar{x})}{\sum_i (x_i - \bar{x})^2}$$

where $N$ is the number of spatial units indexed by $i$ and $j$; $x$ is the variable of interest; $\bar{x}$ is the mean of $x$; $w_{ij}$ is a matrix of spatial weights with zeroes on the diagonal (i.e. $w_{ii} = 0$); and $W$ is the sum of all $w_{ij}$.

For the distributed geographic features or objects, Moran's $I$ Index indicates the following spatial patterns (Figure 9.16). (i) If $I = -1$, it is a perfect dispersion or perfect clustering of dissimilar values of the feature class. (ii) If $I = 0$, it is no autocorrelation or in a perfect randomness distribution. (iii) If $I = +1$, It indicates perfect clustering of similar values the feature class. In GIS practice, we are searching the $I$ value is close to +1 in order to find the clusters of feature value distributions.

In ArcGIS Pro, the p-value and Z-score are computed in replacing the $I$ value. The p-value is a probability. For the pattern analysis tools, it is the probability that the observed spatial pattern was created by some random process. When the p-value is very small, it means it is very unlikely (small probability) that the observed spatial pattern is the result of random processes, so you can reject the null hypothesis. In other words, you can conclude that there is a cluster of geographic objects. Z-scores are standard deviations. If, for example, a tool returns a Z-score of +2.5, you would say that the result is 2.5 standard deviations (Table 9.1; ESRI 2020).

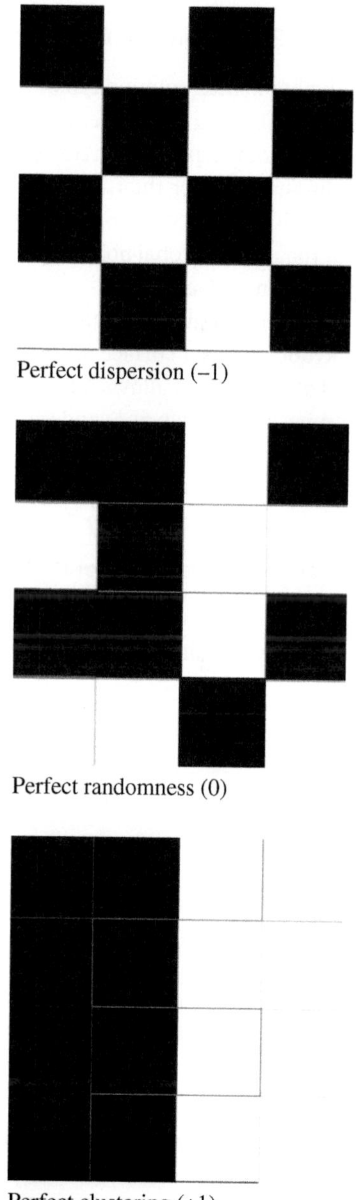


**Figure 9.16** Distribution patterns of the objects judged by Moran's *I* Index.

**Table 9.1** Confidence levels determined by the p-value and Z-score.

| Z-score (SD) | p-Value | Confidence level |
|---|---|---|
| $< -1.65$ or $> +1.65$ | <0.10 | 90% |
| $< -1.96$ or $> +1.96$ | <0.05 | 95% |
| $< -2.58$ or $> +2.58$ | <0.01 | 99% |

## Global Polynomial Interpolation

Polynomial interpolation is a method to estimate or predict values in between the sampled known data locations. In essence, the linear interpolation across the geographic space is the simplest method to interpolate variable values, that is, to draw straight lines between the known data locations and consider the function as the combination of those straight lines. In order to reduce the prediction errors, polynomial functions are used to connect the sampled known data locations for interpolations. Global polynomial interpolation fits one polynomial model or function to the entire prediction surface.

A first-order global polynomial fits a single plane through the entire data; a second-order global polynomial fits a surface with a bend in it, allowing surfaces representing valleys; a third-order global polynomial allows for two bends; and so forth (Figure 9.17). However, when a surface has a varying shape, such as a landscape that slopes, levels out, and slopes again, a single global polynomial will not be suitable for value predictions. Multiple polynomial planes would be able to represent the surface more accurately (ESRI 2020).

## Local Polynomial Interpolation

While global polynomial interpolation utilizes one polynomial model for the entire surface, local polynomial interpolation fits many polynomial models, each within specified overlapping

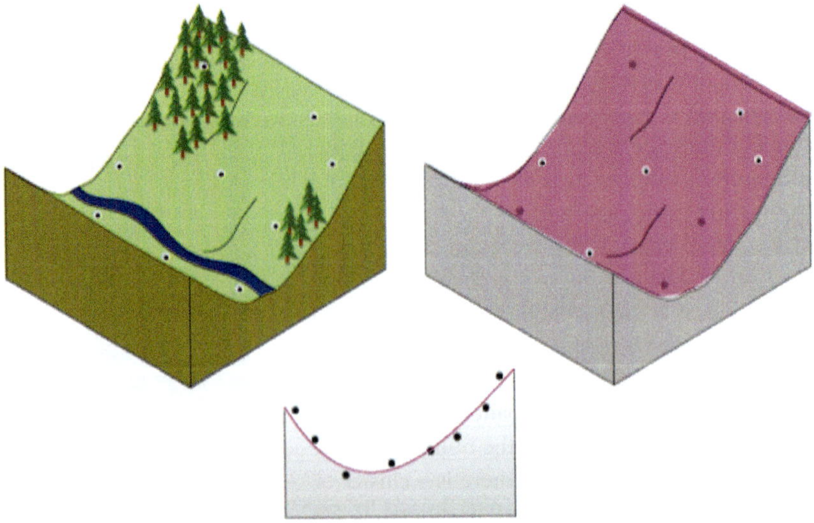

**Figure 9.17** Surface interpolation by second-order global polynomial. *Source:* From Environmental Systems Research Institute, Inc.

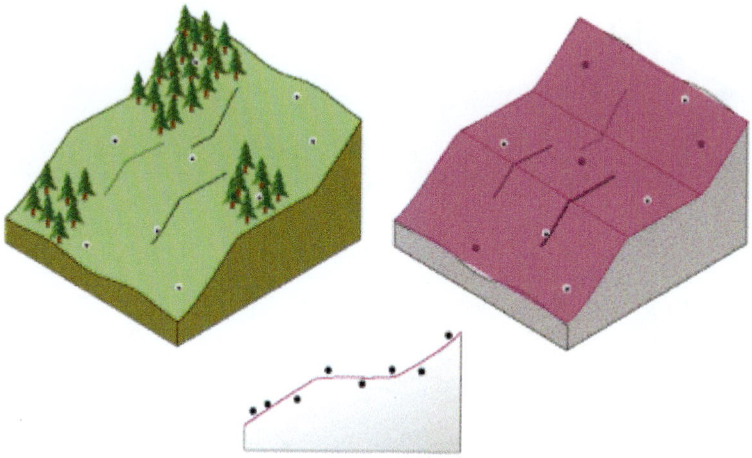

**Figure 9.18** Local polynomial interpolation by multiple functions. *Source:* From Environmental Systems Research Institute, Inc.

neighborhoods. The search neighborhood can be defined by using the size and shape, number of neighbors, and sector configuration. Local polynomial interpolation fits the specified order, such as zero, first, second, third, and so on, polynomial using known data locations only within the defined neighborhood (Figure 9.18). The neighborhoods used for local predictions can overlap, and the value used for each prediction is the value of the fitted polynomial function at the center of the neighborhood.

## Interpolation of Radial Basis Functions

Radial basis functions (RBFs) are a series of exact interpolation techniques to fulfill the requirement that the interpolation surface must pass through each of the measured sample values. There are five different basis functions: (i) thin-plate spline, (ii) spline with tension, (iii) completely regularized spline, (iv) multiquadric function, and (v) inverse multiquadric function. RBFs are conceptually similar to fitting a rubber membrane through the measured sample values (Figure 9.19). while minimizing the total curvature of the surface. The basis function you select determines how the rubber membrane will fit between the values. It is generally considered that RBFs can improve the accuracy of value predictions. In this case, the prediction surface is created using different functions to bend and stretch the predicted local values so that the prediction surface passes through all of the measured values.

## Hotspot Interpolation – Simple Application of SAC

An area of hotspot can be defined as an area that has higher concentration of events or objects compared to the expected number given a random distribution of the objects or events. In essence, the hotspot analysis is to identify clusters in the SAC operations. In practice, given a set of weighted features, the Hot Spot Analysis can identify statistically significant hot spots and cold spots using the local General G index statistic. Hot spot analysis looks at each feature in a neighboring environment. Therefore, a feature with high value may not be a hot spot. A feature that has high value and

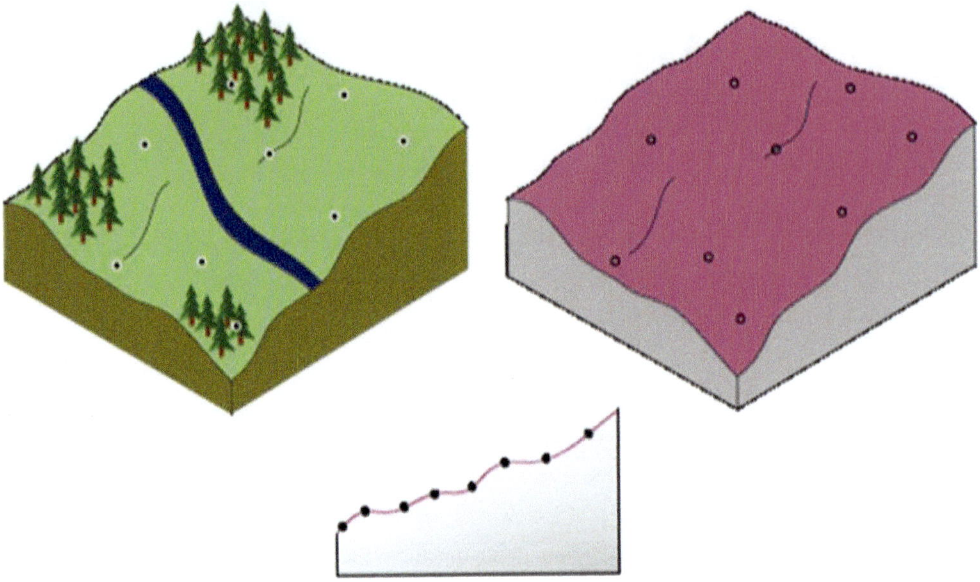

**Figure 9.19** Radial basis functions (RBFs) interpolation. *Source:* From Environmental Systems Research Institute, Inc.

is surrounded by features with high values is a hot spot. In contrast, a feature that has low value and is surrounded by features with low values is a cold spot.

Hotspot analysis is widely applied in many areas of studies, such as fire hazards, emergency rescue management, and public health and epidemiological research. Some major applications in public health practice and research applications are: (i) hotspot analysis of violent trauma occurrences in a city; (ii) spatial relation analysis of hotspots of violent trauma events or victims to the physical environment and the socioeconomic deprivation level of the neighborhood in a region; (iii) hotspot analysis to high potential regions that where future reassortment between human influenza A subtype H3N2 and avian influenza subtype H5N1 may generate a novel influenza virus (Wisniewski et al. 2019).

### Kriging Interpolation – Application of SAC

Kriging is an advanced geostatistical interpolation method that generates a prediction surface from a scattered set of observed or measured points with z-values or attribute values. Unlike other interpolation methods, Kriging function conducts an interactive investigation of the spatial behavior of the phenomena or objects represented by the z-values before users select the best prediction method for generating the output interpolation surface. Kriging, also called Gaussian process regression, is a method of interpolation based on Gaussian process governed by prior covariances.

The Kriging interpolation method was originally developed by South African mining engineer Danie Krige in his master's thesis research. The theory then was improved by French mathematician Georges Matheron (Krige 1951; Matheron 1962). The theoretical base of Kriging is the SAC. Kriging assumes that the distance between sample points reflects a spatial correlation. In other words, autocorrelation means the single variable correlates with no other variables except spatial locations. The following is the Kriging model.

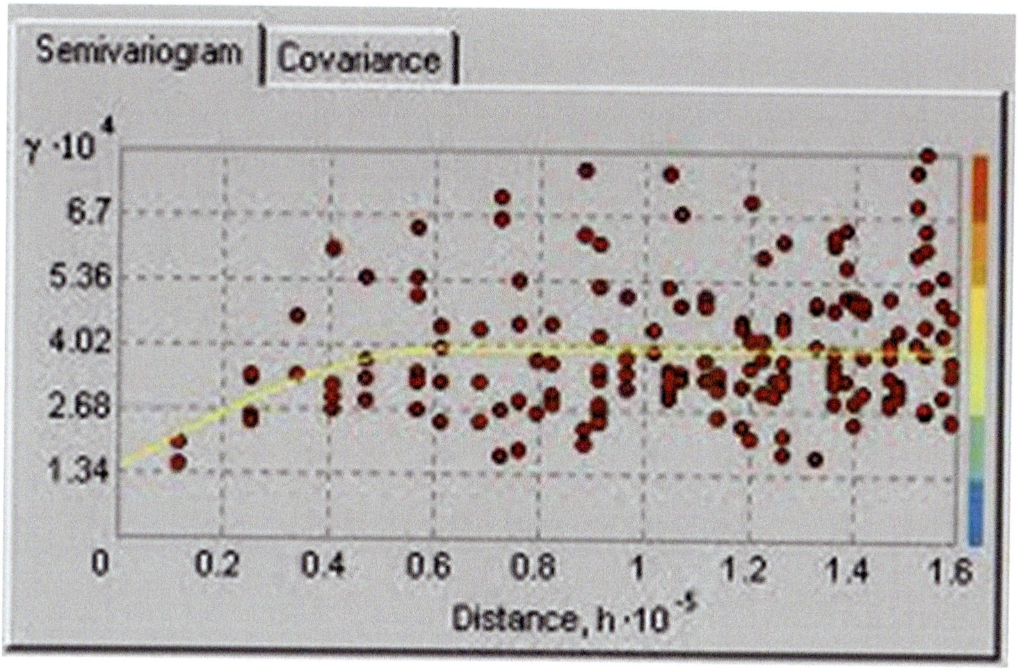


**Figure 9.20** Example of semivariogram in Kriging interpolation.

$$\hat{Z}(s_0) = \sum_{i=1}^{N} \lambda_i Z(s_i)$$

where, $Z(s_i)$ is the measured value at the $i$th location, $\lambda_i$ is an unknown weight for the measured value at the ith location, $s_0$ is the prediction location, and $N$ is the number of measured values.

In practice, Kriging analysis utilizes a variogram to conduct computations of variance in prediction. A variogram or sometimes called a "semivariogram" is a visual depiction of the covariance detected between each pair of points in the sampled data (Figure 9.20). For each pair of points in the sampled data, the gamma-value, or "semi-variance," which is a measure of the half mean-squared difference between their values, is plotted against the distance, or "lag," between these sampled feature locations. The "experimental" variogram is the plot of observed values, while the "theoretical" or "model" variogram is the distributional model that best fits the data. Variogram models can be selected from a variety number of mathematical functions, including linear, spherical, exponential, and power models. Users can make the choice of a variogram model to visually fit the observed or measured sample data. Figure 9.20 is an example of semivariogram model fitting for the prediction of air pollution distributions in Beijing, China, in the summer of 2008.

Fitting a model, or spatial modeling in Kriging analysis, is also known as structural analysis. In spatial modeling of the structure of the measured points, we use the empirical semivariogram to compute the distance decay of predicted values for all pairs of locations. Figure 9.21 shows the pairing of one point (the red point) with all other measured locations. This process continues for each measured point (ESRI 2020).

Kriging interpolation analysis can be further identified in the following major types.

1. **Simple Kriging:** It assumes stationarity of the first moment over the entire domain with a known mean. In other words, the predicted values will vary around a known mean value in the simple Kriging analysis. Stationarity is a condition that shows whether the data has a

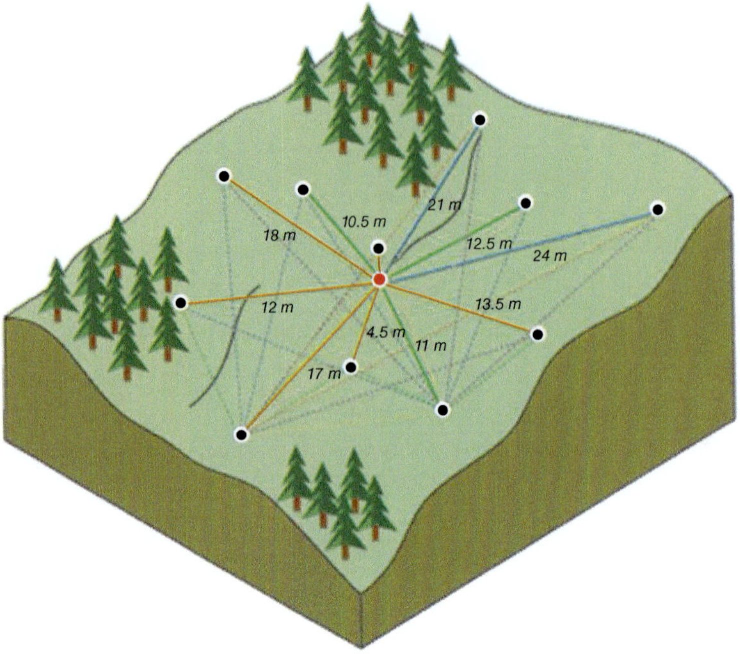


**Figure 9.21** Kriging the impacting weights by distance decay. *Source:* From Environmental Systems Research Institute, Inc.

constant mean and variance in each location. Stationarity is a decision or judgment; it is not a hypothesis.

2. **Ordinary Kriging:** It assumes a constant but unknown mean target value with predicted values varying around it.
3. **Universal Kriging:** It assumes there is a trend of changes for the predicted surface. This relaxes the assumption of stationarity by allowing the mean of the values to change in a deterministic way, such as a second-order polynomial in different locations.
4. **Empirical Bayesian Kriging:** Empirical Bayesian Kriging (EBK) is a geostatistical interpolation method that automates the most difficult aspects of building a valid kriging model. Other Kriging methods in Geostatistical Analyst require a user to manually adjust parameters to obtain accurate or acceptable results. While EBK automatically calculates these parameters through a process of sub-setting observation data and simulations (ESRI 2020).

   EBK also differs from other kriging methods by accounting for the error introduced by estimating the underlying semivariogram. Other kriging methods calculate the semivariogram from known data locations and use this single semivariogram to make predictions at unknown locations. Taking the estimation of semivariogram uncertainty into consideration, EBK can more accurately estimate the standard errors of prediction.

A case study of applying universal kriging (UK) is presented here. Universal kriging method was applied by a team of researchers to interpolate sampling air particulate pollution data (Tang et al. 2010). The field investigations were conducted during the summers and winters in 2007, 2008, and 2009 in Beijing, China. The randomly sampling sites of air pollution are shown in Figure 9.22. An example of an empirical semivariogram is shown in Figure 9.23. The UK interpolation of air particle pollution distributions in the summer of 2008 from 0.3–0.5 to 3.0–5.0 μm are shown in Figure 9.24.

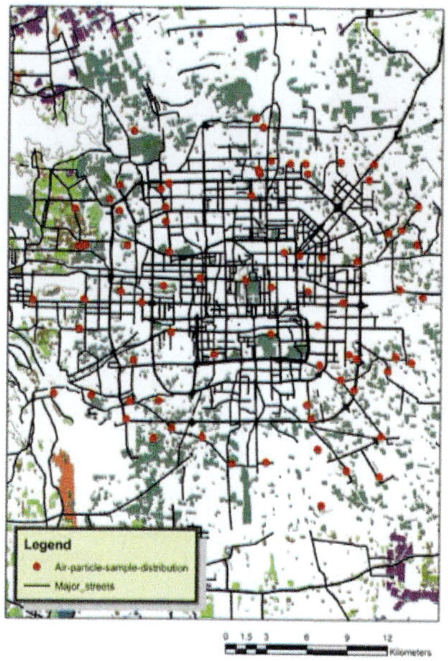


**Figure 9.22** Air particulate pollution sampling sites in Beijing, China.

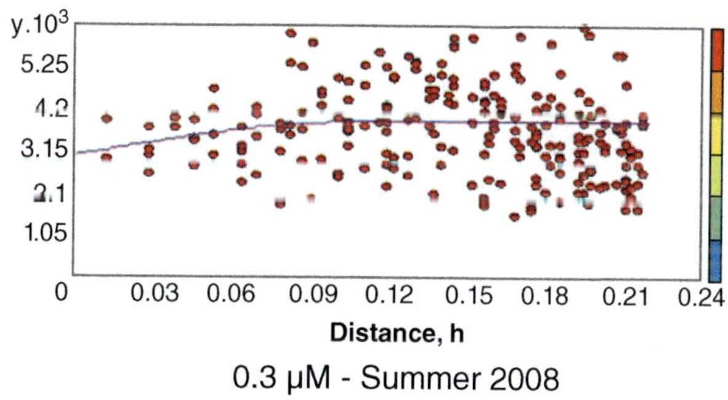


**Figure 9.23** Example of semivariogram of UK prediction of air particulate pollution.

## Multi-variable Spatial Statistical Analysis

Multi-variable spatial statistical analysis that is illustrated in this textbook are those exploratory spatial statistical operations that involve two or more variables in addition to the locational variable. Typically, we have independent or exploratory variable and dependent variable that are both related to their spatial locations. Multi-variable spatial statistical analysis is often called spatial regressions in practice. Regression is a statistical method used in finance, investing, and other disciplines to determine the strength and character of the relationship between one dependent variable (usually denoted by Y) and a series of other variables (known as independent variables, such as X1, X2, X3, ...).

Linear regression, which is also called simple regression or ordinary least squares (OLS), is the most commonly used analytical regression method. Linear regression establishes the linear relationship between two variables based on a line of best fit. Linear regression is thus graphically

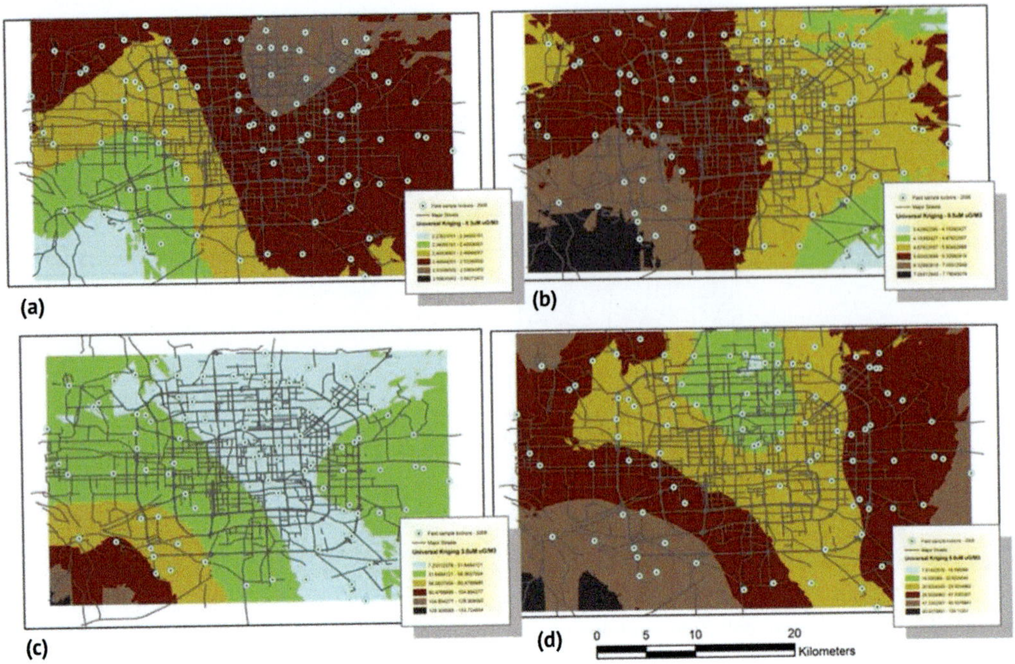


**Figure 9.24** UK modeled air particle pollution distributions of summer 2008: (a) 0.3 μM; (b). 0.5 μM; (c) 3.0 μM; and (d) 5.0 μM.

depicted using a straight line with the slope defining how the change in one variable impacts a change in the other. The following is the linear regression model.

$$Yi = \beta 0 + \beta 1 \cdot Xi + \varepsilon i$$

A spatial regression model in general is the specification combining a spatially autoregressive dependent variable among the set of explanatory variables and spatially autoregressive disturbances (Raymond et al. 2005).

Digital geographic analysis intends to answer the questions not only "Where?," but also "Why?." The following are some important examples of "Where?"

1. Are there places in the United States where people are caught by disease of cancer and are dying at a higher rate than others?
2. Where are the hot spots for crime, 911 emergency calls, or fires?
3. Where do we find the highest proportion of traffic accidents in a city?

Related to each of the above "where?" questions, the next logical question for these types of analyses is to answer the question "why?." For instance, (i) Why are there places in the United States where people are persistently caught by cancer disease? What might be causing this? (ii) Can we model the characteristics of places that experience a lot of crime, 911 calls, or fire events to help reduce these incidents (Figure 9.25)? (iii) What are the factors contributing to highest traffic accidents? Are there policy implications or mitigating actions that might reduce traffic accidents in those particular high-accident areas? One of the ultimate methods to answer these questions is to analyze and model the spatial relationships using multivariable statistical techniques, including OLS regression or geographically weighted regression (GWR).

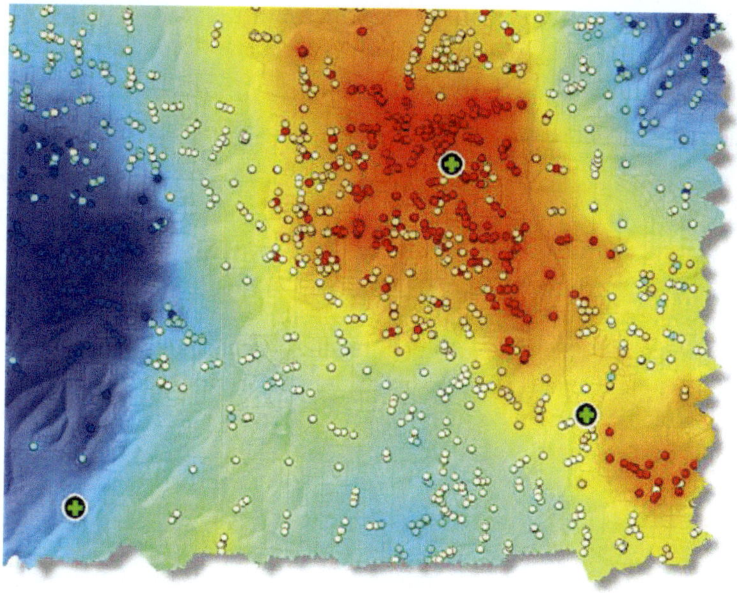

**Figure 9.25** 911 emergency call data showing call hot spots (red), call cold spots (blue), and locations of the fire/police units responsible for responding (green crosses). *Source:* Adapted from ESRI; ESRI 2020.

In essence, understanding the spatial relationships is the goal for us when we conduct spatial regressions. Spatial regression can help us to explain the factors behind observed spatial patterns of geographic objects of features. To understand why people are persistently caught by cancer disease and dying in certain regions in New York State, for instance, we may examine the proximity of residential areas to abandoned old industrial sites across the state (Tang and Anderson 2010).

## Ordinary Least Squares (OLS) Spatial Regression

OLS, also called linear regression, is the most popular regression technique. OLS is also a starting point and the most basic operation for all spatial regression analyses. OLS provides a global model of the dependent variable of geographic features or events that we are trying to understand or predict. It creates a single regression equation to represent all the distributed features. The input feature class can be either points, lines, or polygons. The output feature class contains dependent variable estimates and residuals.

In the ArcGIS Pro environment, two additional regression functions are developed. These are general linear regression and exploratory regression. General linear regression can be used to fit continuous (OLS), binary (logistic), and count (Poisson) models. Exploratory regression evaluates all possible combinations of the input candidate explanatory variables (more than one independent variables), looking for OLS models that best explain the dependent variable within the context of user-specified criteria through weight matrix file input.

## EBK Regression Prediction

EBK Regression Prediction is a geostatistical interpolation method that utilizes EBK with explanatory variable in raster data format as the input. This approach combines kriging with regression analysis to make predictions that are more accurate than either regression or kriging can achieve on their own. The output is one raster prediction surface and one vector diagnostic feature class.

## Geographically Weighted Regression (GWR)

GWR is an important, unique, and local regression method for exploring spatial heterogeneity of spatial data relationships of two or more variables. GWR evaluates local situations or environments of the variables that we are trying to understand or to be used to predict the dependent variable by fitting a regression equation to every geographic object or feature in the dataset. The GWR process facilitates many regression models in the study region based on geographic locations. GWR constructs these separate equations by incorporating the dependent and explanatory variables of the features falling within the neighborhood of each target feature. The shape and extent of each neighborhood analyzed is based on the Neighborhood Type and Neighborhood Selection Method parameters. The simple GWR model at one geographic location (i) is as follows.

$$Yi = \beta 0 + \beta 1X1i + \beta 2X2i + \cdots + \beta nXni + \varepsilon i$$

where $\beta 0$, $\beta 1$..., $\beta n$ are the parameters and $\varepsilon i$ is the error term.

GWR was first introduced in 1996 in geographical research (Brunsdon et al. 1996; Fotheringham et al. 2002). This method works based on the simple yet powerful idea for estimating local models using subsets of field observations or measurements that are centered on each of the focal geographic locations. GWR has been instrumental in highlighting the existence of potentially complex spatial relationships. Worldwide applications of GWR suggest that GWR is a useful tool in interpreting geographic processes and a relatively simple and effective tool for spatial interpolation (Figure 9.26).

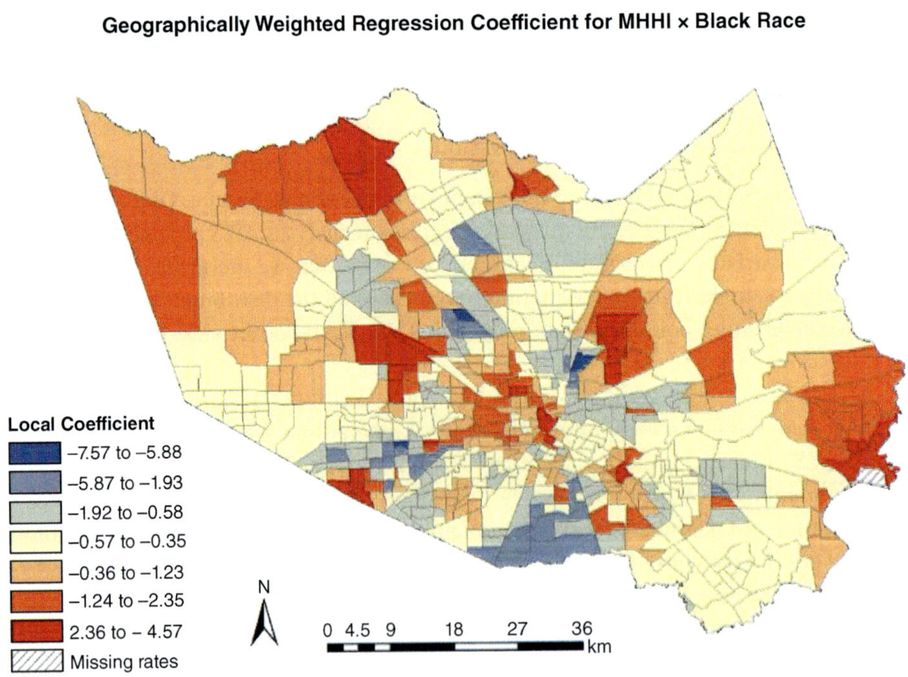


**Figure 9.26** Local $R^2$ of household income of black race at the neighborhood level – Harris County, Texas. *Source:* Ford, 2016/*PLoS One*/CC BY 4.0.

In the ArcGIS Pro environment, the input of GWR analysis is a feature class that contains observed values of dependent variables and exploratory variables. Three model types can be selected, and these are: Continuous (Gaussian), Binary (Logistic), or Count (Poisson). A neighborhood (also known as a bandwidth) for each local regression equation is set up by users. The output for Continuous (Gaussian) modeling, which is the most popularly used method is one prediction feature class and one coefficient raster layer. The output feature class contains fields for observed and predicted dependent variable values, condition number (COND), local $R^2$, explanatory variable coefficients, and standard errors. The local R-squared value ranges from 0 to 1 and represents the strength of the correlations of the local model of the variables. The COND is a measure of the stability of the estimated coefficients. COND s above approximately 1000 indicate instability in the model; this is usually caused by explanatory variables that are highly correlated with each other (ESRI 2020).

In spatial regression analysis, statistical p-value and $R^2$, or local $R^2$ are calculated. P-value is a statistical test to compute a probability for the coefficients associated with each independent variable. The null hypothesis for this statistical test states that a coefficient is not significantly different from zero means explanatory variables do not impact the dependent variable (Figure 9.27). A small p-value, for instance, a coefficient with a p-value of 0.01, is statistically significant at the 99 percent confidence level. Therefore, the associated variable is an effective predictor.

$R^2/R$-squared and adjusted $R^2$ are both statistics derived from the regression equation to quantify model performance. The value of $R^2$ ranges from 0 to 100% (0 to 1.0). If the model fits the observed dependent variable values perfectly, $R^2$ is 1.0. If $R^2/R$ -squared is close to 0, that means the model prediction should be rejected.

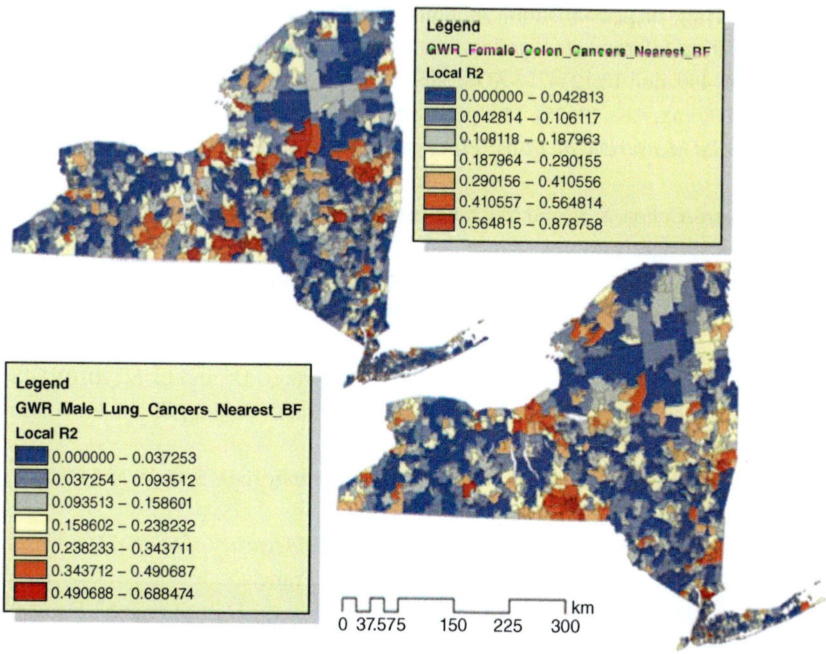


**Figure 9.27** GWR of cancer mortalities and brown field (old industrial site) proximity. *Source:* Adapted from Tang and Anderson, 2010.

## References

Brunsdon, C., Fotheringham, A.S., and Charlton, M.E. (1996). Geographically weighted regression: a method for exploring spatial non-stationarity. *Geographical Analysis* 28 (4): 281–298.

ESRI. (2022). ArcGIS Pro Documentation (The Science of Where). https://pro.arcgis.com/en/pro-app/latest/get-started/quick-access-toolbar.htm.

Ford, M.M. and Highfield, L.D. (2016). Exploring the spatial association between social deprivation and cardiovascular disease mortality at the neighborhood level. *PLoS One* 11 (1): e0146085. doi: 10.1371/journal.pone.0146085.g007.

Fotheringham, A.S., Brunsdon, C., and Charlton, M.E. (2002). *Geographically Weighted Regression: The Analysis of Spatially Varying Relationships*. Danvers, MA 01923, Hoboken, NJ 07030: John Wiley & Sons.

Getis, A. (1999). Spatial statistics. School of Geosciences white paper, The University of Edinburgh. https://www.geos.ed.ac.uk/

Goodchild, M. (2004). "The validity and usefulness of laws in geographic information science and geography." *Annals of the Association of American Geographers* 94 (2): 300–303. doi: 10.1111/j.1467-8306.2004.09402008.

Griffith, D.A. (2005). Spatial autocorrelation. In: *Encyclopedia of Social Measurement*, Page 01–16. https://www.sciencedirect.com/topics/computer

Kappo, O., Orewole, A., and Adedoja, O. (2016). Spatial distribution of terrorists attacks in selected local governments in Borno State, Nigeria. *International Journal of Scientific Research in Computer Science, Engineering and Information Technology* 01 (02): 33–45.

Krige, D.G. (1951). "A statistical approach to some basic mine valuation problems on the Witwatersrand". *J. of the Chem., Metal. and Mining Soc. of South Africa* 52 (6): 119–139. doi: 10.10520/AJA0038223X_4792.

Matheron, G. (1962). Traité de géostatistique appliquée. Editions Technip, Paris, 1962–1963.

Rao, S.S. (2018). Statistical inference for spatial statistics defined in the Fourier domain. *Annals of Statistics* 46 (2) 469–499. doi: 10.1214/17-AOS1556.

Raymond J.G.M. and Florax, P.N., (2005). Misspecification in linear spatial regression models. In: *Encyclopedia of Social Measurement*. https://www.sciencedirect.com/topics/medicine-and-dentistry/spatial-regression

Salas-Parra, R.D., Castro-Ochoa, K.J., and Machado-Aranda, D.A. (2023). Basic science statistics. In: *Translational Surgery*, Pages 195–199. Cambridge, MA 02139: Elsevier Publisher Inc.

Tang, T. and Anderson, C. (2010). Spatial statistical modeling of the pollution impact of old industrial sites on colon and lung cancer incidents in New York State, USA. *IEEE Proceedings – Geoinformatics* IEEE Xplore: Issue, 15–17

Tang, T., Zhao, W., Gong, H., Li, X., Zang, K., Zhao, W.H., Bernosky, J.D., and Li, S. (2010). GIS spatial analysis of population exposure to fine particulate air pollution in Beijing, China. *Environmental Geosciences* 17 (1): 1–16.

Tobler, W.R. (1979). Cellular geography. In: *Philosophy in Geography* (eds. S. Gale, and G. Olsson), 379–86. Dordrecht: Reidel.

Tobler, W.R. (1989). Frame independent spatial analysis. In: *The Accuracy of Spatial Databases* (eds. M.G. Goodchild, and S. Gopal), 115–22. London: Taylor and Francis.

Tobler. W. (1970). "A computer movie simulating urban growth in the Detroit region". *Economic Geography* 46 (Supplement): 234–240.

Wisniewski, C.S., Jones, E.P., Weeda, E.R., Pilch, N.A., and Picone, M.F. (2019). Medical literature evaluation and biostatistics. In: *Clinical Pharmacy Education, Practice and Research*, pp. 143–162. Cambridge, MA 02139: Elsevier Publisher Inc.

## Tutorial in ArcGIS Pro

Copy the tutorial data folder "Chapter09-Tutorial" to your working directory.

Open ArcGIS Pro and create a new ArcGIS Pro project file that is called "MyGIS_Learning09" with the "Map" template. Save the project file into the designated working folder, C:\ or D:\...... Chapter09-Tutorial.

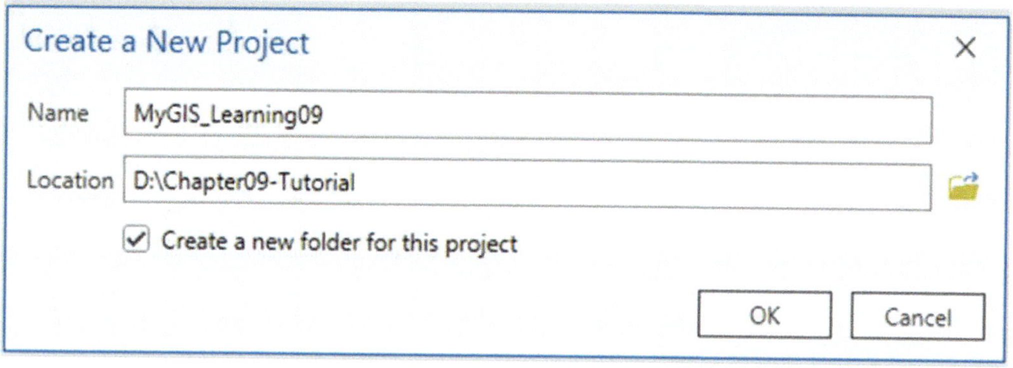


Now, highlight and right click on the Folders on the Catalog Pane to get the contact menu. Then, click on "Add Folder Connections" to add the entire Chapter09-Tutorial folder on your working drive.

1. **New York State 150-year Temperature and Precipitation Distribution Analysis Project**

In the data folder, we downloaded two Excel data tables. One is called "NYS_Annual_Mean Temperature_Total_Precipitation_49_Stations" (.xlsx). It is the aggregation of 49 weather monitoring and measuring stations, most of them operated since the 1800s. As the end result, we summarized two major climate factors: annual mean temperature and total precipitation in this table. We are going to convert this Excel table to a stand-alone geodatabase table first. Then, we are going to produce the weather stations point feature class using coordinate geometry (COGO) and the latitude and longitude information contained in the data table. After the dataset preparation, we are going to use the geostatistical interpolation methods to interpolate the sample point type of data to the spatial data that covers the entire New York State.

At the Catalog Pane, expand Folders ->Chapter09-Tutorial->NYS_Data.gdb (Geodatabase).

Right click on the geodatabase to import the table. The "Table to Geodatabase" tool opens on the Geoprocessing Pane. For "Input Table," navigate to your Chapter09-Tutorial folder to load the Sheet1$ of "NYS_Annual_Mean_Temperature_Total_Precipitation_49_Stations" (.xlsx) file. ArcGIS Pro only works to translate one sheet at a time from Excel to Geodatabase. Make sure the "Output" geodatabase is NYS_Data.gdb. Click on Run to process the conversion.

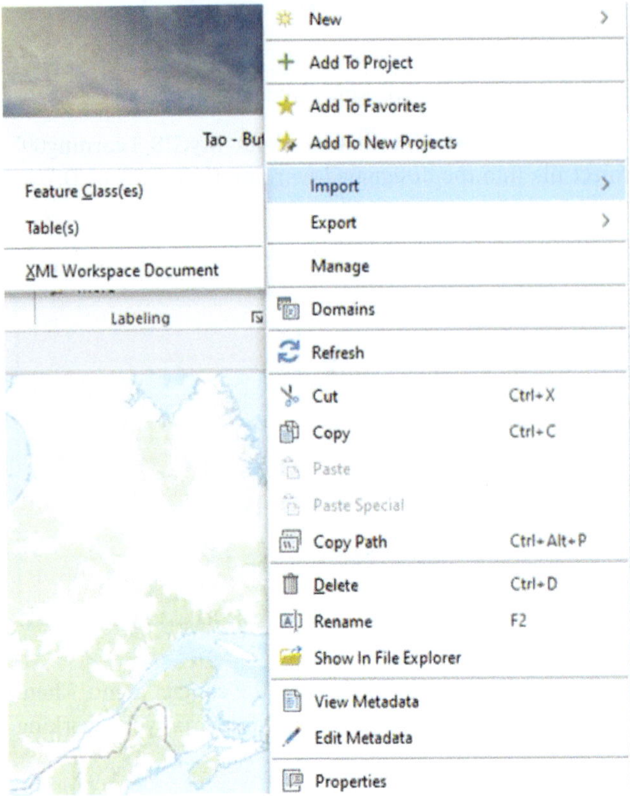


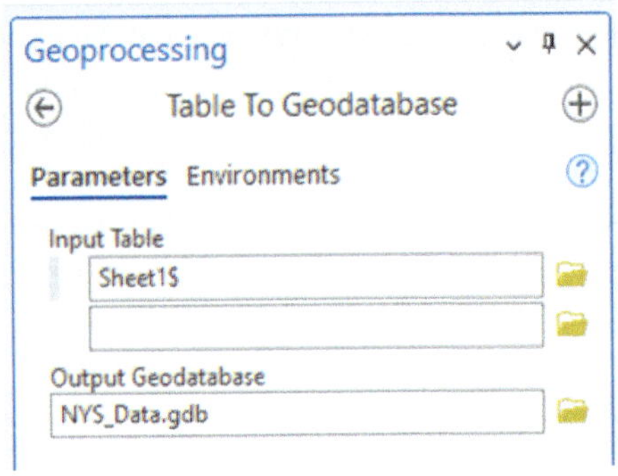


Switch to the Catalog Pane, you notice that the stand-alone table "Sheet1_" is added in the geodatabase. Change the name of this table to "NYS_Weather_Stations." Add the table to the Table of Contents. Open the data table, it should look like this.

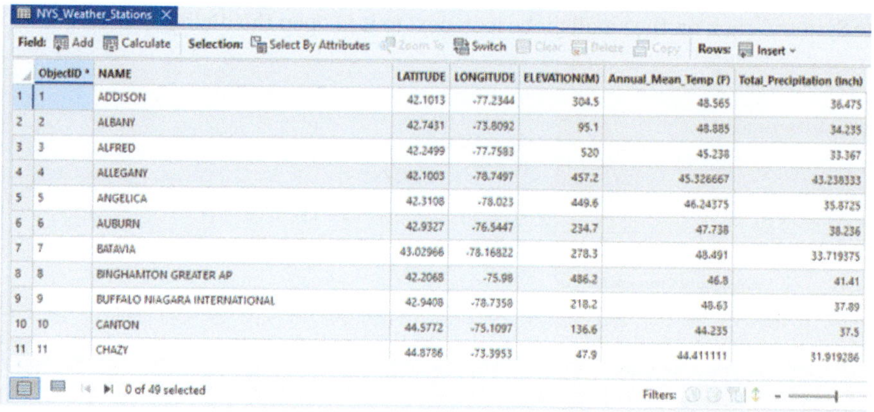

NYS_Weather_Stations

Field: Add Calculate Selection: Select By Attributes Zoom To Switch Clear Delete Copy Rows: Insert

| | ObjectID * | NAME | LATITUDE | LONGITUDE | ELEVATION(M) | Annual_Mean_Temp (F) | Total_Precipitation (inch) |
|---|---|---|---|---|---|---|---|
| 1 | 1 | ADDISON | 42.1013 | -77.2344 | 304.5 | 48.565 | 36.475 |
| 2 | 2 | ALBANY | 42.7431 | -73.8092 | 95.1 | 48.885 | 34.235 |
| 3 | 3 | ALFRED | 42.2499 | -77.7583 | 520 | 45.238 | 33.367 |
| 4 | 4 | ALLEGANY | 42.1003 | -78.7497 | 457.2 | 45.326667 | 43.238333 |
| 5 | 5 | ANGELICA | 42.3108 | -78.023 | 449.6 | 46.24375 | 35.8725 |
| 6 | 6 | AUBURN | 42.9327 | -76.5447 | 234.7 | 47.738 | 38.236 |
| 7 | 7 | BATAVIA | 43.02966 | -78.16822 | 278.3 | 48.491 | 33.719375 |
| 8 | 8 | BINGHAMTON GREATER AP | 42.2068 | -75.98 | 486.2 | 46.8 | 41.41 |
| 9 | 9 | BUFFALO NIAGARA INTERNATIONAL | 42.9408 | -78.7358 | 218.2 | 48.63 | 37.89 |
| 10 | 10 | CANTON | 44.5772 | -75.1097 | 136.6 | 44.235 | 37.5 |
| 11 | 11 | CHAZY | 44.8786 | -73.3953 | 47.9 | 44.411111 | 31.919286 |

0 of 49 selected Filters:

Add the "Counties" feature class from the NYS_Data.gdb to the mapping window as the reference map.

Now, high-light and right click on the NYS_Weather_Stations table on the Table of Contents. Then, select the "Display X, Y data" icon on the contact menu. The "Display X, Y data" dialog opens.

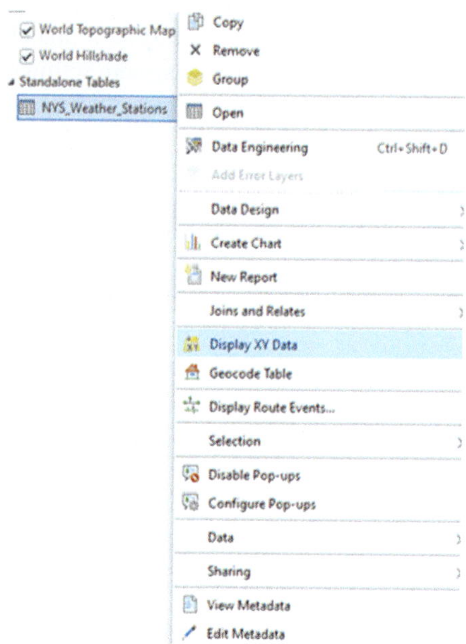


Make sure the Input table is the NYS_Weather_Stations. For "Output Feature class," navigate to your NYS_Data.gdb and save it as "NYSWeather_Stations." Make sure the following.

| | |
|---|---|
| X Field | LONGITUDE |
| Y Field | LATITUDE |
| Z Field | ELEVATION(M) |

For "Coordinate System," click on the small globe icon next to it to open the dialog.

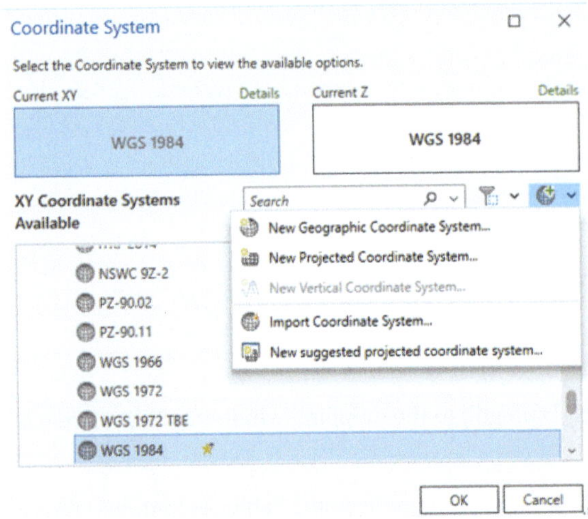


Again, click on the small globe plus icon on this dialog to import the horizontal coordinate system from the New York State "Counties" map layer. Then, for the "Z Coordinate System" match it with WGS_1984 as well. Although this practice still yields the same coordinate system settings, it is a necessary step to align all the data layers. Click OK to process the data.

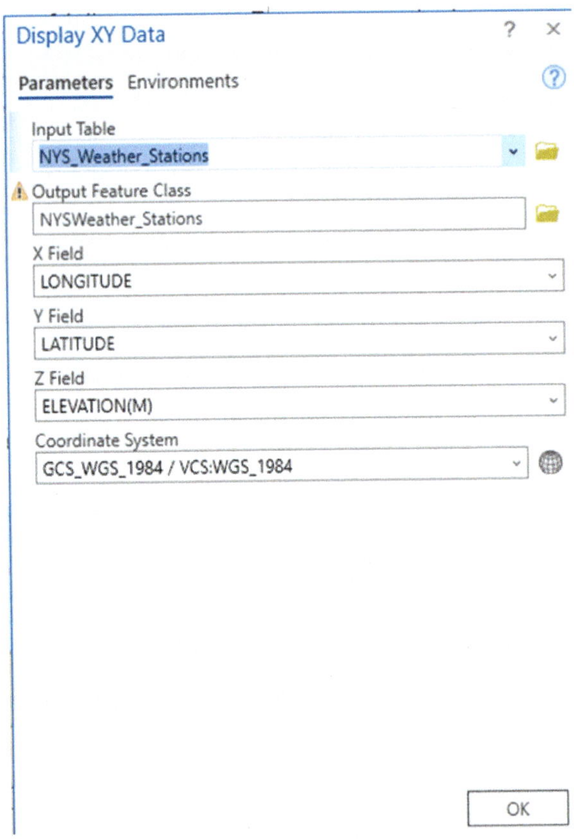

Now, we have the 49 original weather station data added in point feature class. You can open the attribute table of this feature class to view both the map and the characteristics.

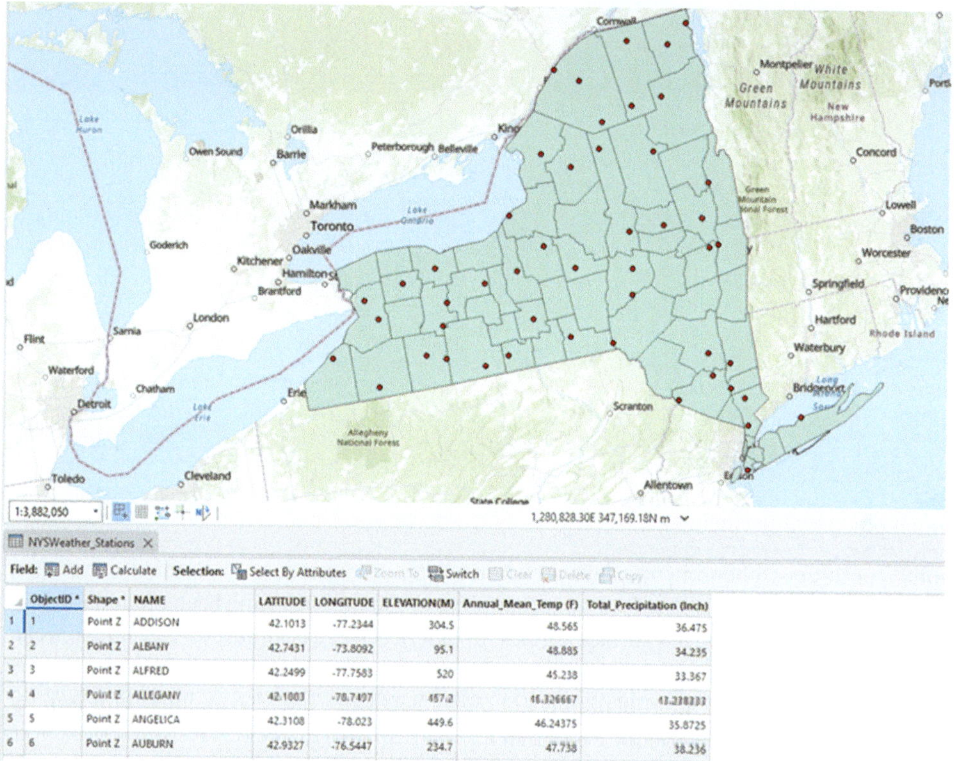


Conventionally, we make the thematic maps of the weather stations and try to understand the distribution patterns of annual mean temperature (°F) or total annual precipitation (inches).

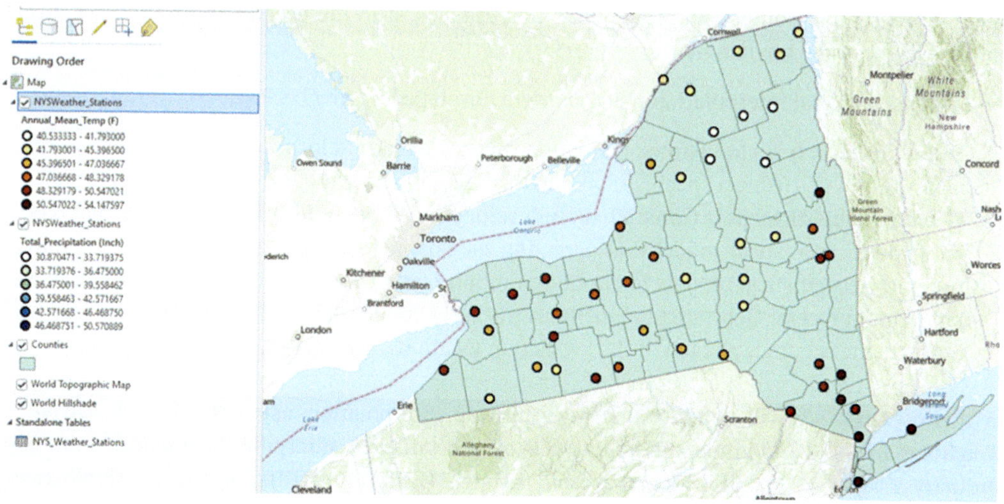

1. Annual mean temperature (°F)

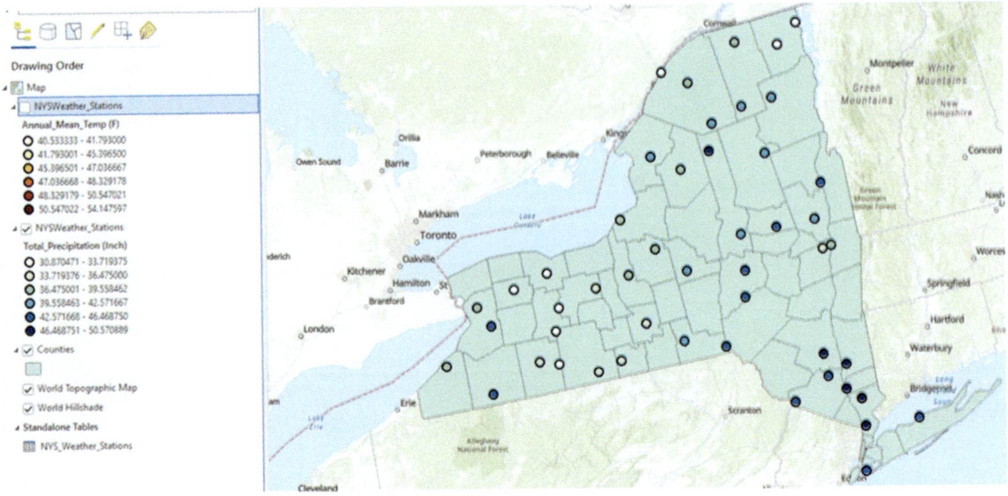


2. Annual precipitation (inches)

Geographic continuous phenomena, such as temperature, precipitation, or humidity, are often measured and monitored at discrete locations. Sampling in the field has long been a standard field technique in Geographic or Earth's surface studies. In practice, there is almost no way humans can measure a particular phenomena or object everywhere in the targeted study areas on the Earth's surface. Therefore, the spatial interpolation is not only a necessary but also an important step for modern Earth's surface-related scientific research.

Spatial interpolation uses measured values from sample points to create a raster surface that contains predicted values for the geographic phenomenon or object. Using spatial interpolation, we can predict or estimate the values of a spatially continuous phenomenon at all other locations in a study area using a set of sampling data. Each grid cell in a prediction surface is predicted using neighboring sample points. Sampling points that are closest to the prediction location are given greater weight in the predictions.

Two groups of spatial interpolation methods were developed in ArcGIS Pro.

1. **Deterministic interpolations:** Prediction values are made at locations based on surrounding measured values using a specific mathematical model where the user defines the parameters.
2. **Geostatistical interpolations:** Prediction values at unknown locations are calculated based on a specific statistical model that depicting the spatial structure that the measured values present. Geostatistical models incorporate randomness of the sampling points into the computational process. Geostatistical methods not only create a prediction surface, but also can estimate the accuracy at each of the prediction locations.

In this exercise, we will conduct one deterministic interpolation using the Inverse Distance Weighting method and three different types of geostatistical interpolation methods, namely Ordinary Kriging (OK), Empirical Bayesian Kriging (EBK), and EBK Regression Prediction. Inverse distance weighted (IDW) interpolation predicts grid cell values using a linearly weighted model of a set of sample points, in which the weight model is a function of inverse distance.

IDW method is an effective tool to interpolate prediction surfaces for locationally dependent variables. Ordinary kriging (OK) is the most widely used kriging method worldwide. OK assumes the constant mean is unknown.

EBK is a geostatistical method that helps the user to build a valid kriging model for accurate prediction without manually adjusting parameters. EBK automatically calculates these parameters through a process of subsetting and simulations. EBK Regression Prediction is a geostatistical interpolation that utilizes the EBK method and a raster explanatory (independent) variable that are known to affect the value of the predicting phenomenon or object to improve the accuracy of the prediction. This method combines kriging process with regression analysis to make predictions that are more accurate than either a kriging or regression alone can achieve.

1. IDW Interpolation

Activate the Analysis menu on the main menu. Click on the "Geostatistical Wizard" icon in the Workflows group 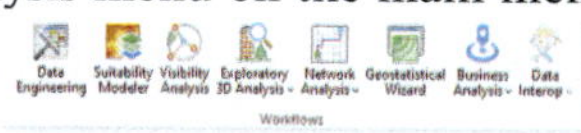 . The "Geostatistical Wizard" dialog opens.

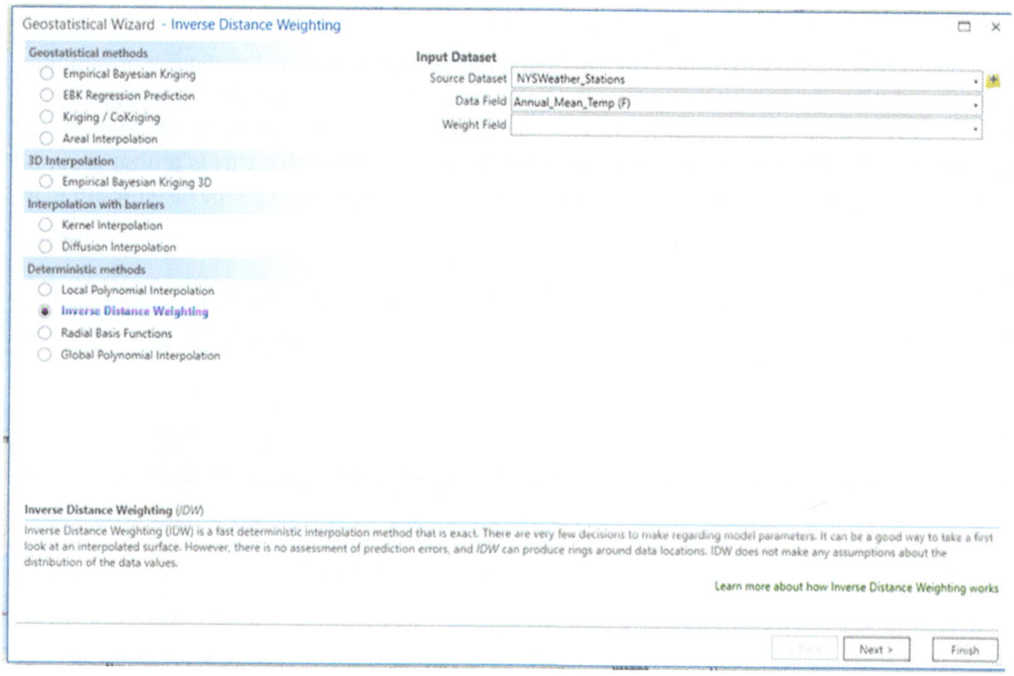


Select "Inverse Distance Weighting" in the Deterministic Method group as the approach.

Make sure the "Input Dataset" on the right-hand side is as following.

| | |
|---|---|
| Source Dataset | NYSWeather_Stations |
| Data Field | Annual_Mean_Temp (F) |

Leave the Weight Field blank and click on the "Next." The second page is presented.

On the form of "General Properties" at the right-hand side of the dialog, change the "Power" to 5 and "Sector Type" of search to 4 Sectors with 45°.

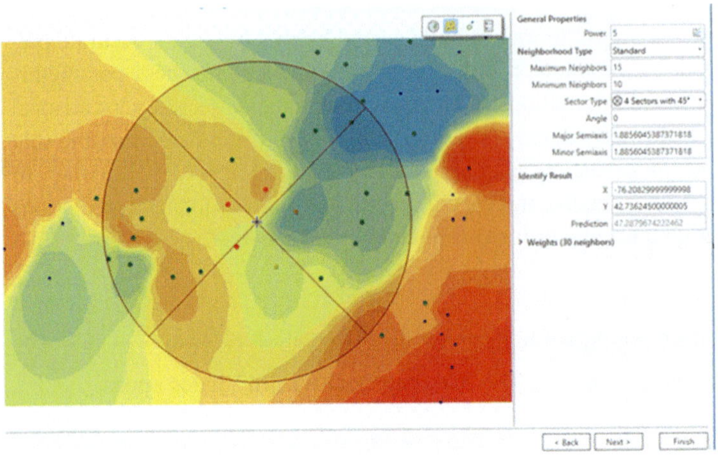


Click on the "Next." We will inspect the prediction model fit, errors, and distribution of sampling data by clicking on the tabs and reading the error table. We noticed that the highest prediction error is 3.759 and the distribution is not an ideal normal pattern. In summary, the Root-Mean-Square (R2) is 1.53 that is much higher than 1.0. We can conclude that this is not an accurate interpolation. Click on "Finish" and "OK" for the summary table, we created the raster result map.

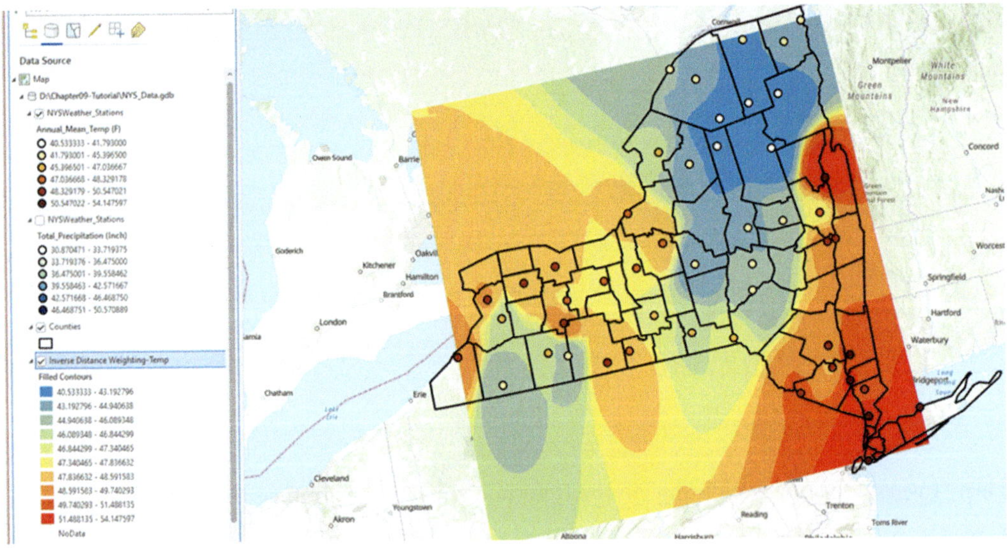


This is only a temporary map layer. Change the map layer name to "Inverse Distance Weighting-Temp" in the Table of Content. We can convert this layer to a raster dataset by right-clicking on it to obtain the contact menu and choose "Export Layer."

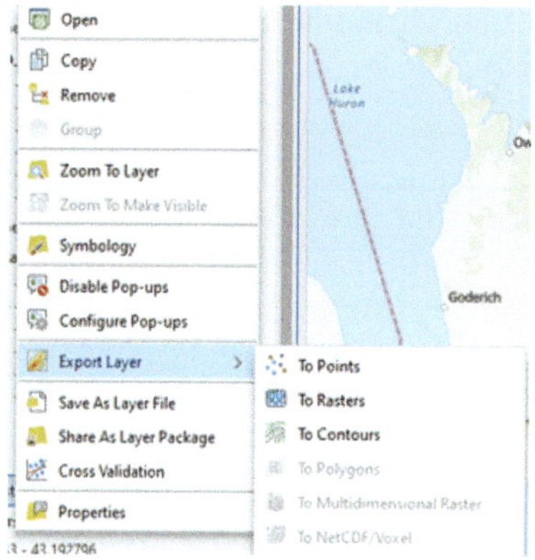


Repeat the above steps for creating the IDW interpolation map layer of the total precipitation.

| | |
|---|---|
| Source Dataset | NYSWeather_Stations |
| Data Field | Total_Precipitation(Inch) |

This time, we select the "Power" of local impact as 3.

The R2 in this case is 2.93. This is not an accurate prediction.

We now just produce this map layer for comparisons. Rename the map layer to "Inverse Distance Weighting-Precip."

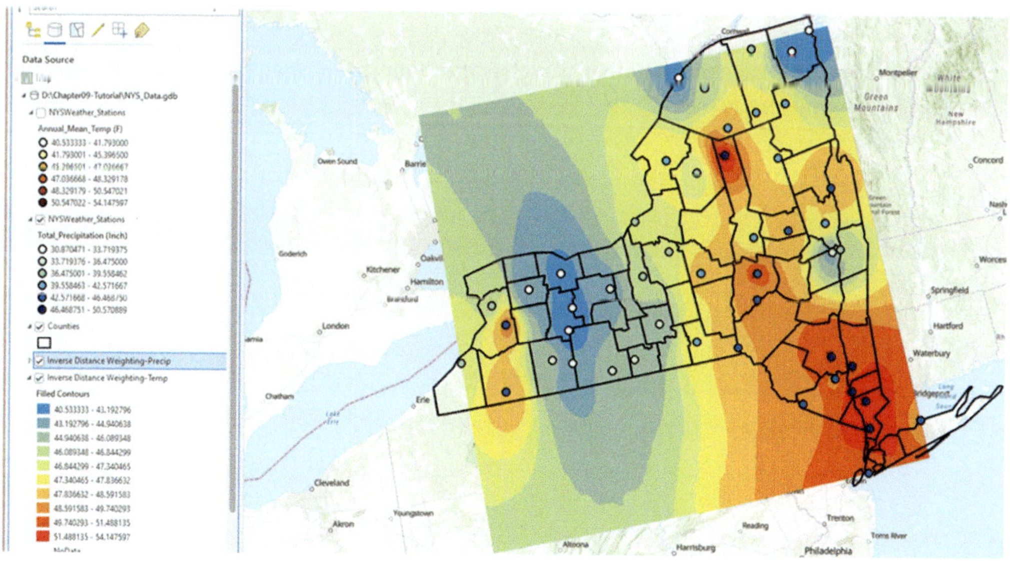


2. Ordinary Kriging

Now, let's open the Geoprocessing Wizard on the Analysis ribbon again to process the data use OK in the Geostatistical Methods group.

Select "Kriging\CoKriging" in the Geostatistical Methods group. Kriging process only use one set of the data input. So, we only fill-in the form of "Input Dataset 1" as follows.

| | |
|---|---|
| Source Dataset | NYSWeather_Stations |
| Data Field | Annual_Mean_Temp (F) |

Click on "Next." On this page of the dialog, we select the "Prediction" map layer of the OK. We can keep the other parameters as the default.

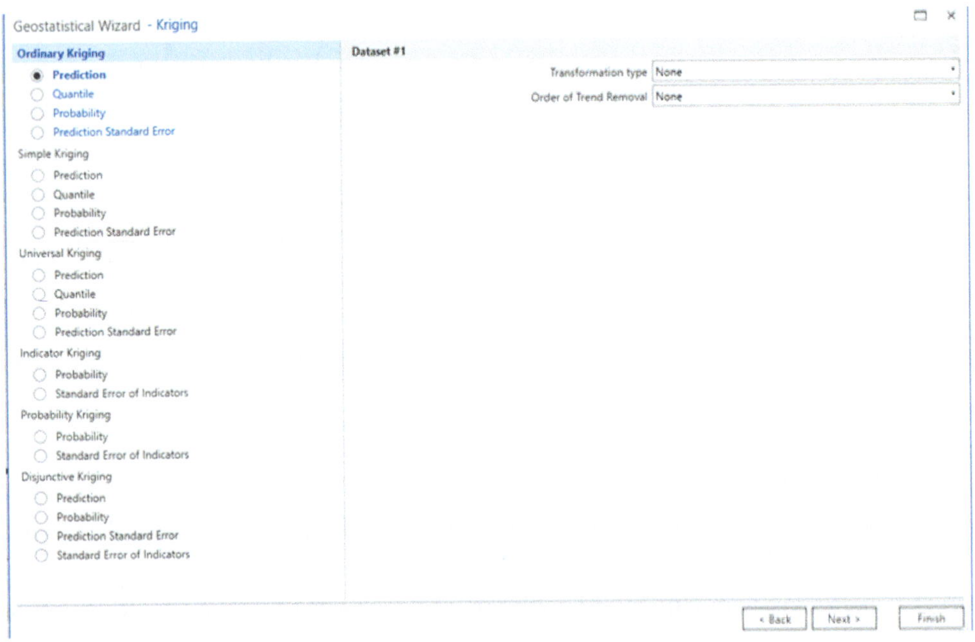

Click on "Next." This is the page of Semivariogram modeling.

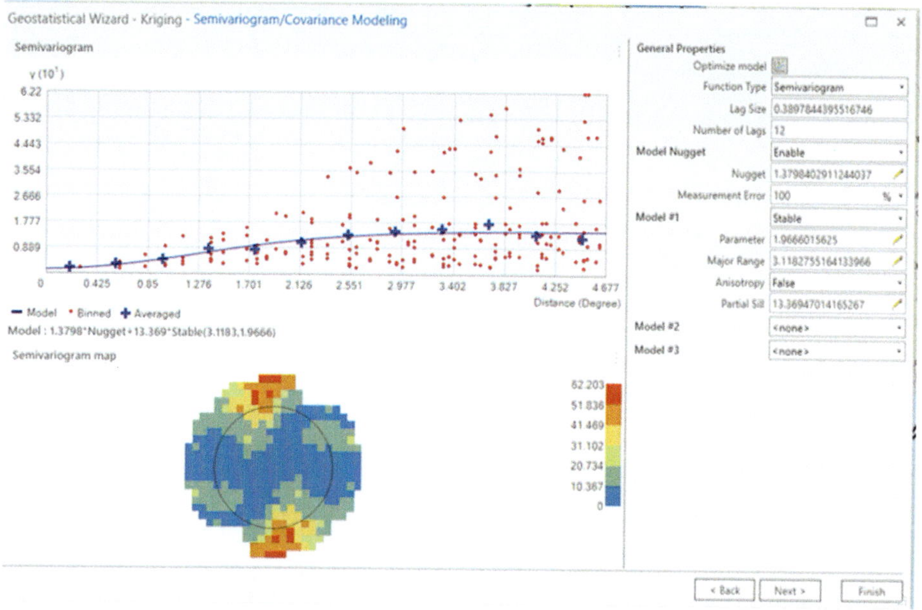


The distribution of sampling values in this case is not in an ideal pattern. Click on "Next."

This is the searching neighborhood page. Change the "Maximum Neighbors" to 6 and "Sector Type" of search to 4 Sectors with 45°.

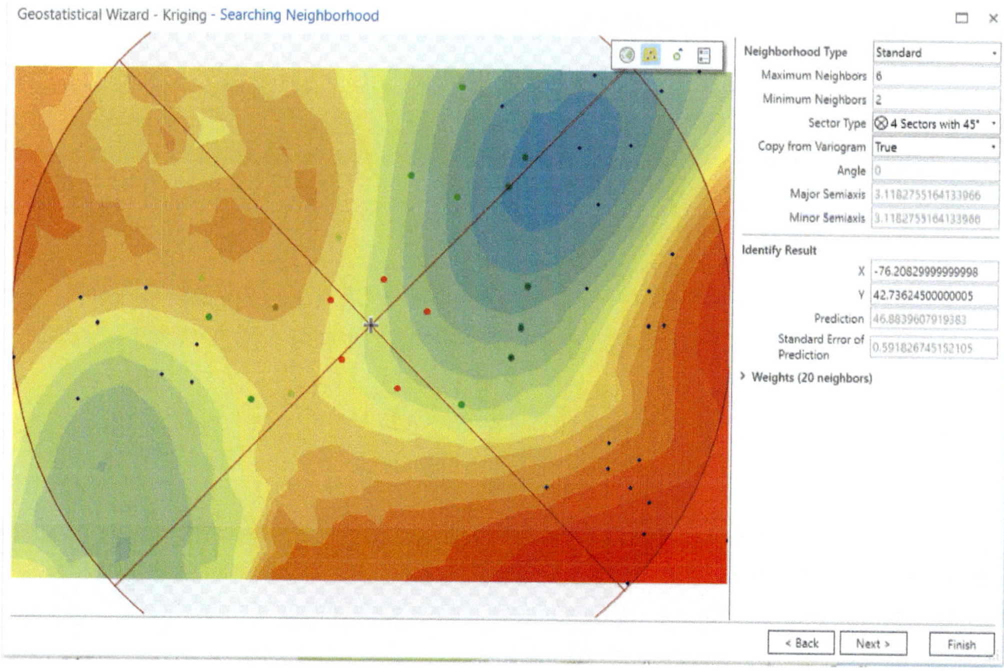

Click on "Next." This is the page of cross validation.

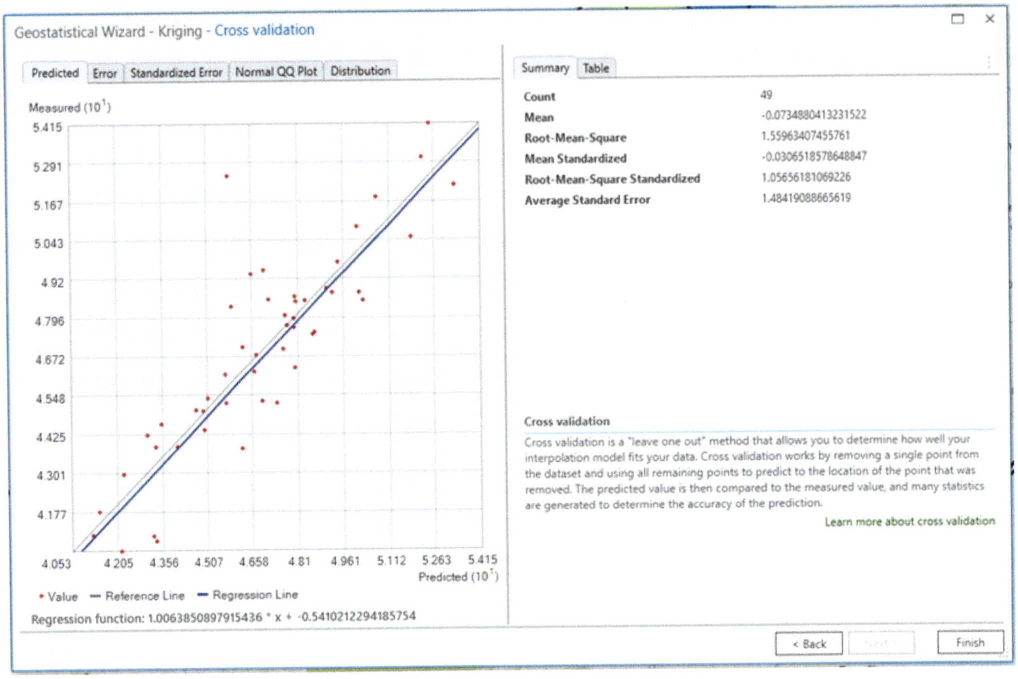


In summary, the R2 in this case is 1.06. It is much improved from that of the predication by IDW. Now, finish the process to create the prediction map layer. Rename the map layer as "Kriging-Temperature."

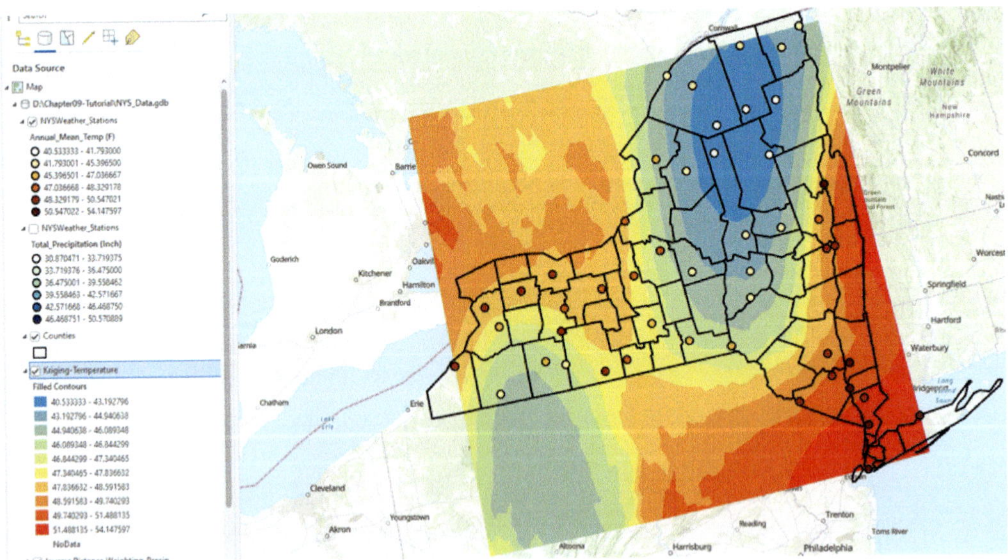

Repeat the above steps for creating the OK interpolation map layer of the total precipitation.

The Semivariogram is as following.

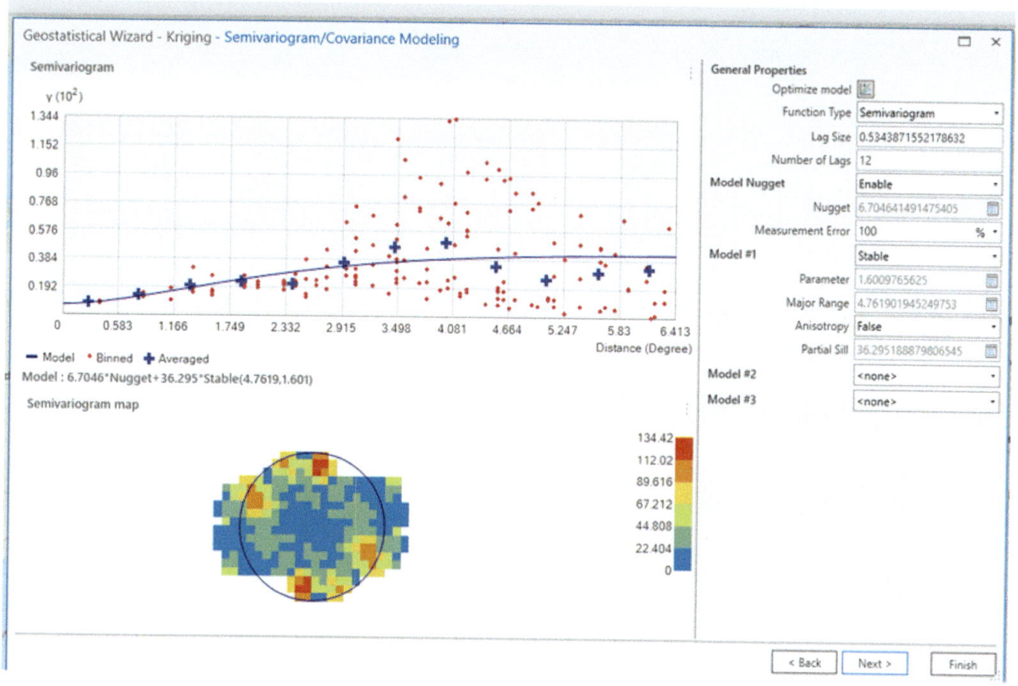


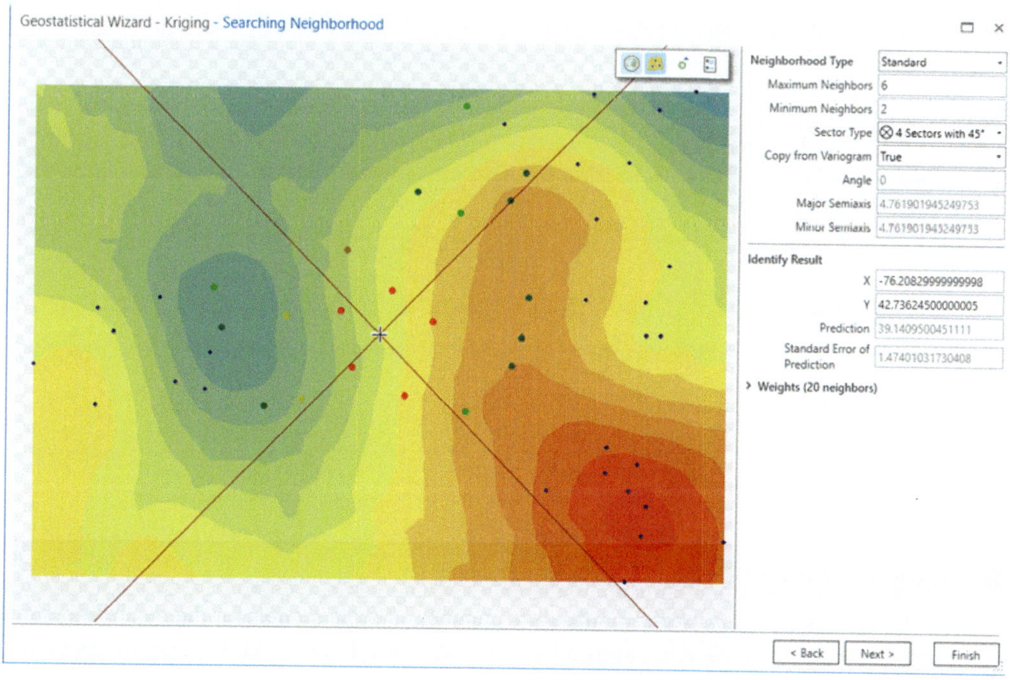

The R2 in this case is 1.001. Again, it is much improved from that of the predication by IDW. Click on "Finish and "OK" for the report to create the prediction map layer.

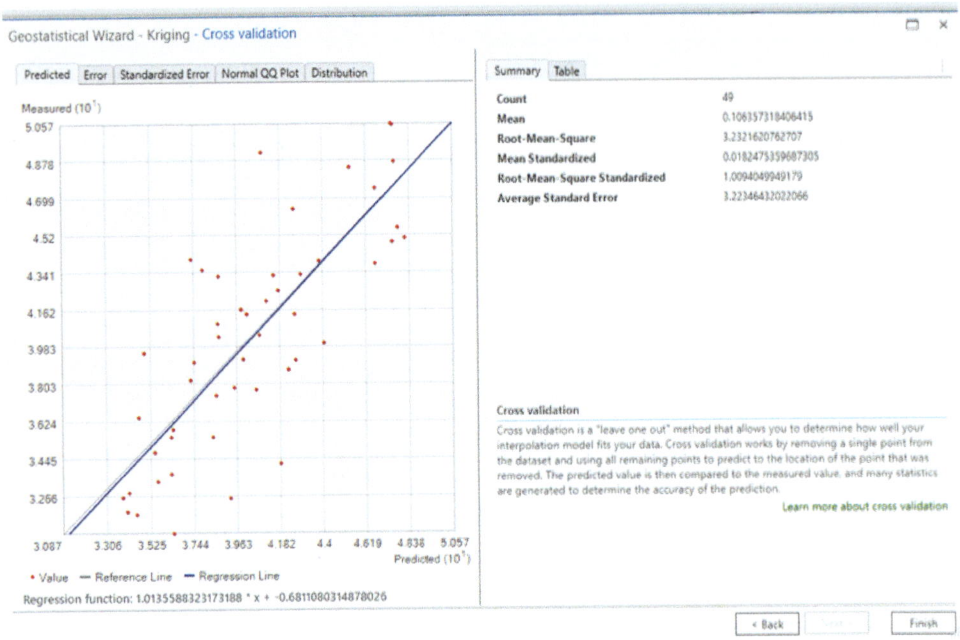


Rename the new map layer as "Kriging-Precipitation."

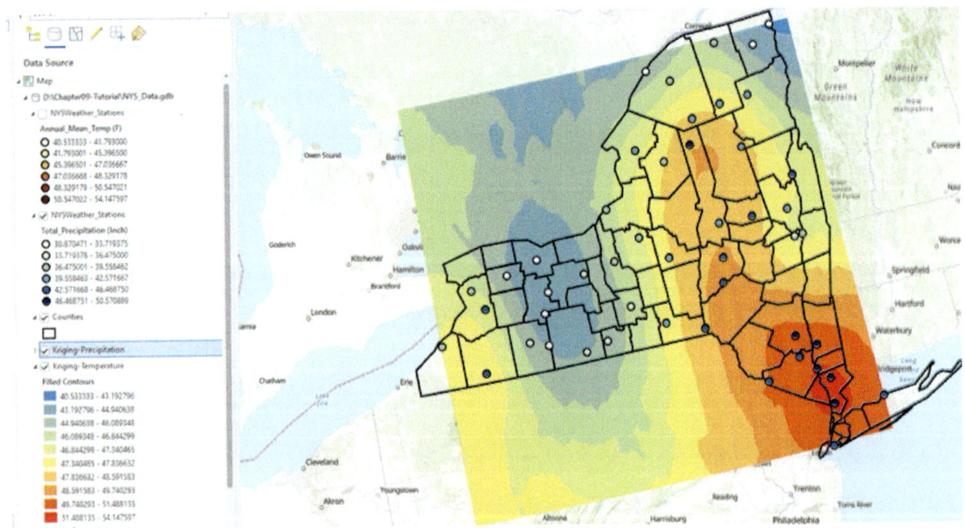


3. Empirical Bayesian Kriging

EBK method divides the study area into subsets to create local-scale models for the mean and error components in searching for a more accurate prediction result. In addition, EBK simulates many semivariogram models to present the uncertainty level in the empirical semivariogram model.

The ideal workflow to perform geostatistical interpolation consists of the following three steps:

1. Conduct exploratory spatial data analysis:_use maps and charts to better understand a given phenomenon or geographic object.
2. Create a raster prediction surface: use a geostatistical interpolation method to create a prediction surface of the given phenomenon or geographic object. In this case, we will use EBK interpolation.
3. Cross-validate the results: compare predicted values to measured values to determine whether the prediction surface is valid. More importantly, this is a process to understand the limitations of your prediction.

In best practice, we use exploratory interpolation to generate various interpolation results and then compare and rank those results using customizable criteria based on cross-validation statistics. Each interpolation result receives a ranking and a summary of the cross-validation statistic that we use. In this case, we will perform exploratory interpolation to determine the best interpolation method for predicting values at unmeasured locations for Mean Annual Temp. We will use a comparison method of a single criterion for the objective of the highest prediction accuracy.

In the Geoprocessing pane, search and open the **exploratory interpolation** tool.

For Input Features, choose NYSWeather_Stations.

For Value Field, choose Annual_Mean_Temp__F_.

For Output Cross Validation Table, navigate to your NYS_Data.gdb and save the new name as "NYS_ExploratoryIn." For Output geostatistical layer with highest rank, navigate to your Chapter09-Tutorial folder and save as "NYS_Exploratory1."

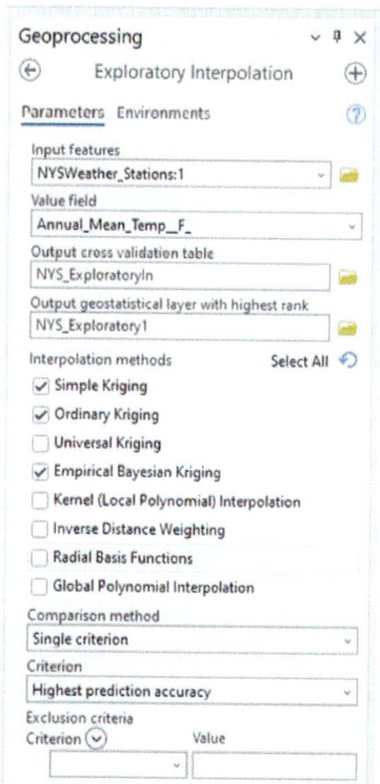

In this case, for interpolation methods, only check Simple Kriging, OK, and EBK. Keep other criteria as default. Run the tool to obtain results.

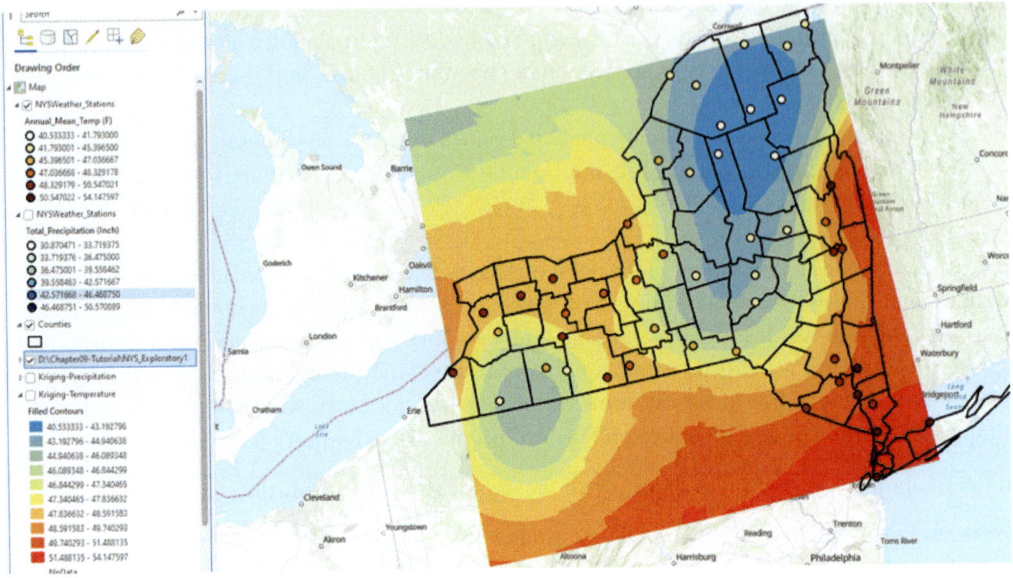


We would like to examine the output cross-validation table.

From the Table of Contents pane, under Standalone Tables, right-click NYS_ExploratoryIn to open it.

NYS_ExploratoryIn

| | Object ID * | Model Description | Rank | Included | Root Mean Square Error | Mean Error | Mean Standardized Error | Root Mean Square Standardized Error | Average Standard Error | Maximum Absolute E |
|---|---|---|---|---|---|---|---|---|---|---|
| 1 | 1 | Empirical Bayesian Kriging - Advanced | 1 | Yes | 1.563475 | -0.023496 | -0.00524 | 1.055977 | 1.517595 | 6.844 |
| 2 | 2 | Ordinary Kriging – Optimized | 2 | Yes | 1.566532 | -0.044075 | -0.008279 | 1.060134 | 1.492931 | 6.728 |
| 3 | 3 | Ordinary Kriging – Default | 3 | Yes | 1.569243 | -0.049474 | -0.010458 | 1.045271 | 1.511027 | 6.739 |
| 4 | 4 | Simple Kriging – Optimized | 4 | Yes | 1.622567 | -0.126366 | -0.044851 | 0.882329 | 1.804646 | 6.715 |
| 5 | 5 | Empirical Bayesian Kriging - Default | 5 | Yes | 1.648041 | -0.030333 | 0.000703 | 0.999757 | 1.630396 | 7.081 |
| 6 | 6 | Simple Kriging – Trend and transformation | 6 | Yes | 1.798304 | -0.086704 | -0.040282 | 0.940583 | 1.888352 | 6.743 |
| 7 | 7 | Simple Kriging – Trend | 7 | Yes | 1.798304 | -0.086704 | -0.040282 | 0.940583 | 1.888352 | 6.743 |
| 8 | 8 | Simple Kriging – Default | 8 | Yes | 2.764936 | -0.441151 | -0.345172 | 6.004948 | 0.743429 | 8.304 |

Reading this result, we recognized that the EBK – Advanced is ranked number 1. So, we should choose this interpolation method. After inspecting the cross-validation result, close the table.

We could use the Geostatistical Wizard in Workflows under the Analysis menu on the main menu to perform the EBK analysis. However, two shortcomings still exist comparing the Geostatistical Wizard with using the specific tool. Geostatistical Wizard cannot readily apply a feature class as a mask to produce a boundary-appropriate raster map result. It also cannot produce a permanent raster dataset in a geodatabase as the end result. GIS analysts need to export the map layers to form raster datasets in a geodatabase. Therefore, we will practice the EBK process using a tool in the Geoprocessing Pane in this practice.

Click on the back arrow on the Geoprocessing Pane and search for the EBK tool. Open the tool and fill in the criteria.

For Input Features, choose NYSWeather_Stations.

For Z Value Field, choose Annual_Mean_Temp__F_.

In interpolation analysis of raster surfaces, the sampling data to predict the surface is called Z value. It could be the elevation or altitude. But it is in most case are other phenomena or objects, such as temperature, precipitation, humidity, family income, or total population. Toward this end, the 3D visualization module – local scene can be used to spatially visualize any geographic object by its 2D locations.

For Output raster, navigate to your NYS_Data.gdb and save it as "Temp_EBK." This will save a permanent raster map layer in the geodatabase. Then, the Output geostatistical layer turns to optional.

Expand the Additional Model Parameters. Change to following.

| | |
|---|---|
| Maximum number of points in each local model | 20 |
| Local model area overlap factor | 1 |
| Number of simulated semivariograms | 50 |

For Search Neighborhood Parameters and Output Parameters, keep all the defaults.

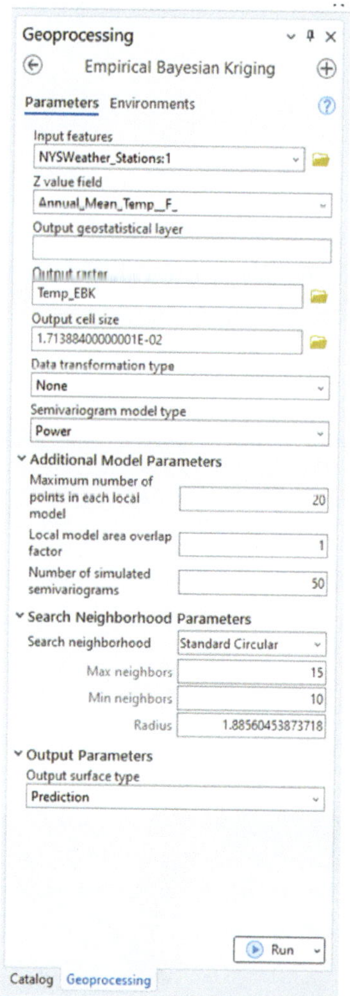

Now, click on the Environments tab on the top to get to the environment settings of this geoprocessing. Change the Mask to "Counties" (means NYS counties map layer). Coincident Points to "Include all coincident data."

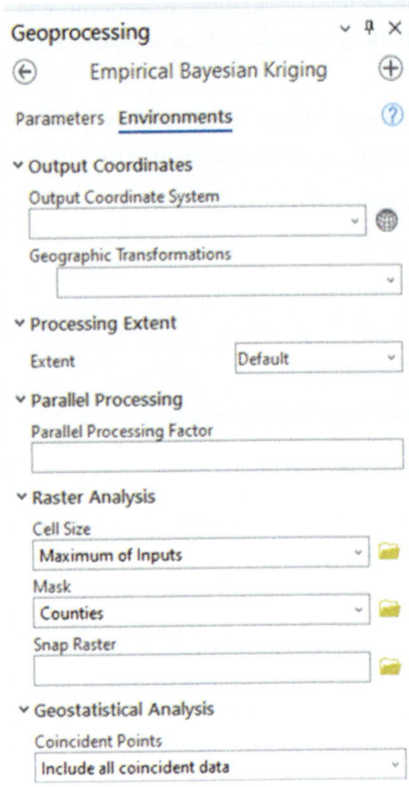


Now, flip back to Parameters page and click on Run to process it.

Turn off all the other map layers except "Annual_Mean_Temp (F)" and Temp_EBK. Use the Temperature color ramp with 8 categories to symbolize the Temp_EBK dataset.

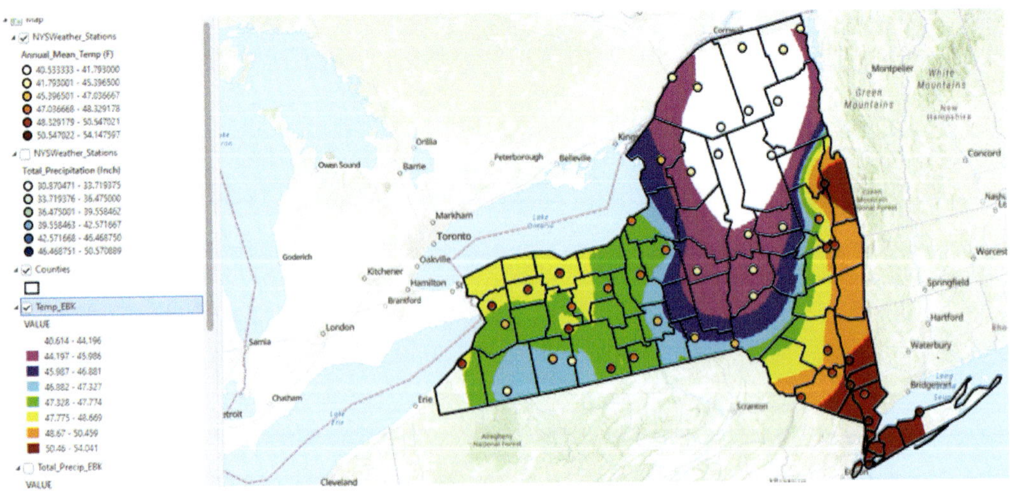

Follow the above steps to create “Total_Precip_EBK” as well using the “Total_Precipitation (Inch)” as the Z value. Again, turn off all the other map layers except “Total_Pricipitation (Inch)” and Total_Precip_EBK. Use the Precipitation color ramp with 8 categories to symbolize the Total_Precip_EBK dataset. Close both the Symbology and the Geoprocessing Panes.

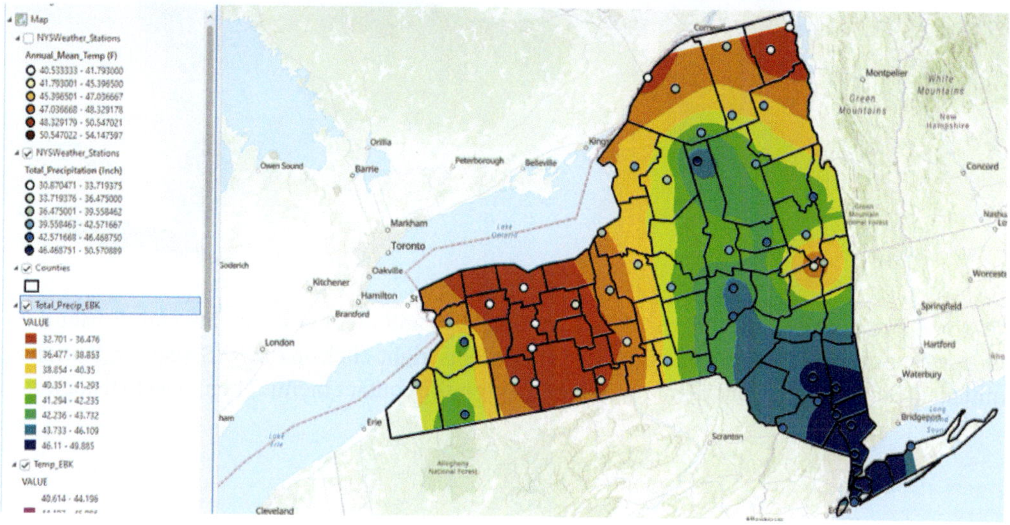


Save the project file.

4. EBK Regression Prediction

Like EBK spatial modeling, EBK regression interpolation also creates local-scale models for mean and error. In addition, the local mean component of EBK regression model utilizes variables and ordinary least squares (OLS) regression to adjust the prediction. The explanatory variables in this approach must be parameters that are related to the predicting phenomenon or object and must be in raster data format. This approach allows the model to change in different areas to account for more local effects. For example, the relationships between the explanatory variables and the prediction phenomenon may vary by aeras of locations. Involving OLS regression, EBK regression model can predict changes of the object more accurately by areas and locations.

If it is necessary, open the MyGIS_Learning09 project with ArcGIS Pro. In the Maps category on the Catalog Pane, add a new mapping frame and name it as “EBK_Regression_Prediction.” On the Table of Contents, copy both NYSWeather_Stations- Annual_Mean_Temp (F) and NYSWeather_Stations- Total_Precipitation (Inch) map layers from the “Map” window to Also, add the NYS_DEM raster dataset from your NYS_Data.gdb in the “EBK_Regression_Prediction” window.

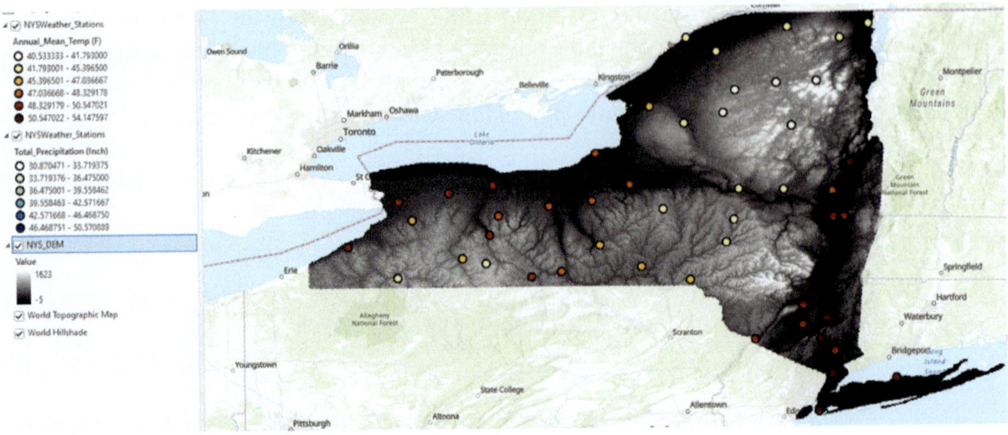


We would like to draw some charts of the sampling data and identify its quality and distribution before we make the predictions. On the Table of Contents, right-click on the NYSWeather_Stations-Annual_Mean_Temp (F) and point to Create Chart on the contact menu. Then, select Histogram.

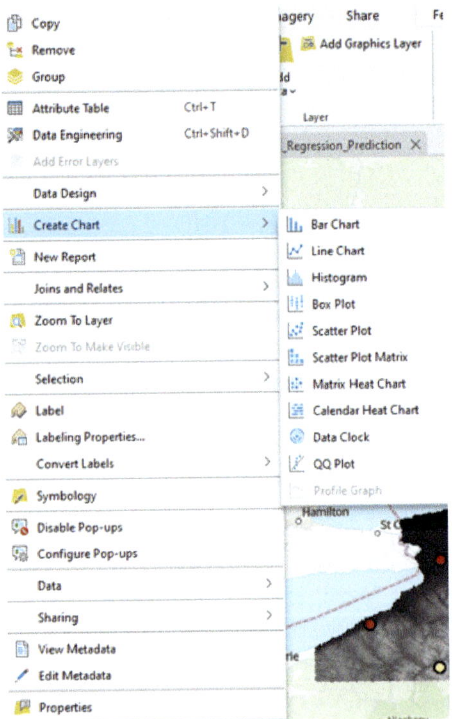


In the Chart Properties pane, for Number, choose Annual_Mean_Temp (F). For the Bins, select 10. On the Table of Contents, right-click NYS_DEM and choose Zoom To Layer. This way, the charted values are connected to the points on the digital map.

Click and select the Bin with the highest value range on the chart. You notice that the coincident point features on the map are also selected and showing in light blue.

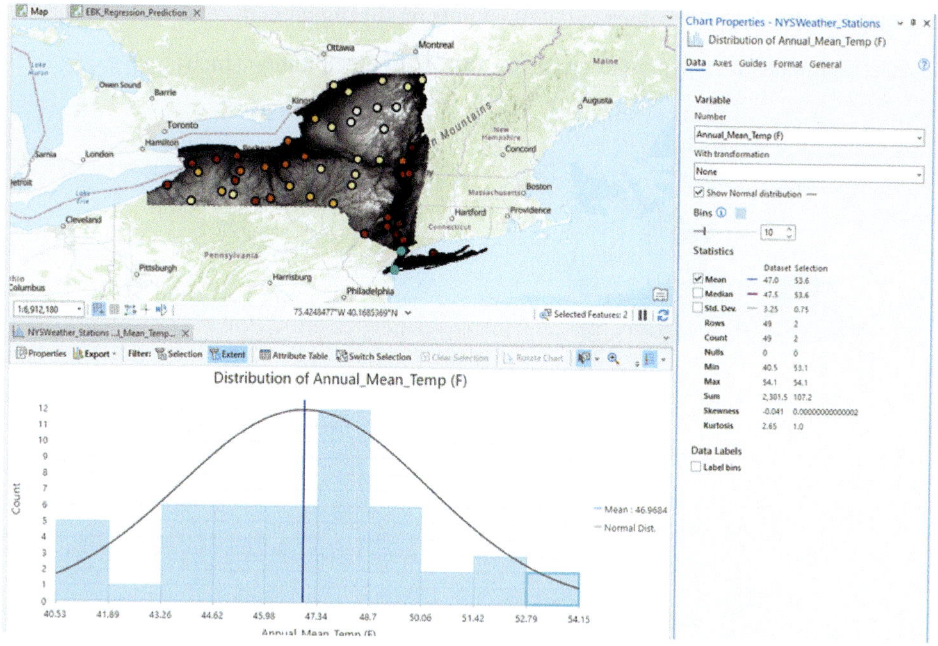


Using this way, you can find the where the highest and the lowest values are located on the digital map. To clear the selection, you can either click Clear in the Selection group on the Map tab or click on the blank space on the chart. In general, the sampling data of annual mean temperature is not an ideal normal distribution but can be used for predictions. Clear selection and close the histogram chart and the Chart Properties pane.

Next, let's create the QQ plot chart for the annual mean temperature using the same method. In the Chart Properties pane, for Compare the Distribution Of, choose Annual_Mean_Temp (F).

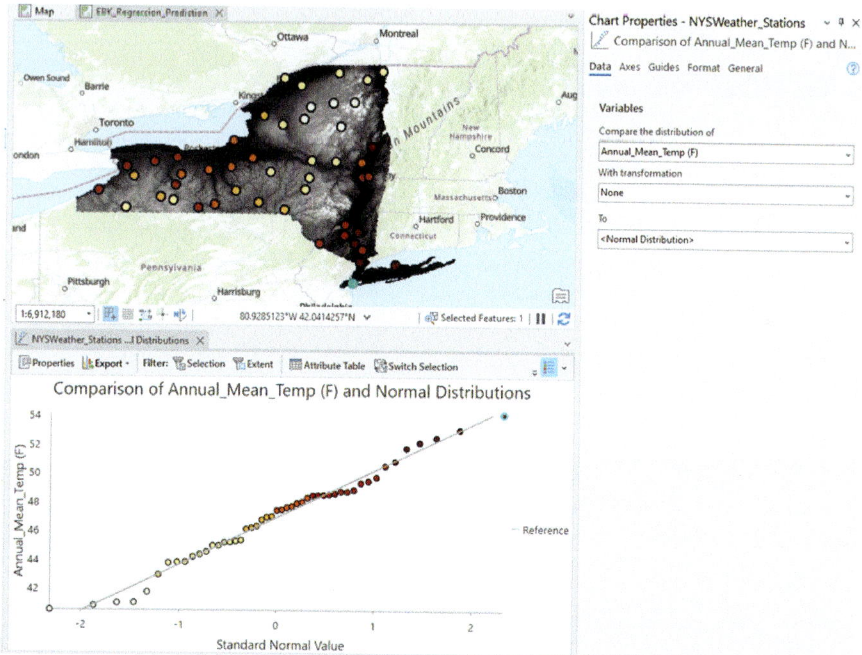

Again, the points on the chart are connected to the points on the digital map.

Close this chart and make the two similar charts for the Total_Precipitation (Inch).

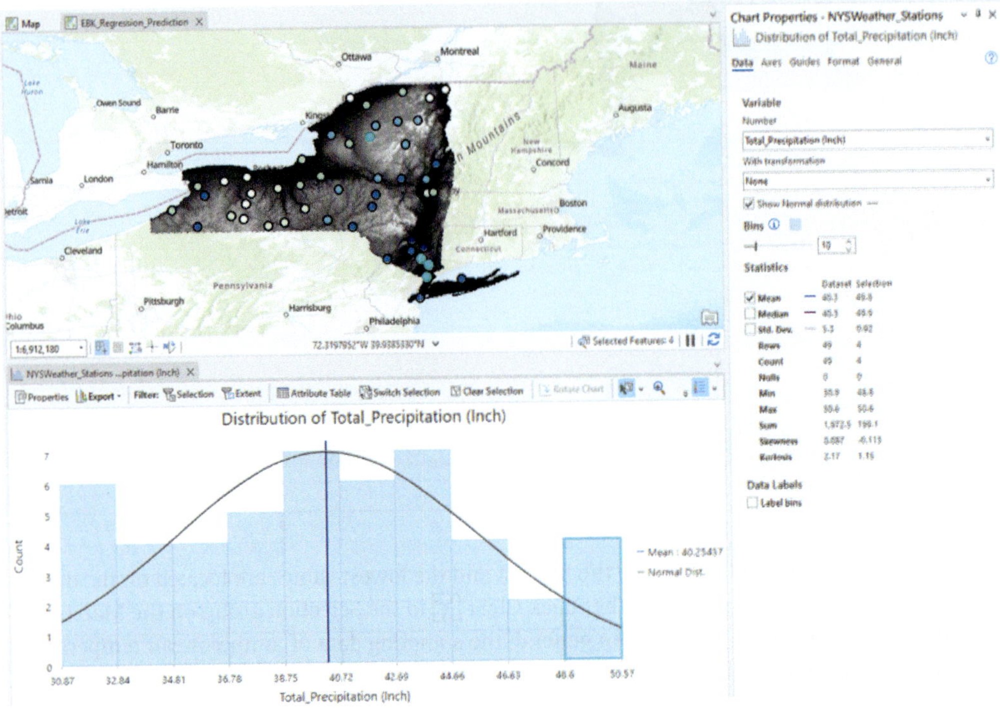


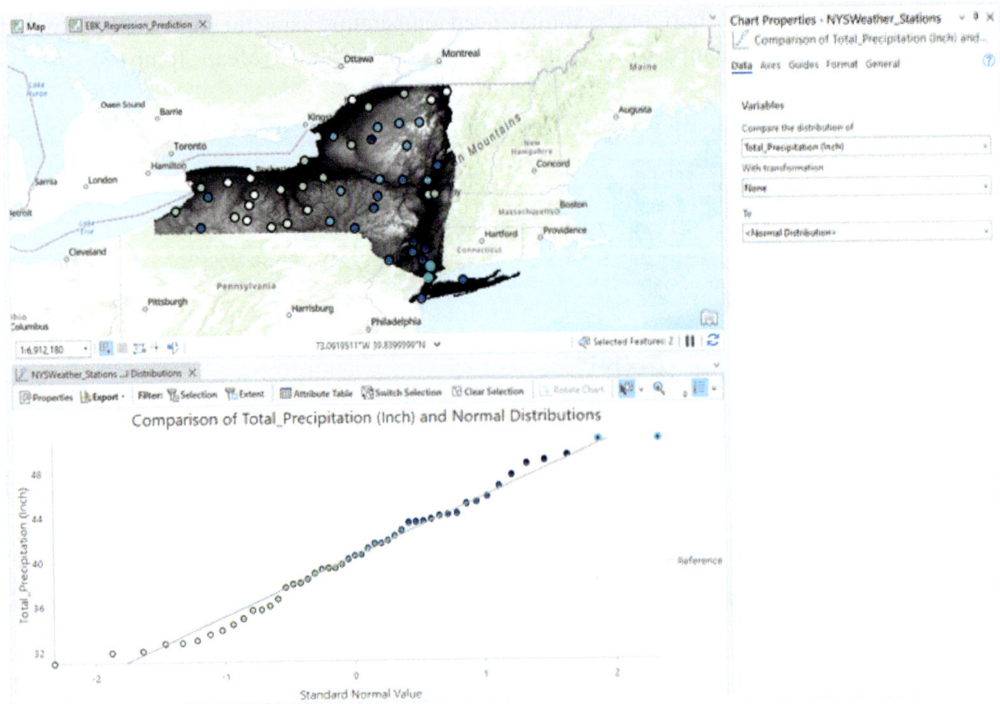

Clear selections and close the chart.

Now, let's use the "Extract Values to Points" tool to evaluate an explanatory variable (elevation) for use with the EBK regression prediction method. On the Analysis tab of the main menu, click Tools to open the Geoprocessing Pane. In the Geoprocessing Pane, search for **Extract Values**.

Open the Extract Multi Values To Points tool.

For Input Point Features, choose NYSWeather_Stations. For Input Rasters, choose NYS_DEM.

For Output Field Name, accept the default.

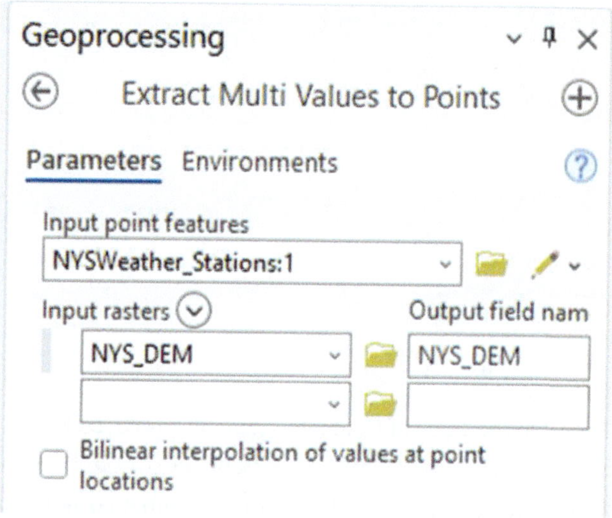


Click Run to process the data.

On the Table of Contents, right-click NYSWeather_Stations- Annual_Mean_Temp (F), point to Create Chart, and choose Scatter Plot. In the Chart Properties pane, set the following parameters:

For X-Axis Number, choose NYS_DEM.

For Y-Axis Number, choose Annual_Mean_Temp (F).

The chart shows that the elevation has a linear relationship with Annual_Mean_Temp (F). The relationship is negative. That means the higher the elevation, the lower the annual mean temperature. Using elevation as the regression factor will improve the accuracy of the prediction.

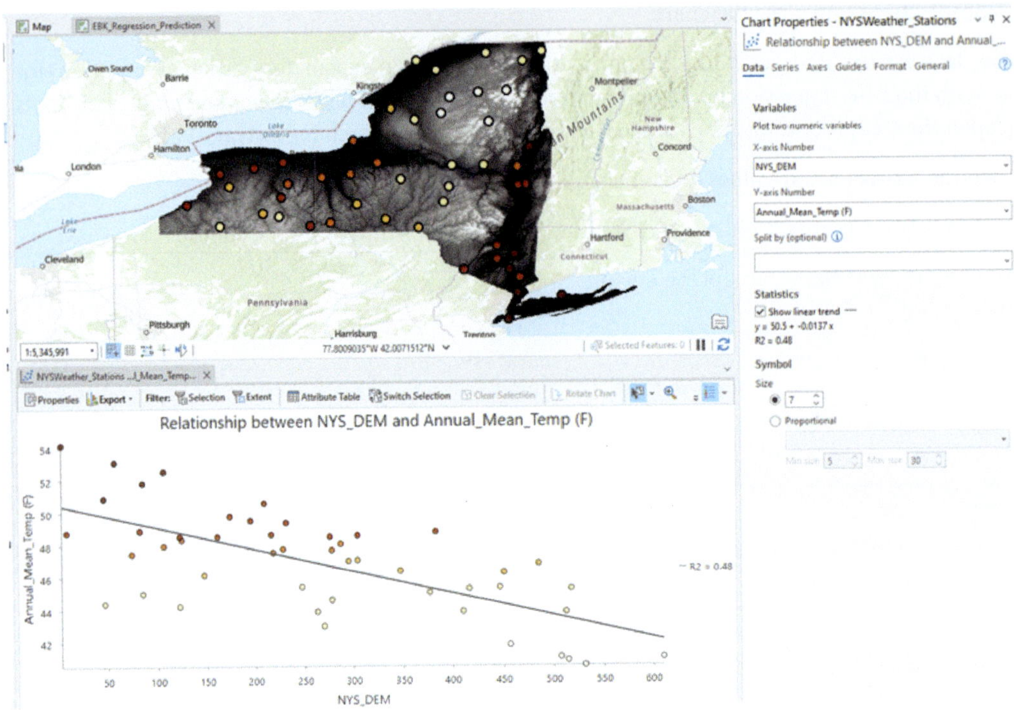


Close the chart and the Geoprocessing Pane. On the Table of Contents, right-click NYS_DEM and choose Zoom To Layer.

Now, let's create prediction surfaces with EBK regression predictions for both annual mean temperature and annual total precipitation for the New York State. Open Geoprocessing Pane from the Analysis tab on the main menu. Search **EBK Regression**. Open the EBK Regression Prediction tool.

For the Input dependent variable features, select the NYSWeather_Stations. For Dependent variable field, select Annual_Mean_Temp_F_. For Input explanatory variable raster, select NYS_DEM.

For Output prediction raster, navigate to your NYS_Data.gdb to save a new dataset as "Temp_EBK_Regression." Adopt the default for Output geostatistical layer and other parameters in the dialog. Click on the Environments tab of the tool. For Mask, choose NYS_DEM.

Then, flip back to the page of Parameters. Click run to process the prediction surface.

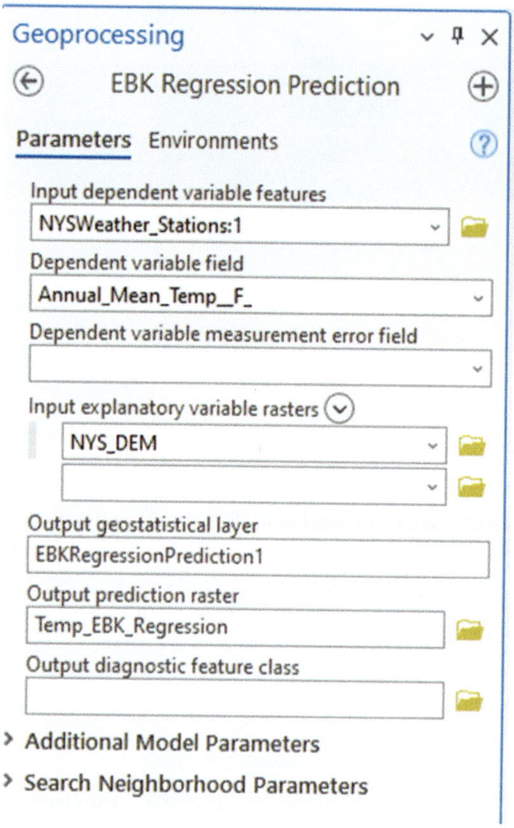


This entire process may take somewhere from 5 to 15 minutes. The result is as follows.

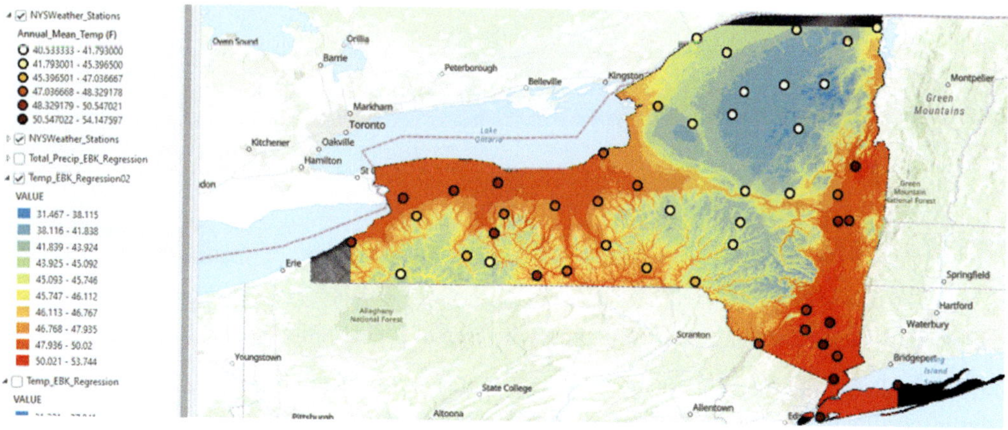


Use Symbology Pane and color ramp of Temperature and eight categories of classification, we obtain the following raster map.

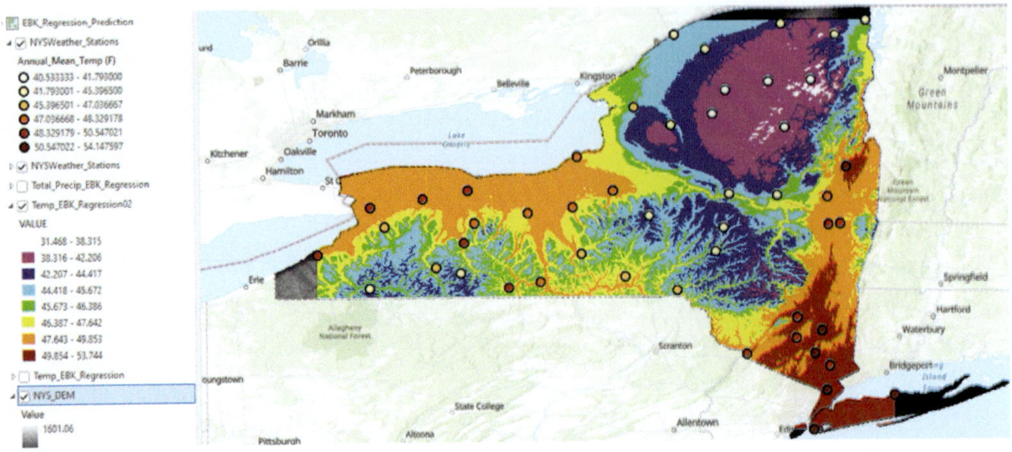


Next, we will use the EBK Regression Prediction tool to create the prediction surface of annual total precipitation for the New York State as well. In this case, for the Input dependent variable features, select the NYSWeather_Stations. For Dependent variable field, select Total_Precipitation_Inch_. For Input explanatory variable raster, select NYS_DEM again.

For Output prediction raster, navigate to your NYS_Data.gdb to save a new dataset as "Total_Precip_EBK_Regression." On the Environments tab of this tool, For Mask, select NYS_DEM again. Run the tool to create the prediction surface.

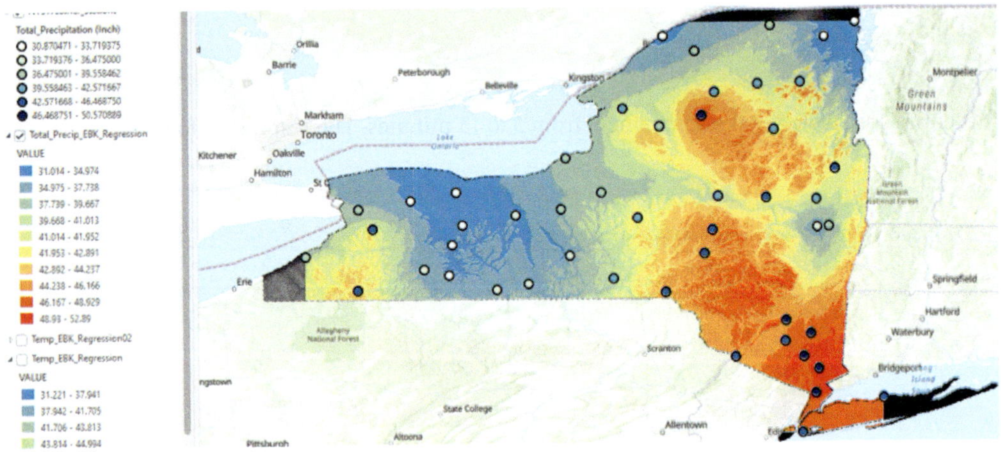


Thematically symbolize this result using the color ramp of Precipitation and eight categories.

You accomplished the New York State 150-year climate conditions study project. Congratulations! After inspection of final maps, you can close the Geoprocessing Pane and save the project. Close the ArcGIS Pro.

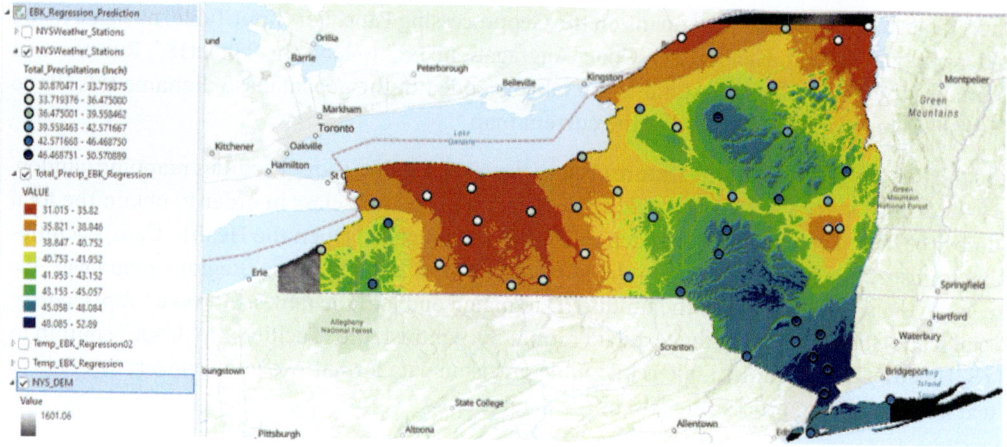


5. New York State Health Care Facility Planning Project

The objective for this project is to analyze the spatial relationship between health care facilities and total population distribution as well as median household income across the entire New York State. This approach will help us to identify some discrepancies and to suggest improvements for the health care facility network in the state. Health care facilities, according to the definition of NYS Department of Health, are "hospitals, hospital extension clinics, diagnostic and treatment centers, licensed home care services agencies, residential health care facilities, long term home health care programs, and adult health care homes." In this project, we will reaggregate data the using Areal Interpolation method to predict the total population in zip code or postal code regions from that was field survey collected in US Census block group regions. Then, we will explore both ordinary least square regression and geographically weighted regression (GWR) to study the impact of total population and median household income on health care facility distributions.

Open the MyGIS_Learning09 project with ArcGIS Pro if it is closed.

On the Catalog Pane, add a new mapping frame "Map1." Rename the mapping frame to "NYS_Health_Care_Facilities." First, we need to import the tabular data in the "Health_Facility_Zip_Code_Aggragation.xlsx" Excel table into NYS_Data geodatabase as a stand-alone table. Right click on NYS_Data.gdb on the Catalog Pane, then select the Import -> Table on the contact menu.

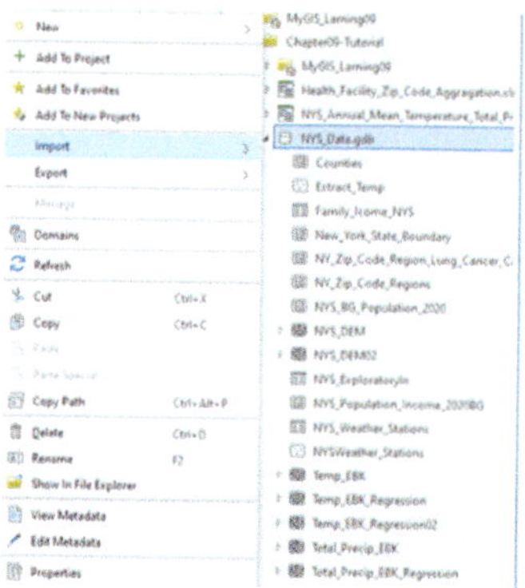

The “Table to Geodatabase” tool opens on the Geoprocessing Pane. For Input Table, navigate to the folder of your “Health_Facility_Zip_Code_Aggragation.xlsx” to load the “Sheet1$.” Run the tool to process the data. “Sheet1_” stand-alone table is added in the geodatabase. Rename this file to “Health_Care_Facilities” and add to the current map.

Add the NY_Zip_Code_Regions feature class to the NYS_Health_Care_Facilities mapping window as well. Now, we need to join the table to the geographic feature class in order to obtain the total number of health care facilities by each of the zip code regions. Open the Health_Care_Facilities table to inspect its fields. Also, open the attribute table of NY_Zip_Code_Regions map layer. We are going to use the zip code as the unique ID to match and join the tables. However, Zip_Code or Postal (code) field are both in the text data format. By contrast, the “Facility_Zip_Code” column in NYS_Health_Care_Facilities table is in double precision data format. We must convert this field to a text field in order to process the join of the databases.

Close the NY_Zip_Code_Regions table and leave Health_Care_Facilities table open. Add a new field called “Zip_Code02” and select “Text” for the Data Type in the table using the Field View tab. For the Length of the Text (Length), replace 255 with 5. Click the save icon in the Fields menu on the main menu to save the changes.

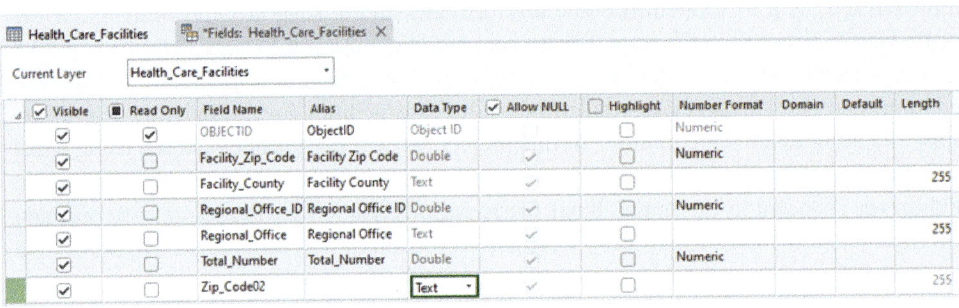


Close the Field View tab and go back to the Table View tab. We need calculate the values of Zip_Code02 field in order to convert the zip code to text format. Right-click on the title bar of the field to get the contact menu. Then, select Calculate Field. Fill in Zip_Code02 = Facility_Zip_Code. Click on Ok to process the data.

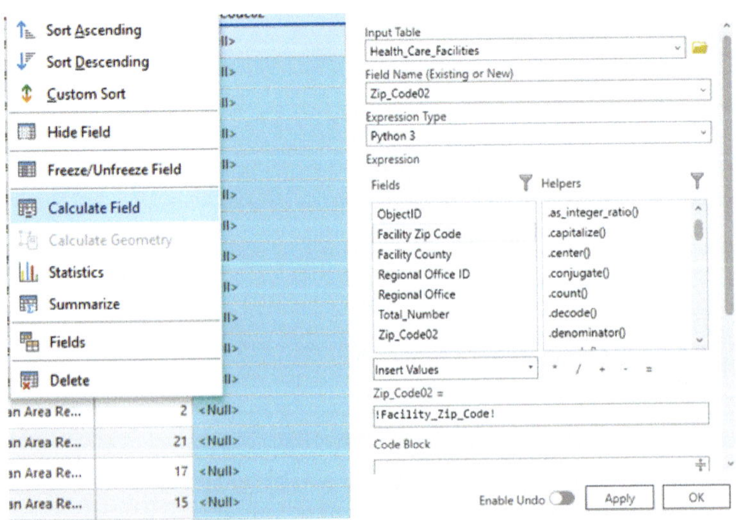

Now, we can join the two tables. Close this table. Right-click on NY_Zip_Code_Regions feature class on the Table of Contents. On the contact menu, select Join and Relate, then Add Join.

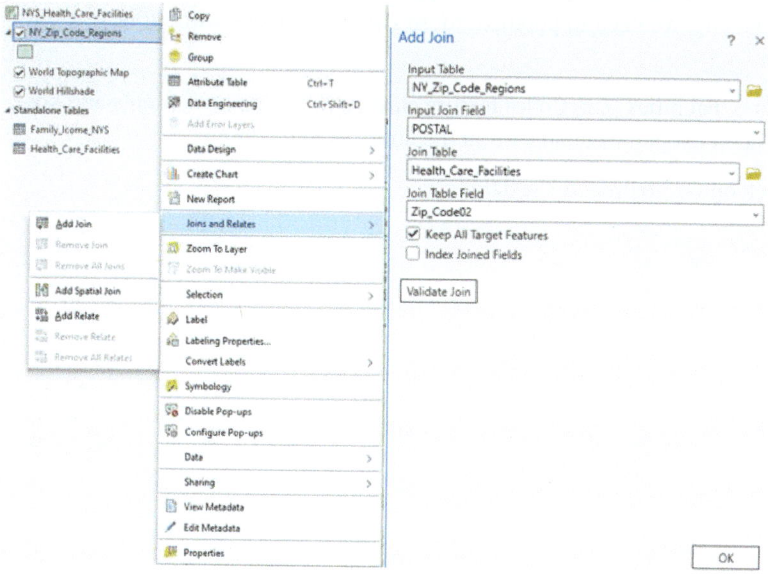


For Input Join Field, select POSTAL. For Join Table, select Health_Care_Facilities. For Join Table Field, select Zip_Code02. Process the function to join the tables. Open the attribute table now, you notice that the fields of total number of facilities and the Zip_Code02 are added. However, the attribute join is only temporary now within this project. We can remove the join any time we would like to. In order to create a permanent feature class with the join, we need to export the feature class to save it in a different title. Close the attribute table. Export the feature class using the contact menu of NY_Zip_Code_Regions via the Data menu. For Output Feature Class, navigate to your NYS_Data.gdb and name the new feature class as "NYS_H_Care_Facilities_Zip_Codes." Click OK to create the feature class.

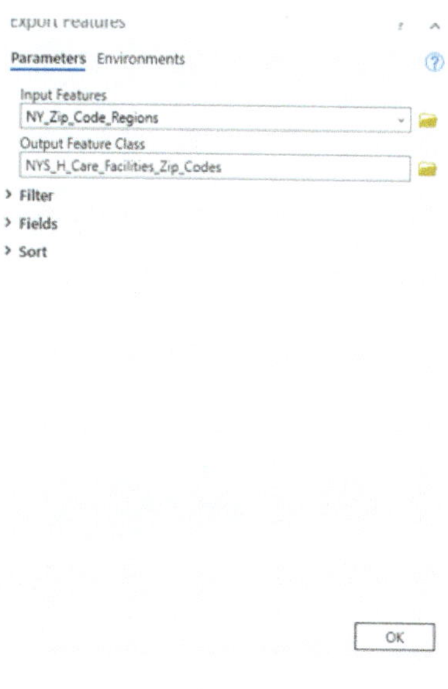

This feature class is a permanent one that contains the health care facility distributions by zip code across the New York State. Remove the NY_Zip_Code_Regions feature class and the Health_Care_Facilities table from the NYS_Health_Care_Facilities mapping frame. Since it is not every zip code region has a health care facility, we need to populate those zip code regions without facility with number "0."

Highlight the NYS_H_Care_Facilities_Zip_Codes feature class in the Table of Contents and click on "Select by Attributes" icon to open its dialog. Build the query as following.

Total_Number => 0. Click on OK to process the data.

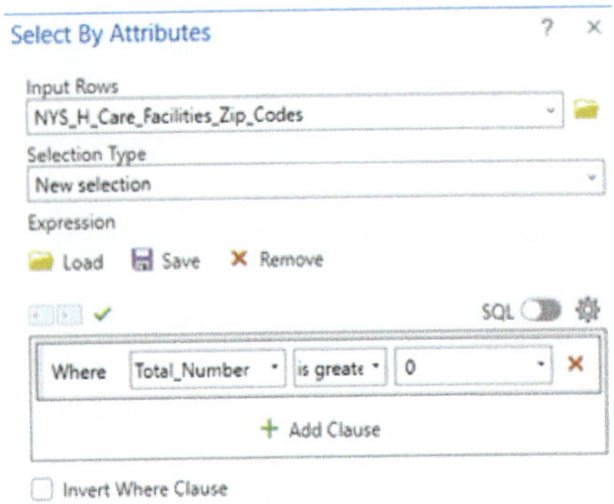


Open the feature class attribute table as well now. Click on the Switch icon Switch on the title bar of the table. This will switch the selected items. In this case, the records with no facility zip code region will be selected. While the <Null> value of Total_Number is selected, right-click on the title of Total_Number column to get to the contact menu. Then, select Calculate Field. In the dialog of Calculate Field, type 0 for "Total_Number =" field. You notice that all these records are changed to 0.

Clear the selection and close the attribute table. Thematically map the Total_Number field using Graduated Colors and the color ramp of Purple-Blue-Green (Continuous) and six classes.

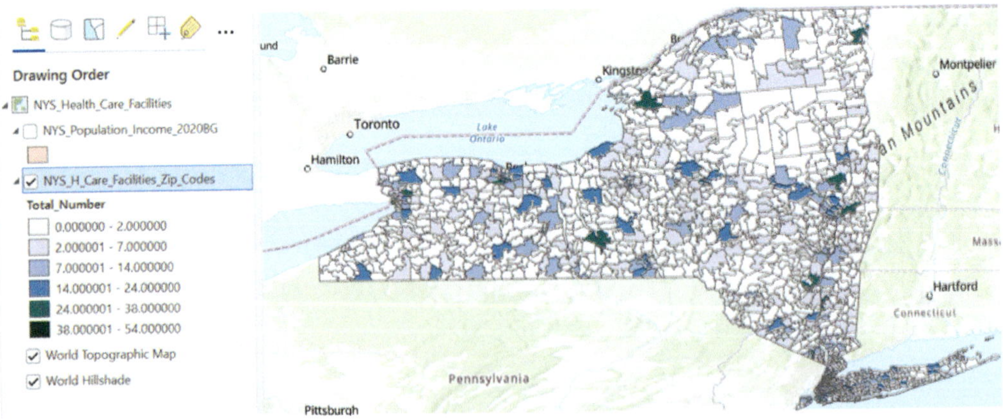


Save the project.

The objective of our project is to study the relationships between health care facilities and total population and median household income by the zip code regions. However, we do not have the population and household income data readily incorporated with our NYS_H_Care_Facilities_Zip_Codes feature class. We are going to use the Areal Interpolation method to transfer the total population and median household income data from that which is based on US Census block-groups to the zip code regions.

In order to make the areal interpolation process less time-consuming and easy to handle, we simplified NYS_Population_Income_2020BG feature class by dissolving many polygon boundaries to create a new feature class called NYS_BG__2020_Pop_In_Dissolve. Add NYS_BG__2020_Pop_In_Dissolve feature class to the NYS_Health_Care_Facilities mapping window. Polygon datasets can represent spatially continuous or discrete data. For a continuous phenomenon, such as mean soil pH value or mean relative humidity, the values are averaged for each polygon area and assumed to be distributed uniformly. For a discrete phenomenon, such as population counts or crime incidents within polygons, the objects are represented as a ratio to the total population or count per unit of each polygon area at a certain time.

Areal interpolation attempts to estimate the underlying grid-cell unit process that was either aggregated or averaged into the source polygons and creates a prediction surface with details beyond the input polygon boundaries. As a result, the prediction surface can be applied to any other polygon dataset to generate more accurate estimations. Therefore, areal interpolation can be used to reaggregate data from a source polygon dataset to a target polygon dataset. In practice, it is significant that a dataset can be reaggregated to downscale from coarse-scale polygons to fine-scale polygons or vice versa. It can also be used to fill in missing data for polygon feature classes.

First, we will visually explore the datasets of two variables, namely "Sum_Total_Population_20" (total population of 2020) and "MEAN Median H Income" (Median household inoomc of 2020), using the thematical mapping method.

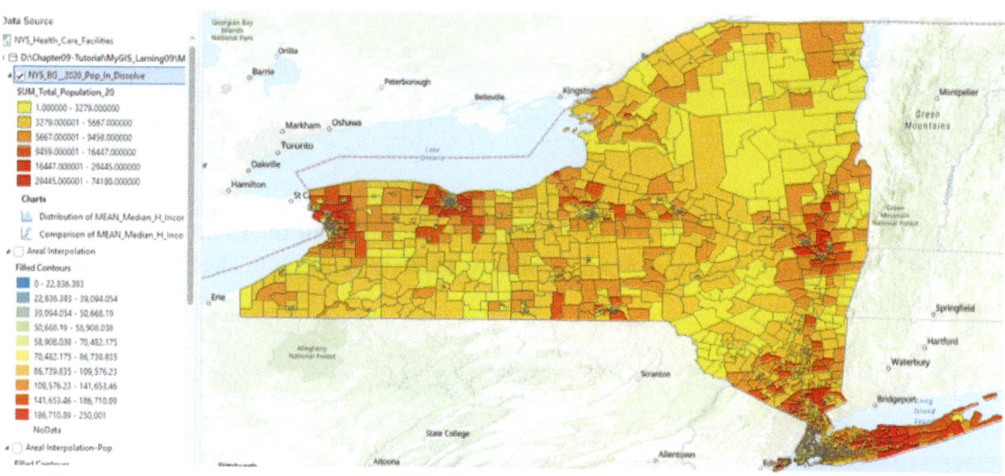

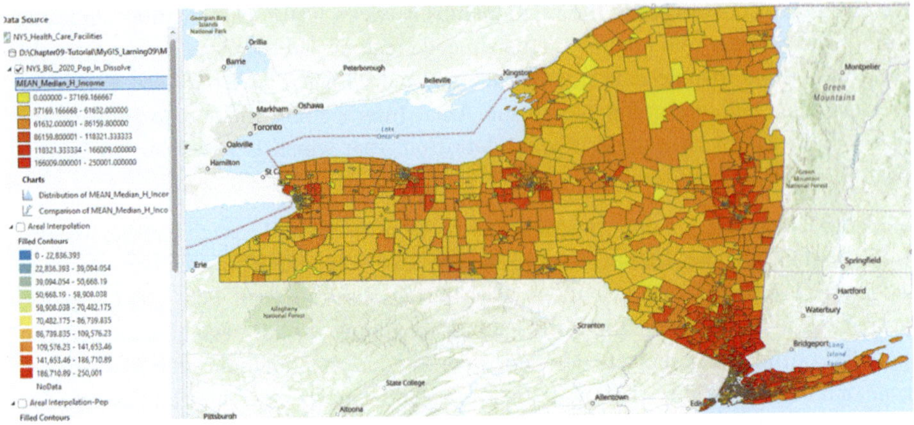


Secondly, we create histogram and QQ plot charts to evaluate the distributions of the values.

In this case, we choose the variable of "MEAN_Median_H_Income" (Median household income of 2020). In the Table of Contents, right-click NYS_BG__2020_Pop_In_Dissolve, point to Create Chart, and choose Histogram on the contact5 menu.

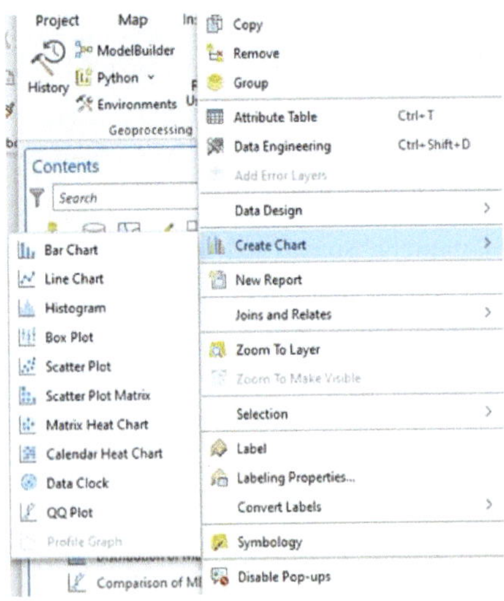


A histogram chart window and the Chart Properties pane appear. In the Chart Properties pane, for Number, click the down arrow and choose MEAN_Median_H_Income. The histogram chart is presented like the following:

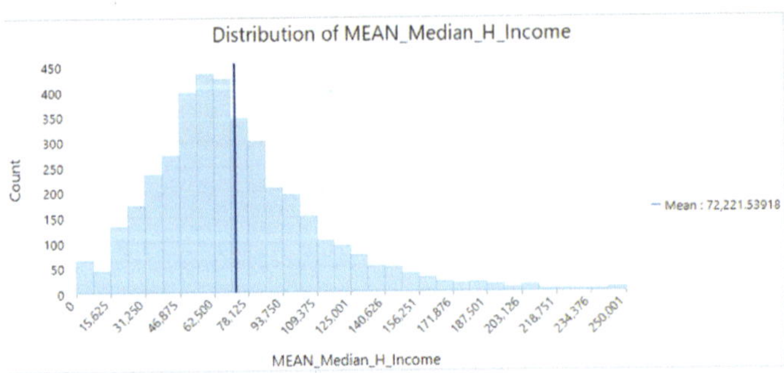

Close the histogram chart and its properties pane.

Using these steps to create a QQ plot for MEAN_Median_H_Income as well. In the Chart Properties pane this time, for Compare the Distribution Of, choose MEAN_Median_H_Income.

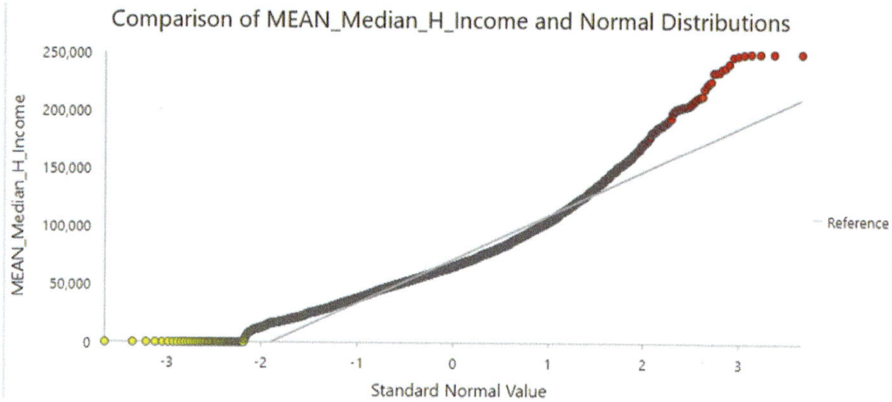


Generally speaking, these two sampled datasets are not ideally in normal distribution. But it can be used for predictions or estimations.

Now let's fit the data to the prediction models to create two prediction surfaces. We are going to use Geostatistical Wizard and areal interpolation method to create two prediction surfaces. In the Table of Contents, select NYS_BG__2020_Pop_In_Dissolve polygon feature class. From the Analysis tab on the main menu, in the Workflows group, click Geostatistical Wizard. The dialog opens. In the Geostatistical Wizard pane, under Geostatistical Methods, choose Areal Interpolation. Under Input Dataset 1, set the following parameters:

For Type, verify that Average is specified. For Source Dataset, verify that NYS_BG__2020_Pop_In_Dissolve is specified. For Value Field, choose Sum_Total_Population_20. Click Next.

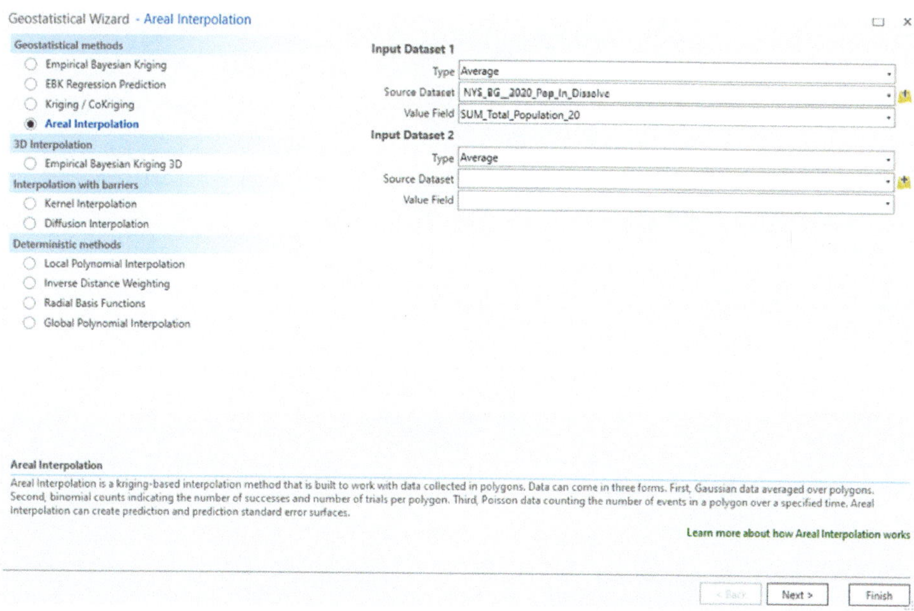

On this page, change the Lattice Spacing to 0.004 and click to enter key on the keyboard. The process to fit in a model may use 5 to 10 minutes. Then, change the Function type to Semivariogram and Confidence Level to 95%

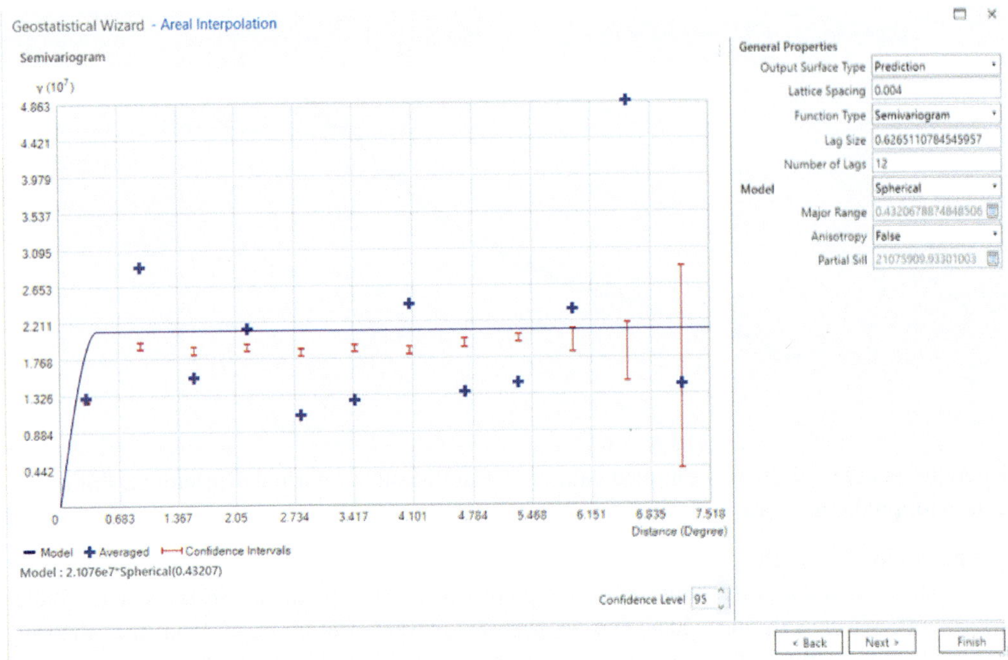


Click on Next, we have the following page.

Keep all the default on this page, we can finish this process to produce the prediction surface layer. Meanwhile, we need to understand that the statistical confidence of this prediction is very low. The prediction surface is as follows:

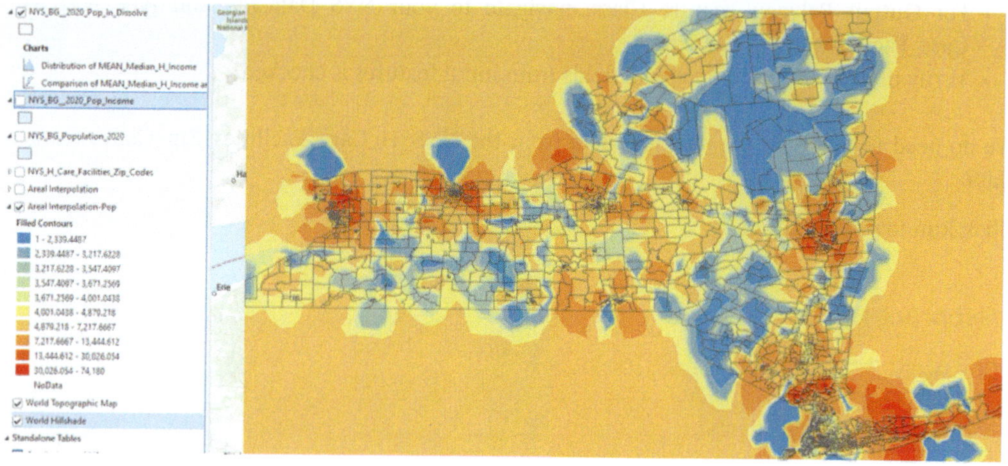


In the Table of Contents, change the name of prediction surface to "Areal Interpolation-Pop."

Use the same steps but change the Value Field for the Areal Interpolation to "MEAN_Median_H_Income" to produce another prediction surface for household income. Change the name of the newly created layer to "Areal Interpolation-Income."

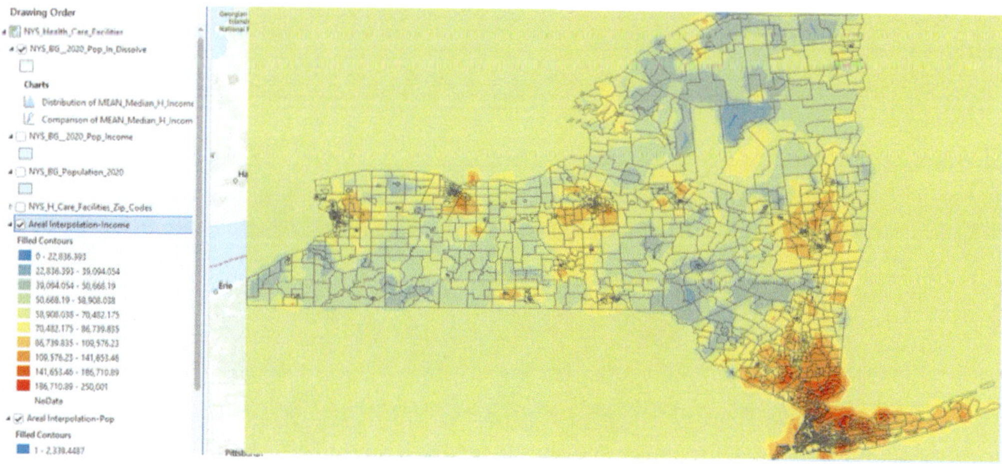


Now, let's apply the prediction surface to the target polygon feature class. In the Table of Contents, turn off all the layers except the NYS_H_Care_Facilities_Zip_Codes.

In the Table of Contents, right-click Areal Interpolation-Pop layer, point to Export Layer, and choose To Polygons. Geoprocessing Pane opens with the Areal Interpolation Layer to Polygons tool.

In the Areal Interpolation Layer to Polygons pane, set the following parameters:

- For Input Areal Interpolation Geostatistical Layer, make sure it is Areal Interpolation-Pop.
- For Input Polygon Features, choose NYS_H_Care_Facilities_Zip_Codes.
- For Output Polygon Feature Class, navigate to your NYS_Data.gdb and type NYS_H_Care_Facilities_Zip_Code_Pop.
- Verify that the box for Append all fields from input features is checked.

We do need to maintain the attribute data from the NYS_H_Care_Facilities_Zip_Codes feature class.

Click Run to process the data.

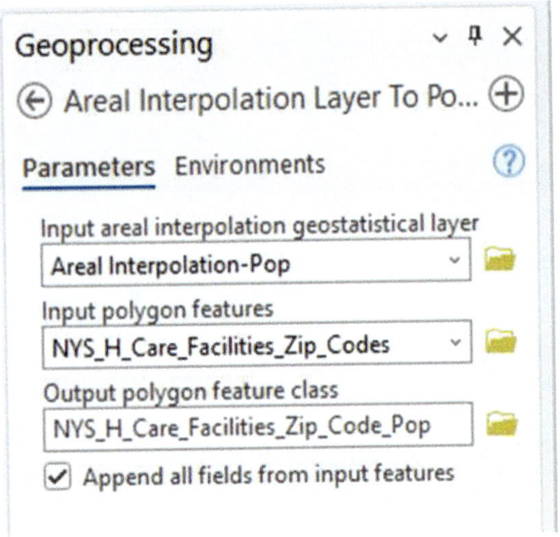


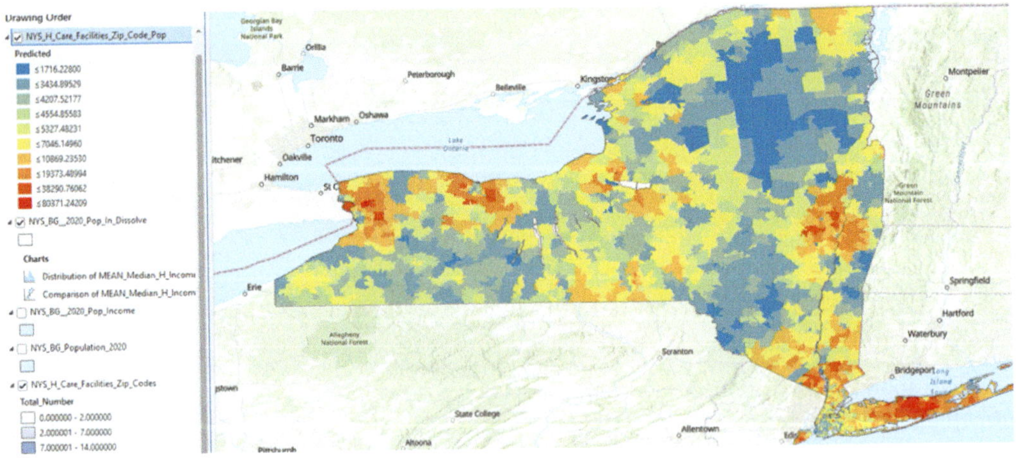

The predicted field is the estimated total population in the zip code regions in 2020. We need to be aware that we do have the total population of 2000 in the original database. However, that is not the population in the 2020 Census survey. Now, we made the predictions using the original 2020 Census survey in zip code geographic regions. Close the Geoprocessing Pane.

Change the thematic mapping back to single symbol for NYS_H_Care_Facilities_Zip_Code_Pop feature class. Open attribute table of it and change the field name from "Predicted" to "Total_Pop_2020_P." Also, change the Alias to "Total_Pop_2020." Click on the save icon on the Fields ribbon on the main menu to save the changes.

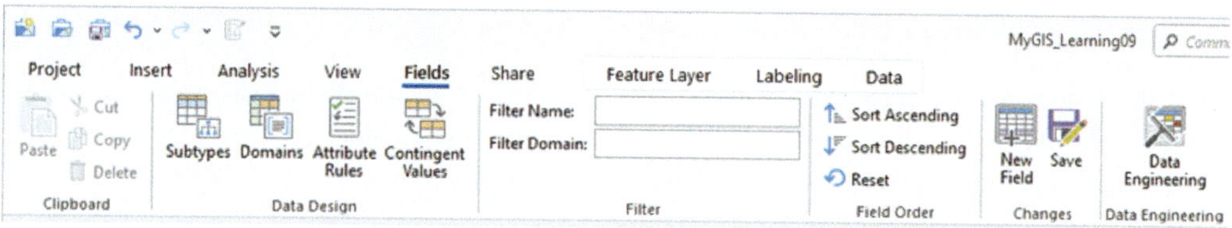


Now let's add the value estimations of median household income to the NYS_H_Care_Facilities_Zip_Code_Pop feature class. Following the above steps again

Right-click on Areal Interpolation-Income in the Table of Contents to export the layer to polygons. In the opened Areal Interpolation Layer to Polygons tool at Geoprocessing Pane, fill in the following:.

For Input Areal Interpolation Geostatistical Layer, make sure it is Areal Interpolation-Income.

- For Input Polygon Features, choose NYS_H_Care_Facilities_Zip_Codes again.
- For Output Polygon Feature Class, navigate to your NYS_Data.gdb and type NYS_H_Care_Facilities_Zip_Code_Income.
- Verify that the box for Append all fields from input features is checked.

Again, we do need to maintain the attribute data from the NYS_H_Care_Facilities_Zip_Codes feature class. Run the tool to create the new feature class.

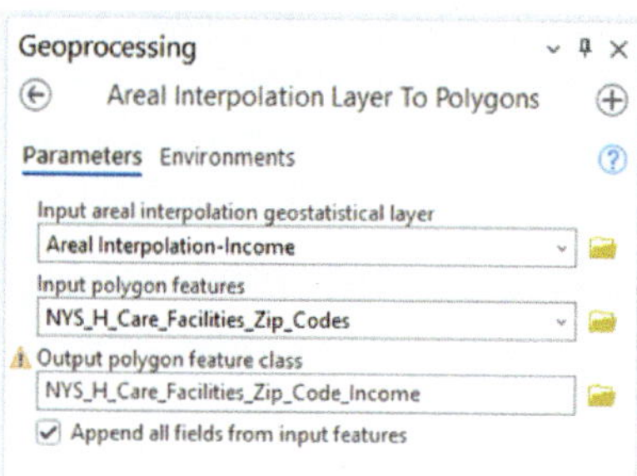

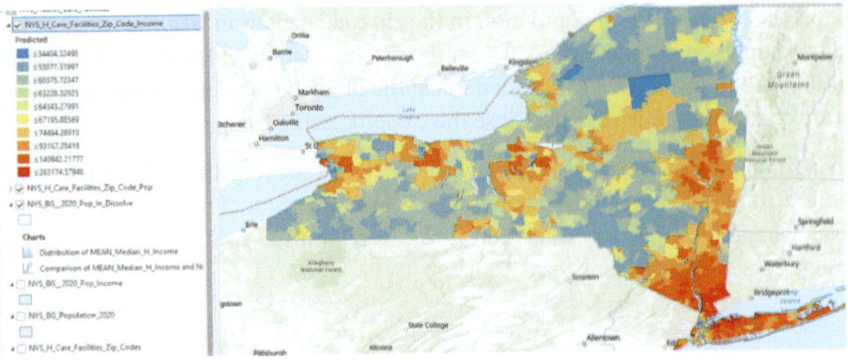

This is the output feature class map of household income. We accomplished the areal interpolations for both total population and household income from Census block-group geography to the zip code regions. Close the Geoprocessing Pane. Save the project and close ArcGIS Pro.

**Multi-variable Analysis of OLS and GWRs:**

If MyGIS_Learning09 project is closed, open it with ArcGIS Pro. Otherwise, continue. Create a new mapping frame on the Catalog Pane under Maps category. Rename the newly created mapping frame as "Regressions_H_Care_Facilities."

The spatial statistical interpolations that we have been exploring up to this stage are only methods to determine value distributions of one variable with its locations. In other words, our focus in these approaches is to predict values of one variable using the sampled or field measured values in the study region. These analyses allow GIS analysts to discover where some phenomenon has occurred. The next level of question in GIS analysis that we need to ask is why do things occur where they do? In the following exercise, we will analyze the impacts of multiple factors on a particular phenomenon using spatial regression models. Understanding where something occurs is beneficial, but understanding why something occurs can help direct remediation efforts to solve the problem. Regression analysis can help us to understand which factors contribute to a phenomenon and determine the extent to which each factor contributes to that phenomenon. For instance, the distribution of crime might be impacted by education level, income level, alcohol or drug use, unemployment rate, etc.

Add the NYS_H_Care_Facilities_Pop_Income_Zip_2020 feature class from your \......\ Chapter09-Tutorial\NYS_Data.gdb into current map (Regressions_H_Care_Facilities).

On the main menu (ribbon), click the Analysis tab. Click Tools to open the Geoprocessing Pane. In the Find Tools field, type ordinary, and then click the OLS tool to open it.

Set the following parameters:

- For Input Feature Class, choose NYS_H_Care_Facilities_Pop_Income_Zip_2020.
- For Unique ID Field, choose Unique_ID.
- For Output Feature Class, navigate to your NYS_Data.gdb and type Results_OLS_POP (D:\ Chapter09-Tutorial\NYS_Data.gdb\Results_OLS_POP).
- For Dependent Variable, choose Total_Number.

It means the total number of medical facilities in a zip code region.

- For Explanatory Variables, check the Total_Population_2020 box.

Click Run.

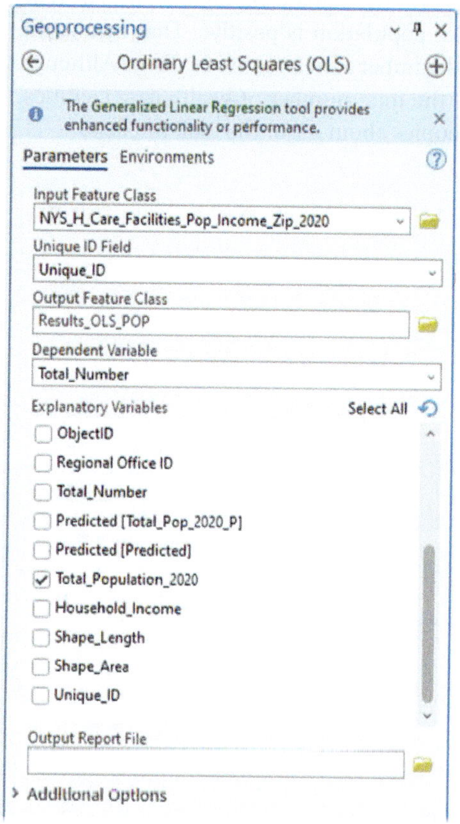


The standard residuals map is produced for the regression of the total number of medical facilities and total population.

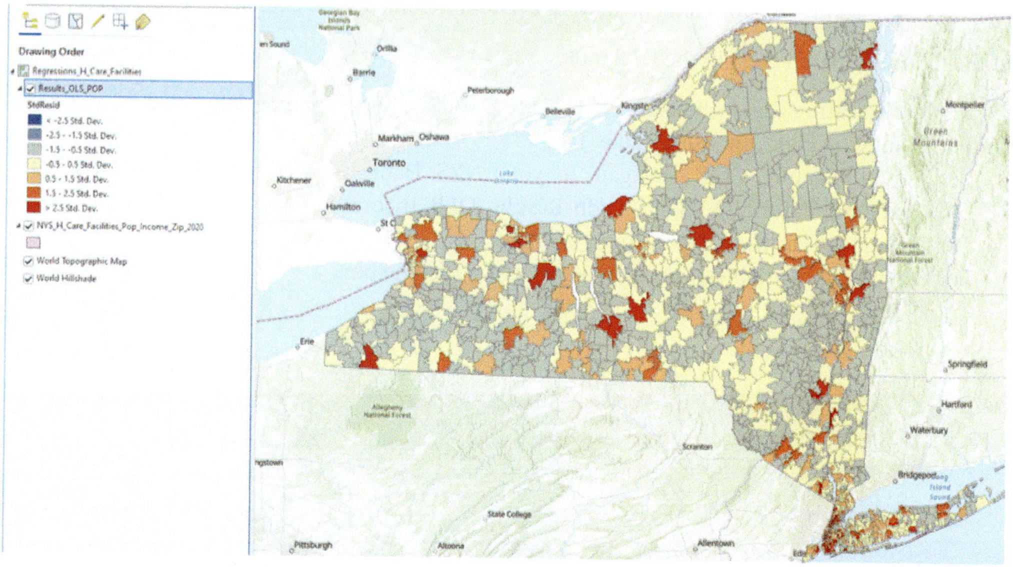

Open the detailed tool process report window; there is an asterisk next to the probability. For both dependent and explanatory variables. Therefore, this regression is statistically significant. The relationship between health care facility numbers and total population is positive. That means the high the population in a zip code region, the high the total number of health care facilities. Although the total population is an important factor to influence the total number of health care facilities, the adjusted R2 is only 0.002972. That means it only occupies about 3% of the total impacts.

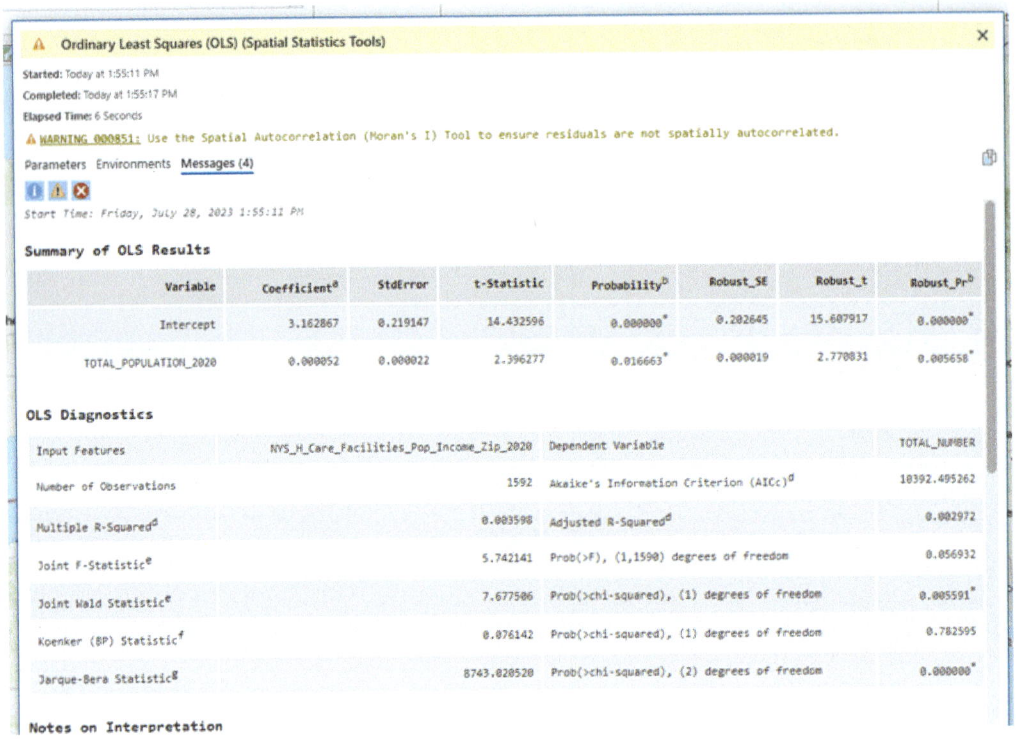
Ordinary Least Squares (OLS) (Spatial Statistics Tools)

Started: Today at 1:55:11 PM

Completed: Today at 1:55:17 PM

Elapsed Time: 6 Seconds

WARNING 000851: Use the Spatial Autocorrelation (Moran's I) Tool to ensure residuals are not spatially autocorrelated.

Parameters Environments Messages (4)

Start Time: Friday, July 28, 2023 1:55:11 PM

Summary of OLS Results

| Variable | Coefficient[a] | StdError | t-Statistic | Probability[b] | Robust_SE | Robust_t | Robust_Pr[b] |
|---|---|---|---|---|---|---|---|
| Intercept | 3.162867 | 0.219147 | 14.432596 | 0.000000* | 0.202645 | 15.607917 | 0.000000* |
| TOTAL_POPULATION_2020 | 0.000052 | 0.000022 | 2.396277 | 0.016663* | 0.000019 | 2.770831 | 0.005658* |

OLS Diagnostics

| | | | |
|---|---|---|---|
| Input Features | NYS_H_Care_Facilities_Pop_Income_Zip_2020 | Dependent Variable | TOTAL_NUMBER |
| Number of Observations | 1592 | Akaike's Information Criterion (AICc)[d] | 10392.495262 |
| Multiple R-Squared[d] | 0.003598 | Adjusted R-Squared[d] | 0.002972 |
| Joint F-Statistic[e] | 5.742141 | Prob(>F), (1,1590) degrees of freedom | 0.056932 |
| Joint Wald Statistic[e] | 7.677506 | Prob(>chi-squared), (1) degrees of freedom | 0.005591* |
| Koenker (BP) Statistic[f] | 0.076142 | Prob(>chi-squared), (1) degrees of freedom | 0.782595 |
| Jarque-Bera Statistic[g] | 8743.020520 | Prob(>chi-squared), (2) degrees of freedom | 0.000000* |

Notes on Interpretation

Now, let's conduct an OLS for using median household income as the explanatory variable.

Set the following parameters in the OLS tool:

- For Input Feature Class, choose NYS_H_Care_Facilities_Pop_Income_Zip_2020.
- For Unique ID Field, choose Unique_ID.
- For Output Feature Class, navigate to your NYS_Data.gdb and type Results_OLS_POP (D:\Chapter09-Tutorial\NYS_Data.gdb\Results_OLS_Income).
- For Dependent Variable, choose Total_Number.
- For Explanatory Variables, check the Household_Income box.

Click Run.

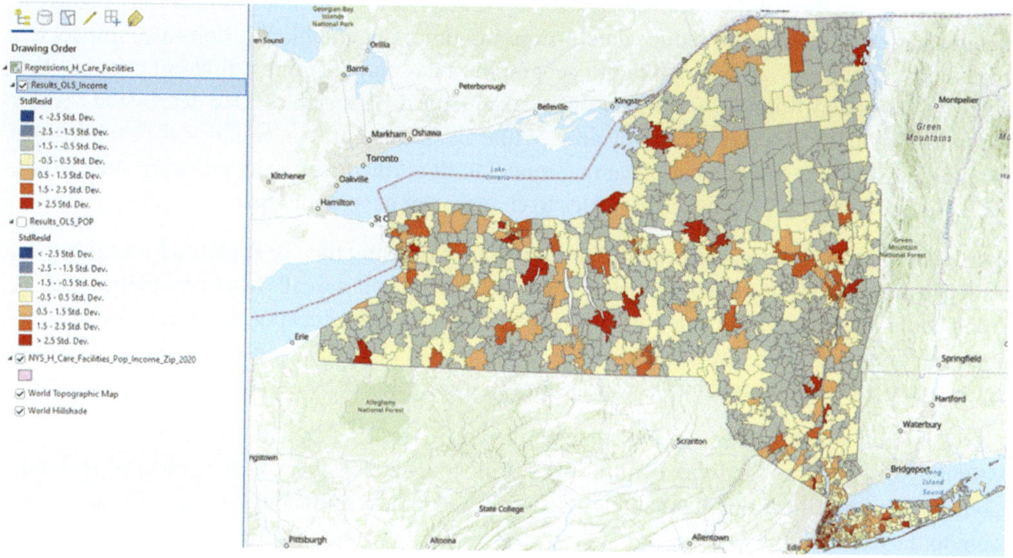


The tool process report shows that the regression is statistically significant and the impact of household income on the total number of heal care facilities is positive. However, the adjusted $R^2$ is much smaller, and the AICc is bigger than last one. AICc is a relative fitness measure for the regressions. Close the Geoprocessing Pane.

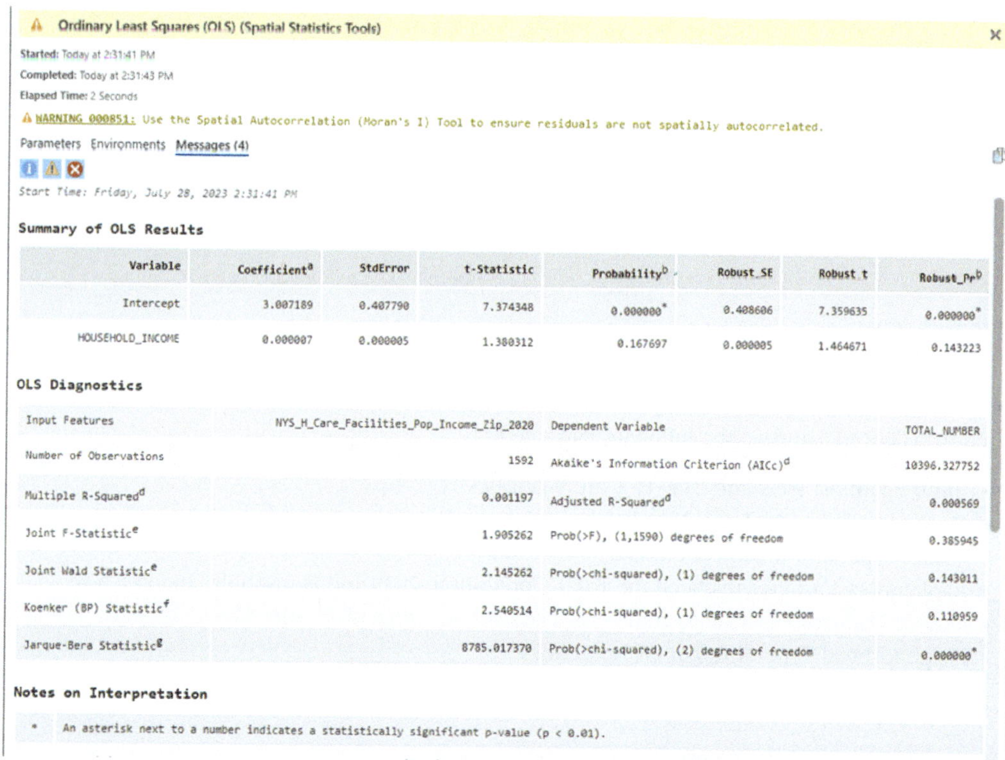

Ordinary Least Squares (OLS) (Spatial Statistics Tools)

Started: Today at 2:31:41 PM

Completed: Today at 2:31:43 PM

Elapsed Time: 2 Seconds

WARNING 000851: Use the Spatial Autocorrelation (Moran's I) Tool to ensure residuals are not spatially autocorrelated.

Parameters Environments Messages (4)

Start Time: Friday, July 28, 2023 2:31:41 PM

**Summary of OLS Results**

| Variable | Coefficient[a] | StdError | t-Statistic | Probability[b] | Robust_SE | Robust_t | Robust_Pr[b] |
|---|---|---|---|---|---|---|---|
| Intercept | 3.007189 | 0.407790 | 7.374348 | 0.000000* | 0.408606 | 7.359635 | 0.000000* |
| HOUSEHOLD_INCOME | 0.000007 | 0.000005 | 1.380312 | 0.167697 | 0.000005 | 1.464671 | 0.143223 |

**OLS Diagnostics**

| | | | |
|---|---|---|---|
| Input Features | NYS_H_Care_Facilities_Pop_Income_Zip_2020 | Dependent Variable | TOTAL_NUMBER |
| Number of Observations | 1592 | Akaike's Information Criterion (AICc)[d] | 10396.327752 |
| Multiple R-Squared[d] | 0.001197 | Adjusted R-Squared[d] | 0.000569 |
| Joint F-Statistic[e] | 1.905262 | Prob(>F), (1,1590) degrees of freedom | 0.385945 |
| Joint Wald Statistic[e] | 2.145262 | Prob(>chi-squared), (1) degrees of freedom | 0.143011 |
| Koenker (BP) Statistic[f] | 2.540514 | Prob(>chi-squared), (1) degrees of freedom | 0.110959 |
| Jarque-Bera Statistic[g] | 8785.017370 | Prob(>chi-squared), (2) degrees of freedom | 0.000000* |

**Notes on Interpretation**

- An asterisk next to a number indicates a statistically significant p-value (p < 0.01).

Close the report window and geoprocessing Pane. Save the project and close the ArcGIS Pro.

Now, let's practice the GWR on a subset dataset in Western New York State. Making the geographic region smaller will help us to reduce data processing time. Geographic locations also impact the regressions of multiple variables. Different geographic areas may encounter different relations between the dependent variable and many different independent variables. OLS regression creates one model for all the features in a study area. While GWR calculates an equation for each feature using nearby features rather than using one model to fit all the features in a dataset. As a result, GWR allows relationships between variables to change over geographic space.

Many examples in practice can present the significance of using the GWR method. For instance, GWR can differentiate the major impact on the number of 911 phone calls in a metropolitan region, which could be strongly connected to large population and low education level in some districts or to high unemployment rate in some others. GWR also can identify cancer mortalities in a study region that could be impacted by water pollution in some areas or by air pollution in other areas.

If MyGIS_Learning09 project is closed, open it with ArcGIS Pro. Otherwise, continue. Add Westeren_NY_H_Care_Facilities feature class from the Catalog Pane to Regressions_H_Care_Facilities window. Check off all the other map layer except the Regressions_H_Care_Facilities and zoom to the extent of this layer.

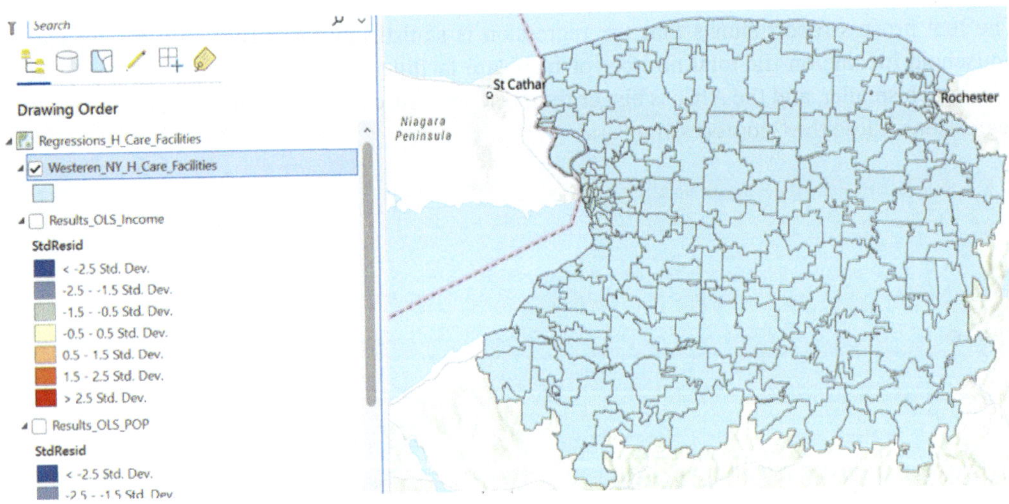


Open the Geoprocessing Pane from Analysis menu on the main menu and search for GWR.

Open the GWR tool and set the following parameters:

- Input Features: Westeren_NY_H_Care_Facilities
- Dependent Variable: Total_Number
- Model Type: Continuous (Gaussian)
- Explanatory Variable(s): check the Total_Population_2020 and Household_Income box
- Output Features: navigate to your NYS_Data.gdb and type in Results_GWR_Western_NY
- Neighborhood Type: Number Of Neighbors
- Neighborhood Selection Method: Golden Search

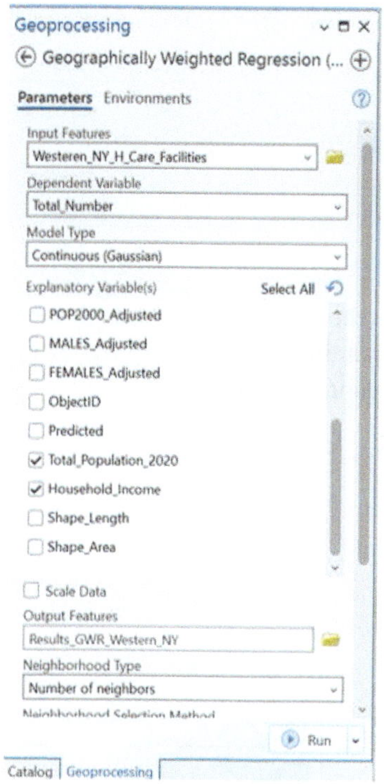


Click Run to process the data. The standardized residual map of GWR is produce.

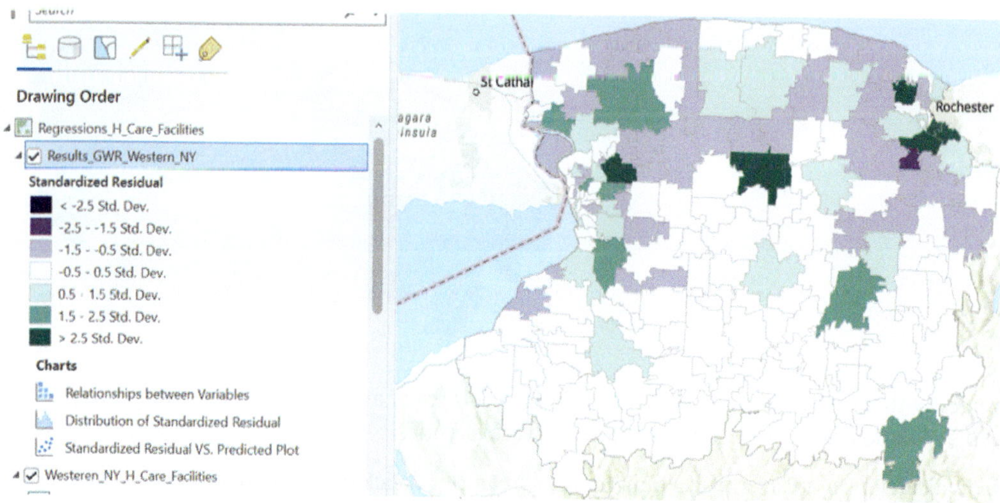


In the Geoprocessing Pane, open the detail report of the GWR tool processing. The AICc value is 1237, and R2 is 0.1814. $R^2$ indicates that it is not a good statistical model but can be used for analyzing relations. Meanwhile the AICc value increased in this case.

**Model Diagnostics**

| | |
|---|---|
| R2 | 0.1814 |
| AdjR2 | 0.0923 |
| AICc | 1237.7807 |
| Sigma-Squared | 30.1084 |
| Sigma-Squared MLE | 27.1698 |
| Effective Degrees of Freedom | 176.8703 |

Now, let's look into the impacts of the two independent variables to the total number of health care facilities in the zip code regions. In the Table of Contents, ensure that the Results_GWR_Western_NY layer is selected, and then open the Symbology Pane from the contact menu.

In the Symbology pane, set the following parameters:

Primary Symbology: Graduated Colors

Field: Coefficient (Total_Population_2020)

Classes: 6

Color Scheme: Yellow-Green (Continuous)

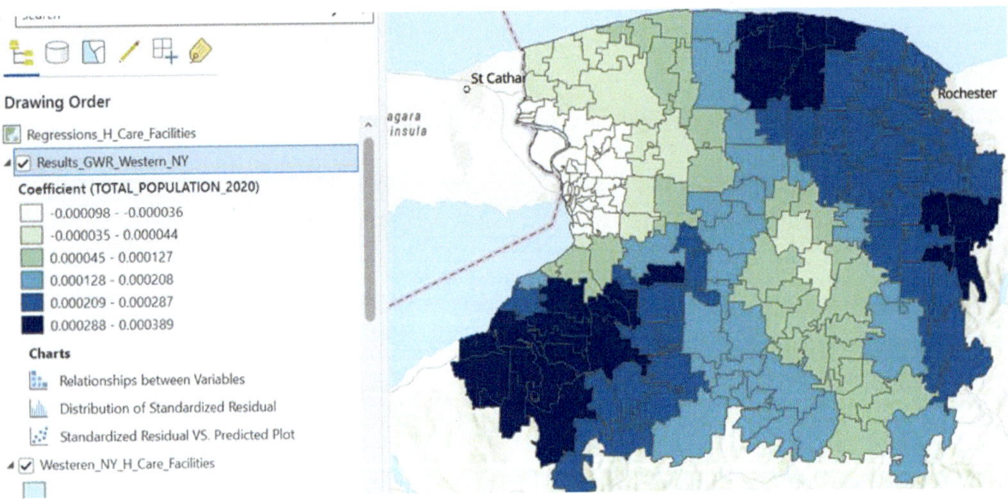


We notice that southwest and northeast regions in this study area demonstrate stronger impact of total population on the total number of health care facilities.

To duplicate this map layer in the Map frame, we can right-click on this map layer (Results_GWR_Western_NY) and copy it on the contact menu. Then, we go to the mapping frame level, highlight and right-click on Regressions_H_Care_Facilities in the Table of Contents to paste it. Now, we have two exactly the same map layers. Open the Symbology Pane and change the top layer presentation field to Coefficient (Household_Income). Then, the map shows the following:

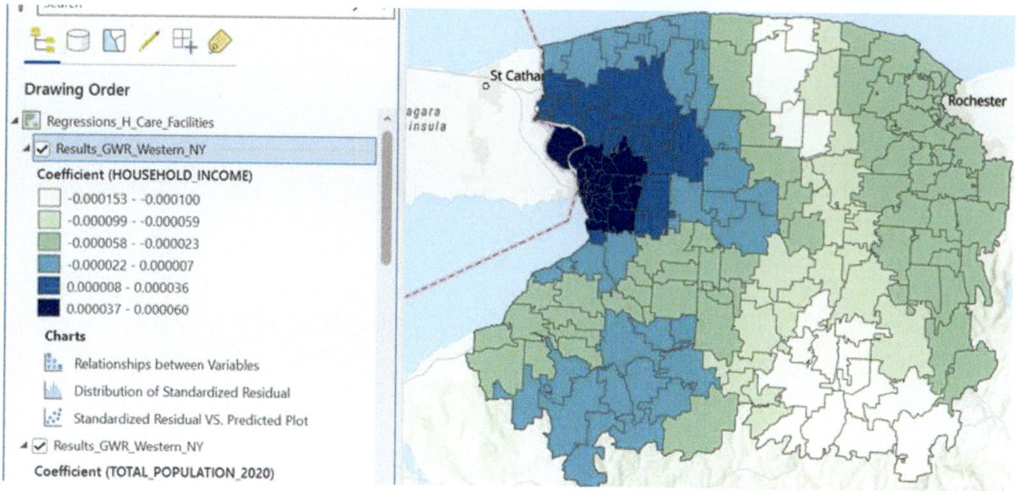


This map shows that median household income has a much stronger impact on the total health facility numbers in urban areas than that in the real areas in Western New York.

Close the Geoprocessing Pane. Save the project and exit the ArcGIS Pro.

# 10

# Network Analysis

## Concepts and Theories

Network is defined as a system of interconnected elements that consists of edges and nodes. In mathematics, a network consists of a set of graphs connecting lines and points. A network can conceptually refer to a structure representing a group of objects or people and relationships between them. In geographic information systems (GIS), analytical environment, and geographical studies, we recognize there are both natural and human-made network systems. For instance, the Mississippi River System or Buffalo River System are natural network systems, while, the United States Interstate Highway System, the National Grid high-voltage power transportation lines, or New York City Subway Transit System are human-made network systems.

According to Curtin (2018), network analysis holds a prominent place in GIS. Network models are the earliest data models that were implemented in GIS. Network analysis is not only broadly useful across domain areas but also has great potential for future development in GIS. Given the fact that extraordinarily broad set of applications in digital geographical analysis which can take advantage of structure of network data, the use of networks in GIS has been grown. Network system visualization and network analysis are likely to remain a dominant platform for spatial analysis in the foreseeable future.

While network analysis in GIS can be applied in various areas of studies, such as studies of road networks, river networks, utility networks, and more, it is the similarity of physical structure and digital data representations rather than the diversity in applications that truly provides the value for research and practices. Networks represent a fundamental spatial domain on which many phenomena can be located and over which many spatial events or processes can be simulated (Curtin 2018).

Network systems are widely used for transportation and utility network modeling in GIS. Topological connectivity on a network system, such as streets and highways, is inherently important for geographic objects (human or goods) to travel over it. Components of a network, such as edges (lines) and junctions (nodes or points), must be interconnected to allow navigation over the entire network system. Additionally, these components have properties that control navigations on the network as well. The fundamental database structure for network analysis is the vector data format although the raster datasets can contribute to the network analysis as the environmental factors. Meanwhile, the network database is unique and different than those regular feature classes of points, lines, and polygons. Network systems are embedded in the unique structure of edges and junctions. In the ArcGIS Pro software environment, line or linear feature classes can be directly

*Fundamentals of Geographic Information Systems (GIS) Analysis: Applying ArcGIS Pro*, First Edition. Tao Tang.

converted to the network systems of edges and junctions. However, line features themselves are not the network components.

According to the flow or transportation directions and the application focus, the network systems that we build in database, analyze, and conduct modeling, can be further identified into two types. These are (i) utility and river network systems, and (ii) transportation (full functioning) network systems. Utility and river networks are network systems of directed flow or traffic. Transportation networks are undirected network systems, which support multiple directional movements of objects or goods.

## Utility or River Network Analysis

A utility network system is a directional or directed transportation system. That means the traveling agent or object, for example, water, sewage, or electricity, flows along the network based upon certain rules built into the network and generally moves toward one certain direction. The path that the water or another flowing agent will take on the network is predetermined by users. The flowing or traveling path and direction can be changed to a utility network, but not by the agent itself. The engineer who controls the network system can change the rules of the network by opening some valves or gates and closing others in order to change the direction of the network flow.

A utility network system is mainly used in GIS environment to manage utility and telecom networks in the real world. Utility network dataset and analytical package provide a comprehensive framework of functionalities for the modeling of utility systems such as electric, gas, water, storm water, wastewater, and telecommunication. Utility network in ArcGIS Pro is designed to model all the components that make up a particular system, such as wires, pipes, valves, zones, devices, and circuits, and enables users to build real-world behaviors of the network into the network features in the digital world that they model (ESRI 2022). In essence, digital network modeling and analysis is to build a digital twin to simulate the real-world system.

Utility network packages in GIS software environment can help users to conduct the following:

1. Create and edit features and objects on a digital network that is used to model every type of utility equipment.
2. Discover how features and objects in the network system are connected.
3. Trace how resources, such as gas, water, and electricity, flow through a particular network system.
4. Visualize how all the dynamic devices of a utility are currently configured and operational.
5. Analyze how the network system is affected by external real-world hazards or natural events such as storms, outage incidents, or equipment failure on the network.

In real-world practice, since utility network systems carry their own unique geographic characteristics, these network systems are treated differently than the regular vector data feature classes in GIS processes from data creation, database configuration, to data analysis and solving the problems, as well as modeling with different scenarios. In the ArcGIS software platform, users must decide the deployment types before creating and configuring a utility network. The utility network system in GIS can be built and accessed with either multiple users (enterprise) or single-user method.

A multiple-user (enterprise) deployment provides the most capabilities of the utility network through a service-based architecture using ArcGIS Pro Enterprise License. This model enables large-scale, multiuser management and full network display, editing, and analysis capabilities through internet maps, internet apps, mobile device apps, and ArcGIS Pro on the desktop. A single-user deployment provides the full analytic capability of the utility network in the GIS system while hosted on a file or mobile geodatabase in local computational environment. In this single-user model, the users interact with the utility network through ArcGIS Pro on the desktop or a mobile device. In this chapter, we focus on the method of single-user deployment based on file geodatabase in ArcGIS Pro.

The major designing of the entire utility network database structure in ArcGIS Pro is to copy and simulate the real-world systems, such as electricity transportation, natural gas transportation, household water supply, or stormwater management in a river system. Users and readers of this textbook are encouraged to create and configure a utility network manually by using basic vector feature class editing tools or using the utility network package tools to convert and reorganize the line or polyline features in a feature dataset that is contained in a file geodatabase in the ArcGIS Pro environment (ESRI 2022).

A utility network can be created and configured manually in ArcGIS Pro using a suite of geoprocessing tools available in the "Utility Network Tools" toolbox, along with other functions or tools in ArcGIS Pro environment. The following steps are a strategic overview of the processes of creating and configuring data structures of a utility network for modeling and simulation purposes (ESRI 2022).

1. Using the "Create Utility Network tool," the utility network data structure can be created inside an existing feature dataset in an ArcGIS Pro geodatabase. During the creation of a utility network, the structure network is also created in the background. The structure network classes are shared across all the domains of the networks (ESRI 2020). Typically, in the file geodatabase feature dataset where we are going to create a utility network, we need to have one line or linear feature class that represents the original linear features such as pipelines, that we surveyed and input into the GIS, and one polygon or area feature class marks the entire service territory or region that we would like to build the utility network system. In this case, the "Service Territory Feature Class" only serves as a boundary marker of the geographic area to establish the utility network system. The value of service territory feature class parameters must exist in the input feature dataset and must be z- value and m- value coordinates enabled.
2. Add domain networks: Utility network systems are not only in the structural distributions of pipelines or channels, but also carry specific services, such as water utility network, natural gas utility network, or electric power utility network, etc. In the interactive field between local government agencies and private companies, these utility network lines are also considered commodities or assets. The concept of domain network was developed for a specific commodity or service type, such as water pipelines, gas pipelines, transportation cables. Generally speaking, the following are the major assets or domains that we study in utility network analysis:
   a. Electricity line
   b. Gas pipeline
   c. Water
   d. Sewer
   e. Stormwater
   f. Telecommunication line

   Normally, when one utility network system is created, one domain network type is attached to it. We can add more domain network systems if the utility system in the study area or territory

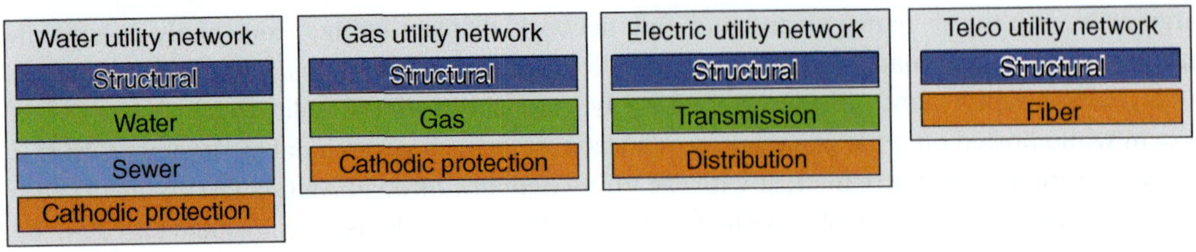


**Figure 10.1** Domain setting up for different utility network needs. *Source:* From Environmental Systems Research Institute, Inc.

has multiple types of utility services, such as gas and water. Separation of utility network structure with domain networks enables one organization, such as a local government to use the same spatial network model to simulate different commodity supplies, either water or natural gas. Figure 10.1 is a chart showing the major categories of domains (Figure 10.1; ESRI 2022).

3. Set up feature classification for network classes.
Each of the features assigned to construct a utility network must be classified into subtypes and domains or asset groups. By default, the ArcGIS Pro provides two fields when we create a utility network, "Subtype" and "Asset Type." Subtypes are the categories of features on a utility network model while "Asset Type" is the attribute that records the real-world name and behavior of a particular feature on a utility network model. Each network class, except SubnetLine or the nested pipeline in a major network, has a requirement that a coded value attribute domain must be assigned on the Asset type field for each subtype (asset group). This is enforced when enabling the network topology on a utility network system. Network properties such as topological connectivity and association rules, network feature categories, and terminal configurations are defined for network classes and in the database of network using the asset group and asset type fields.
The following is an example of network feature classification in a natural gas pipeline or gas domain network:
Real-world network device feature class, "Valve" is defined as one of values of the subtypes the Asset group (sub-domain) field in the utility network data structure. An attribute domain named GasDeviceValveAssetType is created and assigned to the Asset type field for the Valve subtype. It contains the allowable values for a more detailed level of classification for the valve features (Table 10.1).

4. Create and assign terminal configurations for a utility network.
Complex utility networks always have more than one input and output. For example, an electrical transformer device often has two high voltage-side inputs and three low voltage-side outputs. These inputs and outputs are called terminals and can be defined for different types of network devices and junction objects. Terminal stands for changing the utility

**Table 10.1** Allowable values of subtypes in a domain.

| Code | Description |
|---|---|
| 1 | Main |
| 2 | Shutoff |
| 3 | Station |

*Source:* Adapted from ESRI.

network from one to the other, in this case it is from high-voltage transportation system to low-voltage transportation system.

In ArcGIS Pro, terminal configurations are initially added as a property of the utility network layer in a feature dataset using the "Add Terminal Configuration tool." The terminal configuration is set on the network device feature class or Junction Object in the attribute table using the "Set Terminal Configuration tool" at the asset group and asset type levels. For instance, a terminal configuration of a "Transformer" is added to the utility network layer to represent the different terminals of high-voltage electricity or low-voltage electricity. The terminal configuration is then assigned to the network device feature class in a domain network on the Transformer subtype value and the Step-down Asset type value.

Therefore, the advantage of terminal setup on a complex utility network is to constrain the flow directions in trace operations to define valid connectivity associations. This will enhance the work of problem tracing on the utility network system. For a network device or junction object to represent a subnetwork as a subnetwork controller in a partitioned domain network, it must be set with a terminal configuration.

**5.** Create utility network rules.
Two types of utility network rules need to be set up.

In a utility network, network rules are established for network components both in feature class and database locations at the established asset group, asset type, and terminal levels, whichever is applicable. Network rules control the connections to our associations with other components or features of the utility network. These rules can be added to a utility network individually using the "Add Rule tool" or in groups using the tools of "Import Rules" and "Export Rules." Attribute rules in the database allow users to define behaviors related to attributes and geometry of a utility network component. Attribute rules are user-defined rules that are individually used to enhance a particular utility system and help enforce data integrity.

**6.** Create and assign utility network categories.
Utility network categories are a method in ArcGIS Pro to group or tag a collection of network components or features at the asset group and asset type levels. The utility network in ArcGIS Pro is capable of providing network categories, such as subnetwork controllers, taps, or even in an attribute table. These network categories are required for performing certain operations in the network structure established in ArcGIS Pro. For instance, the subnetwork controller network category must be assigned to certain device or junction components or features at the asset group and asset type levels in the main utility network system before the component is set as a subnetwork controller.

Multiple network categories can be added to the utility network and set on the domain network established with it at the asset group and asset type levels. One of the major advantages of using network categories is to enable tracing operations as part of the terminator configuration or as part of the output filter that is used in the entire utility network.

**7.** Create and assign utility network attributes.
Since topological rules are established in the utility network when it is created, network attributes added to the network can be authorized to take the same topological rules. This procedure can be accomplished using the "Add Network Attribute" tool for a specific field and the "Set Network Attribute" tool for one entire utility network class. Utility network attributes stored in the network topology class can be used during tracing operations as well. For instance, in the tracing process, network attribute filters can be used to define the traversability of the trace and filter results. To stop the traversal in a tracing process, utility network attributes can also be used as terminators.

**8.** Assign association roles in a utility network.
Association role of a utility network component is identified into two types. These are structures or containers on a utility network. A structural attachment is a component that flows or transports the materials or signals on the network. While a container is a material or signal storage location for other network features, one utility network component may only serve one of these roles. Association roles are assigned using the "Set Association Role" tool. Setting the association roles of container or structure is limited to component classes and their attribute tables that can support these types of relationships per the utility network feature restrictions.

**9.** Set edge connectivity policies in a utility network.
Edge connectivity policies can be set on network line components or feature classes and the attribute tables of edges on the utility network using the "Set Edge Connectivity" tool. Setting the edge connectivity at the asset group and asset type levels of the utility network allows the snapping rules in the editing environment of ArcGIS Pro to behave differently depending on the connectivity policy that the users would like to enforce. This also enables detail controls for midspan or internal connectivity on edge components and their attribute tables. The Edge Connectivity parameter in the "Set Edge Connectivity" tool is used to control the location on the line or edge that point features or junction objects can connect to it, for example, AnyVertex or EndVertex to split the flows.

**10.** Add tiers and set the subnetwork definition on a utility network.
One or more tiers can be added to the domain network that comes with a utility network when it is established to form a hierarchy or collection of individual partitioned subnetworks. Each utility subnetwork consists of one or more devices representing the subnetwork controller, junctions, and lines for the materials flow and signal transportation. Prior to creating the tiered utility network, the components that serve as subnetwork feature class must be configured. This includes configuring the Asset group and Asset type attributes, setting up terminal configurations, etc. The tier on a utility network is to divide a domain network into logical groupings of subnetworks. During the creation of a tier, basic properties are defined, such as domain network, tier group name, and topology type (partitioned domain networks only). After the tier is created, properties must be defined against it using the "Set Subnetwork Definition" tool. This process will help the users to set up valid subnetwork controllers, devices, and line features for the entire network system.

GIS analysts in utility network studies and management consider all the features of a supply network of goods and services, such as pipeline, electricity transportation lines, or water pumping stations, as network assets or commodities. The purpose and main goal for conducting utility network analysis is to accurately model real-world assets for management and maintenance. Toward this end, the GIS digital utility network model must be capable of being used to manage real-world utility networks. This is the reason why GIS experts working at utility networks have been spending a lot of energy to build their own database structures that deviate from the conventional vector GIS approaches.

Implementation method of separating structure and domain networks in utility network enables GIS analysts to fit a variety of real-world applications, such as fresh water, oil, or electricity into one structural model in a geographic area both logically and in a regional view for simulating individual resource delivery purposes. The focused task in this process is to change the domain network. The utility network data model is designed to be highly scalable in ArcGIS Pro by limiting the number of base feature classes and incorporating a rich classification method that is built into the schema of each component or feature class. All of the base feature classes for the utility network contain two attributes, asset group (ASSETGROUP) and asset type (ASSETTYPE) that incorporate the use of subtypes and attribute domains assigned at the asset group level. The ASSETGROUP field is used for major classification, while the ASSETTYPE field is used for minor classification of utility network objects (ESRI 2022).

## Domain Network Imbedded in the Utility Network Structure

In the utility supply and management industries, we encounter many totally different natures in delivering the resources to the people, such as electricity, natural gas, or fresh water. The common characteristic is that we all use the utility network systems to deliver these goods or resources. However, both the nature of the goods and the commodities or asset types that we use to deliver them are quite different. Therefore, it is very important to identify the domain network from the structure network for utility network management. Meanwhile, redundancy of these utility networks in a specific geographic space at a specific region exists in close proximity in the real world. Therefore, it makes perfect sense for local governments, such as a county or a municipality, to use one digital structural network model in GIS to manage a variety of resource supplies, such as fresh water or natural gas, and to be delivered to citizens. That means one structural utility network can host several domain networks in practice.

In ArcGIS Pro, when a utility network structure is established, a default domain network is also created with it. In essence, domain network represents a specific use of the network system. For example, the water utility network can contain the base structure network with fresh water or sewer water domains. After the utility network structure is established, domain networks can be created using the "Add Domain Network" tool in the "Utility Network Package Tools" box. If business needs or the purpose of deliveries changes, domain networks can be added or changed to an established utility network at any time. Multiple delivery tasks can be simulated on the same utility network system in a region. For example, an additional domain network can easily be added to an existing utility network of a municipality that manages household water supply to be expanded to a natural gas supply domain.

Different utility networks in the real world are interdependent on each other. The utility network modeling in GIS enables the real-world situation to come to life for simulations by allowing capabilities such as connectivity between features from different domain networks running on the same utility network system. The newly developed digital simulation model now is called the digital twin. For instance, we can have an electric domain network that feeds the electricity to a water pump on a fresh water supply domain on the network. The flow rate of electricity domain network we obtained then can be utilized to illustrate and predict the freshwater consumption on the water supply network in the same region.

## Utility Structure Network

The utility structure network is shared among all domain network databases within the utility network. Conceptually, utility network components or features within the structure network do not have any resources such as water or electricity flowing through them but serve to support other network features that are part of delivering a commodity. For example, a vault of utility network, which is considered a structural feature, can contain valves and regulators that are features of a domain network database.

The structure network allows users to expand the foundation of digital structural model into a wide range of real-world practices and services such as network inspections, or business intelligence deliverables about all aspects of the utility network in industries. In essence, the goal of this approach is to build a digital twin to simulate the real-world activities of utility supply management. In this case, the structure network promotes nonnetworked objects on the ground into network analysis and management. Items like poles of the electric network, pads, ducts, trenches, stations, and casings are now all in the network management scheme and can be projected in 3D digital space, such as ArcGIS Pro local scene with their z-values. The key objective of a structure network is to support the association of structural attachments to the established utility network. This is the most efficient way to identify which domain network devices or transportation lines (edges) are attached to a structure feature, such as a pole of electric power lines that may cause a problem to another domain network.

### Hierarchical Classification of Utility Network System by Tiers

In the real-world situation, utility systems are often organized into operational groups for the delivery of a utility resource, such as fresh water, natural gas, or electricity. In the case of electricity transportation, the system is organized into transmission with high voltages, distribution with medium voltages, and local distribution with low voltages. All these are subsystems of the entire electric transportation system that we can call tiers. Natural gas and household water supply systems have tiers too. These are the subsystems for the gathering of natural gas or fresh water, transporting of natural gas or water over long distances, and distribution of natural gas or fresh water to the consumers in their residences or locations. Then, in GIS utility network design, domain networks within the utility network can be divided into tiers to segregate and manage all these subnetworks. Network model of tiers in a hierarchical rank reflects how the utility network delivers a particular resource, such as electricity or fresh water and can also represent characteristic or attribute ranges of a particular resource during the delivery process such as water pressure, electricity voltage, etc.

Real-world utility networks are complex systems. Tier structured utility network modeling in GIS accommodates detail spatial database setup for simulations and digital twin model for day-to-day management. For example, an electricity distribution system can be divided into multiple tiers such as sub-transmission, medium-voltage, and low-voltage levels of subnetwork systems. Tiers as parts of the entire network can be isolated from one another in digital twin situation or real-world situation, such as a valve isolation zone in a pressure-based system of fluid or water medium transportation, to avoid destruction of the entire system in an emergency. Tiered system can also conduct simulation to answer the "what if?" questions before disaster occurs. Tier and domain network concepts accommodate and enforce all unique rules and create templates on the utility network for responsive and appropriate integration of all subsystems. Accommodating rich rules and procedures for each of the commodities to be transported in a region, a comprehensive utility network structure developed in GIS can remain in place when users change distribution and transmission scheme in the system. This strategy also supports communicating or tracing across domains, which is a powerful function for the industries and users to quickly find the issues and formulate best solutions on the network system.

Currently, the tier system in the utility network is defined into two different types in ArcGIS Pro. These two types are partitioned successive tiers or hierarchical nested tiers. Tier definition in the domain network control determines relationships between that network tier to the rest of tiers in a domain network and decides the topology types, such as connectivity that tier use.

Definition of a partitioned tier – Partitioned tiers of subnetwork are sequential subnetwork systems, mainly used in electrical and telecommunication utilities. Partitioned domain networks support topology types of both mesh and radial tires for subnetworks. In partition systems, tiers reside inside of a domain network of the main utility network structure.

Definition of a hierarchical tier – Hierarchical network systems allow the users to model subnetworks that are nested, such as natural gas or household water supply networks. Hierarchical domain networks only support mesh tier topology type for subnetworks. Hierarchical domain networks require tiers to be contained within tier groups.

Each of the tiers can be further partitioned to a group of subnetworks. These subnetworks are a subset of a tier that represents a group of network components that are connected to the same subnetwork controller. Transportation load-based controllers or liquid flow controllers can be used to create subnetworks or circuits across domains. For electric utility networks, a controller could be a circuit breaker or transformer in the network. The advantage of this design is that the management crew of the network system can easily identify, isolate, and sectionalize a particular problem area to reduce the impact to minimum for the customers of the utility network. In the case of natural gas pipeline network systems, we can use pressure valves or regulator stations to isolate all the downstream network assets within the subnetwork and safeguard operating conditions of other areas on the network. Finally, GIS analysts managing the networks can apply the containment rules of topology in a mapping platform to design and assemble different subnetwork systems in modeling the real-world situations. The purpose of all these approaches is to design and implement a better foundation for practical modeling, analyzing, and simulating a real-world utility system to establish a digital twin model.

In essence, utility network systems either natural or human-made have been in human-living environment for a long time. Many of these systems, such as electricity transportation lines, are important daily resource supply facilities and are basic necessities to our society. GIS, among many different digital and computational software, is the most suitable tool to be used to model, manage, and analyze these systems. To build a utility network data structure in GIS is to recreate the real-world system or it is called to create the digital twin of the real-world system. A utility network system in GIS can simulate the different real-world scenarios in supporting the requirements to deliver resources to customers safely, reliably, economically, and on time. GIS-enabled utility network can also serve as the structure in database to record the assets of a utility and provide information about its past, present, and planned future operations. GIS utility network can also be browsed at different computational environments where it is needed, from mobile devices in the field to the desktop computers in operation control centers.

GIS-based utility network management is the most effective and natural way to represent a real-world network system and its assets:

1. GIS makes it more effective for customer service, field collection of data and inspection of devices and assets for utility network operations with its visualization and digital thematic mapping capabilities.
2. GIS platform can create network diagrams of the utility network that allow the users to readily check connectivity of devices and transportation lines. This will help with the logical synthesizing of the real-world network system.
3. GIS platform can help to implement complex assemblies of devices and lines in a network system applying topological rules.
4. GIS can facilitate visualizing a selected pressure zone or electric circuit as a tier-like operation subsystem.

GIS-enabled utility network model provides a large variety of analytic workflows in real-world operations, such as the tracing functions on the network:

1. After a disaster event such as a Hurricane storm, GIS utility network data structure can help to perform network inspections.
2. Accurately calculate the capacity of the resource delivery of the network to the customers in the service region. For instance, GIS analysts can create a load summary and compare the

total transportation capacity and total customer demands for a specific utility network, such as fresh water supply or electricity supply (Tang and Keyser 2007).

3. Conduct tracing on the network upstream or downstream from a given location. For example, GIS network model can help the water utility management to determine which valves should be shut off when a pipeline is broken in order to prevent water resource loss and keep other portions of the network running.
4. GIS-enabled utility network model can accommodate multiple utility systems under the same overall domain in a management region. Tracing within one utility network can help the GIS analyst to identify problems across other network systems. For instance, tracing an outage on the electrical network can also yield the findings of issues affect the delivery of another resource, such as gas or water. GIS network data structure provides GIS analysts with more power to diagnose the locations where the problems are caused and quickly formulate solutions.

Designing and implementing a utility network in GIS can ensure quick information delivery within a large dataset. Converting original feature class-based data to the network data models maximizes the operational efficiency by eliminating operational involvement of feature classes and reducing slow queries of feature relationships. Spatial data attribute domains and subtypes are utilized to classify each of the utility network feature classes. This approach allows for detail modeling down to the single utility feature level. GIS-enabled utility network can also rapidly traverse network features that are connected or attached to structural features.

GIS-enabled utility network models can be assembled by a set of subnetworks that a particular resource flows through at a given time. Each of these subnetworks is a dynamic unit and can be altered by the opening and closing of switch devices and valves. In a GIS-enabled utility network model, we define the start points as the sources of a subnetwork and the end points as sinks. The entire subnetwork can be traced downstream or upstream of the flow. In the GIS simulation environment, when a switch or valve is changed, the extent of a subnetwork is recalculated. Categories and endurabilities of network devices are also defined according to real-world engineering design to ensure protective measures to the facility in detecting situations such as overloading current or excessive pressure that can potentially threaten the subnetwork system during the digital twin operations and simulations.

Utility systems in the real-world are complex and dynamic. GIS-enabled utility network models provide a variety of methods to visualize and explore the utility assets and connectivity of assets in real-world situations. The map view of utility network in ArcGIS Pro shows the overall picture of the real-world utility network in a region. GIS analysts can set the visibility by map scales and layers to show progressively more network features as we zoom into the mapping area of the utility network system. Utility networks in GIS can also be visualized and analyzed both in the map view, which presents the regional distributions of the pipelines or electricity transportation lines, and in the diagram view, which shows a schematic representation of the utility features and connections. In ArcGIS Pro, utility network analysis is a totally different approach than that of the general or fully functioning network analysis. Utility network by its name only models and simulates one-directional movement of resources, goods, or services.

A real-world water main pipeline study for household water supply and demand applying the utility network model is presented here. A team of students and faculty at the SUNY-Buffalo State University conducted a research project to study the fresh water supply for the residents who live in the towns of Amherst and Clarence, Erie County, NY (Tang and Keyser 2007). This project analyzes the impact of the household demand on water supply and water transportation capacity by the infrastructure of pipeline networks in the two towns. The combined geographic area of the two towns is 106.7 square miles (Figure 10.2). The population density of the Town of Amherst is 2190/mile$^2$ and that of the Town of Clarence is 488/mile$^2$ during the study time.

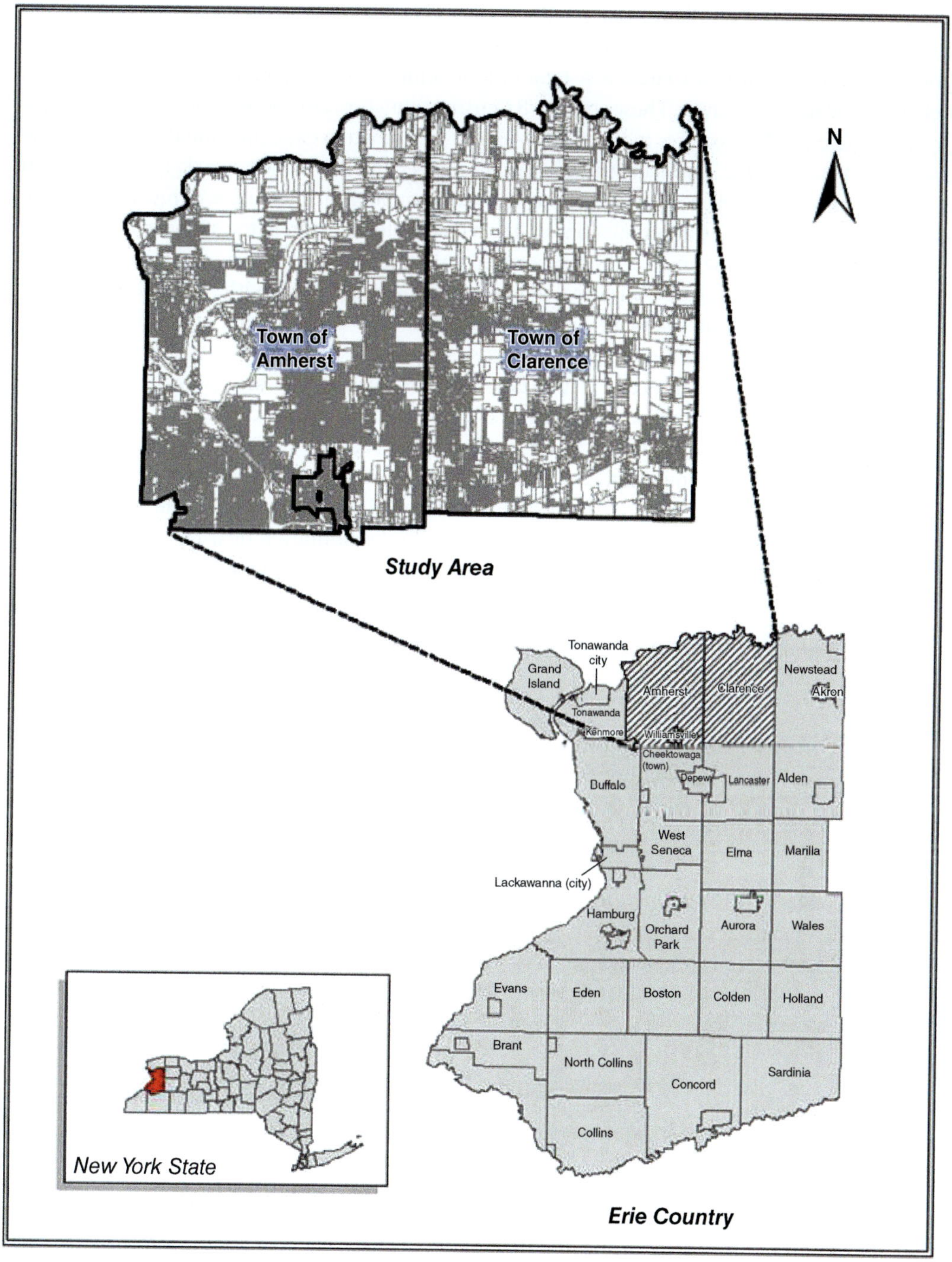


**Figure 10.2** The study area of household water supply and demand using utility network model.

There are a total of 493 miles of water main pipelines and two elevated water storage tanks in the Town of Amherst. There are a total of 211 miles of water main pipelines and one elevated water storage tank in the Town of Clarence. A total of four comprehensive spatial or nonspatial datasets were gathered for GIS analysis. These are: (i) US Census population distribution data by block groups; (ii) water main pipeline features in the two towns and pipe junction data; (iii) boundary files of land ownership parcel for the Towns of Amherst and Clarence; and (iv) the average daily consumption of household water supply per person per day derived from published literature of previous studies. This is used to calculate the accumulations of household water demand in the study area.

Utility network module in ArcGIS was applied to simulate water supply capacity and water demand quantity accumulation with real-world water main pipeline width. The two utility network models, namely "water demand and accumulation" and "water supply capacity" were overlayed to identify the possible discrepancy locations.

The results indicate that the current utilization of the capacity of water supply is 51% at the entrance point or source of this pipeline network. The maximum supply capacity is approximately 29 million gallons per day according to the main transportation pipe conditions. The total demand of household water from the communities in the two towns is 15.1 million gallons per day (Figure 10.3).

The simulation of utility network model shows there is insufficient household water supply in the Southeast part of the Town of Amherst owing to the limitation of trunk pipeline width (Figure 10.4). Insufficient water supply capacities also exist in Northwest (Figure 10.5) and

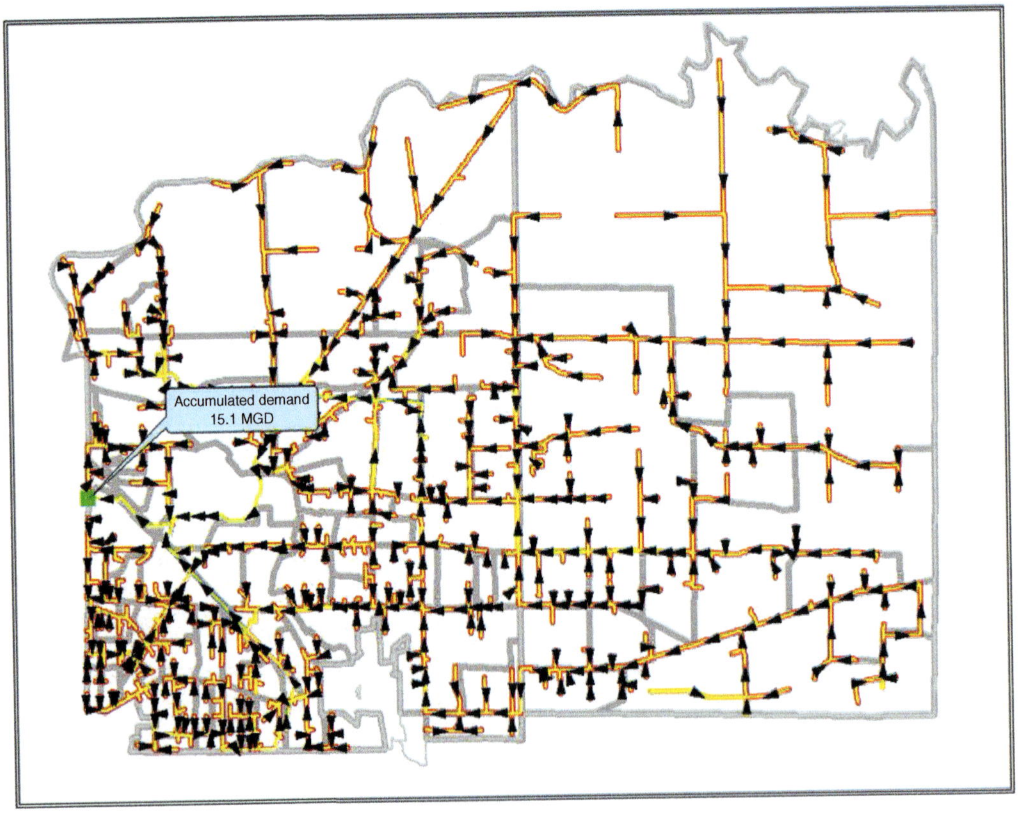


**Figure 10.3** Utility network model for the household water demand accumulation in the towns of Amherst and Clarence.

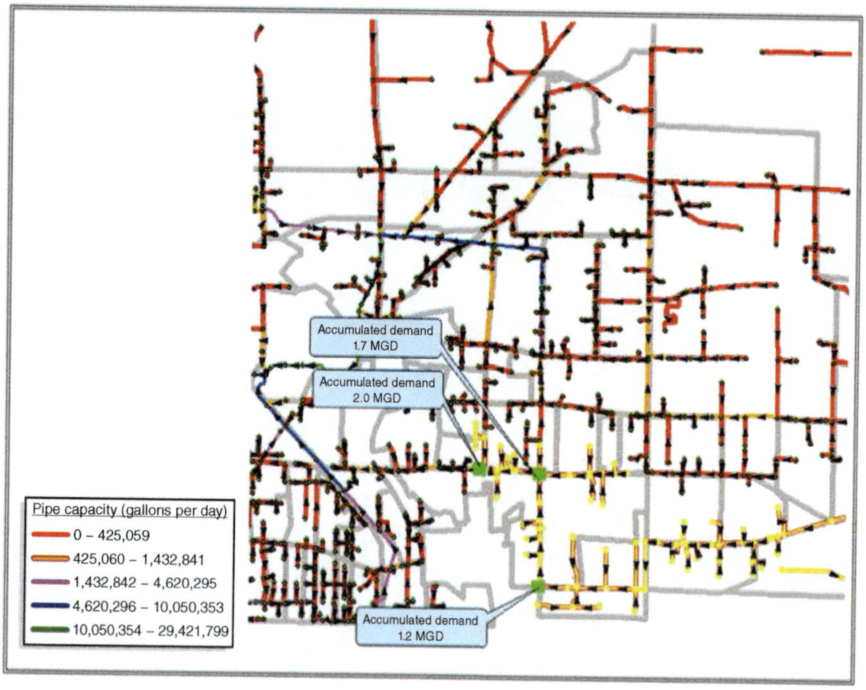


**Figure 10.4** Utility network model shows the accumulated water supply demand (2.0 million gallons per day) larger than supply (~0.5 million gallons per day) in Southeast Amherst.

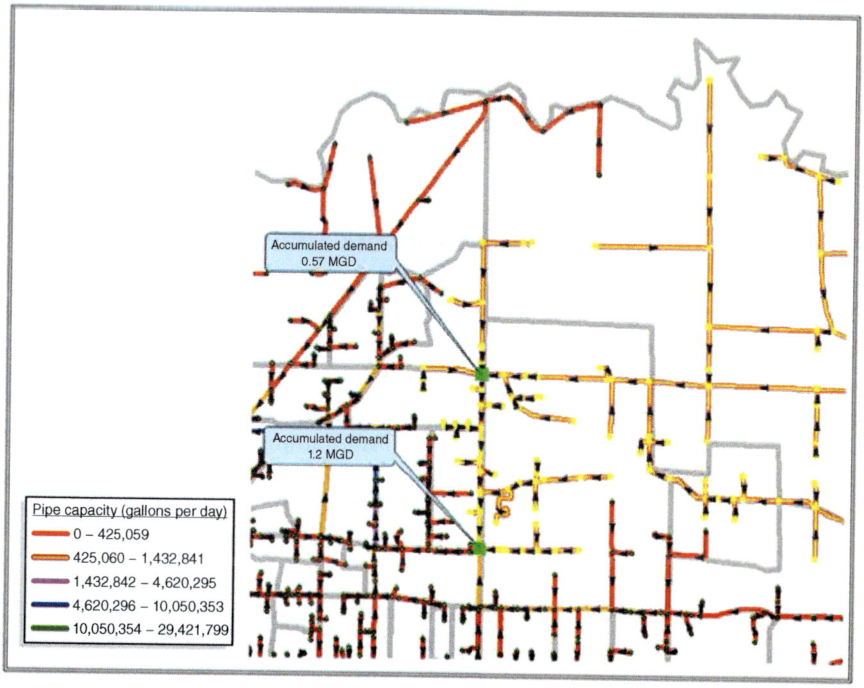


**Figure 10.5** Utility network model shows the accumulated water supply demand (1.2 million gallons per day) larger than supply (~0.85 million gallons per day) in Northwest Clarence.

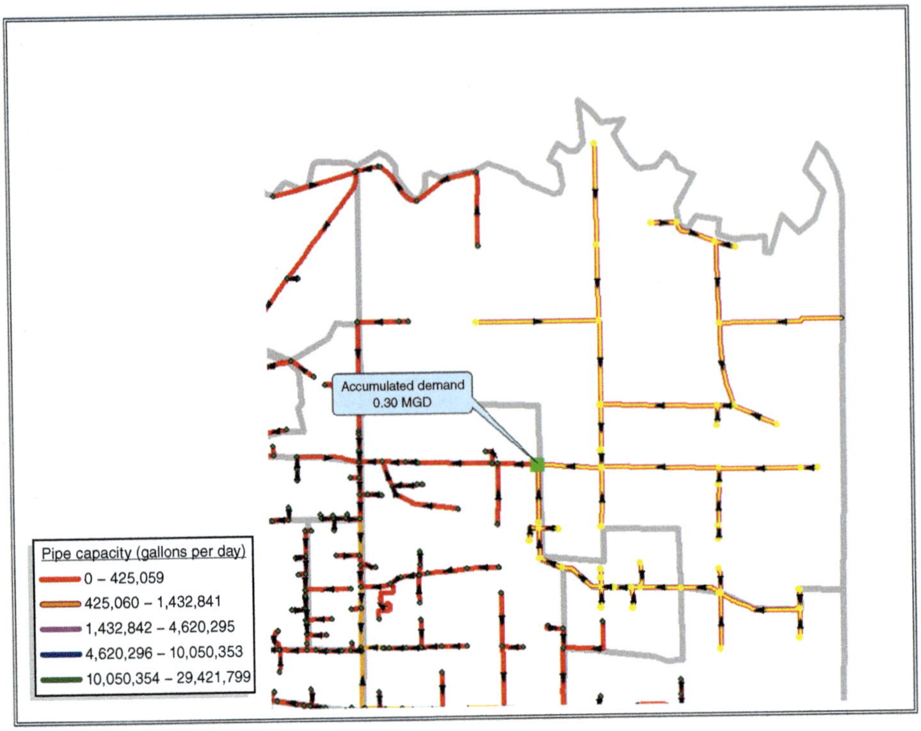


**Figure 10.6** Utility network model shows the accumulated water supply demand (0.3 million gallons per day) larger than supply (~0.13 million gallons per day) in Northeast Clarence.

Southwest (Figure 10.6) parts of the Town of Clarence. These water supply discrepancies might be induced by new residential area constructions and water supply pipeline extensions. The results of this project demonstrate that GIS analysis using the utility network is very effective to predict goods and service distributions and accumulation.

## Fully Functioning or Transportation Network Analysis

Network analysis in GIS indicates operations that deal with two-dimensional or multidimensional movements of goods and services on a structured network. This type of network system is also called a transportation network or a fully functioning network. Network systems have long been used to solve problems in transportation industries or studies. Therefore, the network analysis in GIS generally developed for transportation applications. A strong subdiscipline called transportation GIS or GIS-T has been developed. In this chapter, we will focus on discussion of network analysis for transportation and some issues beyond transportation analysis.

Fundamentally, a network is an interconnected set of points or nodes and lines or edges. In the structure model of GIS network, three kinds of feature elements can be identified: (i) Edges are the links that connect to other elements, such as junctions and over which agents travel; (ii) junctions are the point locations that connect edges and facilitate navigation from one edge to another; and (iii) turns are stored real-world information that impacts agent movement between two or more edges.

Network Analyst extension in ArcGIS Pro is the software platform to process and analyze the network data and to build fully functioning network models. ArcGIS Pro Network Analyst can effectively simulate the movement of goods, logically organize and coordinate the vehicles on the network model. Therefore, the software extension can conduct intelligent transportation network analysis. Network analysis in GIS can help GIS analysts to develop strategic routing plans and make intelligent decisions for the shortest routes as well as other operational needs. Based on the structural network model established in GIS, users can create and implement daily transportation route plans with daily customer data feeding into the system. This will help the users to solve vehicle routing problems and perform post-route spatial network analysis. Approaches of GIS-enabled network analysis not only save time and money but also increase efficiency for an organization or company by integrating the Network Analyst extension and the established structural network model with existing workflows.

When the structural model of a network is created in GIS, the topological rule of connectivity is stored between the source features and a series of connected edges and junctions. For example, in a street network dataset created using the Network Analyst extension, an intersection with traffic light is stored both in the street line features or edges and point features where these lines intersect that is called junctions. Because the connectivity rules and structural model are established, we can perform different routing scenarios through the entire network or part of it.

Street or highway networks, road intersections or junctions are identified and digitally coded in the network database in transportation design and implementation engineering fields (Figure 10.7). For instance, the following is some major standard coding and controlling scheme of intersections: (i) 1-way stop, (ii) 2-way stop, (iii) 3-way stop, (iv) 4-way stop, (v) 4-way warning signal (One street yellow, one street red blinking), (vi) 4-way traffic signal (red, green, yellow), and (vii) Dead end. The following is a scheme of vehicle turning type and intersection coding: (i) Right turn, (ii) Left turn, (iii) U-turn, (iv) Yield, and (v) Go through (right of way).

In a GIS-enabled network system, all the features must be stored in the same feature dataset and coordinate system in a geodatabase. A feature dataset is used to store network topology information, and it groups related features and feature classes together. Network topology is established as an advanced feature class that makes data interactions in the network possible. Network topology in a network feature dataset also regulates what is acceptable behavior given the inherent rules of each network type. For instance, tracing is one relevant advanced feature that is available with

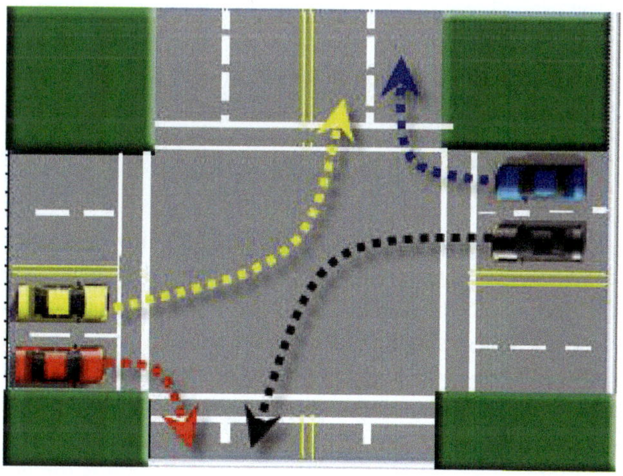

**Figure 10.7** Vehicle turning at an intersection (Drivers 2022)

an established network system. Tracing on a network system can be conducted either through geoprocessing tools that are in the geoprocessing toolbox or accomplished by using the "Network Analysis Wizard" in the Analysis ribbon. The functionality of tracing process returns the shortest route according to existing connectivity and traversability of network structural model. Typically, tracing operation involves at least one starting or origin point and one ending or destination point identified on the network. Users can also input barriers that prevent the route planned to be going through a certain area of the network.

In network analysis, impedance or cost is accounted for in each of the network features or components. Impedance is the amount of resistance that a component offers to current flow or transported goods on the network. Impedance can be further identified on the transportation network as edge impedance and junction impedance. Edge impedance is the cost of traversing the network edge segments, and junction impedance is the cost of traversing a junction on the network. Junction impedance is normally quantified by the results of the studies of vehicle traffic turns in transportation engineering works. Certain attributes in the GIS database can be used in measuring or modeling impedances, such as travel or transit time on a street or changing time of traffic signal light, etc. These are also called the costs.

In order to perform network analyses, we need to create a network dataset in GIS first. The Create Network Dataset tool or function in the Network Analyst toolbox in ArcGIS Pro can be used to create a new network dataset model in an existing file geodatabase feature dataset. This dataset should contain the line feature class, such as "Streets" or "Highways." In this geoprocessing, feature class of "Streets" or "Highways" is used as the source feature class to build the network structure. The topological rule of connectivity is enforced in this process. Creation of connectivity in a network dataset is mainly based on geometric interactions of line endpoints or nodes, line vertices, and points in the original feature class. Therefore, network systems are established in GIS using linear features and feature classes, but a network system is a separate entity from the original digitized or input linear features (ESRI 2022).

Based on the network datasets created in GIS, we can conduct a variety of network analysis. The Network Analyst extension in ArcGIS Pro practically facilitates GIS users to conduct the analyses from finding the shortest paths and planning the drive time, to identifying closest facilities, to choosing the best location for a specific facility, and to finding the best routes for a fleet of vehicles. In order to solve each of these comprehensive issues and generate both graphic and tabular results, Network Analyst extension in ArcGIS Pro creates a network analysis layer for each of the specific tasks.

A total of six major comprehensive tasks or issues of network analysis in GIS are identified.

### Routing or Point-to-point Analysis

This task is to find the best path for a traveling agent or transported object to go through from one location to another or to visit several locations. If the task contains more than two stops in this visit of locations on the network, the GIS analysts can determine the order of specified locations to obtain the best route. Alternatively, Network Analyst extension in ArcGIS Pro can determine the best sequence to visit the locations using the traveling salesperson problem (TSP).

The best route is defined by the specific requirements and environmental conditions. The best route can be the quickest, shortest, or most scenic route, depending on the choice of objectives and conditions of impedance variable. For instance, if the impedance is time, then the best route is the fastest traveling route. By contrast, if the impedance variable is the cost of pay toll, then local roads are a better route than toll-paid interstate highways. In general, the best route can be

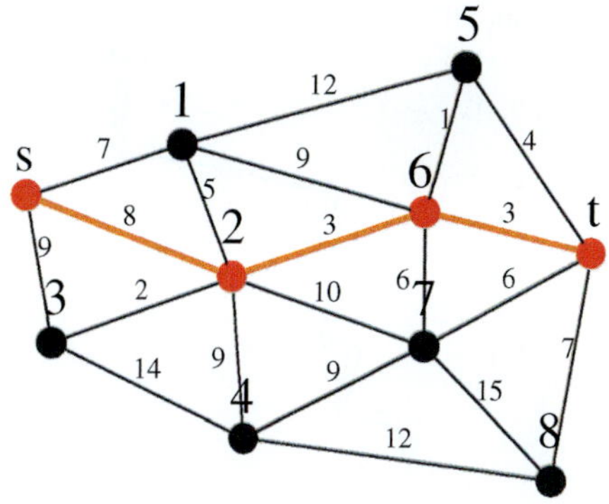


**Figure 10.8** The shortest traveling route from location S to location T on a network of routes.

defined as the route that has the lowest impedance toward the objective to achieve. Figure 10.8 is a diagram that shows the computation of the minimum network cost traveling from location S to location T.

TSP is the computational algorithm that seeks optimum route to answer "for a given list of cities and the distances between each pair of cities presented on the network map, which route is the shortest possible that the salesperson visits each city exactly once and returns to the origin city?"

## Closest Facility

The task of finding the closest facility is to measure all the impedances of traveling between locations of inquirers and facilities to determine which one is the nearest to one another. GIS analysts can also specify the maximum facilities to find and whether the direction of traveling is toward or away from the facilities nearby in this approach. The returning results should be presented on the network map, and the layer displays the best routes between inquirers and facilities. Meanwhile, the report should also show travel costs and driving directions. Examples of closest facility problem applications are finding the closest hospital to an accident, the closest police cars to a crime scene, and the closest store to a customer's home address.

In ArcGIS Pro, users can specify constraints or cutoff threshold if the cost or impedance is beyond which the search is no longer needed anymore. For instance, we can set up the task of search closest facility of hospitals from the site of an accident as within a 15-minute driving distance. Any hospitals that take longer than 15 minutes to reach are not included in the results. In this example, the hospitals are facilities, and the accident is the inquirer. Different than a Google "Close by" search, this process in GIS normally analyzes multiple closest facilities and incidents simultaneously to generate a digital map (Figure 10.9). Therefore, the application of closest facility function can be conducted in a project that searches for the closest subway stations for all the residents who live within one mile buffer zone along the subway system (ESRI 2022).

## Service Area or Location Analysis

The task of creating service areas is to help evaluate accessibility of various facilities and services by a targeted or a home location. The results of service areas should also show how accessibility varies with impedance values as well. Once service areas are created, we can use these regions to analyze how much land, how many people, or how much of anything else is within each of the service regions across the study area. In this type of analysis, GIS analysts can also select whether the direction of traveling is toward or away from the targeted locations. GIS Business location analysis utilizes this function to judge if the targeted retail store location is good or not.

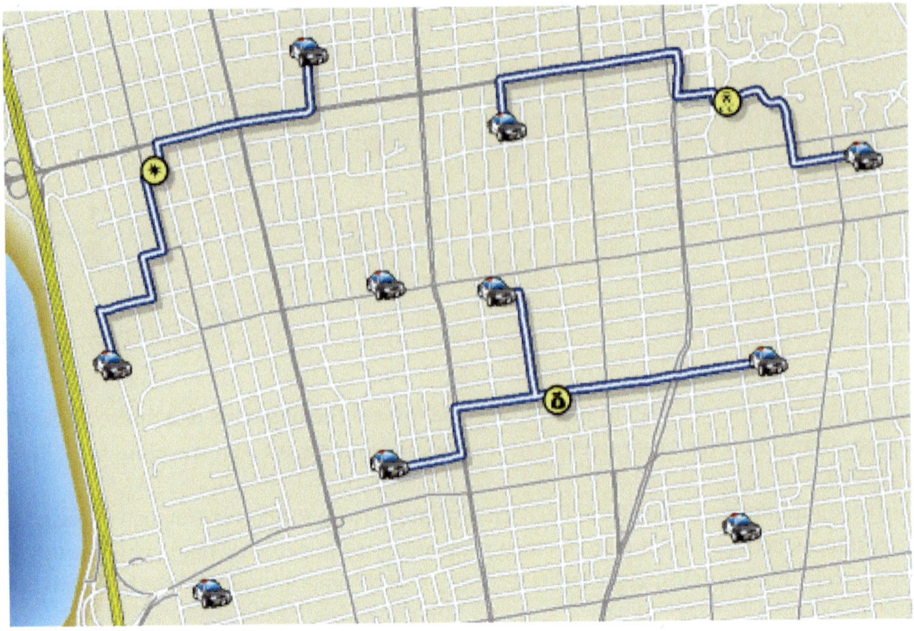

**Figure 10.9** Digital map of the closest facilities to the selected locations. *Source:* ESRI 2022/ Environmental Systems Research Institute, Inc.

Accessibility measures how easy it is to go to a service site from a targeted or a home location. Accessibility evaluation helps answer basic questions, such as, "How many people live within a 15-minute drive radius from a movie theater?" or "How many customers live within a one-mile walking distance from a convenience store?" Multiple irregular rings within service areas show how accessibility changes with an increase in impedance. For example, we can search and show how many hospitals there are within 5-, 10-, and 15-minute drive times of schools (Figure 10.10).

In real-world practice, service areas are actually drive-time areas that correspond to the distance. Drive-distance is impacted by a specific amount of time as well as other criteria. People may have different thoughts on geographic distance or time to travel on different purposes. For instance, people living in an area may not be willing to drive 30 minutes each way to buy groceries weekly. In contrast, the same people may think it is fine for them to drive more than 30 minutes each way going to either a favorite restaurant or a shopping mall. In the modern human built living environment, people are more concerned about the physical impedance and transportation time as major factors on daily activities. Examples in real-world applications for service area analysis with the direction toward the facilities are hospitals, shopping centers, COVID-19 virus infection testing centers, or government motor vehicle registration centers. Examples for those analyses that the direction is away from the facilities are delivery restaurants, such as Pizza Hut, shipping centers or hubs of FedEx or UPS, etc.

One of the most popular applications of the service area analysis is for the study and assessment of new business location sites. Identifying the service areas of currently operating business locations in the same category of services or goods supply is the first step to find the best new business location in the study area (Figure 10.11). Generally speaking, geographic service area analysis needs to be conducted first prior to many other studies for the industrial or economic development in a region.

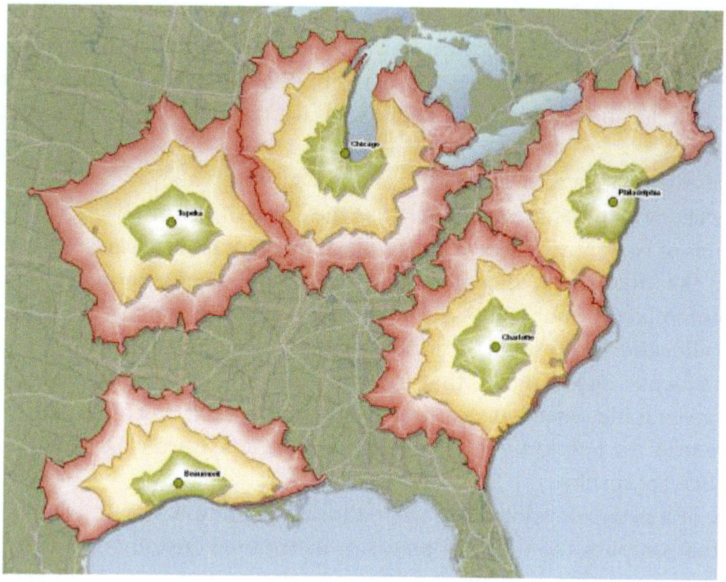

**Figure 10.10** Differentiated driving time map to specific types of service centers, such as hospital or school. *Source:* ESRI 2022/Environmental Systems Research Institute, Inc.

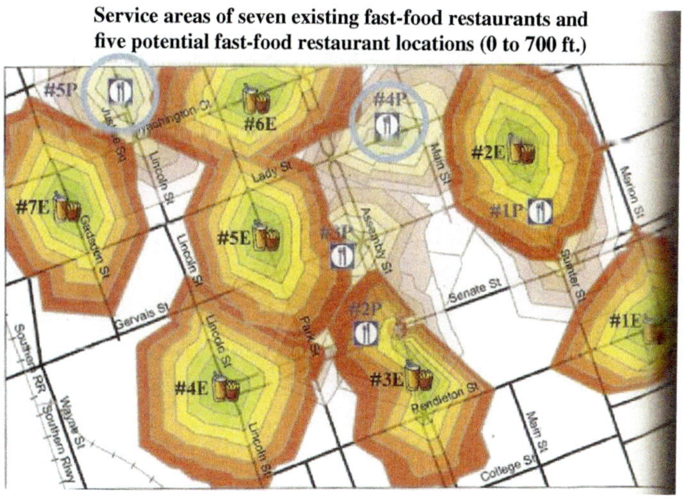


**Figure 10.11** Service area analysis for restaurants. *Source:* Jensen and Jensen, 2007.

## Location-allocation Analysis

Location is one of the fundamental geographic concepts that impacts daily human life. Location analysis is often considered the most important problem solving leading to the success of a private company or a public organization. A private company can profit from a good location with favorable environments, whether it is a small restaurant with mainly local customers or a multinational conglomerate of industries of manufacturing with distribution centers, or a worldwide retail chain of outlets. Meanwhile, when good locations are selected for service oriented public facilities, such as schools, hospitals, libraries, fire stations, and emergency response services (ERS) centers, these facilities can provide high-quality service to the community at a low cost. Therefore, location studies can help us reduce the costs and increase accessibility to goods and services.

Location analysis alone is a more "self-centered" approach simply because those targeted geographic objects or features are established as the focus of the analysis. The importance of interrelationships of two or more sets of geographic objects or feature classes is not emphasized in this case. On the other hand, allocation is not the locations of primary geographic objects, such as distribution facilities of goods, schools, or ERS centers that are involved in the study. The concept of allocation represents the locations of other geographic objects or features, such as customers, students, and patients, that are highly related or that serve as the purpose for those primary geographic objects or features relying on to exist. Location-allocation analysis in essence advanced the network GIS analysis from single feature class centered to multiple feature interactive approaches.

The objective of a newly developed location-allocation analysis in GIS is to locate the facilities in a way that supplies the demand points or distributed customers most efficiently and minimize the cost. In practice, given facilities, which are represented by geographic locations, that provide goods and services and a set of demand points, which are represented by geographic locations of consumers, location-allocation analysis attempts to find the most cost-effective solution to distribute the goods or the services from the facilities to the consumers. Location-allocation, as the name suggests, is a twofold problem that simultaneously locates facilities and allocates demand points to the facilities. Therefore, the focus on resolving this type of issue is neither the facilities nor the demand points, but rather the spatial interrelationship between the two (Figure 10.12). In the business planning and management

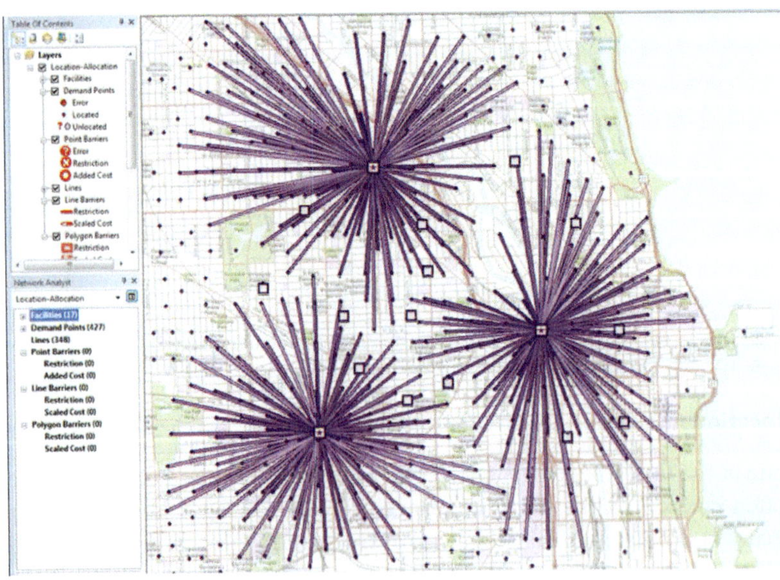


**Figure 10.12** Pipeline presentation map of the distribution centers to customer locations within a particular driving time or distance in a location-allocation analysis. *Source:* ESRI (2022).

perspective, new facility or facilities can be built in the region of study or operations according to the analytical results of location-allocation.

Different facilities that provide different types of goods and services may have different requirements and environmental needs. Therefore, the best locations of different types of facilities or businesses can be quite different. For instance, the emergency response time and ambulance traveling distance are the two major requirements for locating an ERS center. Typically, the ideal situation for locating ERS centers is to make it possible for ambulances to reach the most people within a defined driving time frame, such as 15 minutes. Then, the driving time can be converted to distance on the current GIS network structural model to simulate the reach abilities in the real world. By contrast, when we locate a manufacturing plant, whether it is a production facility for automobiles, appliances, or packaged food, impedance variables, such as time or cost of transportation, could be offset by other factors such as labor cost in an area, or required delivery time by retail outlets in the market region. In many of these cases, distance or transportation time may be considered minor factors for decision-making over many other factors. Location-allocation analysis is the effective method that can answer the question "Where should the manufacturing plant or distribution facilities be located to minimize overall transportation costs or serve the customers in a region faster?"

The following are seven location-allocation problem types:

1. Minimizing impedance or cost
2. Maximizing coverage
3. Maximizing capacitated coverage
4. Minimizing facilities
5. Maximizing attendance or demands
6. Maximizing market share in competition at a market region
7. Target market share in competition at a market region

## Analysis Applying Cost Matrix of Origin-destination

Cost matrix of origin-destination (OD) is the database of various routes on a network structure in a geographic region. The analysis applying cost matrix of OD can find and calculate the least-cost paths on the network from a single or multiple origins to a single or multiple destinations. When a cost matrix of OD is configured for analysis, GIS analysts can define the maximum distance to search and the number of destinations to find. In Figure 10.13, the cost matrix of OD found the least-cost paths from each origin to the four nearest destinations. The output geometric type in this case is straight lines.

Although the cost matrix of OD solver doesn't output line segments on the digital map in GIS that follow the network, the accumulated values stored in the attribute table of these route features reflect real distance on the network, not the straight-line distance on the map. The results of the cost matrix of OD analyses are often used as input for other spatial analyses as well. The network costs of all OD pairs of cost matrix analysis are evaluated in those spatial operations for making decisions. For instance, modeling network loading with network costs in detail can predict the movement of people in a human-built environment, such as people traveling on streets and highways, railroads, or pedestrian paths.

In practice, this type of network analysis produces a large database of the distances and driving times between all the paired locations. In an emergency dispatching situation, the responders can quickly find requested spatial information of paired locations. This approach makes daily emergency response operations much efficient and fast.

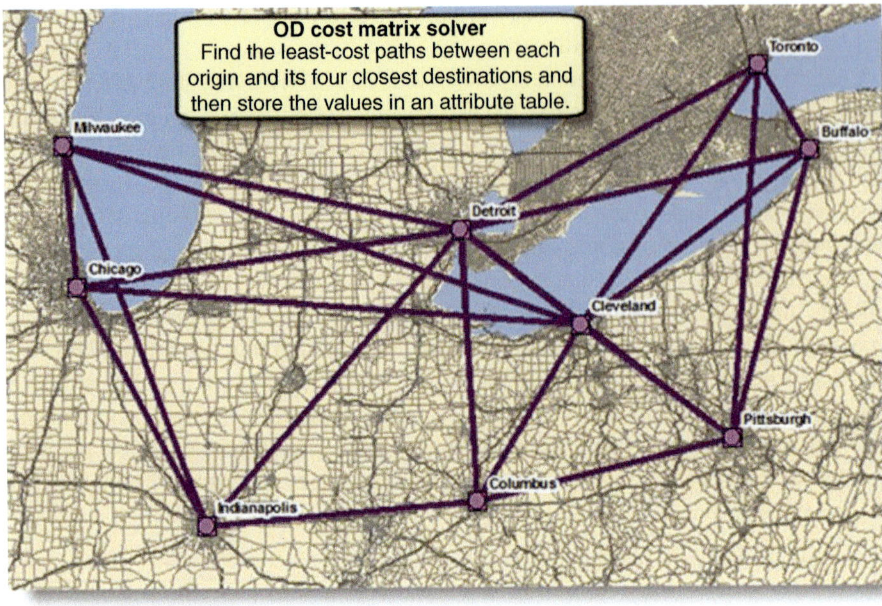


**Figure 10.13** Digital map of cost matrix of OD. *Source:* ESRI (2022).

## Vehicle Routing Problem

The objective of vehicle routing problem (VRP) analysis is to find the best routes for a fleet of vehicles on a GIS enabled transportation network with a specific geographic pattern of customer order or demand locations. The daily work of a transportation fleet management team for delivery vehicles is to plan vehicle routings and make the order to the drivers. In essence, VRP analysis is to obtain the best way for assigning a fleet of vehicles and to sequence and schedule the vehicle visits to a group of customers for goods and service deliveries. VRP analysis can provide a high level of customer service by scheduling delivery or pick-up time windows to the customers while keeping the overall operating and investment costs of operations as low as possible. The challenges of VRP analysis are that GIS analysts need to plan the routes with available resources, such as a limited number of delivery vehicles and within the time limits imposed by driver work shifts, driving speed limits on the road, and time availability of customers to receive the goods and services. (Figure 10.14).

The VRP can be considered a combination of several TSPs that one vehicle is one TSP. Solving a TSP issue is to find the shortest possible route that an individual can take among all the customers at different locations on the network system. In this case, each driver and his or her routing order is one TSP while impacts of other TSPs in the VRP analysis need to be considered as well. Since each of the planned stops on each of TSPs may require the vehicle to stop within a certain time window, many challenging factors from real-world situation, such as the network system traffic condition, and speed limit on different edges of the network, need to be considered as well. In essence, VRP analysis is an interaction analysis of geographic locations, distances, and times.

Figure 10.14 shows the example of delivering goods to grocery stores from a central warehouse location in Western New York. In this case, a fleet of three trucks is available at the warehouse. The warehouse operates only within a certain time window, from 8:00 a.m. to 5:00 p.m. All delivery

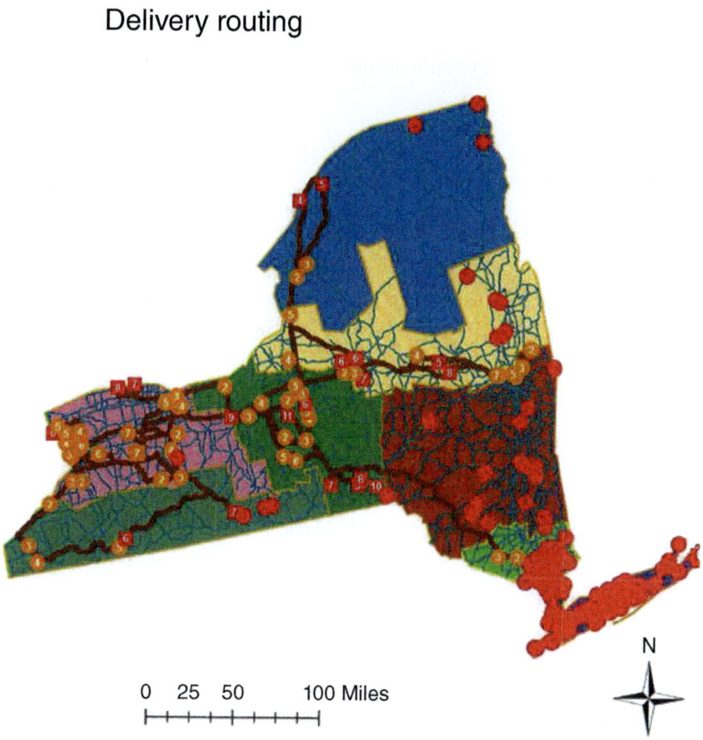


**Figure 10.14** Example of VRP analysis in New York State.

trucks must return to the warehouse during this time period. Each truck has a limited capacity of 15,000 pounds to carry. Each store has a demand for a specific amount of goods in pounds that need to be delivered, and each store has requested a couple of time windows that deliveries should be made. Furthermore, each of the drivers can work only 8 hours per day, requires a break for lunch, and is paid for the delivery quantities and driving time to the stores. The goal is to obtain an itinerary for each of the drivers with the route order of the day, so the deliveries can be made for all the customer requirements. Meanwhile, the total time schedule of a particular route by one driver should be minimized to comply with labor laws. (ESRI 2022).

Applications for GIS-based VRP analysis are very popular in the field of vehicle fleet management. Some examples in the real world are as follows. A large furniture store may need to use several trucks to deliver furniture to homes. A fleet of garbage collection trucks may need to be managed in service for a particular region. A fleet of Amazon delivery vans may need to be routed for a more efficient service to a geographic area.

## Time Dependence of Network Analysis

All the above network analyses deal with the factors of locations, distances, and times. Many of these, such as the VRP, involve in accounting of time or time period. Therefore, the best solution may change as the time and date changes. All these analyses are time dependent. It is always the case that we need to find the best route for a given time of day. For instance, the answers for locating the best place to preposition an ambulance for quick reaction to the emergency situations at 6:00 a.m., 5:00 p.m., or 12:00 a.m. are totally different according to experiences of daily practice. More studies are needed in digital geographic analysis for the interactions of locations, distances, and time.

### Trace Network and Its GIS Database

A trace network is a simple component and connectivity network model for GIS users to visualize, manage, and analyze network type of datasets in ArcGIS Pro. We can consider that a trace network is a simplified version of utility network. A trace network uses a set of connected edges, junctions, or both in conjunction with network attributes in the attribute table for those features to model the flow of resources through the network system.

Feature classes in a feature dataset of a geodatabase in ArcGIS Pro provide the source junctions or edges that the GIS analysts can use to create a trace network. Connectivity within the trace network is based on the geometric coincidence locations (x, y, z) of the features and their data sources. Network topology rules are established when building a trace network. A network topology is required to enable advanced analytic functions such as the ability to set flow direction, run a trace on the network, or generate network diagrams. The "Enable Network Topology" tool is used to create the network topology and establish connectivity in the trace network as well.

## References

Curtin, K.M. (2018). GIS methods and techniques – network analysis. In: *Comprehensive Geographic Information Systems*. Elsevier Publisher.

Drivers, E.D. (2022). *Making Right and Left Turns*. https://driversed.com/driving-information/.

ESRI. (2022). *ArcGIS Pro Help Menu*. https://pro.arcgis.com/en/pro-app/latest/help/.

Jensen, J.R. and Jensen, R.R. (2013). *Introductory Geographic Information Systems*, 400 pages. Pearson Publisher.

Tang, T. and Keyser, G. (2007). Spatial analysis of household water supply and demand in a distributed geographic network in the Towns of Amherst and Clarence, New York. *Middle States Geographer* 40: 133–141.

## Tutorial in ArcGIS Pro

Copy the tutorial data folder "Chapter10-Tutorial" to your working directory.

Open ArcGIS Pro and create a new ArcGIS Pro project file that is called "MyGIS_Learning10" with the "Map" template. Save the project file into the designated working folder, C:\ or D:\...... Chapter10-Tutorial.

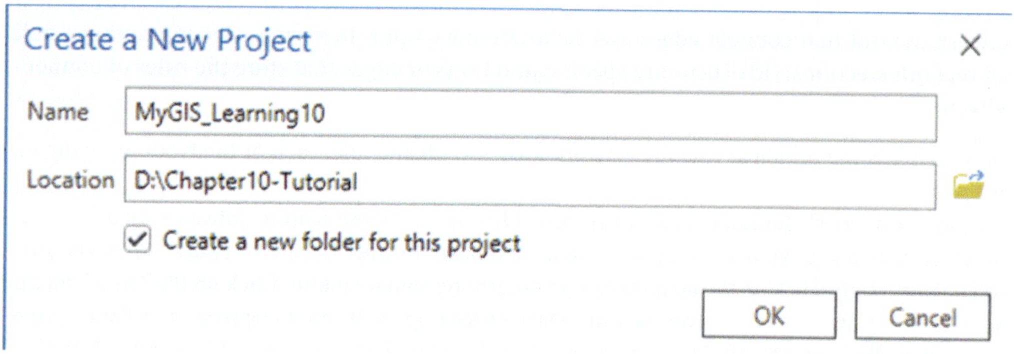


Now, highlight and right-click on the Folders on the Catalog Pane to get the contact menu. Then, click on "Add Folder Connections" to add the entire Chapter10-Tutorial folder on your working drive.

Network analysis in GIS is traditionally categorized into two major approaches: (i) utility network analysis that is directed to one-directional flow or transportations such as water pipelines, oil transportation pipelines, high-voltage electricity transportation lines, and natural rivers; (ii) general network analysis which is directed to two-directional transportation on the network database, such as parcel delivery route planning, traffic analysis, and market competition and suitable location analysis. In the ArcGIS Pro environment, line feature class does not automatically constitute any of the datasets that fulfill each of these two approaches. GIS analysts must either create the network or utility network datasets to facilitate the analysis and approaches or convert the line feature classes to the network or utility network databases in order to conduct the analyses.

1. Creating a network database

Expand the "Urban_Network_Analysis.gdb" geodatabase in the Chapter10-Tutorial folder. This is a geodatabase we prepare for conduct some network analysis using the City of Buffalo as an example. In ArcGIS Pro, line feature classes, such as street centerlines, highways, watermain pipelines, and high-voltage power transportation lines are not necessarily ready to use network dataset the GIS Analysts can use to model the real-world network systems. Creating a network dataset is a necessary first step for various analytical functions, such as creating routes from transportation data, locating facilities or service areas, conducting market area studies, and routing vehicles for delivery. Another way to comprehend the differences between line feature classes and a network dataset is that line feature classes are just collections of linear features to present them in GIS database, while a network dataset is a data model or data structure that the analysts can use it to conduct modeling processes.

A network dataset contains not only the locations and attributes of streets and highways, but also includes information about how they are related or connected to each other. A network system is comprised of network elements. Normally, these elements are created when we convert a source feature class, such as street centerlines, to a network dataset. In ArcGIS Pro, there are mainly three major elements. (i) Edges are connected line segments in a network that are created directly by the source linear features. Edges are connected by junctions and are directional. A network dataset set up for general network analysis always have two-directional traveling mode unless defined in other ways. (ii) Junctions are point features created in a network dataset that connect edges and facilitate navigation from one edge to another, such as street intersections. (iii) Turns are specific junctions or edges that store the rules of connectivity of two or more edges.

Connectivity is the principal rule for network analysis. Meanwhile, each of the types of elements carries some attributes. In particular, the attributes define the impedance or cost for edges and junctions. A network dataset is always imbedded in a geodatabase feature dataset with a line feature class in it. First, we need to create a feature dataset in the "Urban_Network_Analysis.gdb" geodatabase. Right-click on the geodatabase to obtain the contact menu. Click on the "New" menu and "Feature Dataset." The Create Feature Dataset tool opens in the Geoprocessing Pane Name the new feature dataset "City_Transportation." For Coordinate System, import it from the Buffalo_Streets feature class in this geodatabase.

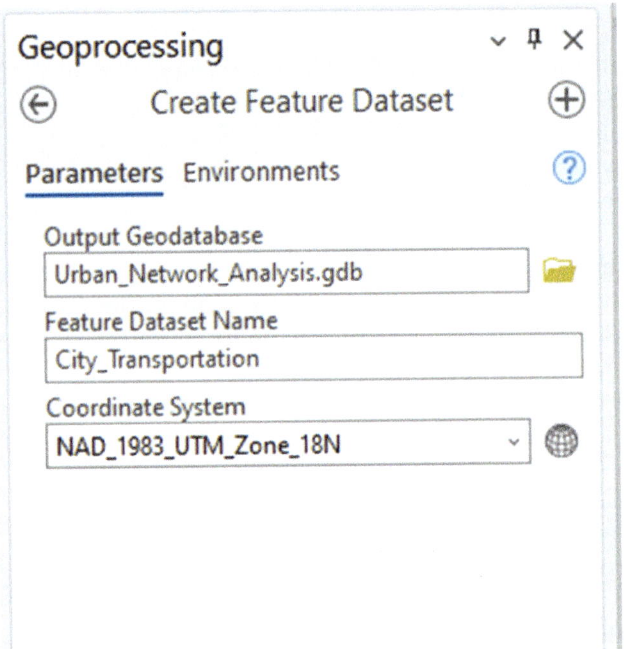


Run the tool to create the feature dataset. Close the Geoprocessing Pane. Drag and move the Buffalo_Streets feature class into the City_Transportation feature dataset.

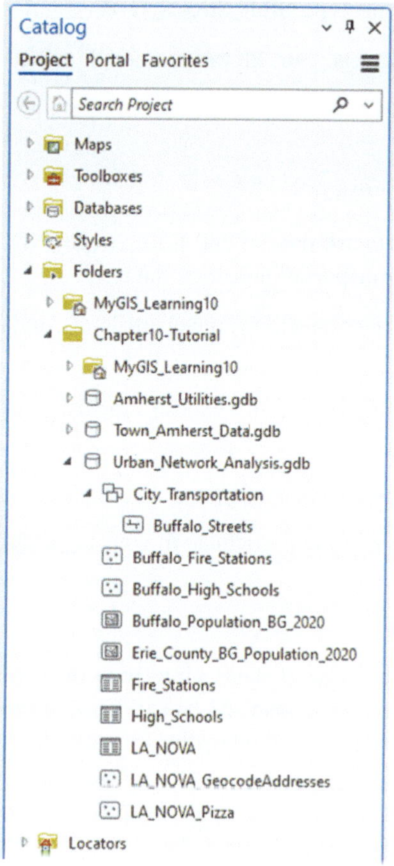


Now, we can create a network dataset in this feature dataset.

Right-click the City_Transportation feature dataset in the geodatabase to obtain the contact menu, point to New, and choose Network Dataset.

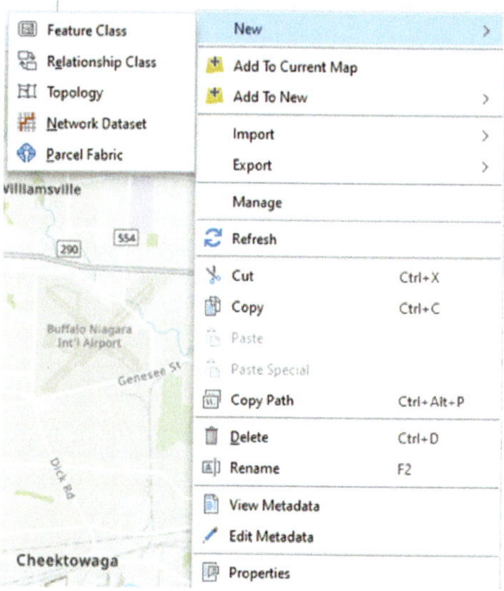

In the Create Network Dataset Pane, for Network Dataset Name, type "Buff_Streets_ND."

For Source Feature Classes, check the box for Buffalo_Streets. For Elevation model, select "Elevation fields."

Click Run.

After the geoprocessing is completed, close the Geoprocessing Pane.

In the Catalog Pane, we notice that the Buff_Streets_ND network dataset and Buff_Streets_ND _Junctions point feature class have been added to the City_Transportation feature dataset.

Next, we will review the network dataset properties and set up connectivity, travel attributes, and directions.

In the Catalog pane, right-click the Buff_Streets_ND network dataset and choose Properties to open the Network Dataset Properties dialog box.

In the Network Dataset Properties dialog box, on the left, click Source Settings.

The Sources tab displays all the source feature classes that currently participate in your network dataset.

Click the Vertical Connectivity tab.

Vertical connectivity has settings that allow us to refine how the network dataset interprets the way that streets are connected. This dataset has elevation values in the fields F_ZLevel and T_ZLevel, so the network dataset will know whether streets intersect or not. We will set up the proper vertical connectivity policies for your network source.

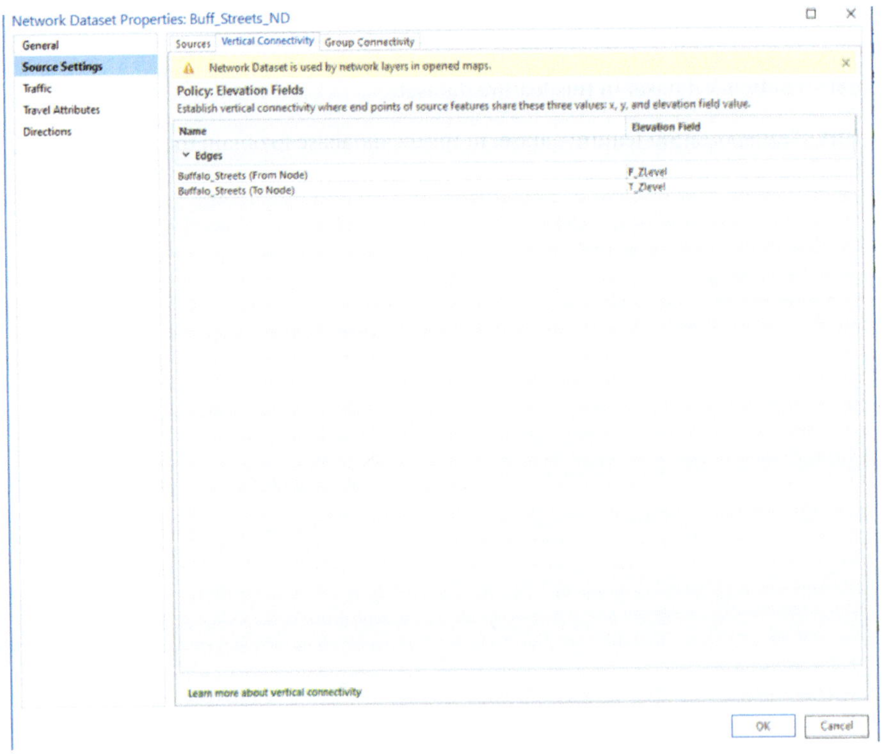

For Buffalo_Streets (From Node), click <None> and, from the drop-down list, choose F_ZLevel

For Buffalo_Streets (To Node), click <None> and choose choose T_ZLevel

On the left, click Travel Attributes.

On the Travel Modes tab, we can set up travel modes to perform analysis for modes such as travel by driving, truck, walking (pedestrian), and so on. Travel modes can be configured to add additional impedances, such as no U-turns and slower top speeds for trucks. Next, you will create a travel mode for driving.

On the right, click the Menu button  and choose New.

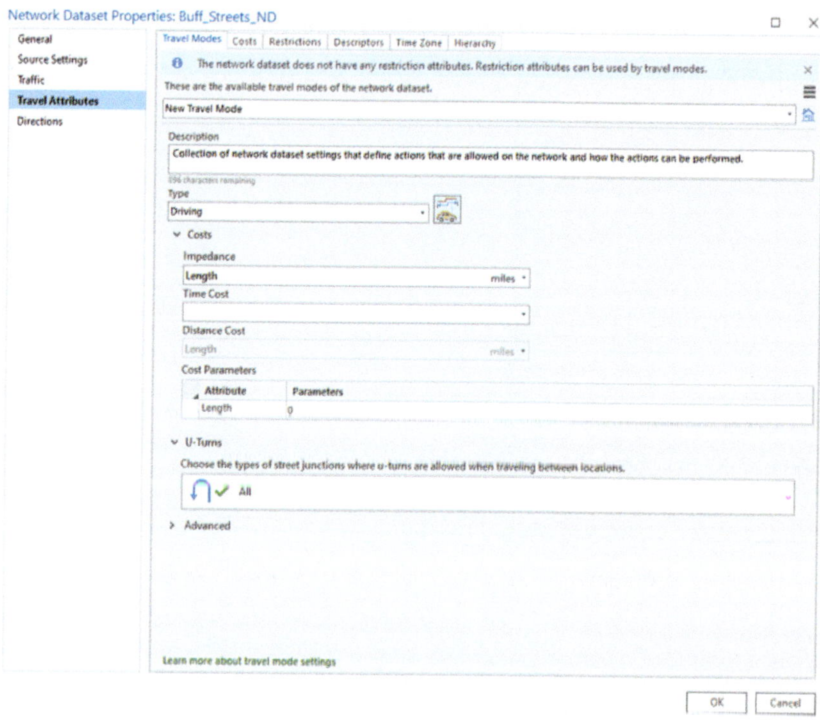


Notice that Length is designated as a cost attribute. We will leave the default settings for Driving travel mode.

Click the Costs tab.

Under Properties, for Units, click the down arrow and choose Miles.

ArcGIS Pro Network Analyst always finds common fields in the source feature classes that we assign to create corresponding network attributes. We use the evaluators to assign these attribute values that we use.

In the Evaluators section, we can see evaluators set from the Buffalo_Streets feature class. We are going to change the value for each evaluator from the Shape field to the Leng_MI field so that meaningful values are assigned to the Length attribute. Leng_MI represents the length of the edge (street segment) in miles.

Under Edges, for Buffalo_Streets (Along), click [Shape] in the Value field.

Click [Shape] again to display the Field Script Settings button, and then click the Field Script Settings button.

In the Field Script dialog box, for Result, delete the existing text and type **[Leng_MI]**.

Click OK.

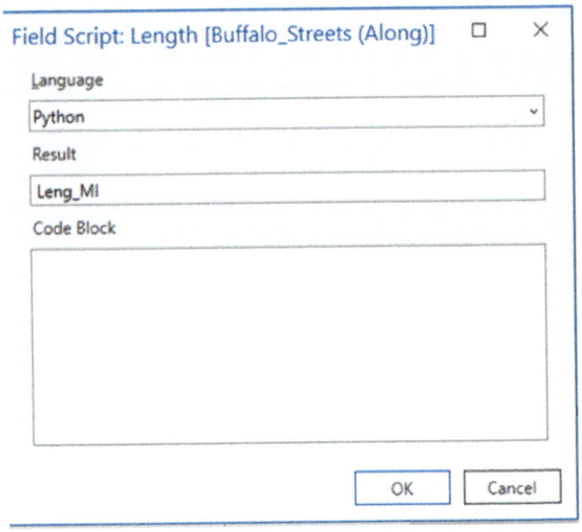


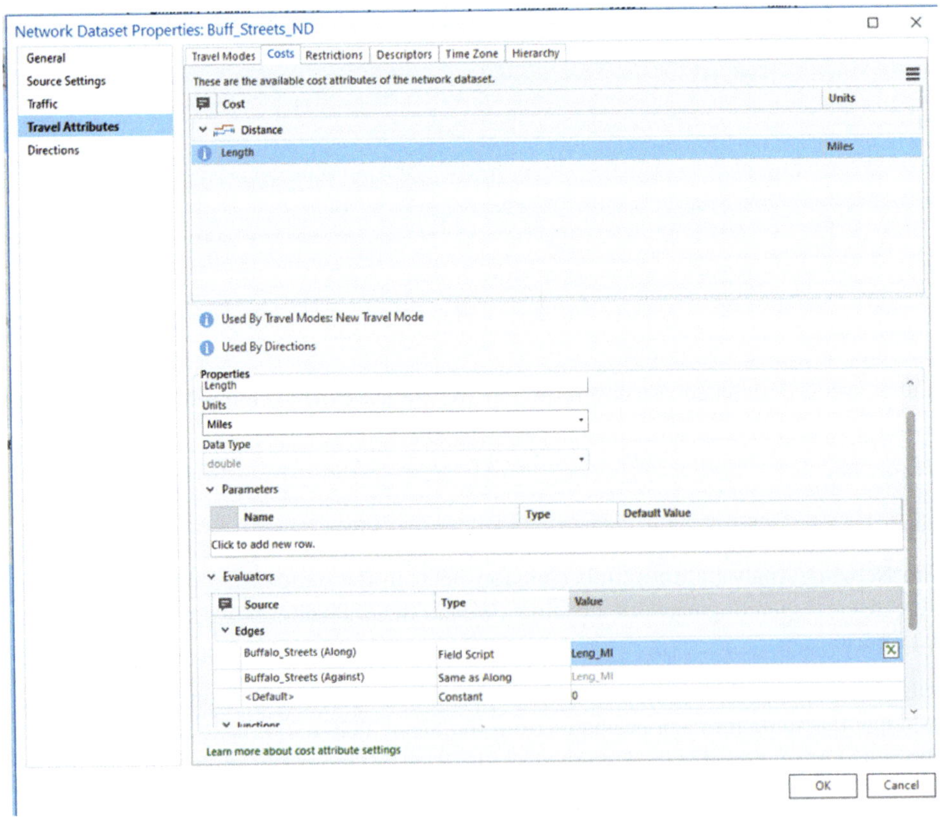

You have assigned a cost attribute using Length in miles.

On the left part of the digital form, click Directions.

On the General tab, check the box for Support Directions.

Notice that the Field Mappings tab is marked with red dashed underlining, indicating an invalid state.

Click the Field Mappings tab.

Directions require at least one edge source to have a Base Name field mapping set. Only the Base Name is required. All the other fields are optional.

Under Primary Name, for Base Name, click <None> and, from the drop-down list, choose StreetName.

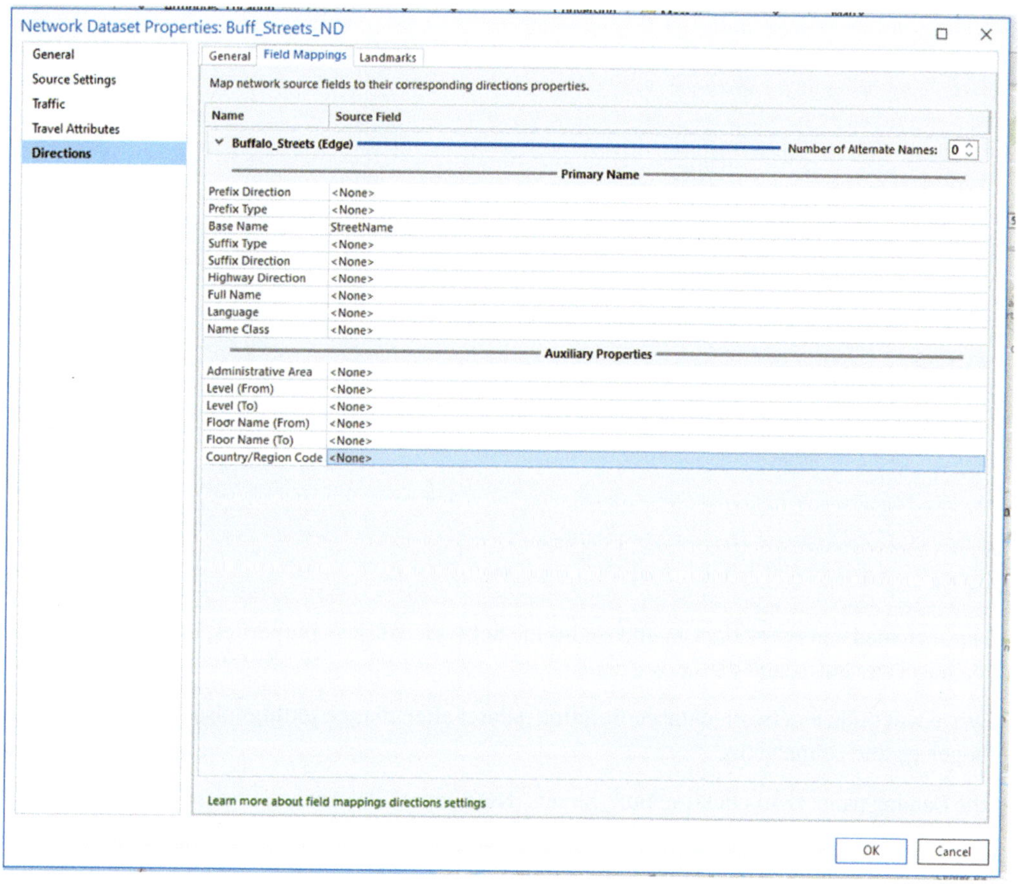


Notice that the Field Mappings tab is no longer in an invalid state.

Click OK in the Network Dataset Properties dialog box.

In the Catalog pane, right-click the Buff_Streets_ND network dataset and choose Properties to open the Network Dataset Properties dialog box again.

Under Summary, notice that the information and inputs for the network dataset that you are about to create have been updated.

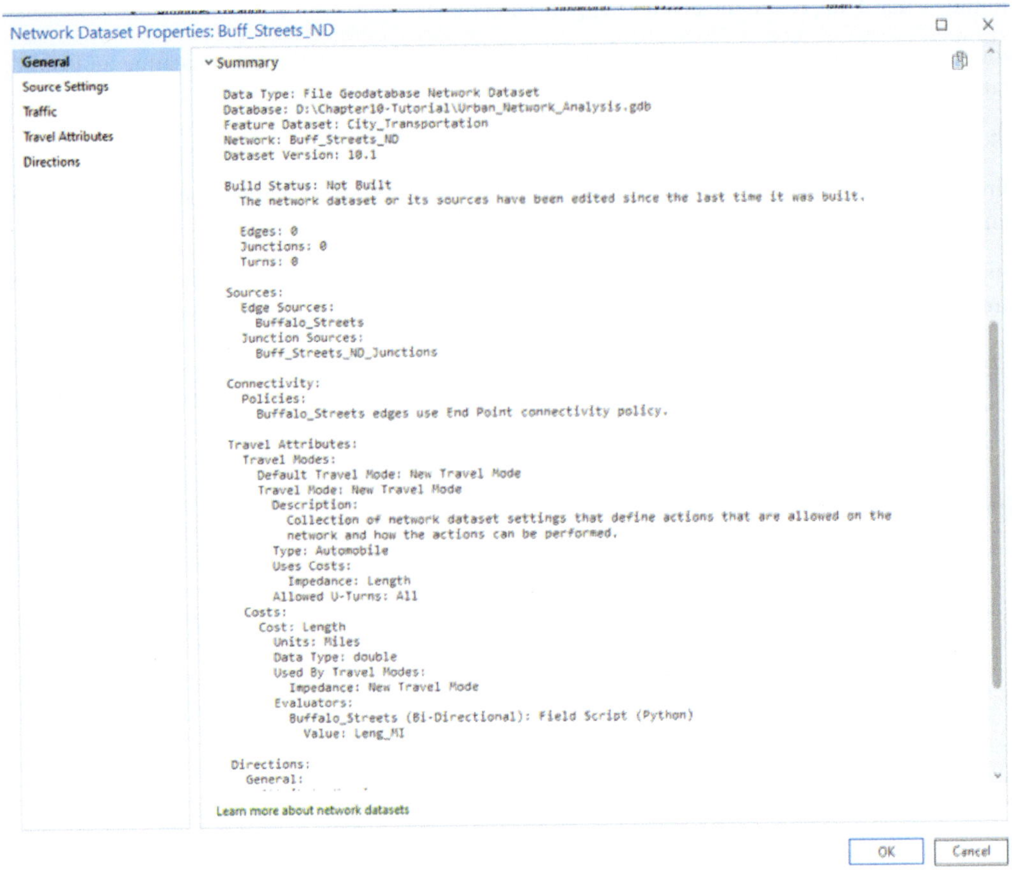


Click OK.

We have created a network dataset and set up the network dataset's properties, including connectivity, travel attributes, and directions.

Now, we will build a network dataset. Building dataset after change settings also enforce the dataset topology and connectivity.

In the Catalog pane, right-click the Buff_Streets_ND network dataset and choose Build.

The Build Network tool determines which network elements are connected and populates the attributes of the network dataset. You must build the network before you can perform any network analysis on it. You also have to build if you edit features in an existing network dataset or you copy and paste (move) it from one workspace to another.

In the Geoprocessing Pane, confirm that Input Network Dataset is set to Buff_Streets_ND.

Click Run.

After the process is accomplished, close the Geoprocessing Pane.

Rename the "Map" (mapping) frame to "Buffalo City Activities" in order to make it more informative.

If the network dataset was not loaded in the mapping frame, you can add it.

After performing these initial steps, you would now be ready to begin analysis.

Save the project and close ArcGIS Pro.

**2.** Creating the optimal delivery route for some shipping company

Open the MyGIS_Learning10 project with ArcGIS Pro. Remove the current network dataset in the Buffalo City Activities mapping frame.

Then, add the Buff_Streets_ND network dataset from the City_Transportation feature dataset in Network_Analysis.gdb. The reason for doing this is to ensure that we have a fresh new network dataset to start the delivery route planning.

The edges display as lines on the map now.

The next step is to set the network data source to reference the Transportation_ND network dataset. On the Analysis ribbon, in the Workflows group, we can find the Network Analysis Icon.

Click the down arrow of Network Analysis Icon, verify that Buff_Streets_ND is listed as Network Data Source.

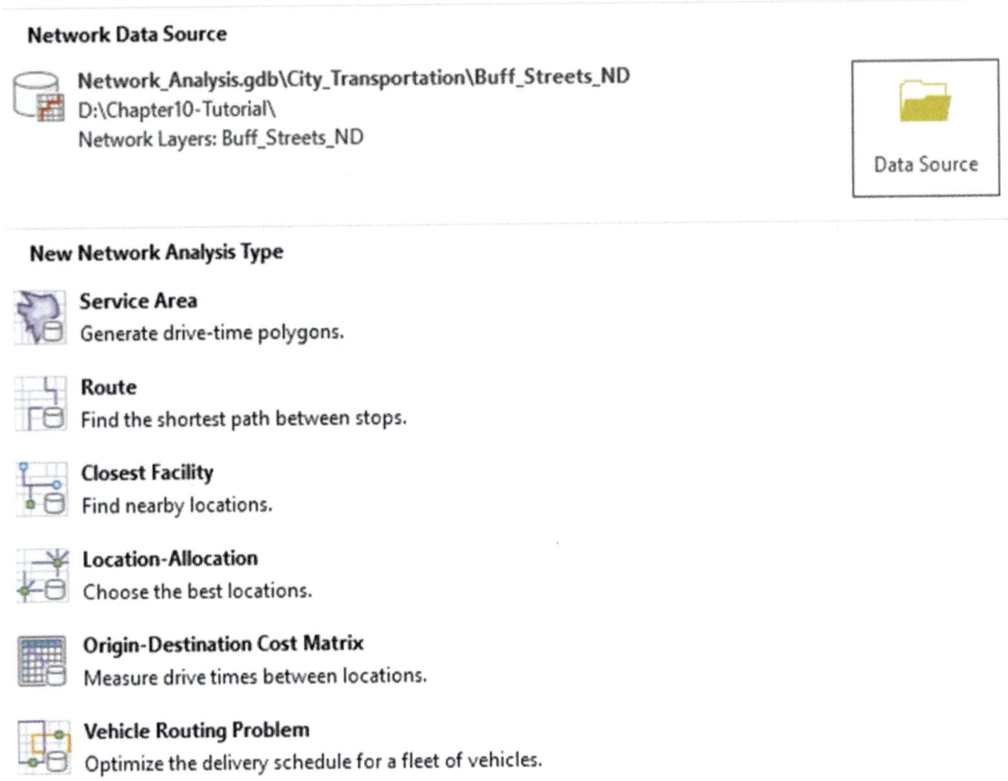


We can follow these steps to change the data source if Buff_Streets_ND is not listed as the data source in this case:

- Click Data Source.
- In the Select Network Data Source dialog box, click Network_Analysis.gdb\T City_Transportation\ Buff_Streets_ND.
- Click OK.

Choose Route from the Network Analysis drop-down list to make an empty route layer. You notice that the symbols of the new route layer is created in the Table of the Contents.

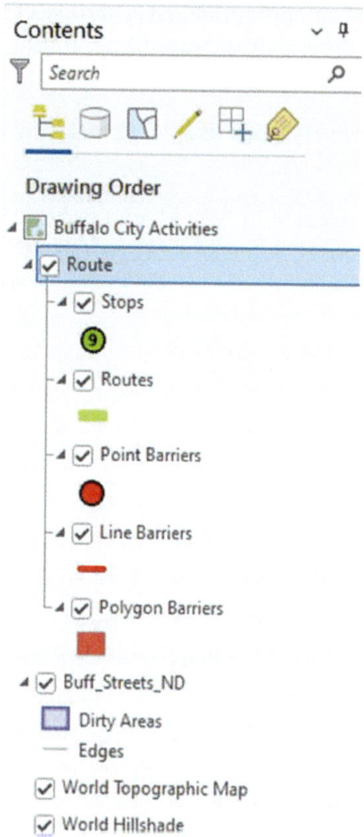


There is a total of six network database related problem solving and analytical approaches on the drop-down list. Each of these approaches is called a solver in ArcGIS Pro.

1. **Service Area:** GIS Analysts can use the network dataset and pre-defined facilities and criteria to create service area polygons.
2. **Route:** GIS Analysts can create the routes of travel or delivery based on the specific criteria and the network dataset, such as the most efficient route or shortest route.
3. **Closest facilities:** This process finds one or more facilities that are closest to a requested location based on criteria of travel time and\or distance. The results will show the best routes and driving directions between the requested location and the facilities that meet the criteria.
4. **Location-allocation:** This process helps to find the best locations for new facilities to serve a set of demand or using locations. Comparing with existing locations, the new facility locations may be trying to cover the most amount of demands, minimize costs or travel time of the demands, or maximize market share.
5. **Origin-destination cost matrix:** This operation finds and measures the least-cost paths along the network from multiple origins to multiple destinations. The best path on the given network dataset is discovered for each of the origin-destination pairs.
6. **Vehicle routing problem:** This approach optimizes a set of routes using a fleet of vehicles for the delivery services of goods or people.

The logic or workflow for each of these approaches is to make an appropriate layer on top of the pre-made network dataset. Then, populate the contents and criteria of the specific approach. Afterwards, we can solve the problem to reach the solutions.

In this case, we need to input a specific feature class or a set of delivery stops in order to make the delivery routes.

Select the Route in the Table of Contents. Route layer must be selected in light blue in order to obtain the ribbons that are related to this layer. On the main ribbon, click the Route Layer tab.

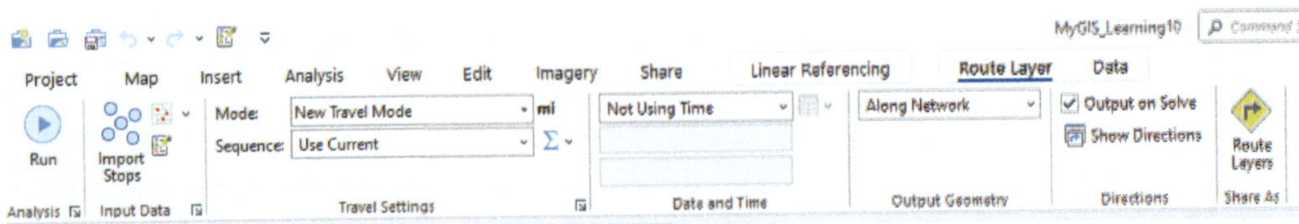


In the Input Data group, click Import Stops.

The Add Locations dialog box opens.

In the Add Locations dialog box, leave Input Network Analysis Layer set to Route and leave Sub Layer set to Stops.

For Input Locations, click the (Browse button) - Folder Icon next to it.

Navigate to your ......\Chapter10-Tutorial\Network_Analysis.gdb and select the Delivery_Locations point feature class.

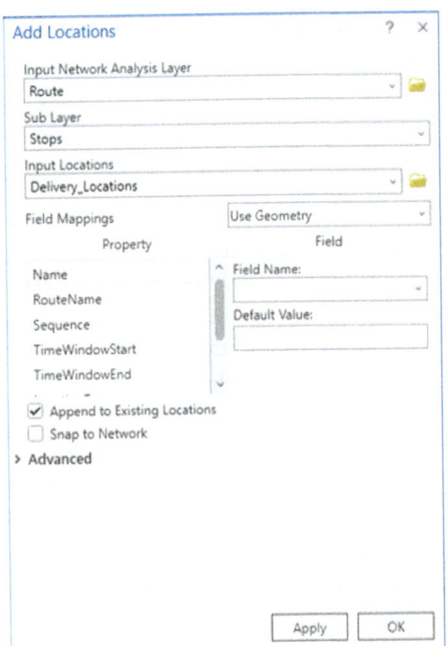

Click OK.

You notice the values of the delivery locations are populated and drawn on the map.

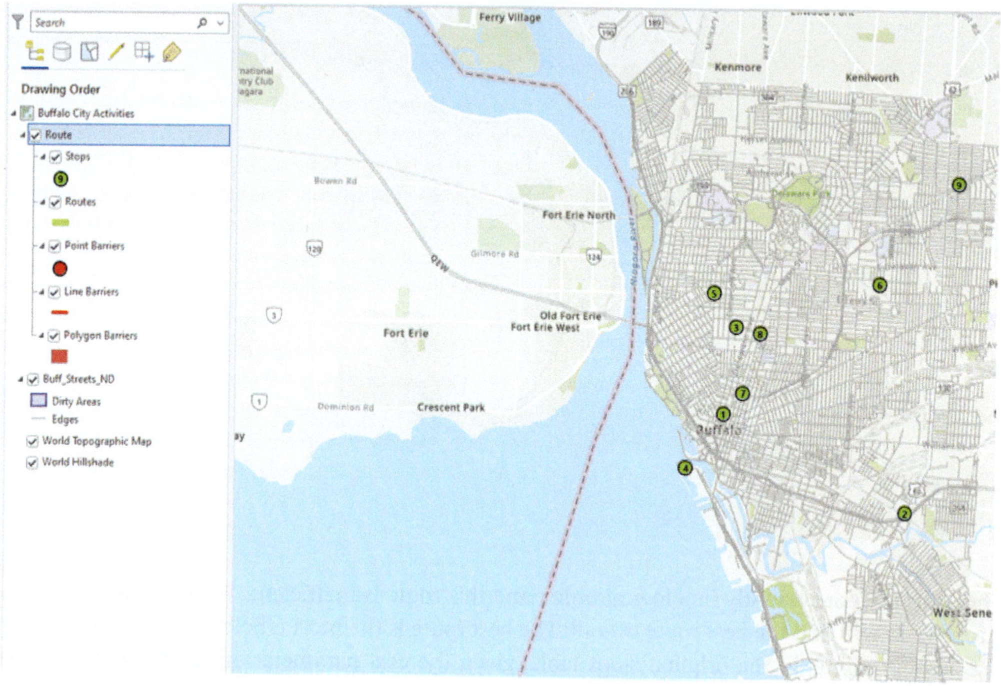


We have imported the stops now, and we are ready to create routes. You will notice that the stops on the map are numbered. The numbering is inherited from the data source, and it represents the order in which the stops were imported into the point feature class.

On the Route Layer tab, in the Travel Settings group, leave Mode set to "New Travel Mode" (driving; we set up the distance in miles of driving when we made the network dataset). However, we did not change the default name and Sequence set to Use Current.

In the Analysis group, click Run.

You notice the new routing map is produced.

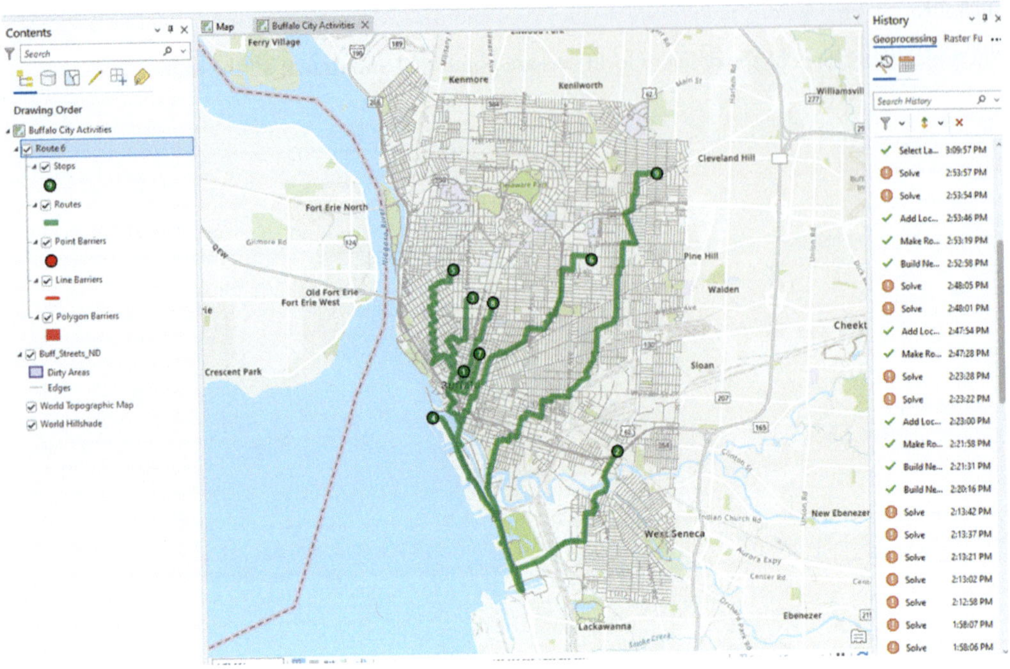


The stops are not currently in a logical order, and this route is inefficient. We are going to create a route that will find the best route overall. The best route is the most efficient way of connecting the delivery points in the original stops table, given the cost parameter or other parameters in the original data source (such as distance or time). This method is useful if the start and stop points do not matter. For example, a delivery truck has already been loaded with the required (demanded) goods and can start making deliveries anywhere. This is the famous "traveling salesperson problem" (or TSP) issue, and it is a common optimization problem modeled in planning and transportation logistics.

On the Route Layer tab, in the Travel Settings group, click the Sequence down arrow and choose Find Best. In the Analysis group, click Run.

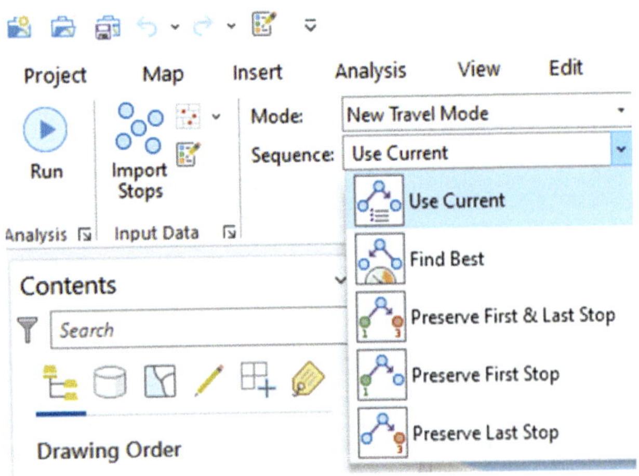

You notice the change of the delivery sequence on the map. Owing to the nature of the original delivery stop layout, there are no big changes in route distances and directions in this case.

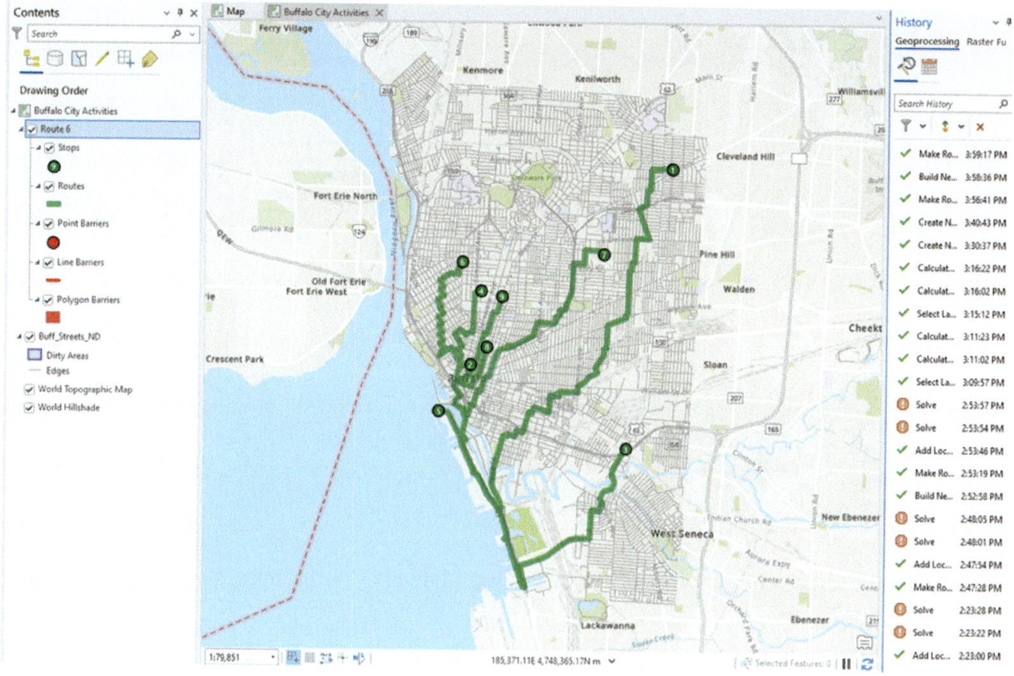


Next, we are going to evaluate and examine the turn-by-turn driving directions that are created for this route. On the Route Layer tab, in the Directions group, click Show Directions.

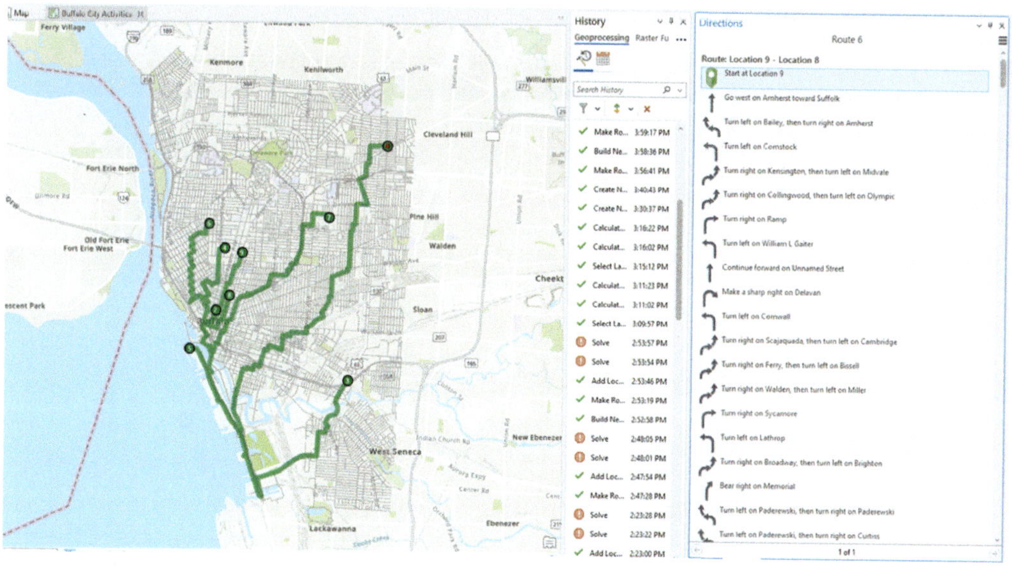

The directions appear in the Directions Pane on the right.

In the Directions Pane, click on any of the steps in the driving directions, a red dot would appear along the route. Users can check both on the digital map and on the instructions.

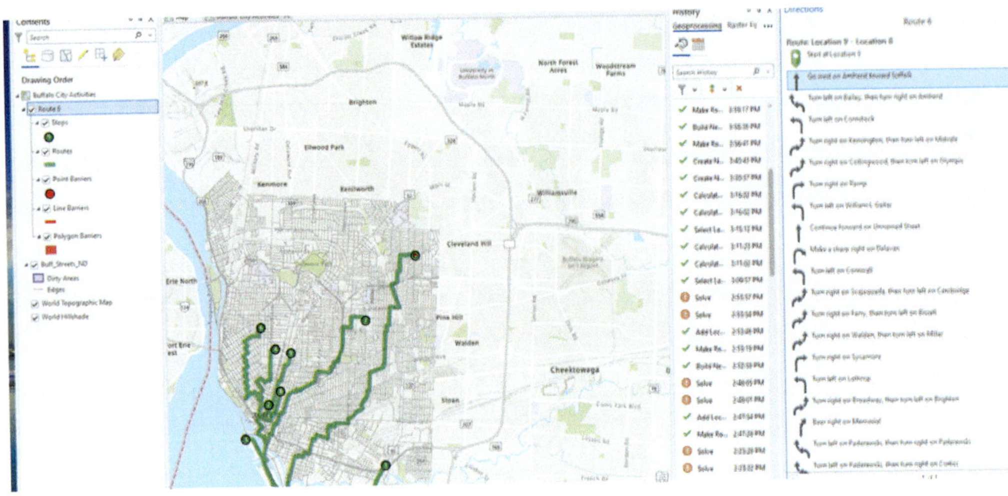

We can also try some other choices of the "traveling (delivery)" sequences, such as preserve last stop.

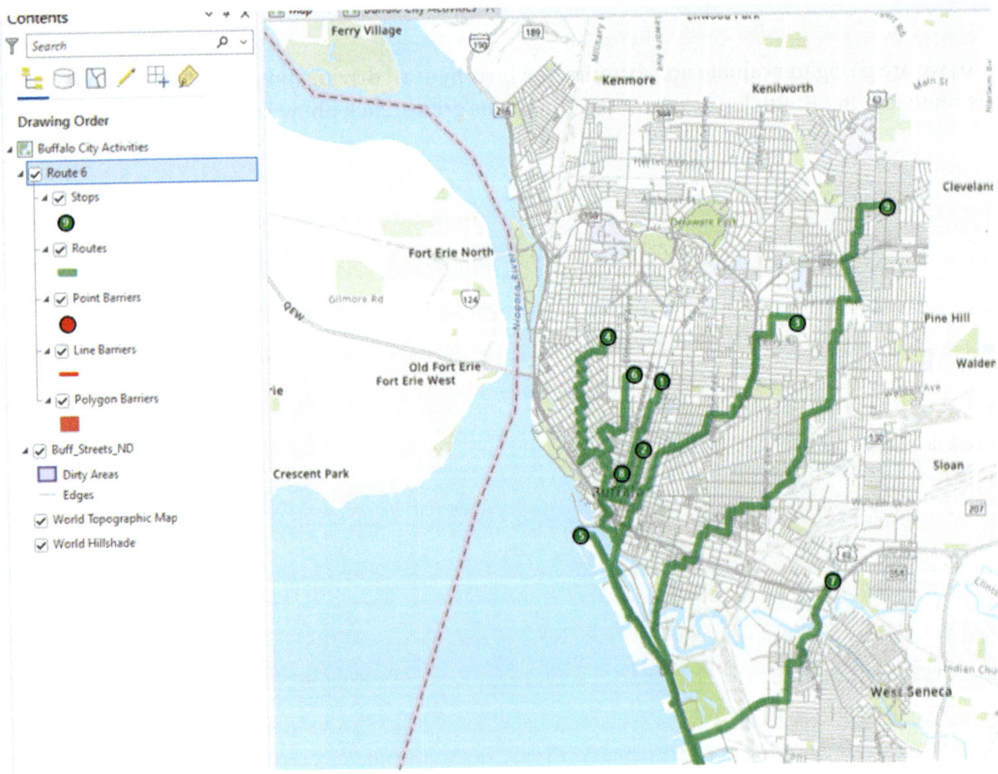

In the real-world, a route can be impeded by an obstruction or a barrier. In this case, we are going to create a polygon barrier feature to model the closed roads in the Highway 33 area.

If necessary, start ArcGIS Pro and open the MyGIS_Learning10 project.

In the Table of Contents, right-click Route and choose Zoom To Layer.

On the ribbon, click the Edit tab.

In the Features group, click Create.

The Create Features Pane opens.

In the Create Features pane, under Route: Polygon Barriers, click Polygon Barriers.

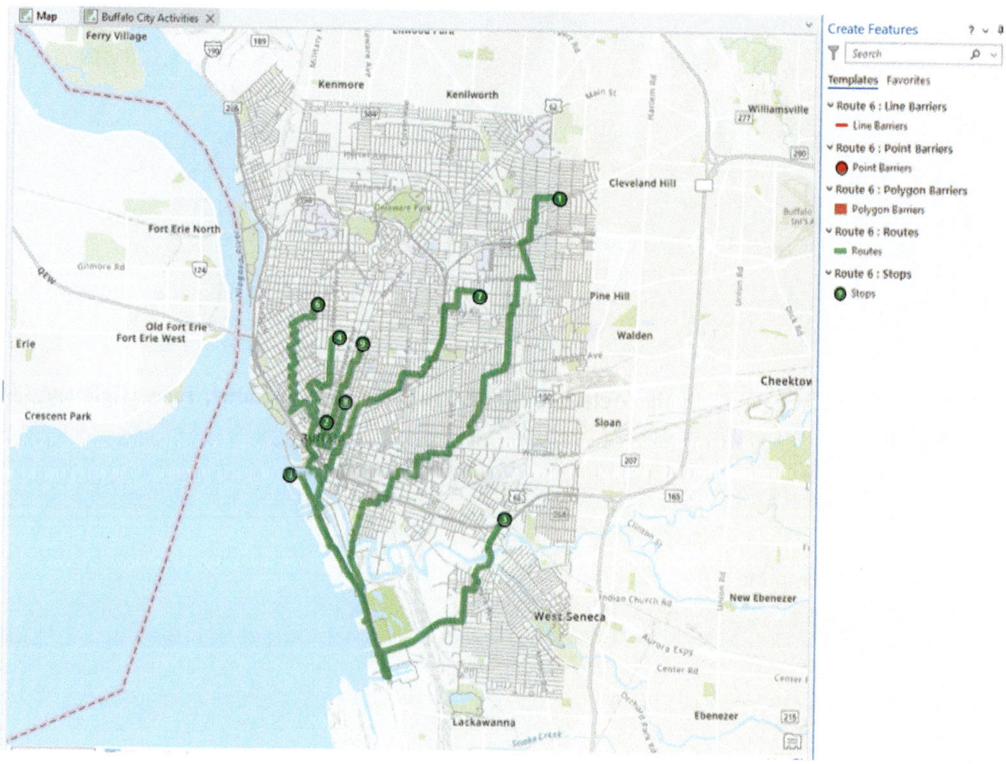


Draw a polygon around the Highway 33 like following. Double-click the last vertex to finish the drawing.

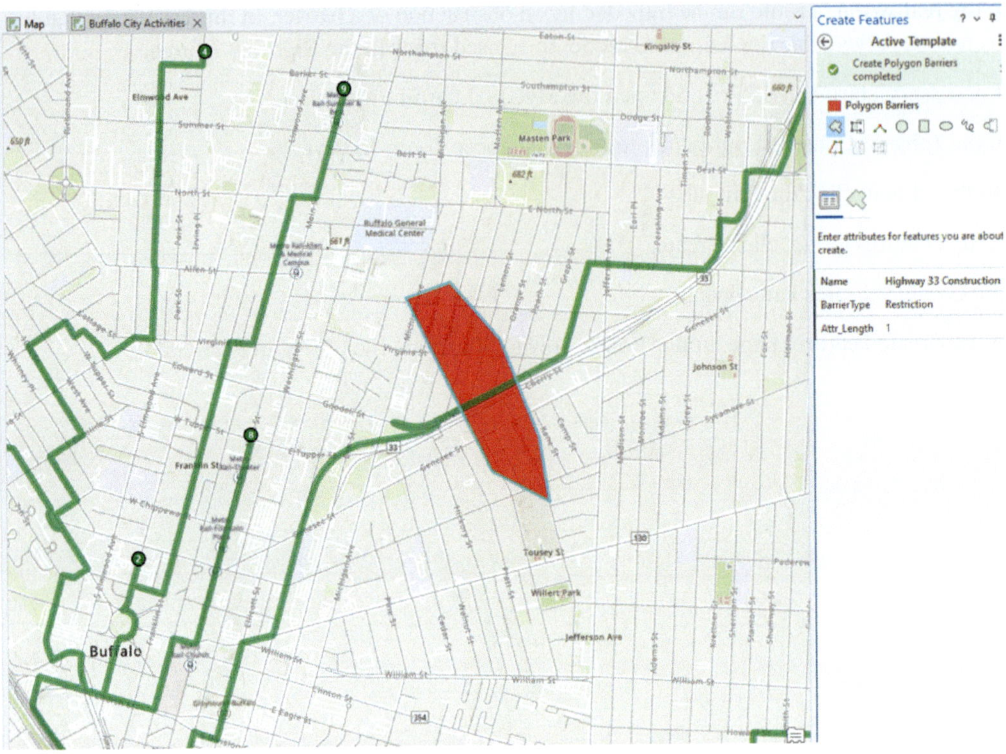


In the Create Features pane, in the Active Template parameters, for Name, type Highway 33 Construction.

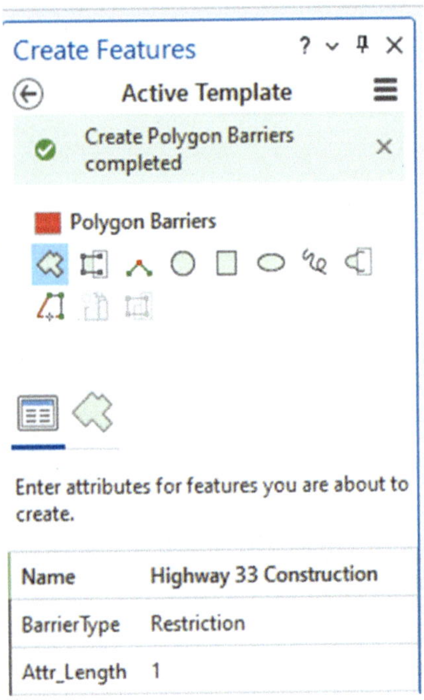

Now we created a barrier, we are going to reanalyze the best route with this construction.

On the Route Layer tab, in the Travel Settings group, verify that Sequence is set to Find Best.

In the Analysis group, click Run. The route map is revised.

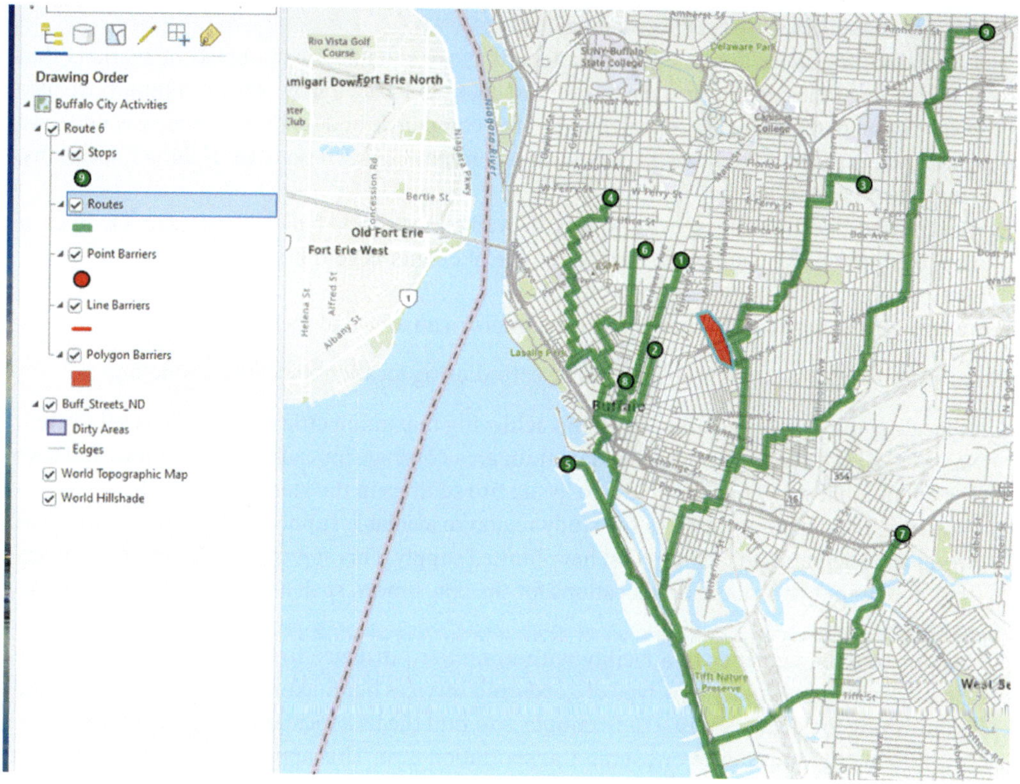


Congratulations! You accomplished the routing planning for network analysis.

Save the project and close the ArcGIS Pro.

**3.** Location–Allocation Analysis:

If the project MyGIS_Learning10 is closed, open it with ArcGIS Pro. Otherwise, continue.

GIS analysts can use location-allocation analysis in ArcGIS Pro to determine the optimal location for planned facilities, such as fast-food stores, bank branches, warehouses, fire stations, or hospitals for a given set of demand locations or the locations where people require the specific services.

On top of the previously made network dataset for our project, two-layer types are required for location-allocation analysis: **facilities and demand points.** Point, line, and area barriers can also be applied in the analysis. A location-allocation model layer allows us to efficiently select the optimal facilities from a set of potential locations for allocating a set of theoretical demands to each of the facilities in the services.

In practice, **facilities** represent existing or possible sites, as well as similar types of facilities in competition with the above. Location-allocation chooses the best candidate facilities to allocate demand to in the most efficient way according to the problem type and criteria we specify. **Demand points** are typically the locations representing the people to request the goods and services that the company's facilities provide.

Facilities can be classified into three types.

1. **Required facilities** are existing facilities.
2. **Candidate facilities** are potential locations that are suitable for the needs or markets that we are going to locate. For instance, if we are going to locate a new pizza restaurant, we also need to find parcels that are for sale, within the company's budget limit, properly zoned, and some other criteria. Candidate facility locations might be based on a set of known locations, research, or suitability analysis.
3. **Competitor facilities** are specific to certain problem types and typically represent facilities of rival businesses that compete for the same type of clients as your business.

**Demand points** represent the number of consumers in an area and their locations.

There are four major types of problems can be solved using location-allocation modeling.

1. Allocating maximum demand values and achieving maximum efficiency of goods supply and services. GIS can help to model the maximum area coverage for each of the facilities and save the operational cost. One example is to locate fire stations in the study region.
2. Matching the demands raised in the study region to available supply of facilities. In this case, we allocate demands to facilities that have limited supply. This approach only optimum allocation. It does not choose the best locations for the consumers, such as allocating K-12 students to schools.
3. Maximizing the demands to a facility with minimized distance to attending. This approach locates facilities with the same type of goods and services but maximizing the attendance and diminishing the distance cost. One example is to find the best location of a grocery store.
4. Minimizing the impedance or overall transportation cost. This approach locates or allocates facilities so that overall transportation cost is minimized. Impedance in the real world could be the time or monetary costs. For instance, we can allocate stores to a set of distribution centers to save budget and/or time to deliver goods.

Now, let us practice location-allocation analysis. In this exercise, we are going to help the La Nova Pizzeria company in Buffalo, NY, to locate two new stores in service to the demands of customers in the City of Buffalo area. The company currently has one store in the city that delivers pizza and chicken wings to serve the entire city. It would like to add two more store locations to reduce the delivery driving time and in getting the products to customers faster.

**Step 1: Making Location-allocation Layer**

In the Catalog Pane, create a new "Map" and change the title of the new map to "La_Nova_Pizza." Navigate to your ......\ Chapter10-Tutorial \ Business_Allocation.gdb \ City_Transportation geodatabase feature dataset. Add the "Streets_ND" network dataset to the"La_Nova_Pizza" mapping frame.

On the main menu, click the Analysis ribbon. In the Geoprocessing group, click the Tools icon. The Geoprocessing Pane opens. In the Geoprocessing Pane, search for location allocation. Click Make Location-Allocation Analysis Layer (Network Analyst Tools) to open it.

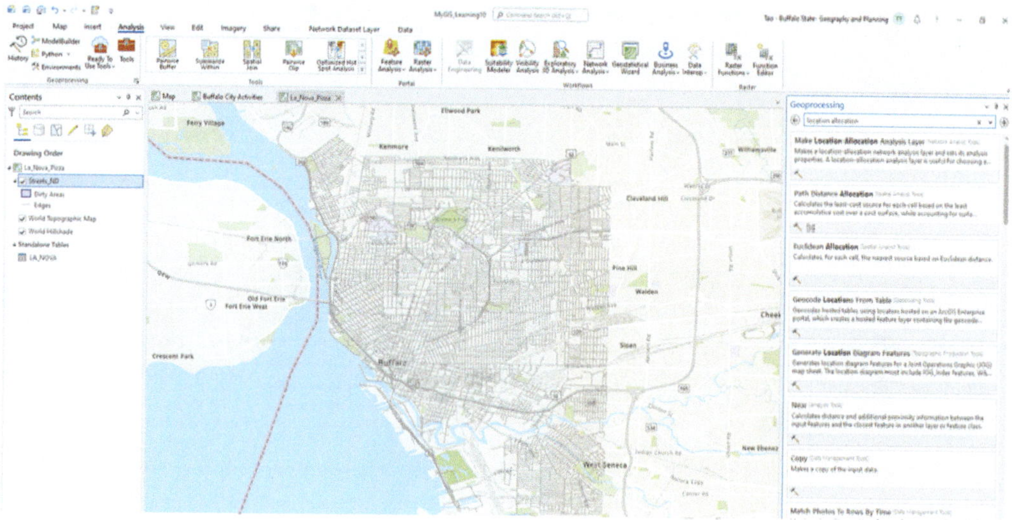

In the Geoprocessing Pane, for Network Data Source, click the down arrow and choose Streets_ND. The restaurant associates will be delivering pizza and chicken wings to the customer locations So, they will be traveling away from facilities.

For Travel Direction, choose Away from Facilities.

We would like each of the drivers to drive the least amount as possible. Expand Problem Type. For Problem Type, click the down arrow and choose Minimize Impedance.

For Number Of Facilities To Find, type 3.

For Cutoff, type in "0.1"

Leave the default values for the rest of the fields.

On the bottom right of the Geoprocessing Pane, click Run.

You notice the new group of Location-Allocation layer symbols is added in the Table of Contents.

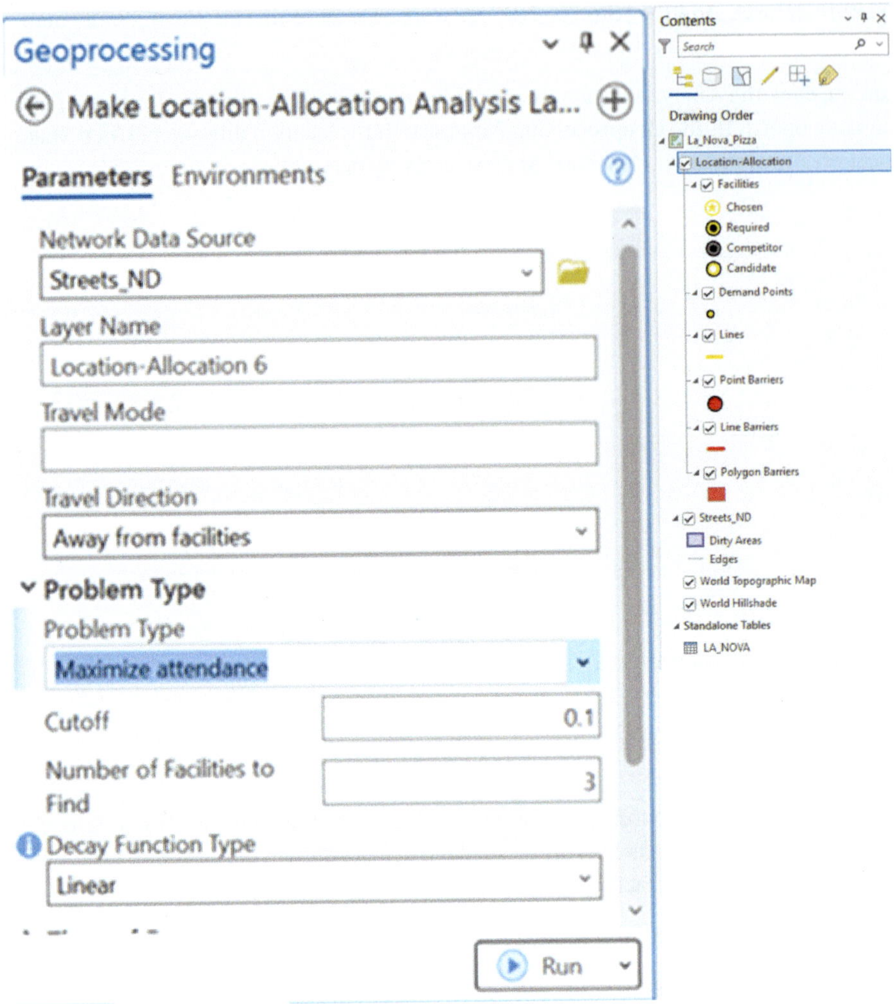


We now have a Location-Allocation Layer ribbon on the main menu. Click the Location-Allocation Layer ribbon and view the location-allocation options.

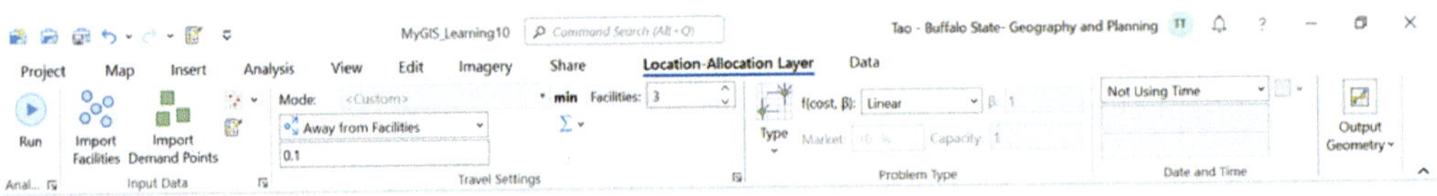


We are going to use these options to complete this exercise.

**Step 2: Add facilities**

We have created a container for holding the components of location-allocation problem. We are going to add the data to this model. We will start with facilities.

First, we will add the existing La Nova restaurant. If necessary, on the ribbon, click the Location-Allocation Layer tab. In the Input Data group, click Import Facilities. The Add Locations dialog box opens.

For Input Network Analysis Layer, verify that Location-Allocation is set. This layer is the group layer that we just created.

Set the following parameters:

- Sub Layer: Choose Facilities.
- Input Locations: Navigate to your ......\ Chapter10-Tutorial \ Business_Allocation.gdb \ City_Transportation geodatabase feature dataset and choose LA_NOVA_Pizza (point feature class).
- Field Mappings: Under Property, choose FacilityType.
- Default Value: Choose Required.

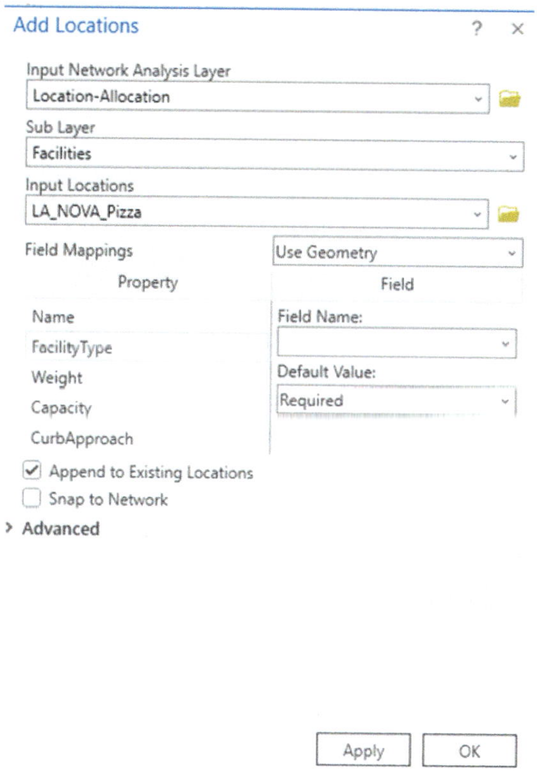


Click OK to run the tool.

You notice the current restaurant is added on the spatial model as a required facility.

### Step 3: Add demand points

Next, we are going to add demand points.

From the Location-Allocation Layer ribbon, in the Input Data group, click Import Demand Points to open the Add Locations tool.

For Sub Layer, choose Demand Points.

For Input Locations, choose Consumers_Demand_Points.

Click OK.

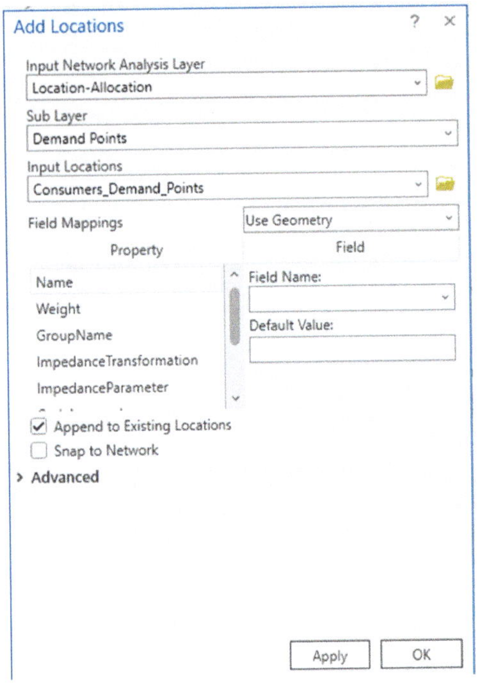

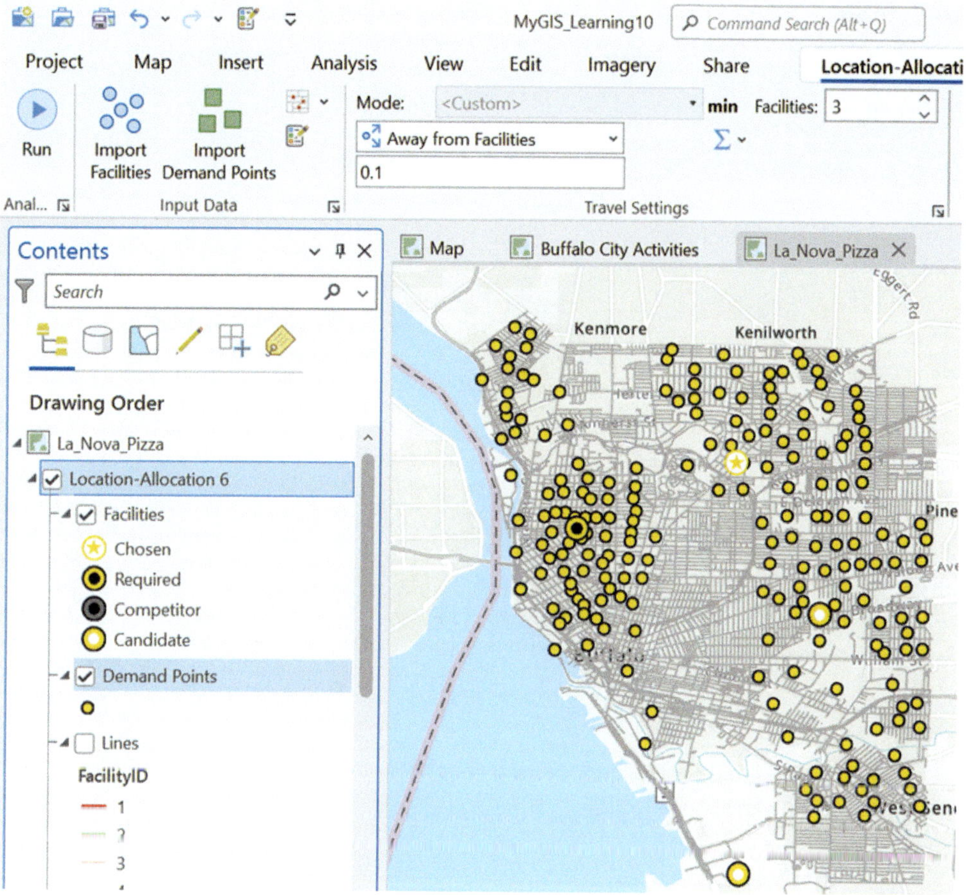


Now, we are ready to solve the problem and find the other two strong candidates for the new restaurant locations.

**Step 4: Solve and view results**

On the main menu, click the Location-Allocation Layer ribbon.

On the left side of the ribbon, in the Analysis group, click the Run button to run the model.

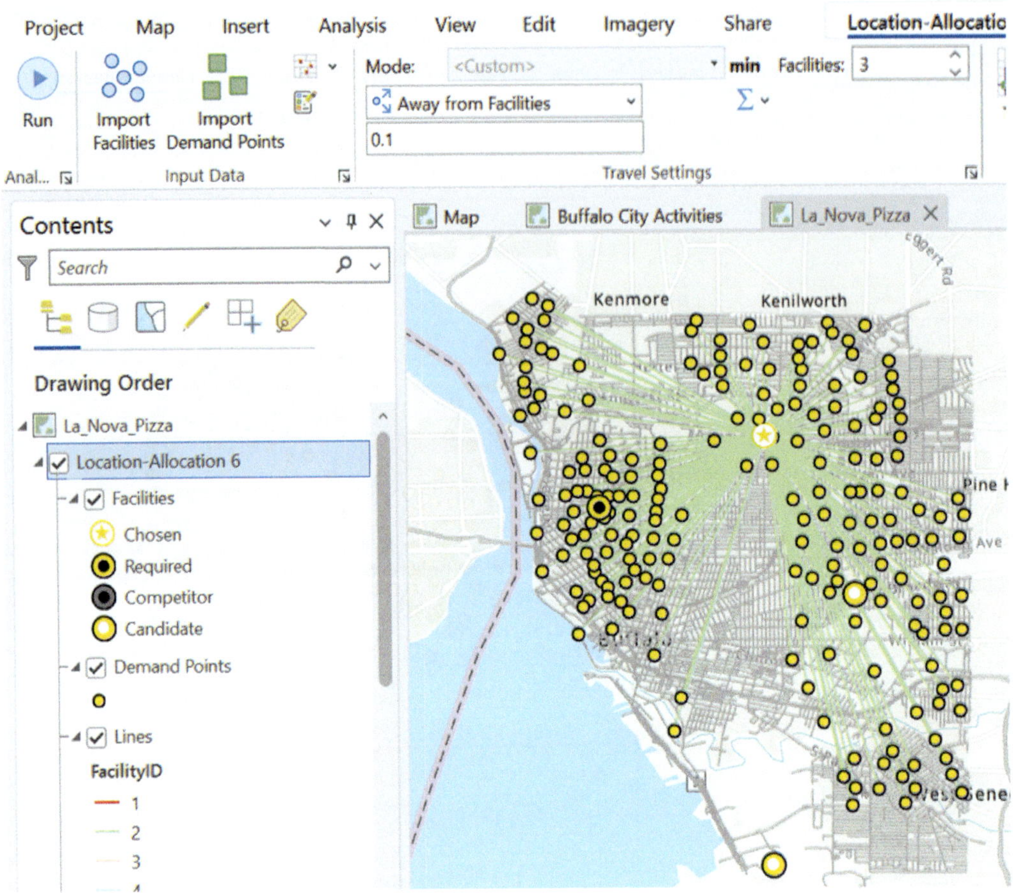


This result shows the best candidate among the three locations for serving the customers.

In order to review the impacts of both current location and candidate locations, we switch from the approach from "**Minimized Weighted Impedance (P-Median)**" to "**Maximize Market Share**" at the Problem Type group on the Location-Allocation Layer ribbon.

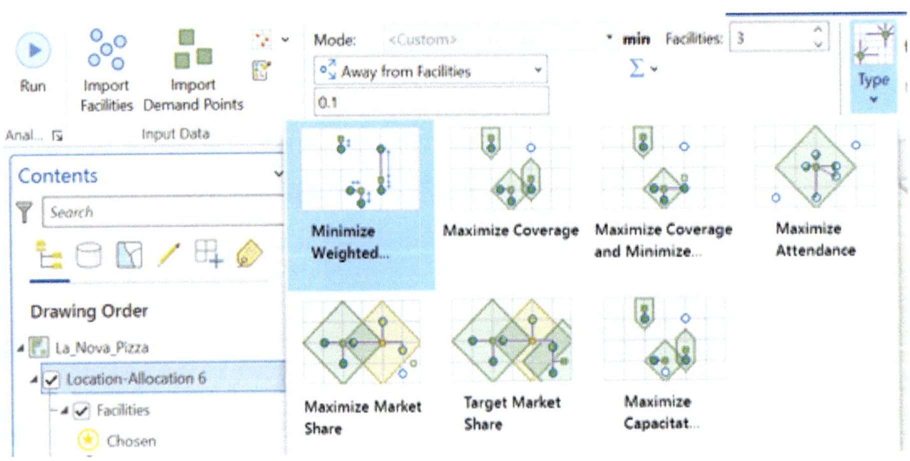

Run the spatial model again at the Location-Allocation Layer ribbon. You will see the maximum market competition among all these existing and planned facilities. We can use the Symbology Pane to differentiate the spike lines from different facility locations.

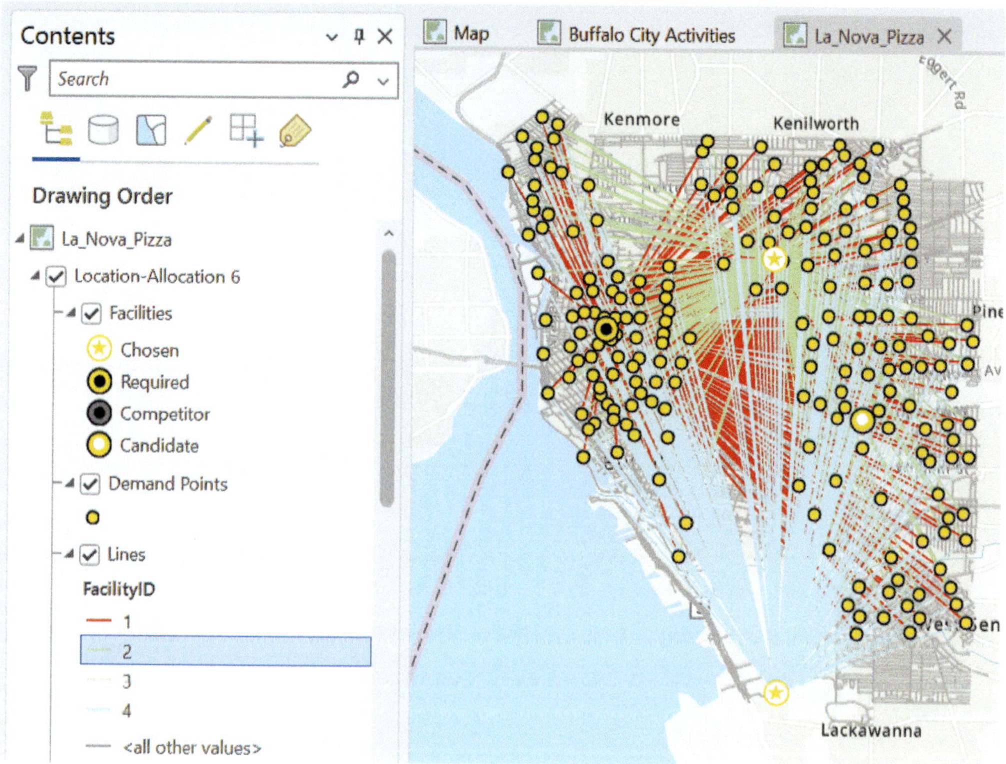


We accomplished this location-allocation project for the La Nova Pizzeria company.

Save the project and close the ArcGIS Pro.

# 11

# Spatial (Geographic) Modeling in GIS and Issues of GIS Software Enhancement

## Concepts and Theories

Modeling as a technical terminology was initiated in art and sculptures. Modeling was originally illustrated as the process of producing a sculptured form of real-world objects or utilizing the illusive methods of volume on a two-dimensional (2D) surface by shading and other means to mimic real-world situations. Modeling is also widely adopted in mathematics, physical sciences, and computer science by using mathematical equations and computational processes to mimic real-world processes and predict the possible results in the future if certain conditions are met in the prediction environment. "Modeling, in the context of Geographic Information Systems (GIS), occurs whenever operations of the GIS attempt to emulate the geographic processes in the real world, at one point in time or over an extended period" (Goodchild 2005).

## GIS Spatial Analysis and GIS Spatial Modeling

The purpose of GIS spatial analysis is to answer questions like "what happens, and where does it happen?". The GIS analytical process is mainly for connecting, illustrating, and visualizing the geographic objects or feature classes to their locations. GIS spatial modeling, by contrast, is to construct models to predict potential outcomes based on either the current environment or sampled values of geographic objects (O'Sullivan and Unwin 2003; de Smith et al. 2013). The spectrum of spatial analysis and spatial modeling can be identified as the following four categories (O'Sullivan and Unwin 2003; Tomlin 2012).

1. **Classical spatial analysis:** various processes of selecting, combining, and reformatting GIS datasets and using primitive operations such as map overlay and data extraction to create new datasets to answer specific spatial analytical questions.
2. **Spatial data analysis:** using exploratory and descriptive statistics to understand and quantify spatial trends of geographic objects.
3. **Spatial statistical analysis:** using statistical methods or models to compare and contrast GIS data of the feature classes. Spatial statistics is used for analyzing spatial relationships of geographic objects in ways classical methods cannot easily do. We can use statistical methods to analyze spatial relationships, distributions, processes, and patterns of geographic features. Towards the statistical prediction end, spatial statistical analysis is also part of the spatial modeling.

*Fundamentals of Geographic Information Systems (GIS) Analysis: Applying ArcGIS Pro*, First Edition. Tao Tang.

4. **Spatial modeling:** constructing models to predict potential outcomes either across geographic space or in a time sequence, or both. Spatial modeling techniques usually make use of statistical models, conventional cartographic modeling techniques like map overlay, data extraction, or the network models.

Although spatial modeling utilizes the majority of spatial statistical methods, in particular the inferential spatial statistics, the objective or the goal of the spatial modeling process is to predict what will possibly be happening in the future. It in general involves a time sequence or time differences between the input values and output results by nature. Meanwhile, the ideal situation for spatial modeling is that the established model can be used to scenario different inputs for different outputs to answer the question, "What if?" For instance, spatial modeling is up to predict if we have this amount of precipitation (rainfall) in this specific region with all the other environmental factors are the same, how much flooding will happen? And where will it be flooded? By contrast, the objective for inferential spatial statistical analysis is to utilize sample data of the object and its distributions to predict or find the entire distributions of the object in the real world. For instance, the Kriging spatial prediction method was invented to find the entire extent of a particular mineral reserve on the ground utilizing a sampling dataset that we can obtain. Since the predicting object has been exist in the natural world for a long time, we only need to estimate its existence locations.

Spatial modeling in essence is an analytical methodology that produces information or data values based on spatial relationships of geographic objects or features. One of the key elements of spatial modeling is the ability to predict or forecast the condition of a geographic region based on certain spatial criteria and relationships of geographic features. One of the classic examples of GIS modeling methodology is the current spatial data usage between an invasive plant species and prediction on areas of potential invasiveness, which is based on known attributes or habitat information, such as soil pH and moisture requirements, rainfall, relative humidity, etc. (Zhang and Tang 2014; Tang and Zhang 2018).

Similar to GIS spatial analysis, spatial modeling can also be performed on both vector and raster data structures. Many modeling processes, in particular those of spatial statistical modeling processes, start with vector data feature classes but produce the prediction or resulting datasets in the grid cell raster surfaces. Spatial modeling can be recognized as one of two categories based on the time scale or time dimension (Tang and Zhang 2018). Static modeling that deals with the prediction of the outcomes at one point in time only. The typical example of static spatial modeling is the GIS implementation of the universal soil loss equation (USLE) model. Many of the spatial statistical model predictions, such as geographically weighted regression (GWR), are static modeling processes. Dynamic modeling, on the other hand, represents a process that modifies or transforms some geographic features on the Earth's surface through time (Goodchild 2005). Applying both raster data of digital elevation models (DEMs) and network structure models for water flow in a river system to predict the flooding event is an example of dynamic spatial modeling.

## Spatial Statistical Modeling in GIS Environment

Since the fundamental requirement to differentiate spatial analysis and spatial modeling is the prediction or estimation of new outcomes, many methods of spatial statistical or geostatistical analysis that we discussed in Chapter 9 can also be classified as spatial statistical modeling. In this chapter, we mainly focus on prediction and modeling perspectives on the spatial statistical models. Applying spatial statistical modeling, GIS analysts need to predict the following three aspects: (i) predicting the intensity or magnitude of occurrence of a geographic object at different geographic locations, (ii) predicting the values of a geographic object or a feature class at different geographic locations, and (iii) predicting the spatial relationships of two or more geographic objects or feature classes with their corresponding geographic locations. These three types of predictions are generally called exploratory spatial statistical modeling, spatial autocorrelation, and spatial regression (Anselin 2005).

### Exploratory Spatial Statistical Modeling

Exploratory spatial statistical modeling is the statistical method used in the prediction for either the existence of a geographic object or the intensity of occurrence of a geographic object or a feature class. In the GIS environment, an example of the former is geographic mean center analysis, and an example of the latter is geographic feature density analysis, including, kernel density, point density, and line density. The mean center is the average geographic coordinate location of all the features in the study region. In a 2D geographic environment, this is the location of means of x-, y-, and in three-dimensional (3D) environment, this is the location of means of x-, y-, z-coordinates. Mean center is very useful for tracking changes in the distribution or for comparing the distributions of different types of geographic features.

Density analysis takes field-sampled or measured values or known quantities of a geographic object or a feature class across the geographic area of the study to predict and classify the changes of intensity of values by different locations. A variety of methods have been developed in GIS. *Kernel density* calculates a magnitude-per-unit area from point or line features using a kernel function to fit a smoothly tapered surface to each point or polyline. In the case of point features, Kernel density calculates the density of point features around each output raster grid cell. In the case of line features, Kernel density is the sum of linear features in the neighborhood of each output raster grid cell. *Point density* calculates a magnitude-per-unit area from point features that fall within a neighborhood around each grid cell. Conceptually, a neighborhood is defined around each raster cell center, and the number of points that fall within the neighborhood is totaled and divided by the area of the neighborhood. *Line density* calculates a magnitude-per-unit area from line (polyline) features that fall within a radius around each grid cell. Conceptually, a circle is drawn around the center of each raster grid cell using the pre-set search radius. The length of the portion of each line that falls within the circle is multiplied by its value of "population field" if it is applied. Otherwise, the proportion of line occupation in the circle would be the density value of the grid cell.

Hot-spot analysis is another type of density modeling that identifies hot spots and cold spots. Hot spot analysis using the Getis-Ord Gi* statistic to calculate for each feature in a feature class. The resulting $z$-scores and $p$-values inform the users where the features with either high or low values cluster spatially in the study area. The characteristic of this prediction is that a feature with a high value may not be a statistically significant hot spot. To be a statistically significant hot spot, a feature will have a high value and be surrounded by other features with high values as well. The local sum for a feature and its neighbors is compared proportionally to the sum of all features; when the local sum is very different from the expected local sum, and when that difference is too large to be the result of random chance, a statistically significant $z$-score is produced. Hot spot areas and cold spot areas are identified.

### Spatial Autocorrelation

Spatial autocorrelation refers to the degree to which one geographic feature in a feature class is similar to other nearby geographic features. "Auto" means self, and "correlation" means association. Spatial autocorrelation utilizes the statistical measures of close objects that are similar to other close objects to predict values of a geographic object at all the locations in the study area. The sampled or field-measured values of the geographic object at known locations serve as the input feature class. The spatial autocorrelation function measures spatial autocorrelation based on both feature locations and feature attribute values simultaneously. Given a feature class and an associated attribute, spatial autocorrelation evaluates whether the pattern expressed is clustered, dispersed, or random. Spatial autocorrelation deals with one geographic variable or one map layer only.

Practically, Kriging is the prediction model to generate estimated values using the concept of spatial autocorrelation. Kriging is an advanced geostatistical method that generates an estimated surface from a scattered set of points with attribute or $z$-values. Kriging assumes that the distance or direction between sample points reflects a spatial correlation that can be used to explain variation in the surface. The Kriging function fits a mathematical (statistical) model to a specified number of sampling points within a specified radius to determine the output value for each of the locations.

**Spatial Regression**

The purpose of spatial regression is to model, examine, and explore spatial relationships of two or more geographic variables. Regression modeling in statistics is for estimating the relationships between one dependent variable and one or more independent variables. Regression model theoretically can be linear or nonlinear, but the most common form of regression analysis is linear regression. Spatial regression methods are added functions in spatial or geographic terms to the statistical regression models. A variety of modeling functions or methods have been developed in GIS. We will discuss a few of them in this chapter.

Ordinary least squares regression (OLS): this method performs global OLS linear regression to generate predictions or to model a dependent variable in terms of its relationships to a set of explanatory variables by geographic locations. Each of the prediction locations accounts for its own variation of the general model.

Geographically weighted regression: GWR is a spatial regression method that uses different regression models at different geographic locations. GWR evaluates a set of local models for the dependent variable and predicts by fitting a regression equation to every feature in the input feature class. GWR constructs these separate equations by incorporating the dependent and explanatory variables of the features falling within the neighborhood of each target feature. The shape and extent of each neighborhood analyzed is based on the neighborhood type and neighborhood selection method parameters.

Forest-based classification and regression: creates models and generates predictions using an adaptation of the random forest algorithm, which is a supervised machine learning method developed by Leo Breiman and Adele Cutler. Predictions can be performed for both categorical variables (classification) and continuous variables (regression). Explanatory variables can take the form of fields in the attribute table of the training features, raster datasets, and distance features used to calculate proximity values for use as additional variables. In addition to validation of model performance based on the training data, predictions can be made to either features or a prediction raster dataset of grid cells.

### Raster Gird Cells and Distributed Network Systems as the Database and Framework for GIS Modeling

The characteristic of raster grid cell structure is that it decomposes the geographic space of a study region by the standard units of grids. This quality presents advantages over vector data. Not only is it suitable to model continuous geographic objects, but it also can demonstrate value transferring or value changes from one grid cell to the next in a logical sequence or during a time period. In addition, the modeling results or predictions represented by raster surfaces are capable of illustrating new attribute values in detail at various locations of the study region. In GIS-based environmental modeling, raster grid cells are most popularly used (DeMers 2002). Some significant examples are: (i) hydrological flow and watershed delineation modeling, (ii) multiple spatial variable map algebra modeling, such as GIS-based universal soil loss or soil erosion (USLE) modeling, (iii) air pollution or water pollution diffusion modeling, (iv) atmospheric condition change modeling, such as using data from the last 100 years to predict temperature change in the future in a study region, and (v) infectious disease transferring and dispersal modeling, such as predicting the spreading of COVID-19 virus in a county or state.

In practice, one of the major advantages of using raster grid cells to model environmental factors than that of using vector data, such as polygons or lines, is its greater precision and accuracy of predicted locations. Raster grid cells, or tessellation data, are more position-oriented data, while vector data are more theme-oriented data (DeMers 2002; Tomlin 2012). Each of the grid cells carries both the predicted value of the attribute and the precise location of the cell. For instance, it is more suitable and effective to use grid cells to model either fertilizer pollution or soil particle loss owing to cultivation practices in farms.

In this chapter, we would like to illustrate two examples of spatial modeling that are based on raster grid cells. One is suitability modeling, and the other is the cellular automata (CA) modeling. *Suitability modeling* is to identify the best locations to site things, or conduct a particular operation, or find areas to be preserved for specific reasons. This process requires multiple variables or themes of grid cells as the input. Mathematical or statistical manipulations can be applied to each of the input raster layers using the map algebra functions. The output raster layer carries the suitable scores that differentiate the best locations or areas for the objective or purpose.

For example, we can use a suitability model to determine the best location to build a shopping center, banana farm, or ski resort. Determination of the best locations to site or preserve some geographic objects is based on the criteria of scientific or social science studies and environmental matches of these criteria at each location. Each of the criteria can be physically realized in GIS as a raster map layer, for example, the slope of terrain, distance to a road or stream, soil pH values, and land use type. Input feature classes are the criteria. Through computing and weighting the criteria, a suitability raster map layer is created. The suitability map layer identifies the relative scores of rankings on locations for the projected purpose. Then, the GIS analysts can summarize the results for the decision-makers to make the final decision.

*Cellular Automata (CA) modeling* is a type of dynamic spatial prediction. A CA consists of three major parts: grid cells, neighborhood environment, and transition rules. Conceptually, CA models implement Tobler's First Law of Geography into simulation modeling. Nowadays, CA are mostly used to model and simulate how extended spatial or geographic objects evolve over time.

CA are a class of models that we can use to explore how local actions generate global patterns through well predefined and specified rules. In CA modeling, decisions are made locally by each cell, which are often arranged on a regular lattice, and the patterns that emerge. Regional spatial patterns emerge through the cell changing status based on specific transition rules and the states of their surrounding cells. For instance, urban growth or deforestation are not coordinated centrally but arise from the bottom up and therefore can be simulated using a CA model (Crooks 2016).

Network spatial data structures or data models in GIS are naturally suitable for modeling scenarios. Prior to installing the components of a network in the real world, we can easily model or simulate the outcomes on either a transportation network or a utility network with different inputs. The newly created professional term in GIS and other related disciplines, such as building information management system (BIM), call this type of process the "digital twin." In essence, the ultimate goal of a digital twin is to model or simulate the real-world situations in the digital and computerized environment either to maximize the effectiveness and efficiency of human engineering design or to minimize the disastrous situation when nature imposes on us. Meanwhile, if the network spatial data structure is utilized to analyze what is happening now for the daily management of its physical counterpart, which is just an analysis.

A network is a system of interconnected linear features and node or junction (point) features that are used to solve problems for transportation, public facility management, accessibility studies, and other human mobility applications. In addition to the geometry and shape, network elements have attributes that control navigation within the network system. The network system of directional and non-directional structures was discussed in Chapter 10. From the modeling point of view, both these two types of distributed networks can be used to predict commodity, signal flow, or goods transportation status.

Network modeling and simulation is a practical approach using vector network structures to solve real-life problems. Utility networks, or directional networks, can be used to model and track flows of liquid commodities, such as fresh water or oil, or electricity or digital signals, such as TV signals. Transportation or nondirectional network structure can be applied for predicting or estimating (i) the best vehicle traveling route, (ii) closest facilities, and (iii) best locations to serve the customers in a region. Details of these modeling approaches are discussed in Chapter 10.

In the natural environment, combining both raster datasets and vector line feature classes, stream networks can be delineated based on a DEM using the output from the Flow Accumulation function. Flow accumulation in its simplest form is the number of upslope cells that flow into each cell. By applying a threshold value of water accumulation, a stream network can be delineated. The newly created vector stream network prediction can be validated using a digitized river system feature class.

## GIS Modeling in Social Economic Environment

In social science and economic studies on the Earth's surface, in particular in urban planning, GIS modeling methods are widely used. Simulations and predictions are popularly applied to answer the what-if questions. One of the major modeling approaches in urban geography or urban planning is to predict the urban growth or urban sprawl using the CA model (Betty and Xie 2005). Urban sprawl in a geographic region is presented as the increase in built-up and paved areas with an impact on the loss of agricultural land, natural vegetation covered space, or ecologically sensitive habitats (Liu and Phinn 2001; Falah et al. 2020). Urban sprawl development is accomplished in three ways, which are ribbon sprawl, leapfrog sprawl, and low-density (radial) sprawl (Falah et al. 2020).

Human consumptive use of land is highly affected by the provision of public transportation and utility infrastructures such as roads and streets, water and sewer lines, electricity power, and telecommunication. The ribbon sprawl is developed along the main transportation network that joins urban activity centers. Land areas without direct access to the transportation and utility network likely remain rural or undeveloped. The leapfrog sprawl is a discontinuous pattern of urbanization in which the developed lands are widely separated from each other, and it costs more than two other ways. Spatial or geographic systems grow in two ways: expansion and compaction. The expansion process leads to more space to be occupied. Compaction increases the density of the geographic objects, either buildings or the density of the population in the urban growth issue.

CA modeling is a spatial and computational method that is capable of simulating a growth process by describing a complex system through a set of simple rules (Mantelas et al. 2008). CA modelling is one of the suitable ways for urban sprawl modelling (Feng et al. 2011). The nature of the CA model is that it follows rules established at the local grid cell level to facilitate development or land value change of adjacent cells mimicking real-world urban land development situations. The behavior of CA-based models is mainly determined by the neighborhood relationships, which makes these models capable of showing dynamic changes over time. In the CA, geographic space is defined as grid cells (Moghadam and Helbich 2013).

Discussing the variety of mathematical equations that are used for computing the land use value changes at the grid cell level, Falah et al. (2020) summarized the general transition rules as follows for the urban sprawl modeling using CA:

1. Do not change the status of the grid cell at simulation periods if the status of a grid cell is urban land use.
2. Do not change the status of the grid cell at simulation periods if the status of a grid cell is constrained.
3. The probability of a grid cell for transformation to urban status will increase if affecting factors are closer to the cell.
4. The probability for transformation to urban status will increase if grid cells have more urban status in the neighborhoods.
5. The status of the grid cell will be transformed to urban if the calculated value of a cell has the maximum value among the other cells in the analytical window (Moghadam et al. 2013).

The second modeling example we would like to discuss in this chapter is the general retail location and service area modeling. Human communities need both commercial and public services. From grocery shopping to purchasing an electronic device to use to seeing a doctor or going to a public library, these are all our daily activities. In the modern retail industries or local public services, the locations of these facilities are very important in serving the people who live in a geographic region. There are two folds of this type of modeling process. The first is to find the best potential locations for new retail facilities. The second is trade or service area modeling of customer demands or service needs. Both of these approaches are being more and more fulfilled by using the vector network data models that we discussed in Chapter 10. In this chapter, we mainly would like to summarize this retail and service modeling based on network data structure and provide some examples.

In the project to spatially model the trade areas of existing and new retail facilities, we can utilize the Generate Threshold Drive Times function to generate drive-time trade areas. This process uses the street network data model and expands outward from each of the facilities until the threshold criterion is reached. The size of the output polygon is determined by the value provided in the threshold values parameter for the selected threshold variable. Overlapping of the trade area polygons presents the real-world business competition of the potential customers. The general rules for selecting the best locations of the new facilities under the condition of the threshold traveling time (drive or walk) are: (i) minimum overlapping of service areas and competition, (ii) maximum potential customers or population density in the trade area chosen, and (iii) household income level is appropriate to the goods or services that these facilities provide.

The service area modeling can help us to answer the following types of questions related to the willing driving time, such as: "What areas are covered within a 5-minute driving time or 3-mile driving distance of my stores?" "What areas are within 4 minutes of our fire stations?" Creating a service (trade area) area is like buffering a point. When you buffer a point, you specify a straight-line

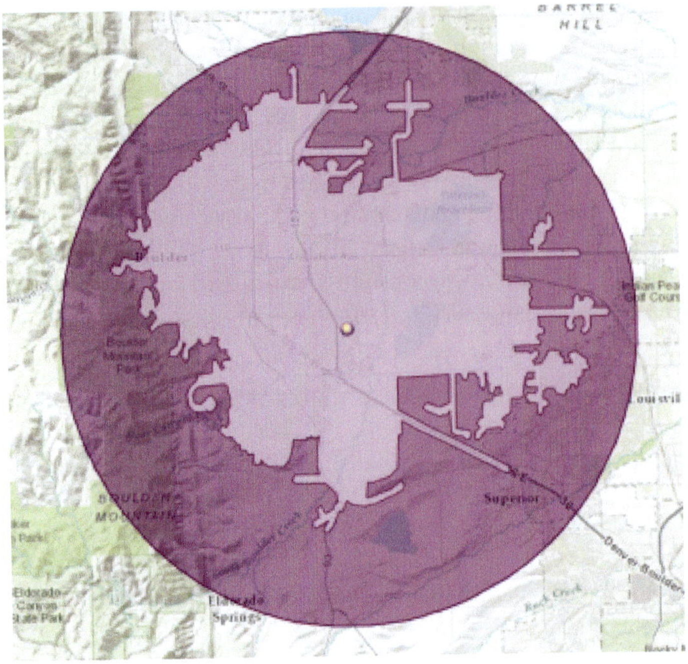


**Figure 11.1** 5-mile on road network driving distance in comparison to 5-mile buffer. *Source:* ESRI 2022/ Environmental Systems Research Institute, Inc.

distance, and a circle is created to show the area within that distance. When you create a service area around a point, you also specify a distance, but unlike a buffer, it represents the maximum distance that can be traveled along a network, such as a road network (Figure 11.1). The result is a service area covering the roads that can be reached within the distance or the time that you specified.

Location-allocation modeling using network datasets can help us to solve many business or industry-related problems. The geographic location of an object is the most significant factor for the decision-making regarding the use of that feature. The success of business operations, industrial facilities, and government services is based on the correctness and effectiveness of the location selections. Right location selections can help keep fixed and overhead costs low and accessibility high.

Given facilities that provide goods and services to the people and a set of demand points that consume them, the goal of the location-allocation modeling is to locate the facilities in a way that supplies the demand points most efficiently. As the name suggests, location-allocation is a twofold problem that simultaneously locates facilities and allocates demand points to the facilities. This is an interactive approach to modeling without considering that one variable is dependent on the others.

## GIS Modeling in the Natural Environment

The ultimate purpose of GIS modeling in the natural environment is to predict outcomes of geographic, ecological, or morphological processes. Although these processes on the Earth's surface are not cyclic in nature, they work in the framework of 3D geographic space and the time scale. GIS provides a distributed digital platform that researchers and GIS analysts can use to model and predict the possible outcomes with field-measured sampling data.

In this chapter, we are going to focus on the discussions of hydrologic or water resource modeling and agent-based modeling (ABM) for land use and land cover in GIS as major examples of applications in the natural environment. A hydrologic model is a simplification of a real-world system from precipitation to runoff that aids in understanding, predicting, and managing water resources in a region. Hydrologic models can be used to study both the water flow and water flow pollutions. GIS can provide a digital distributed platform to conduct hydrological modeling. The hydrologic modeling functions in GIS provide methods to describe the physical components of terrain surface in a study region. GIS analysts can utilize these functions in identifying the ultimate water draining points – the sinks in a watershed, determining the water flow direction, calculating the water flow accumulation, delineating the watersheds, and creating stream networks. The following is the result of GIS-based hydrological modeling using the DEM (Figure 11.2).

The purpose of hydrological modeling is to predict water discharge at each of the locations for the given precipitation event. DEM of grid cells can be used to identify water flow directions and generate water flow accumulations (Maidment 2005; Djokic 2015). Figure 11.3 is a diagram showing a method for computing the water flow directions using DEM.

Accumulations of water flow for each of the grid cells can be quantified by calculating the discharges of how many grid cells are flowing to that particular grid cell. When an artificial threshold of the total number of grid cell discharges is determined, the linear vector feature class of stream segments can be generated. Figure 11.4 shows the computation of water flow accumulation.

In the GIS environment, a unique stream segment ID can be assigned to each of the digital river segments. Based on the elevation and slope of grid cells, water catchments, and watershed can be delineated digitally (Figure 11.5).

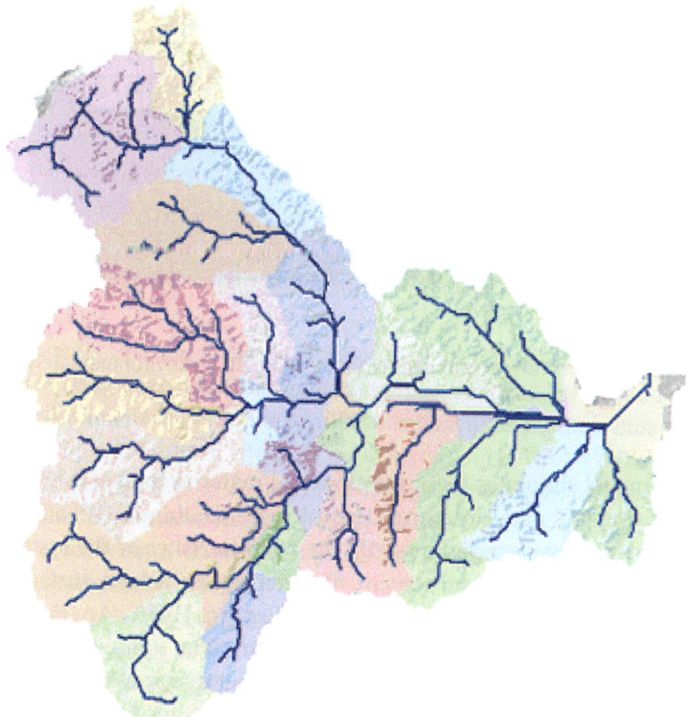

**Figure 11.2** River flow modeling using DEMs.

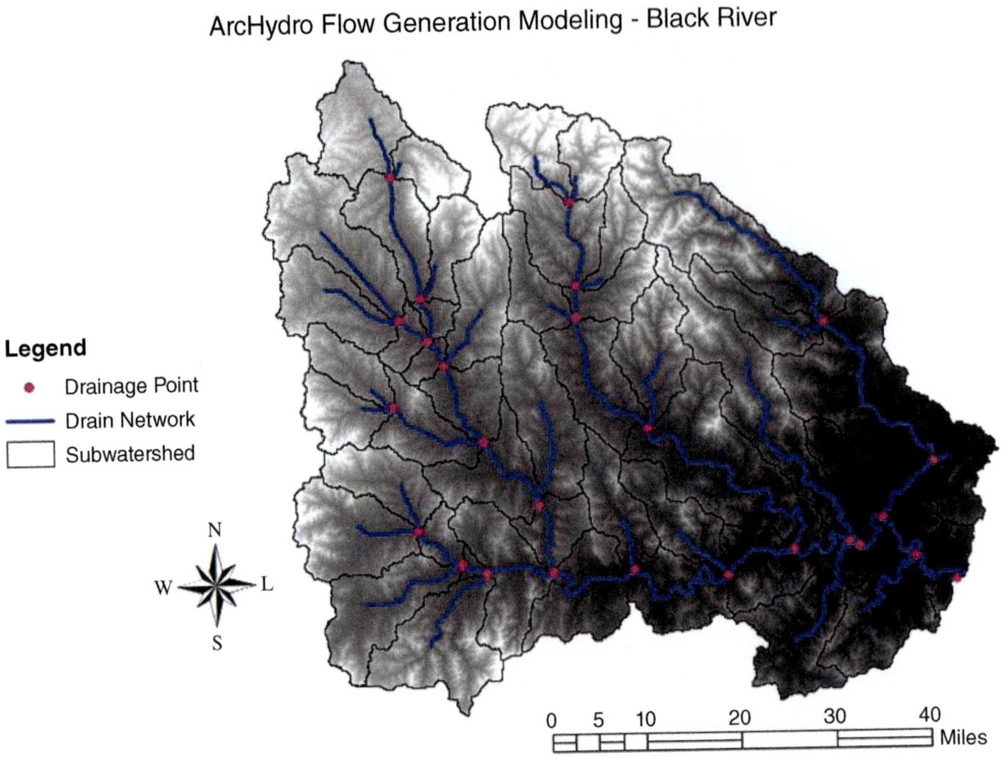


**Figure 11.6** ArcHydro modeling of water flow accumulations in the Black River Drainage Basin.

research combines the entomological and epidemiological factors and geo-spatial environmental factors to predict and simulate the regions and intensity of Malaria infections.

Location-specific mosquito abundance digital maps that play an important role in malaria control activities by guiding future resource allocation for malaria control and identifying hotspots of infections for further investigation. The best modeling practices in an effort to achieve improved efficacy, cost-effectiveness, ecological soundness, and sustainability of mosquito control for malaria were discussed. In this modeling process, computer source code and flowcharts of the spatial ABM were produced, and landscape generation functions that can simulate landscapes with varying spatial heterogeneity of different types of natural environment conditions, including aquatic habitats and houses that the mosquitos may grow.

Another example is to apply agent-based GIS modeling for predicting or simulating the COVID-19 epidemic in the Huangpu District, Shanghai, China (Dong et al. 2022). This research collected and sorted out the geographical coordinates, several feature classes of building and capacity of all types of buildings in the study area, including residential buildings, hospitals, schools, nursing homes, public places and leisure places. Figure 11.7 is the digital map shows the feature classes of the different types of buildings.

Meanwhile, the population data of the residents in the study area of 265,425 families and 662,030 persons were obtained from the newly published Chinese census database (seventh national census). In the model, each of the 662,030 persons is treated as one mobile agent, who can travel from one location to another. The individual movable agents were grouped into the age groups in

**Figure 11.7** Building locations for agent-based simulation modeling. *Source:* Dong et al., 2022/MDPI/CC BY 4.0.

order to simplify the hypothesis process of mobile directions and destinations. People in the group of age 60 or higher in a family mainly stay at home or go to nursing homes. All the grouped agents aged 5–17 are assigned to all kindergartens, middle schools, and high schools, and the number of students is allocated according to the size of the school. Similarly, both universities and vocational colleges assign agents of the age group of 18–24 according to the size of the school. Meanwhile, workplace distribution is generated from occupational statistic studies, along with the number of workers in hospitals, schools, and nursing homes based on their sizes. Figures 11.8 and 11.9 is the digital map showing the agent's original locations and the age group occupation among the total population (Dong et al. 2022).

The agent-based GIS modeling applications over a time period helped field hospitals and medical doctors to predict the possible outcomes. Agents requiring treatment can change their treatment status based on their treatment status and clinical observations, and the initial treatment of the patient can be obtained from clinical data based on the probability of normal hospitalization and ICU hospitalization, combined with the patient's age. Modeling prediction results during a testing time period were validated by the hospital-collected field treatment data. Four aspects (indicators) of the output data from the model were compared with the real-world data in the Huangpu District (the study area). (i) total number of confirmed COVID-19 cases during the simulation, (ii) total number of deaths during the simulation, (iii) average number of new cases per week during the simulation, and (iv) weekly deaths during the simulation. All four factors of predictions are

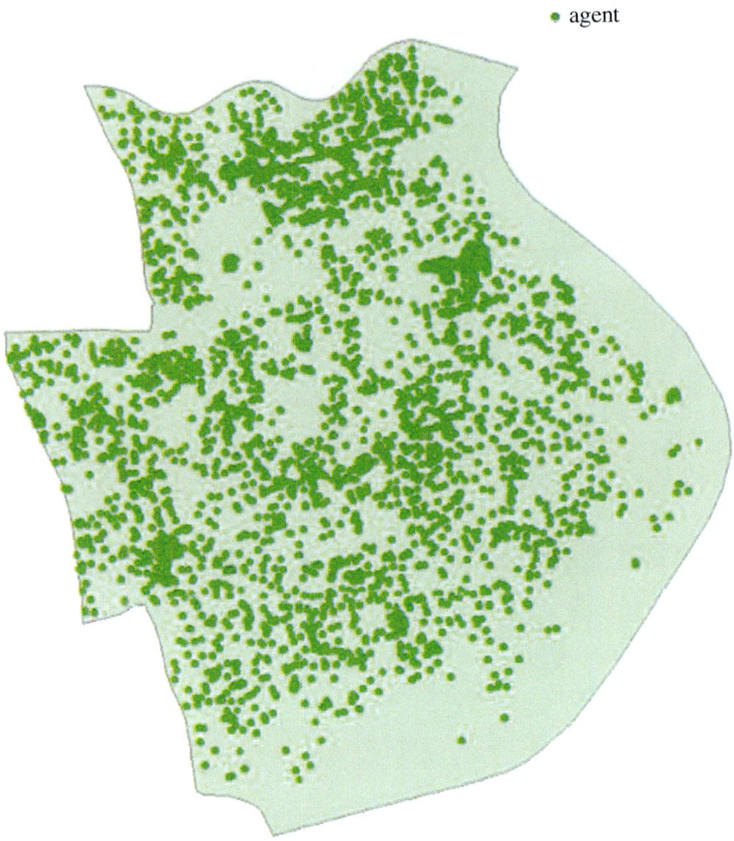


**Figure 11.8** Agent assignment map in the agent-based simulation modeling. *Source*: Dong et al., 2022/ MDPI/CC BY 4.0.

reasonably matched with field-collected datasets and the trends. Figure 11.10 graphs are some of the field data validation test results.

## Issues of GIS Software Enhancement and Programming

Software is a major component in the GIS process although GIS is naturally data-intensive or extensive. The reason that we need to enhance the software is simply because some of the data processing functions do not exist on the current software platform, such as ArcGIS Pro or QGIS. Or a specific business or company needs a particular type of application tool to process its datasets to increase the effectiveness of the workflow. GIS software development or enhancement is the process of designing, building, and maintaining GIS apps. These apps enable users to capture, store, analyze, and visualize spatial data. These apps can either attach to a software platform, such as ArcGIS Pro, or stand independently.

GIS software development involves careful planning, testing, and programming and coding to develop reliable and scalable software. GIS software solutions can be beneficial to various industries, from logistics of transportations to improving efficiency of workflow to vehicle fleet

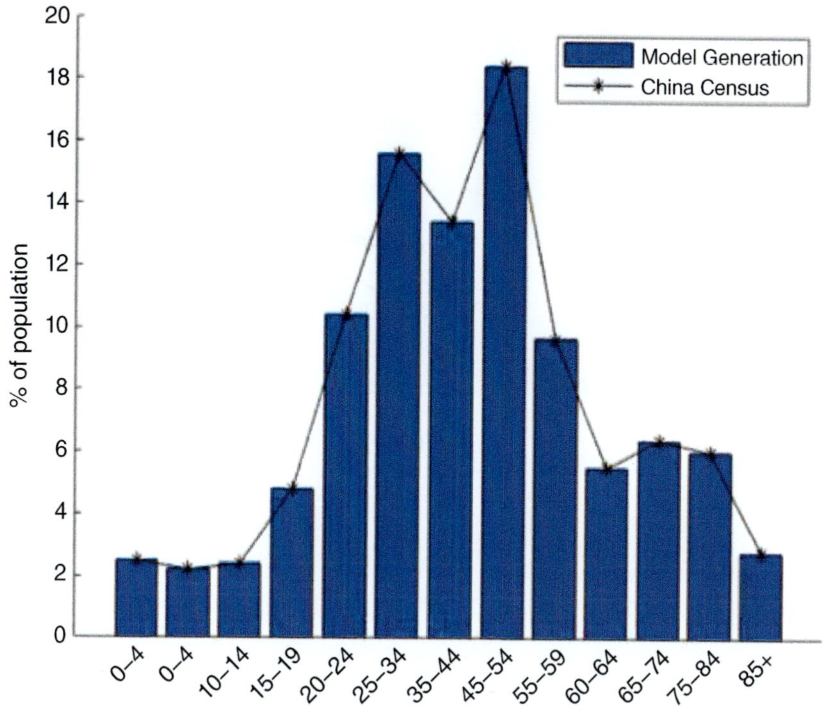


**Figure 11.9** Age characteristics of the residential population. *Source*: Dong et al., 2022/MDPI/CC BY 4.0.

movement management, as well as many other possibilities. The key objectives of GIS software development in practice are data acquisition, data conversion, data analysis, and data visualization (Wieczorek and Delmerico 2009).

Computer programming languages are the backbone of software development. GIS software developers need to use one of these languages to build the specific applications for specific industrial needs or specific tasks. The widely used programming languages in GIS are Python and JavaScript. Python has been used for making functions and tools in ArcGIS Pro. Python has libraries like ArcPy (for ArcGIS) and GeoPandas, making it crucial for spatial data analysis, automation, and geospatial processing. JavaScript is mainly used as an essential language for web-based GIS applications. Libraries of JavaScript like Leaflet and OpenLayers are utilized for interactive maps and spatial data visualization online.

In GIS software development practice, there are six widely agreed stages.

1. Gathering and understanding the requirements
   Understanding the users' needs and goals as well as the constraints of the function or tool applications is the first step for the software development. Thorough research for the needs and objectives to achieve using this planned tool or function needs to be accomplished before the next step.
2. Designing the algorithm and architecture
   Design the architecture and internal workflow of the tool or function. Gathering and logistically putting together the components of the tool or function, such as mathematical models, etc. The design of the GIS app must maintain its scalability, flexibility, and portability.

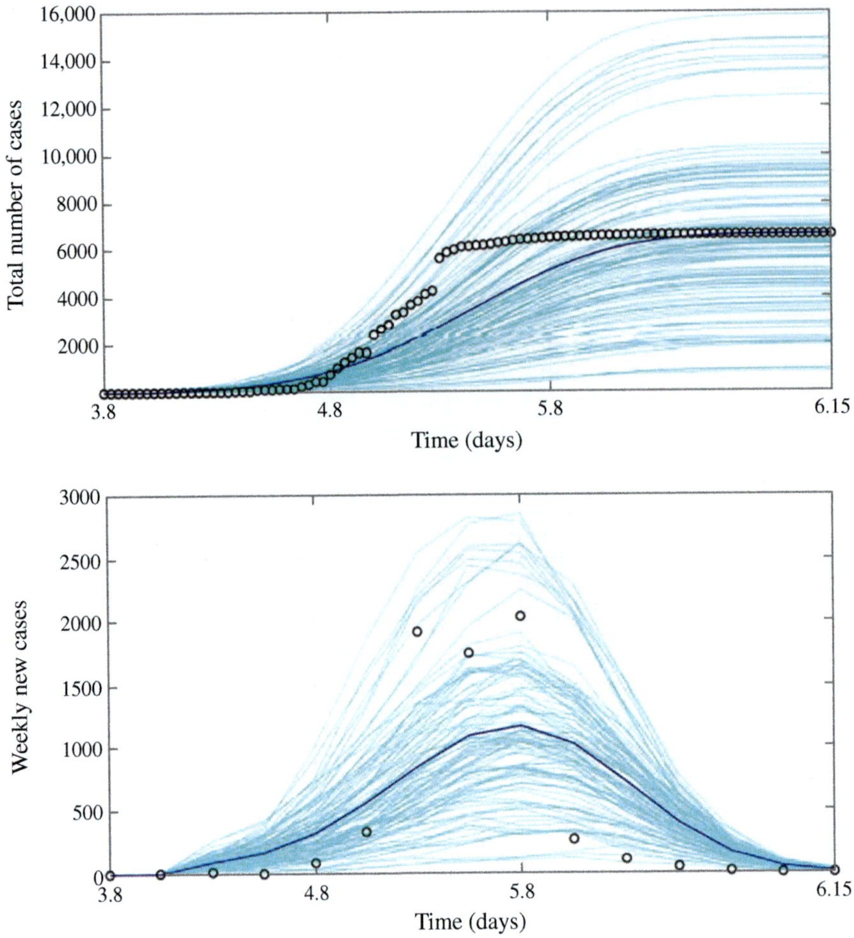


**Figure 11.10** Matching the agent-based modeling predictions to field-measured data. *Source*: Dong et al., 2022/MDPI/CC BY 4.0.

3. Development and programming
   This phase involves writing the program using a particular language. The ideal situation is that the people to write the code understand the GIS applications and the goals.
4. Software testing
   This is a step to confirm that the developed software meets the user requirements and functionalities. There are three major types of testing for software apps in general, and GIS app in particular. (i) functional testing, (ii) performance testing, and (iii) user acceptance testing.
5. Software deployment
   The purpose of developing a GIS app is for the specific users or anyone using the GIS to use. Therefore, deployment method and extent are very important. We should be careful planning and setting up configurations to secure the intended accessibility and functionalities of the software.
6. Software maintenance
   Maintenance means ensuring that the tools or functions continue to perform as planned and meet the changing needs of the users. This process also includes bug fixing, updating, and enhancements of the application.

GIS software developers should use the industry-standard tools and programming language to ensure compatibility in the computational environment; should document the programs, input and output requirements, and should adhere to established conventions for naming, formatting, and documenting the code of the programs (Rogers et al. 2024). GIS software developers also need to take notes for user feedback and use them to improve the software tool or function.

## References

Anselin, L. (2005). Spatial statistical modeling in GIS environment. In: *GIS, Spatial Analysis, and Modeling* (ed. D.J. Maguire, M. Betty, and M.F. Goodchild), 480 pages. Redland, CA, USA: ESRI Press.

Arifin, S.M.N., Madey, G.R., and Collins, F.H. (2016). Spatial agent-based simulation modeling in public health: design, implementation, and applications for malaria epidemiology, 288 pages. Hoboken, NJ, USA: John Wiley & Sons Inc.

Batty, M. and Y. Xie, 2005. Possible Urban Automata. Environment and Planning B: Urban Analytics and City Science. 24 (2): 175–192.

Crooks, A. (2016). Cellular Automata, in Douglas Richardson Editor in Chief: The International Encyclopedia of Geography. New York, NY, USA: John Wiley & Sons, Ltd. https://onlinelibrary.wiley.com/doi/book/10.1002/9781118786352.

DeMers, M. (2002). *GIS Modeling in Raster*, 203 pages, New York, NY, USA: John Wiley & Sons, Inc.

de Smith, M., Goodchild, M., and Longley, P. (2013). *Geospatial Analysis*, 4e. Winchelsea Press. http://www.spatialanalysisonline.com/HTML/index.html.

Djokic, D. (2015). Hydrologic and Hydraulic Modeling with ArcGIS. https://proceedings.esri.com/library/usercont/proc15/tech-workshops/tw_382-228.pdf.

Dong, T., Dong, W., and Xu, Q.L. (2022). Agent simulation model of COVID-19 epidemic agent-based on GIS: a case study of Huangpu District, Shanghai. *International Journal of Environmental Research and Public Health* 19 (10242): 1–19. https://www.mdpi.com/1660-4601/19/16/10242.

ESRI. (2022). ArcGIS Pro Help Menu. https://pro.arcgis.com/en/pro-app/latest/help/.

Falah, N., Karimi, A., and Harandi, A.T. (2020). Urban growth modeling using cellular automata model and AHP (case study: Qazvin city). *Modeling Earth Systems and Environment* 6: 235–248.

Feng, Y., Liu, Y., Tong, X. et al. (2011). Modeling dynamic urban growth using cellular automata and particle swarm optimization rules. *Landscape and Urban Planning* 102 (3): 188–196.

Goodchild, M.F. (2005). GIS, spatial analysis, and modeling overview. In: *GIS, Spatial Analysis, and Modeling* (ed. D.J. Maguire, M. Betty, and M.F. Goodchild), 480 pages. Redland, CA, USA: ESRI Press.

Liu, Y. and Phinn, S.R. (2001). Developing a cellular automaton model of urban growth incorporating fuzzy set approaches. *International Journal of Computers, Environment and Urban Systems* 27: 637–658.

Maidment, D.R. (2005). Hydrologic modeling. In: *GIS, Spatial Analysis, and Modeling* (ed. D.J. Maguire, M. Betty, and M.F. Goodchild), 480 pages. Redland, CA, USA: ESRI Press.

Mantelas, L., Prastacos, P., and Hatzichristos, T. (2008). Modeling urban growth using fuzzy cellular automata. In: *11th AGILE International Conference on Geographic Information Science*. University of Girona, Spain.

Moghadam, H.S. and Helbich, M. (2013). Spatiotemporal urbanization processes in the megacity of Mumbai, India: a Markov chains-cellular automata urban growth model. *Applied Geography* 40: 140–149.

O'Sullivan, D. and Unwin, D. (2003). *Geospatial Information Analysis*. Hoboken, NJ, USA: John Wiley & Sons.

Rogers, S.R., Foust, B.G., and Nelson, J.R. (2024). Standard use of Geographic Information System (GIS) techniques in honeybee research 2.0. *Journal of Apicultural Research* 4–88. doi: 10.1080/00218839.2024.2357977.

Roy, J.R. and Thill. J.C. (2004). Spatial interaction modelling. Papers in Regional Science, 83 (1): 339–361.

Tang, T. and Zhang, J.Z. (2018). Time scale sensitive sensor applications in collecting and analyzing geographic event data. *Annals of Geographic Information Science* 24 (4): 241–253. doi: 10.1080/19475683.2018.1520299.

Tomlin, C.D. (2012). *GIS and Cartographic Modeling*. Redlands, CA, USA: ESRI Press. 420 pages.

Wieczorek, W.F. and Delmerico, A.M. (2009). Geographic Information Systems. *Computational Statistics* 1 (2): 167–186. doi: 10.1002/wics.21.

Zhang, J.Z. and Tang, T. (2014). Spatial and temporal changes of parking lot land use at SUNY – Buffalo State Applying Unmanned Helicopter Remote Sensing. *Middle States Geographer* 47: 92–101.

## Tutorial in ArcGIS Pro

Copy the tutorial data folder "Chapter11-Tutorial" to your working directory.

Open ArcGIS Pro and create a new ArcGIS Pro project file named "MyGIS_Learning11" with the "Map" template. Save the project file in the designated working folder, C:\ or D:\...... Chapter11-Tutorial.

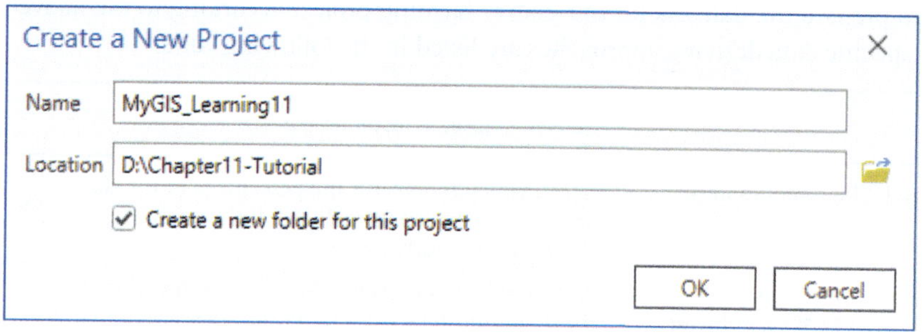


Now, highlight and right-click on the Folders on the Catalog Pane to obtain the contact menu. Then, click on "Add Folder Connections" to add the entire Chapter11-Tutorial folder from your working drive.

One of the practices of spatial modeling is conducting suitability analyses or modeling for specific objectives or goals. Suitability modeling is a type of analysis that locates places that are suitable or favorable for a certain geographic phenomenon or event, such as a new business in a city, a new infrastructure project, or an area for specific agricultural productions. In this case, we are going to conduct a project to locate suitable areas for new cherry orchards in the US part of the Niagara-on-the-Lake region. Spatial modeling processes are driven by effective goals and criteria. The goal of the modeling is to determine the purpose of analysis and the areas of study. The criteria of modeling define the model variables and the values considered suitable for each variable.

Part A: Suitable Cherry Farming Location Project

The workflow of suitability spatial modeling in general can be identified into six steps.

1. Defining the goal and supporting criteria. In our case, the goal of this project is to find and locate suitable areas in the study region for new cherry farming. Consulting with local cherry farming experts, we identified five major criteria currently. These are flat or gentle slope land; slope face to south, southeast, or southwest directions; close to the rivers of water resources for possible irrigation needs; close to express way or highways to facilitate product transportation to the markets; and current land us and land cover are not urban or developed, not an orchard, or an area with historical significance.
2. Deriving datasets that represent the model variables that are defined by the criteria. In our example, both slope and aspect can be created from the digital elevation models (DEM). Distance raster surfaces can be derived from vector feature classes of rivers and express ways in the study area. Detailed land use and land cover suitability raster surface can be derived from the land use and land cover polygon feature class.
3. Transforming the values in each derived datasets to a common suitability scale using one of the modeling methods, unique categories, range of classes, and continuous functions.

4. Combining the transformed datasets, which represent the suitability criteria, into a single suitability surface that meets the analysis goal.
5. Locating the suitable areas for the geographic phenomenon or event using the suitability surface. In this process, both the shape and the area size of the suitable patches need to be set up.
6. Finally, the GIS analyst needs to visually evaluate the suitability surface and the located areas to ensure that the model goal has been met.

First, we are going to prepare the datasets for our cherry farming project according to the above five (5) criteria. The specific data deriving approaches are listed in the following table:

| Criteria | Derived dataset in raster format |
| --- | --- |
| Flat or gentle slope land: Slope should lower than 20% | Slope, which represents the percentage of the slope and can be derived from digital elevation models (DEM) in the study region. |
| Direction of slope face (aspect): south (157.5°–202.5°), southeast (112.5°–157.5°), southwest (202.5°–247.5°) | Aspect, which represents the compass direction that the downhill slope faces in positive degrees from 0° to 360°. Can also be derived from DEM. |
| Close to the rivers of water resources: should be less than two miles. | Distance_To_Rivers, which represents distance from rivers in meters. |
| Close to expressway for quick transportation to the markets: should be less than four miles. | Distance_To_Express_Way, which represents distance from highways in meters. |
| Current land use and land cover: No urban developed land use, no historical heritage site, no competition to current orchards. | LandUse_LandCover_Suitability, which represents current land use or land cover with prior expert judgment suitability scores. |

Since we have prepared the raster datasets for these five criteria in the geodatabase Source_Data.gdb, now we can add these map layers to the MyGIS_Learning11 project.

In the Catalog Pane, change the mapping frame name from "Map" to "Cherry-Farming."

Add the following raster datasets from Source_Data.gdb to the Cherry-Farming window.

1. Niagara_Slope
2. Niagara_Aspect
3. LandUse_LandCover_Suitability
4. Distance_To_Rivers
5. Distance_To_Express_Way

Create a suitability model:

Now, we are going to create a weighted suitability model using the Suitability Modeler. On the Analysis tab of the main menu, in the Workflows group, click Suitability Modeler.

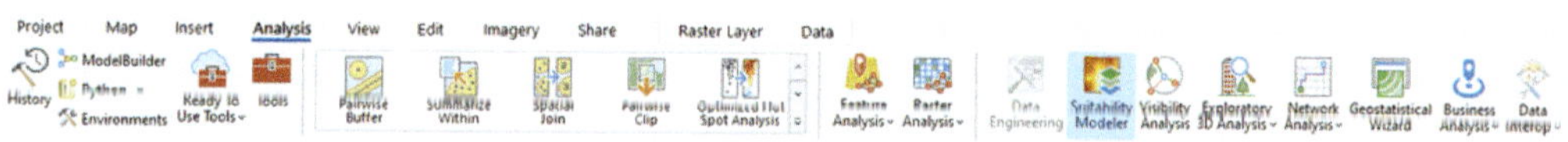


The Suitability Modeler Pane opens.

In the Suitability Modeler Pane, for Model Name, type Cherry-Farming Model.

For the Set Suitability Scale, keep the default scale of 1 to 10.

For Output Suitability Raster, keep the default location and name.

We can choose simply multiply (Multiplier) the criteria map layers or weight by percentage (Percent) using the Suitability Modeler. For Weighted by, select Percent.

On the Suitability Modeler tab, in the Suitability Analysis group, check off the Auto Calculate.

- Auto Calculate will automatically calculate the suitability scores for each of the criteria when you add in the model. Since we are going to perform the criterion transformation one by one and review its result, we turn this function off.

In the Suitability Model group, click Save.

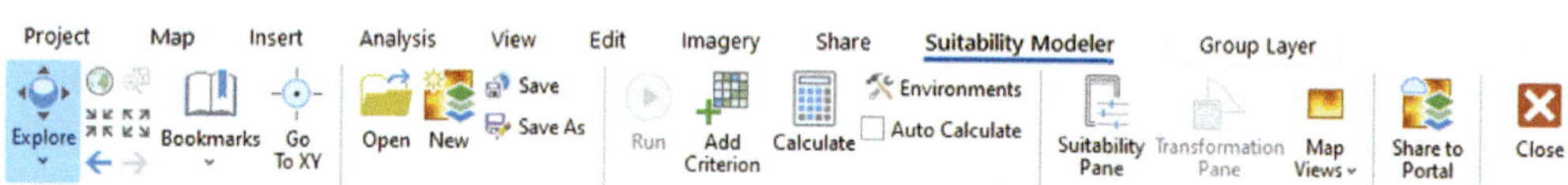

Save the project.

Expand the Spatial Analyst category on the Catalog Pane, You notice that MyGIS_Learning11.sam is added. In this folder, the Cherry-Farming Model is saved.

In the future, you need to right-click on the model name here to open the model in the GIS project.

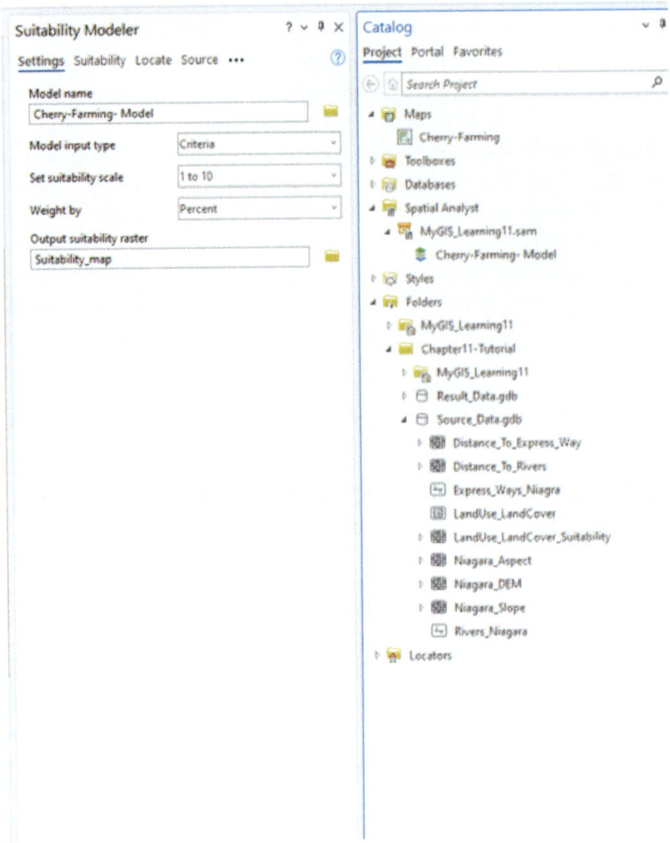


Save the project again.

Add the derived data:

Now, let's add the derived data to the Cherry-Farming Model and the weight percentage of suitability impact for each of the criteria.

In the Suitability Modeler Pane, click the Suitability tab.

Under Criteria, to the right of Input Rasters, click the Add button.

In the Select Input Rasters dialog box, navigate to......\ Chapter11-Tutorial\Source_Data.gdb.

In the Source_Data.gdb, select Niagara_Slope on the list, and then click OK.

You notice that the dataset is added in the suitability model.

Repeat these steps to add Niagara_Aspect, LandUse_LandCover_Suitability, Distance_To_Rivers, and Distance_To_Express_Way too.

We have a total of five variables in the Suitability Modeler now.

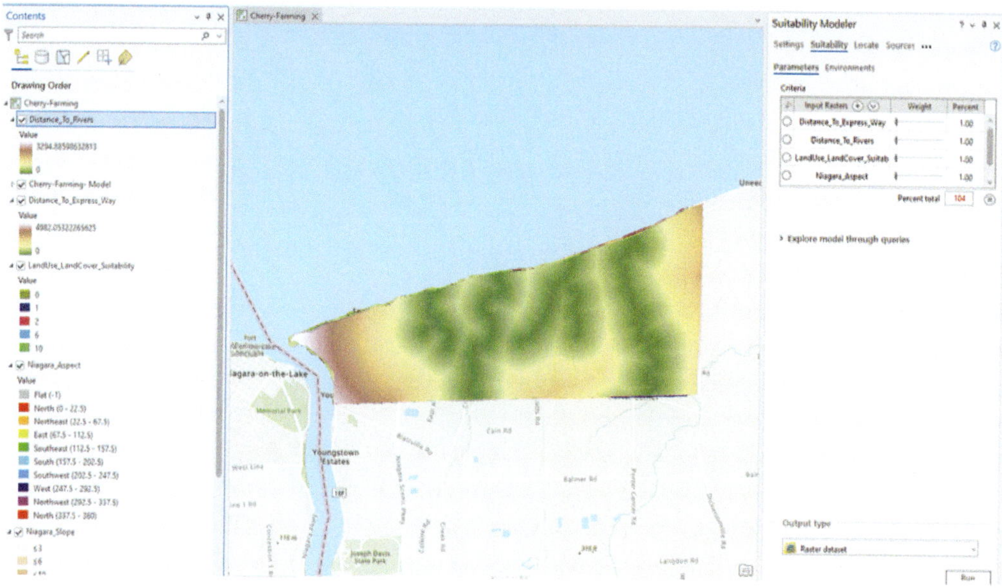

Previously, we discussed with local cherry farming experts and evaluated the weights of each of the five major criteria as follows:

| Variable | Percent |
|---|---|
| Niagara_Slope | 25 |
| LandUse_LandCover_Suitability | 25 |
| Niagara_Aspect | 20 |
| Distance_To_Rivers | 20 |
| Distance_To_Express_Way | 10 |

Enter these percentage weights to the Criteria of the Suitability Modeler.

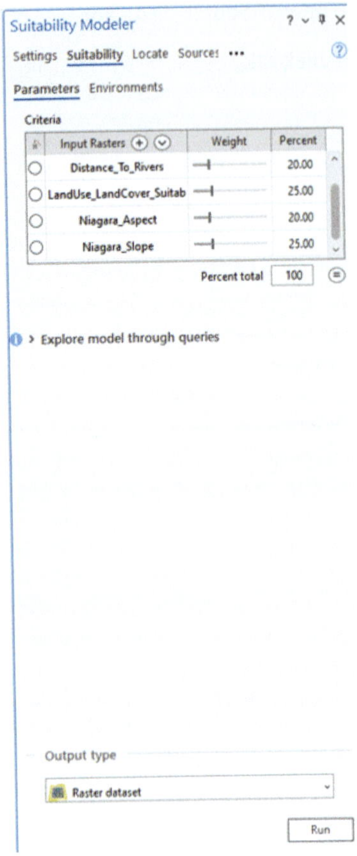


On the Suitability Modeler tab, in the Suitability Model group, click Save.

Save the project.

During the following process, if you would like to take a break, you can always save the Suitability Model (Cherry-Farming Model) on the Suitability Modeler tab. Then, close the Suitability Modeler Pane and save and exit the ArcGIS Pro project. When you are ready to continue, you can open the project in ArcGIS Pro, expand the Spatial Analyst database on the Catalog Pane, and right-click on the saved suitability model (Cherry-Farming Model) to open it again.

Transform data for a weighted suitability model:

In the Suitability Modeler, three approaches can be used to transform the values of a model variable to a common suitability scale. These are categories, ranges of values, and mathematical functions. The data type of the model variable determines which transformation approaches can be used. The general guidelines are shown in the following table.

| Model variable data type | Transformation method | Description |
|---|---|---|
| Discrete data with nominal or ordinal values | Categories | Using a table to transform each category that is represented in the derived raster surface to a score on the suitability scale. For example, cells that are categorized as agricultural land use (x-value) are assigned a suitability score of 2 (y-value). |

| Model variable data type | Transformation method | Description |
|---|---|---|
| Continuous data with interval or ratio values | Ranges of values | Using a table to transform ranges of values that are represented in the derived raster surface to a score on the suitability scale. For example, an x-value of 12 m falls within the range of 10–20 m, so that cell is assigned a suitability score of 2 (y-value). |
| | Mathematical functions | A mathematical function is used to transform the values that are represented in the derived surface to a score on the suitability scale. For example, an x-value of 12 m is input into a linear function to calculate a suitability score of 2 (y-value). |

When we use mathematical transformation functions, we follow the three main principles.

1. Small variable values are preferred: each function rescales the smallest variable values to the highest score on the suitability scale. For instance, the closest distance to rivers for irrigation needs, the closest distance to a highway for transportation of cherry product, or the smallest slope or flat land for the orchards in our case. The functions in this category include Small, MSSmall, Linear (negative slope), and Logistic Decay.
2. Large variable values are preferred: each function rescales the largest variable values to the highest score on the suitability scale. For instance, the criterion of a new residential area development in an urban region with at least 500 m from a superhighway to avoid noise pollution can be transformed using the functions in this category. The functions in this category include Large, MSLarge, Linear (positive slope), Logistic Growth, Exponential, lLogarithm, and Power.
3. Specific variable values are preferred: These functions assign preference to specified variable values or ranges of specified variable values. For instance, the south aspects or slope face (112.5°–247.5°) in our case. The functions in this category include Symmetric Linear, Gaussian, and Near.

Now, let's transform each of the five criteria to the suitability scores on the unified scale. If the Suitability Modeler Pane and Cherry-Farming Model are not open, open them from the catalog Pane. In the Suitability Modeler Pane, click the Suitability tab, and then click the Parameters tab, if necessary.

From the Suitability Modeler ribbon on the main menu, in the Suitability Analysis group, make sure the Auto Calculate box is unchecked or turn off this option (uncheck it). This way, the model data will not automatically update every time that a parameter is changed in the Transformation Pane.

In the Suitability Modeler Pane, to the left of Niagara_Slope, click the circle. The circle will turn green. The Transformation Pane opens below the map view, and the Transformed Niagara_Slope layer and Suitability Map layer are added to the Table of Contents in the Cherry-Farming Model group. A green circle next to an input raster means that the raster is active in the Transformation Pane.

In the Table of Contents, check off the Suitability Map.

In the Transformation Pane, if necessary, click the Continuous Functions tab and then set the following parameters:

- For Function, choose Small.
- For Midpoint, type 10, which is a 10% slope.
- For Point Spread, type 1 to decrease the steepness of the curve.
- For Upper Threshold, type 20, which is a 20% slope.
- For Value Above Threshold, type 1 to assign the lowest suitability to locations that are more than 20% slopes.

Click Calculate icon  on the Suitability Modeler tab in the main menu.

On the Suitability Modeler tab, in the Suitability Model group, click Save.

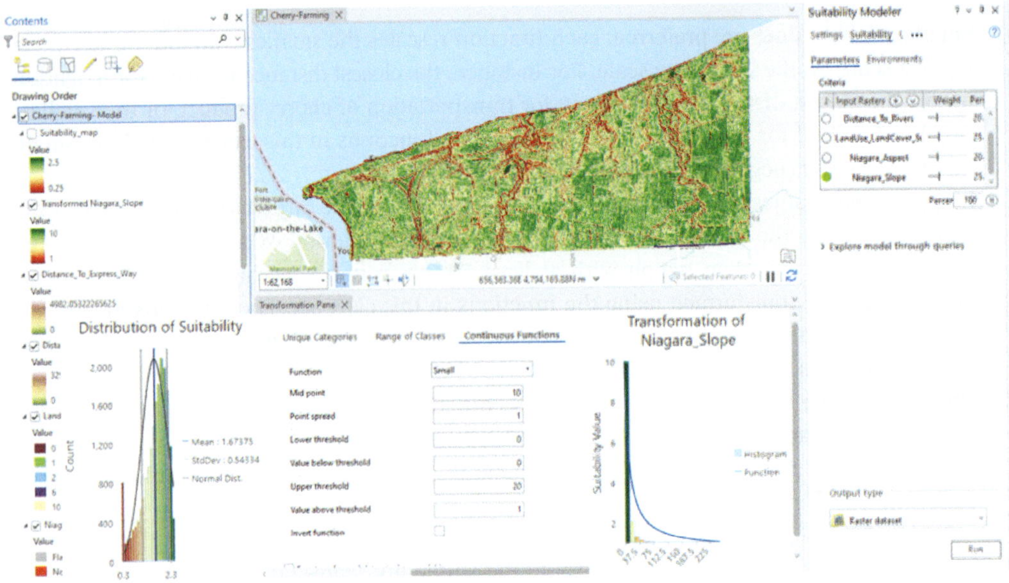


Next, let's transform the Aspect layer. In the Suitability Modeler Pane, to the left of Niagara_Aspect, click the circle, and it turns green. You notice the changes on the Suitability Modeler Pane, the Transformation Pane, and the mapping window.

In the Transformation Pane, click the Continuous Functions tab, if necessary, and then set the following parameters:

- For Function, choose SymmetricLinear.
- For Minimum Value, type 112.5°, which is the start of the southeast direction.
- For Maximum Value, type 247.5°, which is the end of the southwest direction.

Keep the default for the Lower threshold and Upper threshold.

- For the values above and below the threshold, type 3 to assign the lowest suitability if the slope angle does not face to southern directions.

Click Calculate icon  on the Suitability Modeler tab in the main menu.

On the Suitability Modeler tab, in the Suitability Model group, click Save.

Save the project.

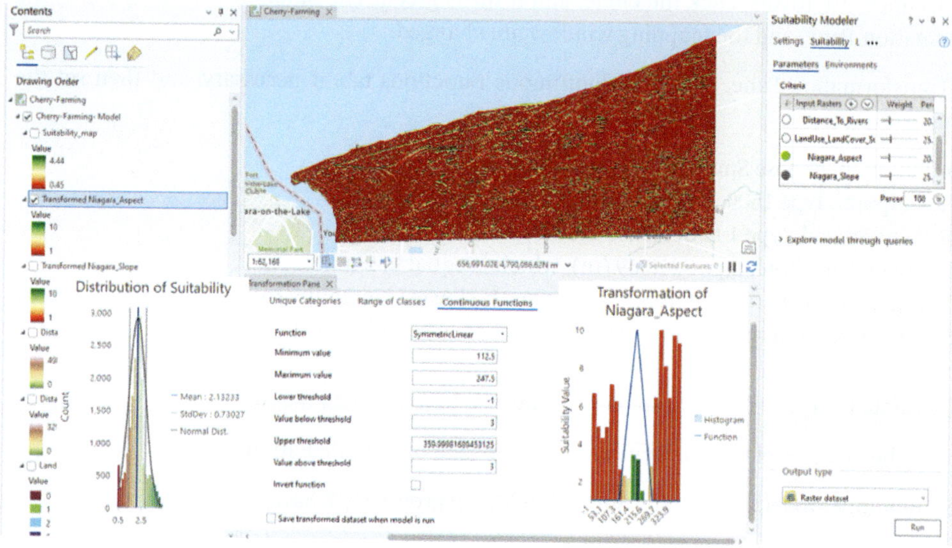


Next, we are going to transform the land use and land cover layer.

In the Suitability Modeler Pane, to the left of LandUse_LandCover_Suitability, click the circle and it turns to green. The Suitability Modeler Pane, the Transformation Pane, and the mapping window are changed. Owing to the method used for data preparation, we can use either unique categories or range of classes to transform this dataset with the same result. We can use the Range of Classes to do the modeling. In this process, we convert the original scores that experts provided to the unified scores here.

For the Field, accept the default of "Value." Click Calculate icon  on the Suitability Modeler tab in the main menu.

On the Suitability Modeler tab, in the Suitability Model group, click Save. Save the project.

Next, we are going to transform the Distance to the Rivers. In the Suitability Modeler Pane, to the left of Distance_To_Rivers, click the circle, and it turns green. The Suitability Modeler Pane, the Transformation Pane, and the mapping window are changed.

In the Transformation Pane, click the Continuous Functions tab, if necessary, and then set the following parameters:

- For Function, choose Small.
- For Midpoint, type 1609.35 (m), which is equal to 1 mile.
- For Point Spread, type 1 to decrease the steepness of the curve.
- For Upper Threshold, type 3218.69 (m), which is equal to 2 miles.
- For Value Above Threshold, type 2 to assign the lowest suitability to locations that are more than one mile from rivers.

Click Calculate icon  on the Suitability Modeler tab in the main menu.



You notice the changes of transformation map after each of the calculations.

On the Suitability Modeler tab, in the Suitability Model group, click Save.

Save the project.

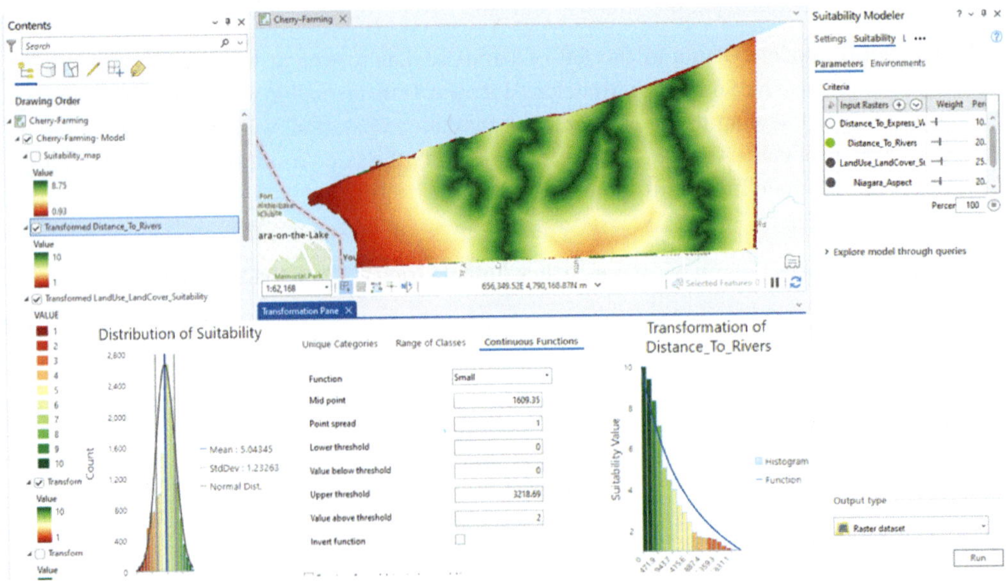



Finally, we transform and evaluate the distance to express the way raster map layer.

In the Suitability Modeler Pane, to the left of Distance_To_Express_Way, click the circle and it turns to green. The Suitability Modeler Pane, the Transformation Pane, and the mapping window are changed. In the Transformation Pane, click the Continuous Functions tab, if necessary, and then set up the following:

- For Function, choose Small.
- For Midpoint, type 3218.69, which is equal to 2 miles.

- For Point Spread, type 1 to decrease the steepness of the curve.
- For Upper Threshold, type 6,437.38, which is equal to 4 miles.
- For Value Above Threshold, type 2 to assign the lowest suitability to locations that are more than four miles from express way.

Click Calculate icon (Calculate) on the Suitability Modeler tab in the main menu. On the Suitability Modeler tab, in the Suitability Model group, click Save.

Save the project.

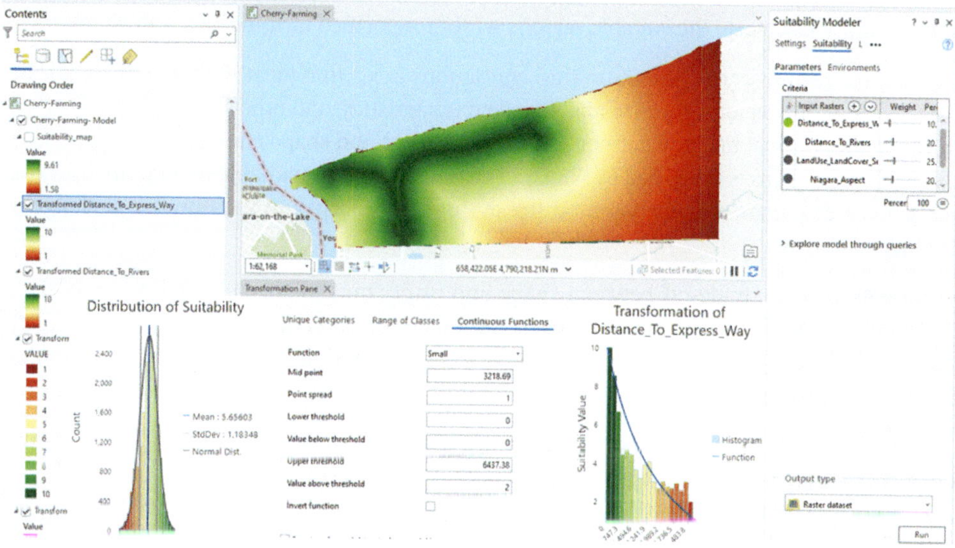


Close the Transformation Pane. Save the project again.

Now, we are going to create a full-resolution suitability map. In the Table of Contents, verify that Suitability Map is turned on, and then turn off and collapse any other visible layers (except for the base-map).

In the Suitability Modeler Pane, click the Suitability tab, and then click the Environments tab. Under Raster Analysis, for Cell Size select "Same as Niagara_Slope."

On the Suitability tab, click the Parameters tab, and then reopen the Output type to select again (make sure) Raster dataset (Create new raster data with full resolution……).

Click Run. The Suitability_map raster layer is produced in the MyGIS_Learning11.gdb (the default geodatabase of the project). The map is also projected in the Cherry-Farming mapping frame.

Save the model and close the Suitability Modeler Pane Save the project and exit if you would like to take a break.

Summarize results and locate the suitable regions with the specified farming area and shape.

Locating suitable regions is the process to find contiguous areas that meet the modeling goal. To locate the best suitable regions, the size and shape characteristics, and other regional criteria that support the modeling goal must be defined. Based on the size and shape characteristics, a pool of candidate suitable regions is identified by ArcGIS Pro. Then, the most suitable regions are selected from the pool of candidate regions based on the evaluation criteria and the number of the geographic objects needed.

In the summary process, trade-off between shape and utility must be made based on input data of the suitability raster map. The extent of each candidate region is determined by the trade-off between maintaining the specified geographic area and shape and maximizing utility (suitability scores). There are three approaches.

1. 100% of geographic regional shape: the region extent is determined by the specified geometry only. For example, it may be important to maintain a square-shaped region when locating a field for an agricultural production.
2. 0–100% of geographic regional shapes: the region extent is determined by a combination of maintaining the specified geometry and maximizing the utility. For example, when modeling animal habitat areas, a 50% trade-off between shape and suitability scores may be the best because animal territories are neither perfectly geometric nor entirely irregular in shape.
3. 0% geographic shape: the region extent will be deviated from the specified shapes that only incorporating connected grid cells that have the highest suitability scores. For example, it may be important to maximize the real growing environment if we are locating a plant species that grows in any area with favorable conditions.

If you close the Cherry-Farming Model, in the Catalog Pane, expand Spatial Analyst, and then expand MyGIS_Learning11.sam. Right-click the model and choose Open.

In the Suitability Modeler Pane, click the Locate tab, and then set the following parameters:

- For Input make sure it is "Suitability_map."

Navigate to your "MyGIS_Learning11.gdb" geodatabase to load the input.

- For Output Raster, type Suitable_Regions.
- For Number of Regions, type 10 to meet the model goal of locating ten optimal regions of new Cherry Farming project.
- For Region Shape, choose Square.
- For Region Orientation, leave the default value of 0.

- For Shape/Utility Trad-eoff (%), leave the default value of 50.
- For Area Units, choose Acres.
- For Total Area, type 300.
- For Region Maximum Area, type 50.
- For Evaluation Method, leave the default option of Highest Average Value so that the located regions have high suitability throughout the area.
- For Distance Units, choose Yards.
- For Minimum Distance Between Regions, type 100.

In the Suitability Modeler Pane and the Locate tab, switch to the Environments tab. For Raster Analysis – Cell Size, select "Maximum of input."

Switch back to the Parameters tab and click on the Run to process the data.

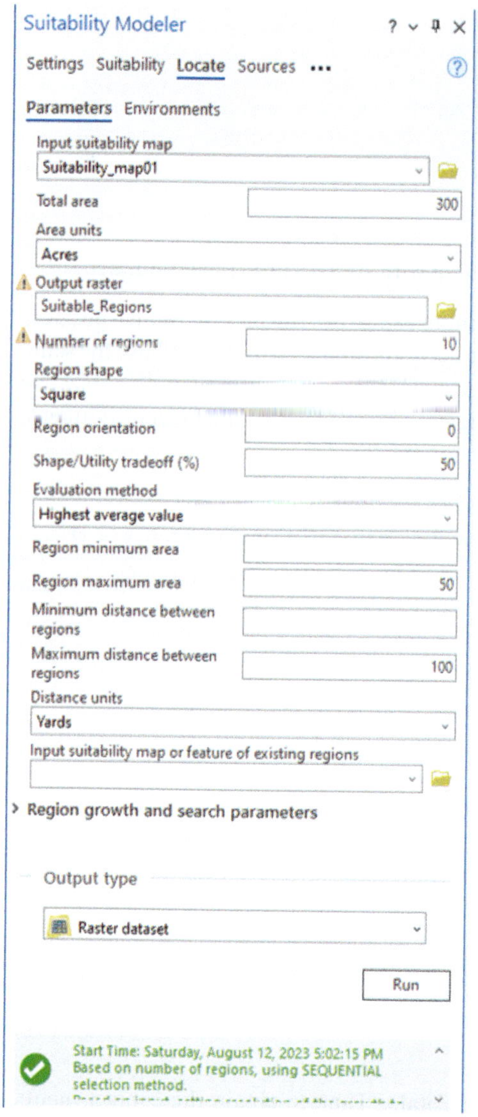

After the process is completed, you notice the Suitable_Regions map is added in the Cherry-Farming mapping frame. Close the Suitability Modeler Pane. Zoom to the entire map layer of Suitable_Regions. You have now accomplished the new cherry farming GIS best location modeling project. Congratulations!

Save the Cherry-Farming Model. Save the project and exit ArcGIS Pro.

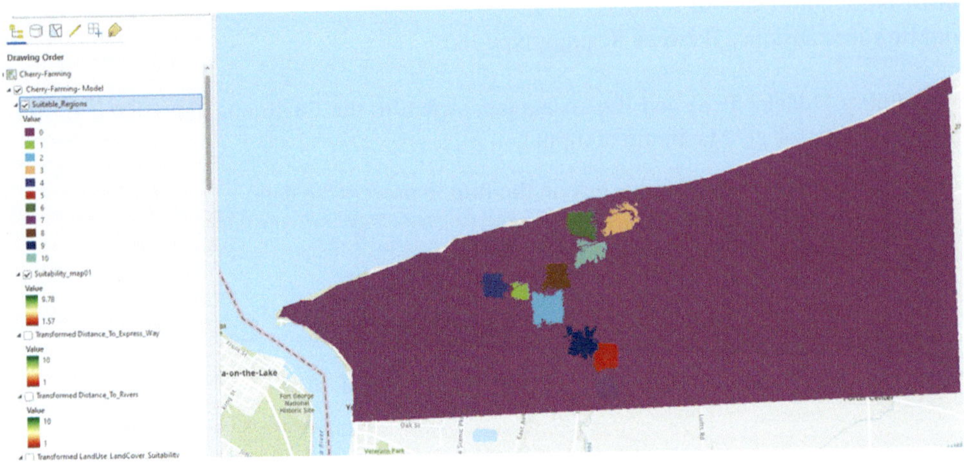


Part B: Streamline Geoprocessing and Workflow Planning with ArcGIS Pro "Model-Builder."

In this exercise, we are going to practice workflow planning using the ArcGIS Pro "Model-Builder" platform and conduct visual programming strategies to accomplish the goals. We will use the same dataset that we used in Chapter 6. Basic spatial analyses in a vector data environment. The objective of this exercise is still the same to use the geoprocessing functions of vector data to identify parcels of land ownership in the County of Erie, NY that falls into the flood zones.

Open ArcGIS Pro and create a new ArcGIS Pro project file that is called "MyGIS_Modeling" with the "Map" template. Save the project file into the designated working folder, C:\ or D:\...... Chapter11-Tutorial.

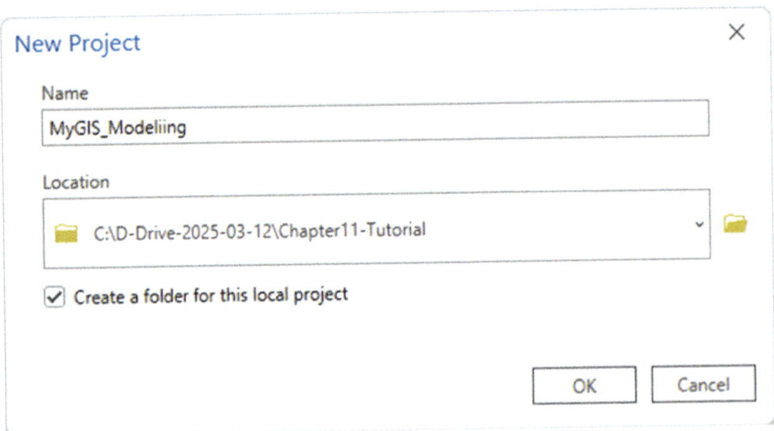


Now, highlight and right-click on the Folders on the Catalog Pane to obtain the contact menu. Then, click on "Add Folder Connections" to add the entire Chapter06-Tutorial folder from your working drive. If you do not have Chapter 06 on your working drive, you need to copy that folder to your working directory that contains Chapter11-Tutorial.Then, add the folder.

Add the three original feature classes or map layers from the Catalog Pane (Erie_County_Spatial_Data.gdb) to the mapping frame. These are: (1) Watersheds_Erie_Niagara, (2) FEMA_Flood_Zones, and (3) Parcels.

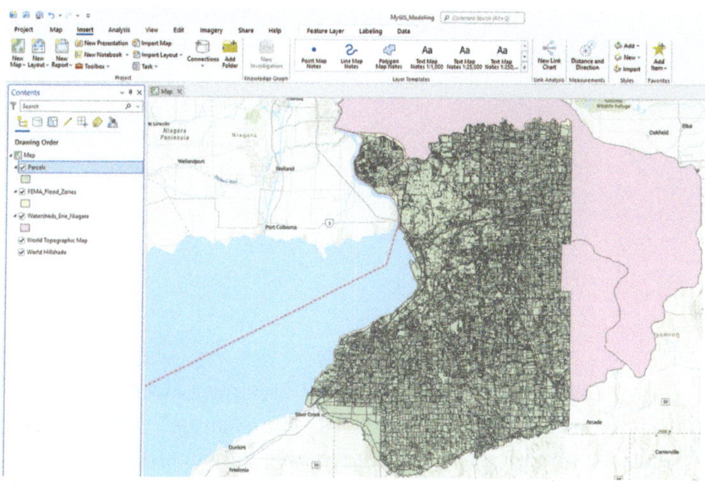

Now, expand the Toolbox category on the Catalog Pane. Highlight and right-click on MyGIS_Modeling.atbx to obtain the contact menu. Click on "New," then "Model" to create a new model in the toolbox that you created with the project. The Model window is open, and the Model tab is activated. Using the save as of "ModelBuilder" ribbon on the top to save the model as a new name called "Parcels_in_FloodZone."

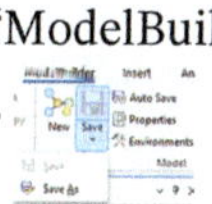

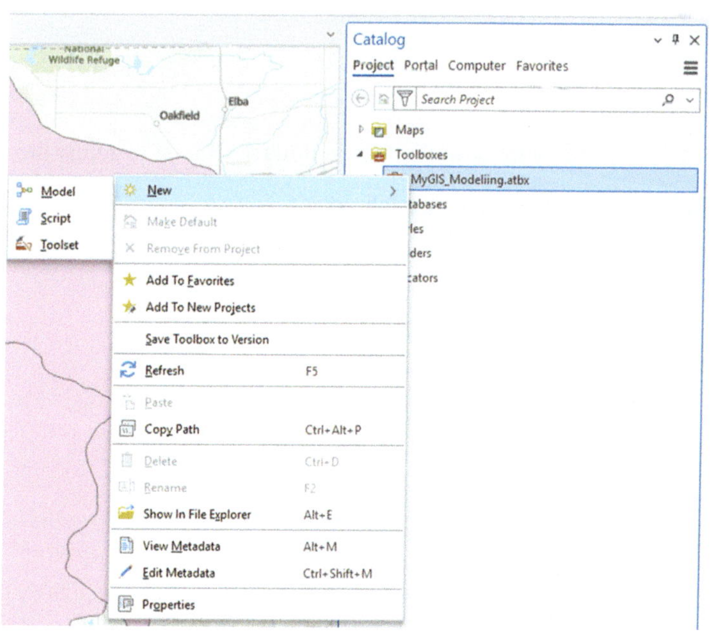

First, we need to clip the parcel land ownership map layer by the watersheds of Erie and Niagara Counties. Open the Geoprocessing Pane and search for the "Clip" tool. Drag and drop the Clip tool from Geoprocessing Pane to the "Parcels_in_FloodZone" model builder window. Then, drag and drop the "Parcels" map layer from the Table of "Contents" on the left-hand side of the window.

Drag a rubber band from the "Parcels" blue oval to the rectangular "Clip" tool. When the dialog opens, select "Input Features or Dataset" for the "Parcels."

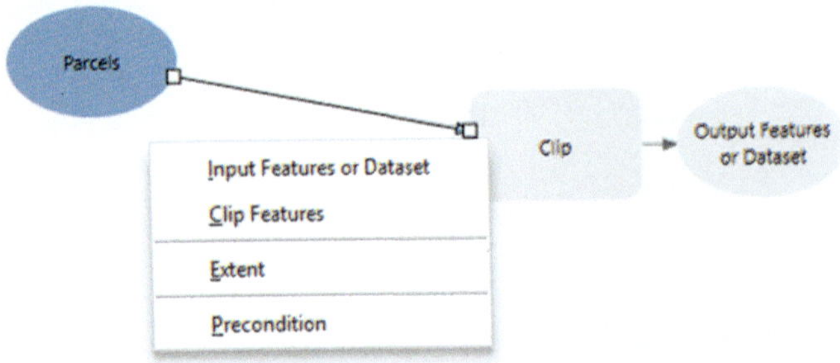


Drag and drop the "Watersheds_Erie_Niagara" to the model builder window as well. Using the rubber band method above to connect to the "Clip" tool as well. Select "Clip feature" this time. You notice that the tool or function is colored now. That means the required parameters are fulfilled and the function or tool is ready to run.

You can always to right-click on either the tool or output dataset to change settings or file names. In this case, we can right-click and open the dialog to change the output file as "Parcels_Clip01." You notice it is saved in the default geodatabase, MyGIS_Modeling.gdb.

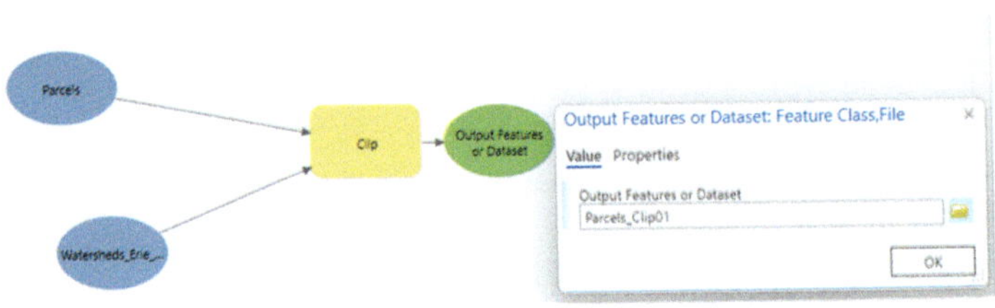


We can always use auto layout or fit to window tools on the "ModelBuilder" ribbon to adjust the layout of the model. 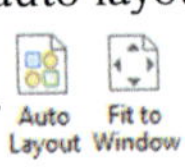 The final view of this step is as follows.


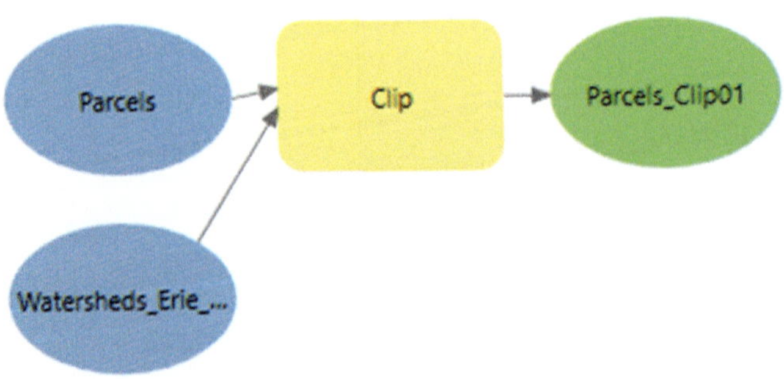

The blue ovals are original input datasets, the brown rectangular shapes are the tools and functions used, and the green ovals are the results of each of the steps of geoprocessing.

Now, let's add all the functions and tools from the Geoprocessing Pane and connect them with the inputs and outputs to form a processing model for the entire GIS project. Remember, the output of one step can serve as the input to the next step.

**Step 1:** Add the "Identity" tool from the Geoprocessing Pane and connect output of "Parcels_Clip01" from the previous step as the input for this step.

Add another copy of "Watersheds_Erie_Niagara" feature class as the "Identity features."

Change the output file name to "Parcels_Watershed01." Click on the auto layout icon on the top menu to reorganize.

**Step 2:** Add the "FEMA_Flood_Zones" feature class from the Table of "Contents" below the current workflow chain in the ModelBuilder window. Search and add the "Select" tool on the Geoprocessing Pane. Connect "FEMA_Flood_Zones" feature class to the "Select" tool, you notice the color change.

However, the query in the "Select" tool is not set up yet. We must open the tool dialog to set it up. Set up a query as "Zone" = "AE" or "Zone" = "AH" as follows.

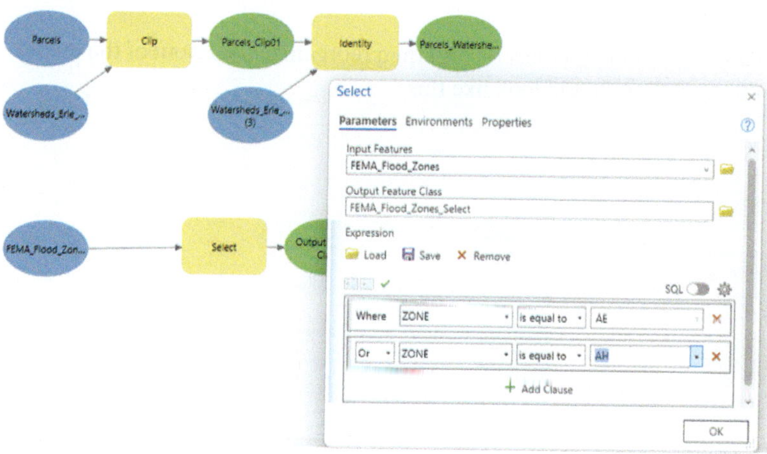


Change the output file name to "FEMA_Flood_Zones_AE_AH01."

**Step 3:** Since FEMA requires the flood zones to be accounted for 200 feet from the original zone boundaries, we need to create the buffer zone. Add the "Buffer" tool from the Geoprocessing Pane. Connect the output file "FEMA_Flood_Zones_AE_AH01" from the previous step to the "Buffer" tool as the input features. Then, open the buffer setting dialog to set the following.

- **a)** Output Features Class: FEMA_Flood_Zones_AE_AH_Buffer01
- **b)** Linear Unit: US Survey Feet
- **c)** Distance (value or field): 200

Click OK to change the settings.

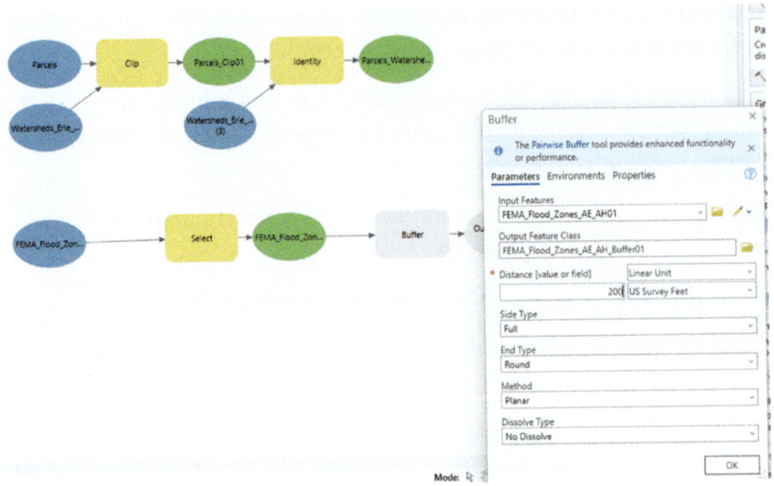


**Step 4:** Now, we can intersect the FEMA flooding zones map layer with the parcels in the watershed map layer to obtain the final result of this project, parcels in the flooding zones.

Add the "Intersect" tool from the Geoprocessing Pane to the window of "Parcels_in_FloodZone" model. Connect both "Parcels_Watershed01" and "FEMA_Flood_Zones_AE_AH_Buffer01" as Input Features.

Name the output file as "Parcels_in_Flood_Zones01" and then reorganize the layout of the model. The completed model of the workflow should look like this.

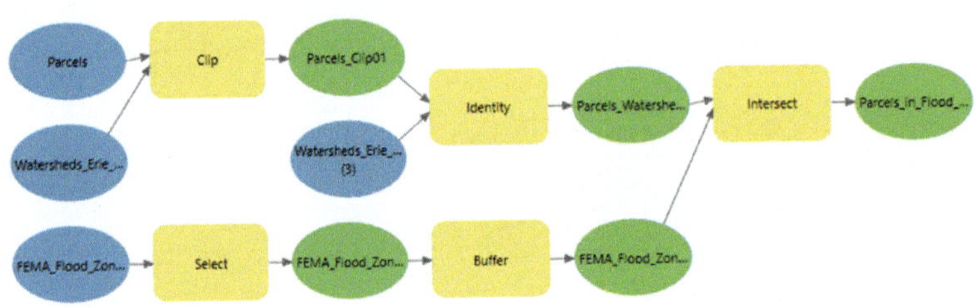


Save this model. Now, we can click on the "Validate" icon on the ModelBuilder ribbon to validate the model. If there is no any "bug" or mistakes, we can click on the run icon to run the model.

Run this model to obtain the results.

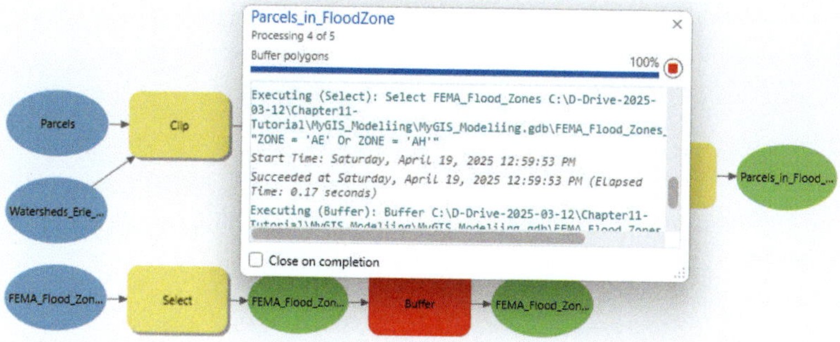


We can activate the "Map" window to add the result feature class or map layer from the geodatabase as follows. You may need to refresh your geodatabase.

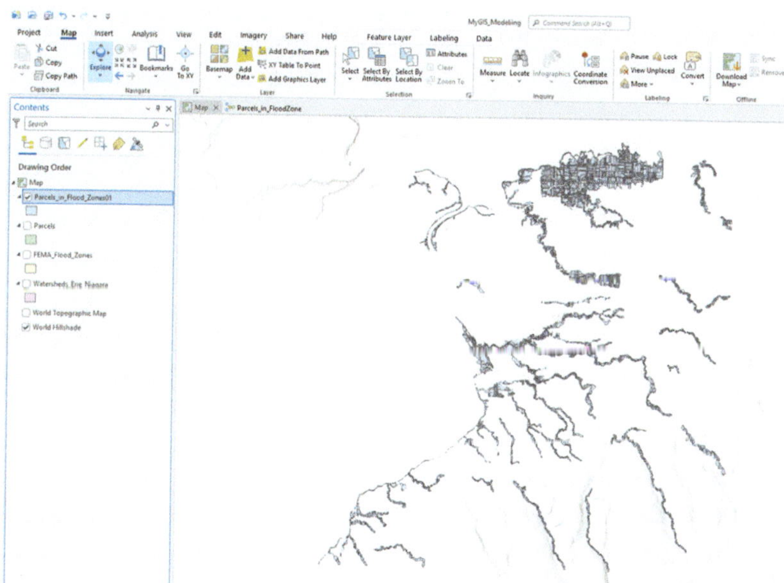


Save the model again. Close the Model-Builder window. When you need to open this model again, you can right-click and choose "Edit" on the contact menu.

Model building in ArcGIS Pro is a kind of visual programming. ArcGIS Pro models streamline the workflow. Model files can be copied and used in different study areas simply just changing the local original inputs.

Congratulations for accomplishing an ArcGIS Pro model!

# Part V

# Geographic Information Presentations and Digital Mapping

# 12

# Digital Mapping, Map Presentations, Cartography, and Geographic Information Sharing – A Summary

## Concepts and Theories

Major tasks of geographic information systems (GIS) are collecting, inputting, analyzing, and presenting spatial data. Digital maps are a major form that GIS systems produce to present and communicate the spatial data and analytical results of spatial data. Digital mapping is the process mainly performed in GIS by which a collection of data is compiled and formatted into a virtual image on a computer screen. GIS-produced maps are all in digital format originally unless they are printed as a hard copy or converted to digital images. The skills and techniques presented in this chapter can be applied in most of the previous chapters for spatial data visualizations and presentations. Since the focus of this book is spatial data analysis and processing, we present these skills and techniques in the last chapter of the book as the "capstone" type of strategy enhancements.

The professional or academic discipline that across the science, engineering, and art design for mapmaking is called cartography conventionally. One of the most recent definitions of cartography discipline states "......the technique of mathematically displaying spatial relations, emphasizing the importance of measurable and quantifiable aspects in creating scientific maps representing physical spaces" (Turnbull 2009). Cartography is an ancient discipline that was established in the early stage of human civilizations on the Earth's surface. The rapid development of GIS technology since the 1960s has relegated the autonomy of cartography. Cartography has since become a tool of GIS in which map production is but one of the functions in GIS processes (Silayo 2002). Meanwhile, the cartographical functionalities imbedded in GIS systems nowadays are much more efficient and powerful for mapmaking tasks. It has revolutionized the cartographic profession, which makes every GIS practitioners much more confident in making high-quality maps and maps for publications.

A map is a bird's-eye drawing of the Earth's surface, either for the entire Earth or part of it. Maps show the locations and shapes of geographic objects or features. Maps were originally drawn on paper. Each map carries a map scale that is the ratio of the distance measured on the map corresponding to the real-world distance. A digital map is a map that is originally produced using a computer software, most popularly GIS software, and presented on computer screen first. Conventionally, maps are classified into two categories: general reference maps and thematic maps. General reference maps, such as topographic maps in quadrangles, are used mainly to present locations of geographic objects. Thematic maps are used mainly to present one or more geographic variables or attributes (McMaster et al. 2005). In the thematic map representations, one variable means one theme, and this is the reason we call this type of map the thematic map. In essence, most of the digital maps we present in GIS are thematic maps since they carry the particular topics or attributes.

*Fundamentals of Geographic Information Systems (GIS) Analysis: Applying ArcGIS Pro*, First Edition. Tao Tang.

Both from geoprocessing perspective and data presentation perspective, maps are a central component in GIS. In ArcGIS Pro, the "Maps" window is designed as one of the major windows on the software platform. The "Maps" window displays layers that represent spatial data. In the digital environment, maps of Earth's surface environments can be presented in two-dimensional (2D) or three-dimensional (3D). The 3D mapping window that is parallel to a "Maps" window in ArcGIS Pro is called a "Scene" (Local Scene or Global Scene). Layers, or map layers, are collections of geographic data. Layers reference a data source; the "Maps" window in ArcGIS Pro interprets data as spatial, the properties and attributes of the spatial data specify how the layer draws on a map, a scene, or a layout. Generally speaking, a map layer shows one attribute or one theme at a time from the layer's attribute table. Meanwhile, from a vector GIS database point of view, one map layer is also called one feature class in a geodatabase. A feature class or a map layer is represented as points, lines, or polygons (areas) in a vector data structure. In raster data structure, one map layer in the "Maps" window normally represents one grid cell surface. In GIS, we need to use symbols to visualize the spatial data or geographic features.

## Principles of Symbolizing the Spatial Data on Digital Maps

GIS analysts need to present feature classes or map layers or raster datasets from time to time in the "Maps" window on a software platform. Graphics of features in points, lines, or polygons can be easily altered by their shape, size, orientation, or color to be distinctive and prominent to readers. Primary visual variables on maps were identified (Robinson et al. 1995; McMaster et al. 2005; Dent et al. 2009). These are shape, size, orientation, color, label, and space arrangement.

1. *Shape*: Shape is the specific graphic characteristic of the feature class to be presented. The shape of features may mimic the real-world shape that the features represent or are used by symbols. The shape of point or line features can be visually impacted by size and dimension combinations. For polygon features, shape is mainly determined by original geometry. In practice, we would like to symbolize different feature classes or map layers of the same geometric category, say points or lines, with different shapes of symbols.
2. *Size*: Size of symbols is the representation of the dimension of the shape, such as width, height, or area. In GIS, the symbol size of point or line features can be changed or classified according to the attribute values to be presented. Or GIS analysts can make different sizes of symbols for individual features.
3. *Orientation*: Orientation for point features is the direction of the symbols. For the line or polygon features, this is the elongation of mark patterns. It can help the readers to identify different features or feature classes.
4. *Color*: Color differences of symbols definitely help readers to identify different types of features. However, we need to present more features, feature classes, or categories of grids than the colors we can use, or human eyes can recognize. Colors can be further identified with differences of hue, value or tone, saturation, or chroma. We will discuss all these in details when we discuss the principles of color.
5. *Label*: Labels are generally the text put on the digital map layers to present the identity of the features. Labels have their own size, font, width, color, etc. In the GIS environment, labels of geographic features can be manually put on or presented from an attribute in the attribute table. Characteristics of labels of one feature class can help the readers to identify it from many others on the digital map.

6. *Space arrangement or arrangement*: For polygon features, arrangement represents the pattern that the markers are used. It can be regularly spaced pattern, or randomly spaced pattern for instance. For line features, the symbols can be dashed lines, dotted lines, or solid lines. For symbol of point features, the arrangement of a single symbol of color and pattern could be alternated.

In summary, the purpose of symbolizing different feature classes or raster datasets on a digital map in GIS is to effectively present analytical results or spatial data to the readers and decision-makers. More intelligent designs of digital map compositions can lead to more effective communications to digital map readers.

## Principles of Color and Color Schemes on Digital Maps

GIS analysts utilize different colors on the digital maps to present different geographic features or feature classes. In essence, colors are reflections of solar radiation or any other type of visible electromagnetic radiation by a specific area of a surface. In our case, this surface is the mapping surface. Human eyes of the map readers can normally recognize different colors. Therefore, we can communicate geographic distributions of features on a digital map using different colors. The nature of colors can be illustrated by the additive color theory. The white solar beam, or the reflections of the solar beam, or any type of visible electromagnetic radiation, are made up of three additive primary colors, red, green, and blue (RGB). Other colors, such as yellow, cyan, and magenta, can be produced by combining any of two additive primary colors. For instance, combining red and green colors will produce yellow.

Details of a specific color presentation are also based on the percentage of combinations for each of the primary colors. Therefore, GIS analysts encounter hundreds of varieties of detailed coloring names when we construct the digital maps. The differentiation of color details is based on the specific category or name of the color, the intensity and vibrancy of the color, and the lightness of the color (Dent et al. 2009). The major components of a color are hue, saturation, and value.

1. *Hue*: hue is the category or name that we give to various colors, such as red color, green color, orange-red, or yellow-orange. Figure 12.1 is a 12-category color wheel that shows the different details of the colors (Figure 12.1).

   Theoretically, a color wheel can contain an almost infinite number of hues. However, limitations of human eye recognition of colors are the maximum capacity for GIS analysts to explore among various colors. In the GIS environment, we normally apply up to 24 hues.
2. *Saturation*: color saturation refers to the intensity of color in an image or a digital map. As the saturation increases, the colors appear to be purer. As the saturation decreases, the colors appear to be more washed-out or pale. Saturation sometimes is also called "chroma." In practice, the saturation is the percentage of a specific color content in comparison to the natural gray background. The more percentage of the color, the higher the saturation. Figure 12.2 shows the differences in saturations of different colors (Figure 12.2).
3. *Value*: color value is the actual shade of lightness. In GIS and IT industries, we also use the terms "lightness" and "color value" interchangeably. Color values are important in providing precise choices in the complex color scale in the digital and computational environment in general and digital mapping in GIS in particular.

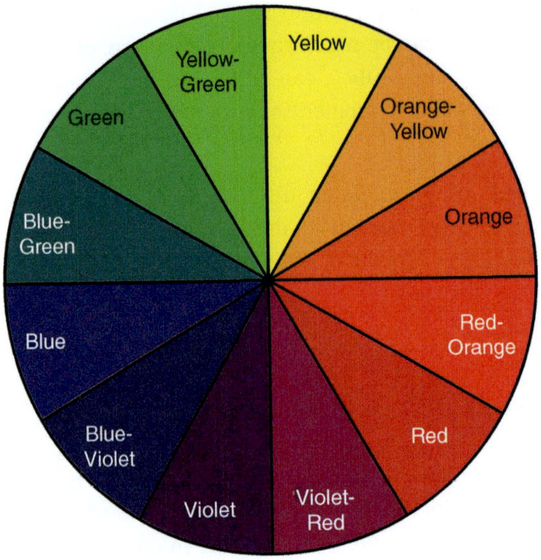


**Figure 12.1** A 12-category color wheel for art and fashion design. *Source:* Blanarovich 2017/ Christina Blanarovich.

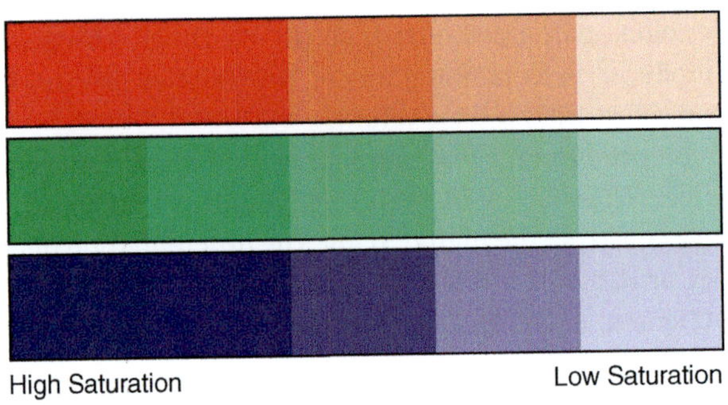


**Figure 12.2** Color saturation diagram. *Source:* Adapted from ESRI 2022.

In the GIS environment, color is a fundamental aspect for symbolizing that are not only used to draw geographic features, but also text and graphics on maps, scenes, and layouts. The color palette is available and imbedded in the "Symbology" geoprocessing dialog. GIS users can choose any specific color or a color combination to present the geographic objects or features, or continuous surface mad by grid cells. We also use color schemes to present geographic objects in GIS.

A color scheme is a range of related colors that can be applied to a collection of features, raster datasets, graphic elements, or symbol components. Color schemes contain at least two colors; the transition between pairs of colors can be either a continuous gradient or a discrete boundary with no blending from one color to another (ESRI 2022).

ArcGIS Pro differentiates the following five types of color schemes (ESRI 2022).

1. **Continuous color schemes**
   Continuous color schemes show a smooth gradation of two or more color combinations (Figure 12.3). Each component color is defined by a major color hue or name. The transitions

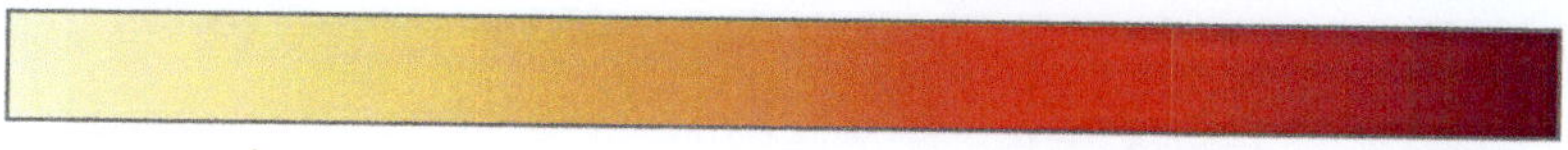

**Figure 12.3** Example of a continuous color scheme. *Source:* ESRI 2022/Environmental Systems Research Institute, Inc.

from one color to another and their positions in representing a thematic value of features from the attribute table are along the scheme and defined as a percentage of the total scheme length from left to right. A 100 or 0% of a particular color is always located at each endpoint. The end point of color representation can be switched as needed, but their proportional positions representing the thematic values cannot be changed.

2. **Discrete color schemes**

   Discrete color schemes consist of two or more colored blocks with borders and no gradation between the color blocks (Figure 12.4). Discrete color schemes can be used to create sequential or divergent color schemes on a map. Each of the color blocks contains a set of specific hue, saturation, value, and color combinations without changes in that block.

3. **Random color schemes**

   Random color schemes are defined solely by a minimum and a maximum hue, saturation, and value (HSV) of colors (Figure 12.5). "The scheme is compiled of discrete hues that are generated randomly so that their H (hue), S (saturation), and V (value) settings lie between the H, S, and V values of the minimum and maximum colors, respectively. A single transparency value is applied to the entire scheme" (ESRI 2022).

   Random color schemes do not contain any selectable components or color and hue transitions. The representation of attribute or thematic values of the geographic features is also randomly assigned. Individual color values can be modified using the Color Scheme Editor.

4. **Multipart color schemes**

   Multipart color schemes consist of any combination of two or more continuous or discrete color subschemes. Subschemes are the same as their stand-alone color schemes, except that continuous color schemes in a multipart color scheme contain only two percentage settings (Figure 12.6).

5. **Bivariate color schemes**

   Bivariate color schemes are 2D in color transition. They are designed for use with bivariate colors symbology only. This mainly occurs in raster datasets. This type of color schemes combines the color values of two discrete color schemes to create a multiclass grid of colors (Figure 12.7). Unlike discrete color schemes, bivariate color schemes can be rotated, but their sequential order cannot be rearranged dynamically.

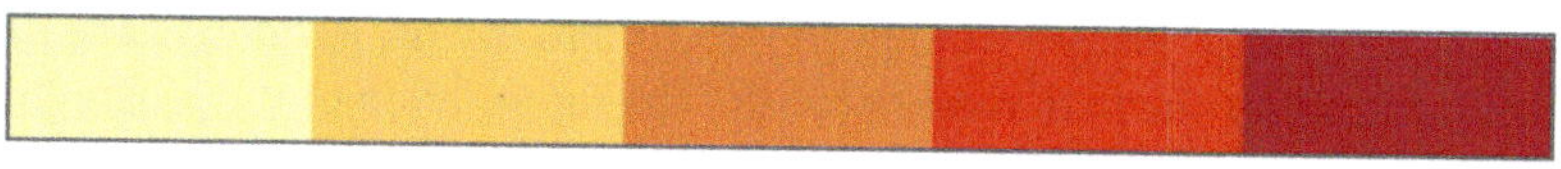

**Figure 12.4** Example of a discrete color scheme. *Source:* ESRI 2022/Environmental Systems Research Institute, Inc.

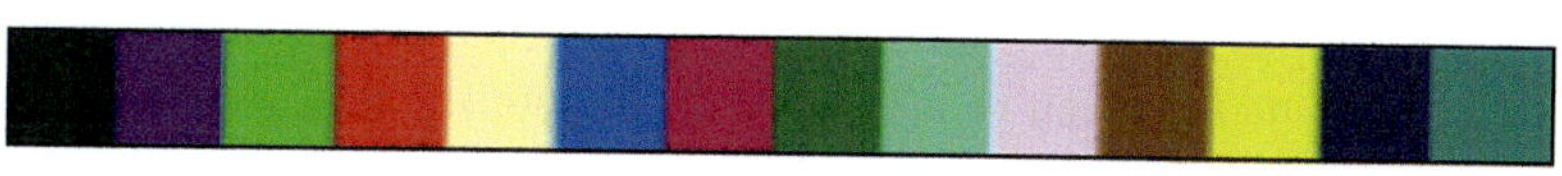

**Figure 12.5** Example of a random color scheme. *Source:* ESRI 2022/Environmental Systems Research Institute, Inc.

**Figure 12.6** Example of a multipart color scheme. *Source:* ESRI 2022/Environmental Systems Research Institute, Inc.

**Figure 12.7** Example of a bivariate color scheme. *Source:* ESRI 2022/Environmental Systems Research Institute, Inc.

In summary, it is significant that GIS analysts apply appropriate colors to present different feature classes or raster datasets in a map combination in communicating the spatial information with digital map readers and decision-makers. Normally, this goal is achieved by two common types of approach. One is to symbolize and color the geographic features by the colors that we see in the real world, and the other is to symbolize and color them according to a meaningful sense of communication. For instance, representing a river feature class in a map composition in blue color in GIS is much more appropriate than using a red color since water flowing in a river is blue. Feature classes of interstate highways and streets can be presented in black or red color depending on the specific map composition circumstance. GIS analysts need to consider other geographic feature classes or raster grid cells in the map composition to determine what type of color and symbol to be used in this case.

## Attribute Data Types and Thematic Mapping in GIS

Characteristics of a set of geographic objects or a feature class in GIS are called attributes. The attributes are stored in the tribute tables that come with the feature classes or grid cell datasets. In vector data structure, one geographic feature class or one map layer can have many attributes in the attribute table. By contrast, an attribute table of a grid cell surface of one raster dataset or map layer can only carry one variable or one set of values. In practice, when we present map layers and make map compositions in a "Maps" window in ArcGIS Pro, we are actually using the attribute data in the attribute table of the map layer to present the information of the geographic features. This, in essence, is a process of thematic mapping in GIS.

Since the attribute table of a vector feature class can contain a variety of fields or columns, GIS analysts can use each of these fields or columns to make a digital thematic map, where one field equals to one mapping theme in our case. Based on the previous discussion in Chapter 4, attribute data in GIS can be identified as nominal data, ordinal data, interval data, or ratio data conceptually. In practice, vector attribute fields can be recognized as numerical fields, text fields, object identifiers, geometry, etc. Mainly either the numerical fields or the text string fields in the attribute table of a geographic feature class are used to make thematic maps. Related to the difference between

**Table 12.1** Guidelines of applying symbol strategies for qualitative and quantitative thematic mapping in ArcGIS Pro.

| | Qualitative | Quantitative |
|---|---|---|
| Point | Preferred: hue, shape<br>Less preferred: orientation, arrangement | Preferred: size, value, lightness<br>Less preferred: perspective height, size |
| Line | Preferred: hue, shape<br>Less preferred: arrangement | Preferred: size, spacing<br>Less preferred: perspective height, value, lightness |
| Area | Preferred: hue, shape<br>Less preferred: orientation, arrangement | Preferred: value, lightness, saturation, size<br>Less preferred: perspective height, hue |
| 2.5D | None recommended | Preferred: perspective height, lightness, value<br>Less preferred: saturation |
| 3D | Preferred: orientation, arrangement, shape<br>Less preferred: hue | Preferred: lightness, value, saturation<br>Less preferred: size, spacing |

*Source:* ESRI 2022/Environmental Systems Research Institute, Inc.

text attributes and numerical attributes, we identify qualitative symbology or quantitative symbology of the digital maps in GIS. GIS analysts need to make sensible choices in mapping the feature attributes in order to effectively communicate their ideas with map readers and decision-makers. For instance, if we would like to show how cities that are represented by point symbols differ in their population size, we need to choose to change the size of the symbol used to represent the points. Larger symbols represent larger magnitudes, and this is how our eyes and brains process the meaning of large symbols compared to smaller ones. In contrast, if we would like to show the difference between a railroad and an interstate highway, changing the size or width of the lines may not immediately show the difference. Instead, we use different symbols and shapes of the lines to show the difference of the two features. Meanwhile, in the task of symbolizing the line features to differentiate the interstate highway from local streets on the map layer, different size or width application is appropriate. Table 12.1 has the ArcGIS Pro guidelines for applying symbol strategies for qualitative and quantitative thematic mapping in GIS (ESRI 2022).

## Symbolization of Vector Data on Digital Maps

Geographic objects stored in a vector GIS database are organized by feature classes of points, lines, and polygons. When we visualize the vector feature classes in a digital mapping window or "Maps" window, they are projected as map layers. In this process, one feature class in the GIS database is in general converted into one map layer. Therefore, we consider a feature class, a map layer, or a feature layer are the same in this discussion. When GIS data is contained in a feature map layer, we can change the way data is represented by symbolizing the feature map layer. In the symbolizing process, GIS analysts need to consider the Vector Map Symbol Properties.

1. *Symbol style*: The symbol style list on the GIS platform contains a collection of a variety of symbols to be used to display the geographic features. GIS analysts make the choice of symbol style based on the meaning and objective of the presentation. GIS analysts can add their own symbols to the existing list on most GIS platforms too.

2. *Symbol display frequency*: This is the control of the number of symbols that are displayed on a vector map layer. The major objective to reduce the display frequency of geographic objects or features on a digital map is to reduce the crowdedness and overwhelming situation for map readers. In practice, the symbol display frequency issue is mainly related to the map scale and the size of the geographic region to be presented. In the GIS environment, users are provided with full control and flexibility of the display frequencies of symbols.
3. *Line symbol properties*: The line symbol properties control the line features or the outside lines of a filled symbol for polygon features. Line symbol properties normally contain line style, line color, and line width.
4. *Fill symbol properties*: The fill symbol properties control the color and pattern of the area inside a filled symbol of polygon features. Fill symbol properties can be used to create point or line symbol filling color and pattern as well. The major components of fill symbol properties are fill pattern, foreground color, background color, etc.
5. *Symbol color scaling*: Symbol color scaling is the method or options for gradationally coloring the symbols. In other words, it is the method to produce color schemes for symbolizing geographic features in GIS.

In ArcGIS Pro, we normally use the "Symbology" menu to open the "Symbology" dialog in performing the task to symbolize geographic features. The following are the major symbolization methods that we use in ArcGIS Pro environment for feature map layers.

1. *Single symbol*: this method applies the same symbol to all features in a map layer. This method is used for drawing a map layer that has only one category of features, such as county boundaries for polygon features or a symbol for all the hospitals in a region.
2. *Unique values*: this method symbolizes qualitative categories of values in the attribute table of a feature class. It applies different symbols to each category of features within the map layer based on one or more attribute fields. For polygon features, one of the random color schemes can be selected to represent values if a lot of categories of values are presented.
3. *Graduated colors*: this method is used to show a quantitative difference between mapped features by varying the color of symbols. It shows quantitative differences in feature values with a range of colors on a digital map layer. Attribute data field is classified into ranges that are each assigned a different color from a color scheme to represent the range. For instance, if our classification scheme has six classes, six different symbol colors are assigned.

   The size of the symbols stays the same, but the color changes with different attribute value ranges. In the case of polygon features, this type of map is called a choropleth map. A choropleth map is a map that uses differences in shading, coloring, or the placing of symbols within predefined geographic areas to indicate the average values of a property or quantity in those areas. GIS analysts can select either a continuous color scheme or a discrete color scheme to present value changes of the feature map layer. In practice, most people choose a continuous color scheme.

   The limitation of using different colors and color quantities to represent feature attribute value changes is that we cannot select many categories of value ranges. Otherwise, the readers of digital maps will be confused by lacking in differences of the colors. It is suggested that the upper limit of the attribute value ranges should be seven to eight.
4. *Bivariate symbol colors*: This method utilizes two graduated colors to show quantitative differences in feature values between two fields. It also shows the quantitative relationship between two fields or variables in a feature map layer. Bivariate symbol colors are a combination of two graduated colors of symbology, each variable or field is classified, and each class is assigned a color.

5. *Unclassed colors*: This method shows quantitative differences in feature values with a range of colors, but not broken into discrete classes. While a graduated color scheme distributes data into discrete classes with unique symbols, unclassed color symbology distributes a color scheme evenly across features.
6. *Graduated symbols*: this method uses the size of symbols to show a quantitative difference among the mapped features (Figure 12.8). Attribute data is classified into ranges that are each then assigned a symbol size to represent the range. For instance, if your classification scheme has five classes, five different symbol sizes are assigned. The color of all the symbols stays the same.
7. *Proportional symbols*: this method is used to show relative differences in quantities of a particular attribute among features. It represents quantitative values of geographic objects as a series of unclassed, proportionally sized symbols. Proportional symbology is similar to the symbology of graduated symbols in that both draw symbols relative size to the magnitude of a feature attribute. But where graduated symbols distribute features into distinct classes, proportional symbols represent quantitative values as a series of unclassed symbols. The size of each feature is based on the specific attribute value of each feature. When symbols are applied to point or line features, the feature's size is modified directly. When symbols are applied to polygon features, a proportionally sized point symbol draws at the center of the polygon.
8. *Dot density*: this method draws quantities as point symbols distributed within polygon features. This method is only available for polygon features in ArcGIS Pro. With the dot density method, we can symbolize the data that is not classified. Instead, quantitative values for one or more fields are represented as a collection of point symbols, typically solid circles, or dots, in each of the polygons presented. Each dot represents a constant number of attribute value,

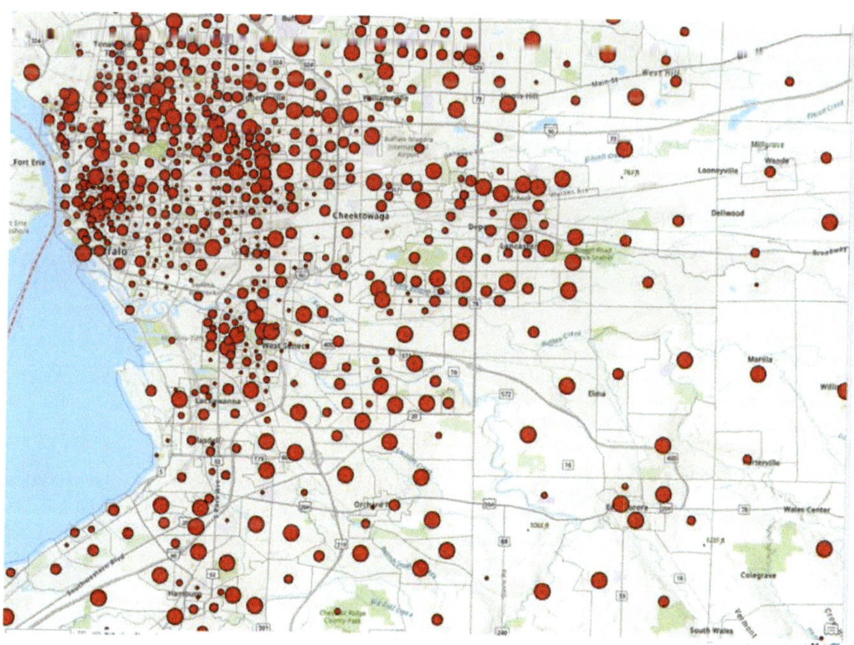


**Figure 12.8** Example of graduated symbols on a map.

such as people or other quantifiable phenomena. The dots are equally sized, even when multiple fields are symbolized together in a layer. The effect is to use the density of dots to present the value distributions.

9. *Chart symbology*: this method draws quantities from multiple attribute fields using chart symbols. Charts can be used as multivariate symbology in ArcGIS Pro to show quantitative differences between attributes, with each part of the chart representing an attribute value that contributes to the overall whole set of values (Figure 12.9). Chart symbology can be used with point, line, or polygon features. For example, we can use pie chart symbols to represent the ethnicity of a district in a city. Each section of the pie chart comprises one of the ethnicities. Each chart symbol can then be sized proportionally by the district's total population.
10. *Heat map symbology*: this method draws or converts the density of points as a continuous patch with a color gradient. It converts point features as a dynamic, representative surface or grid cells to represent the relative density of original point features. Since heat mapping is a representation of point distribution, it works best when there is a large number of point features, especially if their symbols overlap on a digital map.

    Heat map symbology displays the relative density of points as a dynamic raster visualization using a color scheme to indicate density values. Ideally, the color scheme is a smoothly varying set of colors, ranging from cool (sparse density of points) to hot (high density of points). The density definition and the color values are changing as the users zoom in or out on the digital map. The density is calculated using the kernel density method.

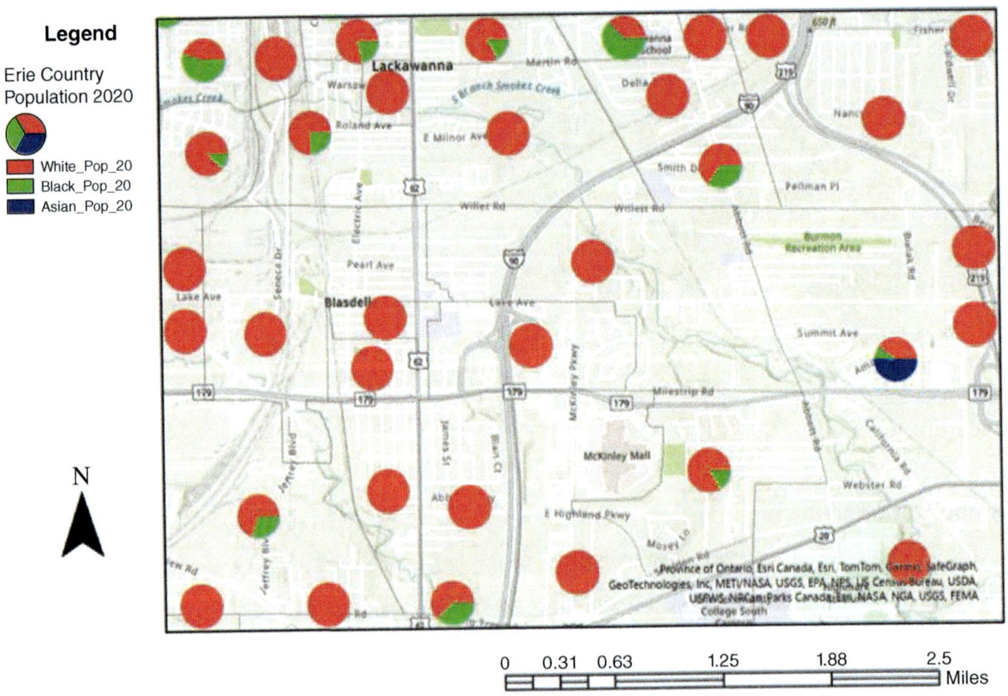


**Figure 12.9** Example of a map using chart symbols to present population distribution by ethnic groups.

## Map Scale and Abstraction

In the GIS environment, we can change the geographic extent in the mapping window or "Maps" window on the fly. This process is technically called zoom-in or zoom-out. Ideally, the mapping symbols and colors should change while the map scale and viewing area in the mapping window are changing. Small map scale portraits large geographic area, large map scale portraits a small geographic area. Feature classes of polygons at large map scale that display the situations in a local region should be abstracted to pints when it is zoomed out at a small scale on a country or a continental level. When zooming out, line feature classes, such as streets and highways, should be generalized, and detailed streets should be graphically eliminated to reduce the cluster effects and increase the clarity.

In essence, digital map layers or map compositions are abstract drawings of geographic objects from the real world. The process of transforming geographic data, which represents the actual human and physical environment, into a map is called map abstraction. The systematic process of removing detail from the geographic to be mapped either in a digital or hard copy format is called map generalization. Map generalization is particularly important when we change the map scales and display areas in a mapping window. Four major components of map generalizations were identified (Robinson et al. 1995; McMaster et al. 2005).

1. *Simplification*: This is a process to determine important features in a feature class and only symbolize and show these individuals according to the attributes of the features. Eliminates the graphic symbols of unwanted details.
2. *Aggregation*: Join together multiple features to form one feature based on their close proximity of locations as map scales change from large to small.
3. *Selection*: This process is an intellectual decision-making process by the GIS analysts. Basically, GIS analysts need to make selections of which geographic feature class or group of features in a feature class are important and need to be kept even if the digital map is changed from large to small map scales.
4. *Exaggeration*: Many significant geographic objects or features are strategically significant or much more important than their proportional size. For instance, the international airport of a city or a region, interstate highway, high-voltage power transportation lines, or railroad transportation lines need to be kept on digital maps no matter what map scale it is. Therefore, GIS analysts need to exaggerate the mapping symbols to make these features always visible.

Several new approaches have been developed in ArcGIS Pro for map generalizations. A few are automatic conflict resolution and generalization workflow set-up, scale-based symbolization classes, setting of display of feature map layers by a certain map scale or scale range, and function of feature generalizations.

### Symbolization of Raster Data on Digital Maps

Raster data in GIS is made by grid cells. The default raster dataset in ArcGIS Pro is called ArcGIS grid. The objects to be symbolized and presented on digital maps in this case are grid cells rather than the geographic objects or features directly. Two major characteristics are identified when we symbolize raster grid cells. These are attribute-driven and color-drawing centered.

1. *Attribute driven*: Owing to the structure characteristics, one raster map layer or dataset carries only one attribute. In the GIS environment, the raster attribute table is also called a value

attribute table. Different attributes, even if they are related to each other, such as precipitation (rainfall) and relative humidity (RH), can only be stored and presented with different sets of grid cells or different raster surfaces or map layers. In the mapping window of a GIS software platform, the default color symbols are made by their carried unique attribute. GIS analysts can change and modify colors and color schemes of the grid cell surfaces in the mapping window.

2. *Color drawing centered*: Since each of the grid cells in raster data is actually a small square or rectangle polygon in shape, a color symbol is the center component to present value differences of the attribute of a raster dataset or map layer. It is similar to symbolizing polygon features in vector data; the size and shape of symbols are not important in raster data presentations. While it is also different than vector polygon symbolization, color and color scheme or color ramp, but not spatial pattern type of symbols, are mainly used for grid cell presentations.

According to the technology and the purpose that the raster datasets were created for, we can recognize there are two types of raster data: thematic raster or imagery raster data. Thematic raster data are those datasets that carry an attribute that is different than the collected reflections of solar radiation or electromagnetic radiation. By contract, imagery raster data are those remotely sensed images that carry the reflection intensities of the ground geographic features of either solar radiation or electromagnetic radiation.

In the GIS environment, both thematic and imagery raster datasets are displayed by colors and color ramps. Color scheme and color ramp are two different terms that are used interchangeably for the same concept in GIS. The RGB and the coloring rules that we used for vector data symbolization are also emphasized in raster data presentation. Distinctive colors of hue, saturation, and value will lead to better presentations of raster surfaces or map layers to the audience. Originating from remote sensing science and image processing, color and contrast stretch techniques are widely used to present raster data in GIS.

If the raster data represent continuous data, we can apply a color contrast stretch to it based on the statistics of the raster dataset. Color contrast stretch increases the visual contrast of the raster display. For example, we can apply a stretch when the raster display appears dark or has little contrast. In an imagery data case, the dataset may not contain the entire range of values that our computer can display. Therefore, we stretch the imagery data values to the entire range. This will result in a distinctive imagery map, and some geographic objects or features may become easier to be interpreted.

Color contrast stretch of raster data, in essence, is to stretch the grayscale of color or black to white presentations of the values of grid cells or pixels stored in a computer database. The most popular method to store color imagery data in a computer is to use an 8-bit image file. An 8-bit color graphic file stores digital image information in a computer's memory or in an image file, so that each pixel is represented by 8 bits or 1 byte. The maximum number of colors that can be displayed at any given time is 256 or $2^8$. In the black to white situation, Value 0 means the pixel or grid cell is totally black. By contrast, 255 means the pixel or grid cell is totally white. Values of 0 to 255 are also used to range each of the primary colors, RGB. For instance, 0 of red fill means there is no red color in this cell, while 225 of red fill means the maximum red in the grid cell. The following is an example comparing histograms of color tones before and after the stretch (Figure 12.10). The standard stretches used in GIS software environments are Minimum–Maximum, Percent Clip, Standard Deviation, Histogram Equalize, and Gamma Stretch.

Meanwhile, raster data represents the entire mapping area by a huge amount of grid cells or pixels. In order to facilitate a relatively faster retrieval of the data to present on a digital map

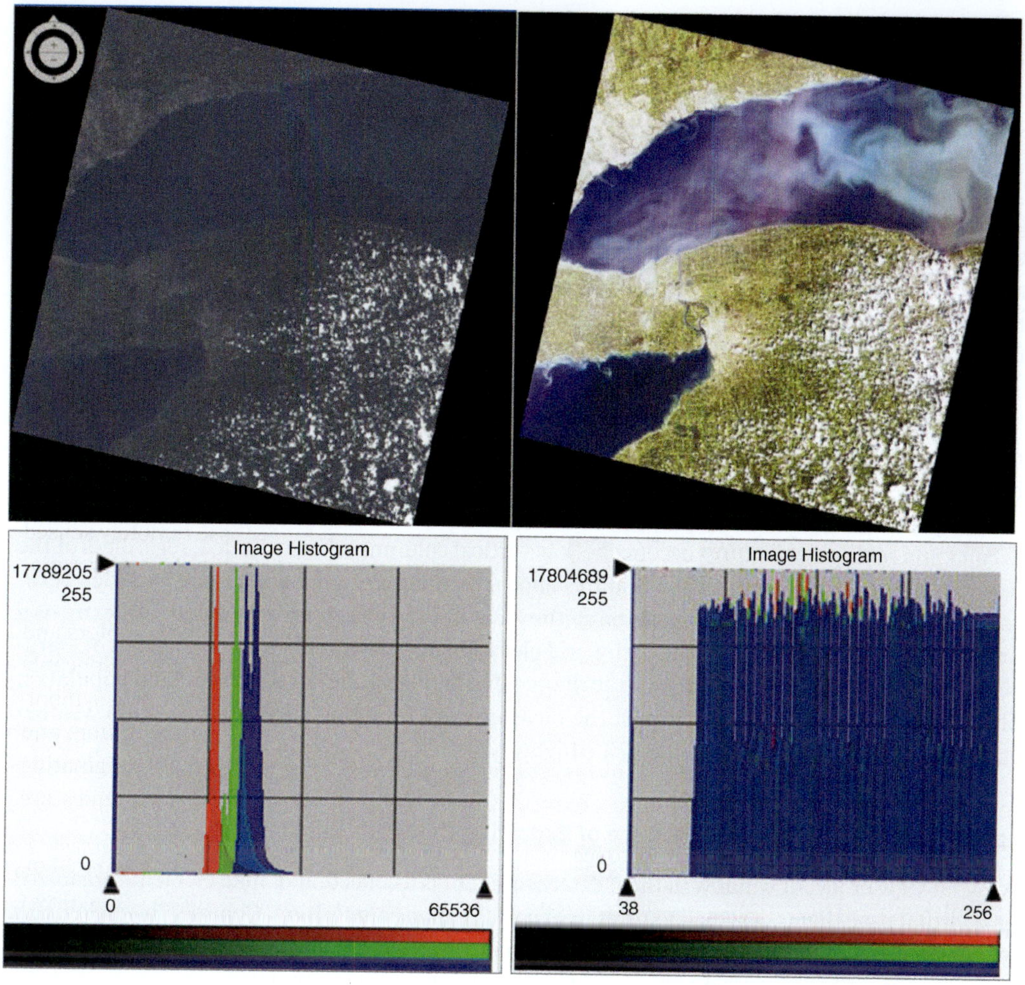


**Figure 12.10** Example of Histogram Equalize color contrast stretch on a raster satellite imagery map. *Source:* Dr. Tao Tang.

composition, a pyramid or quadtree structure is built when raster data is input into GIS the first time. The best way to improve and reduce the time it takes to display a large raster dataset is by creating raster pyramids. Pyramids are copies of the original raster dataset that have been successively resampled in decreasing levels of resolution, fourfolds, with the same file name as the raster dataset. Several other raster grid cell compression methods have been developed to facilitate more effective computational storage and display as well.

## Symbolization in 3D Scenes

A 3D object scene layer represents objects such as buildings that are explicitly modeled in three dimensions. We can visualize a 3D object scene layer with textures embedded into the 3D features. A 3D object scene layer is created from a multi-patch feature layer or a 3D object feature layer in ArcGIS Pro. A 3D object scene layer is only supported in the 3D category of a scene.

We can symbolize a 3D object scene layer with a single symbol, unique values, unclassed colors, graduated colors, or graduated symbol symbology, and by varying symbology using the attributes in its attribute table. Face culling is available for a 3D object scene layer. Modify the layer's face culling property to see through the front or back of the object depending on which side of the object faces the viewers.

A 3D vector map layer symbolization using a single symbol, unique values, unclassed colors, graduated colors, or graduated symbol symbology, and by varying symbology, are similar to those applied in the 2D "Maps" window. Face culling is to modify the layer face culling property and allows you to see through the front, back, or both faces of 3D objects, depending on which side of the objects the users are viewing. The function is enabled for layers containing area-based data, including extruded polygons, multi-patches, TIN triangles, and raster data.

A 2D feature class can be extruded into a 3D model for viewing and map presentations in the "Scene" window in ArcGIS Pro. Extrusion is the process of stretching a flat, 2D feature class or map layer vertically to create a 3D object in a scene. For example, we can extrude building polygons by a height value to create 3D building shapes in the "Scene" window in ArcGIS Pro. Only points, lines, and polygons feature classes support extrusion (ESRI 2022).

Point and multipoint features become hollow vertical columns when extruded, regardless of the point symbol applied. Extruded line features appear as walls. Polygon features are extruded vertically to create blocks. In 3D visualization in the ArcGIS Pro environment, the values of extrusion or the Z-value do not need to be the real-world elevation or height of features. The values of extrusion can be any numerical values in the attribute table of the feature class, such as total population of the regions, air pollution quantity of the measurement locations, precipitation of the weather stations, or traffic volume on the interstate highways.

## Layout View of GIS Data and Interface of Exporting Digital or Hard Copy Maps

A layout view or layout window in the GIS environment is a collection of map elements organized on a virtual page that is designed to print or export the map composition. Common map elements include one or more map frames that each contain an ordered set of map layers, a scale bar, a north arrow, a map title, descriptive text, and a legend. Inserts of views on databases, other graphics, and digital images can also be added. GIS has revolutionized the processes of making maps. In a digital mapping window, GIS analysts can change the map scale on the fly and are able to show the geographic objects or features at different regional extents, such as at a county level, state level, national level, or continental level, easily and straightforwardly. Symbols and colors of feature map layers can be changed quickly, and distribution patterns of geographic objects can be presented right away. However, the traditional cartographic or mapping elements, map design, and layout are still a necessity for printing hard copies and exporting digital maps.

In GIS, layout view or window is the formal way of making map compositions. The "Maps" window only shows map layers, and it is a utility for GIS analysts to conduct symbolization and presentation of geographic features. Layouts contain all of the components of a final map as graphical elements. Some of these graphical elements may be dynamically linked to the "Maps" window and other windows in the ArcGIS Pro project, such as data frames or charts. Dynamically linked elements, such as data frames and charts, change in the layout when the source changes in other windows. The dynamic link between the mapping and layout windows can be removed by the GIS analyst.

Only in the layout window are insert functions of the map elements available. Layouts can be set up as landscape (wide) or portrait (tall) orientation. Map composition of layout can be exported in digital format, published on the internet, or printed. The page size varies depending on the specifications for the output. What you see on the layout is what you will get when you print or export the map composition to the same page size.

Digital map compositions can be exported using the "Export Map" dialog on the "Share" ribbon in ArcGIS Pro. There are 12 export file types available in ArcGIS Pro, including both vector and raster formats. The vector formats are AIX, EMF, EPS, PDF, SVG, and SVGZ, which support a mixture of vector and raster data. The raster formats are BMP, JPEG, PNG, TIFF, TGA, and GIF. These file exporting functions are solely raster export formats and automatically rasterize any vector data in your map or layout.

AIX files are used with the ArcGIS Maps for Adobe Creative Cloud extension. BMP files are native Windows raster images. EMF are useful for embedding in Windows documents, such as MS-Word, because the vector portions of the EMF can be resized without loss of quality. EPS files use the PostScript page description language to describe vector and raster objects. PostScript is the publishing industry standard for high-end graphics files, cartography, and printing. GIF files are of a legacy raster format that is used for graphics on the web. JPEG files are compressed image files. JPEG files support 24-bit color and have been a popular choice for use on the internet because JPEG file size is much smaller than many other image formats. PDF files are designed to be consistently viewable and printable across different platforms. PDF files are commonly used for distributing documents on the internet, and the format is now an official ISO standard for document interchange. PNG is a versatile raster format that can be displayed in internet browsers and inserted into other documents. It supports high-bit-depth color and uses a lossless compression that is suitable for presenting maps. SVG (Scalable Vector Graphics) is an XML-based file format that has been specifically designed for viewing on the web. SVG can contain both vector and raster information. TIFF files are the best choice for importing into image editing applications and are also a common raster data format in GIS environment. However, TIFF files consume large computer storage space and are not supported to be directly incorporated on the internet.

## Sharing Geographic Information through Internet and Online Map Presentations

Modern GIS software platforms include both local computer-based packages, such as ArcGIS Pro and GIS packages on the internet, such as ArcGIS Online. Many functions, datasets, and analytical results can be shared or transferred across the local and internet-based platforms. Many functions built in ArcGIS Pro facilitate data and map sharing and transporting them to the internet. Meanwhile, internet web pages are very good media to publish GIS analytical results and maps.

ArcGIS Online is an internet cloud-based suite of GIS software. ArcGIS Online can be used anytime, anywhere worldwide. Hundreds to thousands of people can interact with the data and maps published online at the same time.

In summary, one of the major trends in the near future for the computationally enabled digital mapping is to create 3D or four-dimensional (4D), which includes time-dimension and time scale (Tang and Zhang 2018), environmental scenes on the Earth's surface. This type of presentations can be either incorporating map projection strategy or mimicking the real-world situation without map projections at all. The real-world environment on the Earth's surface is four-dimensional (4D; x, y, z, and time). We have been developing the capable computational technology to make a digital twin of the Earth's surface environment in recent years. Or in other words, the digital mapping will shine with help from a true virtual reality.

## References


Blanarovich, C. (2017). Engagement session styling by NYC engagement photographer. https://christinazen.com/2017/08/engagement-session-styling-nyc-engagement-photographer/.

Dent, B.D., Torguson, J., and Hodler, T. (2009). *Cartography: Thematic Map Design*, 6e. New York, NY; Boston, MA, USA: McGraw Hill. 336 pages.

ESRI. (2022). ArcGIS Pro help manual. 2022. https://pro.arcgis.com/en/pro-app/latest/help/.

McMaster, R.B., Slocum, T.A., and Kessler, F.C. (2005). *Thematic Cartography and Geographic Visualization*, 2e. Upper Saddle River, NJ, USA: Prentice Hall. 518 pages.

Robinson, A., Morrison, J., Muehrke, P. et al. (1995). *Elements of Cartography*, 6e. New York, NY: Wiley. 674 pages.

Silayo, E.H. (2002). Cartography in a GIS environment. *The International Archives of the Photogrammetry, Remote Sensing and Spatial Information Sciences* XXXIV (Part 6/W6): 106–112.

Tang, T. and Zhang, J. (2018). Time scale sensitive sensor applications in collecting and analyzing geographic event data. *Annals of Geographic Information Science* 24 (4): 241–253. https://doi.org/10.1080/19475683.2018.1520299.

Turnbull, D. (2009). Science and scientism, cartography. In: *International Encyclopedia of Human Geography*, 37–42. Melbourne, VIC, Australia: Elsevier. https://doi.org/10.1016/B978-008044910-4.00066-3.

## Tutorial in ArcGIS Pro

Copy the tutorial data folder "Chapter12-Tutorial" to your working directory.

Open ArcGIS Pro and create a new ArcGIS Pro project file with the "Map" template and call is as "MyGIS_Learning12." Save the project file into the designated working folder, C:\ or D:\......Chapter12-Tutorial.

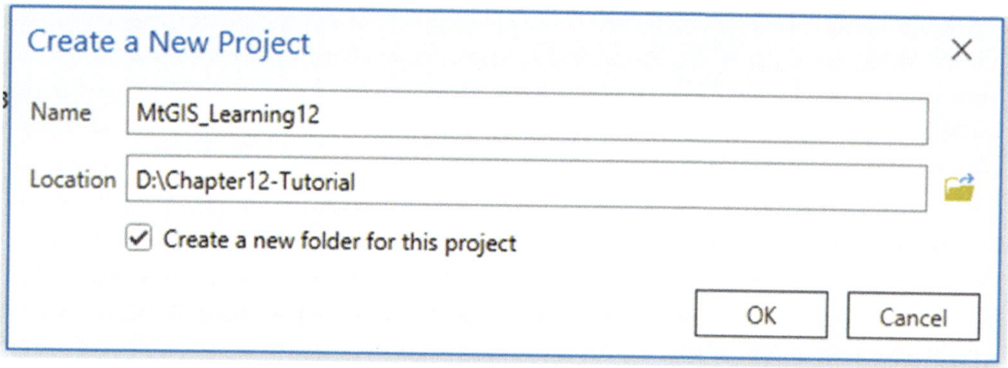


Now, highlight and right-click on the Folders on the Catalog Pane to obtain the contact menu. Click on "Add Folder Connections" on the contact menu and add the entire Chapter 12-Tutorial folder from your working drive.

Designing and constructing a digital map in GIS always constitutes three steps. First, we need to define the purpose of making this map. Each of the digital maps that we construct in GIS should have a theme. This is the reason that specific map presentations and mapping projects are also called thematic mapping or making thematic maps. The second step is to gather the data layers, either in vector or raster format or both to support the purpose. The third step is to author or construct the map. Cartography, or making maps using symbols and labels by their own contents is a subdiscipline in GIS. In this tutorial, wc can only introduce the techniques very briefly.

On the Catalog Pane, expand the Chapter 12-Tutorial folder, then the NYS_Mapping_Data.gdb (geodatabase). Add the following map layers to the "Map" window: 1 "County_Boundaries," 2 "Major_Transportation_Routes," 3 "NYS_Public_Schools."

When we add the map layers from database, the symbols and colors are not properly presented in most cases. First, we will make some simple improvements by changing the names of map layers. Then, we are going to change the symbols and colors of the features on the map. We will also practice adding metadata to the map and changing the base-map layer for a specific digital map construction purpose.

In the Table of Contents, highlight and right-click on the **County_Boundaries** map layer to get the contact menu. Then, click on the Properties to open its dialog. In the Layer Properties dialog, while the General tab is highlighted on the left-hand side, change the Name (of the map layer) to "Population by County." Click OK, we will see the name change of the "County_Boundaries" map layer.

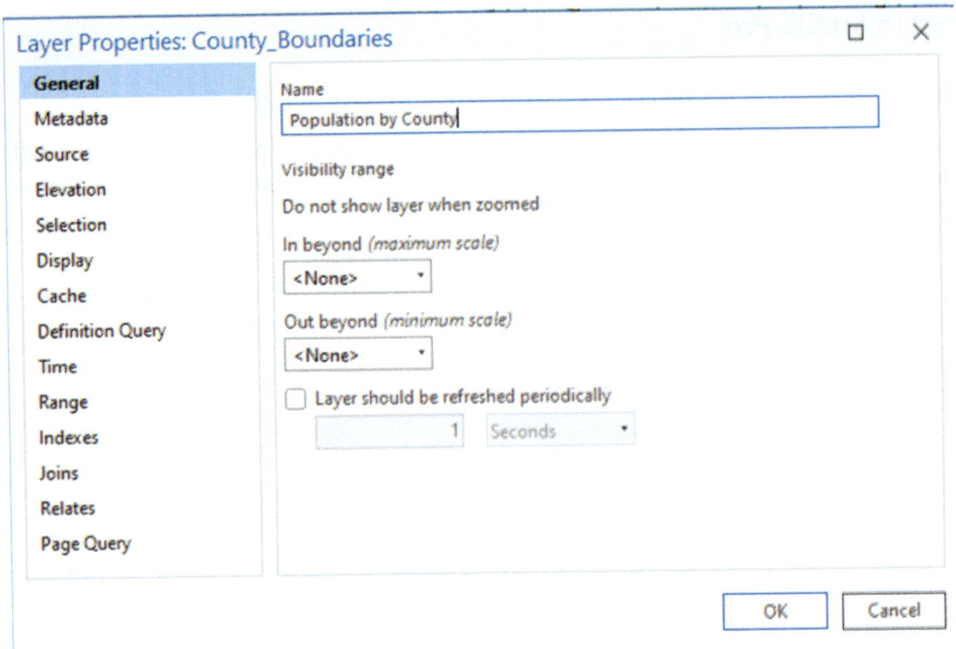


Using the same method above, change the "Major_Transportation_Routes" map layer to Major Roads, and the "NYS_Public_Schools" map layer to New York State Public Schools. Changing the names of map layers in a map composition helps the readers to understand the presentations of the GIS Analysts. Meanwhile, it doesn't change the original file name in the geodatabase.

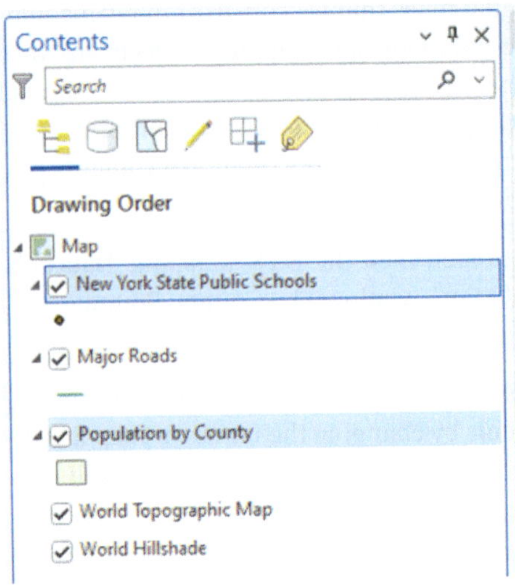


On the **Quick Access Toolbar**, click **Save** icon to save the project file.

1. Symbolize the population by county map layer.

In the Table of Contents, uncheck the check boxes for the New York State Public Schools and Major Roads layers to turn them off. Leave the Population by County layer on.

We need to study the attribute table of the map layer in order to pick up appropriate themes in making map presentations. In the Table of Contents, right-click the Population by County layer and click Attribute Table to Open the Table.

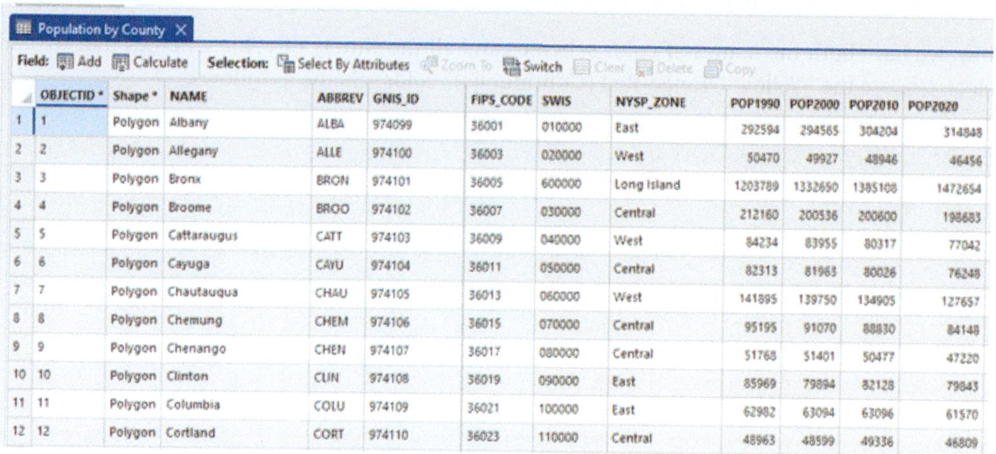
Population by County

Field: Add Calculate Selection: Select By Attributes Zoom To Switch Clear Delete Copy

| | OBJECTID * | Shape * | NAME | ABBREV | GNIS_ID | FIPS_CODE | SWIS | NYSP_ZONE | POP1990 | POP2000 | POP2010 | POP2020 |
|---|---|---|---|---|---|---|---|---|---|---|---|---|
| 1 | 1 | Polygon | Albany | ALBA | 974099 | 36001 | 010000 | East | 292594 | 294565 | 304204 | 314848 |
| 2 | 2 | Polygon | Allegany | ALLE | 974100 | 36003 | 020000 | West | 50470 | 49927 | 48946 | 46456 |
| 3 | 3 | Polygon | Bronx | BRON | 974101 | 36005 | 600000 | Long Island | 1203789 | 1332650 | 1385108 | 1472654 |
| 4 | 4 | Polygon | Broome | BROO | 974102 | 36007 | 030000 | Central | 212160 | 200536 | 200600 | 198683 |
| 5 | 5 | Polygon | Cattaraugus | CATT | 974103 | 36009 | 040000 | West | 84234 | 83955 | 80317 | 77042 |
| 6 | 6 | Polygon | Cayuga | CAYU | 974104 | 36011 | 050000 | Central | 82313 | 81963 | 80026 | 76248 |
| 7 | 7 | Polygon | Chautauqua | CHAU | 974105 | 36013 | 060000 | West | 141895 | 139750 | 134905 | 127657 |
| 8 | 8 | Polygon | Chemung | CHEM | 974106 | 36015 | 070000 | Central | 95195 | 91070 | 88830 | 84148 |
| 9 | 9 | Polygon | Chenango | CHEN | 974107 | 36017 | 080000 | Central | 51768 | 51401 | 50477 | 47220 |
| 10 | 10 | Polygon | Clinton | CLIN | 974108 | 36019 | 090000 | East | 85969 | 79894 | 82128 | 79843 |
| 11 | 11 | Polygon | Columbia | COLU | 974109 | 36021 | 100000 | East | 62982 | 63094 | 63096 | 61570 |
| 12 | 12 | Polygon | Cortland | CORT | 974110 | 36023 | 110000 | Central | 48963 | 48599 | 49336 | 46809 |

In this case, the "POP2020" column or field stands, the total population of the US Census 2020 survey. We can use this most recent field-surveyed data to symbolize the counties. Close the attribute table. Right-click the Population by County layer and click Symbology on the contact menu.

In the **Symbology** pane, make sure the **primary symbology** icon on the top is selected. Click on the drop-down arrow and click on **Graduated Colors**.

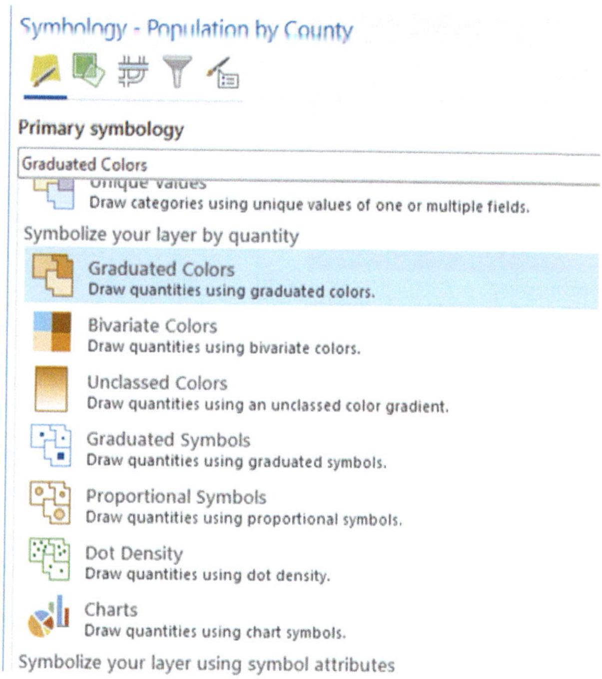

For Classes, select 6. For Method (feature classification method), select Quantile.

Then, expand the Color Scheme drop-down list.

Check on both "Show names" and "Show all."

Now, we have both the names of color schemes and all the variety of color schemes that we can select. Choose the "Yellow-Orange-Red (Continuous)" color scheme as follows:

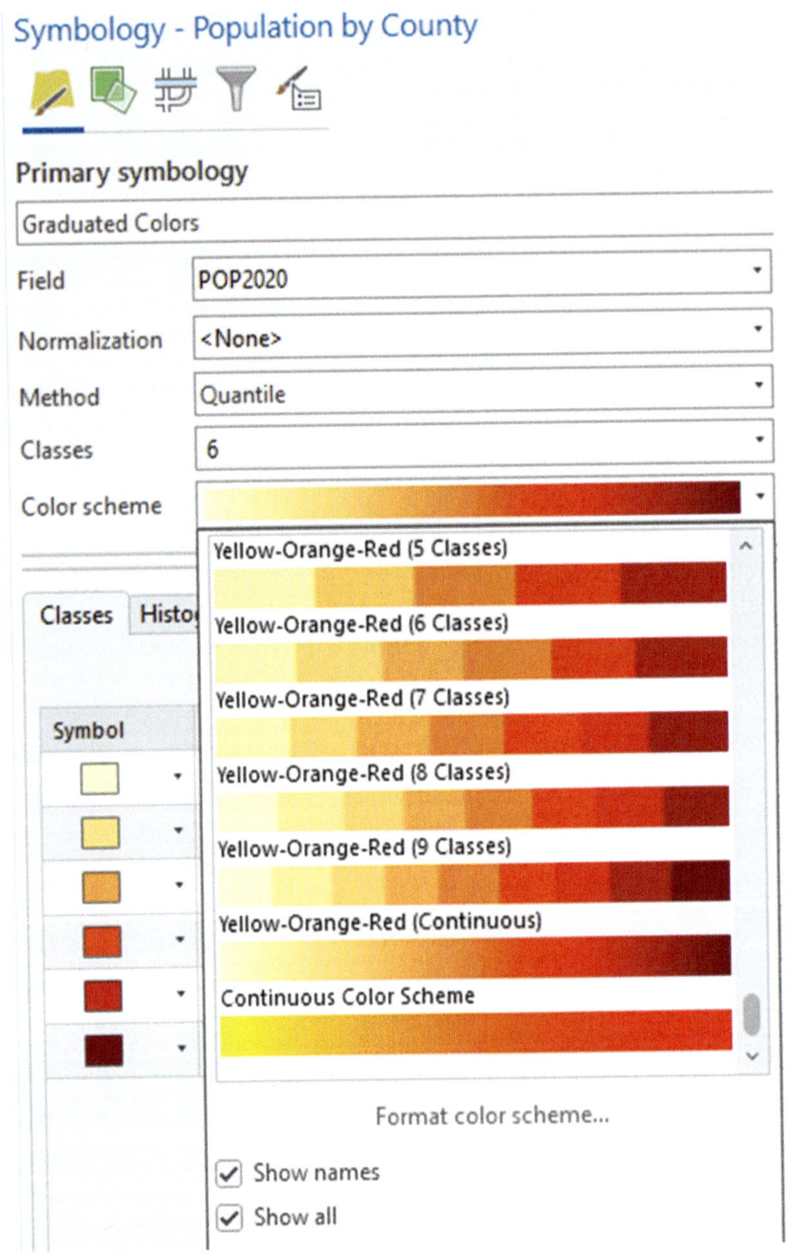

Now, you notice the change of the map as follows:

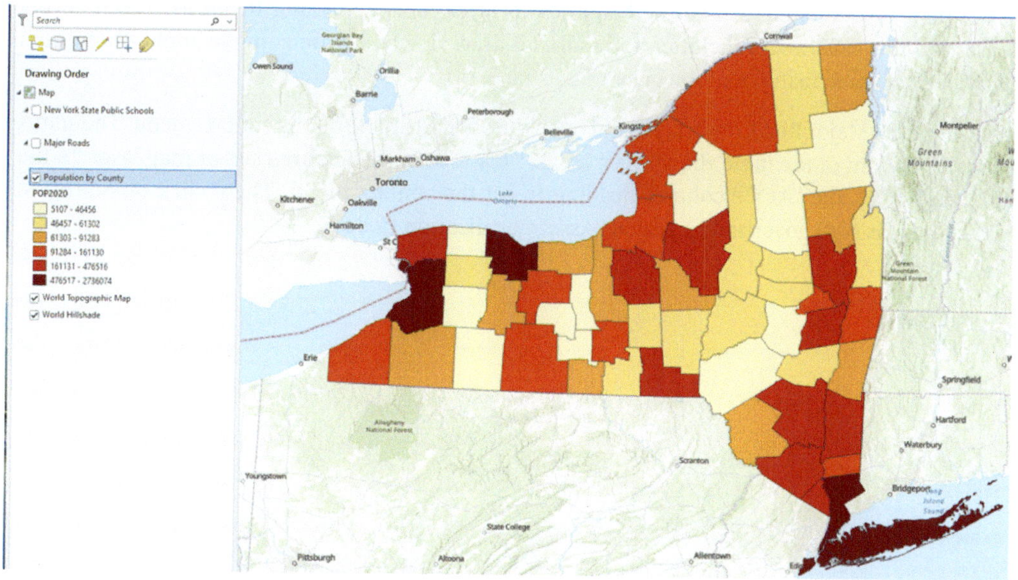


Click on the sub-title of the Population by County map layer, "POP2020" and replace (type in) it by "Total population of 2020."

Data classification methods are used for classifying numerical fields in the attribute table of a feature class or map layer for graduated symbology, graduated colors, or graduated symbols. In ArcGIS Pro, there are a total of seven (7) classification methods available now. These are:

1. *Defined interval*: user define interval range and size. A series of classes with the same value range will be produced. The number of classes are based on the interval size and maximum data size and is determined automatically.
2. *Equal interval*: using equal interval to divide the range of attribute values into equal-sized subranges. This allows the user to specify the number of intervals, and the class breaks based on the value range are automatically determined.
3. *Quantile*: in a quantile classification, each class contains an equal number of features. A quantile classification is specifically good for linearly distributed data. Quantile assigns the same number of data values to each class.
4. *Natural breaks (Jenks)*: in natural breaks classification (Jenks), classes are based on natural groupings inherent in the database. Class breaks are created in a way that best groups similar values together and maximizes the differences between classes.
5. *Geometrical interval*: the geometrical interval classification scheme creates class breaks based on class intervals that have a geometric series. The algorithm creates geometric intervals by minimizing the sum of squares of the number of elements in each class. This ensures that each class range has approximately the same number of values in each class and that the change between intervals is consistent.
6. *Standard deviation*: this method shows the user how much a feature's attribute value varies from the mean. The mean and standard deviation are calculated automatically.
7. *Manual interval*: user manually defines intervals and select their own classes, to manually add class breaks and to set class ranges that are appropriate for the dataset.

**2.** Symbolize the major roads map layer.

In the Table of Contents, turn on the Major Roads layer. Since the Major Roads map layer is very dense, and this causes visual clutter and makes it draw slowly, we are going to apply a definition query to filter some features of the layer out.

Right-click the Major Roads map layer and open the attribute table on the contact menu. The status bar at the bottom of the table shows over 133,221 features in the feature class or map layer. In the table, right-click the "ACC" column heading to get the contact menu. Then, click on Statistics. ACC stands for NYS government-issued "Asset Certification Code."

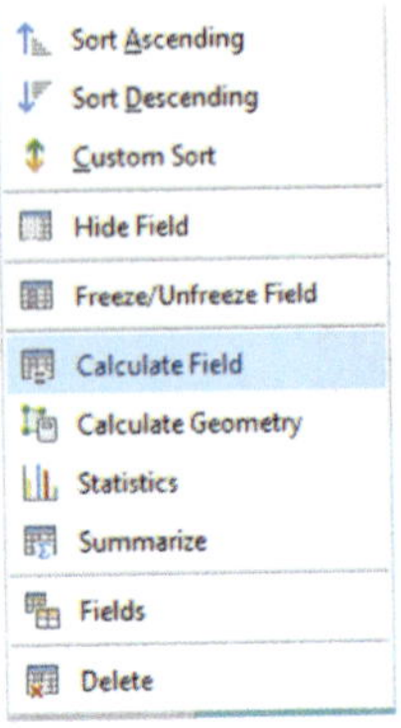


A bar chart opens showing the counts for each of the major road type (not every category is graphed). The vast majority are in the "Major Road" category in this case.

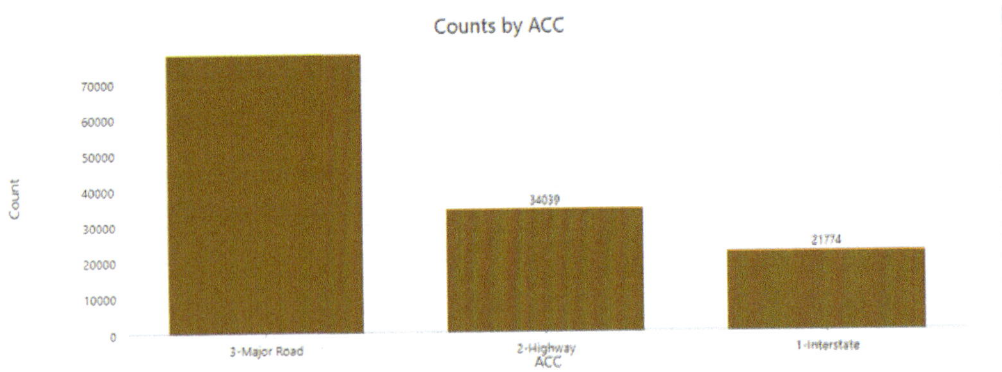


Close the chart window and the attribute table.

We are going to filter out some of these street and highway line segments using a definition query. In the Table of Contents, right-click the Major Roads layer to open its contact menu and select properties.

On the Layer Properties dialog box, click the Definition Query tab. Click New definition query. In the **Query 1** expression dialog, click the **Where** drop-down arrow and click **ACC**. Click the condition drop-down arrow and click includes the values. Click the values drop-down arrow and check the following boxes:

1. **Interstate**
2. **Highway**

The query window should look like this.

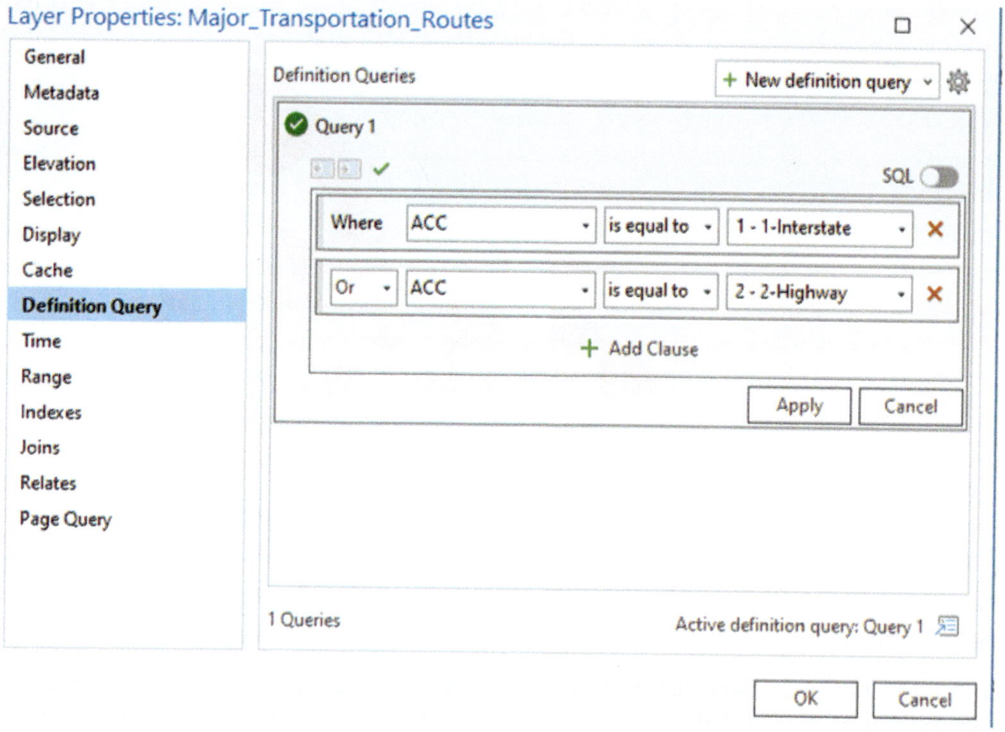


Or look like this in SQL environment.

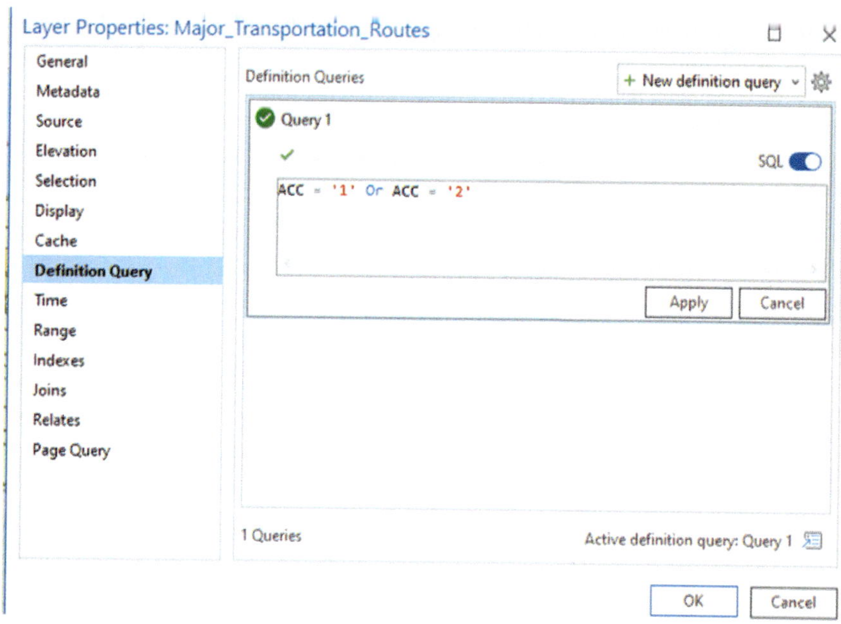

Click OK on the Layer Properties dialog. The filtered Major Roads layer presents. These are mainly interstate highways.

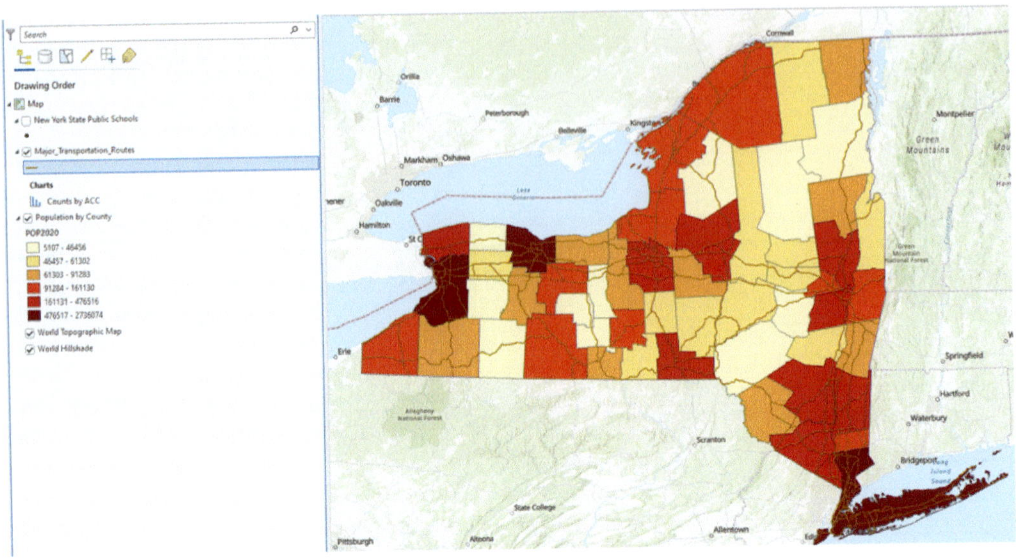


There are only 55,813 road features in this layer now (checking in attribute table).

In the Table of Contents, right-click the Major Roads layer and click Symbology on the contact menu. In the Symbology Pane, click the Primary Symbology drop-down arrow and click Unique Values.

Select and confirm that the Field 1 drop-down list is set to SPEED. On the Classes tab, click Add all values [icon]. Click More and uncheck Show all other values. In the list of values, click the row containing the value of "65" to select it – but don't click the trunk symbol itself. Click Move selected values up or drag it up until the row is on the top. Drag the other symbols according to the speed value from high to low for this map layer from the top down except "85." Remove the rows of "85," "all the other values," and "5" to "25" by right-click to get the contact menu while the row is highlighted.

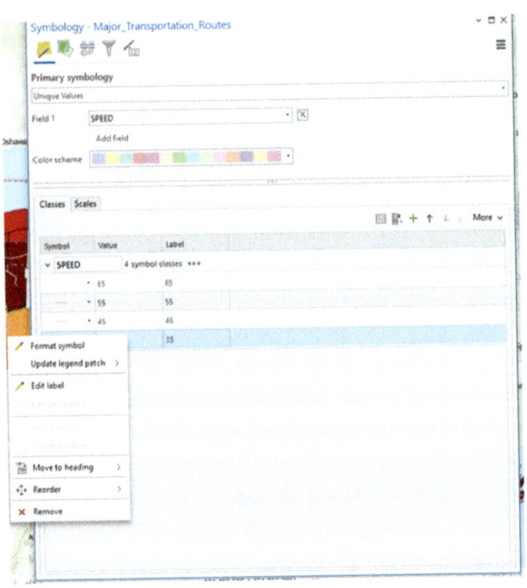

The result should look like this.

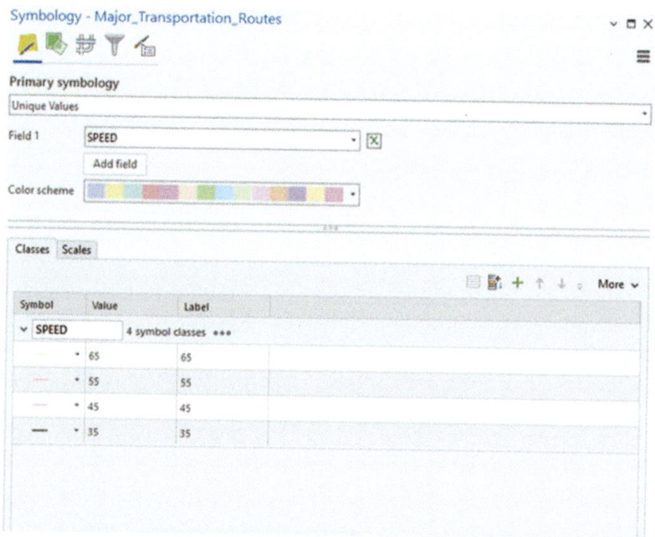


Your symbol colors may be different. The four road symbol classes are now ordered by speed limit. This order is reflected in the Table of Contents.

Now, we only have four major categories that we would like to use for mapping. Close the Symbology Pane.

In the Table of Contents – not the Symbology Pane – under the Major Roads layer, click the symbol for "65." The Symbology pane shows options for formatting line symbols. At the top of the pane, click the Gallery tab if necessary. If necessary, type highway and press the Enter key in the search box

In the search results, under ArcGIS 2D, hover over the second Highway symbol (Limited Access). Click the symbol. At the top of the pane, click the Properties tab. Set the Line width value to 3 pt and press the Enter key. Click Apply. In the Contents pane, click the other Road Type symbols in turn and change their values in the Symbology pane using the table below as a guide.

| Value | Symbol | Color and tone | Line width |
|---|---|---|---|
| 65 | Highway Limited Access | Light Blue | 4 pt |
| Expressway | Highway Limited Access | Light Blue | 2.5 pt |
| Highway | Highway (3D Billboards) | Cretan Blue | 3 pt |
| Turnpike | Highway Ramp (3D Billboards) | Cretan Blue | 1.5 pt |

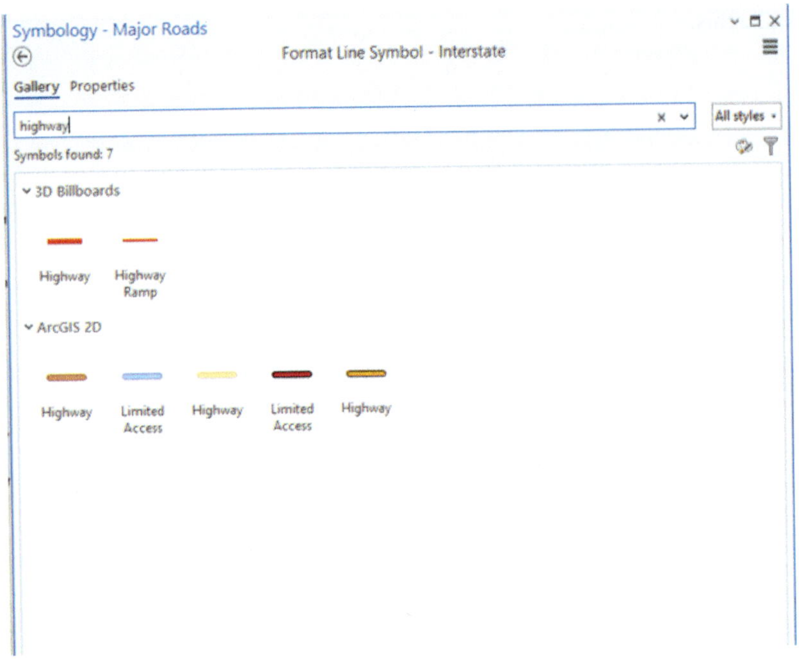


The map now looks like this.

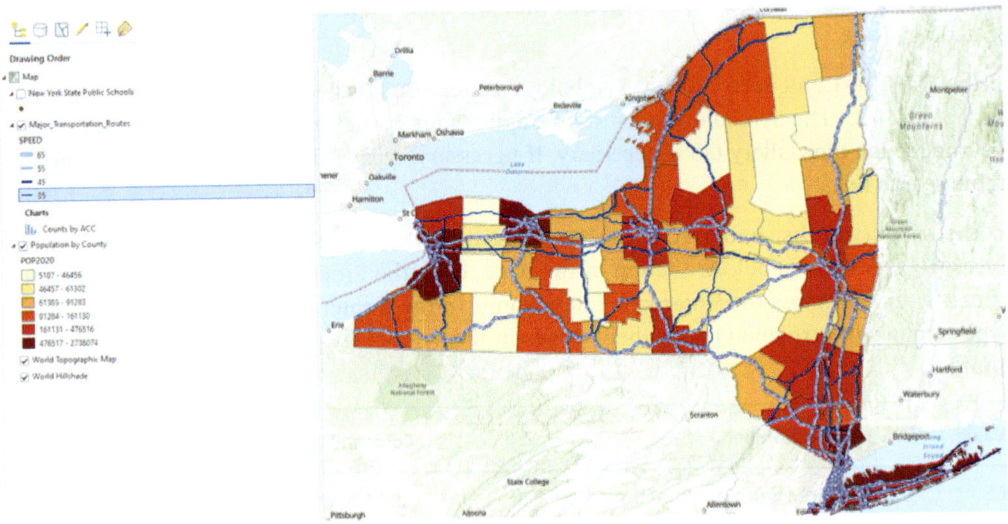


Open the Symbology Pane again while the "Major_Transportation_Routes" layer is highlighted.

Change the labels as the following:

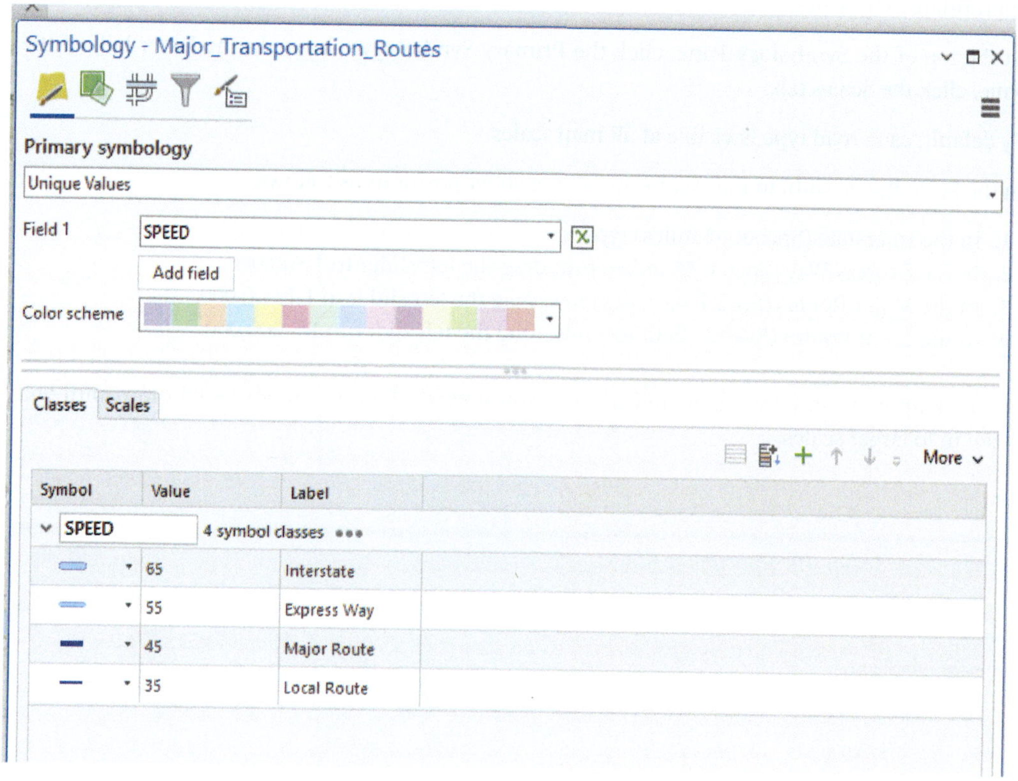


You notice the labels on the Table of Contents are changed following these names. We can also change the name of this map layer or erase the underbars between the words of the title. The map layer on the Table of Contents looks like this now.

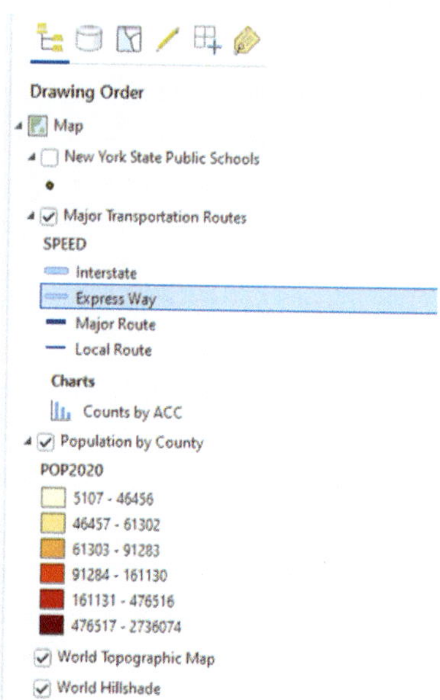

Now we will set scale ranges for the symbol classes so the major transportation routes draw at appropriate map scales.

At the top of the Symbology Pane, click the Primary Symbology tab. In the lower half of the pane, click the Scales tab.

By default, each road type is visible at all map scales.

In the Scale Range (min to max) column, set the slider positions as follows:

1. In the Interstate (Speed: 65 miles) type row, drag the left slider to 1:100,000.
2. In the Express Way (Speed: 55 miles) row, drag the left slider to 1:500,000.
3. In the Major Routes (Speed: 45 miles) row, drag the left slider to 1:1,000,000.
4. In the Local Routes (Speed: 35 miles) row, drag the left slider to 1:5,000,000.

On the map now, only Interstate and trunk roads appear. The other roads won't draw until you zoom in to larger scales.

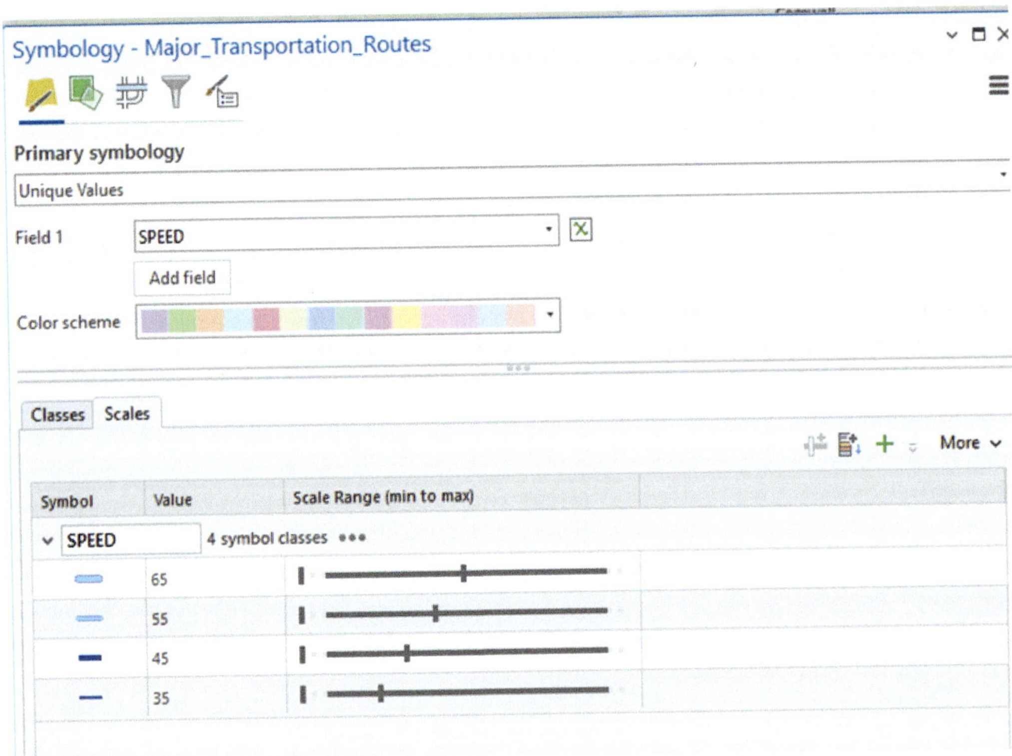

The map layout looks like this.

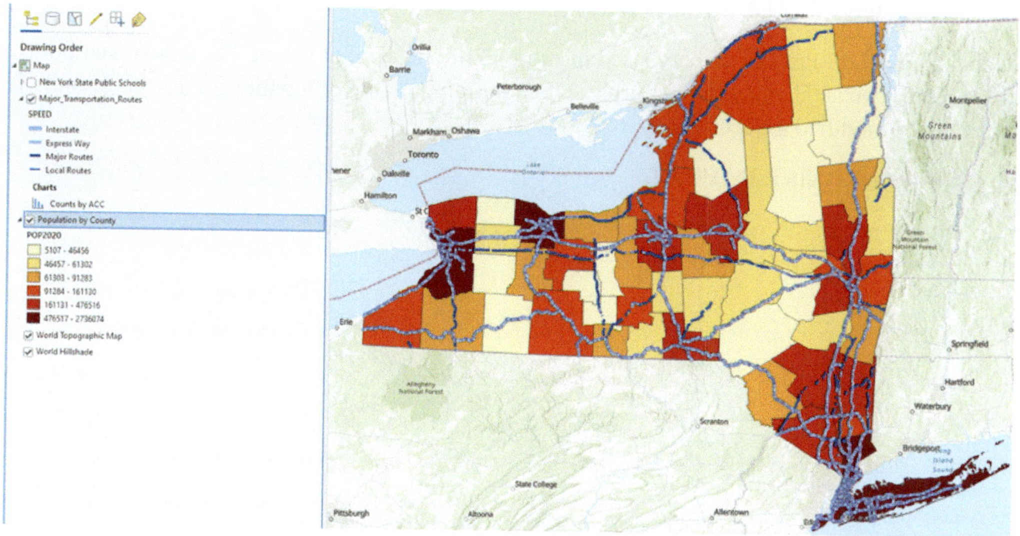


**3.** Symbolize the New York State Public Schools feature layer.

We are going to customize NYS public schools with transparency and aggregate symbols that are close together into single symbols uses the feature clustering method. The size of the symbol will be presented in proportion based on the cluster size.

In the Table of Contents, turn on the New York State Public Schools layer. Open the Symbology Pane, and under the Primary Symbology tab, click on the symbol to open the symbol gallery.

Type in "school" to search school symbols in the gallery.

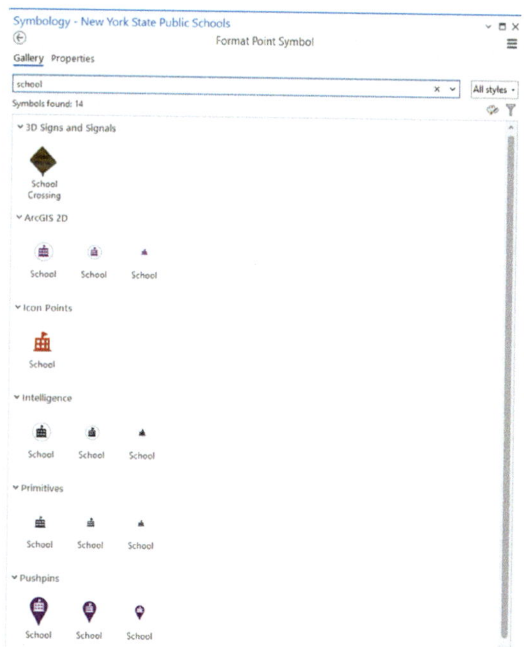

Under ArcGIS 2D, click the medium-sized School icon (its category is POI-Medium).

1. Click the Properties tab and click the Layers tab under it.

Symbols in ArcGIS Pro can be composed of more than one symbol layers. The School symbol is composed of two symbol layers – a purple school building on top of a white circle. These symbol layers can be modified independently.

2. Click the white circle to select it.

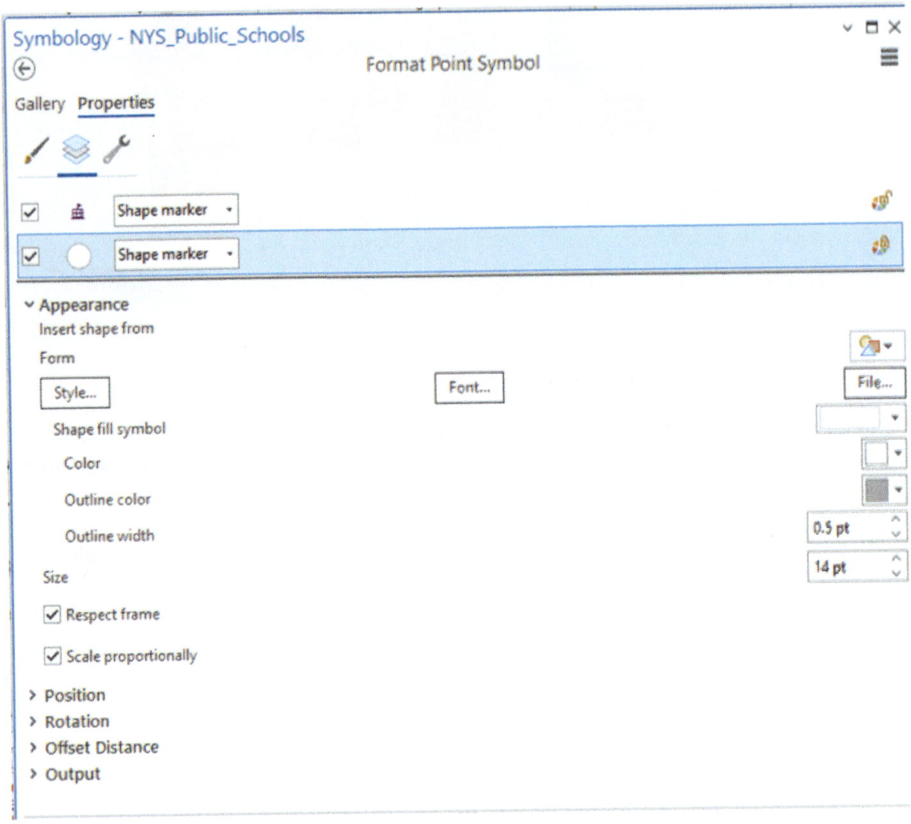


Expand the Appearance section if necessary. Click the Color button and click Color Properties. On the Color Editor dialog box, change the Transparency setting to 50% and press the Enter key.

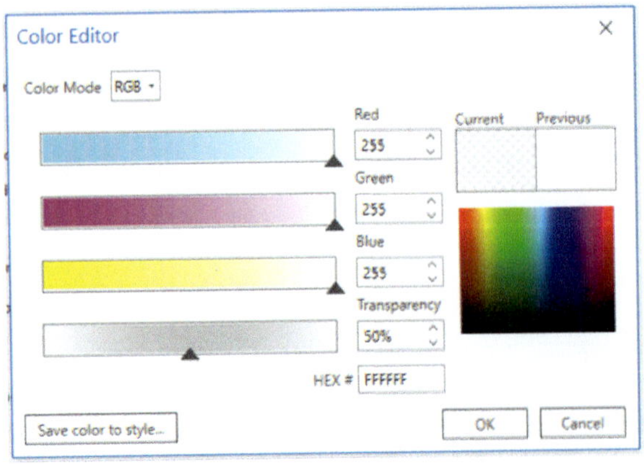

Click OK on the Color Editor and click Apply in the Symbology Pane.

In the Table of Contents, ensure that the New York State Public Schools layer is selected. On the ribbon at the top, click the Feature Layer tab. In the Drawing group, click Aggregation and click Clustering.

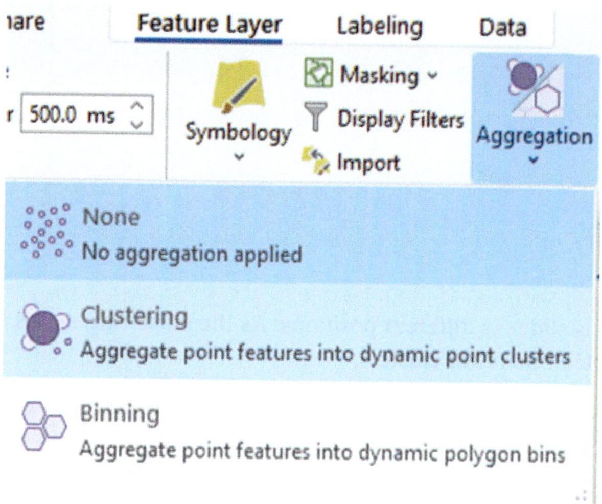


On the map, clusters of schools are now represented by proportionally sized symbols. The clusters update dynamically as you zoom in and out on the map. We can conduct some zoom-in experiments.

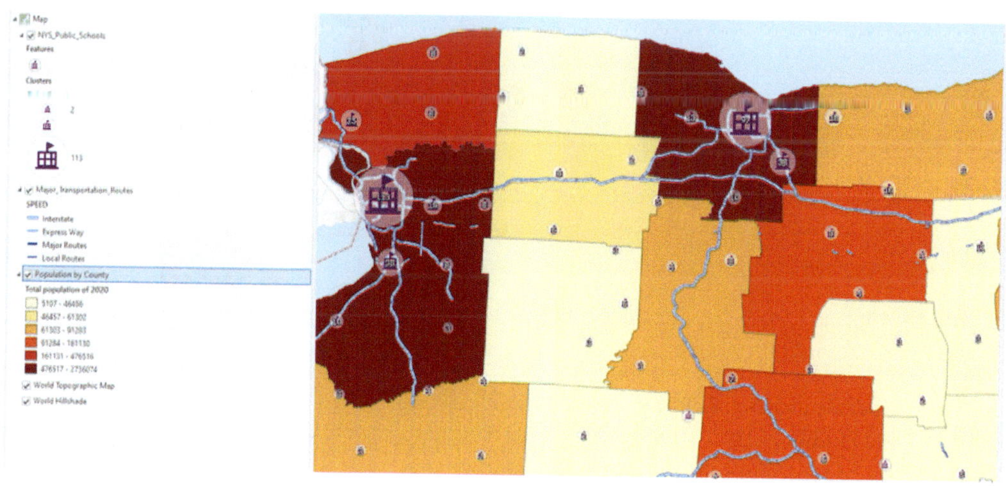

Return to the entire map extent and open the Symbology Pane. At the top of the Symbology Pane, click the Clusters tab and click the Cluster settings tab under it.

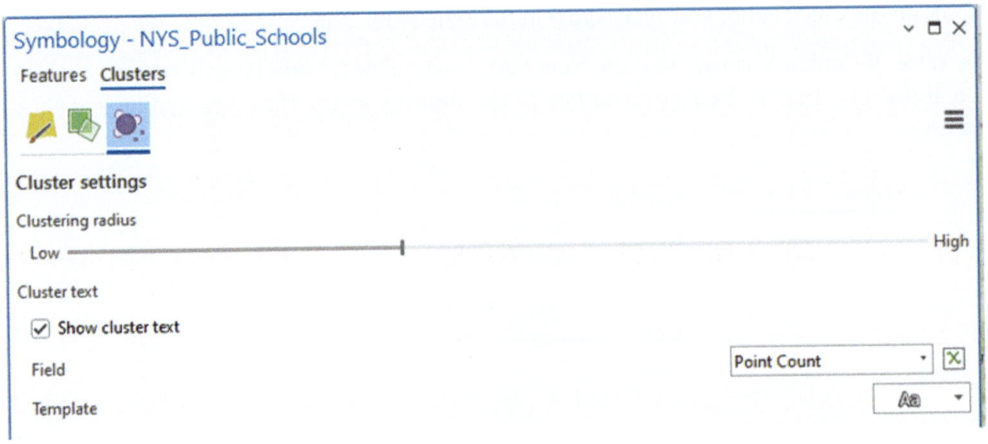


Experiment by moving the clustering radius slider to different positions. As the clustering radius becomes higher, more symbols are aggregated into fewer clusters.

Uncheck the Show cluster text check box to remove the count number of schools from the cluster symbols. Move the slider to a position about roughly 50% of the distance from the low end.

Close the Symbology Pane. Your map may look like this.

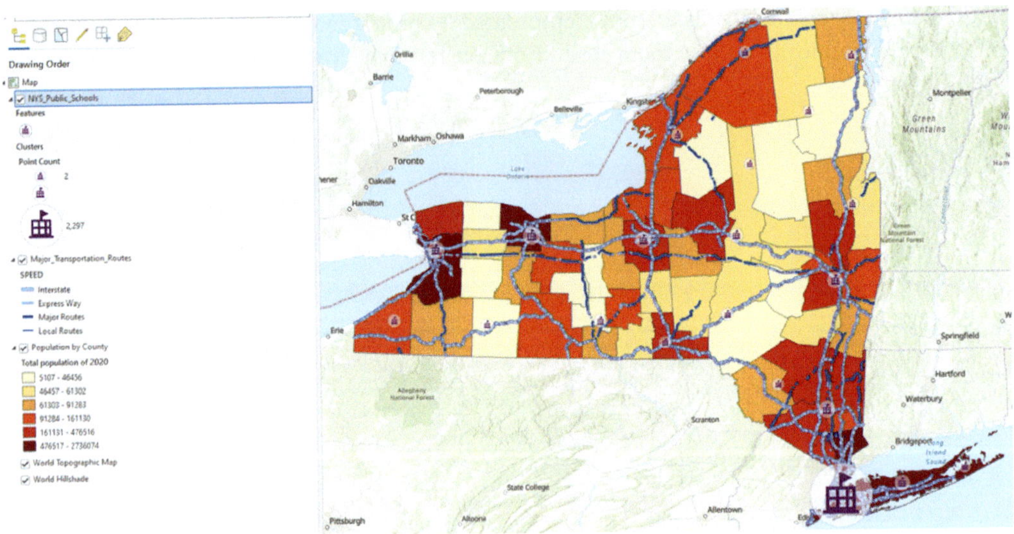


4. Label the population by county map layer.

We are going to label counties on this map composition. However, if we label all of them, it is going to be overwhelmingly crowded. Therefore, we are going to put the county labels on the map.

Ensure that the Population by County layer is selected in the Table of Contents. On the ribbon of the main menu at the top, click the Labeling tab. In the Label Class group, click SQL Query.

The Label Class Pane opens. We are going to construct a SQL query to filter out the counties with population below 50,000 for labeling.

In the Label Class Pane, click new expression. In the expression box, click the Where drop-down arrow and click POP2020 (Population 2020). Click the condition drop-down arrow and click is greater than. Click in the values box and type 500,000. Click Apply.

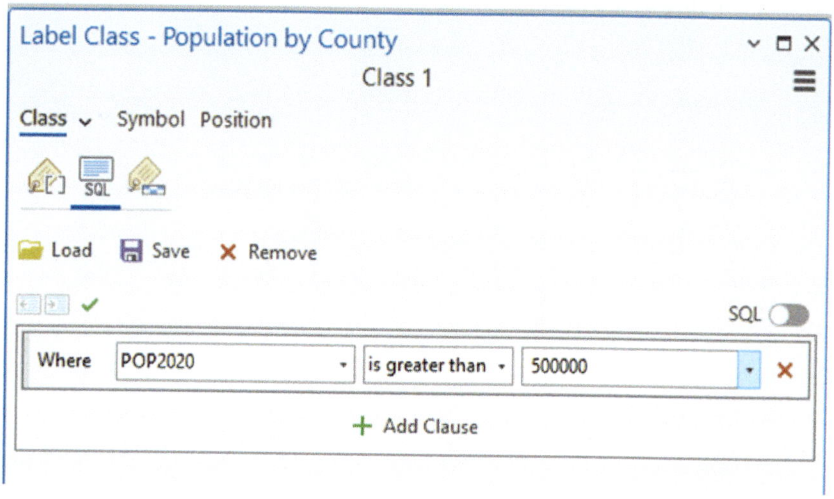


Now, right-click the Population by County layer in the Table of Contents to get the contact menu and click Label. On the map, only counties with populations over 500,000 are labeled.

At the top of the Label Class Pane, click the Symbol tab and confirm that the General tab is selected under it.

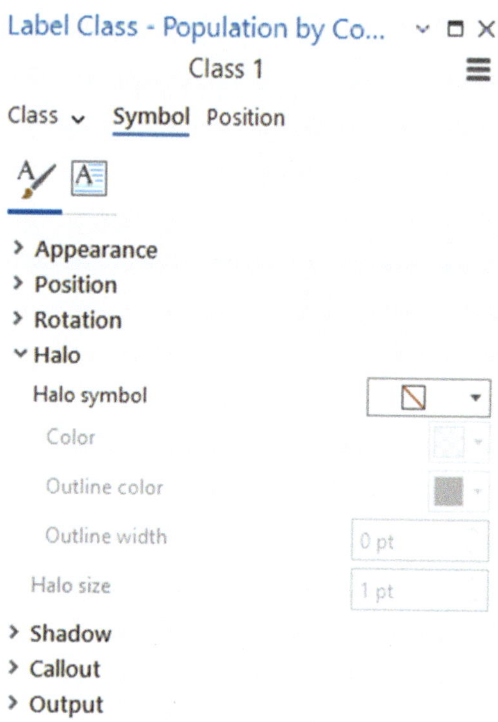

To help the labels stand out from the background, we can add a halo. A halo is a background fill color that separates the text from features on the map.

Expand Halo and click the Halo symbol drop-down arrow. Under Polygon symbols, click the White fill symbol. Confirm that the halo size is 2 pt.

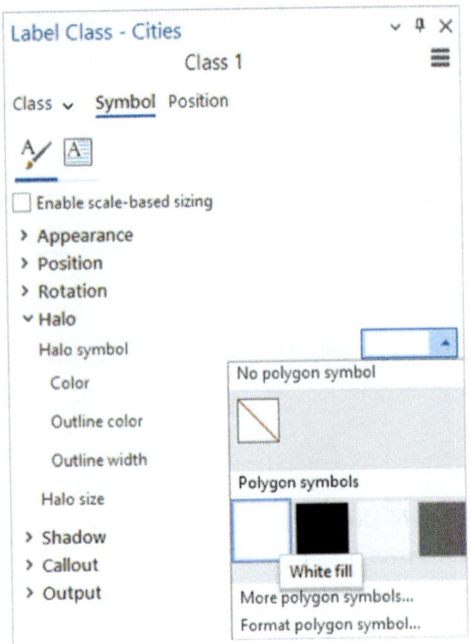


Click Apply. Close the Symbology Pane. The labels are distinguished to show on the map.

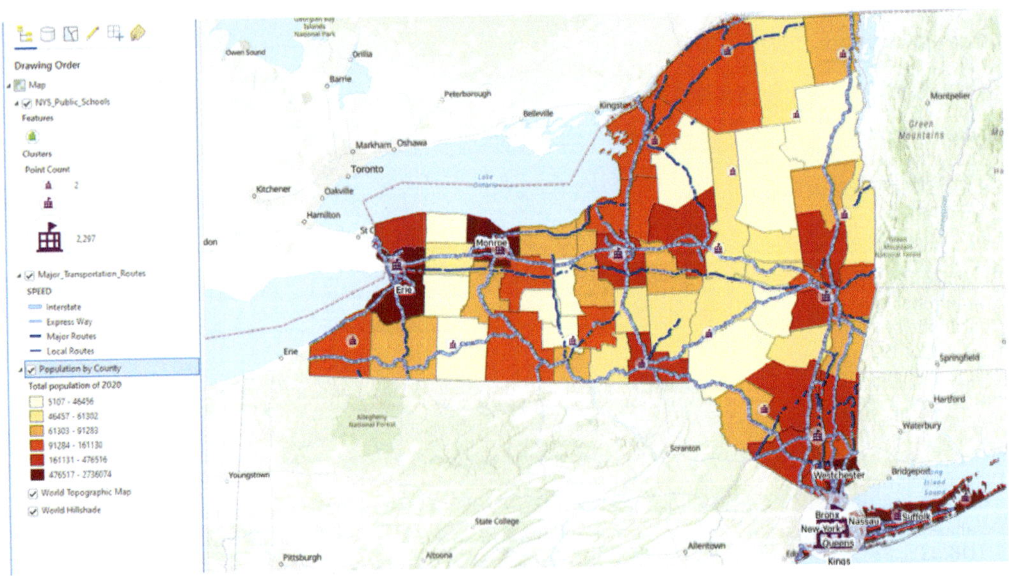


We accomplished the basic map compositions. Save your project and close ArcGIS Pro.

# Index




*Fundamentals of Geographic Information Systems (GIS) Analysis: Applying ArcGIS Pro*, First Edition. Tao Tang.

Printed and bound by CPI Group (UK) Ltd, Croydon, CR0 4YY

17/09/2026

14964334-0002